Supplement E

The chemistry of ethers, crown ethers, hydroxyl groups and their sulphur analogues
Part 2

THE CHEMISTRY OF FUNCTIONAL GROUPS

A series of advanced treatises under the general editorship of
Professor Saul Patai

$$-\overset{|}{\underset{|}{C}}-OH; \quad -\overset{|}{\underset{|}{C}}-SH; \quad -\overset{|}{\underset{|}{C}}-O-\overset{|}{\underset{|}{C}}-; \quad -\overset{|}{\underset{|}{C}}-S-\overset{|}{\underset{|}{C}}-$$

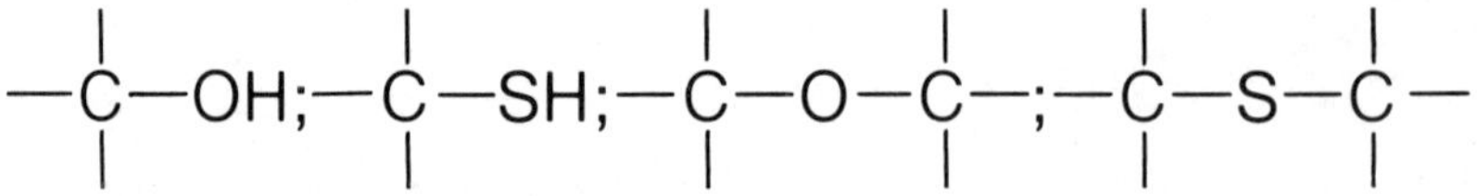

Supplement E

The chemistry of ethers, crown ethers, hydroxyl groups and their sulphur analogues
Part 2

Edited by

SAUL PATAI

The Hebrew University, Jerusalem

1980

JOHN WILEY & SONS

CHICHESTER – NEW YORK – BRISBANE – TORONTO

An Interscience ® Publication

CHEMISTRY

6454-3262

Copyright © 1980 by John Wiley & Sons Ltd.

All rights reserved.

No part of this book may be reproduced by any means,
nor transmitted, nor translated into a machine language
without the written permission of the publisher.

ISBN 0 471 27771 1 (Pt. 1)
ISBN 0 471 27772 X (Pt. 2)
ISBN 0 471 27618 9 (SET)

Typeset by Preface Ltd., Salisbury, Wiltshire.
Printed in the United States of America.

QD305
E7C48
V.2
CHEM

Contributing Authors

M. Bartók — Department of Organic Chemistry, József Attila University, Szeged, Hungary.

R. G. Bergstrom — Department of Chemistry, California State University, Hayward, California, U.S.A.

G. Bertholon — Groupe de Recherches sur les Phénols, C.N.R.S. of France (E.R.A. 600), Université Claude Bernard'Lyon 1, 43 Boulevard du 11 Novembre 1918, 69621 Villeurbanne Cedex, France

E. Block — Department of Chemistry, University of Missouri-St Louis, St. Louis, Missouri 63121, U.S.A.

C. H. Bushweller — Department of Chemistry, University of Vermont, Burlington, Vermont 05405, U.S.A.

R. L. Failes — Department of Chemistry, Macquarie University, New South Wales 2113, Australia.

P. Fischer — Institut für Organische Chemie, Biochemie und Isotopenforschung, Universität Stuttgart, Stuttgart, Bundesrepublik Deutschland.

M. H. Gianni — Department of Chemistry, St Michael's College, Winooski, Vermont 05404, U.S.A.

I. Goldberg — Institute of Chemistry, Tel-Aviv University 61390 Tel-Aviv, Israel.

G. Gottarelli — Faculty of Industrial Chemistry, University of Bologna, Italy.

D. A. Laidler — I.C.I. Corporate Laboratory, Runcorn, England and Department of Chemistry, University of Sheffield, England.

R. Lamartine — Group de Recherches sur les Phénols, C.N.R.S. of France (E.R.A. 600), Université Claude Bernard Lyon 1, 43 Boulevard du 11 Novembre 1918, 69621 Villeurbanne Cedex, France.

K. L. Láng — Department of Organic Chemistry, József Attila University, Szeged, Hungary.

C. L. Liotta — School of Chemistry, Georgia Institute of Technology, Atlanta, Georgia 30332, U.S.A.

Á. Molnár — Institute of Organic Chemistry, József Attila University, Szeged, Hungary.

P. Müller — Département de Chimie Organique, Université de Genève, Genève, Suisse.

9099

P. Pasanen   Department of Chemistry, University of Turku, SF-20500
                       Turku 50, Finland.

M. Perrin   Laboratoire de Minéralogie-Cristallographie, C.N.R.S. of
                       France (E.R.A. 600) Université Claude Bernard Lyon 1, 43
                       Boulevard du 11 Novembre 1918, 69261 Villeurbanne Cedex,
                       France.

R. Perrin   Group de Recherches sur les Phénols, C.N.R.S. of France
                       (E.R.A. 600), Université Claude Bernard Lyon 1, 43 Boulevard
                       du 11 Novembre 1918, 69621 Villeurbanne Cedex, France.

K. Pihlaja   Department of Chemistry, University of Turku, SF-20500,
                       Turku 50, Finland.

J. Royer   Groupe de Physique Moléculaire et Chimie Organique
                       Quantiques, C.N.R.S. of France (E.R.A. 600), Université
                       Claude Bernard Lyon 1, 43 Boulevard du 11 Novembre 1918,
                       69621 Villeurbanne Cedex, France.

B. Samorï   Faculty of Industrial Chemistry, University of Bologna, Italy.

H.-P. Schuchmann   Institut für Strahlenchemie im Max-Planck-Institut für
                       Kohlenforschung, Stiftstrasse 34–36, D-4330 Mülheim a.d.
                       Ruhr, West Germany.

J. S. Shapiro   Department of Chemistry, Macquarie University, New South
                       Wales 2113, Australia.

T. Shono   Department of Synthetic Chemistry, Kyoto University,
                       Kyoto 606, Japan.

C. von Sonntag   Institut für Strahlenchemie im Max-Planck-Institut für
                       Kohlenforschung, Stiftstrasse 34–36, D-4330 Mülheim a.d.
                       Ruhr, West Germany.

P. J. Stang   Chemistry Department, The University of Utah, Salt Lake
                       City, Utah 84112, U.S.A.

V. R. Stimson   Department of Physical and Inorganic Chemistry, University
                       of New England, Armidale 2351, Australia.

J. F. Stoddart   I.C.I. Corporate Laboratory, Runcorn, England and Department
                       of Chemistry, University of Sheffield, England.

C. Van de Sande   Department of Organic Chemistry, State University of Gent,
                       Krijgslaan, 217 (Block S.4), B-9000 Gent, Belgium.

F. Vögtle   Institut für Organische Chemie und Biochemie der Universität,
                       Gerhard-Domagk Strasse 1, D-5300 Bonn, West Germany.

E. Weber   Institut für Organische Chemie und Biochemie der Universität,
                       Gerhard-Domagk Strasse 1, D-5300 Bonn, West Germany.

M. Zieliński   Institute of Chemistry, Jagiellonian University, Cracow, Poland.

Foreword

The present *Supplement E* brings material related to the chapters which appeared in the main volumes on *The Ether Linkage* (1967), on *The Hydroxyl Group* (1971), and on *The Thiol Group* (1974). It is characteristic of the rapid development of organic chemistry that crown ethers, which are the subjects of the first three weighty chapters of this volume, had not even been mentioned in the main volume on ethers, thirteen years ago!

This volume contains several chapters dealing with sulphur analogues of alcohols and ethers. However, the first in a set of volumes (*The Chemistry of the Sulphonium Group*) on various sulphur-containing groups is already in press and further volumes of the set are being planned.

Chapters on 'Thermochemistry' and on 'Cyclic sulphides' were also planned for this volume, but did not materialize.

Jerusalem, June 1980.   SAUL PATAI

The Chemistry of Functional Groups
Preface to the series

The series 'The Chemistry of Functional Groups' is planned to cover in each volume all aspects of the chemistry of one of the important functional groups in organic chemistry. The emphasis is laid on the functional group treated and on the effects which it exerts on the chemical and physical properties, primarily in the immediate vicinity of the group in question, and secondarily on the behaviour of the whole molecule. For instance, the volume *The Chemistry of the Ether Linkage* deals with reactions in which the C—O—C group is involved, as well as with the effects of the C—O—C group on the reactions of alkyl or aryl groups connected to the ether oxygen. It is the purpose of the volume to give a complete coverage of all properties and reactions of ethers in as far as these depend on the presence of the ether group but the primary subject matter is not the whole molecule, but the C—O—C functional group.

A further restriction in the treatment of the various functional groups in these volumes is that material included in easily and generally available secondary or tertiary sources, such as Chemical Reviews, Quarterly Reviews, Organic Reactions, various 'Advances' and 'Progress' series as well as textbooks (i.e. in books which are usually found in the chemical libraries of universities and research institutes) should not, as a rule, be repeated in detail, unless it is necessary for the balanced treatment of the subject. Therefore each of the authors is asked *not* to give an encyclopaedic coverage of his subject, but to concentrate on the most important recent developments and mainly on material that has not been adequately covered by reviews or other secondary sources by the time of writing of the chapter, and to address himself to a reader who is assumed to be at a fairly advanced post-graduate level.

With these restrictions, it is realized that no plan can be devised for a volume that would give a *complete* coverage of the subject with *no* overlap between chapters, while at the same time preserving the readability of the text. The Editor set himself the goal of attaining *reasonable* coverage with *moderate* overlap, with a minimum of cross-references between the chapters of each volume. In this manner, sufficient freedom is given to each author to produce readable quasi-monographic chapters.

The general plan of each volume includes the following main sections:

(a) An introductory chapter dealing with the general and theoretical aspects of the group.

(b) One or more chapters dealing with the formation of the functional group in question, either from groups present in the molecule, or by introducing the new group directly or indirectly.

ix

(c) Chapters describing the characterization and characteristics of the functional groups, i.e. a chapter dealing with qualitative and quantitative methods of determination including chemical and physical methods, ultraviolet, infrared, nuclear magnetic resonance and mass spectra: a chapter dealing with activating and directive effects exerted by the group and/or a chapter on the basicity, acidity or complex-forming ability of the group (if applicable).

(d) Chapters on the reactions, transformations and rearrangements which the functional group can undergo, either alone or in conjunction with other reagents.

(e) Special topics which do not fit any of the above sections, such as photochemistry, radiation chemistry, biochemical formations and reactions. Depending on the nature of each functional group treated, these special topics may include short monographs on related functional groups on which no separate volume is planned (e.g. a chapter on 'Thioketones' is included in the volume *The Chemistry of the Carbonyl Group*, and a chapter on 'Ketenes' is included in the volume *The Chemistry of Alkenes*). In other cases certain compounds, though containing only the functional group of the title, may have special features so as to be best treated in a separate chapter, as e.g. 'Polyethers' in *The Chemistry of the Ether Linkage*, or 'Tetraaminoethylenes' in *The Chemistry of the Amino Group*.

This plan entails that the breadth, depth and thought-provoking nature of each chapter will differ with the views and inclinations of the author and the presentation will necessarily be somewhat uneven. Moreover, a serious problem is caused by authors who deliver their manuscript late or not at all. In order to overcome this problem at least to some extent, it was decided to publish certain volumes in several parts, without giving consideration to the originally planned logical order of the chapters. If after the appearance of the originally planned parts of a volume it is found that either owing to non-delivery of chapters, or to new developments in the subject, sufficient material has accumulated for publication of a supplementary volume, containing material on related functional groups, this will be done as soon as possible.

The overall plan of the volumes in the series 'The Chemistry of Functional Groups' includes the titles listed below:

The Chemistry of Alkenes (two volumes)
The Chemistry of the Carbonyl Group (two volumes)
The Chemistry of the Ether Linkage
The Chemistry of the Amino Group
The Chemistry of the Nitro and Nitroso Groups (two parts)
The Chemistry of Carboxylic Acids and Esters
The Chemistry of the Carbon–Nitrogen Double Bond
The Chemistry of the Cyano Group
The Chemistry of Amides
The Chemistry of the Hydroxyl Group (two parts)
The Chemistry of the Azido Group
The Chemistry of Acyl Halides
The Chemistry of the Carbon–Halogen Bond (two parts)
The Chemistry of Quinonoid Compounds (two parts)
The Chemistry of the Thiol Group (two parts)
The Chemistry of Amidines and Imidates
The Chemistry of the Hydrazo, Azo and Azoxy Groups (two parts)

The Chemistry of Cyanates and their Thio Derivatives (two parts)
The Chemistry of Diazonium and Diazo Groups (two parts)
The Chemistry of the Carbon–Carbon Triple Bond (two parts)
Supplement A: The Chemistry of Double-bonded Functional Groups (two parts)
Supplement B: The Chemistry of Acid Derivatives (two parts)
The Chemistry of Ketenes, Allenes and Related Compounds (two parts)
Supplement E: The Chemistry of Ethers, Crown Ethers, Hydroxyl Groups and their Sulphur Analogues (two parts)

Titles in press:

The Chemistry of the Sulphonium Group
Supplement F: The Chemistry of Amines, Nitroso and Nitro Groups and their Derivatives

Future volumes planned include:

The Chemistry of Peroxides
The Chemistry of Organometallic Compounds
The Chemistry of Sulphur-containing Compounds
Supplement C: The Chemistry of Triple-bonded Functional Groups
Supplement D: The Chemistry of Halides and Pseudo-halides

Advice or criticism regarding the plan and execution of this series will be welcomed by the Editor.

The publication of this series would never have started, let alone continued, without the support of many persons. First and foremost among these is Dr Arnold Weissberger, whose reassurance and trust encouraged me to tackle this task, and who continues to help and advise me. The efficient and patient cooperation of several staff-members of the Publisher also rendered me invaluable aid (but unfortunately their code of ethics does not allow me to thank them by name). Many of my friends and colleagues in Israel and overseas helped me in the solution of various major and minor matters, and my thanks are due to all of them, especially to Professor Z. Rappoport. Carrying out such a long-range project would be quite impossible without the non-professional but none the less essential participation and partnership of my wife.

The Hebrew University
Jerusalem, ISRAEL

SAUL PATAI

Contents

xiv

Contents

CHAPTER **14**

Oxiranes

M. BARTÓK and K. L. LÁNG
Department of Organic Chemistry, József Attila University,
Szeged, Hungary

Abbreviations

AcAc	Acetylacetone
DATMP	Diethylaluminium 2,2,6,6-tetramethylpiperidide
DMF	Dimethylformamide
DMSO	Dimethylsulphoxide
LAH	Lithium aluminium hydride
MCPBA	*m*-Chloroperoxybenzoic acid
NBA	*N*-Bromoacetamide
NBS	*N*-Bromosuccinimide
PAA	Peroxyacetic acid
PBA	Peroxybenzoic acid
PNPBA	*p*-Nitroperoxybenzoic acid
TDAP	Tris(dimethylamino)phosphine
TMC	Tetramethyl carbamide
Ts	*p*-Toluenesulphonyl

I. INTRODUCTION

The earlier literature data on the synthesis and chemistry of oxiranes were reviewed by Dittus[1] in 1965 and by Gritter[2] in 1967. Since then, the work relating to the synthesis and chemical transformations of the oxiranes has been surveyed by numerous authors[3-19]. Only a few of these surveys are of a general nature, the majority dealing with some special area. Some of them discuss experimental results that were published five to six years ago. Accordingly, the present review is based mainly on the conclusions drawn from the experimental data of the most recent period (up to the end of 1977). Of the other results since 1965, only those are mentioned that are of general validity, or which were not dealt with in the previous reviews.

II. SYNTHESIS OF OXIRANES

A. By Oxidation of Alkenes

Direct oxidation of alkenes continues to be the main method of preparing oxiranes both in the laboratory and in industry. Significant new results have been achieved in the development of the procedures of liquid-phase oxidation of alkenes. Efforts have been made to perform this oxidation under the mildest possible experimental conditions, which allows an increase in the selectivity of oxirane formation and also the selective oxidation of more sensitive compounds.

1. Oxidation with peroxy acids

Details on the peroxy acid oxidation of alkenes, the Prilezhaev reaction[18], are to be found in some very good reviews, which deal with the mechanism and stereochemistry of the reaction and its practical modifications[3,6,7,10,11,13,17,18].

The accepted mechanism of alkene oxidation with peroxy acids is that outlined in equation (1). The process involves an addition reaction, where the alkene is the nucleophile and the peroxy acid the electrophile, but binding of the electrophilic species is not followed by binding of an external nucleophilic species.

$$\text{\Large$>$C=C$<$} + RCO_3H \longrightarrow \left[\begin{array}{c} \text{activated} \\ \text{complex} \end{array} \right] \longrightarrow \text{\Large$\triangle$}O + RCO_2H \qquad (1)$$

The fine mechanism of the reaction is still not known in every respect, for it depends on the electrophilic and nucleophilic characters of the two reactants, their stereostructures and reaction conditions such as temperature, solvent, catalyst, etc. All these factors have a considerable influence on the structure and stability of the transition complex, and on the process determining the reaction rate. After wide-ranging kinetic investigations, Dryuk[20] gave the reaction mechanism as in equation (2). This mechanism is supported by studies of the stereochemical course,

$$\text{\Large$\searrow$C$\parallel$C$\nearrow$} + RCO_3H \; \rightleftharpoons \; \begin{array}{c} \text{electron} \\ \text{donor--} \\ \text{acceptor} \\ \text{complex} \end{array} \longrightarrow \begin{array}{c} \text{activated} \\ \text{complex} \\ \\ \text{other} \\ \text{transformations} \end{array} \longrightarrow \text{\Large$\triangle$}O \qquad (2)$$

kinetics and acid catalysis of the reaction, and the side-reactions accompanying it and by the following experimental observations: electron-repelling groups on the alkene increase the reaction rate; the reaction rate is higher for peroxy acids containing electron-attracting substituents; basic solvents decrease the rate of epoxidation. The solvent effect is connected with hydrogen bonds between the peroxy acids and the solvents.

Other investigations[21-28] also deal with the mechanism of the reaction, and with the structure of the transition complex[20,29,30]. Significant conclusions may also be drawn from the results of stereochemical investigations (see below). The 1,3-dipolar cycloaddition mechanism[31-34] has not been confirmed by the recent experimental results.

In contrast with other electrophilic additions, the peroxy acid oxidation is stereochemically *syn*-stereospecific. In the case of cycloalkenes, the C—O bond in the oxirane formed displays axial orientation. With sterically-hindered alkenes, epoxidation occurs from the less-hindered side. The more important stereochemical regularities[13] described earlier for the epoxidation of various types of compounds have been supported by more recent studies; some of these are presented here.

Stereoselectivity to varying degrees has been observed on the peroxy acid epoxidation of some new compound types (equations 3–8)[35-40].

(Ref. 35)   (3)

M. Bartók and K. L. Láng

(Ref. 36)　(4)

n = 1,2

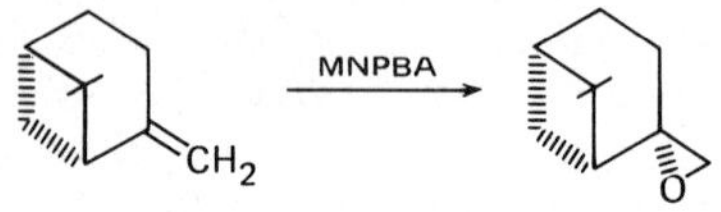

(Ref. 37)　(5)

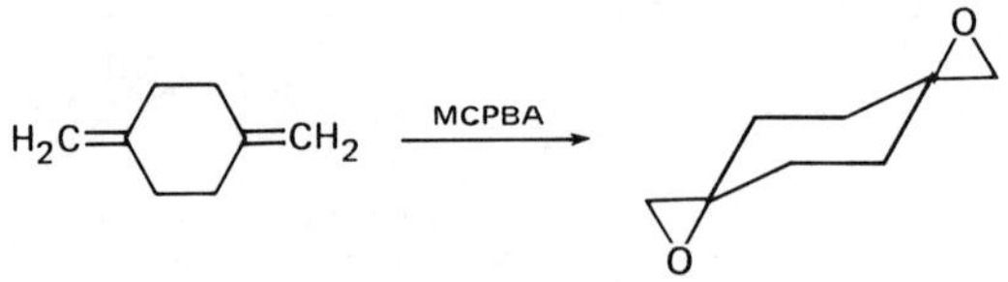

(Ref. 38)　(6)

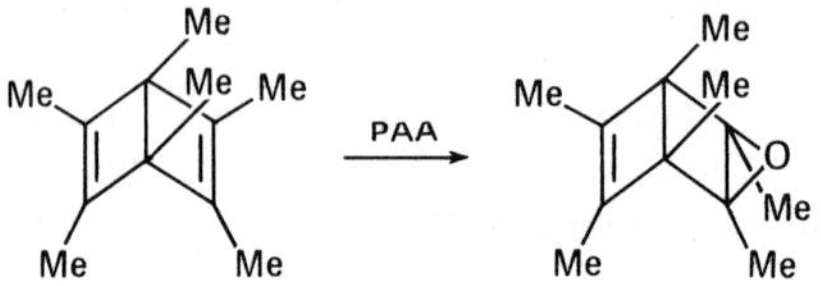

(Ref. 39)　(7)

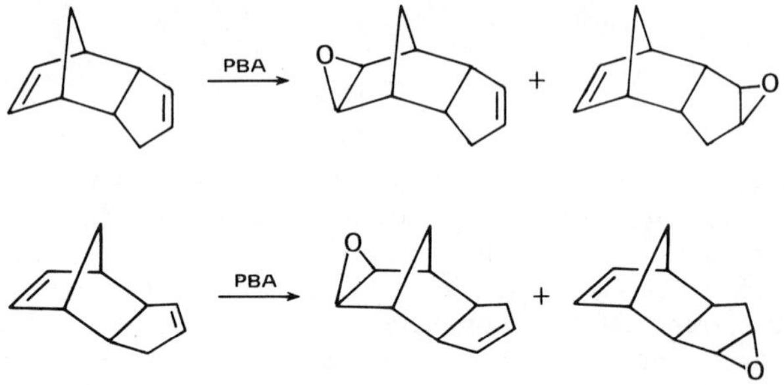

(Ref. 40)　(8)

The epoxidation of olefins containing various functional groups is also stereoselective in many cases, as a consequence of steric, electronic and conformational effects. Examples are given in equations (9)–(15)[41–48].

In recent years studies have been made of other compound types and the stereochemical course of their reactions, e.g. for olefins containing a high number

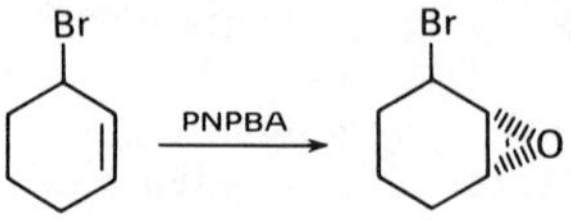

(Ref. 41)　(9)

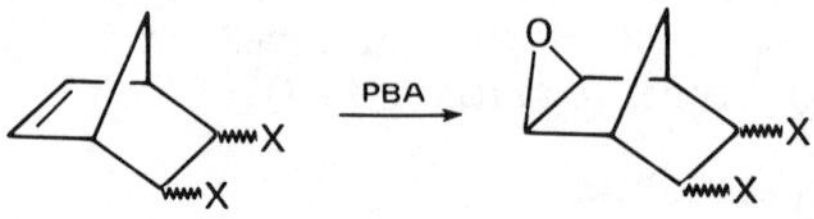

(Ref. 42)　(10)

X = Cl,Br,CO$_2$Me,CN

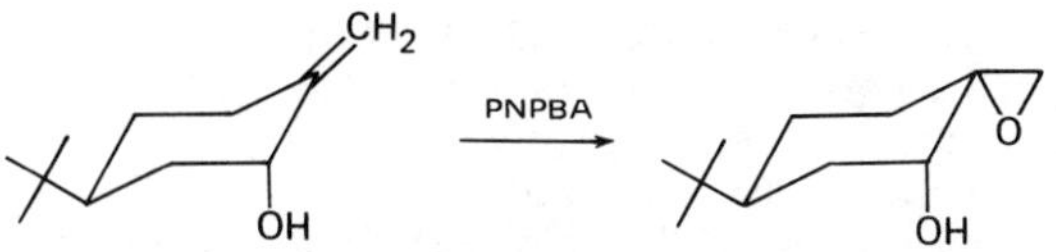

(Ref. 43) (11)

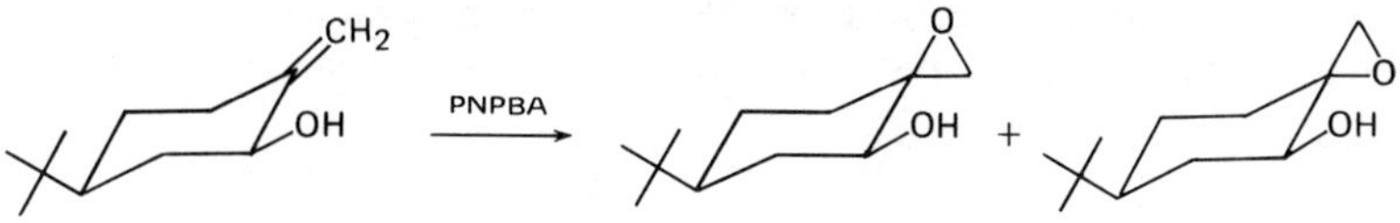

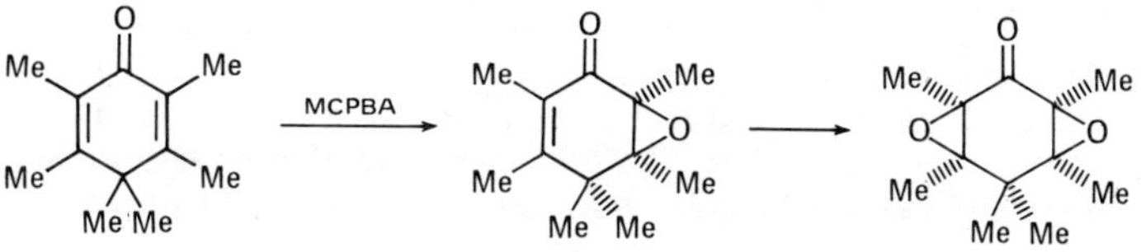

(Ref. 44) (12)

(Ref. 45) (13)

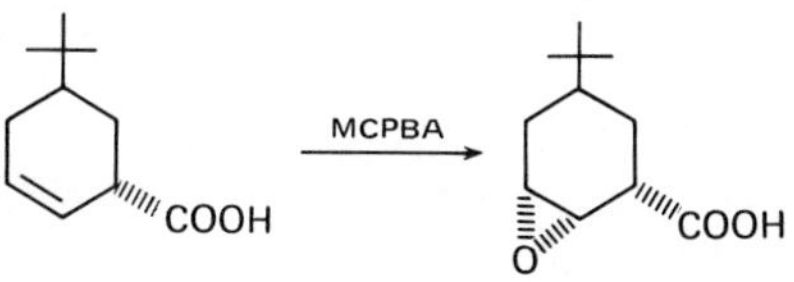

(Refs. 46, 47) (14)

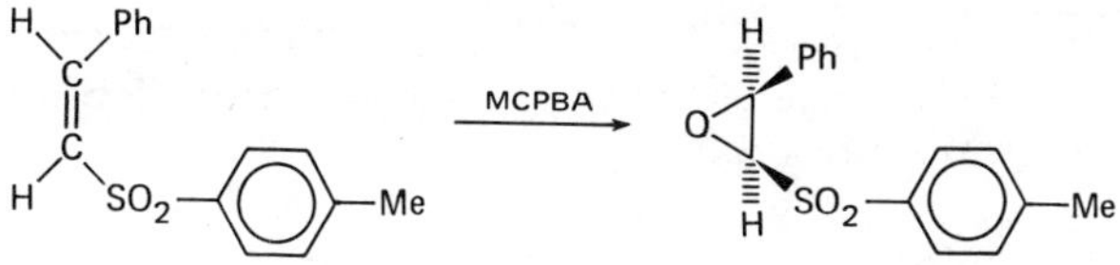

(Ref. 48) (15)

of carbon atoms[49], cyclic alkenes and dienes[40,50-55], aromatic systems[56], unsaturated alcohols and their derivatives[57-62], steroids[63-65], unsaturated carboxylic acids and their derivatives[66,67], olefin propellanes[68,69], phosphine oxides[70] and phospholenes[71].

Enantiostereoisomeric oxiranes may be prepared by epoxidation with chiral peroxy acids[72-79]. A method has been elaborated for the separation of racemic oxiranes, using optically active lanthanide complexes[80].

Peroxy acid oxidation is currently the most frequently employed method of epoxidation in the organic preparative laboratory. It gives very good yields, and may also be used for relatively sensitive compounds, such as unsaturated alcohols[81], terpenes[82], acenaphthene[83] and allenes[84-87], or for the preparation of halogenated oxiranes[88-90].

Of the peroxy acids, MCPBA is most favoured, except for procedures elaborated to meet special needs. Alkenes undergoing reaction with difficulty are epoxidized at higher temperature in the presence of radical inhibitors[91]. Peroxy acid stabilizers increase the yield[92]. In the preparation of acid-sensitive oxiranes or the oxidation of acid-sensitive olefins, an alkaline two-phase solvent system is employed at room temperatures[93,94]. Polymer-supported peroxy acids may be used for the oxidation of some olefins[95,96]. In certain cases *in situ* peroxy acid procedures are used[3,97,98].

New epoxidizing reagents have recently been introduced, e.g. *o*-sulphoperbenzoic acid[63], *p*-methoxycarbonylperoxybenzoic acid[99], [bis(benzoyldioxy)iodo] benzene[100], *O*-benzylmonoperoxycarbonic acid[101], peroxycarboximidic acids formed from nitriles with hydrogen peroxide[102-107], peroxycarbaminic acids[108,109], peroxyacetyl nitrate[110], disuccinyl peroxide[111] and benzeneperoxyseleninic acid[112].

2. Oxidation with hydrogen peroxide

Hydrogen peroxide may be used for epoxidation in the presence of phenyl isocyanate[113]. Hydrogen peroxide as a direct epoxidizing agent can be employed for the epoxidation of electron-poor olefins. The procedures are of great importance, since compounds may thus be epoxidized even when the peroxy acid procedures have proved ineffective.

a. Oxidation with alkaline hydrogen peroxide. The earlier literature has been reviewed by Berti[13]. The essence of the method is illustrated in equations (16) and (17). The mechanism of the process depends on the starting compound. No general and completely clear-cut correlations have yet emerged as regards the stereochemistry or stereoselectivity of this epoxidation.

$$H_2O_2 + OH^- \rightleftharpoons HO_2^- + H_2O \tag{16}$$

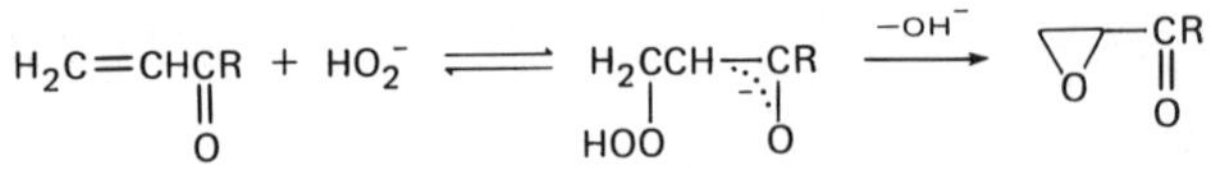

$$\tag{17}$$

The procedure has been employed effectively for numerous types of compounds: α,β-unsaturated ketones[114-117], nitro olefins[118], α,β-unsaturated nitriles[119,120] *endo*- and *exo*-cyclic enones[121-123] and steroids[124-128]. The epoxidation is often of very high stereoselectivity (equations 18–22):

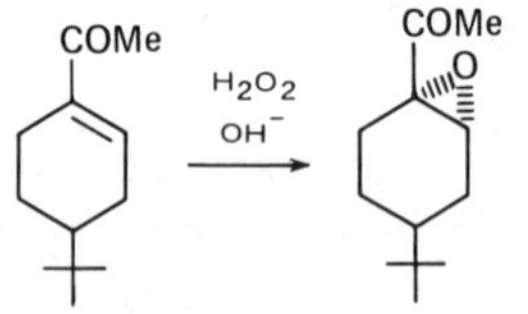

(Ref. 129) (18)

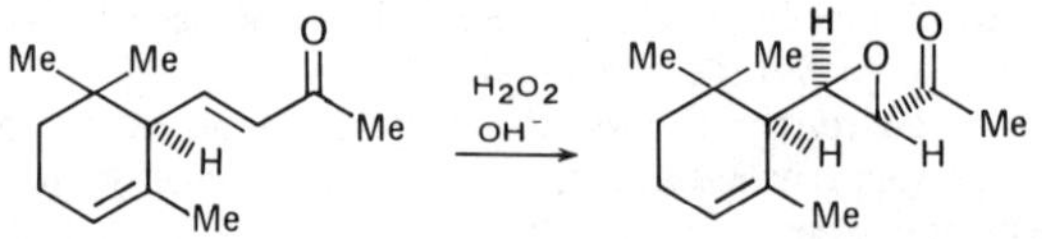

(Ref. 130) (19)

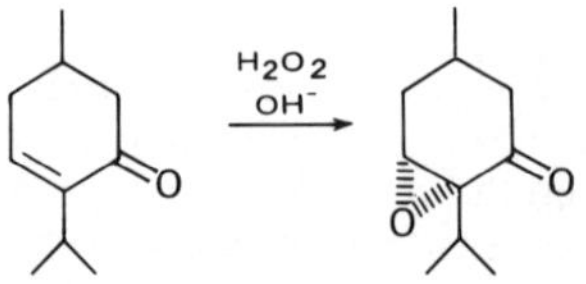

(Ref. 131) (20)

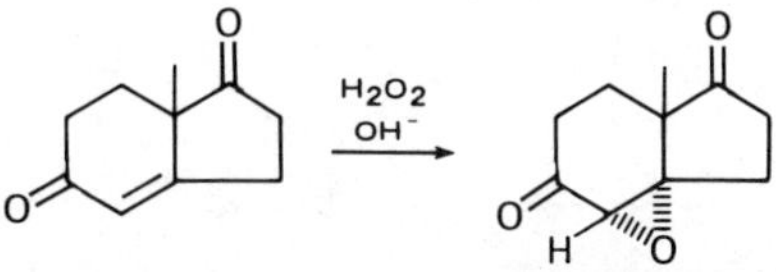

(Ref. 132) (21)

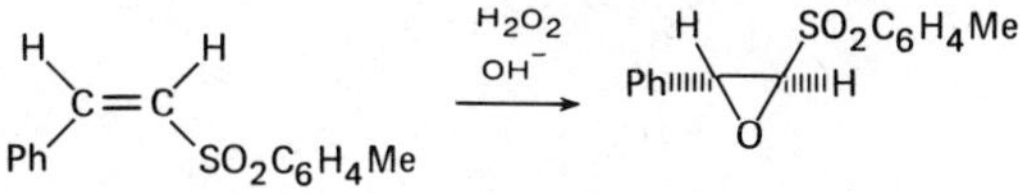

(Refs. 48, 133) (22)

With a chiral phase-transfer catalyst being used as base, optically-active oxiranes may be prepared in excellent yield[134] (equation 23).

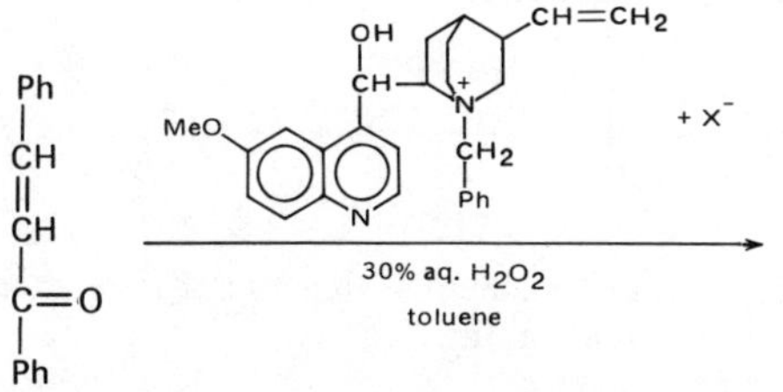

(23)

Hydrogen peroxide is also used in the new procedure of Kametani and co-workers[135].

b. *Oxidation with hydrogen peroxide and catalyst.* Some acids and their various transition metal salts are used as catalysts[10,136-144]. The most frequent catalyst is sodium tungstate ($HWO_4^- + H_2O_2 \rightleftharpoons HWO_5^- + H_2O$), which may behave both as a nucleophilic and an electrophilic reagent, depending on the substrate and the experimental conditions. The epoxidation process is shown in equation (24). Mechan-

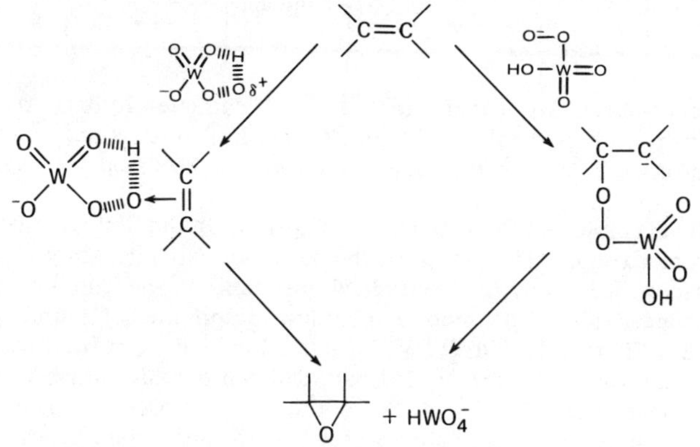

(24)

istic studies confirm this reaction path[142,143,145-149], and at the same time provide information on the stereochemical course of the reaction[150-152] (equations 25 and 26).

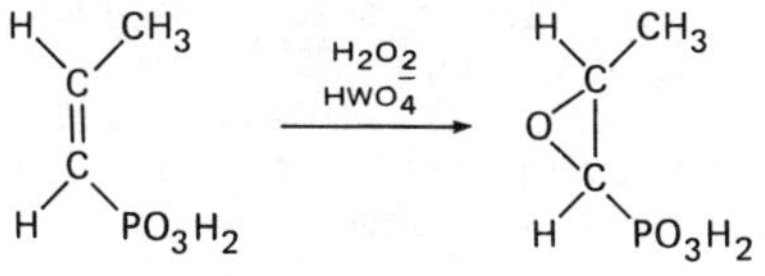

(25)

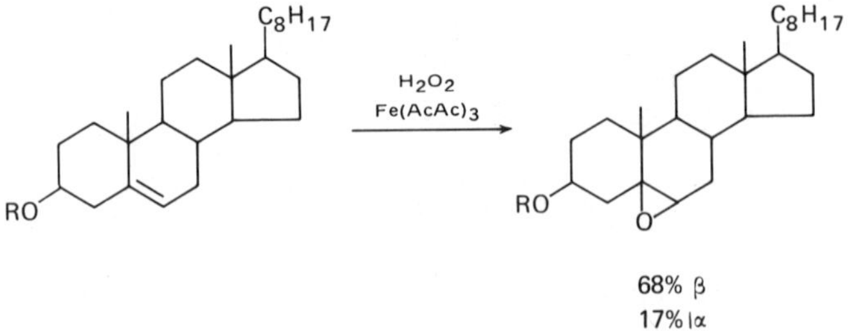

$$(26)$$

68% β
17% |α

Peroxo complexes readily prepared from hydrogen peroxide and MoO_3 can be likewise employed to produce oxiranes[153-156] (equation 27).

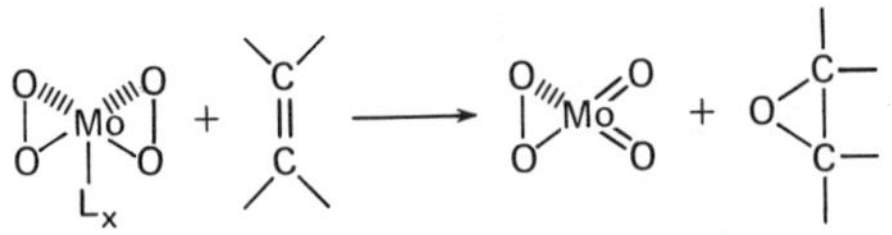

$$(27)$$

Useful conclusions have been reached as regards the mechanism[156-159] and stereochemistry[160] of the epoxidation process.

3. Oxidation with organic hydroperoxides

Epoxidation of olefins with organic hydroperoxides and metal complex catalysts is both a laboratory method and an industrial procedure. Many reviews[10,161-164] and patents[165-170] deal with this topic. The essence of the procedure is given in equation (28). The following organic hydroperoxides are most frequently used for

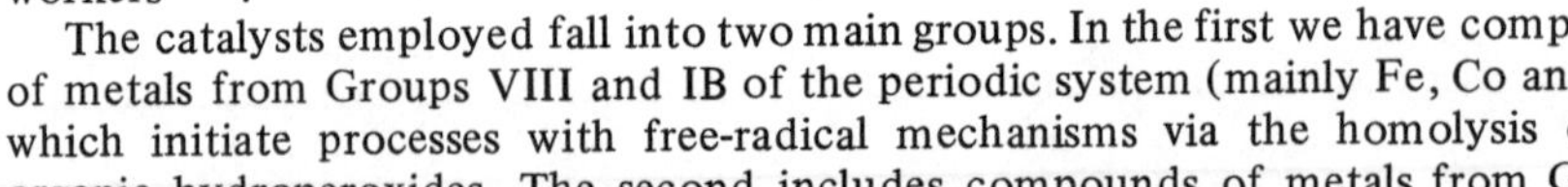

$$(28)$$

epoxidation: t-butyl hydroperoxide[171,172], cumene hydroperoxide[173,174], ethylbenzene hydroperoxide[175,176] and t-amyl hydroperoxide[177]. The effect of the hydroperoxide structure on the epoxidation is discussed by Sheldon and co-workers[178].

The catalysts employed fall into two main groups. In the first we have compounds of metals from Groups VIII and IB of the periodic system (mainly Fe, Co and Cu), which initiate processes with free-radical mechanisms via the homolysis of the organic hydroperoxides. The second includes compounds of metals from Groups IVB, VB and VIB (mainly Mo, W, V, Cr and Ti), which exert their catalytic effects by means of heterolysis of the O—O bond. The various Mo and V complexes have found the widest application[179,180]. In the liquid-phase homogeneous catalytic procedure, the metal compounds used (acetylacetonates, naphthenates, carbonyls, oxalates, chlorides, nitrates, etc. and complexes containing different ligands) dissolve well under the given experimental conditions. For heterogeneous catalysis, catalysts supported on Al_2O_3 and SiO_2[181-183] and catalysts bound to synthetic resin[184,185] are mainly used. Various boron compounds have been similarly applied as catalysts or catalyst components[186-189].

The increasing demands relating to the epoxidation procedure are demonstrated not only by the patents, but also by the research aimed at improving the economic efficiency of the method[190,191]. Very recent investigations[192,193] indicate that with chiral metal complex catalysts the method may be employed to prepare enantiomers.

With a view to gaining a deeper understanding of the mechanism of the epoxidation process, wide-ranging examinations of the following have been carried out: reaction kinetics[172,174,194-207], isotope tracing[159,208], intermediates[195,209-213], transition complexes[156,157,174,178,194,214,215], various spectra[212,216,217], stereochemistry (see later) and solvent effects[179,218]. These indicate that the epoxidation mechanism involves the steps shown in equation (29).

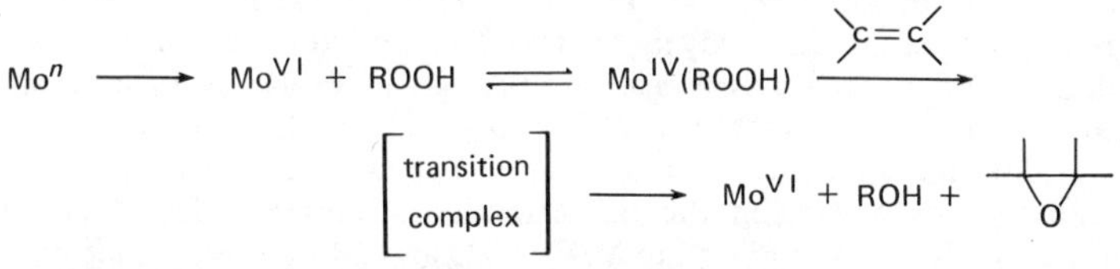

$$Mo^n \longrightarrow Mo^{VI} + ROOH \rightleftharpoons Mo^{IV}(ROOH) \xrightarrow{}$$

$$\begin{bmatrix} \text{transition} \\ \text{complex} \end{bmatrix} \longrightarrow Mo^{VI} + ROH + \overset{\triangle}{\vee}$$

(29)

The mechanism may vary very considerably, depending on the catalyst used, the substrate and the reaction parameters. It is most important to study and understand the coordination of hydroperoxide by the catalyst centre, and the rate-determining oxygen transfer.

Stereochemical examinations have confirmed the stereoselective character of the epoxidation process[164]. From cis-olefins cis-oxiranes are formed, and from trans-olefins trans-oxiranes[174]. The epoxidation of cyclic olefins was also shown to be stereoselective[177]. Besides permitting unambiguous conclusions as to the mechanism of the epoxidation, the stereoselective epoxidation of olefins containing various functional groups is also of great preparative importance[57,219-227] (e.g. equations 30 and 31).

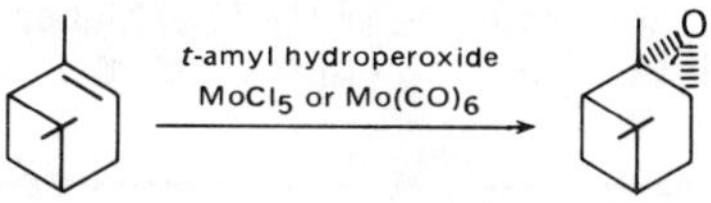

(Ref. 177) (30)

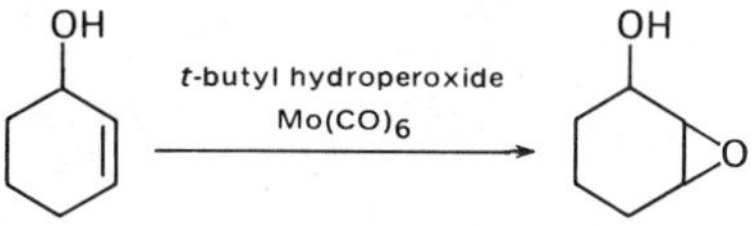

(Ref. 222) (31)

4. Oxidation with oxygen

The literature data relating to the procedures are summarized in some monographs and reviews[10,164,228,229]. Direct olefin epoxidation methods with oxygen can be divided into two main groups: oxidation with oxygen without the application of catalysts, and homogeneous and heterogeneous catalytic epoxidation procedures.

Epoxidation procedures not involving catalysts may be classified on the basis of the step-initiating oxidation. Accordingly, they may be thermal procedures[230-232], photocatalytic procedures[233-237] or radical-catalysed procedures[238,239]. Special mention must be made of the cooxidation procedures[10,236,238,240-242], in which the alkenes are oxidized in the presence of substances prone to radical formation.

If these methods are compared, from the aspect of application, with the methods described previously and those to be discussed below, the following conclusions may be drawn. The selectivity of these direct oxidation procedures is low; only in certain cases does the yield attain 50%[230,243], although an excellent yield has been described by Shimizu and Bartlett[236]. Thus, they are not very satisfactory as laboratory procedures, but may be of industrial importance in the case of simpler olefins.

A very large number of publications have appeared on studies of the mechanisms[232,236,243-248] and stereochemistry[40,235,236,238,245] of the processes. The epoxidation process is a radical chain-reaction. Depending on the reaction conditions, the chain-propagating radical may be the peroxyacyl radical, the alkenylperoxy radical, etc. In some cases the epoxidation is stereoselective[235].

The procedures based on catalysis by metal complexes are results of research in the past decade. Their great advantages are the considerably lower temperature and the improved selectivity, and hence higher oxirane yields may be attained under milder experimental conditions. It is useful to divide into two main groups the complex catalysts employed in the oxidation of olefins[164]. The first (group A) contains the complexes of the Group VIIB, VIII and IB metals (mainly Co, Ni, Mn, Cu, Ir, Rh, Pt and Ru), and the second (group B) those of the Group IVB, VB and VIB metals (mainly Mo, V, W, Cr and Ti). The oxidizing activity of the group A compounds is higher, but at the same time the selectivity is generally low. Reference may be made to some recent experimental data[249-254], while one reaction is given as illustration[249] in equation (32). Certain metal complexes from group B epoxidize

$$
\text{(cyclohexene)} + O_2 \xrightarrow[\text{conv. 20\%}]{\text{RhCl(PPh}_3)_3} \text{(epoxide)} O + \overset{OOH}{\text{()}} + \overset{OH}{\text{()}} + \overset{O}{\text{()}} \tag{32}
$$

$$
 7\% 76\% 2\% 15\%
$$

alkenes with lower activity, but with considerably higher selectivity[255-258] (equation 33). Epoxidation by these methods is the subject of several

$$
CH_3-CH{=}CH_2 + O_2 \xrightarrow[\substack{\text{conv. 8\%} \\ \text{sel. 70\%}}]{\substack{\text{MoO}_2\text{(AcAc)}_2 \\ \text{CH}_2\text{Cl}_2}} CH_3-\overset{O}{\overset{\diagdown\diagup}{CH{-}CH_2}} \tag{33}
$$

patents[259-262]. Work has also been carried out with mixtures of metal complexes from groups A and B[222,249,263,264].

Investigations on the mechanism of epoxidation in the presence of metal complexes have been reported in many papers[257,264-273]. In general, these suggest that the process occurs by a radical chain-reaction, the characters of the key intermediates being fundamentally influenced by the properties of the central metal atom and of the ligands surrounding it, and also by the nature of the substrate.

More recent data on olefins with various heterogeneous catalysts mainly deal with the Ag-catalyst procedure[274-278]. Detailed kinetic studies[279-282] and the stereochemistry of the epoxidation[283] have been reported, as well as the use of new heterogeneous catalysts[261,284-293].

5. Other methods of oxidation

Other methods may be employed, mainly when a very hindered double bond is to be epoxidized, or in the event of special needs. Experimental results described for ozone, chromic acid, permanganate and hypochlorite ion are reviewed by Berti[13].

Oxidation with ozone was found to be stereospecific[13]. Ozone has also been used for the epoxidation of propylene in such a way that intermediates suitable for epoxidation were first prepared from it[294,295].

Chromic acid oxidation may be employed only with tri- and tetra-substituted olefins[296-298]. The mechanism of the process seems to involve a carbonium ion type intermediate[298,299].

Epoxidation of 1 with peracetic acid is not stereoselective, but with Na_2CrO_4, $KMnO_4$ or O_3 high stereoselectivity is observed[52] (equation 34).

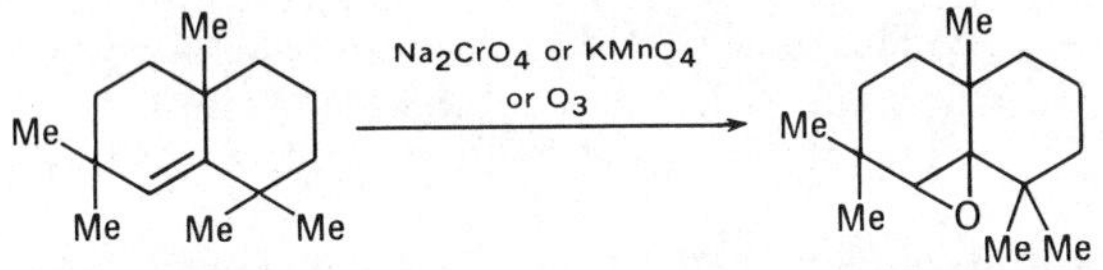

(1)   (34)

Hypochloric acid and its salts can be used primarily for the epoxidation of electron-poor olefins, and very favourably because of the stereospecific nature of the process[48,300]. A cis-oxirane is formed from a cis-olefin. The mechanism of the process may be explained in accordance with equation (35)[300].

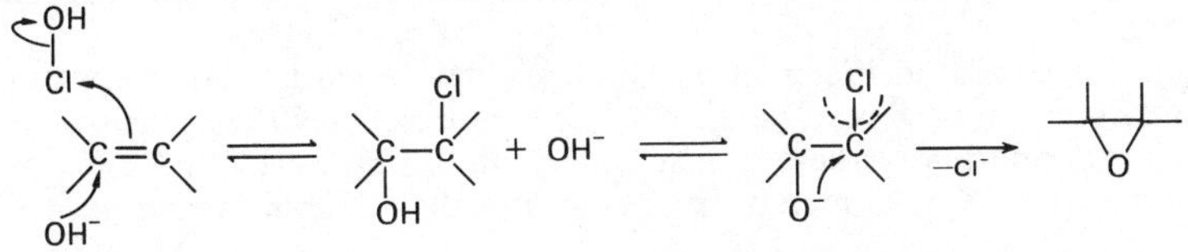

(35)

With this method, 3,4-epoxybutanone-2 can be prepared in very good yield[301], as can phenanthrene 9,10-oxide with a phase-transfer catalyst[302].

Shackelford and coworkers[303] have elaborated a new stereoselective epoxidation method, with an alkaline solution of xenon trioxide. Kruse and coworkers[304] achieved good oxirane yields by applying $NaClO_3$, OsO_4 and $Tl(OAc)_3$ for the epoxidation of C_4 alkenes.

The electrochemical oxidation of olefins has also been used to prepare oxiranes[305].

Five-membered cyclic phosphoranes are transformed almost quantitatively to oxiranes[306] (equation 36).

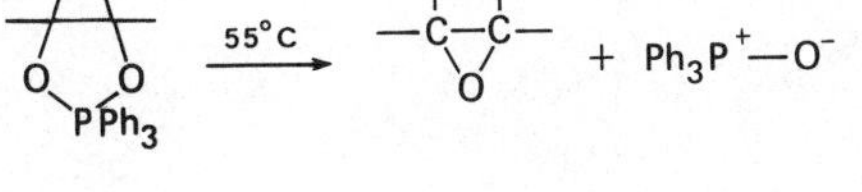

(36)

B. From 1,2-Difunctional Compounds by 1,3-Elimination

2-Substituted alkanols and their esters can be converted to oxiranes by 1,3-elimination via an S_Ni mechanism (equation 37). In the transition state of the

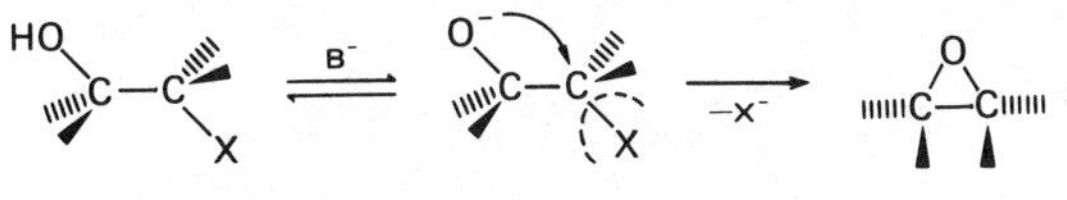

(37)

$X = Cl, Br, I, OSO_2R, OCOR, NR_3^+, N_2^+, OH$

elimination process, the reacting groups are in the antiperiplanar conformation. The oxirane formation is stereospecific. The importance of the individual procedures is very well reflected by the recently published reviews[13,16,17]. Studies on the mechanism[307,308] and stereochemistry of the different reactions have revealed many of their details and the scope of their applicability.

Most papers describe the use of halohydrins, which can be prepared relatively simply and stereospecifically by various procedures: from olefins by the addition of hypohalous acids [usually produced *in situ* e.g. from *t*-butyl hypochlorite[309], *N*-bromosuccinimide (NBS)[35,310,311] or *N*-bromoacetamide (NBA)[36,38,312]], from α-haloketones by reduction[313,314] and from α-halooxo compounds by a Grignard reaction[315]. Epoxycyanides may be obtained from bromoketones by the action of cyanide[316,317]. Iodohydrins may be prepared from olefins in the presence of oxidants[318].

Four chlorohydrin isomers prepared from 3-*t*-butylcyclohexene are transformed stereoselectively to the corresponding *cis*- and *trans*-oxiranes[319]. In conformity with earlier stereochemical studies, variously substituted *trans*(diaxial)-cyclohexane-halohydrins are converted to oxirane derivatives, and the corresponding *cis* compounds to cyclohexanone derivatives in the presence of Ag_2CO_3/celite[320].

The halohydrin route has been used to prepare good yields of α,β-epoxysulphon-amides[321], α-fluorooxiranes[315,322], α-bromooxiranes[314] and optically active oxiranes[323,324].

With NBS, a stereospecific method has been developed for the preparation of vinyloxiranes containing *Z*-configuration double bonds[325]. NBS can also be used in the selective epoxidation of the terminal C=C bond of polyenes[311].

Aromatic oxiranes are mainly prepared by the alkaline reaction of halo-acetates[326-329].

By a modification of the halohydrin method, with the use of tributylethoxytin or tributyl-2-halogenalkoxytin, oxiranes may be prepared in excellent yields[330].

If the iodohydrins can be prepared, high oxirane yields can be achieved[331]. With the modification of the iodohydrin method shown in equation (38), a general

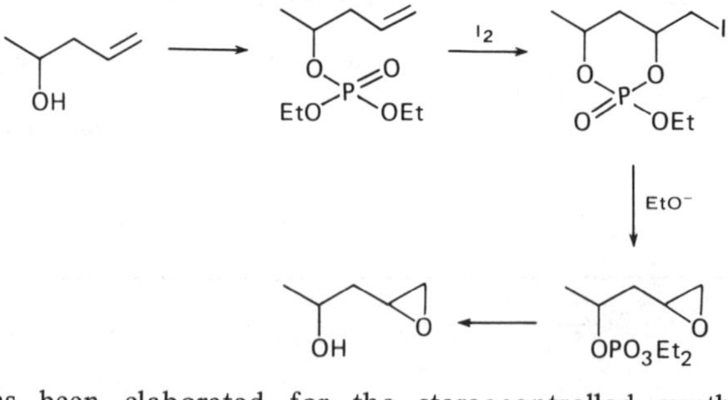

$$(38)$$

procedure has been elaborated for the stereocontrolled synthesis of acyclic oxiranes[332].

In an aprotic solvent, the bicyclo[2.2.1]heptane iodolactone can be converted to an oxirane derivative[333] (equation 39).

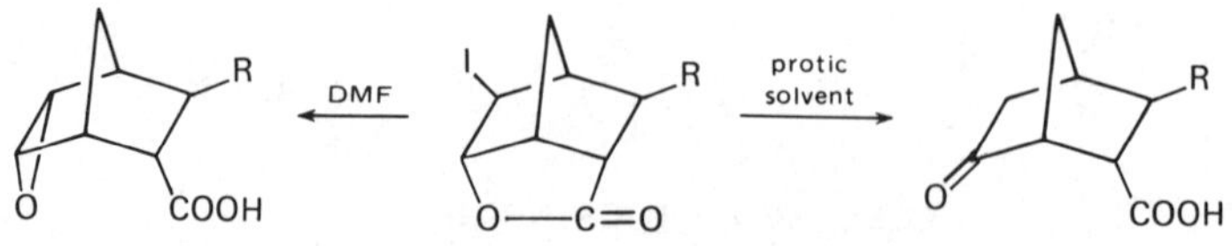

$$(39)$$

A widely used method is to prepare sulphonate esters from 1,2-diols by a generally regioselective reaction, and to transform these to oxiranes under basic conditions. This ring-closure method too is stereoselective[129,334-338] (e.g. equations 40 and 41).

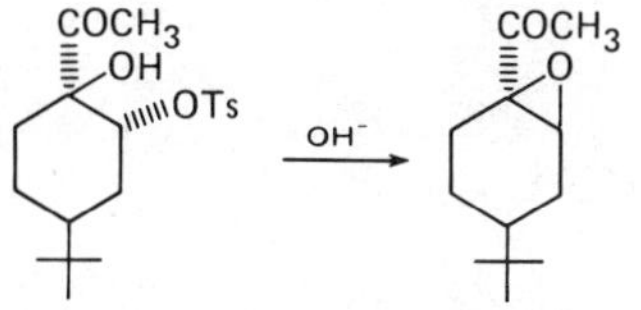

(Ref. 129) (40)

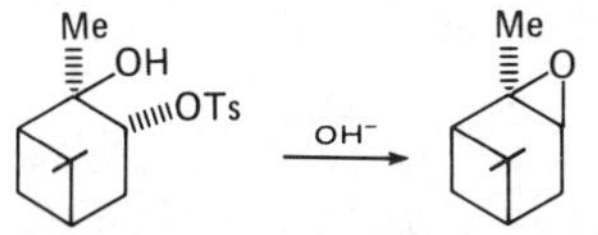

(Ref. 336) (41)

An exception to the *anti* elimination rule was found when the oxirane compound was formed from the *cis*-tosylate[339] (equation 42).

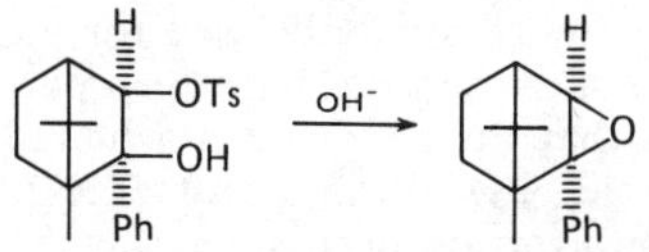

(42)

Cis- and *trans*-2 may be prepared from the corresponding diols (equation 43)[334].

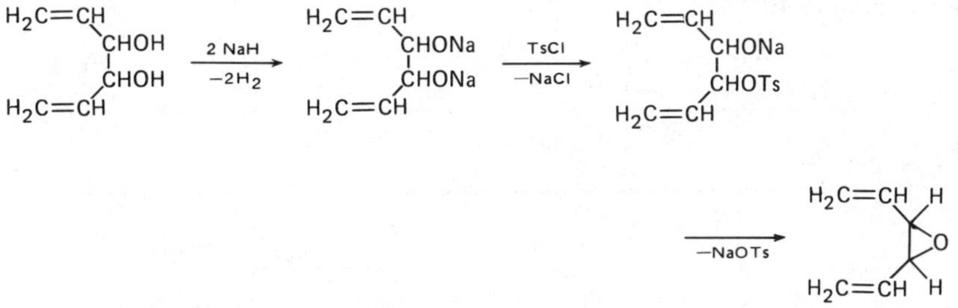

(43)

(2)

Carboxylate anions[340], trimethylammonium ions[341-343] and diazonium ions[331] have also featured as leaving groups for oxirane synthesis.

In the preparation of alkali-sensitive oxiranes, Ag_2O is used for ring-closure of the halohydrins[344-346].

In many cases the 1,3-elimination procedures cannot be replaced by other oxidation methods, due to the sensitivity of the starting substituted olefin[347]. An important application of the halohydrin procedure is for the preparation of oxiranes with configurations opposite to those obtained with the peroxy acid method[35,36, 38,312,348-351] (equations 44–46). The method can be similarly employed for the stereoselective preparation of steroid β-oxiranes[346,352]. A new stereospecific chlorooxirane synthesis has been developed with *t*-butyl hypochlorite as epoxidizing

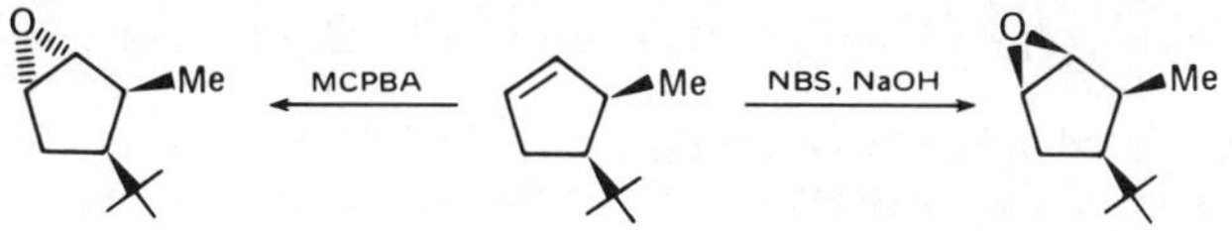

(Ref. 348) (44)

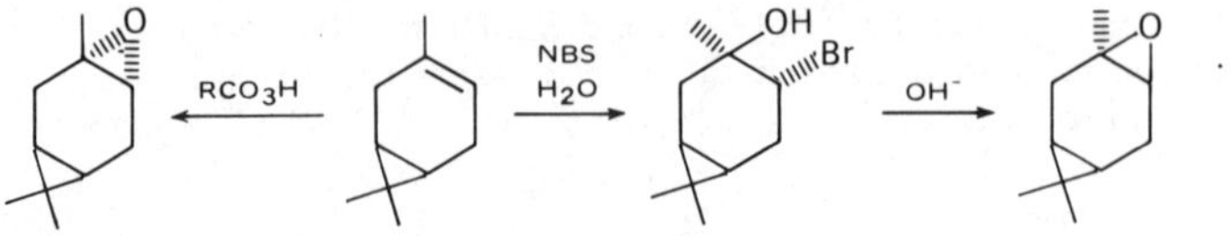

(Ref. 349)   (45)

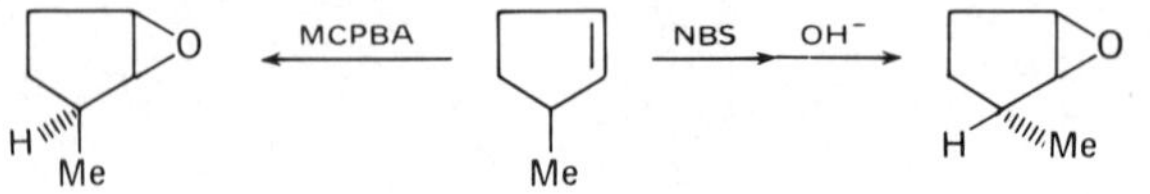

(Ref. 35)   (46)

agent[309]. The reaction proceeds with neighbouring-group participation (equation 47). Steroid chlorooxiranes are formed by a similar reaction mechanism[353].

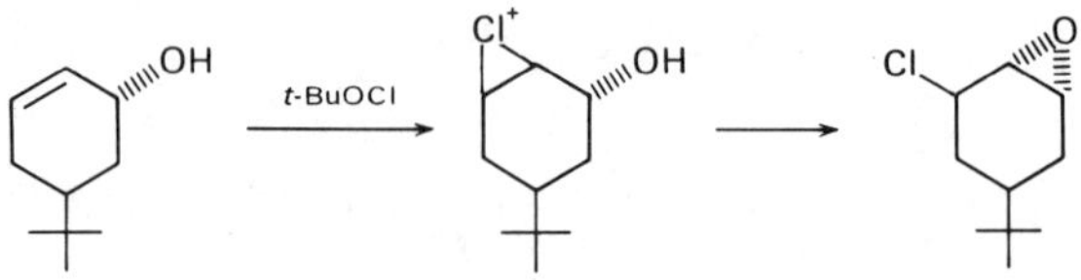

(47)

The 1,3-elimination method can be similarly used for the stereoselective preparation of acyclic oxiranes. Three such methods have been published in recent years; these have the common feature that the synthesis is achieved via cyclic intermediates[332,354,355]. As an example, the synthesis of R,R-2,3-epoxybutane[354] is shown in equation (48). Double inversion occurs, so that the diol and the oxirane have the same configurations. Both oxirane isomers may be prepared from the same diol[355] (equation 49).

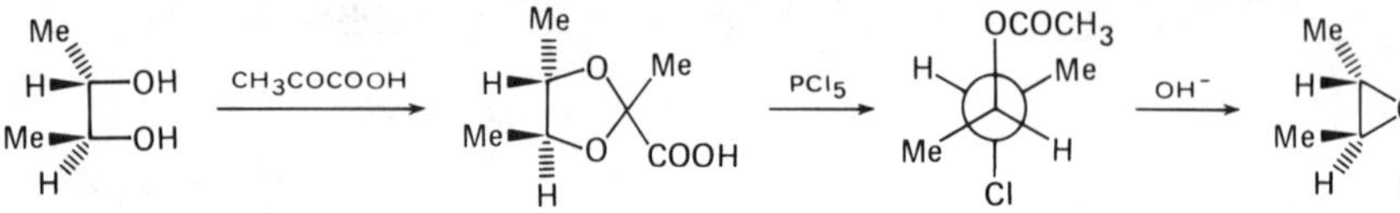

(48)

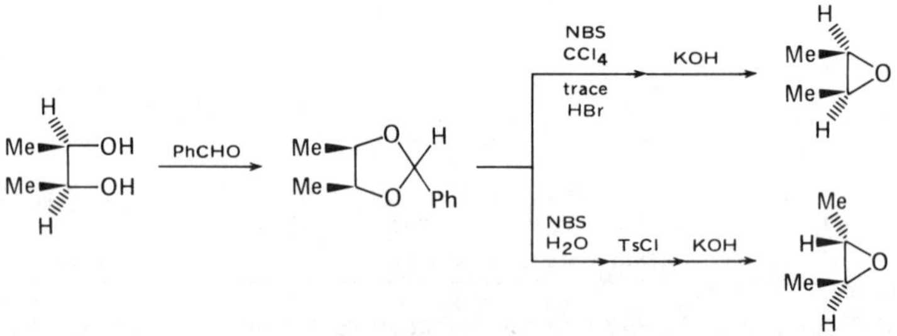

(49)

Another 1,3-elimination is the base-catalysed decomposition of β-hydroxyalkyl-mercurichlorides[356] (equation 50). The reaction is accompanied by the formation of isomeric oxo compounds.

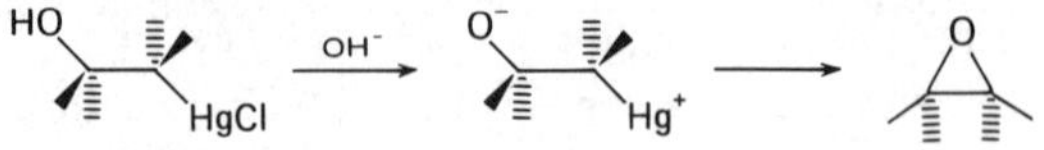

(50)

Oxiranes have been prepared by the thermolysis of 1,2-diol monoesters[357] (equation 51).

Oxiranes may be formed by the dehydration of 1,2-diols. The presence of oxirane as intermediate has been demonstrated in the pinacoline-type rearrangement of

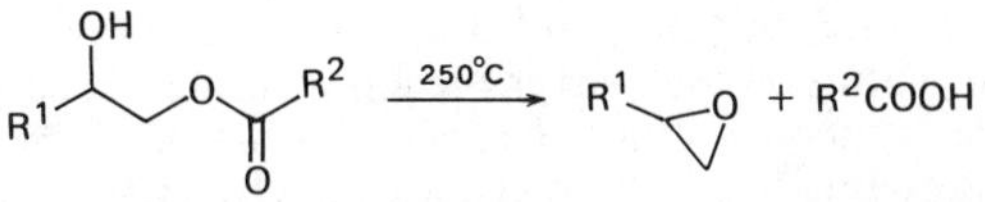

$$R^1 \underset{OH}{\overset{}{\diagdown}} O \underset{O}{\overset{}{\diagup}} R^2 \xrightarrow{250°C} R^1 \triangleleft O + R^2COOH \qquad (51)$$

tetraarylethylene glycol[358]. Formation of the oxirane ring has similarly been proved in the case of diols with a steroid skeleton[359], and on the dehydration of diamantyl glycol in the presence of acids[360].

A one-step synthesis of oxiranes has been achieved in the reaction of diaryl-dialkoxysulphuranes with 1,2-diols[361] (equation 52).

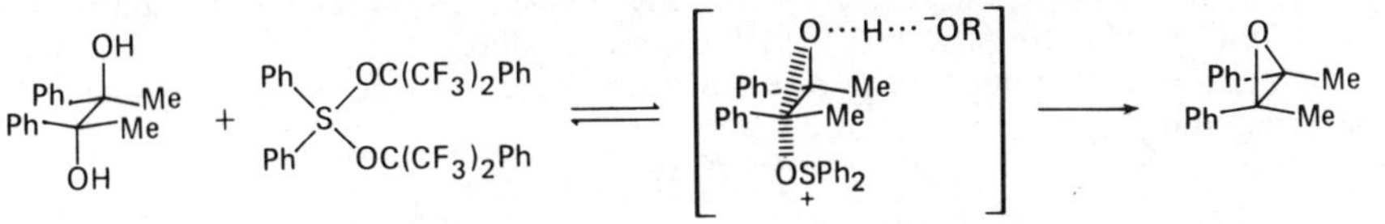

$$\underset{Ph}{\overset{Ph}{\diagdown}}\underset{OH}{\overset{OH}{\diagup}}\underset{Me}{\overset{Me}{\diagup}} + \underset{Ph}{\overset{Ph}{\diagdown}}S\underset{OC(CF_3)_2Ph}{\overset{OC(CF_3)_2Ph}{\diagup}} \rightleftharpoons \left[\begin{array}{c} O\cdots H\cdots^-OR \\ \underset{Ph}{\overset{Ph}{\diagdown}}\underset{OSPh_2}{\overset{Me}{\diagup}}\underset{Me}{\diagup} \\ + \end{array} \right] \longrightarrow \underset{Ph}{\overset{O}{\diagdown}}\underset{Me}{\overset{Me}{\diagup}} \quad (52)$$

Oxiranes may also be prepared with TDAP from *meso*-1,2-diols in the presence of CCl_4[362,363] (equation 53).

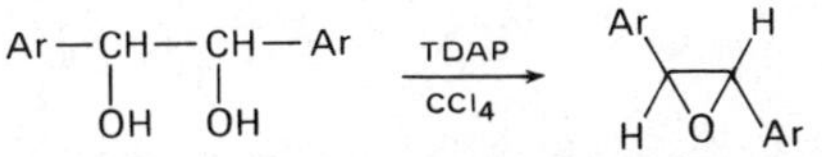

$$\underset{OH}{\overset{}{Ar-CH-}}\underset{OH}{\overset{}{CH-Ar}} \xrightarrow[CCl_4]{TDAP} \underset{H}{\overset{Ar}{\diagdown}}\underset{O}{\overset{}{\triangle}}\underset{Ar}{\overset{H}{\diagup}} \qquad (53)$$

A general method has been developed for the preparation of polycyclic aromatic oxiranes; the final reaction step is the conversion of the corresponding diol to the oxirane by heating with DMF–dimethylacetal[364-366] (equation 54).

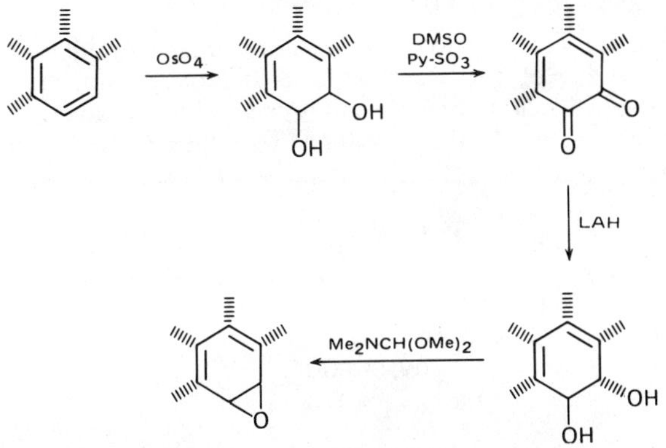

(54)

C. From Carbonyl Compounds

Various nucleophiles react with carbonyl compounds to produce new C–C bonds, and oxiranes are formed. Depending on the nucleophilic reagent, numerous modifications of the procedure outlined in equation (55) have been developed. A number of monographs treat the individual methods from different aspects[1,13,16,17]. Here we shall confine ourselves to a brief survey relating to the procedures, stressing the results of the past few years.

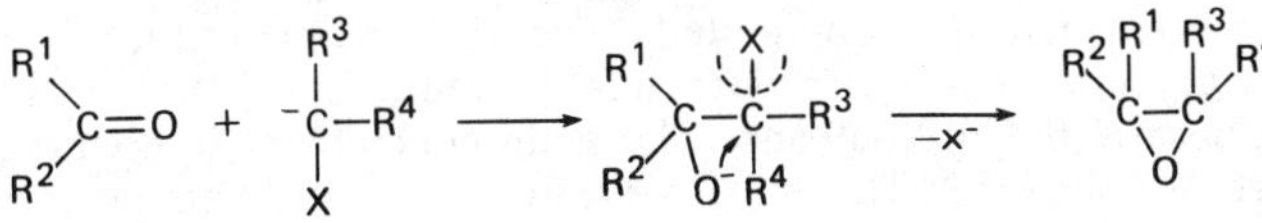

$$\underset{R^2}{\overset{R^1}{\diagdown}}C=O + \underset{X}{\overset{R^3}{\overset{|}{C}}}-R^4 \longrightarrow \underset{R^2}{\overset{R^1}{\diagdown}}\underset{O^- R^4}{\overset{X}{\overset{|}{C}-\overset{|}{C}-R^3}} \xrightarrow{-x^-} \underset{R^2}{\overset{R^1 R^3}{\overset{|}{\underset{O}{C-C}}}}\underset{}{\overset{|}{R^4}} \qquad (55)$$

The most useful method for the preparation of oxiranes containing substituents of an electronegative nature is the Darzens reaction, which proceeds by the above scheme. Besides carbonyl compounds, the following may serve as starting material: α-halocarbonyl compounds[367,368], α-halocarboxylic acid derivatives[369-377], α-halonitriles[378-382], α-halosulphoxides[383,384], α-halosulphones[385,386] and α-halosulphides[387].

The reaction has been studied in detail to establish the effects of various solvents and bases[13]. The phase-transfer catalysis technique has recently been introduced[381,382,386].

Detailed information on studies of the mechanism of the Darzens reaction is to be found in the literature[13]; it is concluded that[372] the formation of the oxiranes can be interpreted as the result of three reaction steps: proton exchange, aldolization and ring-closure (equation 56).

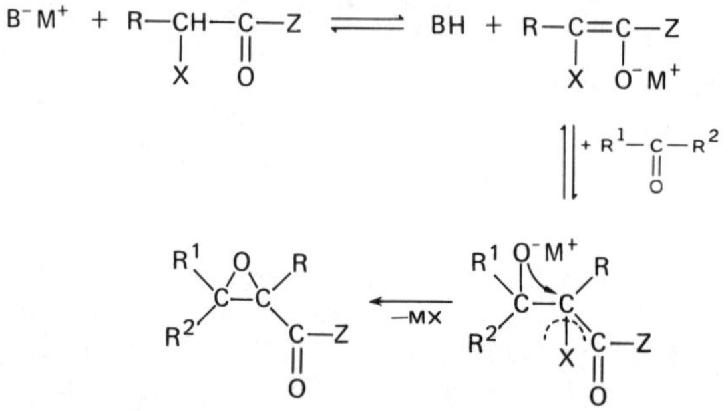

$$ (56) $$

In spite of complex investigations[13,378], a uniform picture has not yet emerged as to the steric course of the reaction. The stereochemistry of the process is influenced by the substituents, the base employed and the solvent.

The Darzens reaction was further developed by White[388] (equation 57).

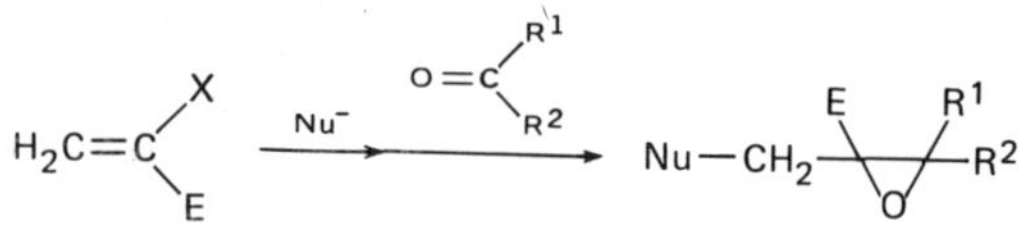

X = CN, Cl
E = CN, Cl, COOEt
Nu = stabilized carbanion

$$ (57) $$

In a manner analogous to the Darzens reaction, 2-methoxyoxiranes and 2-cyano-oxiranes can be prepared from carbonyl compounds with methoxide ion[389] or cyanide ion[316,317] (equation 58).

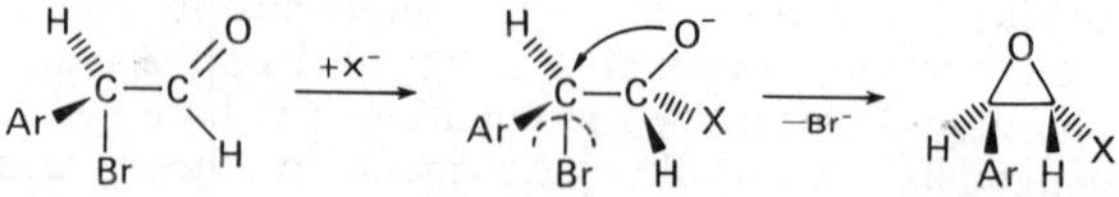

$$ (58) $$

β-Epoxyketones may be prepared in good yield (50—80%) by the dimerization of α-bromoketones in the presence of $Ni(CO)_4$ in DMF[390].

Diazoalkanes with carbonyl compounds give two main products: an oxirane and a carbonyl compound isomeric with this[391] (equation 59). The first step is nucleophilic attack of the diazoalkane. The main conclusions in connection with the reaction are as follows[13]. Of the two parallel reactions, oxirane formation is

$$\begin{array}{c} R^1 \\ \diagdown \\ R^2 \diagup \end{array} C{=}O \ + \ R^3CHN_2 \ \longrightarrow \ \begin{array}{c} R^1 \ \ \ O^- \\ \diagdown \ \ \ \diagup \\ R^2 \diagup \ \underset{R^3}{|} \end{array} C{-}CHN_2^+ \ \xrightarrow{-N_2} \ \begin{array}{c} R^1 \ \ O \\ \diagdown \diagup \diagdown \\ R^2 \diagup \ \ C{-}CHR^3 \\[4pt] R^1COCHR^2R^3 \end{array} \tag{59}$$

generally of subordinate importance, but may predominate with acyclic carbonyl compounds having electron-attracting substituents in the α-position. Equatorial attack of the diazoalkane is favoured in the case of cyclic ketones. In spite of recent new applications[392-395], the procedure is of minor importance for the preparation of oxiranes.

A very good method for the preparation of oxiranes from carbonyl compounds is the Corey synthesis[13,16,17] with sulphonium (3) and oxosulphonium (4) ylides. Recent investigations have led to the proposal of many active methylene transfer reagents, such as 5–9.

$$Me_2\overset{+}{S}{-}\overset{-}{C}H_2 \qquad\qquad Me_2\overset{+}{\underset{\underset{O}{\|}}{S}}{-}\overset{-}{C}H_2$$

$$\qquad (3) \qquad\qquad\qquad (4)$$

$$\begin{array}{c} NR_2^2 \\ \overset{+|}{R^1{-}S}{-}\overset{-}{C}HR^3 \\ (Ar) \ \overset{\|}{O} \end{array} \qquad \begin{array}{c} NTs \\ \overset{\|}{R^1{-}S}{-}\overset{-}{C}R^2R^3 \\ (Ar) \ \overset{\|}{O} \end{array} \qquad Me_2\overset{+}{\underset{\underset{O}{\|}}{S}}{-}\overset{-}{C}H{-}Ar$$

$$(5)^{396-398} \qquad\qquad (6)^{399,\,400} \qquad\qquad (7)^{401}$$

$$Ph_2\overset{+}{S}{-}\overset{-}{\underset{\underset{CH_2}{\diagdown\diagup}}{C}}{-}CH_2 \qquad\qquad Ph_2\overset{+}{S}{-}\overset{-}{C}HR$$

$$(8)^{402} \qquad\qquad (9)^{403}$$

Yields of more than 80% may be attained. The reagents can in general be easily prepared and stored. Because of all these advantages, different variants of the procedure have become widely used[13,402,404-409]. Introduction of the phase-transfer technique means further advantages of application[410,411]. Asymmetric syntheses too may be carried out with optically-active reagents[412-414]. The currently accepted mechanism of the process is shown in equation (60).

$$\begin{array}{c} O \\ \| \\ C \\ R^1 \diagdown R^2 \end{array} + \begin{array}{c} Y^+ \\ | \\ C^- \\ R^3 \diagdown R^4 \end{array} \longrightarrow \begin{array}{c} O^- \ R^4 \\ | \ \ | \\ R^1{-}C{-}C{-}R^3 \\ | \ \ | \\ R^2 \ Y^+ \end{array} \xrightarrow{-Y} \begin{array}{c} O \\ \diagdown \diagup \diagdown \\ C{-}C \\ R^1 | \ \ | R^4 \\ R^2 \ R^3 \end{array} \tag{60}$$

Many authors have dealt with the stereochemistry of the reaction[13,338,396, 399,401,415-417]. The reaction is in general stereospecific; the reagent used has a substantial effect on the stereochemical course. Less bulky reagents (e.g. 3) attack the C=O group from the more sterically hindered side, and the bulkier reagents (e.g. 4, but also the decisive majority of reagents generally) from the less sterically hindered side[13].

Oxiranes can also be prepared from carbonyl compounds with reagents of type $RSCH_2Li$[401,418-421] (equation 61). As in the Corey reaction, the process

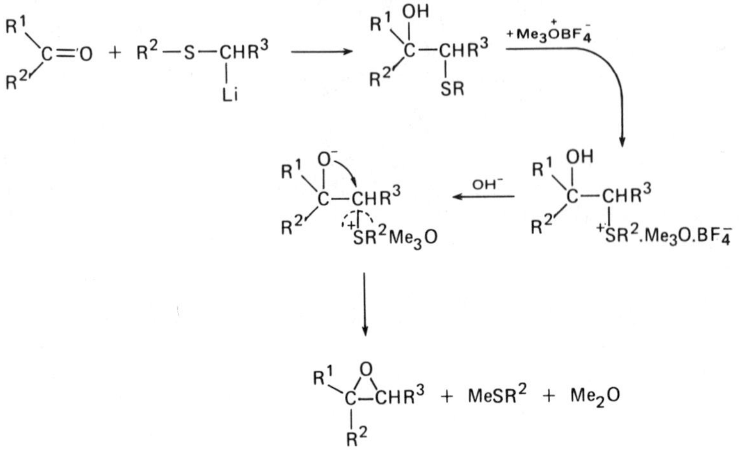

$$R^2{-}S{-}CHR^3 \text{ ... (equation 61)}$$

$$(61)$$

$$\underset{R^2}{\overset{R^1}{C}}{-}CHR^3 \;+\; MeSR^2 \;+\; Me_2O$$

takes place via a betaine intermediate. Yields vary between 50 and 90%[420] (equation 62).

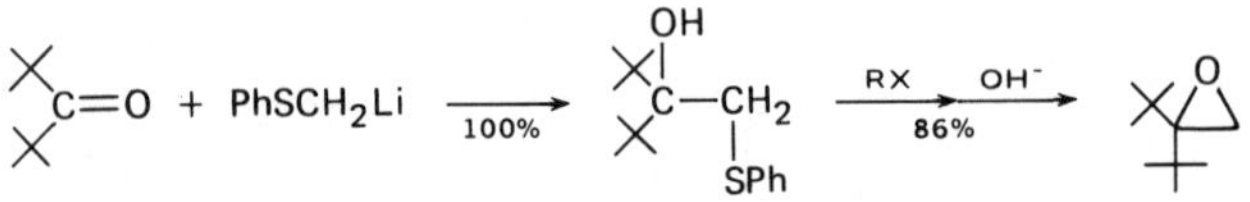

$$(62)$$

A method similar in principle was developed recently[422–427]. The new reagent is the alkylseleno or arylseleno carbanion, comparatively simply prepared from carbonyl compounds (equation 63).

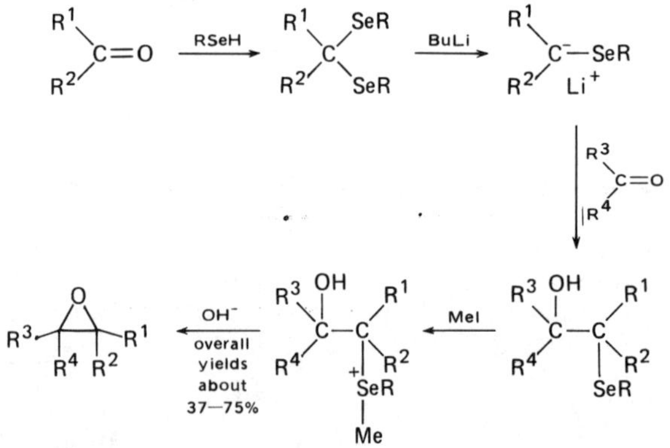

$$(63)$$

— Carbonyl compounds with a geminal bromolithium reagent prepared *in situ* also give oxiranes[428–430] (equation 64). The yield is 60–70%.

Oxiranes are found by the reaction of two moles of an aromatic aldehyde with TDAP[13,431,432].

A new catalytic procedure has been developed for the preparation of α-keto-oxiranes (yield ca. 90%), by the reaction of ketones or keto alcohols with copper(II) methoxides of the type $CuX(OMe)L$ (where $X = Cl^-$, Br^- or ClO_4^-, and L = pyridine, bipyridyl, etc.)[433,434] (equation 65).

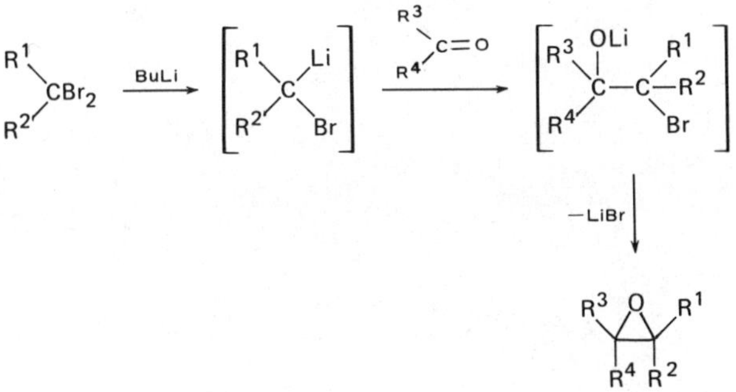

$$\underset{R^2}{\overset{R^1}{>}}CBr_2 \xrightarrow{\text{BuLi}} \left[\underset{R^2}{\overset{R^1}{>}}\underset{Br}{\overset{Li}{C}}\right] \xrightarrow{\underset{R^4}{\overset{R^3}{>}}C=O} \left[\underset{R^4}{\overset{R^3}{>}}\underset{Br}{\overset{OLi}{C-C}}\overset{R^1}{\underset{R^2}{<}}\right]$$

$$\Big\downarrow -\text{LiBr} \qquad\qquad (64)$$

$$\underset{R^4\ R^2}{\overset{R^3\ O\ R^1}{\bigtriangleup}}$$

$$2\,RCOCH_3 + 2\,Cu^{II}(OCH_3)XL \xrightarrow[-2Cu^{II}XL]{-2\,CH_3OH} \underset{Me\ \ O\ \ R}{\bigtriangleup}^{R} \qquad (65)$$

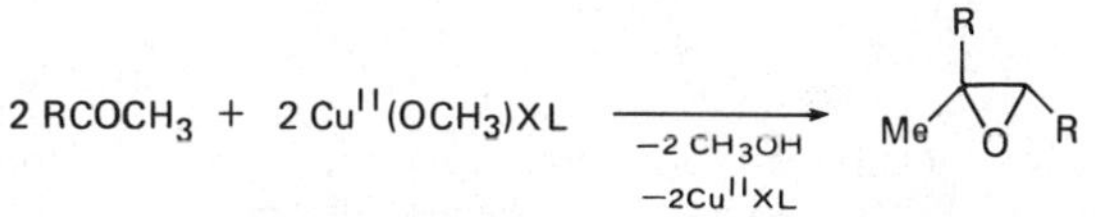

III. REACTIONS OF OXIRANES

A. Deoxygenation

Deoxygenation may be induced with both electrophilic and nucleophilic reagents. The former attack at the oxygen atom of the oxirane, and the latter at the carbon atom linked to the oxygen. The question of which of the two carbon atoms of the oxirane ring is attacked by the reagent is decided by the substituents on them and by the nucleophilic reagent. In certain cases the deoxygenation is stereospecific, so that, depending on the reagents and reaction conditions employed, retention or inversion may occur. On the basis of the results of the past few years[16], this type of reaction has become suitable for the stereospecific preparation of olefins.

1. Deoxygenation with electrophilic reagents

The metals of the first transition series fall into the following sequence as regards their activities in deoxygenation reactions[435]: $V > Cr > Co > Ti > Ni$. The metal atom attacks at the oxygen, and isomeric radicals are formed as intermediates[436]. The metal pair $Zn-Cu$ is also used as a reagent[437,438]. This deoxygenation is not stereoselective, as the rate of rotation about the $C-C$ bond in the intermediate radical is almost the same as the rate of formation of the $C=C$ bond.

With Ti(II) as reagent, prepared from $TiCl_3$ with $LiAlH_4$, the mechanism of the deoxygenation may be outlined as in equation (66)[439].

$$\underset{}{\bigtriangleup}O \xrightarrow{Ti^{II}} \underset{Ti^{III}\ O}{\overset{\cdot}{\bigtriangleup}} \xrightarrow{Ti^{II}} \underset{O}{\overset{\cdot}{\bigtriangleup}}Ti^{II} \longrightarrow \bigtimes + O=Ti^{IV} \qquad (66)$$

$MgBr_2 + Mg/Hg$ may also be used as deoxygenating agents[440]. In deoxygenations with tungsten reagents obtained from WCl_6 with various lithium compounds, stereoselectivity accompanied by retention has been observed in all cases[441]. Metal complexes too may be applied as electrophilic deoxygenating reagents for oxiranes containing electron-attracting substituents[442] (equation 67).

Other electrophilic deoxygenating reagents are cobalt and iron carbonyls[443]. In the case of *cis*- and *trans*-epoxymethyl succinates the deoxygenation is stereoselective, leading to inversion in both cases. In the presence of iron pentacarbonyl tetramethylcarbamide (TMC), oxirane undergoes deoxygenation in accordance with the mechanism shown in equation 68[444]. It can be seen from this scheme that both the central atom and one of the ligands may act as the electrophilic centre of the reagent.

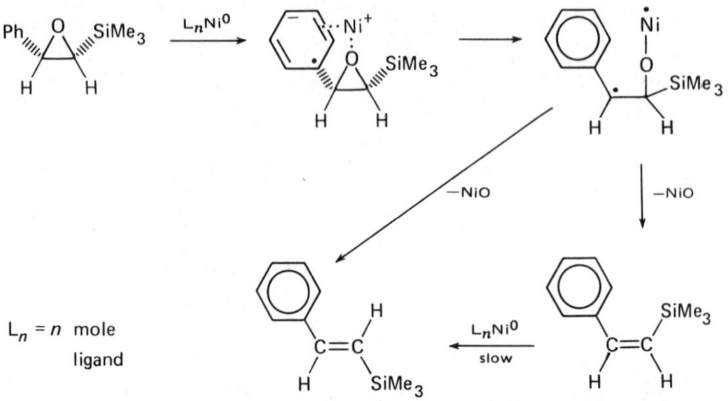

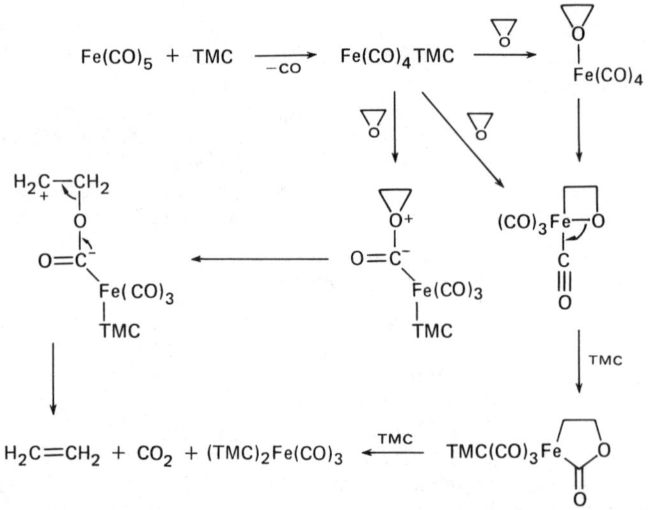

Chemically produced carbon atoms may also be utilized for deoxygenation[445-447] (equation 69). A high degree of stereoselectivity with retention of the configuration has been observed on the deoxygenation of cis- and trans-2,3-dimethyloxiranes with carbon atoms[447].

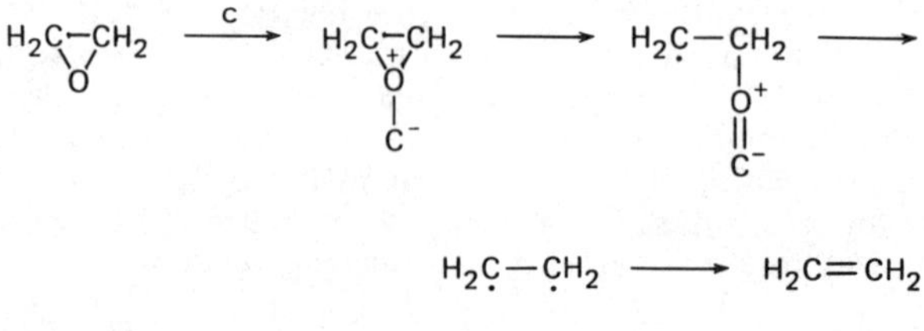

2. Deoxygenation with nucleophilic reagents

One of the most important representatives of this type is the deoxygenation of oxiranes with compounds $R_3P=Y^{448}$ (where Y may be S^{449}, Se^{450} or Te^{451}). In such reactions, first heteroatom exchange occurs, and then the olefins are formed by elimination of the heteroatom of the resulting episulphide, episelenide or epitelluride[448] (equation 70). These deoxygenation methods are stereospecific, with retention of configuration. With sodium O,O-diethyl phosphorotelluroate as reagent, the reaction is explained as in equation (71)[451]. Deoxygenation via heteroatom exchange can also be achieved with KSeCN[452] (equation 72).

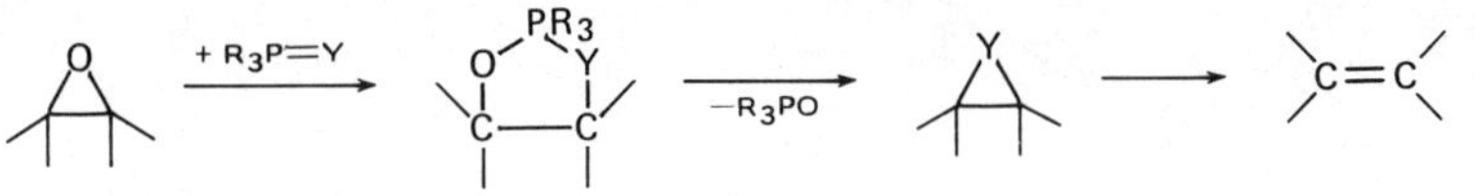

(70)

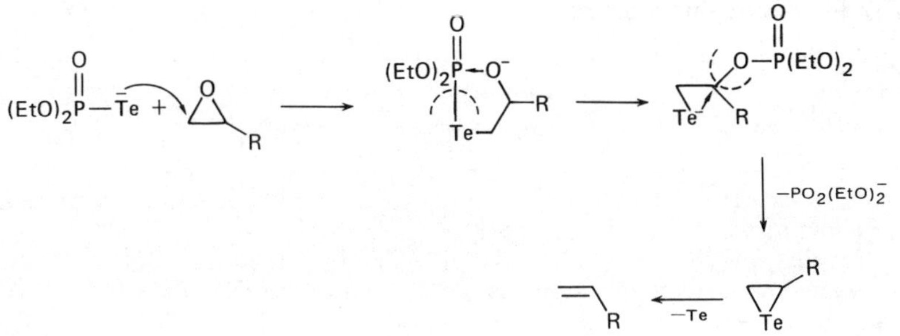

(71)

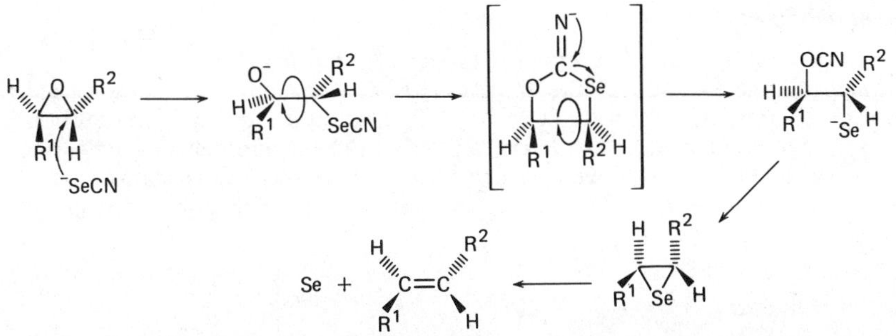

(72)

Ph$_2$PLi too is suitable for deoxygenation[453,454] (equation 73). Since the nature of the method is stereospecific, it is suitable for the isomerization of olefins via oxiranes.

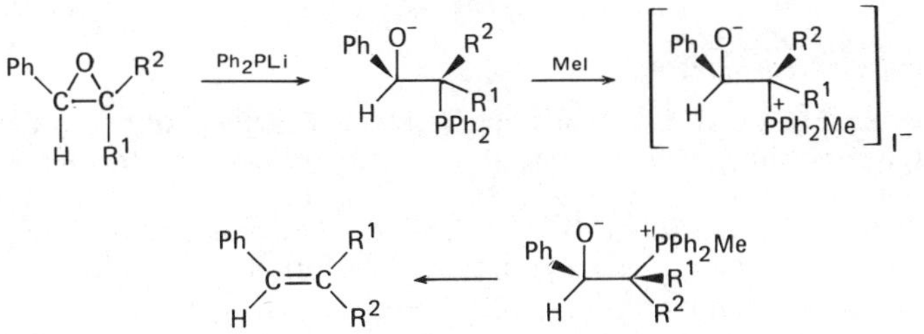

(73)

α,β-Epoxysilanes can be subjected to stereospecific deoxygenation by various methods[455,456]. This procedure is also suitable for the isomerization of olefins, and for the preparation of heteroatom-substituted olefins with epoxysilanes[456]. Inversion occurs if the silyl alcohol formed in the first step is reacted with acid, whereas reaction with base results in retention (equation 74).

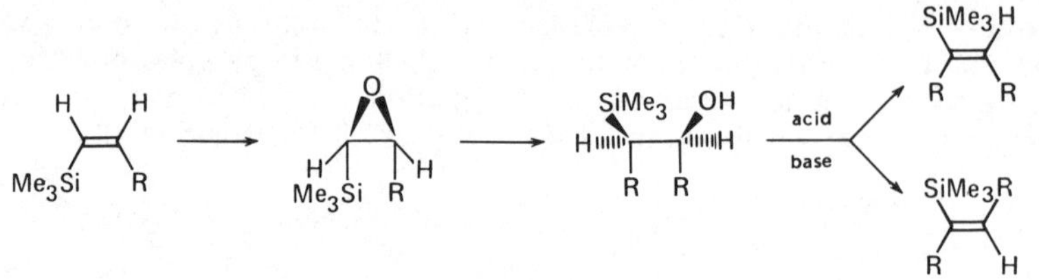

$$(74)$$

Deoxygenation with trimethylsilylpotassium[457] is stereospecific and is accompanied by inversion (equation 75).

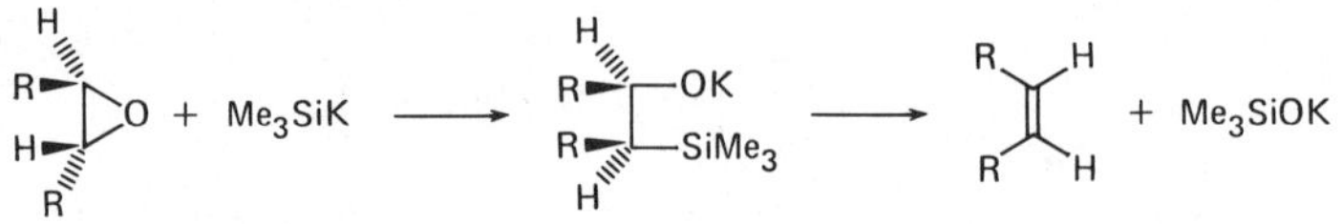

$$(75)$$

If oxiranes are reacted with organolithium compounds, in addition to deoxygenation substituted olefins are formed[458].

The complexes $K_2Fe(CO)_4$, $KHFe(CO)_4$[459] and $C_5H_5Fe(CO)_2Na$[460,461] may serve as nucleophilic deoxygenating reagents. In the latter case the process is accompanied by retention of configuration.

3. Other deoxygenations

Complex oxiranes undergo enzymatic biodeoxygenation[462].

A study has been made of the transformation of cyclohexene oxide on metal complexes of type MY (M = Na, Co, Ni, Cu; Y = ethylenediamine) incorporated into the skeleton of synthetic zeolites[463]. Cyclohexadiene and benzene are formed, as the deoxygenation is followed by dehydrogenation and aromatization. Deoxygenation has also been observed in the catalytic hydrogenolysis of phenyloxiranes[464].

B. Rearrangements

Because of the strained ring, the oxiranes are very reactive compounds, and are capable of many types of rearrangements, discussed in several recent reviews[5,9,12,16,17,465,466]. The main products of the rearrangement of oxiranes are carbonyl compounds and α,β-unsaturated alcohols.

1. Base-catalysed rearrangements

The base-catalysed rearrangements involve either α- or β-elimination. The latter is of great synthetic importance, since it gives allyl alcohol derivatives with good stereo- and regio-selectivity (equation 76). α-Elimination is illustrated in equation (77). The carbenoid intermediate[467,468] is stabilized by transannular C–H insertion. If there is no possibility for this, ketones may be formed. Examples are also to be found of γ,δ and ω-eliminations[12].

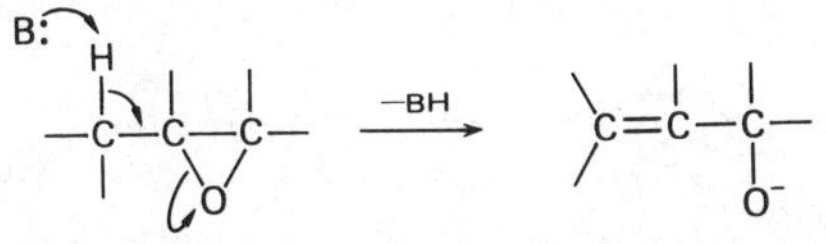

$$\text{(76)}$$

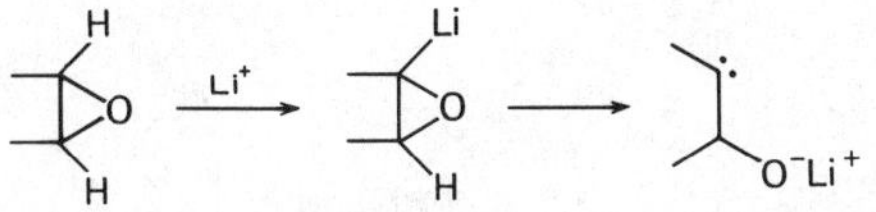

$$\text{(77)}$$

In the case of aliphatic and alicyclic oxiranes, regioselective hydrogen elimination occurs from the least-substituted carbon atom[469,470,470a], with stereoselective formation of the *trans*-olefin[469,471] and in certain instances the occurrence of *cis* elimination[472]. Equation (78) shows a characteristic example of regio- and stereo-selective isomerization[473].

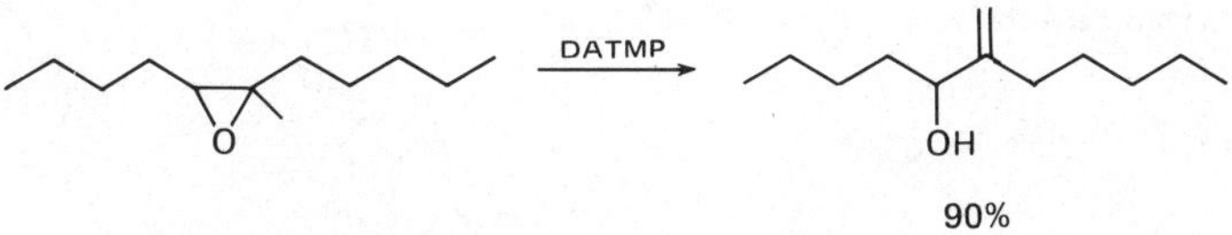

$$\text{(78)}$$

90%

For epoxycyclohexanes the rearrangement to allyl alcohols is maximum with LiNR$_2$ (R = primary alkyl) as reagent; with bulkier bases isomerization occurs to the cyclohexanone[474]. Newer investigations[475] show that at higher temperatures β-elimination and formation of the allyl alcohol is favoured, whereas α-elimination is predominant at lower temperatures. Hence the latter may be suitable for the preparation of bicyclic alcohols. If appropriate reaction conditions are employed, β-elimination can be suppressed[476] (equation 79). Transannular insertion may also

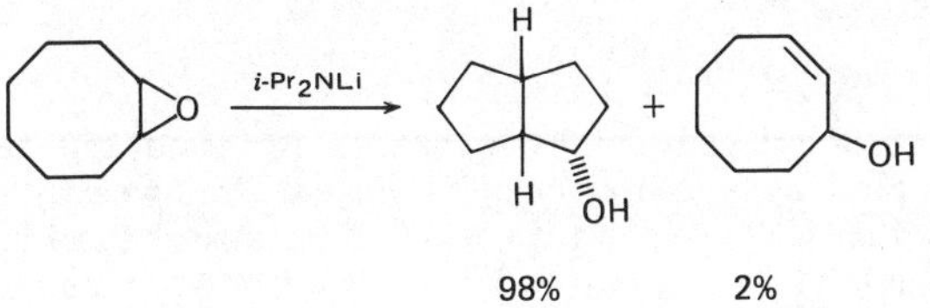

$$\text{(79)}$$

98%   2%

be a convenient preparative tool in the case of compounds that are otherwise difficult to prepare[477] (equation 80). Elimination with ketone formation generally

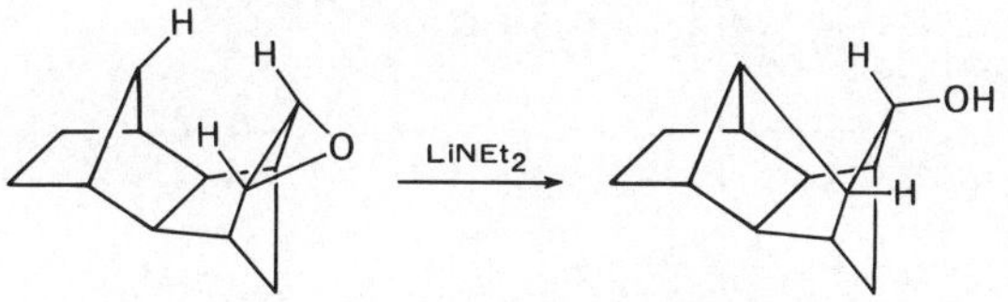

$$\text{(80)}$$

occurs if the β-elimination is excluded and no transannular hydrogen is available[478]. With LiNEt$_2$, γ,δ-unsaturated oxiranes are transformed to cyclopropane derivatives[479] (equation 81). Aryl-substituted oxiranes rearrange to carbonyl compounds on the action of LiNEt$_2$[480] (equation 82). In the case of benzyloxirane, however, very rapid β-elimination takes place[481] (equation 83).

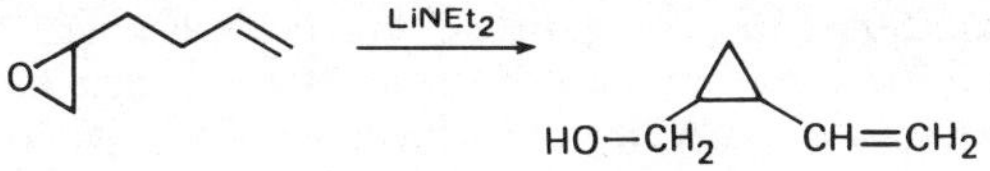

$$\text{(81)}$$

HO—CH$_2$ CH=CH$_2$

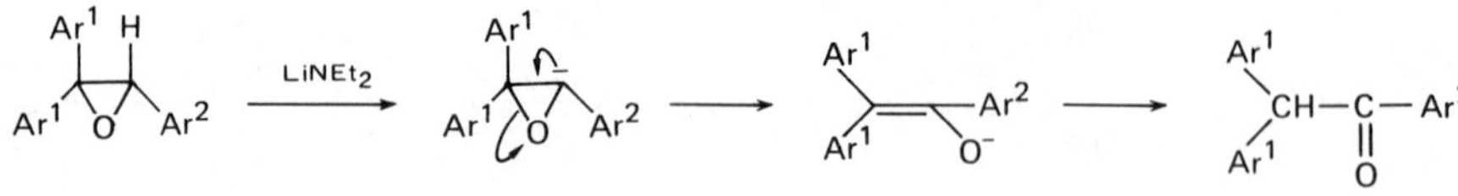

$$(82)$$

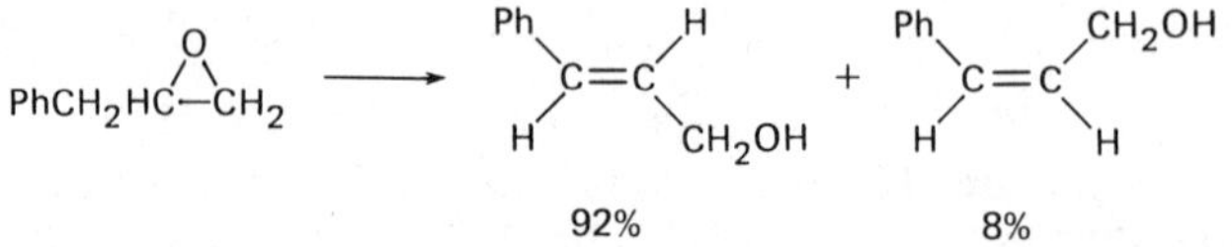

$$(83)$$

92%   8%

Under basic conditions compounds containing a *trans*-hydroxy group in the position α to the oxirane ring tend to be converted to the isomeric α-hydroxyoxirane via intramolecular nucleophilic substitution[9,482,483] (equation 84). The process is known as oxirane migration.

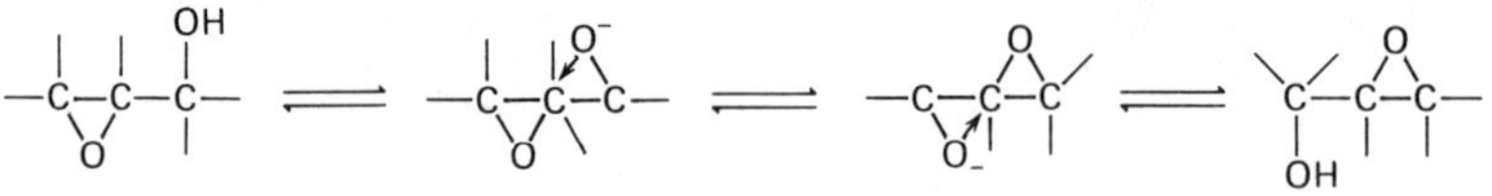

$$(84)$$

The rearrangements of α-epoxyketones have been widely studied[64,484-486]. Compounds in which a methylene or methyne group is bonded to the carbon atom adjacent to the carbonyl group, undergo the Favorskii rearrangement (γ-elimination) under nonpolar conditions, and allyl rearrangement under polar conditions. A different rearrangement yields diketones, which undergo benzylic acid rearrangement.

Rearrangements of other oxirane types, on the action of various basic reagents, have also been studied in detail[82,487-494].

2. Acid-catalysed rearrangements

Oxiranes give carbonyl compounds with both Brönsted and Lewis acids. The initial step is the binding of the electrophilic agent, followed by splitting of the C—O bond; this either leads to the formation of a classical carbonium ion, or the bond-splitting and migration of group R occur in a concerted manner (equation 85).

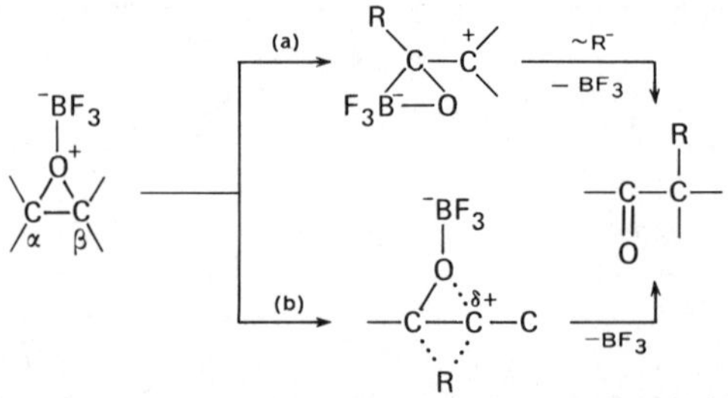

$$(85)$$

The nature and rate of the reaction are influenced by the electrophile and also by the substituents. The stereoselective character of the process is generally not too high. From stereochemical data obtained for oxiranes containing a tertiary carbon atom, the formation of a discrete carbonium ion intermediate has been assumed[495-500] (equation 86). To clarify the mechanism of transformation of oxiranes not containing a tertiary carbon atom, the rearrangements of deuterated derivatives of *n*-hexyloxirane have been investigated[501].

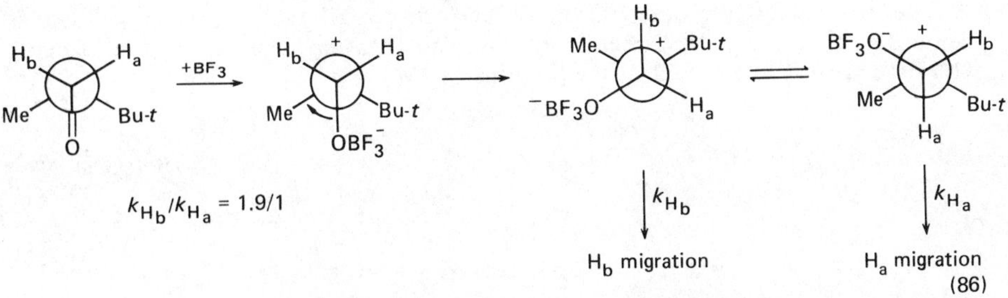

$$k_{H_b}/k_{H_a} = 1.9/1$$

H_b migration   H_a migration
(86)

Many publications have appeared on the isomerizations of alkyl- and aryl-substituted oxiranes also containing various functional groups[500,502-513]. In the rearrangement of oxiranes containing a carbonyl group on the action of Lewis acids, the migration of the functional group may be observed as well[514] (equation 87). In

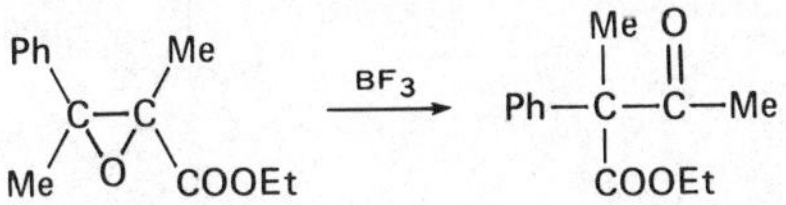

(87)

a study of the Lewis acid-catalysed acyl migration reaction[515], a concerted mechanism was confirmed (equation 88).

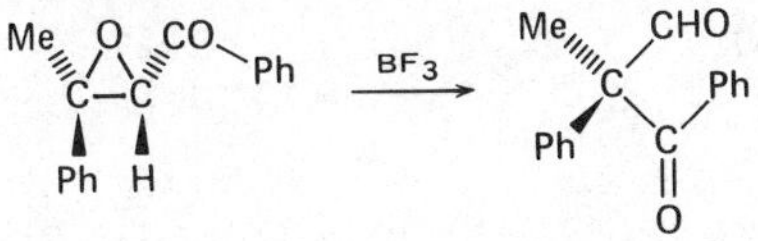

(88)

The isomerizations of the cyclic oxiranes have been examined in detail because of their great variety[345,505,516-524] (e.g. equation 89)[525].

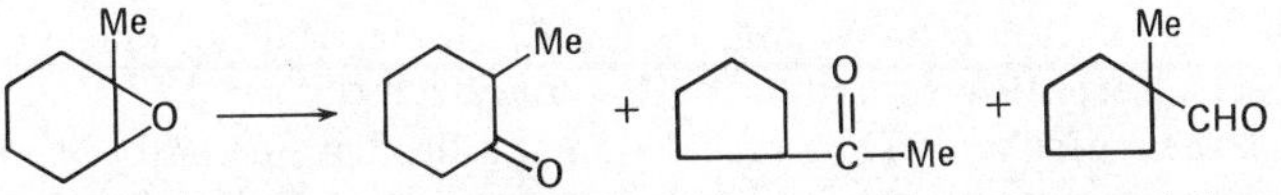

(89)

The individual reaction directions are strongly influenced by the reagent employed, the experimental conditions and by electronic and stereochemical factors[526-532] (equations 90–93).

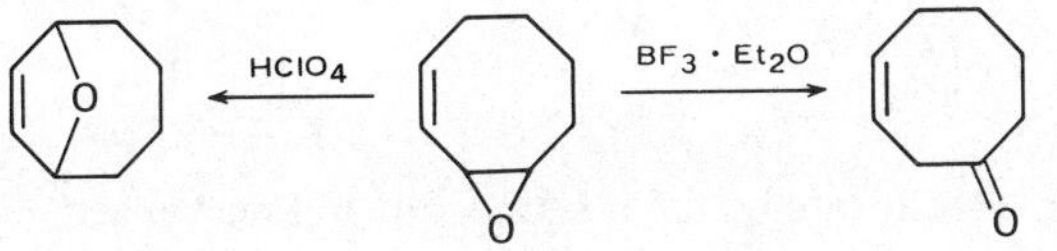

(Ref. 524)   (90)

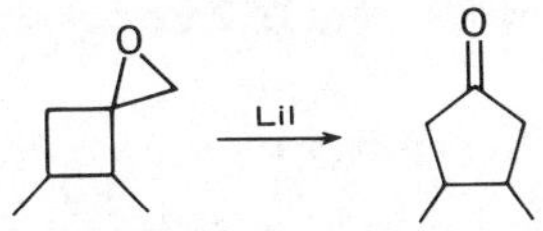

(Ref. 528)   (91)

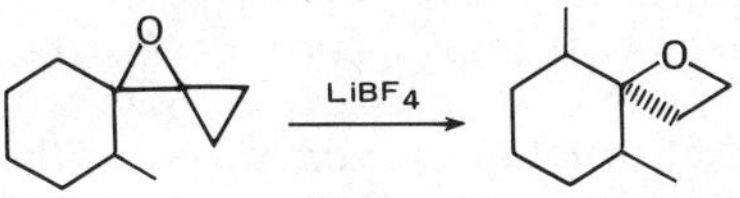

(Ref. 530)   (92)

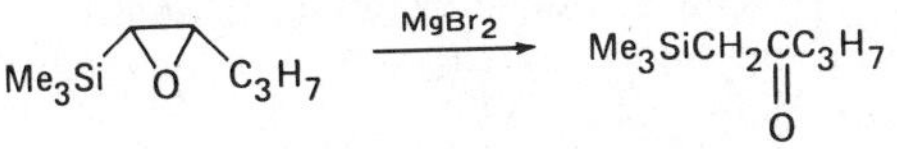

(Ref. 531)   (93)

The acid-catalysed isomerization of cyclopropyloxiranes has been studied in some detail[533-537]. The direction of the isomerization depends on the reactant and the experimental conditions (equations 94—96).

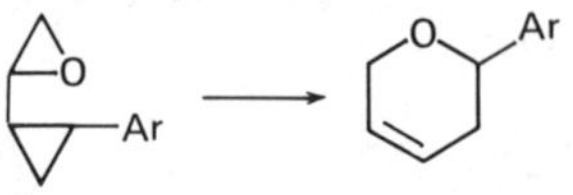

$$ \text{(Ref. 535)} \quad (94) $$

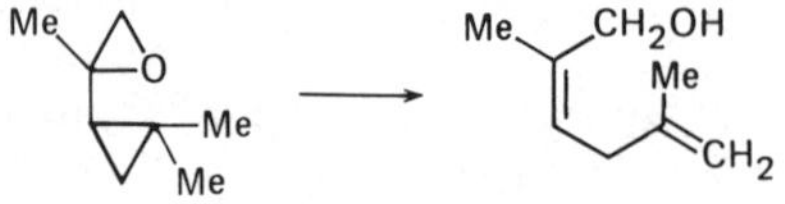

$$ \text{(Ref. 536)} \quad (95) $$

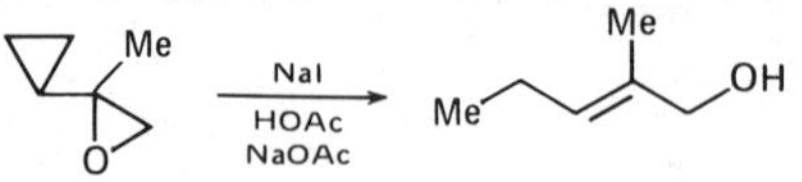

$$ \text{(Ref. 537)} \quad (96) $$

An interesting ring-expansion reaction has been observed for cyclopentanol-oxiranes[538,539] (equation 97).

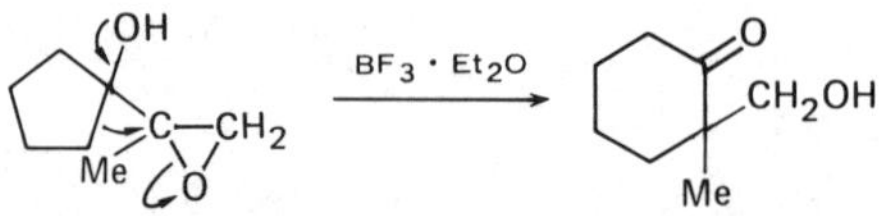

$$ \text{(97)} $$

Detailed studies have also been made of the isomerizations of various steroid oxiranes[517,540-545]. On the action of $BF_3 \cdot Et_2O$ the oxirane ring linked to the steroid skeleton is isomerized to an oxolane[540]. The ring-expansion is attributed to the overcrowding of the oxirane ring. In the BF_3-catalysed rearrangement of 5,6-epoxy steroids, a long-range substituent effect has been observed[544].

Because of their biochemical interest, arene oxides have recently been subjected to very detailed investigation[546]. These compounds isomerize on the action of acids (equation 98). It was proposed[547-549] that the concerted ring-opening and

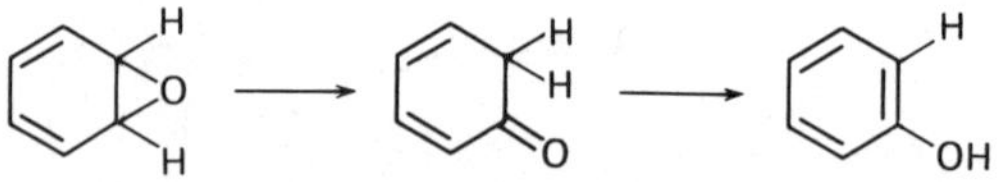

$$ \text{(98)} $$

hydrogen transfer are followed by the dienone—phenol rearrangement. More detailed studies strongly suggest the involvement of a carbonium cation[550].

3. Thermal and photochemical rearrangements

Thermal and photochemical rearrangements of oxiranes involve homolysis of a C–C bond. From a theoretical investigation of the thermal splitting of the C–C bond in the oxirane[551], and on the basis of other studies[552,553], it has been concluded that a biradical structure is more probable than a carbonyl ylide. However, some workers justify the existence of ylide intermediates[554-558]. The formation of the latter was also assumed in the pyrolysis of α-keto-α-cyanooxiranes[559] (equation 99).

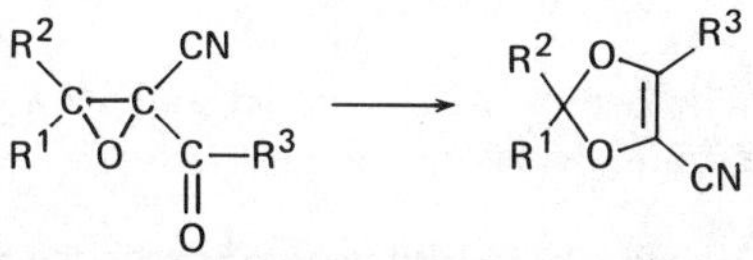

$$(99)$$

Various oxiranes have been studied in detail as regards their thermal and photochemical rearrangements in recent years[44,89,390,534,536,560-564a].

4. Rearrangement on the action of heterogeneous catalysts and metal complexes

Most studies deal with the catalytic activities of various metals, metal oxides, phosphates and zeolites.

The isomerizing activities of the transition metals have been examined on some model compounds[565-570] (e.g. equations 100–102). The formation of carbonyl compounds is a characteristic transformation.

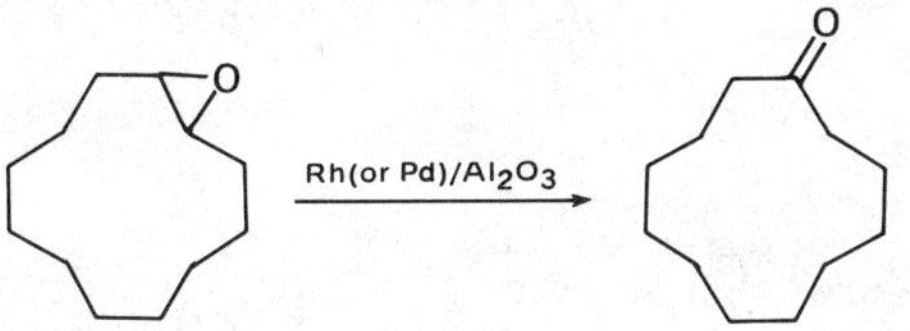

(Ref. 567)   (100)

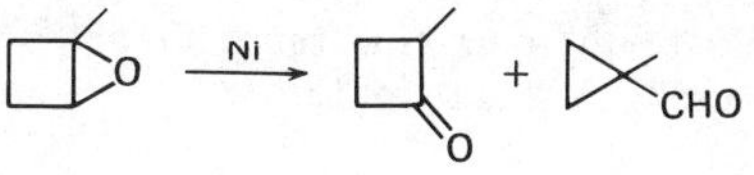

(Ref. 568)   (101)

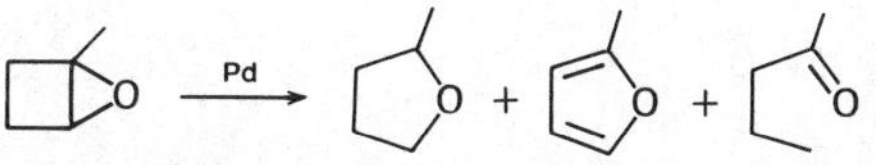

(Ref. 569)   (102)

Wide-ranging examinations have been carried out in an attempt to establish the mechanism of the catalytic reaction[565,568,569,571,572].

On oxide catalysts (Al_2O_3, SiO_2, MgO, TiO_2 and ZnO) oxiranes are isomerized to carbonyl compounds and unsaturated alcohols[573-578,526] (e.g. equation 103).

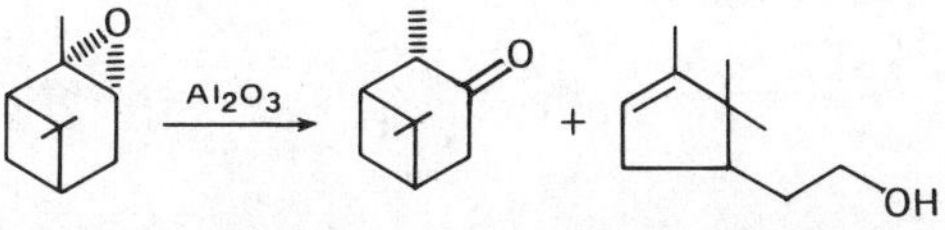

(Ref. 578)   (103)

Investigations relating to the isomerizing effect of phosphates[526,579-583] have extended to the catalyst Li_3PO_4. Using the latter, a general method has been elaborated for the preparation of unsaturated alcohols from oxiranes (equation 104).

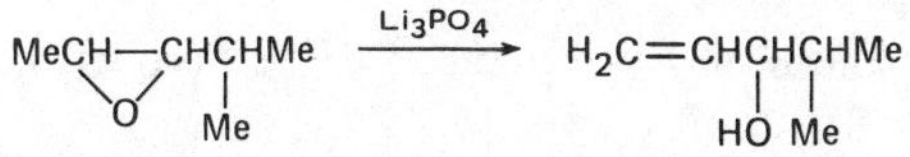

(Ref. 580)   (104)

Modified zeolite types catalyse the isomerization of oxiranes to carbonyl compounds also[567,576,584-586].

Recent studies indicate that certain metal complexes also catalyse the isomerization to carbonyl compounds of oxiranes containing a π-electron system[583-593]. The experimental data obtained so far on the isomerization of aliphatic and alicyclic oxiranes have proved that only pentacyanocobalt complexes are active[594].

5. Other rearrangements

Homoallyl rearrangement occurs with α- and β-pineneoxiranes in the presence of $Et_3N–HF$[595]. Phenyloxirane is isomerized to phenylacetaldehyde on natural graphites[596].

Spirooxiranes containing an amine function undergo isomerization accompanied by ring-expansion[597] (equation 105).

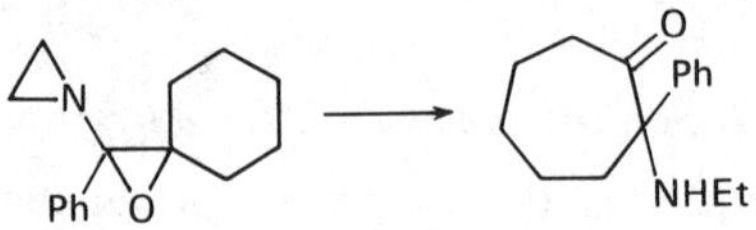

$$(105)$$

The isomerization presented in equation (106) may be used for the synthesis of oxiranes that are otherwise difficult to prepare (e.g. certain steroid oxiranes)[598,599].

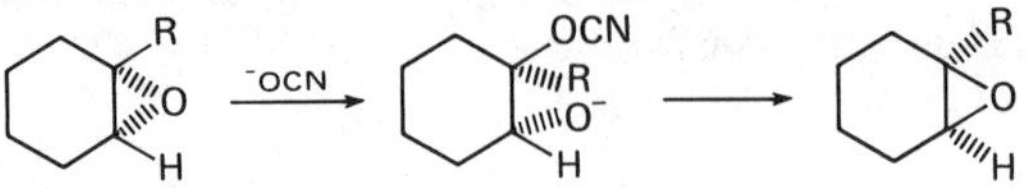

$$(106)$$

C. Oxidation

Oxidations will be emphasized that are also of preparative importance: On the action of HIO_4, oxiranes containing an olefin bond can be transformed in good yield to dialdehydes, the double bond remaining unaffected[600] (equation 107). Phase-transfer agents can also be used for this oxidation[601].

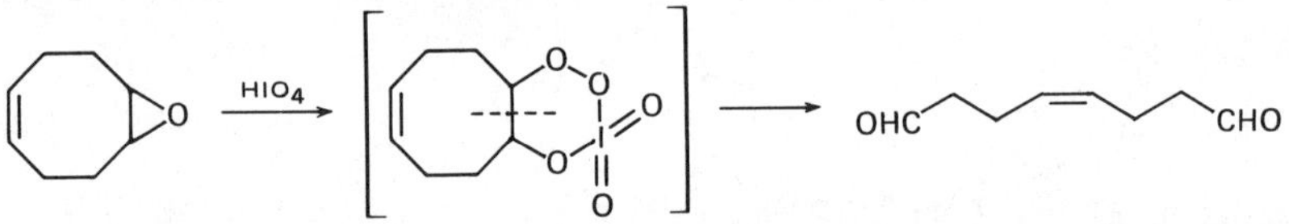

$$(107)$$

Dialdehydes may also be prepared using H_2O_2[602], but oxiranes undergo perhydrolysis also with H_2O_2[603a,b] (equation 108). In the base-catalysed addition of hydroperoxides to oxiranes[604] β-hydroxyperoxides are formed (equation 109).

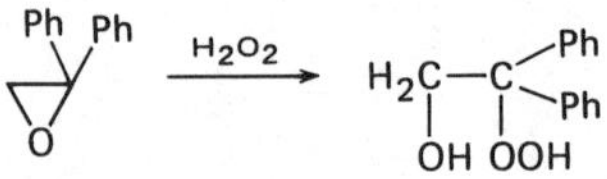

$$(Ref. 603a) \quad (108)$$

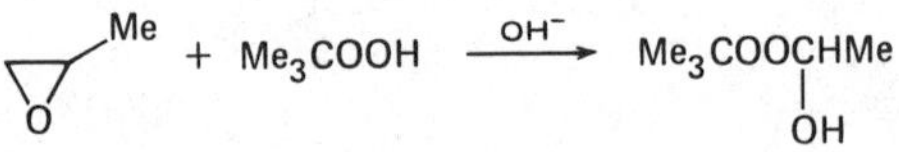

$$(109)$$

On the action of DMSO, α-ketols may be produced[605,606] (equation 110).

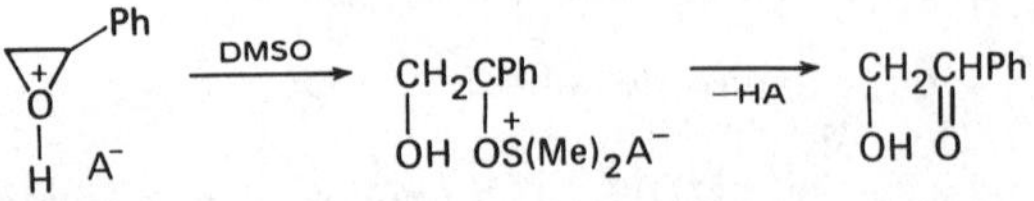

$$(110)$$

Oxiranes containing low numbers of carbon atoms may be oxidized to oxalic acid with HNO_3[607].

D. Reduction

The reduction of oxiranes with various reagents leads to the formation of alcohols. The development in this area is well reflected by the reviews[5,9,17] that have appeared since 1967[2].

1. Reduction with complex metal hydrides

Most of the publications deal with reduction with $LiAlH_4$. Other reagents used are AlH_3, $LiAlH_4$ + $AlCl_3$, $LiBH_4$, $NaBH_4$, $Zn(BH_4)_2$, and their deuterated analogues.

The regioselectivity, stereoselectivity and mechanism of the reaction were studied by Villa and coworkers[505,608-611] who conclude[611] that reduction with a complex metal hydride may proceed either by an intramolecular or an intermolecular mechanism, and that the reduction may also be accompanied by rearrangement (equations 111–113). Whether or not the different individual mechanisms occur is

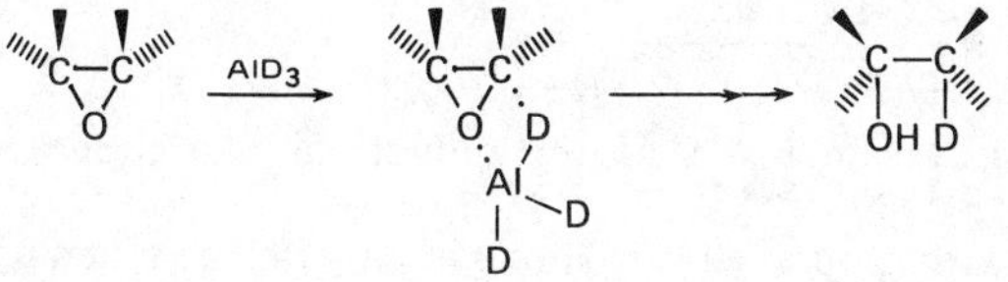

(111)

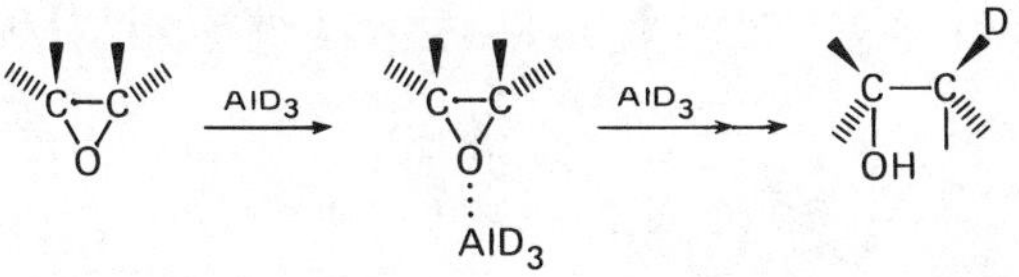

(112)

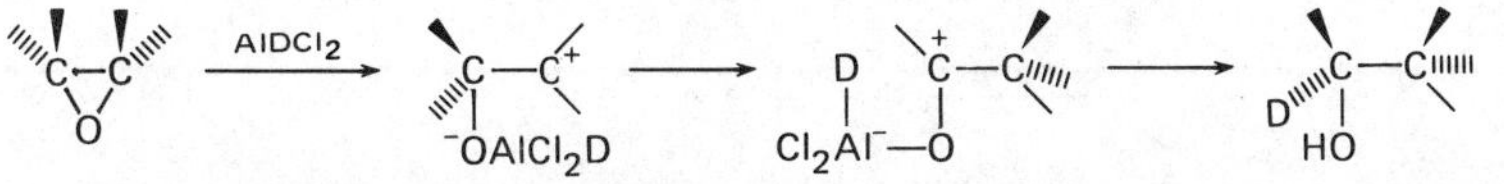

(113)

determined by the steric and electric properties of the oxiranes and by the experimental conditions. Other investigations too[327,612,613] support the following findings. On the reduction of oxiranes with $LiAlH_4$, the H^- ion attacks predominantly on the side opposite to the O; that is, the reduction is accompanied by Walden inversion on the carbon atom which took part in the cleavage. In contrast, the carbon atom not participating in the cleavage retains its original configuration. The extent of the inversion depends on the nature of the transition state. If the lifetime of the carbonium ion formed is relatively long, the product is obtained with retention of configuration.

In the course of the $LiAlH_4$ reduction of oxiranes the H^- ion generally attacks at the least-hindered carbon atom; that is, that carbon atom takes part in the cleavage which has the lowest number of substituents.

Equations (114)–(117) illustrate some of the regio- and stereo-selective reductions of open-chain and alicyclic oxiranes[37,45,51,219,614-616].

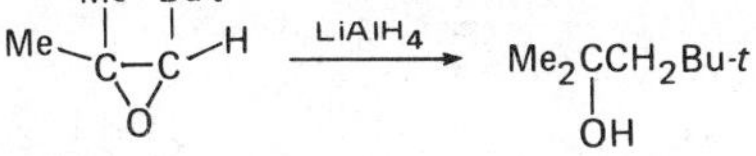

(Ref. 614)   (114)

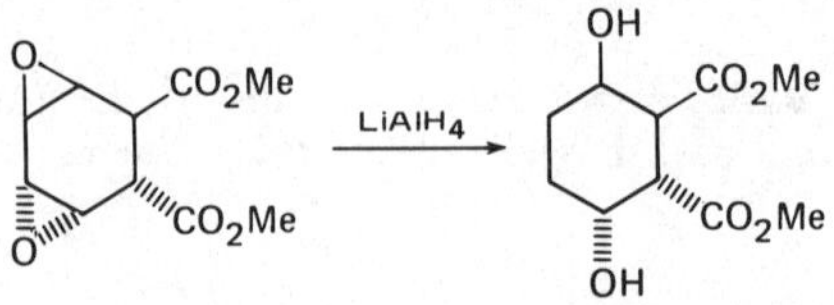

(Ref. 45) (115)

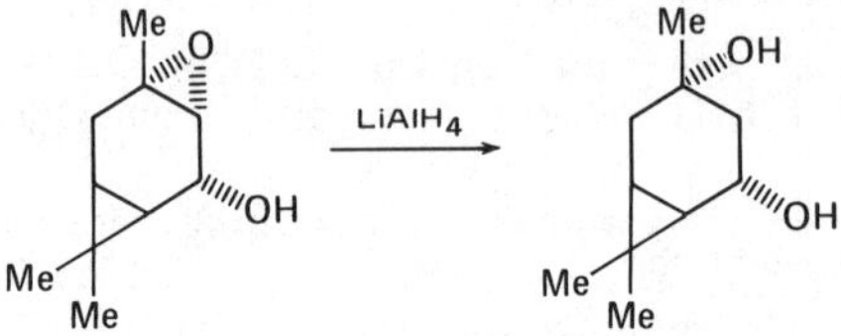

(Ref. 615) (116)

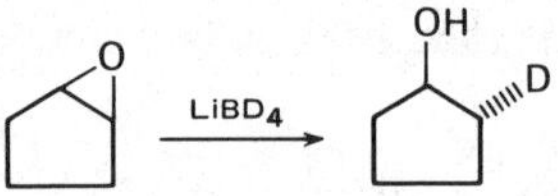

(Ref. 616) (117)

Studies have also been made of the reductions of oxiranes containing other functional groups [87,115,116,470a,617-624].

Oxiranes react with diborane more slowly than with the metal hydrides discussed so far. The oxirane ring is generally opened in the opposite manner to that suggested by the Markownikoff rule[625-627] (e.g. equations 118 and 119). Depending on the

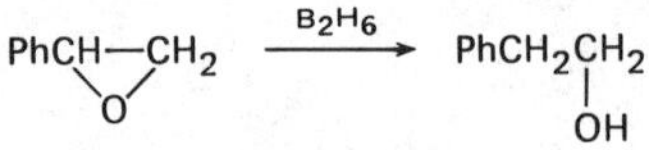

$$PhCH\!-\!CH_2 \xrightarrow{B_2H_6} PhCH_2CH_2\text{—}OH$$

(Ref. 625, 626) (118)

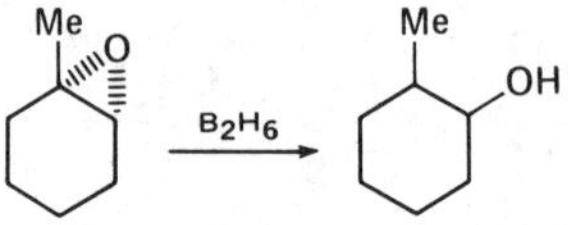

(Ref. 627) (119)

reactant and the experimental conditions, however, the ring-opening may also proceed in accordance with the Markownikoff rule[626,628].

The diborane reduction of α,β-unsaturated oxiranes displays the regioselectivity depicted in equation (120)[629].

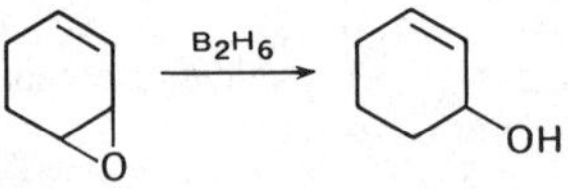

(120)

2. Catalytic hydrogenolysis

Catalytic hydrogenolysis of oxiranes yields alcohols, and many studies deal with the preparation of primary alcohols from olefins, via oxirane intermediates[630-636], and the stereochemistry[222,568,569,637,638] and mechanism[568,569,636] of the hydrogenolysis (equation 121). Among good catalysts are various supported and

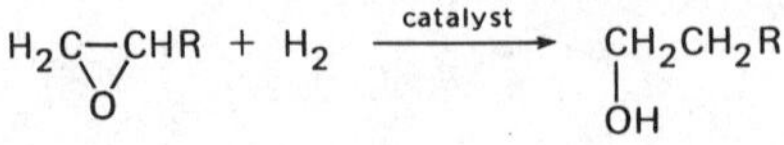

$$H_2C\!-\!CHR + H_2 \xrightarrow{catalyst} CH_2CH_2R\text{—}OH$$

(121)

support-free metal catalysts[222,630-632], metal borates[633], phosphorus-containing metal catalysts[634] and metal-containing zeolites[635]. The configuration of the alcohol formed is strongly influenced by the catalyst, the reactant and the experimental conditions[631,639].

The review by Akhrem and coworkers[5] deals with ring-openings accompanied by retention of configuration. With 1,2-dimethylcyclohexene oxide[638] hydrogenolysis on Raney nickel and Pd(OH)$_2$ results in retention, while on PtO$_2$ it results in inversion (equation 122).

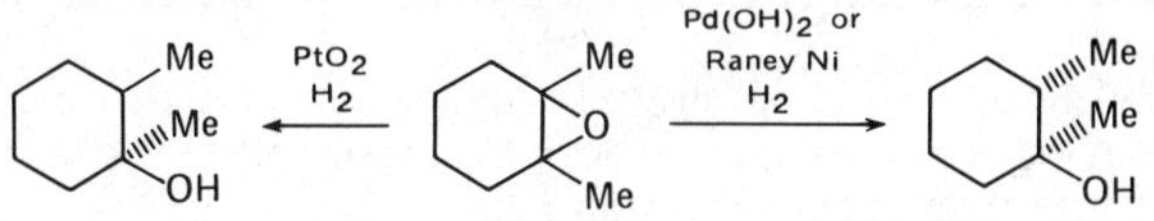

$$\tag{122}$$

Extensive stereo- and regio-selectivities have also been observed in the hydrogenolysis of bicyclic monoterpene oxiranes on a Raney nickel catalyst[637] (e.g. equation 123).

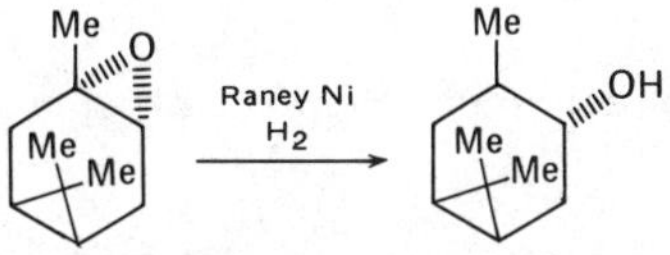

$$\tag{123}$$

Nickel opens the ring on the more sterically hindered, and palladium on the less sterically hindered side[568,569,640]. The selectivities of Raney nickel and Raney copper are likewise not identical[636].

3. Other reductions

Much work has dealt with the application of alkali metals, and mainly lithium, to the reduction of oxiranes to alcohols[39,121,618,641-647]. Liquid ammonia and ethylenediamine are generally used as solvents. These processes (equations 124—127) are usually regio- and stereo-selective.

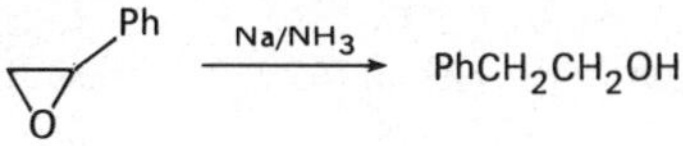

$$\text{(Ref. 641)} \quad (124)$$

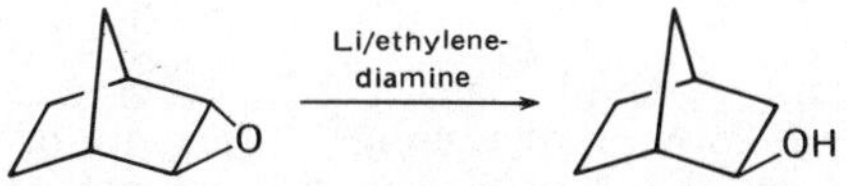

$$\text{(Ref. 646)} \quad (125)$$

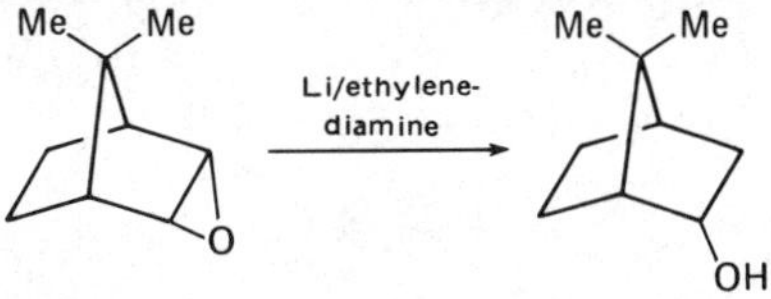

$$\text{(Ref. 647)} \quad (126)$$

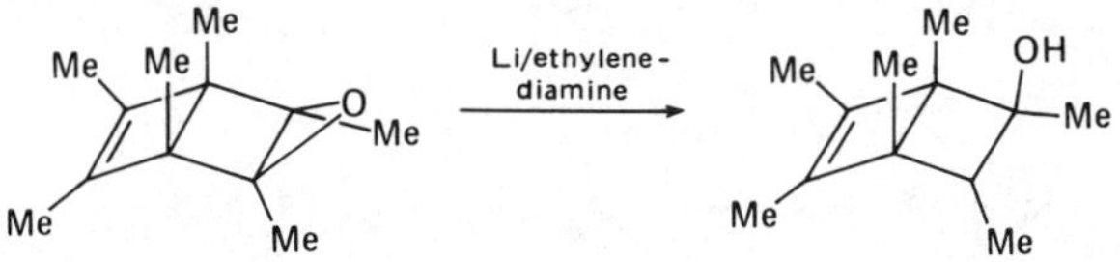

$$\text{(Ref. 39)} \quad (127)$$

The reagents open the oxirane ring on the more sterically hindered side, with retention of configuration. The alkali metal procedures are simple and clean methods for the reduction of sterically hindered oxiranes. A general synthesis has been elaborated for the preparation of 2-ethynylcycloalkanols with this procedure[644].

The regioselectivity is the opposite if the reduction is performed in alcoholic medium, when isopropanol is formed from methyloxirane[643]. (The oxirane ring is similarly cleaved on the less sterically-hindered side in the reduction of steroid oxiranes with Cr^{2+} [648].)

Lithium triethyl borohydride has proved an excellent reagent for the reduction of sterically hindered oxiranes prone to rearrangement[649,650]. The reaction results in 'Markownikoff alcohols' (equation 128).

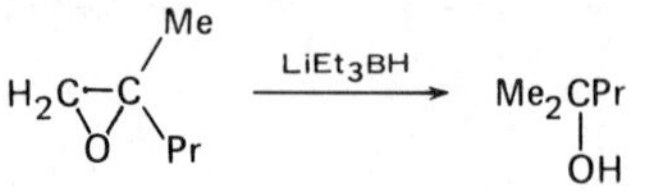

$$H_2C-C\overset{Me}{\underset{O}{|}}Pr \xrightarrow{LiEt_3BH} Me_2CPr \atop \underset{OH}{|} \qquad (128)$$

Aliphatic and aromatic oxiranes are reduced with opposite regioselectivities by 10[651] (equations 129 and 130).

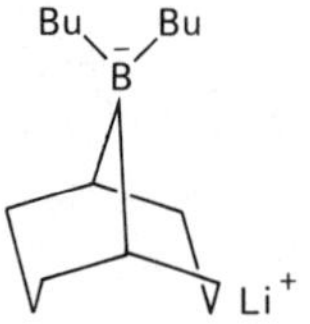

$$PhCH-CH_2 \xrightarrow{\quad 10 \quad} PhCH_2CH_2 \atop \underset{OH}{|} \qquad (129)$$

(10)

$$Me(CH_2)_7CH-CH_2 \xrightarrow{\quad 10 \quad} Me(CH_2)_7CHMe \atop \underset{OH}{|} \qquad (130)$$

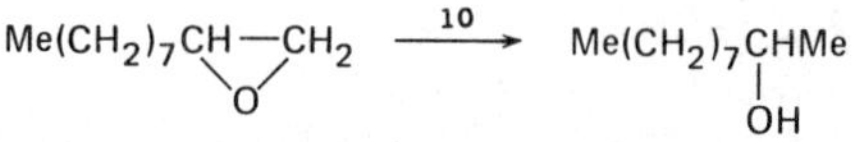

The regioselectivities are opposite in the reductions of α,β-unsaturated oxiranes with i-Bu$_2$AlH and with Ca/NH_3 [642].

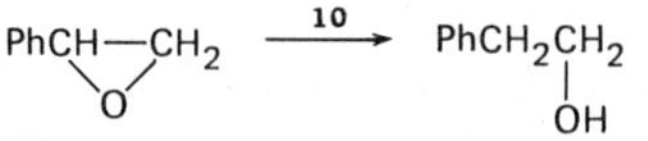

Oxiranes may also be reduced to alcohols with alkoxyaluminium hydrides[620,652] and with aluminium trialkyls[653].

E. Polymerization

Since the monograph by Furukawa and Saegusa[654], the state of development of the various polymerization methods has been well surveyed by a number of reviews up to 1976[655-662]. Hence we shall mention only a few recent characteristic researches[663-668].

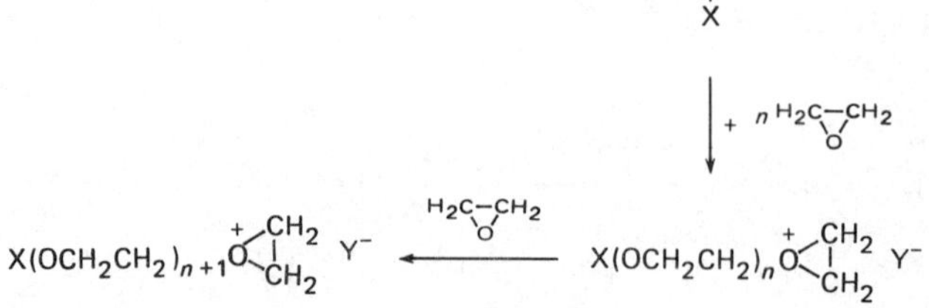

$$H_2C-CH_2 + X^+Y^- \rightleftharpoons H_2C-CH_2 \atop \underset{X}{\overset{+}{O}}\;Y^-$$

$$\downarrow + n\,H_2C-CH_2$$

$$(131)$$

$$X(OCH_2CH_2)_{n+1}O\overset{+}{\underset{CH_2}{\diagdown}}{\diagup}^{CH_2}\,Y^- \longleftarrow X(OCH_2CH_2)_nO\overset{+}{\underset{CH_2}{\diagdown}}{\diagup}^{CH_2}\,Y^-$$

Lewis acid-catalysed cationic polymerization is outlined in equation (131), and the anionic polymerization induced by basic catalysts in equation (132).

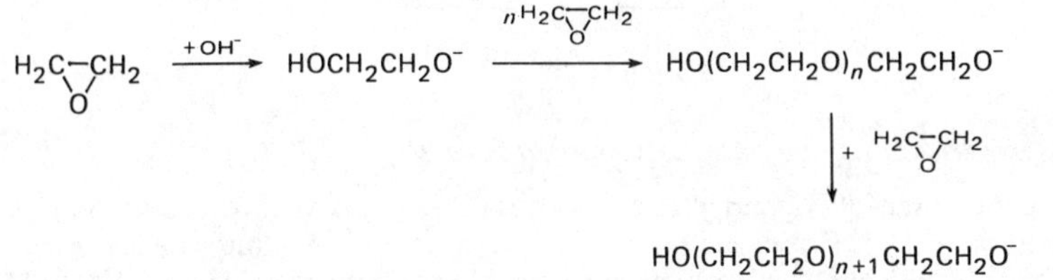

$$(132)$$

Numerous variants exist within the two main groups, and the literature already referred to also deals with radical polymerizations.

F. Formation of Heterocyclic Compounds

Attention is drawn to three reviews connected with this topic[8,17,19].

1. Ring-transformation of three-membered heterocyclic compounds into other three-membered heterocyclic compounds

Most experimental data deal with the transformation of oxiranes to thiiranes. Equation (133) presents an example of the stereospecific reaction[669].

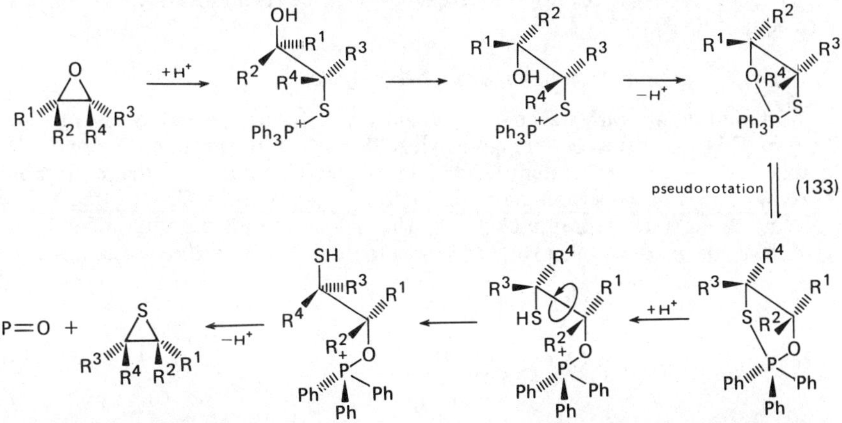

$$(133)$$

Heteroatom exchange occurs with CS_2[670], with 3-methylbenzenethiazole-2-thione in the presence of trifluoroacetic acid[671], and with 1-phenyl-5-mercapto-tetrazole[672]. The yields are high. Oxiranes also react with phosphine selenides in the presence of trifluoroacetic acid[673] (equation 134). The reaction is again stereo-specific.

$$(134)$$

A single-step aziridine synthesis has also been developed[674]; the transformation of oxirane to aziridine occurs by nucleophilic attack of the amidophosphate ester anion on the less-substituted carbon atom, with ring-closure by phosphate elimination (equation 135).

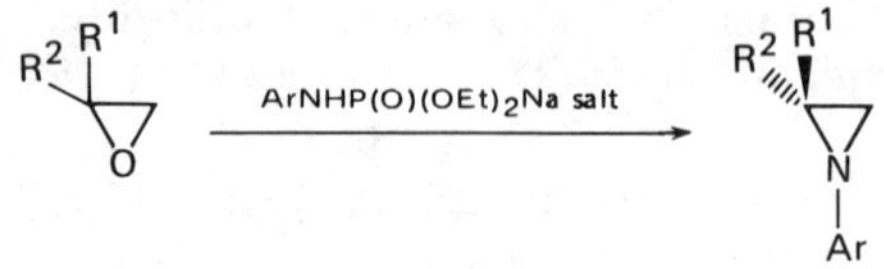

$$\tag{135}$$

2. Ring-expansion to one-heteroatom heterocycles

In the presence of a copper salt, vinyloxirane reacts with diazomethane to give 3-vinyloxetane[675]. Oxocarboxylic acid derivatives[676] and dicarboxylic acid derivatives[677-680] yield γ-lactones with oxiranes (equations 136 and 137).

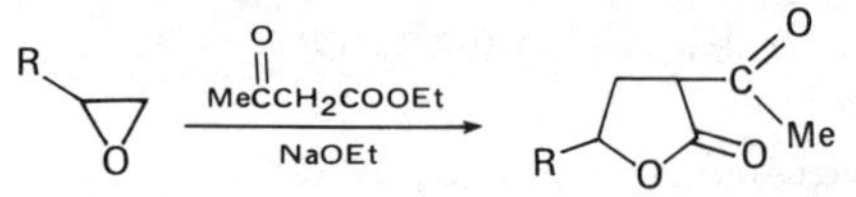

$$\tag{136}$$

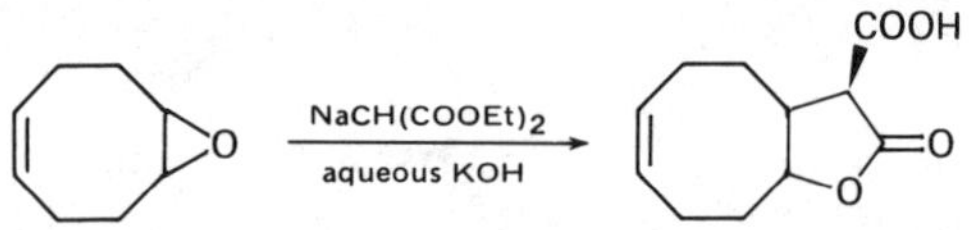

$$\tag{137}$$

On the action of BF_3, certain steroid oxiranes undergo isomerization with ring-expansion to yield oxolanes[540].

By acid catalysis, cyclopropyloxiranes can be isomerized to dihydropyrans (see equation 94).

3. Transformation to two-heteroatom heterocycles

Carbonyl compounds react with oxiranes via acid- or base-catalysed ring-opening to give 1,3-dioxolanes in very good yield[681-689]. For example, (E)- and (Z)-2,3-octene oxides are converted with total stereoselectivity to the corresponding erythro- and threo-acetonide on the action of anhydrous $CuSO_4$, the (Z)-oxide reacting three times more quickly[683]. The (E)- and (Z)-2-methyl-3-phenyloxiranes give the same erythro- (66%) and threo-acetonide (34%) mixture (equation 138).

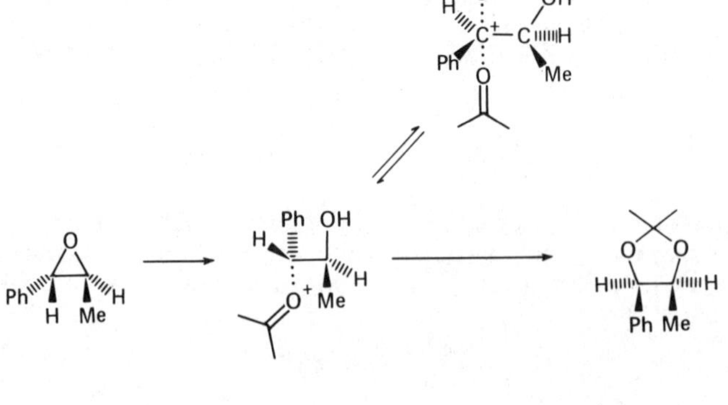

$$\tag{138}$$

In the presence of various catalysts (bases, transition-metal complexes), oxiranes react with CO_2 to form 1,3-dioxolanones[682,688,689] (equation 139).

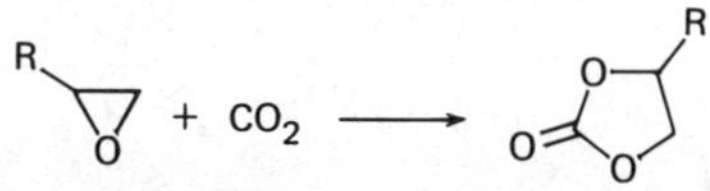

$$(139)$$

Equations (140)–(147) illustrate the preparation from oxiranes of compounds with oxazoline[690-695], oxathiolane[696], oxaphospholane[697] and oxathia-phospholane[698] skeletons.

The transformations presented in equations (140) and (143) are stereospecific. Oxiranes can also be converted in good yield to trithiocarbonates with NaS_2COEt (sodium O-ethyl xanthate)[699], and to oxazolidines with carbodiimide[700] (equations 148 and 149).

Compounds with 1,3-oxazine[701] and 1,4-oxazine[702,703] skeletons can be prepared from oxiranes with various reactants. An example is presented in equation (150). A trioxan ring is formed in equation (151)[704].

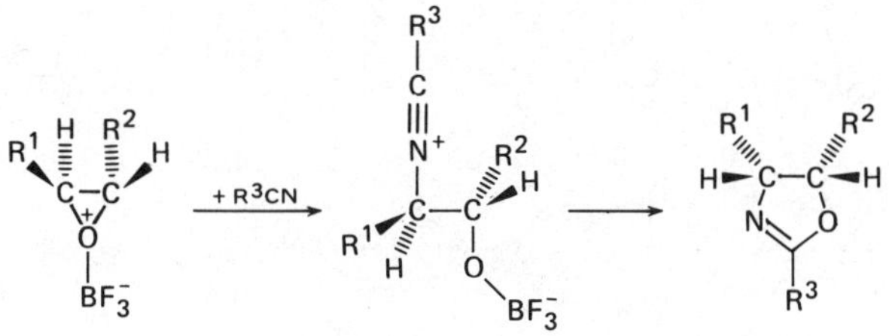

(Ref. 690) (140)

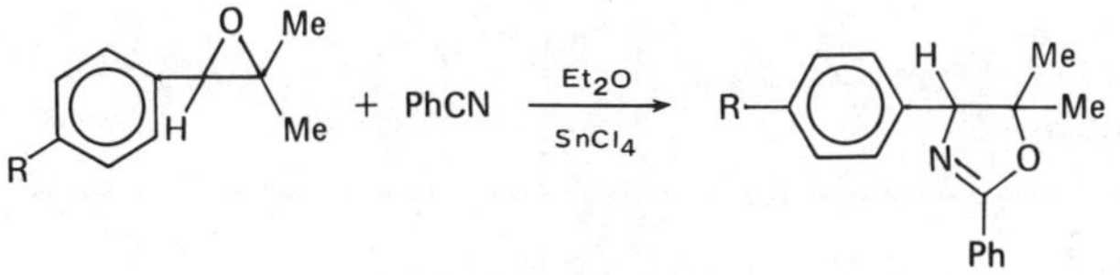

(Ref. 692) (141)

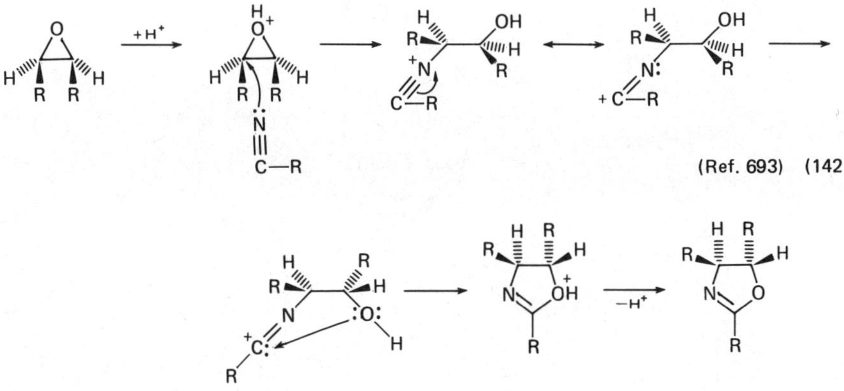

(Ref. 693) (142)

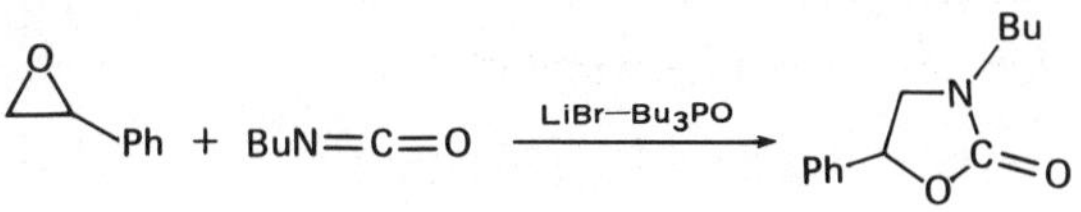

(Ref. 694) (143)

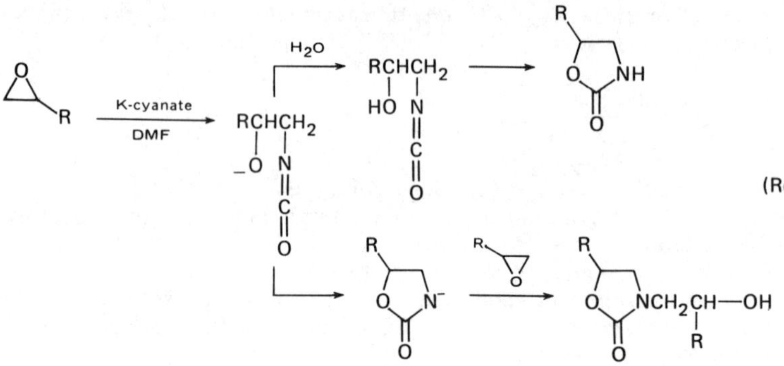

(Ref. 695) (144)

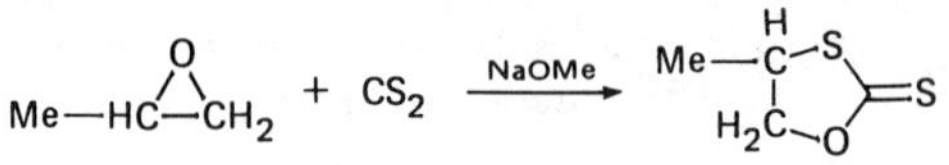

(Ref. 696) (145)

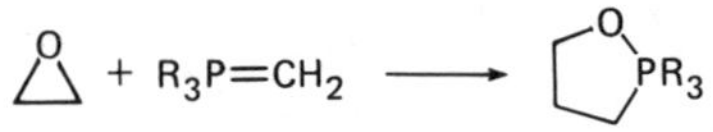

(Ref. 697) (146)

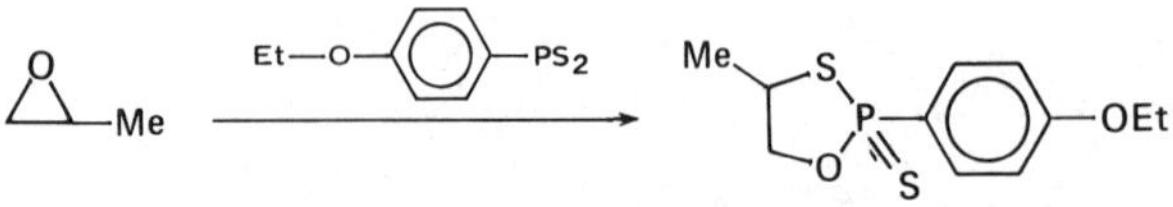

(Ref. 698) (147)

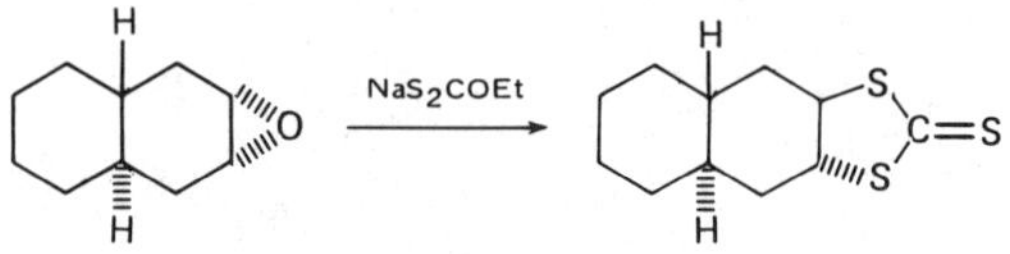

(148)

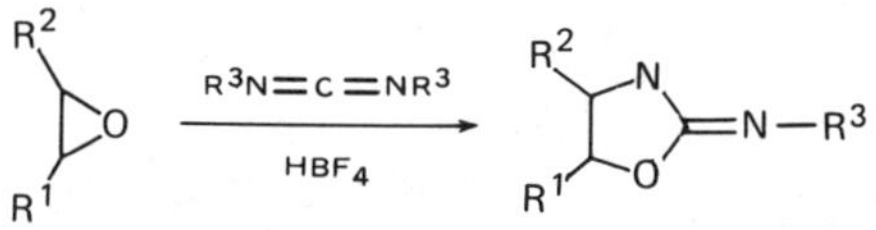

(149)

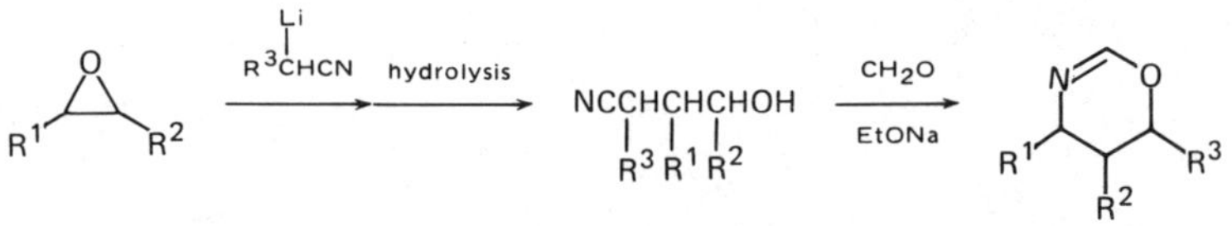

(Ref. 701)

(150)

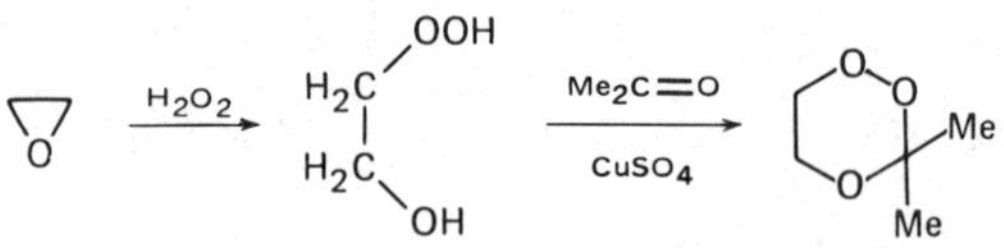

(151)

4. Transformation of oxiranes containing a functional group, by ring-expansion

The ring-expansion of oxiranes to four-membered heterocyclic compounds can be seen in equations (152) and (153).

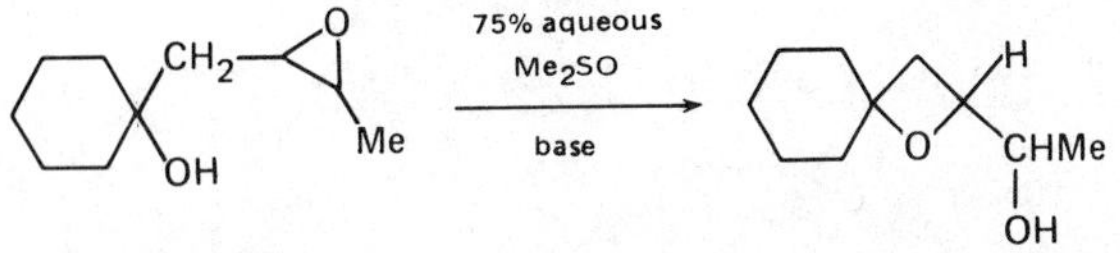

(Refs. 705, 706) (152)

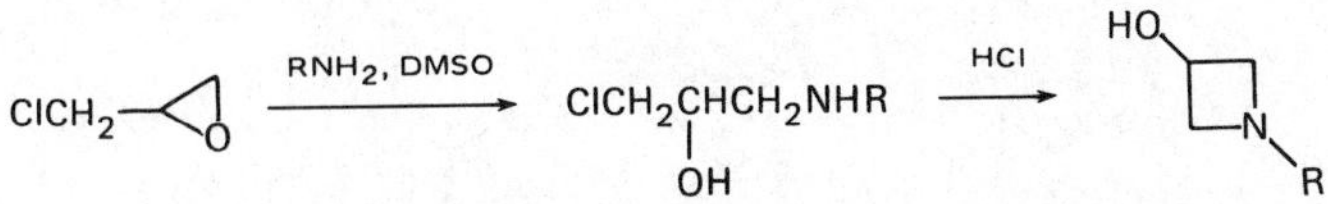

(Ref. 707) (153)

Equations (154)–(168) show the ring-transformations of oxiranes to five-membered heterocyclic compounds. Phenolate neighbouring-group participation has been found in the opening of the oxirane ring[709] (equation 155). By means of 1,3-dipolar cycloaddition[710], dihydrofuran derivatives are formed (equation 156).

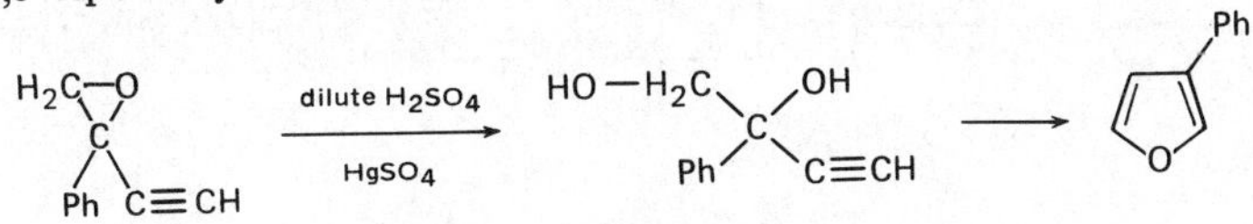

(Ref. 708) (154)

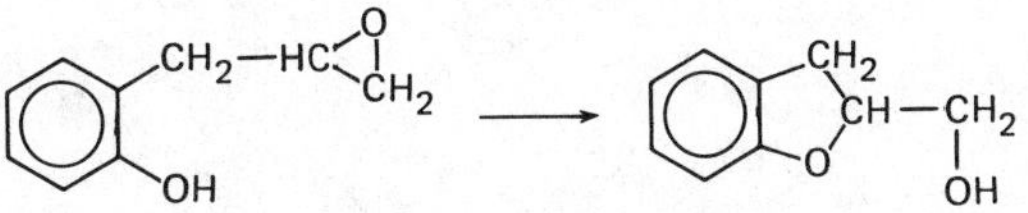

(Ref. 709) (155)

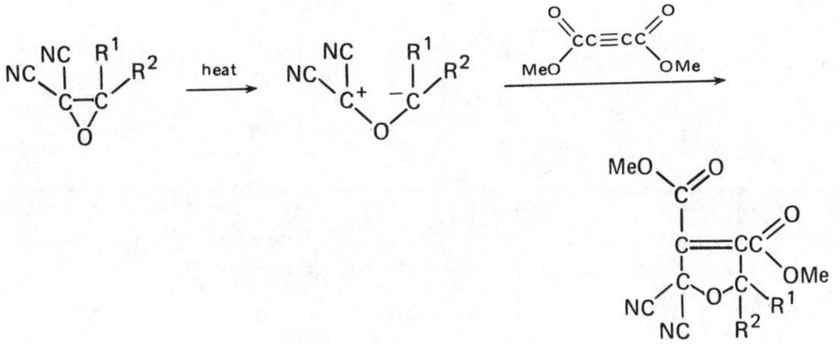

(Ref. 710) (156)

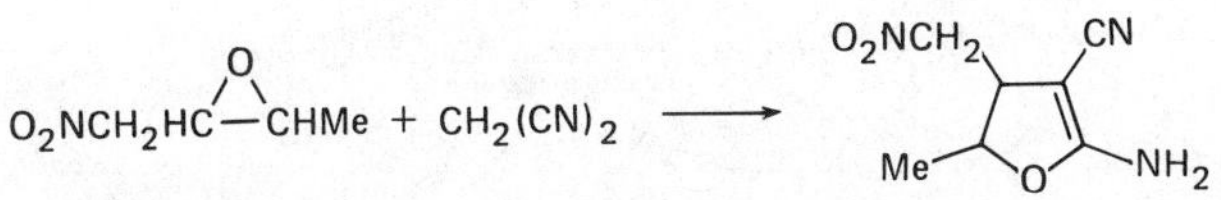

(Ref. 711) (157)

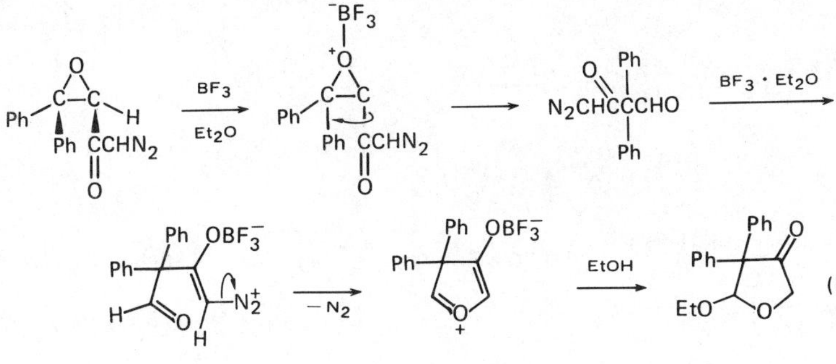

(Ref. 512) (158)

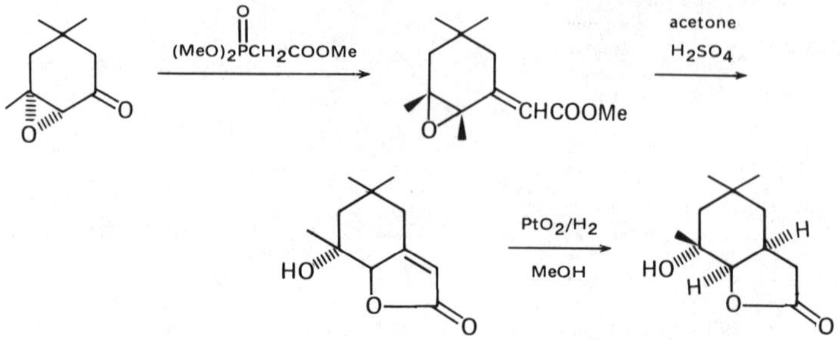

(Ref. 712)   (159)

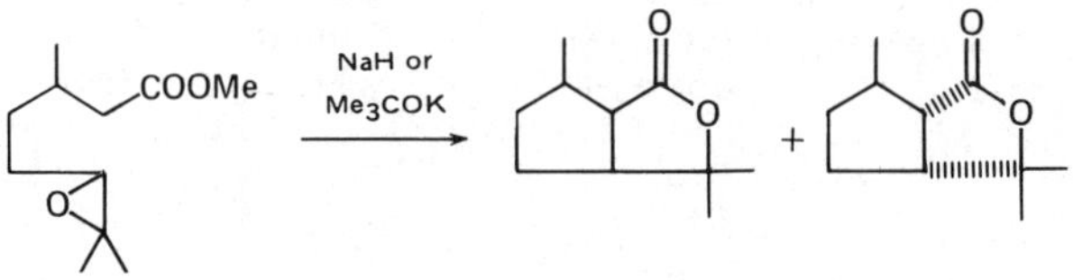

(Ref. 713)   (160)

(Ref. 714)   (161)

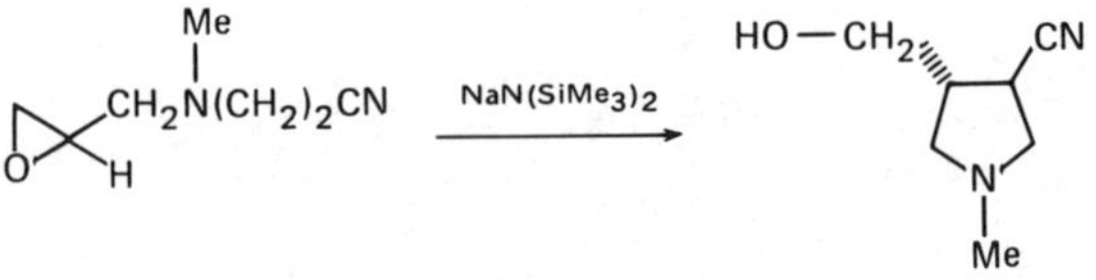

(Ref. 715)   (162)

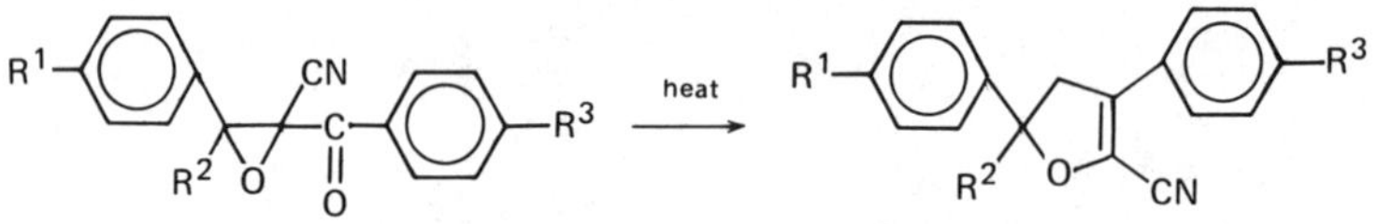

(Ref. 716)

(163)

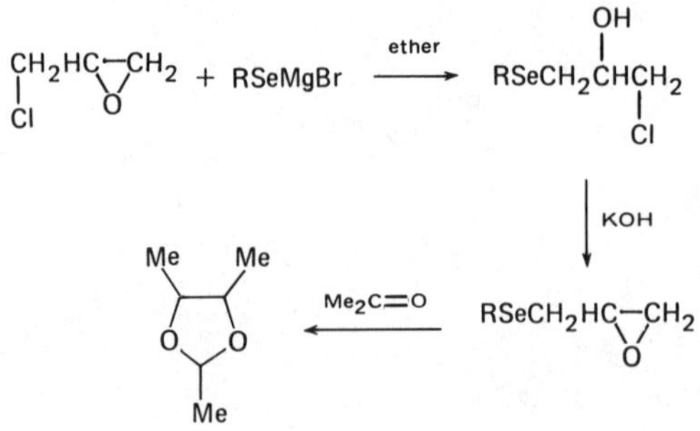

(Ref. 717)   (164)

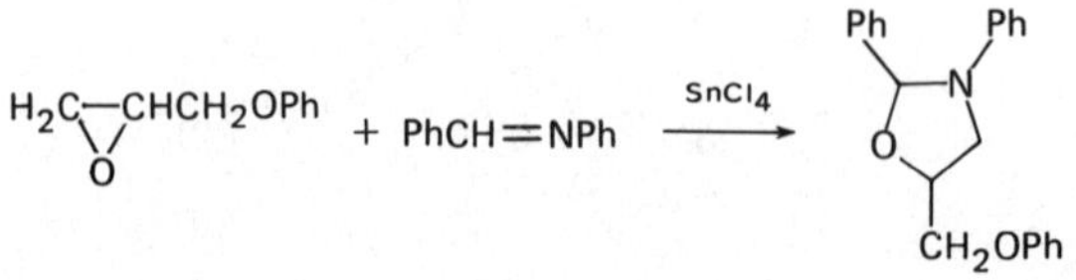

(Ref. 718)   (165)

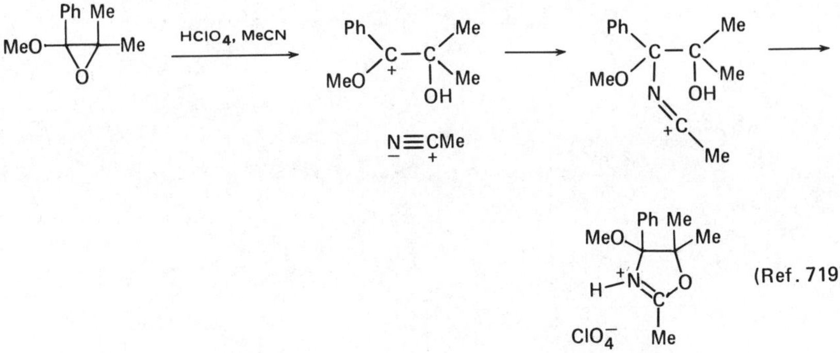

(Ref. 719) (166)

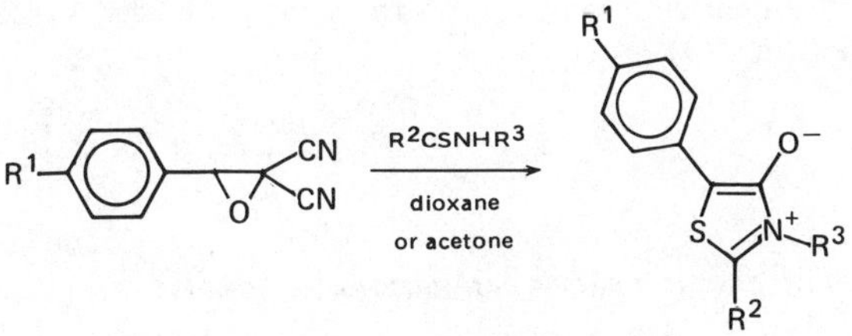

(Ref. 720) (167)

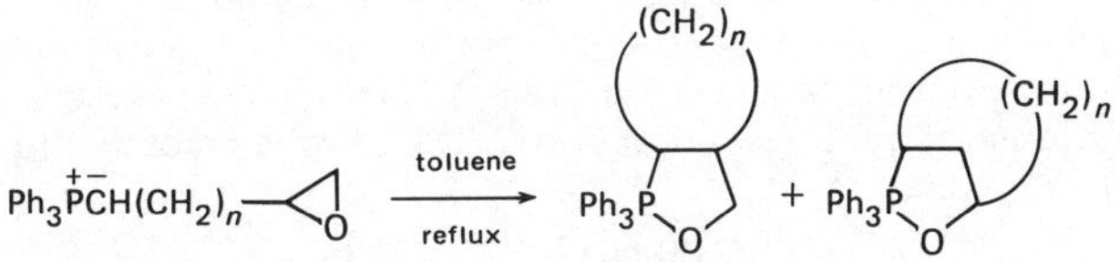

(Ref. 721) (168)

The syntheses presented above generally display very good yields. Additional studies yielded other five-membered[722-730] and six-membered heterocyclic compounds[731-733].

G. Reaction with Organometallic Compounds.

In the past ten years, numerous publications have dealt with the reactions of oxiranes with organometallic compounds. The Grignard compounds, dialkyl-magnesiums, trialkylaluminiums and lithium dialkylcuprates are the most important organometallic reagents.

1. Reaction with Grignard compounds

Organomagnesium compounds were the earliest used organometallic compounds for the transformation of oxiranes to alcohols[1,734-736]. In the case of substituted oxiranes, the reaction generally gives an alcohol mixture (equation 169).

Route (a) shows the normal addition, route (b) occurs on the action of the magnesium halide ($2\,RMgX \rightleftharpoons MgX_2 + MgR_2$), and route (c) is due to metal halide-catalysed isomerization of the oxiranes to carbonyl compounds. The latter two reactions do not take place in the case of MgR_2. Via route (a), cyclopentene oxides yield 2-substituted cyclopentanols. Higher cycloalkene oxides give ring-contraction

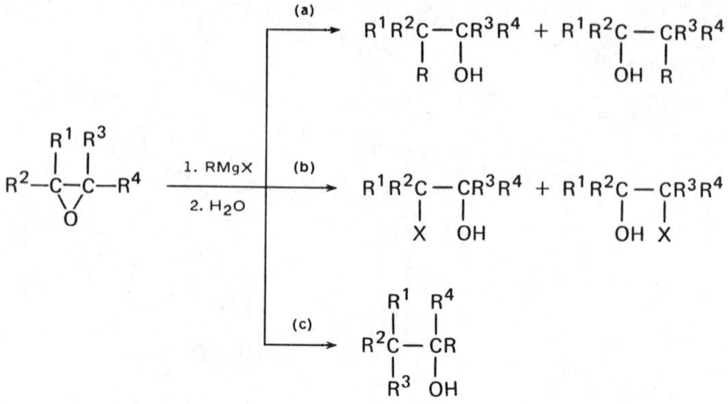

$$(a) \quad R^1R^2C\!-\!CR^3R^4 + R^1R^2C\!-\!CR^3R^4$$
$$\qquad\qquad \underset{R\;\;\;OH}{|\;\;\;|} \qquad\qquad \underset{OH\;\;\;R}{|\;\;\;|}$$

$$\underset{R^1\;R^3}{\overset{R^1\;R^3}{R^2\!-\!\underset{\diagdown O \diagup}{C}\!-\!\underset{}{C}\!-\!R^4}} \quad \overset{1.\;RMgX}{\underset{2.\;H_2O}{\longrightarrow}} \quad (b) \quad R^1R^2C\!-\!CR^3R^4 + R^1R^2C\!-\!CR^3R^4 \qquad (169)$$
$$\qquad\qquad\qquad\qquad\qquad \underset{X\;\;\;OH}{|\;\;\;|} \qquad\qquad \underset{OH\;\;\;X}{|\;\;\;|}$$

$$(c) \quad \underset{R^3\;\;\;OH}{\overset{R^1\;\;\;\;R^4}{R^2C\!-\!CR}}$$

and rearrange to aldehydes, which in turn react with the reagent in the usual manner[737] (equation 170).

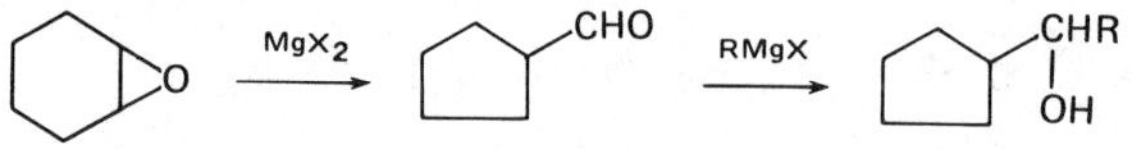

$$\underset{}{\overset{}{\bigcirc\!\!\!\!\diagup\!\!\diagdown O}} \quad \overset{MgX_2}{\longrightarrow} \quad \overset{CHO}{\bigcirc} \quad \overset{RMgX}{\longrightarrow} \quad \underset{OH}{\overset{CHR}{\bigcirc}} \qquad (170)$$

2. Reaction with magnesium alkyls and aluminium alkyls

Both types of organometallic compound react with oxiranes to give alcohols[734,738-742]. Comprehensive work has been carried out on the comparison of the reactivities of the two types of compound and the mechanism of the reactions[743]. With a given oxirane, the two organometallic compounds give alcohols with different structures (equation 171). The stereostructure of the

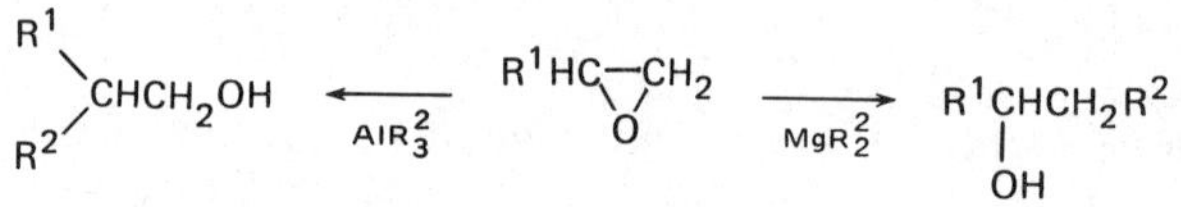

$$\underset{R^2}{\overset{R^1}{\diagdown}}\!\!CHCH_2OH \quad \underset{AlR_3^2}{\longleftarrow} \quad \underset{O}{\overset{R^1HC\!-\!CH_2}{\diagdown\!\!\diagup}} \quad \underset{MgR_2^2}{\longrightarrow} \quad \underset{OH}{\overset{R^1CHCH_2R^2}{|}} \qquad (171)$$

alcohol formed is also determined by the type of organometallic reagent: in the case of dialkylmagnesium, inversion always occurs at the reacting carbon atom. In both cases a two-step process is assumed (equations 172 and 173).

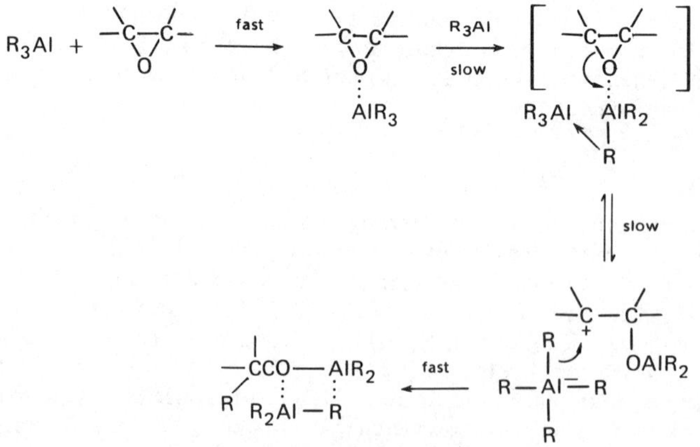

$$R_3Al + \overset{}{\underset{O}{\diagdown C \diagup C \diagdown}} \quad \overset{fast}{\longrightarrow} \quad \underset{\underset{AlR_3}{\vdots}}{\overset{}{\underset{O}{\diagdown C \diagup C \diagdown}}} \quad \overset{R_3Al}{\underset{slow}{\longrightarrow}} \quad \left[\underset{R_3Al \diagdown \underset{R}{\overset{AlR_2}{|}}}{\overset{}{\diagdown C \diagup C \diagdown}} \right]$$

$$\Big\Updownarrow slow \qquad\qquad (172)$$

$$\underset{R\;\;\;R_2Al-R}{\overset{}{-CCO\!-\!AlR_2}} \quad \overset{fast}{\longleftarrow} \quad \underset{R}{\overset{\underset{R}{-C-C-}}{R-Al\overset{=}{}R}}\,\underset{}{\overset{+}{OAlR_2}}$$

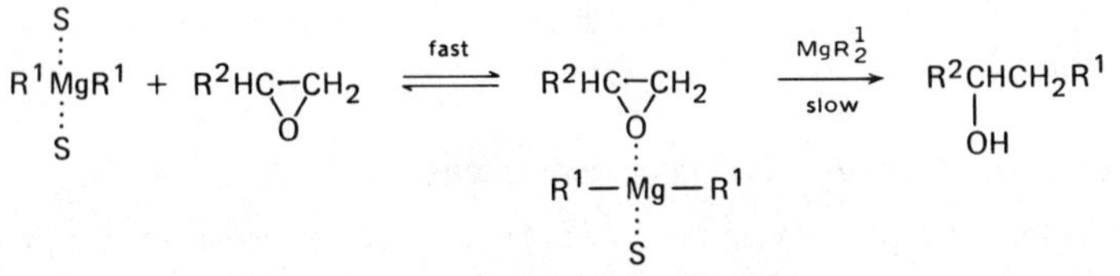

$$R^1MgR^1 + R^2HC\!\!-\!\!CH_2 \xrightarrow{\text{fast}} R^2HC\!\!-\!\!CH_2 \xrightarrow[\text{slow}]{MgR^1_2} R^2CHCH_2R^1 \quad (173)$$

S = solvent molecule

3. Reaction with lithium dialkylcuprates

Organolithium compounds generally react at the less-substituted carbon atom with asymmetrically substituted oxiranes[744-746] (equation 174). Similarly, cyclo-

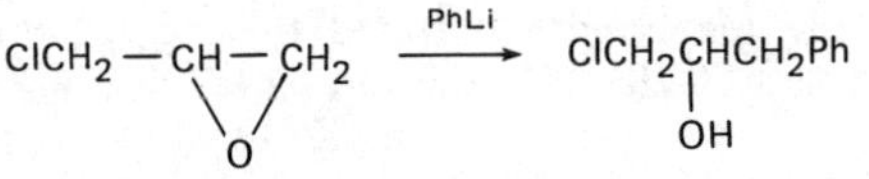

$$ClCH_2\!-\!CH\!-\!CH_2 \xrightarrow{PhLi} ClCH_2CHCH_2Ph \quad (174)$$

hexene oxide or 2,3-dimethyloxirane react with neopentylallyllithium to give the regular addition products[747] (e.g. equation 175).

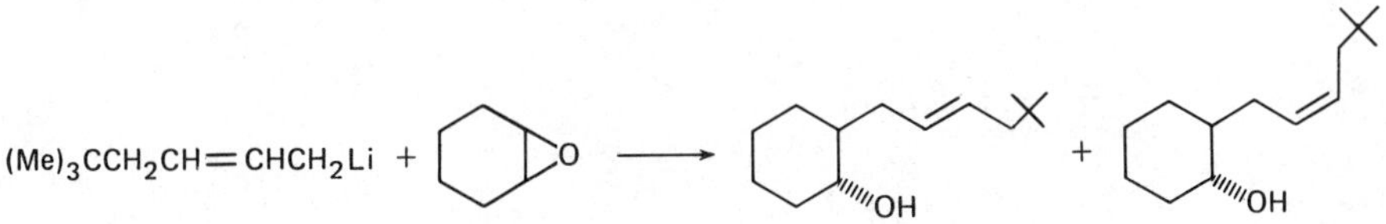

$$(Me)_3CCH_2CH\!=\!CHCH_2Li + \quad (175)$$

Lithium organocuprates are much more effective in their reactions with oxiranes than methyllithium or phenyllithium, and good regioselectivity has been observed[748-750]. The reaction requires much milder conditions than in the case of other organometallic compounds (equation 176). Lithium dimethylcuprate does

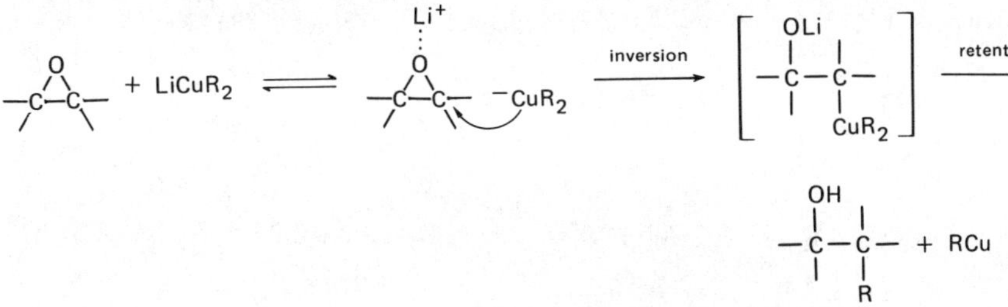

$$(176)$$

not react with tetrasubstituted oxiranes[751]. Oxiranes containing unprotected carbonyl groups react only via their oxirane function. Accordingly, the reaction may be utilized for the α-alkylation of α,β-epoxyketones (α,β-unsaturated ketones)[752,753] (equation 177). In general a large excess of the reagent must be taken, and only one of the alkyl groups is incorporated. If the stoichiometric

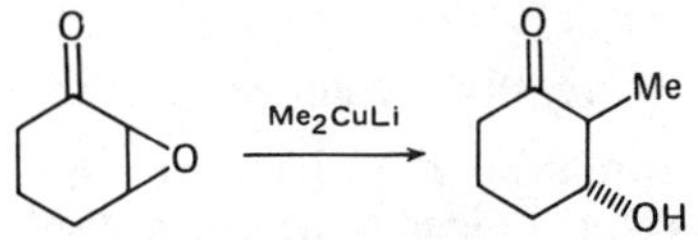

$$\xrightarrow{Me_2CuLi} \quad (177)$$

quantity of R(CN)CuLi is used, the desired alcohol may be obtained in high yield (>90%)[754].

4. Reaction with other organometallic compounds

Dialkylcadmium and dialkylzinc do not react with oxiranes. In the presence of $MgBr_2$, however, dialkylcadmium transforms phenyloxirane to a benzyl alkyl carbinol[755] (equation 178).

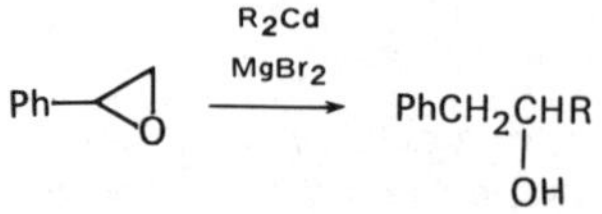

$$Ph-\!\triangleleft_O \xrightarrow[\text{MgBr}_2]{\text{R}_2\text{Cd}} PhCH_2\underset{\underset{\text{OH}}{|}}{C}HR \qquad (178)$$

Trimethylchlorosilane reacts with oxiranes to give 1,2-chlorohydrin trimethylsilyl ethers[756] (equation 179). In the presence of magnesium, bistrimethylsilyloxy derivatives are formed[757]. Trimethylisothiocyanatosilane[758] and trimethylsilyl cyanide[759] react in a similar manner.

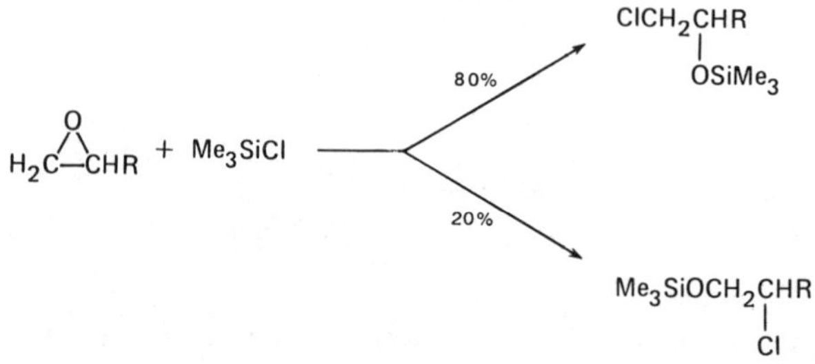

$$
\underset{\text{H}_2\text{C}-\text{CHR}}{\overset{\text{O}}{\triangle}} + Me_3SiCl \longrightarrow
\begin{array}{c}
\overset{80\%}{\nearrow} \quad ClCH_2\underset{\underset{\text{OSiMe}_3}{|}}{C}HR \\[2em]
\underset{20\%}{\searrow} \quad Me_3SiOCH_2\underset{\underset{\text{Cl}}{|}}{C}HR
\end{array}
\qquad (179)
$$

Oxiranes give olefins in stereospecific transformations with lithiumtrialkylsilane and stannate[453,760-762].

Certain organoaluminium compounds react with oxiranes to yield β-hydroxy acetylenes or β-hydroxy olefins[763-768] (e.g. equations 180 and 181).

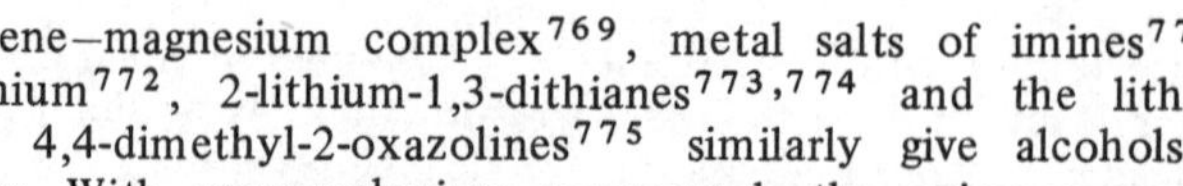

$$\bigcirc\!\!\!\!\!\triangle O + Et_2AlC\equiv CC_6H_{13} \longrightarrow \underset{\text{OH}}{\overset{C\equiv CC_6H_{13}}{\bigcirc}} \qquad (180)$$

$$\underset{\text{MeHC}-\text{CH}_2}{\overset{\text{O}}{\triangle}} + Et_2AlCH=CHEt \longrightarrow MeCH\underset{\underset{\text{OH}}{|}}{C}H_2CH=CHEt \qquad (181)$$

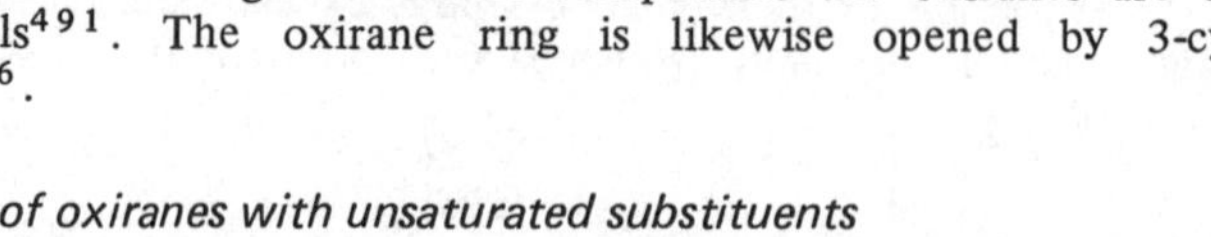

The mircene–magnesium complex[769], metal salts of imines[770,771], polychloroaryllithium[772], 2-lithium-1,3-dithianes[773,774] and the lithium salts of 2-substituted 4,4-dimethyl-2-oxazolines[775] similarly give alcohols on reaction with oxiranes. With organoselenium compounds the oxiranes are converted to allyl alcohols[491]. The oxirane ring is likewise opened by 3-cyclohexenylpotassium[776].

5. Reaction of oxiranes with unsaturated substituents

With organometallic compounds, and particularly lithium alkylcuprates, vinyloxiranes mainly participate in a 1,4-addition, which displays extensive stereoselectivity[494,777-779] (equation 182). The reactions of lithium alkenylcuprates and vinyloxiranes lead to 2,5-dienol systems[780].

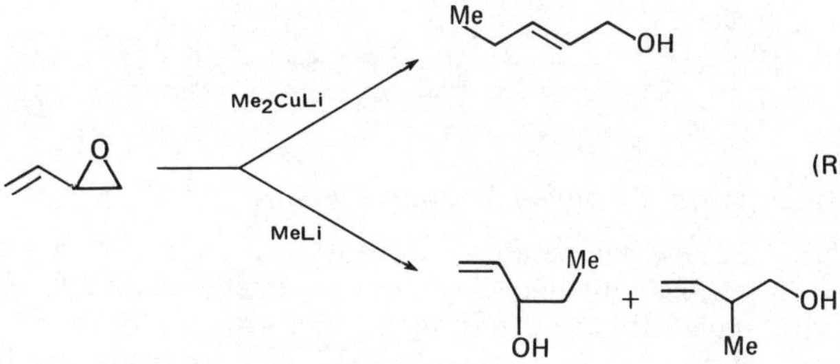

(Ref. 777)   (182)

Comparative investigations have been carried out on the transformations of 1,3- and 1,4-cyclohexadiene monoxides and vinyloxirane with certain types of organometallic compounds[781,782] (e.g. equations 183−186). Cyclopentadiene monoxide gives different products with diethylhexynylaluminium in ether and in toluene[783]

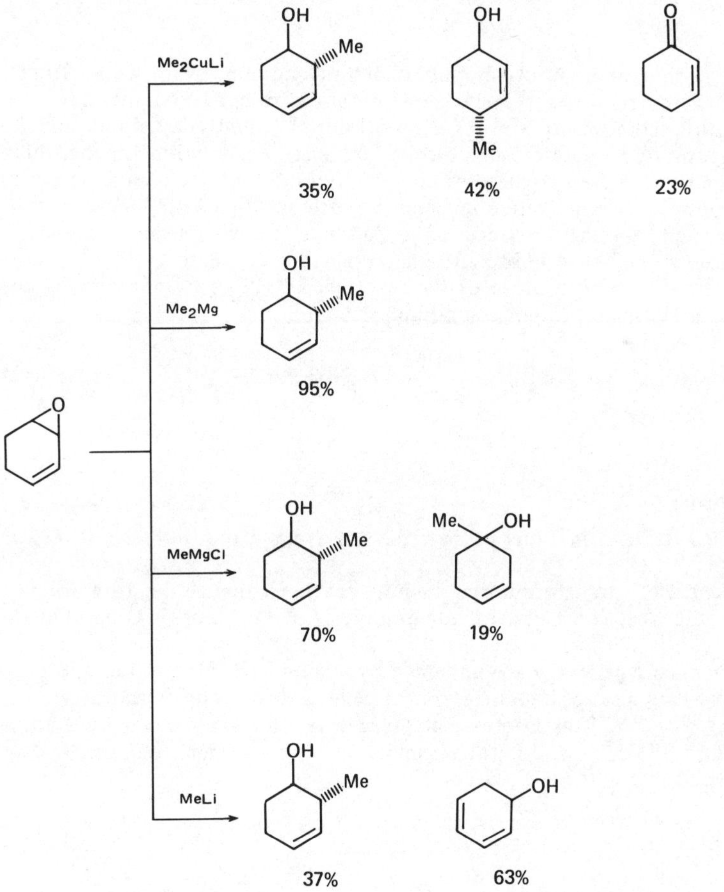

(183)

35%   42%   23%

(184)

95%

(185)

70%   19%

(186)

37%   63%

(equation 187). Cyclooctatetraene monoxide reacts with an alkynyl Grignard compound to give a cycloheptatriene derivative via ring-contraction[784].

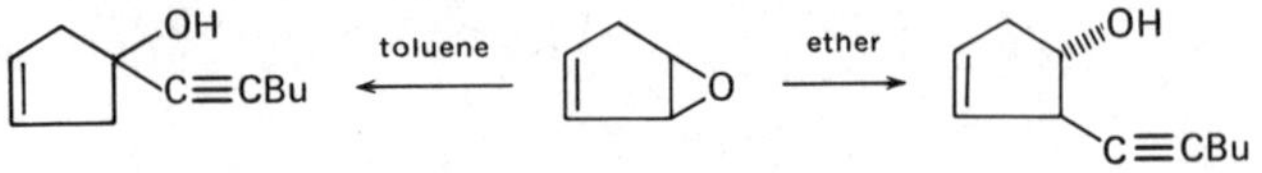

$$(187)$$

6. Reaction of oxiranes containing functional groups

With LiCuR$_2$ at low temperature, α-acetoxyoxiranes give α-alkylketones in moderate yield[785], while α,β-epoxysilanes give a β-hydroxysilane[455,786]. α-Chloroepoxycarboxylic acid esters give rise to a chlorocarbonyloxirane with Grignard reagents[787] (equation 188). α,β-Epoxyketones or open-chain aldehydes can be

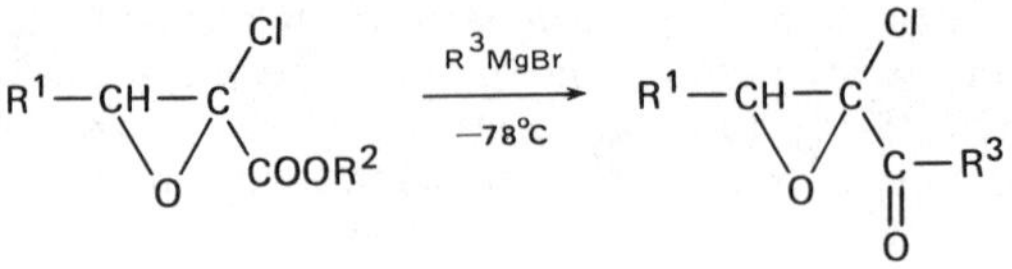

$$(188)$$

30—70%

prepared with Grignard compounds and dialkylmagnesium from cyanooxiranes, depending on their structures[788]. The transformations of cyanooxiranes have been studied with lithium dialkylcuprates[789], alkyllithium[790] and trialkylaluminium[791]. At low temperatures, α-heterosubstituted oxiranes react with organolithium compounds, and the 1,2-epoxyalkyllithium compounds obtained serve as an important nucleophilic oxirane source in organic syntheses[792]. With LiCuR$_2$, with a Grignard compound in the presence of a Cu$^+$ salt, or with trialkylborane[793], alkynyloxiranes can be converted to allene alcohols in good yield[794,795] (equation 189). Studies have also been made of the reactions of chloroxiranes with organomagnesium[796] and organolithium compounds[797-799].

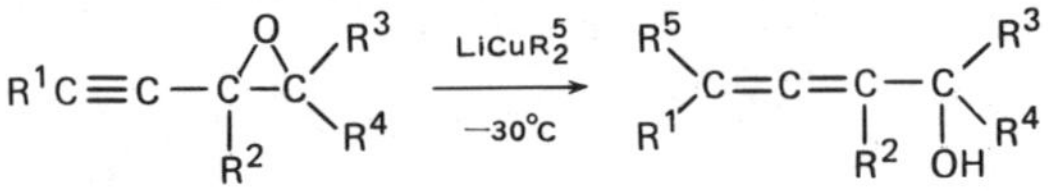

$$(189)$$

H. Photochemistry

Photochemical transformations of oxiranes are treated in a number of reviews and monographs[16,17,800-803].

The photochemical transformations include rearrangements, the formation of carbenes, and other reactions, all involving homolysis of a C—C or C—O bond of the oxirane ring.

Rearrangements are generally accompanied by isomerization (equation 190); this frequently plays only a subordinate role, but it nevertheless occurs with noteworthy stereoselectivity[804,805]. The intermediate carbonyl ylide is formed by disrotational ring-opening[806-809], and is then converted to the isomeric oxirane by ring-closure after rotation about the C—O bond.

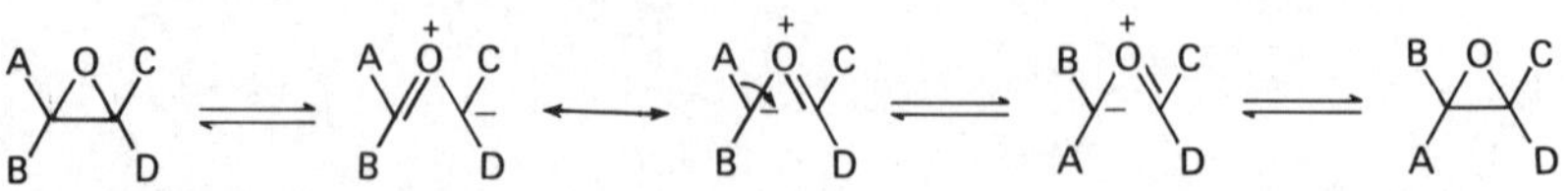

$$(190)$$

Oxiranes containing strongly electron-attracting substituents (e.g. CN, COOEt)
yield carbenes[810-813]. For example, on the photolysis of **11** and **12**, **13** and **14**,

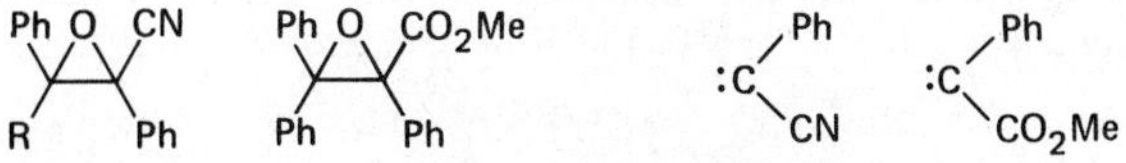

(**11**) (**12**) (**13**) (**14**)

respectively, are formed. The mechanism of carbene formation was studied by
Griffin and coworkers[814], who suggested that it takes place via an ionic mechanism.
On the double photolysis of **15** at low temperature, both ylide and carbene formation were demonstrated. On this basis, the mechanism of equation (191) was
assumed, with the note that the photochemical reaction of **16** may be followed by
concerted or other processes which give rise finally to **17**.

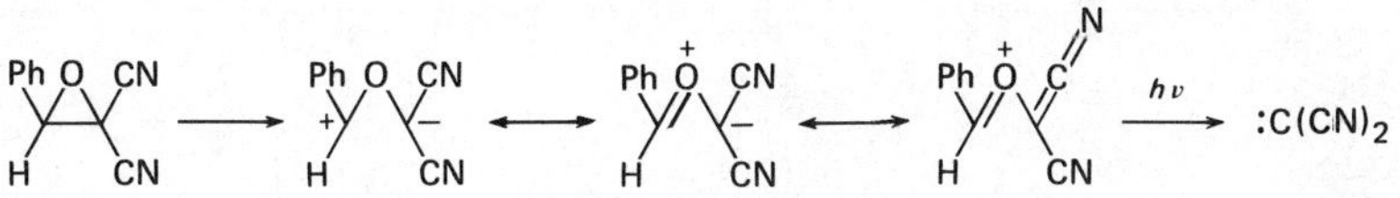

(**15**) (**16**) (**17**)

Although the intermediate may also be an ylide[806,815], the first step in most
photochemical reactions is the homolytic splitting of one of the C—O bonds[815].

On the low-pressure photolysis of propylene oxide, propionaldehyde and acetone
are formed[816] (equation 192). If the pressure is raised, the amount of acetone

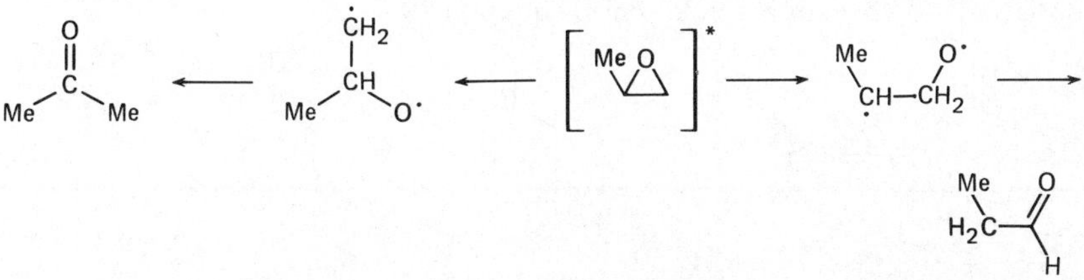

increases, and it emerges from the quenching effect that, under these conditions,
the propionaldehyde and acetone cannot be formed from a common intermediate.

Among photocatalytic transformations of oxiranes containing various functional
groups[552,553,817-837], some characteristic examples are presented in equations
(193)—(198).

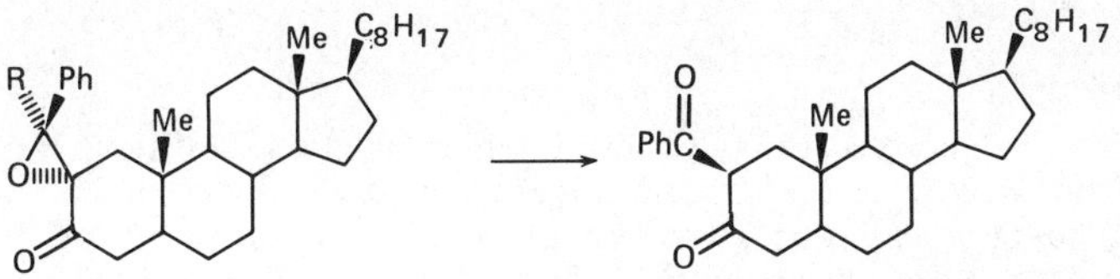

(Ref. 821) (193)

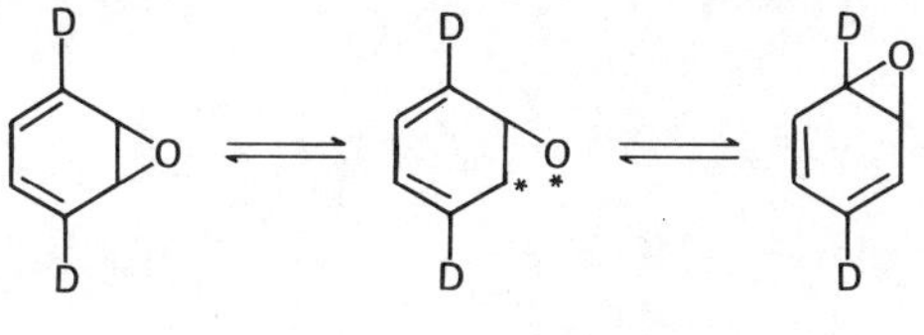

(194)

Spiro-α-carbonyloxiranes are converted to dicarbonyl compounds[820,821] (equation 193). At room temperature benzene oxide is transformed to phenol, while at low temperature oxygen migration around the aromatic ring and ketene formation can also be detected[822] (equations 194 and 195). Equation (196) shows

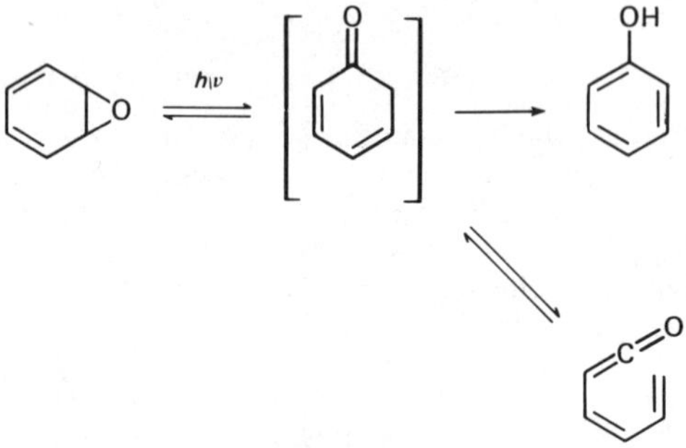

(195)

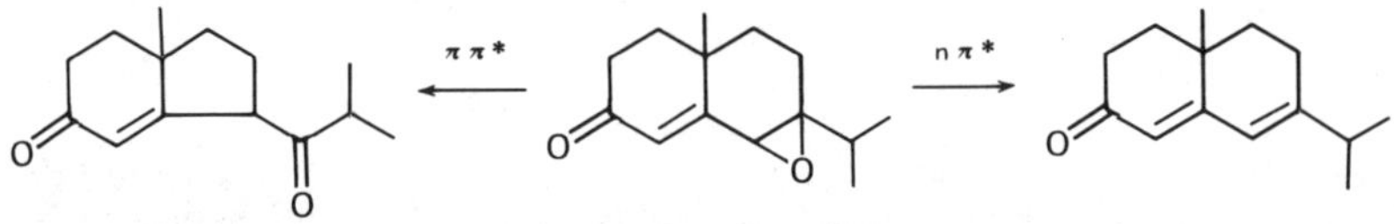

(196)

that the direction of the rearrangement also depends on the mode of excitation[825].

Murray and coworkers[829] proposed a general scheme for the photochemical transformations of β,γ-epoxycycloketones (equation 197).

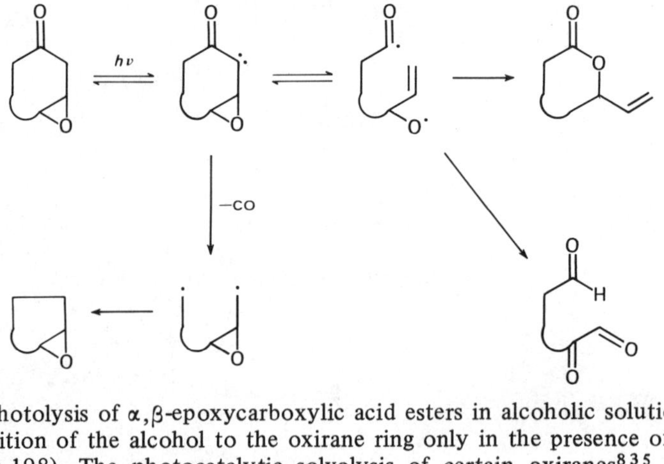

(197)

The photolysis of α,β-epoxycarboxylic acid esters in alcoholic solution[833,834] gives addition of the alcohol to the oxirane ring only in the presence of Fe^{3+} ions (equation 198). The photocatalytic solvolysis of certain oxiranes[835], and their photoreduction on the action of alcohols[836], have also been examined. With NBS or other brominating reagents, α-bromooxiranes and α-bromoketones may be prepared by photochemical means[837].

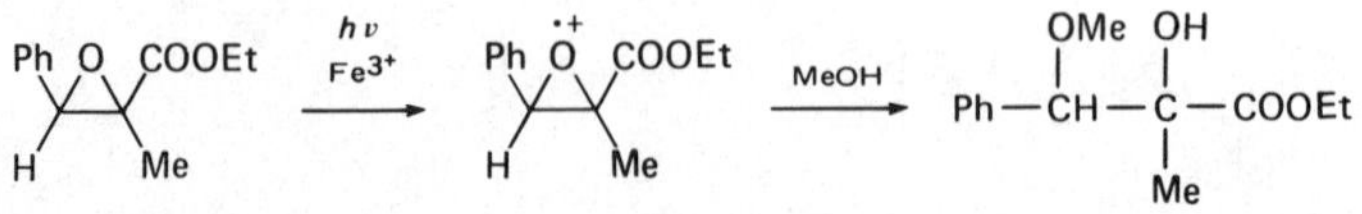

(198)

I. Thermally induced Reactions

Thermally-induced reactions of oxiranes yield rearrangements to carbonyl compounds and unsaturated alcohols, as well as other rearrangements[555-557,838-841].

The kinetics of rearrangement of oxiranes to carbonyl compounds and unsaturated alcohols[843-846] indicate that these are monomolecular homogeneous processes; the intermediate biradicals are converted to end-products via intramolecular rearrangement. The radicals playing the key roles in most of the thermal and photochemical reactions of oxiranes can be detected by ESR and their structures studied[842].

The mechanism of the electrocyclization and isomerization processes is outlined in equation (199). Investigation of the stereochemistry of electrocyclization[334,556],

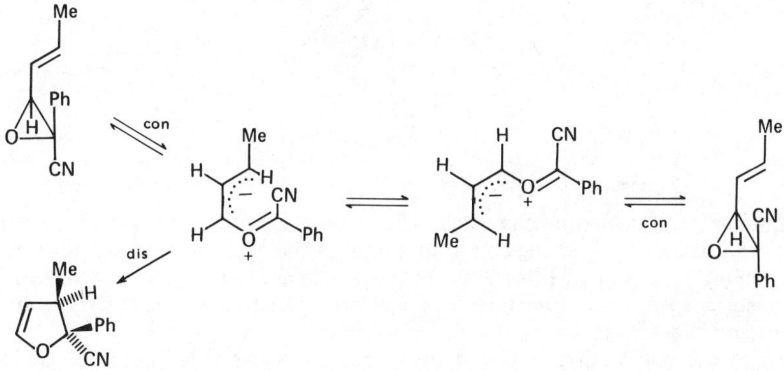

(199)

[840,841] has shown that only cis-dihydrofurans are formed. The first step is cleavage of a C—C bond, showing that the biradical structure is favoured[551]. The ring-opening is conrotational[554,840,847].

Stereospecific formation of dihydrofurans proceeds via disrotational ring-closure of the ylide[557,840]. The isomerization can similarly be explained in accordance with equation (199). The formation of dihydrooxepines from the cis-oxirane is a concerted [3,3]sigmatropic rearrangement, the transition state having a boat conformation[554,847] (equation 200).

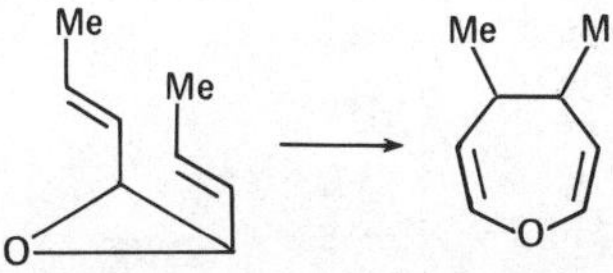

(200)

Ylides formed from oxiranes containing electron-attracting substituents have given a possibility for a new type of dioxolane syntheses too[716] (equation 163).

Much new information has been acquired in connection with the pyrolysis of oxiranes linked to large unsaturated rings[848-850]. Additionally, the radical-induced transformations of oxiranes have been investigated[803,848,851-853]. In conclusion, attention is drawn to the review by Huisgen[854] on the electrocyclic ring-opening reactions of the oxiranes.

J. Ring-opening with Nucleophilic Reagents

The most frequent reactions of oxiranes are those involving opening of a C—O bond, in the course of which 1,2-difunctional compounds may be obtained. The

C—O bond may be opened by direct nucleophilic attack on one of the carbon atoms, or first the oxygen is protonated (or a complex is formed with the electrophilic centre of the reagent) and this is followed by nucleophilic attack on the carbon (equations 201 and 202). The equations also illustrate the stereochemical

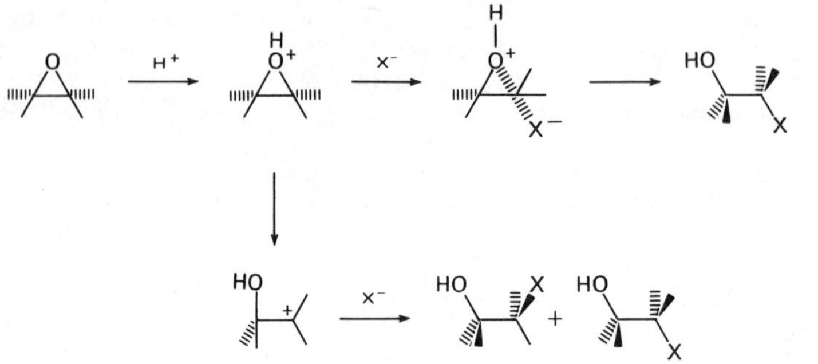

(201)

(202)

X = OH, SH, F, Cl, Br, I, CN, OR, OAr, SR, SAr, O_2R, RCO_2, etc.

consequences of the two mechanisms. The mechanism and stereochemistry depend on the structure of the starting compound and on the experimental conditions.

In general, reactions in basic and neutral media occur by an A2 mechanism, and involve stereospecifically *trans* stereochemistry. There is a particularly abundant literature on the acid-catalysed reactions of the oxiranes.

Most of the publications referred to in recent reviews[9,16,17] or published since deal with factors of a steric, stereoelectronic, polar or conjugative nature, resulting in the regioselectivity and stereoselectivity of the ring-opening. A much-discussed subject is the mechanism of acid-catalysed reactions. The experimental results have been interpreted on the basis of the A2, the A1 or the borderline mechanism.

Comprehensive kinetic studies[855] on the acid catalysis of alkyl-substituted oxiranes in aqueous and non-aqueous media pointed to a competition between the A2 and A1 mechanisms, with the predominance of the former. Anhydrous conditions favour the A1 mechanism, since the halide ion does not play a role in the formation of the transition state. For resolution of the contradictions, a new mechanistic concept is proposed, in which the conjugate acid of the substrate forms a close ion pair (equation 203).

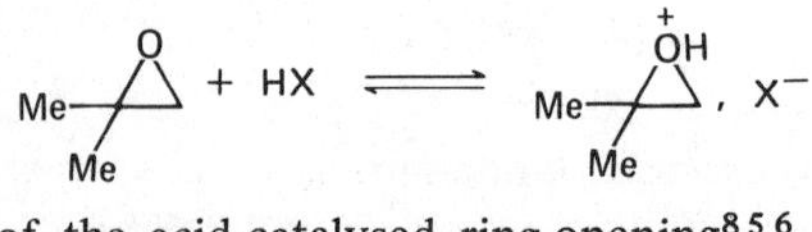

(203)

In another study of the acid-catalysed ring-opening[856] it was concluded that primary and secondary aliphatic oxiranes react by the A2 mechanism, but further investigations are necessary for tertiary and monoaryl-substituted oxiranes.

The stereochemistry of the base-catalysed hydrolysis of aryl-substituted oxiranes points to a concerted S_N2 mechanism. With acid hydrolysis, and S_N1 mechanism is suggested for the *trans*-oxirane, and an S_N2 mechanism for the *cis* isomer[857-859].

Many investigations have recently been carried out on the acid hydrolysis of oxiranes[860-869]. The reaction rate and steric course[860] depend to a large extent not only on the configuration of the substrate, but also on the solvent type (equation 204). In a solvent with a low dielectric constant, mainly *cis* opening

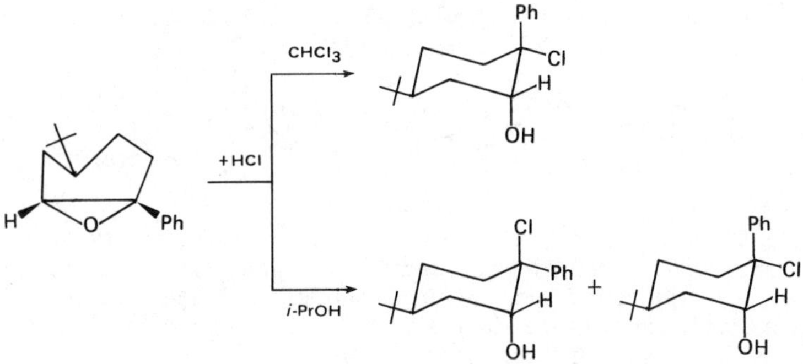

$$(204)$$

occurs, with configuration retention. In water or in alcohols, the stereospecificity is lower. The retention can be ascribed in part to the formation of a solvent-protected ion pair, in which the attack by the anion proceeds internally on the electron-deficient benzyl carbon atom (equation 205).

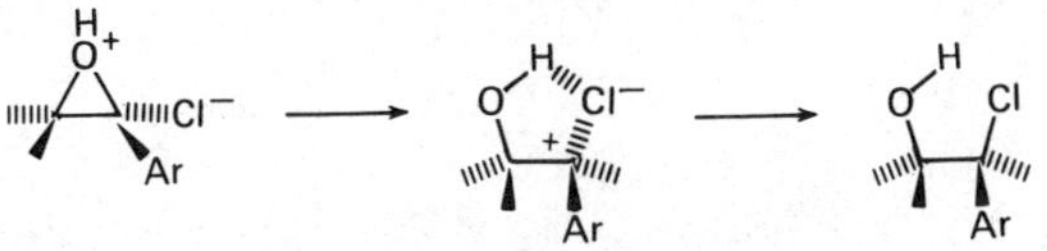

$$(205)$$

In the course of stereochemical studies (equations 206 and 207), it has also been proved that the transition state leading to the *cis* products has a high degree of carbocationic character; the tendency towards the retention product is explained

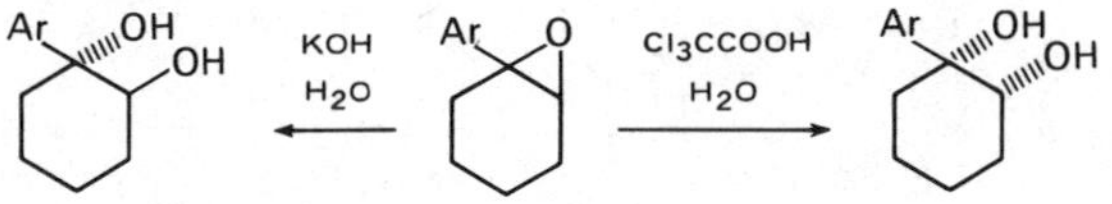

(Ref. 863)   (206)

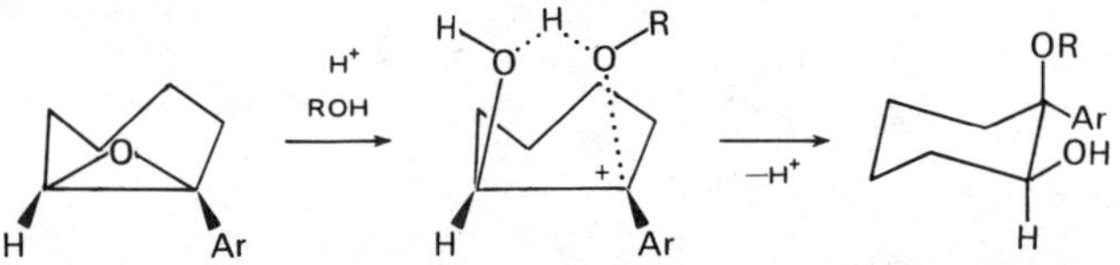

(Ref. 865)   (207)

by the favourable entropy content of the transition state of *cis* addition and by the relatively low enthalpic barrier to the breaking of the benzylic C—O bond. At the same time, almost total antistereoselectivity can be observed in aliphatic and cyclo aliphatic oxiranes[348,866]. The importance of the activation parameters in mechanistic studies is confirmed by recent results on the solvolysis of 1-arylcyclo-hexene oxides[865,867]. Attempts have been made to separate the inductive, conformational and stereoelectronic effects[868]; the conclusion was reached that the inductive effect on the regioselectivity of the reaction plays the determining role, but the other factors are not negligible.

In agreement with the regularities mentioned above, *cis* ring-opening has also been observed with other types of compounds on the action of various electrophilic reagents[5,432]. Neighbouring-group participation is manifested most often in *cis* ring-opening[5,869–872].

The nucleophilic participation of TDAP and DMSO has been demonstrated in acid-promoted ring-opening reactions of oxiranes. Stable phosphonium and sulphonium salts are formed[873-875] (equation 208).

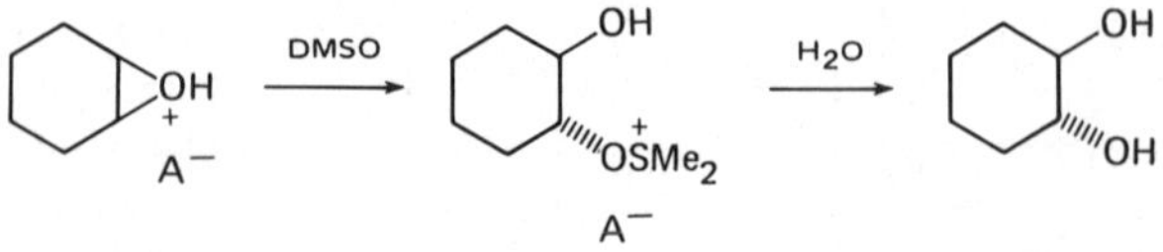

$$ (208) $$

In recent years, interest has grown in polycyclic aromatic oxides, which are regarded as mediators in polycyclic aromatic carcinogenesis. A number of teams have dealt with the various ring-opening reactions of K-region and non-K-region aromatic oxiranes, and with the kinetics of their hydrolyses[546,876-880].

Many studies deal with the stereochemistry[332,881-883] and mechanisms[881, 884-895] of the ring-opening. Others deal with the acid-catalysed[129,680,834, 896-898] or base-catalysed[322,491,899-904] ring-openings of various oxiranes, and with their utilization in synthetic organic chemistry[491,693,834,905-909], including ring-opening reactions with carbanions[680,776,908,910]. A number of new examples are illustrated in equations (209)–(213).

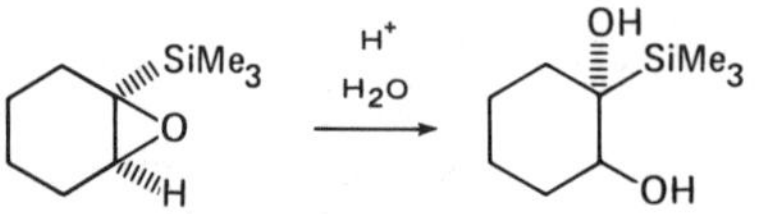

$$ \text{(Ref. 911)} \quad (209) $$

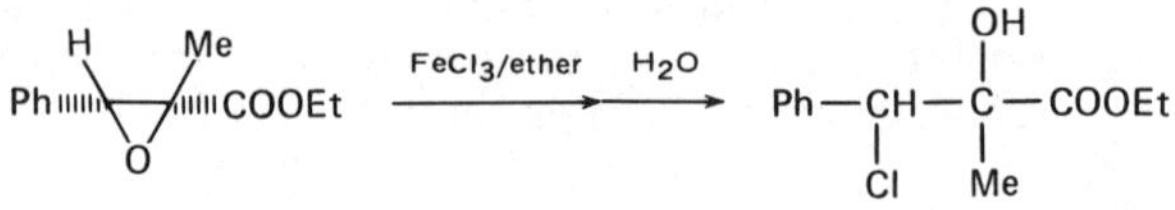

$$ \text{(Ref. 834)} \quad (210) $$

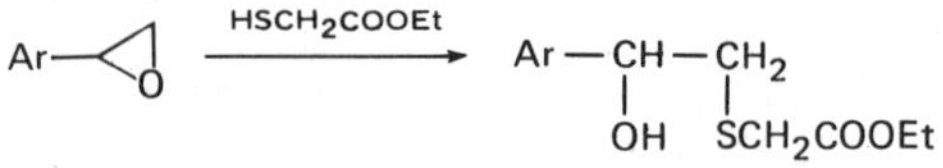

$$ \text{(Ref. 912)} \quad (211) $$

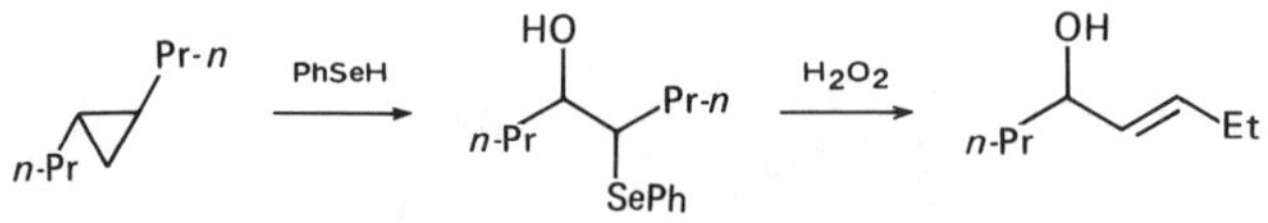

$$ \text{(Ref. 491)} \quad (212) $$

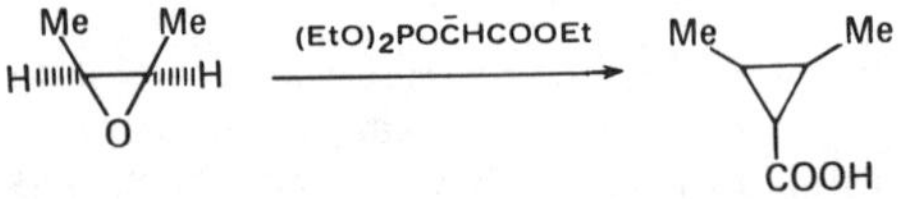

$$ \text{(Ref. 908)} \quad (213) $$

The solvolysis of oxiranes has also been investigated on synthetic ion-exchange resins[913], alumina[914-916] and silica gel[917], and extensive stereoselectivity has been observed in certain cases[915,916].

An interesting ring-opening occurs on the alcoholysis of oxiranes in the dark in the absence of catalysts[918] (equation 214).

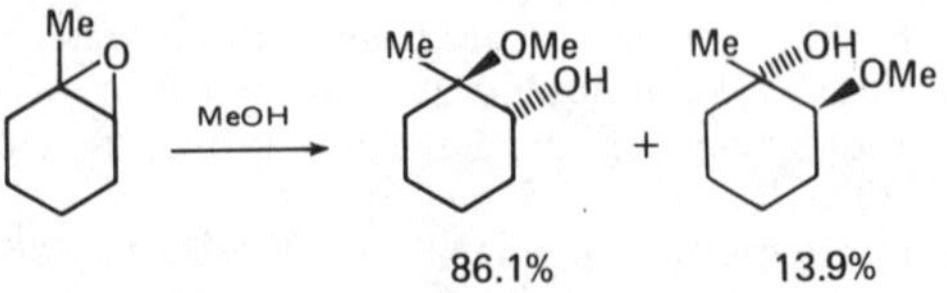

$$ (214) $$

86.1%   13.9%

New investigations have been carried out on the transformations of various oxiranes to yield 1,2-amino alcohols[341,919-929] leading to a deeper understanding of the stereochemistry and the $S_N 2$-type mechanism of the transformation, and to broad synthetic applications. Two examples are presented in equations (215) and (216). Similar studies have led to the recognition of two further modes of anchimeric assistance[931,932].

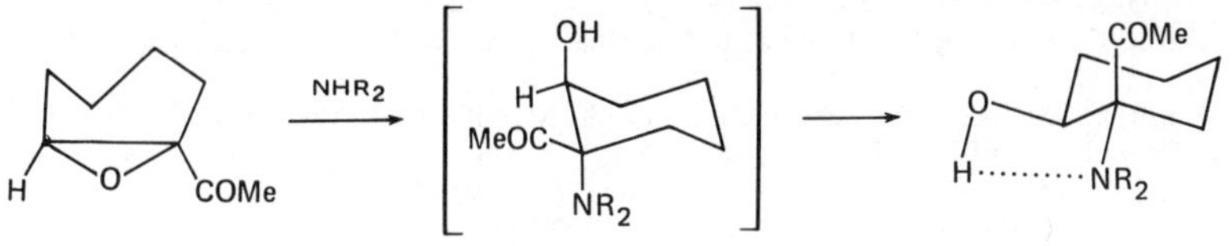

(Ref. 930)

(215)

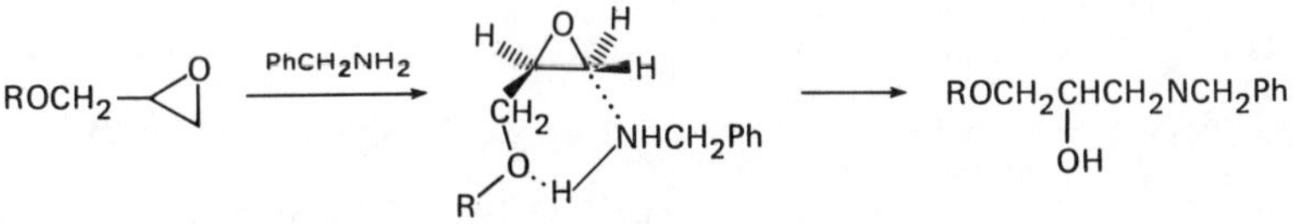

$ROCH_2CHCH_2NCH_2Ph$ (Ref. 931)

(216)

K. Other Reactions

Because of the exceptional reactivity of oxiranes (there is perhaps no reactant towards which oxiranes are immune), it has not been possible to describe a number of special transformations. Of these, some may be listed that are employed in synthetic organic chemistry or in the chemical industry. Recent results confirm that oxiranes may be used effectively for Friedel–Crafts-type syntheses[933,934]; many reactions are known with various organic[935-941] and inorganic[942-950] halogen compounds, organic sulphur compounds[951,952] and organic phosphorus compounds[950,953,954]. The reactions of oxiranes with CO_2[955,956] are also of industrial importance.

IV. REFERENCES

1. G. Dittus in *Methoden der Organischen Chemie (Houben-Weyl)*, Vol. VI/3, Georg Thieme Verlag, Stuttgart, 1965, pp. 367–487.
2. R. J. Gritter in *The Chemistry of the Ether Linkage*, (Ed. S. Patai), John Wiley and Sons, London, 1967, pp. 373–410.
3. D. Swern, *Encycl. Polim. Sci. Technol.*, **6**, 83 (1967).
4. P. V. Zimakov, *Okis' Etilena*, Khimiya, Moscow, 1967, pp. 1–317.
5. A. A. Akhrem, A. M. Moiseenkov and V. N. Dobrynin, *Usp. Khim.*, **37**, 1025 (1968).
6. R. C. Fahey in *Topics in Stereochemistry*, Vol. 3 (Eds. E. L. Eliel and N. L. Allinger), John Wiley and Sons, New York, 1968, p. 294.
7. D. N. Kirk and M. P. Hartshorn, *Steroid Reaction Mechanisms*, Elsevier, Amsterdam, 1968, p. 71.
8. V. N. Yandovskii, V. S. Karavan and T. I. Temnikova, *Usp. Khim.*, **39**, 571 (1970).
9. J. G. Buchanan and H. Z. Sable in *Selective Organic Transformations*, Vol. 2 (Ed. B. S. Thyagarajan), John Wiley and Sons, New York, 1972, pp. 1–95.
10. D. I. Metelitsa, *Usp. Khim.*, **41**, 1737 (1972).
11. J. Rouchaud, *Ind. Chim. Belg.*, **37**, 741 (1972).
12. V. N. Yandovskii and B. A. Ershov, *Usp. Khim.*, **41**, 785 (1972).
13. G. Berti in *Topics in Stereochemistry*, Vol. 7 (Eds. E. L. Eliel and N. L. Allinger), John Wiley and Sons, New York, 1973, pp. 93–251.
14. D. N. Kirk, *Chem. Ind. (Lond.)*, 109 (1973).

15. K. Matsumoto, *Kagaku No Ryoiki*, **27**, 148 (1973); *Chem. Abstr.*, **79**, 42253w (1973).
16. S. G. Wilkinson, *Int. Rev. Sci., Org. Chem., Ser. 2*, **2**, 111 (1975).
17. W. L. F. Armarego in *Stereochemistry of Heterocyclic Compounds*, Part 2, John Wiley and Sons, New York, 1977, pp. 12–36.
18. E. N. Prilezhaeva, *Prilezhaev reaction: Electrophilic oxidation*, Nauka, Moscow, 1974.
19. H. C. Van der Plas in *Ring Transformations of Heterocycles*, Vol. 1, Academic Press, London, 1973, pp. 1–43.
20. V. G. Dryuk, *Tetrahedron*, **32**, 2855 (1976).
21. P. L. Barili, G. Bellucci, B. Macchia, F. Macchia and G. Parmigiani, *Gazz. Chim. Ital.*, **101**, 300 (1971).
22. P. L. Barili, G. Bellucci, G. Berti, F. Marioni, A. Marsili and I. Morelli, *J. Chem. Soc., Chem. Commun.*, 1437 (1970).
23. R. D. Bach and H. F. Henneike, *J. Amer. Chem. Soc.*, **92**, 5589 (1970).
24. R. Kavčič and B. Plesničar, *J. Org. Chem.*, **35**, 2033 (1970).
25. T. Asahara, M. Seno, Y. Shimozato and C. Nagasawa, *Kogyo Kagaku Zasshi*, **73**, 2332 (1970); *Chem. Abstr.*, **74**, 140672y (1971).
26. L. Červený, J. Barton and V. Růžička, *Scientific Papers of the Prague Institute of Chemical Technology*, **C24**, 125 (1976).
27. B. Capon, J. Farquarson and D. J. McNeillie, *J. Chem. Soc., Perkin II*, 914 (1977).
28. M. S. Sytilin, *Zh. Fiz. Khim.*, **51**, 488 (1977).
29. K. M. Ibne-Rasa, R. H. Pater, J. Ciabattoni and J. O. Edwards, *J. Amer. Chem. Soc.*, **95**, 7894 (1973).
30. R. P. Hanzlik and G. O. Shearer, *J. Amer. Chem. Soc.*, **97**, 5231 (1975).
31. H. Kwart and D. M. Hoffman, *J. Org. Chem.*, **31**, 419 (1966).
32. K. D. Bingham, G. D. Meakins and G. H. Whitham, *J. Chem. Soc., Chem. Commun.*, 445 (1966).
33. H. Kwart, P. S. Starcher and S. W. Tinsley, *J. Chem. Soc., Chem. Commun.*, 335 (1967).
34. A. Azman, B. Borstnik and B. Plesničar, *J. Org. Chem.*, **34**, 971 (1969).
35. R. A. Finnegan and P. J. Wepplo, *Tetrahedron*, **28**, 4267 (1972).
36. J. P. Girard, J. P. Vidal, R. Granger, J. C. Rossi and J. P. Chapat, *Tetrahedron Letters*, 943 (1974).
37. Y. Bessière, M. M. El Gaied and B. Meklati, *Bull. Soc. Chim. Fr.*, 1000 (1972).
38. A. G. Causa, H. Y. Chen, S. Y. Tark and H. J. Harwood, *J. Org. Chem.*, **38**, 1385 (1973).
39. L. A. Paquette, S. A. Lang, Jr., M. R. Short and B. Parkinson, *Tetrahedron Letters*, 3141 (1972).
40. T. Sato and E. Murayama, *Bull. Chem. Soc. Japan*, **47**, 1207 (1974).
41. P. L. Barili, G. Bellucci, F. Marioni and V. Scartom, *J. Org. Chem.*, **40**, 3331 (1975).
42. Y. Yanagida, H. Shigesato, M. Nomura and S. Kikkawa, *Nippon Kagaku Kaishi*, 657 (1975); *Chem. Abstr.*, **83**, 113273y (1975).
43. P. Chautemps and J. L. Pierre, *Tetrahedron*, **32**, 549 (1976).
44. H. Hart, H. Verma and I. Wang, *J. Org. Chem.*, **38**, 3418 (1973).
45. S. A. Cerefice and E. K. Fields, *J. Org. Chem.*, **41**, 355 (1976).
46. S. G. Davies and G. H. Whitham, *J. Chem. Soc., Perkin I*, 2279 (1976).
47. S. G. Davies and G. H. Whitham, *J. Chem. Soc., Perkin I*, 572 (1977).
48. R. Curci and F. Di Furia, *Tetrahedron Letters*, 4085 (1974).
49. H. J. Bestmann, O. Vostrowsky and W. Stransky, *Chem. Ber.*, **109**, 3375 (1976).
50. E. Kh. Kazakova, Z. G. Isaeva and Sh. S. Bikeev, *Dokl. Akad. Nauk SSSR*, **226**, 346 (1976).
51. E. Kh. Kazakova, L. N. Surkova, Z. G. Isaeva and Sh. S. Bikeev, *Dokl. Akad. Nauk. SSSR*, **236**, 363 (1977).
52. A. R. Hochstetler, *J. Org. Chem.*, **40**, 1536 (1975).
53. F. Plenat, F. Pietrasanta, M. R. Darvich and H. Christol, *Bull. Soc. Chim. Fr.*, 2227 (1975).
54. H. Prinzbach and Ch. Ruecker, *Angew. Chem.*, **88**, 611 (1976).
55. C. W. Greengrass and R. Ramage, *Tetrahedron*, **31**, 689 (1975).

56. K. Ishikawa and G. W. Griffin, *Angew. Chem.*, **89**, 181 (1977).
57. S. Tanaka, H. Yamamoto, H. Nozaki, K. Sharpless, R. Michaelson and J. Cutting, *J. Amer. Chem. Soc.*, **96**, 5254 (1974).
58. C. W. Wilson and P. E. Philip, *Australian J. Chem.*, **28**, 2539 (1975).
59. M. R. Demuth, P. E. Garrett and J. D. White, *J. Amer. Chem. Soc.*, **98**, 634 (1976).
60. M. G. Hyman, M. N. Paddon-Row and R. N. Warrener, *Synth. Commun.*, **5**, 107 (1975).
61. P. P. Sane, V. R. Tadwalkar and A. S. Rao, *Indian J. Chem.*, **12**, 444 (1974).
62. P. Chamberlain, M. L. Roberts and G. H. Whitham, *J. Chem. Soc., (B)*, 1374 (1970).
63. J. M. Bachhawat and N. K. Mathur, *Tetrahedron Letters*, 691 (1971).
64. P. B. D. de la Mare and R. D. Wilson, *J. Chem. Soc., Perkin II*, 975 (1977).
65. W. G. Salmond and M. C. Sobala, *Tetrahedron Letters*, 1695 (1977).
66. B. A. Chiasson and G. A. Berchtold, *J. Org. Chem.*, **42**, 2008 (1977).
67. L. Pizzala, J. P. Aycard and H. Bodot, *J. Org. Chem.*, **43**, 1013 (1978).
68. W. J. W. Mayer, I. Oren and D. Ginsburg, *Tetrahedron*, **32**, 1005 (1976).
69. K. Weinges and H. Baake, *Chem. Ber.*, **110**, 1601 (1977).
70. Y. Kashman and O. Awerbouch, *Tetrahedron*, **31**, 45 (1975).
71. C. Symmes Jr., and L. D. Quin, *Tetrahedron Letters*, 1853 (1976).
72. R. C. Ewins, H. B. Henbest and M. A. McKervey, *J. Chem. Soc., Chem. Commun.*, 1085 (1967).
73. D. R. Boyd and M. A. McKervey, *Quart. Rev. (Lond.)*, **22**, 111 (1968).
74. R. M. Bowman and M. F. Grundon, *J. Chem. Soc. (C)*, 2368 (1967).
75. F. Montanari, I. Moretti and G. Torre, *J. Chem. Soc., Chem. Commun.*, 135 (1969).
76. D. R. Boyd, D. M. Jerina and J. W. Daly, *J. Org. Chem.*, **35**, 3170 (1970).
77. F. Montanari, I. Moretti and G. Torre, *Gazz. Chim. Ital.*, **104**, 7 (1974).
78. F. Montanari, I. Moretti and G. Torre, *Boll. Sci. Fac. Chim. Ind. Bologna*, **26**, 113 (1968).
79. W. H. Prikle and P. L. Rinaldi, *J. Org. Chem.*, **42**, 2080 (1977).
80. B. T. Golding, P. J. Sellars and Ah Kee Wong, *J. Chem. Soc., Chem. Commun.*, 570 (1977).
81. T. Itoh, K. Jitsukawa, K. Kaneda and S. Teranishi, *Tetrahedron Letters*, 3157 (1976).
82. S. Watanabe, K. Suga, T. Fujita and N. Takasaka, *J. appl. Chem. Biotechnol.*, **24**, 639 (1974).
83. G. P. Petrenko and V. P. Ivanova, *Zh. Org. Khim.*, **8**, 1065 (1972).
84. J. K. Crandall, W. H. Machleder and S. A. Sojka, *J. Org. Chem.*, **38**, 1149 (1973).
85. J. K. Crandall, W. W. Conover, J. B. Komin and W. H. Machleder, *J. Org. Chem.*, **39**, 1723 (1974).
86. J. Grimaldi and M. Bertrand, *Bull. Soc. Chim. Fr.*, 957 (1971).
87. J. Grimaldi and M. Bertrand, *Bull. Soc. Chim. Fr.*, 973 (1971).
88. E. Elkik and M. LeBlanc, *Compt. Rend. (C)*, **299**, 173 (1969).
89. K. Griesbaum, R. Kibar and B. Pfeffer, *Liebigs Ann. Chem.*, 214 (1975).
90. Yu. V. Zeifman, E. M. Rokhlin, V. Utebaer and I. L. Knunyants, *Dokl. Akad. Nauk. SSSR*, **226**, 1337 (1976).
91. Y. Kishi, M. Aratani, H. Tanino, T. Fukayama, T. Goto, S. Inoue, S. Sugiura and H. Kakoi, *J. Chem. Soc., Chem. Commun.*, 64 (1972).
92. N. Isogai, T. Okawa and T. Takeda, *German Patent*, No. 2,629,188; *Chem. Abstr.*, **87**, 5783p (1977).
93. W. K. Anderson and T. Veysoglua, *J. Org. Chem.*, **38**, 2267 (1973).
94. K. Ishikawa, H. C. Charles and G. W. Griffin, *Tetrahedron Letters*, 427 (1977).
95. C. R. Harrison and P. Hodge, *J. Chem. Soc., Chem. Commun.*, 1009 (1974).
96. C. R. Harrison and P. Hodge, *J. Chem. Soc., Perkin I*, 605 (1976).
97. W. Dittmann and K. Hamann, *Chemiker-Zeitung*, **95**, 857 (1971).
98. A.-G. Hoechst, *Dutch Patent* No. 75 08,894; *Chem. Abstr.*, **85**, 192531y (1976).
99. N. Kawabe, K. Okada and M. Ohno, *J. Org. Chem.*, **37**, 4210 (1972).
100. B. Plesničar and G. A. Russell, *Angew. Chem. (Intern. Ed. Engl.)*, **9**, 797 (1970).
101. R. M. Coates and J. W. Williams, *J. Org. Chem.*, **39**, 3054 (1974).
102. S. A. Kozhin and E. I. Sorochinskaya, *Zh. Obsch. Khim.*, **44**, 2350 (1974).

103. Yu. S. Shabarov, S. A. Blagodatskikh and M. I. Levina, *Zh. Org. Khim.*, **11**, 1223 (1975).
104. W. Cocker and D. H. Grayson, *J. Chem. Soc., Perkin II*, 791 (1976).
105. G. Farges and A. Kergomard, *Bull. Soc. Chim. Fr.*, 4476 (1969).
106. T. Mori, K. H. Yang, K. Kimoto and H. Nozaki, *Tetrahedron Letters*, 2419 (1970).
107. R. J. Ferrier and N. Prasad, *J. Chem. Soc. (C)*, 575 (1969).
108. J. Rebek, S. F. Wolf and A. B. Mossman, *J. Chem. Soc., Chem. Commun.*, 711 (1974).
109. E. P. Kyba and D. C. Alexander, *Tetrahedron Letters*, 4563 (1976).
110. K. R. Darnall and J. N. Pitts, *J. Chem. Soc., Chem. Commun.*, 1305 (1970).
111. C. Wilkins, *Synthesis*, 156 (1973).
112. P. A. Grieco, Y. Yokoyama, S. Gilman and M. Nishizawa, *J. Org. Chem.*, **42**, 2034 (1977).
113. M. Naotake, S. Noboru and T. Shigeru, *Tetrahedron Letters*, 2029 (1970).
114. R. D. Temple, *J. Org. Chem.*, **35**, 1275 (1970).
115. P. Chautemps and J. L. Pierre, *Bull. Soc. Chim. Fr.*, 2899 (1974).
116. P. Chautemps, *Compt. Rend. (C)*, **284**, 807 (1977).
117. I. G. Tishchenko, I. F. Revinskii and V. N. Sytin, *Zh. Org. Khim.*, **13**, 347 (1977).
118. H. Newman and R. B. Angier, *Tetrahedron*, **26**, 825 (1970).
119. M. Igarashi, M. Akano, A. Fujimoto and H. Madrikawa, *Bull. Chem. Soc. Japan*, **43**, 2138 (1970).
120. A. Robert and A. Foucaud, *Bull. Soc. Chim. Fr.*, 4528 (1969).
121. J. D. McChesney and A. F. Wycpatela, *J. Chem. Soc., Chem. Commun.*, 542 (1971).
122. B. A. Brady, M. M. Healey, J. A. Kennedy, W. I. O'Sullivan and E. M. Philbin, *J. Chem. Soc., Chem Commun.*, 1434 (1970).
123. J. Carduff and D. G. Leppard, *J. Chem. Soc., Perkin I*, 1325 (1977).
124. F. Khuong-Huu, D. Herlem and M. Beneche, *Bull. Soc. Chim. Fr.*, 2702 (1970).
125. M. E. Kuehne and J. A. Nelson, *J. Org. Chem.*, **35**, 161 (1970).
126. H. B. Henbest and W. R. Jackson, *J. Chem. Soc. (C)*, 2459 (1967).
127. B. Pelc and E. Kodicek, *J. Chem. Soc. (C)*, 1568 (1971).
128. A. A. Akhrem, A. V. Kamernitskii, I. G. Reshetova and K. Yu. Chernyuk, *Izv. Akad. Nauk. SSSR., Ser. Khim.*, 709 (1973).
129. A. A. Akhrem, I. D. Vladimirova and V. N. Dobrinin, *Izv. Akad. Nauk. SSSR, Ser. Khim.*, 1316 (1967).
130. G. Ohloff and G. Uhde, *Helv. Chim. Acta*, **53**, 531 (1970).
131. J. Katsuhara, H. Yamasaki and N. Yamamoto, *Bull. Chem. Soc. Japan*, **43**, 1584 (1970).
132. B. M. Trost and T. N. Salzmann, *J. Chem. Soc., Chem. Commun.*, 571 (1975).
133. B. Zwanenburg and J. Ter Wiel, *Tetrahedron Letters*, 935 (1970).
134. R. Helder, I. C. Hummelen, R. W. P. H. Laane, J. S. Wiering and H. Wynberg, *Tetrahedron Letters*, 1831 (1976).
135. T. Kametani, H. Nemoto and K. Fukumoto, *Heterocycles*, **6**, 1365 (1977).
136. A. T. Menyailo, Kh. E. Khcheyan, M. V. Pospelov, I. E. Pokrovskaya, O. R. Kaliko, T. A. Bortyan, I. K. Alferova and L. D. Krasnoslobodskaya, *Dutch Patent*, No. 73 12,323; *Chem. Abstr.*, **83**, 206084c (1975).
137. Y. Watanabe, T. Nishizawa and J. Kobayashi, *U.S. Patent*, No. 3,806,467 (1974); *Chem. Abstr.*, **81**, 78090w (1974).
138. A. M. Matucci, E. Perotti and A. Santambrogio, *J. Chem. Soc., Chem. Commun.*, 1198 (1970).
139. G. G. Allan and A. N. Neogi, *J. Catal.*, **19**, 256 (1970).
140. G. G. Allan and A. N. Neogi, *J. Phys. Chem.*, **73**, 2093 (1969).
141. H. C. Stevens and A. J. Kaman, *J. Amer. Chem. Soc.*, **87**, 734 (1965).
142. V. N. Sapunov and N. N. Lebedev, *Zh. Org. Khim.*, **2**, 273 (1966).
143. G. G. Allan and A. N. Neogi, *J. Catal.*, **16**, 197 (1970).
144. K. Franz, G. Hauthal, J. Klaassen and S. Busch, *D.D.R. Patent*, No. 122,379 (1976); *Chem. Abstr.*, **87**, 135001t (1977).
145. Z. Raciszewski, *J. Amer. Chem. Soc.*, **82**, 1267 (1960).

146. M. Igarashi and H. Midorikawa, *J. Org. Chem.*, **32**, 3399 (1967).
147. M. A. Beg and I. Ahmad, *J. Catal.*, **39**, 260 (1975).
148. M. A. Beg and I. Ahmad, *J. Org. Chem.*, **42**, 1590 (1977).
149. M. A. Beg and I. Ahmad, *Indian J. Chem. (A)*, **15**, 105 (1977).
150. B. G. Christensen, W. J. Leanza, T. R. Beattie, A. A. Patchett, B. H. Arison, R. E. Ormond, F. A. Kuehl, G. Albers-Schonberg and O. Jardetzky, *Science*, **166**, 123 (1969).
151. E. J. Glamkowski, G. Gal, R. Purick, A. J. Davidson and M. Sletzinger, *J. Org. Chem.*, **35**, 3510 (1970).
152. M. Tohma, T. Tomita and M. Kimura, *Tetrahedron Letters*, 4359 (1973).
153. H. Mimoun, I. Seree de Roch, L. Sajus and P. Menguy, *French Patent*, No. 1,549,184 (1968); *Chem. Abstr.*, **72**, 3345p (1970).
154. H. Mimoun, I. Seree de Roch, P. Menguy and L. Sajus, *German Patent*, No. 1,815,998 (1969); *Chem. Abstr.*, **72**, 90640x (1970).
155. H. Mimoun, I. Seree de Roch, P. Menguy and L. Sajus, *German Patent*, No. 1,817,717 (1970); *Chem. Abstr.*, **72**, 10042n (1970).
156. II. Mimoun, I. Scrcc dc Roch and L. Sajus, *Tetrahedron*, **26**, 37 (1970).
157. H. Arakawa, Y. Moro-oka and A. Ozaki, *Bull. Chem. Soc. Japan*, **47**, 2958 (1974).
158. A. Akhrem, T. Timoshchuk and D. Metelitsa, *Tetrahedron*, **30**, 3165 (1974).
159. K. Sharpless, J. Townsend and D. Williams, *J. Amer. Chem. Soc.*, **94**, 296 (1972).
160. G. A. Tolstikov, V. P. Yur'ev, I. A. Gailyunas and U. M. Dzhemilev, *Zh. Obshch. Khim.*, **44**, 215 (1974).
161. R. Hiatt in *Oxidation Techniques and Applications in Organic Synthesis*, Vol. 2 (Ed. R. Augustine), Marcel Dekker, New York, 1971, p. 133.
162. R. A. Sheldon and J. K. Kochi, *Advan. Catal.*, **25**, 272 (1976).
163. A. Doumaux in *Oxidation*, Vol. 2 (Ed. R. Augustine), Marcel Dekker, New York, 1971, pp. 141–185.
164. J. E. Lyons, *Aspects Homogeneous Catalysis*, **3**, 1 (1977).
165. V. A. Belyaev, A. A. Petukhov, A. N. Bushin, B. A. Plechev, A. B. Feigin, M. I. Farberov, A. G. Liakumovich, G. I. Rutman and S. I. Kryukov, *U.S.S.R. Patent*, No. 466,221; *Chem. Abstr.*, **83**, 28077 (1975).
166. P. E. Bost and M. Costantini, *German Patent*, No. 2,428,559; *Chem. Abstr.*, **83**, 43169t (1975).
167. I. Kende, G. Keresztury and S. Dombi, *Hungarian Patent*, No. 10,383; *Chem. Abstr.*, **84**, 105377r (1976).
168. A. L. Stautzenberger, *U.S. Patent*, No. 3,931,249; *Chem. Abstr.*, **84**, 105379t (1976).
169. M. N. Sheng, J. G. Zajacek and T. N. Baker, III, *U.S. Publ. Pat. Appl.*, No. B 521,324; *Chem. Abstr.*, **84**, 135449h (1976).
170. S. Ozaki, T. Takahashi and I. Sudo, *Japanese Patent*, No. 74 24,003; *Chem. Abstr.*, **83**, 9754v (1975).
171. N. Indictor and W. Brill, *J. Org. Chem.*, **29**, 2074 (1964).
172. E. Gould, R. Hiatt and K. Irwin, *J. Amer. Chem. Soc.*, **90**, 4573 (1968).
173. S. A. Nirova, N. N. Bayanova, S. P. Luchkina, E. P. Krysin and L. G. Andronova, *Neftekhimiya*, **15**, 756 (1975).
174. M. N. Sheng and J. G. Zajacek, *Adv. Chem. Ser.*, **76**, 418 (1968).
175. R. Landau, *Hydrocarbon Proc.*, **46**, 141 (1967).
176. Á. Gedra, L. Sümegi, A. Németh and D. Gál, *Magy. Kém. Folyóirat*, **80**, 368 (1974); *Chem. Abstr.*, **81**, 168849e (1974).
177. V. P. Yur'ev, I. A. Gailyunas, Z. G. Isaeva and G. A. Tolstikov, *Izv. Akad, Nauk. SSSR, Ser. Khim.*, 919 (1974).
178. R. A. Sheldon and J. A. Van Doorn, *J. Catal.*, **31**, 427 (1973).
179. R. A. Sheldon, J. A. Van Doorn, C. W. A. Schram and A. J. De Jong, *J. Catal.*, **31**, 438 (1973).
180. M. N. Sheng, J. G. Zajacek and T. N. Baker, *Amer. Chem. Soc., Div. Petrol. Chem. Prepr.*, **15**, E19 (1970).
181. P. Forzatti, F. Trifiro and I. Pasquon, *Chim. Ind. (Milan)*, **56**, 259 (1974).
182. Kh. E. Khcheyan and I. K. Alferova, *Khim. Promysl.*, **10**, 742 (1977).

183. F. Trifiro, P. Forzatti and I. Pasquon, *Catal. Proc. Int. Symp., 1974*, 509–519 (1975).
184. S. S. Srednev, S. I. Kryukov and M. I. Farberov, *Kinet. Katal.*, **16**, 1472 (1975).
185. G. L. Linden and M. F. Farona, *J. Catal.*, **48**, 284 (1977).
186. P. F. Wolf and R. K. Barnes, *J. Org. Chem.*, **34**, 3441 (1969).
187. R. A. Sheldon and J. A. Van Doorn, *J. Catal.*, **34**, 242 (1974).
188. L. Červený, A. Marhoul, V. Růžička, A. Hora and J. Novak, *Czech. Patent*, No. 156,839; *Chem. Abstr.*, **83**, 131434f (1975).
189. J. S. McIntyre and R. J. B. Wilson, *Canadian Patent*, No. 967,973; *Chem. Abstr.*, **83**, 163981y (1975).
190. C. Y. Wu and H. E. Swift, *J. Catal.*, **43**, 380 (1976).
191. B. N. Bobylev, M. I. Farberov, E. P. Tenenitsyna, D. I. Epshtein and N. V. Dormidontova, *Neftekhimiya*, **16**, 255 (1976).
192. R. C. Palermo and K. B. Sharpless, *J. Amer. Chem. Soc.*, **99**, 1990 (1977).
193. S. Yamada, T. Mashiko and Sh. Terashima, *J. Amer. Chem. Soc.*, **99**, 1988 (1977).
194. G. Howe and R. Hiatt, *J. Org. Chem.*, **36**, 2493 (1971).
195. R. Sheldon, *Rec. Trav. Chim.*, **92**, 253 (1973).
196. T. N. Baker, G. J. Mains, M. N. Sheng and J. G. Zajacek, *J. Org. Chem.*, **38**, 1145 (1973).
197. V. N. Sapunov, I. Margitfalvi and N. N. Lebedev, *Kinet. Katal.*, **15**, 1442 (1974).
198. C. Su, J. Reed and E. Gould, *Inorg. Chem.*, **12**, 337 (1973).
199. M. I. Farberov, G. A. Stozhkova, A. V. Bondarenko, T. M. Kirik and N. A. Ognevskaya, *Neftekhimiya*, **11**, 404 (1971).
200. I. Ya. Mokrousova, L. A. Oshin, M. R. Flid and Yu. A. Treger, *Kinet. Katal.*, **17**, 792 (1976).
201. E. Costa Novella, P. J. Martinez de La Cuesta and E. Rus Martinez, *Ann. Quim.*, **70**, 545 (1974).
202. V. A. Gavrilenko, E. I. Evzerikhin and I. I. Moiseev, *Izv. Akad. Nauk SSSR, Ser. Khim.*, 29 (1977).
203. V. A. Gavrilenko, E. I. Evzerikhin and I. I. Moiseev, *Izv. Akad. Nauk SSSR, Ser. Khim.*, 34 (1977).
204. V. A. Gavrilenko, E. I. Evzerikhin, I. I. Moiseev and I. Sh. Fish, *Izv. Akad. Nauk SSSR, Ser. Khim.*, 1746 (1977).
205. V. N. Sapunov, I. Yu. Litvintsev, R. B. Svitych and N. N. Rzhevskaya, *Kinet. Katal.*, **18**, 408 (1977).
206. B. N. Bobylev, L. V. Mel'nik, M. I. Farberov and L. I. Bobyleva, *Kinet. Katal.*, **18**, 811 (1977).
207. S. A. Kesarev, B. N. Bobylev, M. I. Farberov and L. I. Bobyleva, *Neftekhimiya*, **17**, 576 (1977).
208. A. O. Chong and K. B. Sharpless, *J. Org. Chem.*, **42**, 1587 (1977).
209. V. A. Gavrilenko, E. I. Evzerikhin, V. A. Kolosov, G. M. Larin and I. I. Moiseev, *Izv. Akad. Nauk SSSR, Ser. Khim.*, 1954 (1974).
210. R. Sheldon, *Rec. Trav. Chim.*, **92**, 367 (1973).
211. J. Kaloustian, L. Lena and J. Metzger, *Tetrahedron Letters*, 599 (1975).
212. K. E. Khcheyan, L. N. Samter and A. G. Sokolov, *Neftekhimiya*, **15**, 415 (1975).
213. H. Arakawa and A. Ozaki, *Chem. Letters*, 1245 (1975).
214. F. Trifiro, P. Forzatti, S. Preite and I. Pasquon, *Proceedings 1st Conference on Chemical Uses of Molybdenum, 1973*, 169 (1975).
215. K. B. Yatsimirskii, V. M. Belousov and A. P. Filippov, *Dokl. Akad. Nauk SSSR*, **224**, 1369 (1975).
216. P. Kok and I. P. Skibida, *Zh. Fiz. Khim.*, **48**, 2400 (1974).
217. R. B. Svitych, N. N. Rzhevskaya, A. L. Buchachenko, O. P. Yablonskii, A. A. Petukhov and V. A. Belyaev, *Kinet. Katal.*, **17**, 921 (1976).
218. V. N. Sapunov, I. Yu. Litvintsev, I. Margitfalvi and N. N. Lebedev, *Kinet. Katal.*, **18**, 620 (1977).
219. V. P. Yur'ev, I. A. Gailyunas, L. V. Spirikhin and G. A. Tolstikov, *Zh. Obshch. Khim.*, **45**, 2312 (1975).
220. J. Lyons, *Homogeneous Catalysis-II. Joint Symposium of the Division of Industrial and Engineering Chemistry and Petroleum Chemistry*, 166th Meeting, ACS, Chicago, Illinois, August 1973.

221. K. Sharpless and R. Michaelson, *J. Amer. Chem. Soc.*, **95**, 6136 (1973).
222. J. Lyons in *Catalysis in Organic Synthesis*, (Eds. P. Rylander and H. Greenfield), Academic Press, New York, 1976, pp. 235–255.
223. Ya. M. Paushkin, I. M. Kolesnikov, B. T. Sherbanenko, S. A. Nizova and L. M. Vilenskii, *Kinet. Katal.*, **13**, 493 (1972).
224. G. A. Tolstikov, V. P. Yur'ev, I. A. Gailyunas and U. M. Dzhemilev, *Zh. Obshch. Khim.*, **44**, 215 (1974).
225. U. M. Dzhemilev, V. P. Yur'ev, G. A. Tolstikov, F. B. Gershanov and S. R. Rafikov, *Dokl. Akad. Nauk SSSR*, **196**, 588 (1971).
226. R. Breslow and L. M. Maresca, *Tetrahedron Letters*, 623 (1977).
227. I. A. Gailyunas, E. M. Tsyrlina, N. I. Solov'eva, N. G. Komalenkova and V. P. Yur'ev, *Zh. Obshch. Khim.*, **47**, 2394 (1977).
228. N. M. Emanuel, E. T. Denisov and Z. K. Maizus, *Cepnye reakcii okisleniya uglevodorodov v zhidkoi faze*, Nauka, Moskva, 1965.
229. A. A. Syrov and V. K. Gyskovskii, *Usp. Khim.*, **39**, 817 (1970).
230. A. Creiner, *J. prakt. Chem.*, **38**, 207 (1968).
231. A. Padwa and L. Brodski, *Tetrahedron Letters*, 1045 (1973).
232. H. Hart and P. B. Lavrik, *J. Org. Chem.*, **39**, 1793 (1974).
233. F. Tsuchiya, M. Kuwa and T. Ikawa, *Kogyo Kagaku Zasshi*, **73**, 2655 (1970); *Chem. Abstr.*, **74**, 125287h (1971).
234. H. Takeshita, H. Kanamori and T. Hatsui, *Tetrahedron Letters*, 3139 (1973).
235. L. A. Paquette, C. C. Liao, D. C. Liotta and W. E. Fristad, *J. Amer. Chem. Soc.*, **98**, 6412 (1976).
236. N. Shimizu and P. D. Bartlett, *J. Amer. Chem. Soc.*, **98**, 4193 (1976).
237. E. Vogel, A. Breuer, C. D. Sommerfield, R. E. Davis and L. K. Liu, *Angew. Chem.*, **89**, 175 (1977).
238. F. Tsuchiya and T. Ikawa, *Can. J. Chem.*, **47**, 3191 (1969).
239. C. J. Michejda and D. H. Campbell, *J. Amer. Chem. Soc.*, **98**, 6728 (1976).
240. A. D. Vreugdenhil and H. Reit, *Rec. Trav. Chim.*, **91**, 237 (1972).
241. H. Kropf and H. R. Yazdanbakhch, *Synthesis*, 711 (1977).
242. W. F. Brill and N. Indictor, *J. Org. Chem.*, **29**, 710 (1964).
243. V. M. Parfenov and Z. K. Maizus, *Neftekhimiya*, **11**, 416 (1971).
244. E. A. Blyumberg, M. G. Bulygin and N. M. Emanuel, *Dokl. Akad. Nauk SSSR*, **166**, 353 (1966).
245. D. J. M. Ray and D. J. Waddington, *J. Amer. Chem. Soc.*, **90**, 7176 (1968).
246. T. V. Filippova, E. A. Blyumberg, L. I. Kas'yan, Ya. L. Letuchii and L. A. Sil'chenko, *Dokl. Akad. Nauk SSSR*, **210**, 644 (1973).
247. D. S. Jones and S. J. Moss, *Int. J. Chem. Kinet.*, **6**, 443 (1974).
248. J. J. Havel and C. J. Hunt, *J. Phys. Chem.*, **80**, 779 (1976).
249. H. Arzoumanian, A. Blanc, U. Hartig and J. Metzger, *Tetrahedron Letters*, 1011 (1974).
250. S. Imamura, T. Otani and H. Teranishi, *Bull. Chem. Soc. Japan*, **48**, 1245 (1975).
251. E. Gould and M. Rado, *J. Catal.*, **13**, 238 (1969).
252. C. Sharma, S. Sethi and S. Dev, *Synthesis*, 45 (1974).
253. D. Holland and D. Milner, *J. Chem. Soc., Dalton*, 2440 (1975).
254. A. Fusi, R. Ugo, F. Fox, A. Pasini and S. Cenini, *J. Organomet. Chem.*, **26**, 417 (1971).
255. J. Rouchaud and J. Mawaka, *J. Catal.*, **19**, 172 (1970).
256. V. A. Tulupov and T. N. Zakhar'eva, *Zh. Fiz. Khim.*, **49**, 272 (1975).
257. J. E. Lyons, *Tetrahedron Letters*, 2737 (1974).
258. E. DeRuiter, *Erdöl und Kohle*, **25**, 510 (1972).
259. C. Bocard, C. Gadelle, H. Mimoun and I. Seree de Roch, *French Patent*, No. 2,044,007 (1971); *Chem. Abstr.*, **75**, 151660q (1971).
260. S. B. Cavitt, *U.S. Patent*, No. 3,856,827 (1974); *Chem. Abstr.*, **82**, 98755r (1975).
261. A. V. Bobolev, A. S. Tatikolov, N. N. Lukaschina, I. S. Krainov and N. M. Emanuel, *German Patent*, No. 2,313,023 (1973); *Chem. Abstr.*, **80**, 83867p (1974).
262. R. D. Smetana, *U.S. Patent*, No. 3,637,768; *Chem. Abstr.*, **76**, 99718u (1972).
263. K. Shin and I. Kehoe, *J. Org. Chem.*, **36**, 2717 (1971).
264. A. Fusi, R. Ugo and G. Zanderighi, *J. Catal.*, **34**, 175 (1974).

265. J. E. Lyons and J. O. Turner, *J. Org. Chem.*, **37**, 2881 (1972).
266. R. A. Budnik and J. K. Kochi, *J. Org. Chem.*, **41**, 1384 (1976).
267. T. Itoh, K. Kaneda and S. Teranishi, *Bull. Chem. Soc., Japan*, **48**, 1337 (1975).
268. V. K. Tsykovskii, V. C. Fedorov and Y. L. Moskovich, *Zh. Obshch. Khim.*, **45**, 248 (1975).
269. S. Muto and Y. Kamiya, *J. Catal.*, **41**, 148 (1976).
270. M. E. Pudel' and Z. K. Maizus, *Izv. Akad. Nauk SSSR, Ser. Khim.*, 43 (1975).
271. K. Takao, H. Azuma, Y. Fujiwara, T. Imanaka and S. Teranishi, *Bull. Chem. Soc. Japan*, **45**, 2003 (1972).
272. J. Farrar, D. Holland and D. Milner, *J. Chem. Soc., Dalton*, 815 (1975).
273. M. J. Y. Chen and J. K. Kochi, *J. Chem. Soc., Chem. Commun.*, 204 (1977).
274. V. Bazant, J. Beranek, B. Jiricek and R. Kubicka, *Czech. Patent*, No. 152,768; *Chem. Abstr.*, **82**, 4749s (1975).
275. D. Bryce-Smith, *Chem. Ind. (Lond.)*, 154 (1975).
276. F. Jaminon-Beekman and C. A. M. Weterings, *Dutch Patent*, No. 73 05,029; *Chem. Abstr.*, **82**, 98740g (1975).
277. S. Ozaki, T. Takashaki, A. Tamaoki and T. Kiyoura, *Japanese Patent*, No. 75 32,091; *Chem. Abstr.*, **83**, 9759a (1975).
278. J. Wasilewski and A. Gawdzik, *Przemysl Chem.*, **55**, 357 (1976); *Chem. Abstr.*, **86**, 42901a (1977).
279. M. I. Temkin, *Kinet. Katal.*, **18**, 547 (1977).
280. Sh. L. Guseinov, I. T. Frolkina, L. A. Vasilevich, A. K. Avetisov and A. I. Gel'bstein, *React. Kinet. Catal. Letters*, **6**, 409 (1977).
281. Sh. L. Guseinov, I. T. Frolkina, L. A. Vasilevich, A. K. Avetisov and A. I. Gel'bstein, *Kinet. Katal.*, **18**, 1455 (1977).
282. D. Kamenski and D. Bonchev, *Kinet. Katal.*, **19**, 633 (1978).
283. W. F. Richey, *J. Phys. Chem.*, **76**, 213 (1972).
284. T. Mitsuhata, K. Matsuda and T. Kumasawa, *Japanese Patent*, No. 74 26,603; *Chem. Abstr.*, **82**, 98748r (1975).
285. C. Mazzocchia, R. Del Rosso and P. Centola, *ICP*, **4**, 57 (1976); *Chem. Abstr.*, **86**, 42839m (1977).
286. H. Mimoun, C. Gadelle, C. Bocard, I. Seree de Roch and P. Baumgartner, *French Patent*, No. 2,187,774; *Chem. Abstr.*, **81**, 3752e (1974).
287. J. Rouchaud and J. J. Fripiat, *Bull. Soc. Chim. Fr.*, 78 (1969).
288. A. V. Bobolev, A. S. Tatikolov, N. N. Lukashina, I. S. Krainov and N. M. Emanuel, *U.S. Patent*, No. 3,957,690; *Chem. Abstr.*, **85**, 52324n (1976).
289. J. E. Lyons, *U.S. Patent*, No. 4,021,389; *Chem. Abstr.*, **87**, 22382z (1977).
290. M. Sato, Y. Kobayashi and M. Hirakuni, *Japanese Patent*, No. 75 00,007 (1975); *Chem. Abstr.*, **83**, 43171n (1975).
291. M. Baccouche, J. Ernst, J. H. Fuhrhop, R. Schlözer and H. Arzoumanian, *J. Chem. Soc., Chem. Commun.*, 821 (1977).
292. K. M. Sokova, G. A. Zelenaya and A. N. Bashkirov, *Neftekhimiya*, **16**, 445 (1976).
293. D. Dimitrov, V. Angelov, A. Badev and V. Badeva, *Rev. Chim. (Bucharest)*, **24**, 899 (1973).
294. I. K. Alferova, T. A. Bortyan and M. V. Pospelov, *Neftekhimiya*, **17**, 582 (1977).
295. R. E. Keay and G. A. Hamilton, *J. Amer. Chem. Soc.*, **98**, 6578 (1976).
296. G. S. Aulakh, M. S. Wadia and P. S. Kalsi, *Chem. Ind. (Lond.)*, 802 (1970).
297. P. S. Kalsi, K. S. Kumar and M. S. Wadia, *Chem. Ind. (Lond.)*, 31 (1971).
298. G. E. M. Mousse and S. O. Abdalla, *J. Appl. Chem.*, **20**, 256 (1970).
299. J. S. Littler, *Tetrahedron*, **27**, 81 (1971).
300. A. Robert and A. Foucaud, *Bull. Soc. Chim. Fr.*, 2531 (1969).
301. G. R. Wellman, B. Lam, E. L. Anderson and E. V. White, *Synthesis*, 547 (1976).
302. S. Krishnan, D. G. Hamilton and A. Gordon, *J. Amer. Chem. Soc.*, **99**, 8121 (1977).
303. S. A. Shackelford and G. U. Yuen, *Inorg. Nucl. Chem. Letters*, **9**, 605 (1973); *J. Org. Chem.*, **40**, 1869 (1975).
304. W. Kruse, *J. Chem. Soc., Chem. Commun.*, 1610 (1968); W. Kruse and T. M. Bednarski, *J. Org. Chem.*, **36**, 1154 (1971).

305. Farbenfabriken Bayer A.-G. and Pullman Inc., *French Patent*, No. 2,008,606 (1970); *Chem. Abstr.*, **73**, 66000p (1970).
306. P. D. Bartlett, A. L. Baumstark and M. E. Landis, *J. Amer. Chem. Soc.*, **95**, 6486 (1973).
307. E. N. Barantsevich and T. I. Temnikova, *Zh. Org. Khim.*, **2**, 648 (1966).
308. T. H. Cromartie and C. G. Swain, *J. Amer. Chem. Soc.*, **98**, 545 (1976).
309. B. Ganem, *J. Amer. Chem. Soc.*, **98**, 858 (1976).
310. W. Cocker and D. H. Grayson, *Tetrahedron Letters*, 4451 (1969).
311. R. P. Hanzlik, *Org. Synth.*, **56**, 112 (1977).
312. R. Carlson and R. Ardon, *J. Org. Chem.*, **36**, 216 (1971).
313. G. Berti, F. Bottari, G. Lippi and M. Macchia, *Tetrahedron*, **24**, 1959 (1968).
314. P. Duhamel, L. Duhamel and J. Gralak, *Compt. Rend. (C)*, **269**, 1658 (1969).
315. J. Cantacuzene and J. M. Normant, *Compt. Rend. (C)*, **271**, 748 (1970).
316. J. Cantacuzene and R. Jantzen, *Tetrahedron*, **26**, 2429 (1970).
317. J. Cantacuzene and M. Atlani, *Tetrahedron*, **26**, 2447 (1970).
318. J. W. Cornforth and D. T. Green, *J. Chem. Soc. (C)*, 846 (1970).
319. J. C. Richer and C. Freppel, *Tetrahedron Letters*, 4411 (1969).
320. M. Fetizon, M. Golfier, M. T. Montaufier and J. Rens, *Tetrahedron*, **31**, 987 (1975).
321. W. E. Truce and L. W. Christensen, *J. Org. Chem.*, **36**, 2538 (1971).
322. R. A. Bekker, G. V. Asratyan, B. L. Dyatkin and I. L. Knunyants, *Tetrahedron*, **30**, 3539 (1974).
323. J. Gombos, *Chem. Ber.*, **109**, 2645 (1976).
324. B. T. Golding, D. R. Hall and S. Sakrika, *J. Chem. Soc., Perkin I*, 1214 (1973).
325. J. C. Paladini and J. Chuche, *Bull. Soc. Chim. Fr.*, 192 (1974).
326. D. Dansette and D. M. Jerina, *J. Amer. Chem. Soc.*, **96**, 1224 (1974).
327. N. N. Akhtar and D. R. Boyd, *J. Chem. Soc., Chem. Commun.*, 591 (1975).
328. H. Yagi and D. M. Jerina, *J. Amer. Chem. Soc.*, **95**, 243 (1973).
329. H. Yagi and D. M. Jerina, *J. Amer. Chem. Soc.*, **97**, 3185 (1975).
330. B. Delmond, J.-C. Pommier and J. Valade, *J. Organomet. Chem.*, **35**, 91 (1972).
331. R. Wylde and F. Forissier, *Bull. Soc. Chim. Fr.*, 4508 (1969).
332. P. A. Bartlett and K. K. Jernstedt, *J. Amer. Chem. Soc.*, **99**, 4829 (1977).
333. H. Christol, J. Coste and F. Plenat, *Tetrahedron Letters*, 1143 (1972).
334. J. C. Pommelet, N. Manisse and J. Chuche, *Compt. Rend. (C)*, **270**, 1894 (1970).
335. P. W. Feit, N. Rastrup-Andersen and R. Matagne, *J. Med. Chem.*, **13**, 1173 (1970).
336. Z. Chabudzinski, Z. Rykowski, U. Lipnicka and D. Sedzik-Hibner, *Rocz. Chem.*, **46**, 1443 (1972); *Chem. Abstr.*, **77**, 1403310r (1972).
337. S. Holand and R. Epsztein, *Synthesis*, 706 (1977).
338. J. C. Paladini and J. Chuche, *Bull. Soc. Chim. Fr.*, 187 (1974).
339. J. M. Coxon, M. P. Hartshorn and A. J. Lewis, *Australian J. Chem.*, **24**, 1009 (1971).
340. G. Berti, F. Bottari and A. Marsili, *Tetrahedron*, **25**, 2939 (1969).
341. G. Berti, B. Macchia, F. Macchia and L. Monti, *J. Org. Chem.*, **33**, 4045 (1968).
342. M. Rosenberger, A. J. Duggan, R. Borer, R. Müller and G. Saucy, *Helv. Chim. Acta*, **55**, 2663 (1972).
343. P. Moreau, A. Casadevall and E. Casadevall, *Bull. Soc. Chim. Fr.*, 2013 (1969).
344. J. D. McClure, *J. Org. Chem.*, **32**, 3888 (1967).
345. K. Undheim and B. P. Nilsen, *Acta Chem. Scand.*, **B29**, 503 (1975).
346. M. Parilli, G. Barrone, M. Adinolfi and L. Mangoni, *Tetrahedron Letters*, 207 (1976).
347. C. A. Grob and R. A. Wohl, *Helv. Chim. Acta*, **49**, 2175 (1966).
348. R. Hanselaer, M. Samson and M. Vandewalle, *Tetrahedron*, **34**, 2393 (1978).
349. W. Cocker and D. H. Grayson, *Tetrahedron Letters*, 4451 (1969).
350. J. P. Vidal, J. P. Girard, J. C. Rossi, J. P. Chapat and R. Granger, *Org. Mag. Reson*, **6**, 522 (1974).
351. J. Klinot, K. Waisser, L. Streinz and A. Vystrčil, *Coll. Czech. Chem. Commun.*, **35**, 3610 (1970).
352. T. Nambara, K. Shimada and S. Goya, *Chem. Pharm. Bull. (Tokyo)*, **18**, 453 (1970).
353. B. O. Lindgren and C. M. Svahn, *Acta Chem. Scand.*, **24**, 2699 (1970).
354. M. S. Newman and C. H. Chen, *J. Amer. Chem. Soc.*, **94**, 2149 (1972).

355. D. A. Seeley and J. McElwee, *J. Org. Chem.*, **38**, 1691 (1973).
356. R. A. Kretchmer, R. A. Conrad and E. D. Mihelich, *J. Org. Chem.*, **38**, 1251 (1973).
357. H. R. Ansari and R. Clark, *Tetrahedron Letters*, 3085 (1975).
358. Y. Pocker and B. P. Ronald, *J. Org. Chem.*, **35**, 3362 (1970).
359. M. Fetizon and P. Foy, *Compt. Rend.*, **263**, 821 (1966); *Coll. Czech. Chem. Commun.*, **35**, 440 (1970).
360. H. Wynberg, E. Boelema, J. H. Wieringa and J. Strating, *Tetrahedron Letters*, 3613 (1970).
361. J. C. Martin, J. A. Ranz and R. J. Arhart, *J. Amer. Chem. Soc.*, **96**, 4604 (1974).
362. R. Boigegrian and B. Castro, *Tetrahedron Letters*, 3459 (1975).
363. R. Boigegrian and B. Castro, *Tetrahedron*, **32**, 1283 (1976).
364. S. H. Goh and R. G. Harvey, *J. Amer. Chem. Soc.*, **95**, 242 (1973).
365. H. Choo and R G. Harvey, *Tetrahedron Letters*, 1491 (1974).
366. D. Avnir, A. Grauer, D. Dinur and J. Blum, *Tetrahedron*, **31**, 2457 (1975).
367. N. H. Cromwell and J. L. Martin, *J. Org. Chem.*, **33**, 1890 (1968).
368. N. F. Woolsey and M. H. Khalil, *J. Org. Chem.*, **38**, 4216 (1973).
369. F. W. Bachelor and R. K. Bansal, *J. Org. Chem.*, **34**, 3600 (1969).
370. J. Seyden-Penne, M. C. Roux-Schmitt and A. Roux, *Tetrahedron*, **26**, 2649 (1970).
371. E. B. Castro, J. Villieras and N. Ferracutti, *Comp. Rend. (C)*, **268**, 1403 (1969).
372. D. J. Dagli, P. Yu and J. Wemple, *J. Org. Chem.*, **40**, 3173 (1975).
373. R. F. Borch, *Tetrahedron Letters*, 3761 (1972).
374. J. D. White, J. B. Bremner, M. J. Dimsdale and R. L. Garcea, *J. Amer. Chem. Soc.*, **93**, 281 (1971).
375. J. Villieras, P. Coutrot and J. C. Combret, *Compt. Rend. (C)*, **270**, 1250 (1970).
376. G. Lavielle, J. C. Combret and J. Villieras, *Compt. Rend. (C)*, **272**, 2175 (1971).
377. J. Villieras, G. Lavielle and J. C. Combret, *Bull. Soc. Chim. Fr.*, 898 (1971).
378. G. Kyriakakou and J. Seydon-Penne, *Tetrahedron Letters*, 1737 (1974).
379. B. Deschamps and J. Seyden-Penne, *Compt. Rend. (C)*, **271**, 1097 (1970).
380. P. Coutrot, J. C. Combret and J. Villieras, *Compt. Rend. (C)*, **270**, 1674 (1970).
381. J. Jonczyk, M. Fedorynski and M. Makosza, *Tetrahedron Letters*, 2395 (1972).
382. J. M. McIntosh and H. Khalil, *J. Org. Chem.*, **42**, 2123 (1977).
383. D. F. Tavares, R. E. Estep and M. Blezard, *Tetrahedron Letters*, 2373 (1970).
384. V. Reutrakul and W. Kanghae, *Tetrahedron Letters*, 1377 (1977).
385. P. F. Vogt and D. F. Tavares, *Can. J. Chem.*, **47**, 2875 (1969).
386. A. Jonczyk, K. Banko and M. Makosza, *J. Org. Chem.*, **40**, 266 (1975).
387. D. F. Tavares and R. E. Estep, *Tetrahedron Letters*, 1229 (1973).
388. D. R. White, *J. Chem. Soc., Chem. Commun.*, 95 (1975).
389. J. J. Riehl and L. Thil, *Tetrahedron Letters*, 1913 (1969).
390. F. Yoshisato and S. Tsutsumi, *J. Amer. Chem. Soc.*, **90**, 4488 (1968).
391. G. W. Cowell and A. Ledwith, *Quart. Rev.*, **24**, 119 (1970).
392. R. S. Bly, F. B. Culp and R. K. Bly, *J. Org. Chem.*, **35**, 2235 (1970).
393. A. Giddey, F. G. Cocu, B. Pochelon and Th. Posternak, *Helv. Chim. Acta*, **57**, 1963 (1974).
394. F. M. Dean and B. K. Park, *J. Chem. Soc., Chem. Commun.*, 162 (1974).
395. R. A. Bekker, G. G. Melikyan, B. L. Dyatkin and I. L. Knunyants, *Zh. Org. Khim.*, **12**, 1377 (1976).
396. C. R. Johnson, M. Haake and C. W. Shroeck, *J. Amer. Chem. Soc.*, **92**, 6594 (1970).
397. C. R. Johnson and P. E. Rogers, *J. Org. Chem.*, **38**, 1793 (1973).
398. C. R. Johnson and E. R. Janiga, *J. Amer. Chem. Soc.*, **95**, 7692 (1973).
399. C. R. Johnson and G. F. Katekar, *J. Amer. Chem. Soc.*, **92**, 5753 (1970).
400. C. R. Johnson, R. A. Kirchoff, R. J. Reischer and G. F. Katekar, *J. Amer. Chem. Soc.*, **95**, 4287 (1973).
401. T. Durst, R. Viau, R. van den Elzen and C. H. Nguyen, *J. Chem. Soc., Chem. Commun.*, 1334 (1971).
402. M. J. Bogdanowicz and B. M. Trost, *Tetrahedron Letters*, 887 (1972).
403. E. J. Corey and W. Oppolzer, *J. Amer. Chem. Soc.*, **86**, 1899 (1964).
404. M. Hetschko and J. Gosselck, *Chem. Ber.*, **106**, 996 (1973).

405. H. Braun, G. Huber and G. Kresze, *Tetrahedron Letters*, 4033 (1973).
406. R. S. Matthews and T. E. Meteyer, *Synth. Commun.*, **2**, 399 (1972).
407. B. M. Trost and M. J. Bogdanowicz, *J. Amer. Chem. Soc.*, **95**, 5311 (1973).
408. E. J. Corey and M. Chaykovsky, *Org. Synth.*, **49**, 78 (1969).
409. J. Bryan Jones and R. Grayshan, *J. Chem. Soc., Chem. Commun.*, 741 (1970).
410. Y. Yano, T. Okonogi, M. Sunaga and W. Tagaki, *J. Chem. Soc., Chem. Commun.*, 527 (1973).
411. A. Merz and G. Märkl, *Angew. Chem. (Intern. Ed.)*, **12**, 845 (1973).
412. C. R. Johnson and C. W. Schroeck, *J. Amer. Chem. Soc.*, **90**, 6852 (1968).
413. C. R. Johnson and C. W. Schroeck, *J. Amer. Chem. Soc.*, **95**, 7418 (1973).
414. T. Hiyama, T. Mishima, H. Sawada and H. Nozaki, *J. Amer. Chem. Soc.*, **97**, 1626 (1975).
415. J. Adams, L. Hoffman and B. M. Trost, *J. Org. Chem.*, **35**, 1600 (1970).
416. R. W. LaRochelle, B. M. Trost and L. Krepski, *J. Org. Chem.*, **36**, 1126 (1971).
417. Y. Tamura, S. M. Bayomi, K. Sumoto and M. Ikeda, *Synthesis*, 693 (1977).
418. C. R. Johnson and C. W. Shroeck, *J. Amer. Chem. Soc.*, **93**, 5303 (1971).
419. J. M. Townsend and K. B. Sharpless, *Tetrahedron Letters*, 3313 (1972).
420. J. R. Shanklin, C. R. Johnson, J. Ollinger and R. M. Coates, *J. Amer. Chem. Soc.*, **95**, 3429 (1973).
421. C. R. Johnson, C. W. Shroeck and J. R. Shanklin, *J. Amer. Chem. Soc.*, **95**, 7424 (1973).
422. W. Dumont and A. Krief, *Angew. Chem. (Intern. Ed.)*, **14**, 350 (1975).
423. D. Van Ende, W. Dumont and A. Krief, *Angew. Chem. (Intern. Ed.)* **14**, 700 (1975).
424. A. Anciaux, A. Eman, W. Dumont, D. Van Ende and A. Krief, *Tetrahedron Letters*, 1613, 1615, 1617 (1975).
425. W. Dumont and A. Krief, *Angew. Chem.*, **87**, 347 (1975).
426. D. Van Ende, W. Dumont and A. Krief, *Angew Chem.*, **87**, 709 (1975).
427. D. Van Ende and A. Krief, *Tetrahedron Letters*, 457 (1976).
428. G. Cainelli, A. U. Ronchi, F. Bertini, P. Grasselli and G. Zubiani, *Tetrahedron*, **27**, 6109 (1971).
429. G. Cainelli, N. Tangari and A. U. Ronchi, *Tetrahedron*, **28**, 3009 (1972).
430. M. Becker, H. Marschall and P. Weyerstahl, *Chem. Ber.*, **108**, 2391 (1975).
431. V. Mark, *J. Amer. Chem. Soc.*, **85**, 1884 (1963).
432. G. Flad, R. Sabourin and P. Chovin, *Bull. Soc. Chim. Fr.*, 1347 (1975).
433. C. Neri and E. Perrotti, *British Patent*, No. 1,331,856 (1973); *Chem. Abstr.*, **80**, 27081r (1974).
434. R. Bianchi, C. Neri and E. Perrotti, *Ann. Chim. (Roma)*, **65**, 47 (1975).
435. J. A. Gladysz, J. G. Fulcher and S. Togashi, *J. Org. Chem.*, **41**, 3647 (1976).
436. J. K. Kochi, D. M. Singleton and L. J. Andrenos, *Tetrahedron*, **24**, 3503 (1968).
437. S. M. Kupchan and M. Maruyama, *J. Org. Chem.*, **36**, 1187 (1971).
438. T. Kurokawa, K. Nakanishi, W. Wu, H. Y. Hsu, M. Maruyama and S. M. Kupchan, *Tetrahedron Letters*, 2863 (1970).
439. J. E. McMurry and M. P. Fleming, *J. Org. Chem.*, **40**, 2555 (1975).
440. F. Bertini, P. Grasselli and G. Zubiani, *J. Chem. Soc., Chem. Commun.*, 144 (1970).
441. K. B. Sharpless, M. A. Umbreit, M. T. Nieh and T. C. Flood, *J. Amer. Chem. Soc.*, **94**, 6538 (1972).
442. J. Eish and R. Kyoung, *J. Organomet. Chem.*, **139**, 45 (1977).
443. P. Dowd and R. Kang, *J. Chem. Soc., Chem. Commun.*, 384 (1974).
444. H. Alper and D. Des Roches, *Tetrahedron Letters*, 4155 (1977).
445. P. B. Shevlin, *J. Amer. Chem. Soc.*, **94**, 1379 (1972).
446. P. S. Skell, K. J. Klabunde, J. H. Klonka, J. S. Roberts and D. L. William-Smith, *J. Amer. Chem. Soc.*, **95**, 1547 (1973).
447. R. H. Parker and P. B. Shevlin, *Tetrahedron Letters*, 2167 (1975).
448. F. Mathey and G. Muller, *Compt. Rend. (C)*, **281**, 881 (1975).
449. T. H. Chan and J. R. Finkenbine, *J. Amer. Chem. Soc.*, **94**, 2880 (1972).
450. D. L. J. Clive and C. V. Denyer, *J. Chem. Soc., Chem. Commun.*, 253 (1973).
451. D. L. J. Clive and S. M. Menchen, *J. Chem. Soc., Chem. Commun.*, 658 (1977).

452. J. M. Behan, R. A. W. Johnstone and M. J. Wright, *J. Chem. Soc., Perkin I*, 1216 (1975).
453. E. Vedejs and P. L. Fuchs, *J. Amer. Chem. Soc.,* **93**, 4070 (1971); **95**, 822 (1973).
454. A. J. Bridges and G. H. Whitham, *J. Chem. Soc., Chem. Commun.*, 142 (1974).
455. P. F. Hudrlik, D. Peterson and R. J. Rona, *J. Org. Chem.*, **40**, 2263 (1975).
456. P. F. Hudrlik, A. M. Hudrlik, R. J. Rona, R. N. Misra and G. P. Withers, *J. Amer. Chem. Soc.*, **99**, 1993 (1977).
457. P. B. Dervan and M. A. Shippey, *J. Amer. Chem. Soc.*, **98**, 1265 (1976).
458. J. K. Crandall and L. C. Lin, *J. Amer. Chem. Soc.*, **89**, 4527 (1967).
459. Y. Takegami, Y. Watanabe and T. Mitsudo, *Bull. Chem. Soc. Japan*, **42**, 202 (1969).
460. M. Rosenblum, M. R. Saidi and M. Madhavarao, *Tetrahedron Letters*, 4009 (1975).
461. W. P. Giering, M. Rosenblum and J. Tancrede, *J. Amer. Chem. Soc.*, **94**, 7170 (1972).
462. G. W. Ivie, *Science*, **191**, 4230 (1976).
463. M. V. Gusenkov, V. Yu. Zakharov and B. V. Romanovskii, *Neftekhimiya*, **18**, 105 (1978).
464. S. Yoshihiro, *J. Nat. Chem. Lab. Ind.*, **72**, 101 (1977).
465. I. I. Schiketanz and F. Badea, *Studii Cerc. Chim.*, **18**, 567 (1970); *Chem. Abstr.*, **73**, 109588f (1970).
466. R. N. McDonald, in *Mechanisms of Molecular Migrations*, Vol. 3 (Ed. B. S. Thyagarajan), John Wiley and Sons, New York, 1973, pp. 67–107.
467. A. C. Cope, G. A. Berchtold, P. E. Peterson and S. H. Sharman, *J. Amer. Chem. Soc.*, **82**, 6370 (1960).
468. J. K. Crandall, L. C. Crawley, D. B. Banks and L. C. Lin, *J. Org. Chem.*, **36**, 510 (1971).
469. B. Rickborn and R. P. Thummel, *J. Org. Chem.*, **34**, 3583 (1969).
470. J. K. Crandall and L. C. Lin, *J. Org. Chem.*, **33**, 2375 (1968).
470a. J. P. Montheard and Y. Chrétien-Bessière, *Bull. Soc. Chim. Fr.*, 336 (1968).
471. A. C. Cope and J. K. Heeren, *J. Amer. Chem. Soc.*, **87**, 3125 (1965).
472. R. P. Thummel and B. Rickborn, *J. Amer. Chem. Soc.*, **92**, 2064 (1970); *J. Org. Chem.*, **36**, 1365 (1971).
473. A. Yasuda, S. Tanaka, K. Oshima, H. Yamamoto and H. Nozaki, *J. Amer. Chem. Soc.*, **96**, 6513 (1974).
474. C. L. Kissel and B. Rickborn, *J. Org. Chem.*, **37**, 2060 (1972).
475. R. K. Boeckman, Jr., *Tetrahedron Letters*, 4281 (1977).
476. J. K. Whitesell and P. D. White, *Synthesis*, 602 (1972).
477. J. R. Neff and J. E. Nordlander, *Tetrahedron Letters*, 499 (1977).
478. J. K. Crandall and L. H. Chang, *J. Org. Chem.*, **32**, 435 (1967).
479. M. Apparu and M. Barrelle, *Tetrahedron Letters*, 2837 (1976).
480. A. C. Cope, P. A. Trumbull and E. R. Trumbull, *J. Amer. Chem. Soc.*, **80**, 2844 (1958).
481. R. P. Thummel and B. Rickborn, *J. Org. Chem.*, **37**, 3919 (1972).
482. H. Paulsen, K. Eberstein and W. Koebernich, *Tetrahedron Letters*, 4377 (1974).
483. H. Paulsen and K. Eberstein, *Tetrahedron Letters*, 1495 (1975).
484. H. O. House and W. F. Gilmore, *J. Amer. Chem. Soc.*, **83**, 3980 (1961).
485. R. W. Mouk, K. M. Patel and W. Reusch, *Tetrahedron*, **31**, 13 (1975).
486. D. H. R. Barton and Y. Houminer, *J. Chem. Soc., Perkin I*, 919 (1972).
487. F. Ya. Perveev and L. N. Gonoboblev, *Zh. Org. Khim.*, **5**, 1001 (1969).
488. V. Srinivasan and E. W. Warnhoff, *Can. J. Chem.*, **54**, 1372 (1976).
489. E. W. Warnhoff and V. Srinivasan, *Can. J. Chem.*, **55**, 1629 (1977).
490. J. A. Turner and W. Herz, *J. Org. Chem.*, **42**, 2006 (1977).
491. K. B. Sharpless and R. F. Lauer, *J. Amer. Chem. Soc.*, **95**, 2697 (1973).
492. J. Carnduff and D. G. Leppard, *J. Chem. Soc., Perkin I.* 1325 (1977).
493. P. Y. Bruice, T. C. Bruice, H. G. Selander, H. Yagi and D. M. Jerina, *J. Amer. Chem. Soc.*, **96**, 6814 (1974).
494. G. C. M. Aithie and J. A. Miller, *Tetrahedron Letters*, 4419 (1975).
495. J. M. Coxon, M. P. Hartshorn and B. L. S. Sutherland, *Tetrahedron Letters*, 4029 (1969).
496. B. N. Blackett, J. M. Coxon, M. P. Hartshorn and K. E. Richards, *Tetrahedron*, **25**, 4999 (1969).

497. J. M. Coxon, M. P. Hartshorn and C. N. Muir, *Tetrahedron*, **25**, 3925 (1969).

498. B. N. Blackett, J. M. Coxon, M. P. Hartshorn, B. L. J. Jackson and C. N. Muir, *Tetrahedron*, **25**, 1479 (1969).

499. B. N. Blackett, J. M. Coxon, M. P. Hartshorn and K. E. Richards, *J. Amer. Chem. Soc.*, **92**, 2574 (1970).

500. B. N. Blackett, J. M. Coxon and M. P. Hartshorn, *Australian J. Chem.*, **23**, 2077 (1970).

501. J. M. Coxon and C. Lim, *Australian J. Chem.*, **30**, 1137 (1977).

502. J. M. Coxon, M. P. Hartshorn, A. J. Lewis, K. E. Richards and W. H. Swallow, *Tetrahedron*, **25**, 4445 (1969).

503. G. Kolaczinki, R. Mehren and W. Stein, *Fette, Seifen, Anstrichmittel*, **73**, 553 (1971).

504. A. Ya. Zapevalov, I. P. Kolenko and V. S. Plashkin, *Zh. Org. Khim.*, **11**, 1622 (1975).

505. R. Guyon and P. Villa, *Bull. Soc. Chim. Fr.*, 2593 (1975).

506. D. J. Goldsmith, B. C. Clark, Jr. and R. C. Joines, *Tetrahedron Letters*, 1211 (1967).

507. L. Canonica, M. Ferrari, U. M. Pagnoni, F. Pelizzoni, S. Maroni and T. Salvatori, *Tetrahedron*, **25**, 1 (1969).

508. J. K. Crandall and W. H. Machleder, *J. Heterocyclic Chem.*, **6**, 777 (1969).

509. S. P. Singh and J. Kagan, *J. Amer. Chem. Soc.*, **91**, 6198 (1969).

510. J. Wemple, *J. Amer. Chem. Soc.*, **92**, 6694 (1970).

511. D. J. Dagli, R. A. Gorski and J. Wemple, *J. Org. Chem.*, **40**, 1741 (1975).

512. A. C. Brauwer, L. Thijs and B. Zwanenburg, *Tetrahedron Letters*, 807 (1975).

513. R. A. Gorski, D. J. Dagli and J. Wemple, *J. Amer. Chem. Soc.*, **98**, 4588 (1976).

514. J. Kagan, D. A. Agdeppa, S. P. Singh, D. A. Mayers, C. Bogajian, C. Poorker and B. E. Firth, *J. Amer. Chem. Soc.*, **98**, 4581 (1976).

515. J. M. Domagala and R. D. Bach, *J. Amer. Chem. Soc.*, **100**, 1605 (1978).

516. P. Kropp, *J. Amer. Chem. Soc.*, **88**, 4926 (1966).

517. M. P. Hartshorn and D. N. Kirk, *Tetrahedron*, **21**, 1547 (1965).

518. R. C. Cambie and W. A. Denny, *Australian J. Chem.* **28**, 1153 (1975).

519. B. Gioia, A. Marchesini and U. M. Pagnoni, *Gazz. Chim. Ital*, **107**, 39 (1977).

520. G. H. Boelsma, *Dutch Patent*, No. 73 11,335; *Chem. Abstr.*, **83**, 79413g (1975).

521. L. H. Schwartz, M. Feil, A. J. Kascheres, K. Kaufmann and A. M. Levine, *Tetrahedron Letters*, 3785 (1967).

522. C. Maignan and F. Rouessac, *Bull. Soc. Chim. Fr.*, 550 (1976).

523. C. W. Bird and Y. C. Yeong, *Synthesis*, 27 (1974).

524. N. Heap, G. E. Green and G. H. Whitham, *J. Chem. Soc. (C)*, 160 (1969).

525. R. Rickborn and R. M. Gerkin, *J. Amer. Chem. Soc.*, **90**, 4193 (1968).

526. K. Arata, S. Akutagawa and K. Tanabe, *J. Catal.*, **41**, 173 (1976).

527. J. H. Kennedy and C. Buse, *J. Org. Chem.*, **36**, 3135 (1971).

528. M. L. Leriverend and P. Leriverend, *Compt. Rend. (C)*, **280**, 791 (1975).

529. B. Rickborn and R. M. Gerkin, *J. Amer. Chem. Soc.*, **93**, 1693 (1971).

530. B. M. Trost, M. Preckel and L. M. Leichter, *J. Amer. Chem. Soc.*, **97**, 2224 (1975).

531. P. F. Hudrlik, R. N. Misra, G. P. Withers, A. M. Hudrlik, R. J. Rona and J. P. Arcoleo, *Tetrahedron Letters*, 1453 (1976).

532. P. Coutrot and Cl. Legris, *Synthesis*, 118 (1975).

533. J. A. Donnelly, P. Bennett, S. O'Brien and J. O'Grady, *Chem. Ind. (Lond.)*, 500 (1972).

534. J. A. Donnelly, J. G. Hoey, S. O'Brien and J. O'Grady, *J. Chem. Soc., Perkin I*, 2030 (1973).

535. J. A. Donnelly, S. O'Brien and J. O'Grady, *J. Chem. Soc., Perkin I*, 1674 (1974).

536. J. A. Donnelly and J. G. Hoey, *J. Chem. Soc., Perkin I*, 2364 (1975).

537. H. Nakamura, H. Yamamoto and H. Nozaki, *Tetrahedron Letters*, 111 (1973).

538. C. J. Cheer and C. R. Johnson, *J. Org. Chem.*, **32**, 428 (1967).

539. C. J. Cheer and C. R. Johnson, *J. Amer. Chem. Soc.*, **90**, 178 (1968).

540. G. Berti, S. Catalano, A. Marsili, I. Morelli and V. Scartoni, *Tetrahedron Letters*, 401 (1976).

541. J. M. Coxon, M. P. Hartshorn, C. N. Muir and K. E. Richards, *Tetrahedron Letters*, 3725 (1967).

542. J. W. ApSimon and J. J. Rosenfeld, *J. Chem. Soc.(D)*, 1271 (1970).

543. I. Morelli, S. Catalano, G. Moretto and A. Marsili, *Tetrahedron Letters*, 717 (1972).

544. I. G. Guest and B. A. Marples, *J. Chem. Soc., Perkin I*, 900 (1973).

545. C. W. Lyons and D. R. Taylor *J. Chem. Soc., Chem. Commun.*, 647 (1976).
546. T. C. Bruice and P. Y. Bruice, *Acc. Chem. Res.*, **9**, 378 (1976).
547. D. M. Jerina, J. W. Daly, B. Witkop, P. Zaltzman-Nirenberg and S. Udenfriend, *J. Amer. Chem. Soc.*, **90**, 6525 (1968).
548. D. M. Jerina, J. W. Daly and B. Witkop, *J. Amer. Chem. Soc.*, **90**, 6523 (1968).
549. D. R. Boyd, J. W. Daly and D. M. Jerina, *Biochemistry*, **11**, 1961 (1972).
550. G. J. Kasparek and T. C. Bruice, *J. Amer. Chem. Soc.*, **94**, 198 (1972).
551. K. Yamaguchi and T. Fueno, *Chem. Phys. Letters*, **22**, 471 (1973).
552. D. R. Paulson, F. Y. N. Tang and R. B. Sloan, *J. Org. Chem.*, **38**, 3967 (1973).
553. D. R. Paulson, G. Korngold and G. Jones, *Tetrahedron Letters*, 1723 (1972).
554. H. H. J. McDonald and R. J. Crawford, *Can. J. Chem.*, **50**, 428 (1972).
555. R. J. Crawford, V. Vukov and H. Tokunaga, *Can. J. Chem.*, **51**, 3718 (1973).
556. J. C. Paladini and J. Chuche, *Tetrahedron Letters*, 4383 (1971).
557. J. C. Pommelet, N. Manisse and J. Chuche, *Tetrahedron*, **28**, 3929 (1972).
558. A. Robert and B. Moisan, *J. Chem. Soc., Chem. Commun.*, 337 (1972).
559. B. Moisan, A. Robert and A. Foucaud, *Tetrahedron*, **30**, 2867 (1974).
560. M. C. Flowers and R. M. Parker, *Int. J. Chem. Kinet.*, **3**, 443 (1971).
561. D. L. Garin, *J. Org. Chem.*, **34**, 2355 (1969).
562. G. P. Petrenko and V. P. Ivanova, *Zh. Org. Khim.*, **6**, 2576 (1970).
563. P. F. Hudrlik, C. Wan and G. P. Withers, *Tetrahedron Letters*, 1449 (1976).
564. D. Bethall, G. W. Kenner and P. J. Powers, *J. Chem. Soc., Chem. Commun.*, 227 (1968).
564a. R. A. Kretchmer and W. J. Frazee, *J. Org. Chem.*, **36**, 2855 (1971).
565. G. Sénéchal, J. C. Duchet and D. Cornet, *Bull. Soc. Chim. Fr.*, 783 (1971).
566. M. Bartók, I. Török I. Szabó, *Acta Chim. Acad. Sci. Hung.*, **76**, 417 (1973).
567. F. A. Chernyshkova and D. V. Musenko, *Neftekhimiya*, **16**, 250 (1976).
568. J. C. Duchet and D. Cornet, *Bull. Soc. Chim. Fr.*, 1135 (1975).
569. J. C. Duchet and D. Cornet, *Bull. Soc. Chim. Fr.*, 1141 (1975).
570. G. N. Koshel, M. I. Farberov, T. N. Antonova and I. I. Glazurina, *U.S.S.R. Patent*, No. 513,966 (1976); *Chem. Abstr.*, **85**, 176936d (1976).
571. M. Bartók, *Acta Chim. (Budapest)*, **88**, 395 (1976).
572. F. Notheisz and M. Bartók, *Acta Chim. (Budapest)*, **95**, 335 (1977).
573. F. Humbert and G. Guth, *Bull. Soc. Chim. Fr.*, 2867 (1966).
574. K. Arata, J. O. Bledsoe and K. Tanabe, *J. Org. Chem.*, **43**, 1660 (1978).
575. K. Arata and K. Tanabe, *Chem. Letters*, 321 (1976).
576. K. Arata, S. Akutagawa and K. Tanabe, *Bull. Chem. Soc. Japan*, **48**, 1097 (1975); **49**, 390 (1976).
577. K. Arata, K. Kato and K. Tanabe, *Bull. Chem. Soc., Japan*, **49**, 563 (1976).
578. V. S. Joshi and S. Dev, *Tetrahedron*, **33**, 2955 (1977).
579. M. I. Farberov, A. V. Bondarenko, V. M. Obukhov, E. P. Tepenitsina, B. N. Bonylev, I. P. Stepanova, G. A. Stepanov, A. N. Bushin and V. Sh. Fel'bdlyum, *U.S.S.R. Patent*, No. 429,050 (1974); *Chem. Abstr.*, **81**, 17093lu (1974).
580. S. S. Srednev, S. I. Kryukov and M. I. Farberov, *Zh. Org. Khim.*, **10**, 1608 (1974); **12**, 1885 (1976); **13**, 257 (1977).
581. M. N. Sheng, *Synthesis*, 194 (1972).
582. T. Imanaka, Y. Okamoto and S. Teranishi, *Bull. Chem. Soc. Japan*, **45**, 1353 (1972).
583. Y. Shinohara and A. Niiyama, *Japanese Patent*, No. 72 13,009; *Chem. Abstr.*, **77**, 33922v (1972).
584. P. B. Venuto and P. S. Landis, *Advan. Catal.*, **18**, 259 (1968).
585. T. Imanaka, Y. Okamoto and S. Teranishi, *Bull. Chem. Soc. Japan*, **45**, 3251 (1972).
586. M. R. Musaev, S. D. Mekhtiev, I. K. Magamedov and F. M. Mamedov, *Azerb. Khim. Zh.*, 135 (1970); *Chem. Abstr.*, **75**, 35070y (1971).
587. R. Grigg and G. Shelton, *J. Chem. Soc., Chem. Commun.*, 1247 (1971).
588. R. Grigg, R. Hayers and A. Sweeney, *J. Chem. Soc., Chem. Commun.*, 1248 (1971).
589. J. Blum, *J. Mol. Catal.*, **3**, 33 (1977).
590. D. Milstein, O. Buchman and J. Blum, *Tetrahedron Letters*, 2257 (1974).
591. D. Milstein, O. Buchman and J. Blum, *J. Org. Chem.*, **42**, 2299 (1977).

592. H. Alper, D. Des Roches, T. Durst and R. Legault, *J. Org. Chem.*, **41**, 3611 (1976).
593. R. W. Ashworth and G. A. Berchtold, *Tetrahedron Letters*, 343 (1977).
594. J. Y. Kim and T. Kwan, *Chem. Pharm. Bull*, **18**, 1040 (1970).
595. G. Farges and A. Kergomard, *Bull. Soc. Chim. Fr.*, 315 (1975).
596. J. C. Volta, V. Perrichon and J. M. Cognion, *Carbon*, **16**, 59 (1978).
597. C. L. Stevens and P. M. Pillai, *J. Amer. Chem. Soc.*, **89**, 3084 (1967).
598. K. Jankowski and J. Y. Daigle, *Synthesis*, 32 (1971).
599. K. Jankowski and J. Y. Daigle, *Can. J. Chem.*, **49**, 2594 (1971).
600. J. P. Nagarkatti and K. R. Ashley, *Tetrahedron Letters*, 4599 (1973).
601. T. Okimoto and D. Swern, *J. Am. Oil. Chem. Soc.*, **54**, 867A (1977).
602. Bayer, A.-G., *German Patent*, No. 2,252,719 (1974); *Chem. Abstr.*, **81**, 25084r (1974).
603a. W. Adam and A. Rios, *J. Chem. Soc. (D)*, 822 (1971).
603b. V. Subramanyam, C. L. Brizuela and A. H. Soloway, *J. Chem. Soc., Chem. Commun.*, 508 (1976).
604. H. Kropf, M. Ball, H. Schroeder and G. Witte, *Tetrahedron*, **30**, 2943 (1974).
605. T. Tsuji, *Tetrahedron Letters*, 2975 (1967).
606. T. M. Santosusso and D. Swern, *Tetrahedron Letters*, 4261 (1968).
607. L. Levine, *U.S. Patent*, No. 3,652,669; *Chem. Abstr.*, **76**, 139939a (1972).
608. E. Laurent and P. Villa, *Bull. Soc. Chim. Fr.*, 249 (1969).
609. R. Guyon and P. Villa, *Bull. Soc. Chim. Fr.*, 1375 (1972).
610. R. Guyon and P. Villa, *Bull. Soc. Chim. Fr.*, 2583 (1975).
611. R. Guyon and P. Villa, *Bull. Soc. Chim. Fr.*, 2599 (1975).
612. M. L. Michailović, V. Andrejević, J. Milovanović and J. Janković, *Helv. Chim. Acta*, **59**, 2305 (1976).
613. S. Ushio and S. Tshitsugu, *Tetrahedron Letters*, 981 (1978).
614. E. C. Ashby and B. Cooke, *J. Amer. Chem. Soc.*, **90**, 1625 (1968).
615. B. A. Arbuzov, A. N. Karaseva, Z. B. Isaeva and T. P. Povodyreva, *Dokl. Akad. Nauk SSSR*, **233**, 366 (1977).
616. R. H. Cornforth, *J. Chem. Soc.(C)*, 928 (1970).
617. A. R. Davies and G. H. R. Summers, *J. Chem. Soc.(C)*, 1227 (1967).
618. J. C. Richer and C. Freppel, *Can. J. Chem.*, **46**, 3709 (1968).
619. G. R. Krow and J. Reilly, *Tetrahedron Letters*, 1561 (1975).
620. B. Cooke, E. C. Ashby and J. Lott, *J. Org. Chem.*, **33**, 1132 (1968).
621. P. T. Lansbury, D. J. Scharf and V. A. Pattison, *J. Org. Chem.*, **32**, 1748 (1967).
622. B. C. Hartman and B. Rickborn, *J. Org. Chem.*, **37**, 4246 (1972).
623. H. C. Brown and N. M. Yoon, *J. Amer. Chem. Soc.*, **88**, 1464 (1966).
624. N. M. Yoon and J. Kang, *Taehan Hwahak Hoechi*, **19**, 355 (1975); *Chem. Abstr.*, **84**, 59065n (1976).
625. H. C. Brown and N. M. Yoon, *J. Chem. Soc., Chem. Commun.*, 1549 (1968).
626. P. A. Marshall and R. H. Prager, *Australian J. Chem.*, **30**, 141 (1977).
627. H. C. Brown and N. M. Yoon, *J. Amer. Chem. Soc.*, **90**, 2686 (1968).
628. D. K. Murphy, R. L. Alumbaugh and B. Rickborn, *J. Amer. Chem. Soc.*, **91**, 2649 (1969).
629. M. Zaidlewicz and A. Uzarewicz, *Rocz. Chem.*, **47**, 1433 (1973); *Chem. Abstr.*, **80**, 36736k (1974).
630. S. Mitsui and Y. Nagahisa, *Chem. Ind. (Lond.)*, 1975 (1965).
631. S. Mitsui and I. Imaizumi, M. Hisashige and Y. Sugi, *Tetrahedron*, **29**, 4093 (1973).
632. S. Suzuki, *U. S. Patent*, No. 3,975,449 (1976); *Chem. Abstr.*, **86**, 54967q (1977).
633. M. Amagasa and T. Aoki, *Japanese Patent*, No. 73 31,083; *Chem. Abstr.*, **80**, 26745y (1974).
634. K. Isogai, T. Ogino, T. Hiiro, K. Endo and N. Yokokawa, *Yuki Gosei Kagaku Kyokai Shi*, **34**, 492 (1976); *Chem. Abstr.*, **86**, 4881m (1977).
635. F. A. Chernishkova, D. V. Musenko and L. A. Blandina, *Neftekhimiya*, **14**, 188 (1974).
636. M. Barták and R. A. Karakhanov, *Acta Phys. Chem. Szeged*, **20**, 453 (1974).
637. A. Suzuki, M. Miki and M. Itoh, *Tetrahedron*, **23**, 3621 (1967).
638. Y. Nagahisa, Y. Sugi and S. Mitsui, *Chem. Ind. (Lond.)*, 38 (1975).
639. A. Sohma and S. Mitsui, *Bull. Chem. Soc. Japan*, **43**, 448 (1970).

640. G. Sénéchal and D. Cornet, *Bull. Soc. Chim. Fr.*, 773 (1971).
641. E. M. Kaiser, C. G. Edmonds, S. D. Grubb, J. W. Smith and D. Tramp, *J. Org. Chem.*, **36**, 330 (1971).
642. R. S. Lenox and J. A. Katzenellenbogen, *J. Amer. Chem. Soc.*, **95**, 957 (1973).
643. T. Kojima and K. Katayama, *Japanese Patent*, No. 76 26,808; *Chem. Abstr.*, **85**, 77752j (1976).
644. R. G. Carlson and W. W. Cox, *J. Org. Chem.*, **42**, 2382 (1977).
645. L. A. Paquette, K. H. Fuhr, S. Porter and J. Clardy, *J. Org. Chem.*, **39**, 467 (1974).
646. H. C. Brown, S. Ikegami and J. H. Kawakami, *J. Org. Chem.*, **35**, 3243 (1970).
647. H. C. Brown, J. H. Kawakami and S. Ikegami, *J. Amer. Chem. Soc.*, **92**, 6914 (1970).
648. C. H. Robinson and R. Henderson, *J. Org. Chem.*, **37**, 56 (1972).
649. H. C. Brown, *Govt. Rep. Announce (U.S.)*, **73**, 68 (1973); *Chem. Abstr.*, **80**, 37205y (1974).
650. S. Krishnamurthy, R. M. Schubert and H. C. Brown, *J. Amer. Chem. Soc.*, **95**, 8486 (1973).
651. Y. Yamamoto, H. Toi, A. Sonoda and S.-I. Murahashi, *J. Chem. Soc., Chem. Commun.*, 672 (1976).
652. T. K. Jones and J. H. J. Peet, *Chem. Ind. (Lond.)*, 995 (1971).
653. J. L. Namy and D. Abenhaim, *J. Organomet. Chem.*, **43**, 95 (1972).
654. J. Furukawa and T. Saegusa, *Polymerization of Aldehydes and Oxides*, John Wiley and Sons, New York, 1963.
655. K. S. Kazanskii, A. A. Solov'yanov and S. G. Entelis, *Advan. Ionic Polym., Proc. Int. Symp.*, 77 (1972).
656. I. Benedek and D. Feldman, *Studii Cercet. Chim.*, **21**, 433 (1973); *Chem. Abstr.*, **80**, 83673x (1974).
657. S. G. Entelis and G. Korovina, *Makromol. Chem.*, **175**, 1253 (1974).
658. N. Spassky, P. Dumas, M. Sepulchre and P. Sigwalt, *J. Polym. Sci., Polym. Symp.*, 327 (1975).
659. P. Sigwalt, *Pure Appl. Chem.*, **48**, 257 (1976).
660. T. Tsuruta, *Pure Appl. Chem.*, **48**, 267 (1976).
661. N. S. Enikolopiyan, *Pure Appl. Chem.*, **48**, 317 (1976).
662. E. J. Goethals, *Pure Appl. Chem.*, **48**, 335 (1976).
663. R. J. Kern, *J. Org. Chem.*, **33**, 388 (1968).
664. L. P. Blanchard, C. Raufast, H. H. Kiet and S. L. Malhotra, *J. Macromol. Sci. Chem.*, **A9**, 1219 (1975).
665. L. P. Blanchard, G. G. Gabra and S. L. Malhotra, *J. Polym. Sci., Polym. Chem. Ed.*, **13**, 1619 (1975).
666. J. A. Orvik, *J. Amer. Chem. Soc.*, **98**, 3322 (1976).
667. M. Rodriguez and J. E. Figueruelo, *Makromol. Chem.*, **176**, 3107 (1975).
668. K. S. Kazanskii, A. N. Tarasov, I. E. Paleeva and S. A. Dubrovskii, *Vysokomol. Soed.*, **20**, 391 (1978).
669. T. H. Chan and I. R. Finkenbine, *J. Amer. Chem. Soc.*, **94**, 2880 (1972).
670. G. Lebrasseur, *French Patent*, No. 1,505,715; *French Patent*, No. 1,605,472; *Chem. Abstr.*, **70**, 19908u (1969); **87**, 168005a (1977).
671. V. Calo, L. Lopez, L. Marchese and G. Pesce, *J. Chem. Soc., Chem. Commun.*, 621 (1975).
672. E. Lippman, D. Reifegerste and E. Kleinpeter, *Z. Chem.*, **17**, 60 (1977).
673. T. H. Chan and J. R. Finkenbine, *Tetrahedron Letters*, 2091 (1974).
674. I. Shanah, Y. Ittah and J. Blum, *Tetrahedron Letters*, 4003 (1976).
675. M. Kapps and W. Kirmse, *Angew. Chem. (Intern. Ed.)*, **8**, 75 (1969).
676. L. A. Mukhamedova, T. M. Malyshko, R. R. Shagidullin and N. V. Teptina, *Khim. Geterotsikl. Soedin*, 195 (1969).
677. N. Bensel, H. Marshall and P. Weyerstahl, *Chem. Ber.*, **108**, 2697 (1975).
678. G. Descotes, B. Giroud-Abel and J. C. Martin, *Bull. Soc. Chim. Fr.*, 2466 (1967).
679. A. Chatterjee and D. Banerjee, *Tetrahedron Letters*, 4559 (1969).
680. A. Chatterjee, D. Banerjee and R. Mallik, *Tetrahedron*, **33**, 85 (1977).
681. S. Watanabe, T. Fujita and K. Suga, *Yuki Gosei Kagaku Kyokaishi*, **35**, 290 (1977); *Chem. Abstr.*, **87**, 135271f (1977).

682. F. Nerdel, J. Buddrus, G. Scherowsky, D. Klammar and M. Fligge, *Liebigs Ann. Chem.*, **710**, 85 (1967).
683. R. P. Hanzlik and M. Leinwetter, *J. Org. Chem.*, **43**, 438 (1978).
684. T. I. Temnikova and V. N. Yandovskii, *Zh. Org. Khim.*, **4**, 1006 (1968).
685. N. L. Madison, *U.S. Patent*, No. 3,324,145 (1967); *Chem. Abstr.*, **68**, 39610c (1968).
686. F. G. Ponomarev, N. N. Chernousova and G. N. Yashchenko, *Zh. Org. Khim.*, **5**, 226 (1969).
687. E. E. Gilbert and E. J. Rumanowski, *U.S. Patent*, No. 3,285,936 (1966); *Chem. Abstr.*, **66**, 75986 (1967).
688. R. J. De Pasquale, *J. Chem. Soc., Chem. Commun.*, 157 (1973).
689. H. Koinuma, H. Kato and H. Hirai, *Chem. Letters*, 517 (1977).
690. J. R. L. Smith, R. O. C. Norman and M. R. Stillings, *J. Chem. Soc., Perkin I*, 1200 (1975).
691. T. I. Temnikova and V. N. Yandovskii, *Zh. Org. Khim.*, **4**, 178 (1968).
692. V. N. Yandovskii and T. I. Temnikova, *Zh. Org. Khim.*, **9**, 1376 (1973).
693. R. A. Wohl and J. Cannie, *J. Org. Chem.*, **38**, 1787 (1973).
694. I. E. Herweh and W. I. Kaufman, *Tetrahedron Letters*, 809 (1971).
695. M. E. Dyen and D. Swern, *J. Org. Chem.*, **33**, 379 (1968).
696. P. L. De Benneville and L. J. Exner, *U.S. Patent*, No. 3,409,635; *Chem. Abstr.*, **70**, 20047u (1969).
697. H. Schmidbaur and P. Hull, *German Patent*, No. 2,545,073; *Chem. Abstr.*, **87**, 53438b (1977).
698. R. S. Edmundson, *Chem. Ind. (Lond.)*, 1809 (1967).
699. M. E. Ali, N. G. Kardouche and L. N. Owen, *J. Chem. Soc., Perkin I*, 748 (1975).
700. M. Radau and K. Hartke, *Arch. Pharm.*, **305**, 665 (1972).
701. U. Schoellkopf and R. Jentsch, *Angew. Chem.*, **85**, 355 (1973).
702. K. Jankowski and C. Berse, *Can. J. Chem.*, **46**, 1939 (1968).
703. A. P. Sinekov, F. N. Gladysheva, V. S. Etlis and V. S. Kutyreva, *Khim. Geterotsikl. Soedin.*, **4**, 475 (1970).
704. M. Schulz and K. Kirschke, *Advan. Heterocycl. Chem.*, **8**, 182 (1968).
705. A. Murai, M. Ono and T. Masamune, *J. Chem. Soc., Chem. Commun.*, 864 (1976).
706. A. Murai, M. Ono and T. Masamune, *Bull. Chem. Soc. Japan*, **50**, 1226 (1977).
707. V. R. Gaertner, *J. Org. Chem.*, **32**, 2972 (1967).
708. D. Miller, *J. Chem. Soc. (C)*, 12 (1969).
709. B. Capon and J. W. Thompson, *J. Chem. Soc., Perkin II*, 917 (1977).
710. J. J. Pommeret and A. Robert, *Tetrahedron*, **27**, 2977 (1971).
711. I. F. Sokovishina and V. V. Perekalin, *Zh. Org. Khim.*, **11**, 52 (1975).
712. N. Bensel, H. Marschall and P. Weyerstahl, *Tetrahedron Letters*, 2293 (1976).
713. J. Wolinsky, P. Hull and E. M. White, *Tetrahedron*, **32**, 1335 (1976).
714. F. Ya. Perveev, L. N. Shil'nikova and R. Ya. Irgal, *Zh. Org. Khim.*, **5**, 1337 (1969).
715. R. Achini and W. Oppolzer, *Tetrahedron Letters*, 369 (1975).
716. A. Robert and B. Moisan, *J. Chem. Soc., Chem. Commun.*, 337 (1972).
717. S. E. Kurbanov, I. M. Akhmedov, F. G. Gasanov and D. T. Radzabov, *Dokl. Akad. Nauk Azerb. SSR*, **33**, 23 (1977).
718. S. Hayashi, M. Furukawa, Y. Fujino, H. Okabe and T. Nakao, *Chem. Pharm. Bull.*, **19**, 2404 (1971).
719. N. J. Leonard and B. Zwanenburg, *J. Amer. Chem. Soc.*, **89**, 4456 (1967).
720. M. Baudy and A. Robert, *J. Chem. Soc., Chem. Commun.*, 23 (1976).
721. A. Turcant and M. Le Corre, *Tetrahedron Letters*, 789 (1977).
722. G. El' Naggar and B. A. Ershov, *Zh. Org. Khim*, **5**, 1369 (1968).
723. B. A. Ershov, L. A. Kaunova, Yu. L. Kleiman and G. V. Markina, *Zh. Org. Khim.*, **4**, 1764 (1968).
724. B. Meklati and Y. Bessière-Chrétien, *Bull. Soc. Chim. Fr.*, 3133 (1972).
725. A. Padwa and M. Rostoker, *Tetrahedron Letters*, 281 (1968).
726. D. P. G. Hamon and L. J. Holding, *J. Chem. Soc. (C)*, 1330 (1970).
727. C. Sabate-Alduy and J. Lemarte, *Compt. Rend. (C)*, **270**, 1611 (1970).
728. W. I. Middleton, *J. Org. Chem.*, **31**, 3731 (1966).
729. A. Takeda, S. Wada, M. Fujii and H. Tanaka, *Bull. Chem. Soc. Japan*, **43**, 2997 (1970).

730. I. L. Knunyants, V. V. Shokina and I. V. Galakhov, *Khim. Geterotsikl. Soedin*, 873 (1966).
731. R. Scheffold, P. Bissig, K. L. Ghatah, B. Granwehr and B. Patwardhan, *Chimia*, **29**, 463 (1975).
732. Sh. Fujisaki, Sh. Okano, Sh. Sugiyama, S. Murata and Sh. Kajigaeshi, *Nippon Kagaku Kaishi*, 344 (1975); *Chem. Abstr.*, **83**, 9930z (1975).
733. L. S. Stanishevskii, I. G. Tishchenko, Yu. V. Glazkov and A. Ya. Guzikov, *Zh. Org. Khim.*, **8**, 860 (1972).
734. K. Nützel in *Methoden der Organischen Chemie (Houben-Weyl)* Vol. 13/2a, Georg Thieme Verlag, Stuttgart, 1973, p. 343.
735. S. Hata, Y. Yano, H. Matsuda and S. Matsuda, *Kogyo Kagaku Zasshi*, **71** 704, (1968); *Chem. Abstr.*, **70**, 11251d (1969).
736. A. Schaap and J. F. Arens, *Rec. Trav. Chim.*, **87**, 1249 (1968).
737. I. Matsuda and M. Sugishita, *Bull. Chem. Soc. Japan*, **40**, 174 (1967).
738. H. Lehmkuhl and K. Ziegler in *Methoden der Organischen Chemie (Houben-Weyl)*, Vol. 13/4, Georg Thieme Verlag, Stuttgart, 1970, p. 243.
739. A. J. Lundeen and A. C. Oehlschlager, *J. Organomet. Chem.*, **25**, 337 (1970).
740. J. L. Namy, E. Henry-Basch and P. Freon, *Bull. Soc. Chim. Fr.*, 2249 (1970).
741. D. Abenhaim and J. L. Namy, *Tetrahedron Letters*, 1001 (1972).
742. W. Kuran, S. Posynkiewicz and J. Serzyho, *J. Organomet. Chem.*, **73**, 187 (1974).
743. G. Boireau, D. Abenhaim, J. L. Namy and E. Henry-Basch, *Zh. Org. Khim.*, **12**, 1841 (1976).
744. K. Schöllkopf in *Methoden der Organischen Chemie (Houben-Weyl)*, Vol. 13/1, Georg Thieme Verlag, Stuttgart, 1970, p. 217.
745. G. H. Posner in *Organic Reactions*, Vol. 22, John Wiley and Sons, New York, 1975, p. 287.
746. G. Leandri, H. Monti and M. Bertrand, *Compt. Rend. (C)*, **271**, 560 (1970).
747. W. H. Glase, D. P. Duncan and D. J. Donald, *J. Org. Chem.*, **42**, 694 (1977).
748. C. R. Johnson, R. W. Herr and D. M. Wieland, *J. Org. Chem.*, **38**, 4263 (1973).
749. J. F. Normant, *Synthesis*, 78 (1972).
750. E. J. Corey, K. C. Nicolaou and D. J. Beames, *Tetrahedron Letters*, 2439 (1974).
751. B. C. Hartman, T. Livinghouse and B. Rickborn, *J. Org. Chem.*, **38**, 4346 (1973).
752. E. J. Corey, L. S. Melvin, Jr. and M. F. Haslanger, *Tetrahedron Letters*, 3115 (1975).
753. P. L. Fuchs, *J. Org. Chem.*, **41**, 2935 (1976).
754. R. D. Acker, *Tetrahedron Letters*, 3407 (1977).
755. J. Deniau, E. Henry-Basch and P. Freon, *Bull. Soc. Chim. Fr.*, 4414 (1969).
756. O. Ceder and B. Hansson, *Acta Chem. Scand.*, **B30**, 574 (1976).
757. R. Calas and J. Dunogues in *Organometallic Chemistry Reviews* (J. Organomet. Chem. Library 2) (Ed. D. Seyferth), Elsevier, Amsterdam, 1976, p. 336.
758. A. V. Fokin, A. F. Kolomiet, Yu. N. Studner and A. I. Repkin, *Izv. Akad. Nauk SSSR, Ser. Khim.*, 348 (1974).
759. W. Lidy and W. Sundermeyer, *Tetrahedron Letters*, 1449 (1973).
760. D. D. Davis and C. E. Gray, *J. Org. Chem.*, **35**, 1303 (1970).
761. P. F. Hudrlik and D. Peterson, *J. Am. Chem. Soc.*, **97**, 1464 (1975).
762. E. Vedejs, K. A. J. Snoble and P. L. Fuchs, *J. Org. Chem.*, **38**, 1178 (1973).
763. E. Negishi, Sh. Baba and A. D. King, *J. Chem. Soc., Chem. Commun.*, 17 (1976).
764. J. Fried, C. H. Lin and S. H. Ford, *Tetrahedron Letters*, 1379 (1969).
765. D. B. Malpass, S. C. Watson and G. S. Yeargin, *J. Org. Chem.*, **42**, 2712 (1977).
766. M. Nause, K. Utimoto and H. Nozaki, *Tetrahedron*, **30**, 3037 (1974).
767. J. Fried, J. C. Sih, C. H. Lin and P. Dalven, *J. Amer. Chem. Soc.*, **94**, 4343 (1972).
768. E. J. Negishi in *New Application of Organometallic Reagents in Organic Synthesis* (J. Organomet. Chem. Library 1), (Ed. D. Seyferth), Elsevier, Amsterdam, 1976, p. 113.
769. R. Baker, R. C. Cookson and A. D. Saunders, *J. Chem. Soc. Perkin I*, 1809 (1976).
770. W. E. Harvey and D. S. Tarbell, *J. Org. Chem.*, **32**, 1679 (1967).
771. P. F. Hudrlik and C. N. Wan, *J. Org. Chem.*, **40**, 2963 (1975).
772. N. J. Foulger and B. J. Wakefield, *J. Chem. Soc., Perkin I*, 971 (1974).
773. D. Seebach and E. J. Corey, *J. Org. Chem.*, **40**, 231 (1975).

774. S. Torii, K. Uneyama and M. Isihara, *J. Org. Chem.*, **39**, 3645 (1974).
775. A. I. Meyers, E. D. Michelich and R. L. Nolen, *J. Org. Chem.*, **39**, 2783 (1974).
776. J. Hartmann and M. Schlosser, *Synthesis*, 328 (1975).
777. R. J. Anderson, *J. Amer. Chem. Soc.*, **92**, 4978 (1970).
778. R. W. Herr and C. R. Johnson, *J. Amer. Chem. Soc.*, **92**, 4978 (1970).
779. D. M. Wieland and C. R. Johnson, *J. Amer. Chem. Soc.*, **93**, 3047 (1971).
780. C. Cahiez, A. Alexakis and J. F. Normant, *Synthesis*, 528 (1978).
781. J. Staroscik and B. Rickborn, *J. Amer. Chem. Soc.*, **93**, 3046 (1971).
782. C. B. Rose and S. K. Taylor, *J. Org. Chem.*, **39**, 578 (1974).
783. G. A. Crosby and R. A. Stephenson, *J. Chem. Soc., Chem. Commun.*, 287 (1975).
784. J. Hambrecht, H. Straub and E. Miller, *Chem. Ber.*, **107**, 2985 (1974).
785. R. A. Amos and J. A. Katzenellenbogen, *J. Org. Chem.*, **42**, 2537 (1977).
786. P. F. Hudrlik in *New Application of Organometallic Reagents in Organic Synthesis* (J. Organomet. Chem. Library 1), (Ed. D. Seyferth), Elsevier, Amsterdam, 1976, p. 144.
787. P. Coutrot, J. C. Combret and J. Villieras, *Tetrahedron Letters*, 1553 (1971).
788. J. Cantacuzene and A. Keramat, *Bull. Soc. Chim. Fr.,* 4540 (1968).
789. J. M. Normant, *Compt. Rend. (C)*, **277**, 1045 (1973).
790. J. M. Normant, *Tetrahedron Letters*, 4253 (1973).
791. J. Cantacuzene and J. Normant, *Tetrahedron Letters*, 2947 (1970).
792. J. J. Eisch and J. A. Galle, *J. Organomet. Chem.*, **121**, C 10 (1976).
793. A. Suzuki, N. Miyaura and M. Itoh, *Synthesis*, 305 (1973).
794. P. R. Ortiz de Montellano, *J. Chem. Soc., Chem. Commun.*, 709 (1973).
795. P. Vermeer, J. Meijer, C. De Graaf and H. Scheurs, *Rec. Trav. Chim.*, **93**, 47 (1974).
796. R. Nouri-Bimorghi, *Bull. Soc. Chim. Fr.*, 2812 (1969).
797. R. Nouri-Bimorghi, *Bull. Soc. Chim. Fr.*, 2971 (1971).
798. H. Molines, J. Normant and C. Wakselmann, *Tetrahedron Letters*, 951 (1974).
799. G. Kobrich, W. Werner and J. Grosser, *Chem. Ber.*, **106**, 2620 (1973).
800. G. W. Griffin, *Angew. Chem. (Intern. Ed.)*, **10**, 537 (1971).
801. G. W. Griffin and A. Padwa in *Photochemistry of Heterocyclic Compounds* (Ed. O. Buchardt), John Wiley and Sons, New York, 1976.
802. N. R. Bertoniere and G. W. Griffin, *Organic Photochemistry*, Vol. 3 (Ed. O. L. Chapman), Marcel Dekker, New York, 1973, Chap. 2.
803. A. P. Meleshevits, *Usp. Khim.*, **39**, 444 (1970).
804. J. Muzart and J. P. Pete, *Tetrahedron Letters*, 303 (1977).
805. J. Muzart and J. P. Pete, *Tetrahedron Letters*, 307 (1977).
806. Thap Do Minh, A. M. Trozzolo and G. W. Griffin, *J. Amer. Chem. Soc.*, **92**, 1402 (1970).
807. I. J. Lev, K. Ishikawa, N. S. Bhacca and G. W. Griffin, *J. Org. Chem.*, **41**, 2654 (1976).
808. G. A. Lee, *J. Org. Chem.*, **41**, 2656 (1976).
809. V. Markowski and R. Huisgen, *Tetrahedron Letters*, 4643 (1976).
810. T. I. Temnikova, I. P. Stepanov and L. A. Dotzenko, *Zh. Org. Khim.*, **3**, 1707 (1967).
811. C. P. Panayotis, D. Hilmar, E. Meyer and G. W. Griffin, *J. Amer. Chem. Soc.*, **89**, 1967 (1967).
812. N. R. Bertoniere, S. P. Rowland and G. W. Griffin, *J. Org. Chem.,* **36**, 2596 (1971).
813. A. M. Trozzolo, W. A. Yager, G. W. Griffin, H. Kristinsson and I. Sarkar, *J. Amer. Chem. Soc.*, **89**, 3357 (1967).
814. G. W. Griffin, K. Ishikawa and I. J. Lev, *J. Amer. Chem. Soc.*, **98**, 5697 (1976).
815. R. S. Becker, R. O. Bost, J. Kolc, N. R. Bertoniere, R. I. Smith and G. W. Griffin, *J. Amer. Chem. Soc.*, **92**, 1302 (1970).
816. D. R. Paulson, A. S. Murray, D. Benett, E. Mills, Jr., V. O. Terry and S. D. Lopez, *J. Org. Chem.*, **42**, 1252 (1977).
817. H. Eichenberg, H. R. Wolf and O. Jeger, *Helv. Chim. Acta*, **60**, 743 (1977).
818. M. Tokuda, M. Hatanga, J. Imai, M. Itoh and A. Suzuki, *Tetrahedron Letters*, 3133 (1971).
819. P. M. M. Van Haard, L. Thijs and B. Zwanenburg, *Tetrahedron Letters*, 803 (1975).
820. H. D. Becker, T. Bremholt and E. Adler, *Tetrahedron Letters*, 4205 (1972).
821. J. Muzart and J. P. Pete, *Tetrahedron Letters*, 3919 (1974).

822. D. M. Jerina, B. Witkop, C. L. McIntosh and O. L. Chapman, *J. Amer. Chem. Soc.*, **96**, 5578 (1974).
823. S. P. Pappas and B. La Quoc, *J. Amer. Chem. Soc.*, **95**, 7906 (1973).
824. H. Kristinsson, R. A. Mateer and G. W. Griffin, *J. Chem. Soc., Chem. Commun.*, 415 (1966).
825. A. K. Dey and H. R. Wolf, *Helv. Chim. Acta*, **61**, 626 (1978).
826. K. Maruyama, S. Arakawa and T. Oysuki, *Tetrahedron Letters*, 2433 (1975).
827. R. G. F. Giles and I. R. Green, *J. Chem. Soc., Chem. Commun.*, 1332 (1972).
828. H. Prinzbach and M. Klaus, *Angew. Chem. (Intern. Ed.)*, **8**, 276 (1969).
829. R. K. Murray, Jr., T. K. Morgan, Jr., A. S. J. Polley, Ch. A. Andruszkiewicz, Jr. and D. L. Goff, *J. Amer. Chem. Soc.*, **97**, 938 (1975).
830. H. Hart, C. Peng and E. Shih, *Tetrahedron Letters*, 1641 (1977).
831. H. Hart, C. Peng and E. Shih, *J. Org. Chem.*, **42**, 3635 (1977).
832. J. M. Coxon and G. S. C. Hii, *Australian J. Chem.*, **30**, 161 (1977).
833. J. Kagan, P. Y. Juang, B. E. Firth, J. T. Przybytek and S. P. Singh, *Tetrahedron Letters*, 4289 (1977).
834. J. Kagan, B. E. Firth, N. Y. Shih and C. G. Boyajian, *J. Org. Chem.*, **42**, 343 (1977).
835. M. Tokuda, V. V. Chung, A. Suzuki and M. Itoh, *J. Org. Chem.*, **40**, 1858 (1975).
836. S. N. Merchant, S. C. Sethi and H. R. Sonawane, *Indian J. Chem.*, **B15**, 82 (1977).
837. M. M. Movsumzade and A. L. Shabanov, *Azerb. Khim. Zh.*, **2**, 35 (1973).
838. R. J. Crawford, S. B. Lutener and R. D. Cockcroft, *Can. J. Chem.*, **54**, 3364 (1976).
839. V. Vukov and R. J. Crawford, *Can. J. Chem.*, **53**, 5367 (1975).
840. J. C. Paladini and J. Chuche, *Bull. Soc. Chim. Fr.*, 197 (1974); M. S. Medimagh and J. Chuche, *Tetrahedron Letters*, 793 (1977).
841. W. Eberbach and B. Burchardt, *Chem. Ber.*, **111**, 3665 (1978).
842. A. J. Dobbs, B. C. Gilbert, H. A. H. Laue and R. O. C. Norman, *J. Chem. Soc., Perkin II*, 1044 (1976).
843. M. C. Flowers and D. E. Penny, *J. Chem. Soc., Faraday I*, **70**, 355 (1974).
844. M. C. Flowers and D. E. Penny, *J. Chem. Soc., Faraday I*, **71**, 851 (1975).
845. M. C. Flowers and T. Öztürk, *J. Chem. Soc., Faraday I*, **71**, 1509 (1975).
846. M. C. Flowers, *J. Chem. Soc., Faraday I*, **73**, 1927 (1977).
847. A. Dahmen, H. Hamberger, R. Huisgen and V. Markowski, *J. Chem. Soc., Chem. Commun.*, 1192 (1971).
848. J. K. Crandall and R. J. Watkins, *Tetrahedron Letters*, 1717 (1967).
849. P. Schiess and M. Wisson, *Helv. Chim. Acta*, **57**, 1692 (1974).
850. E. Lewars and G. Morrison, *Tetrahedron Letters*, 501 (1977).
851. E. L. Stogryn and M. H. Gianni, *Tetrahedron Letters*, 3025 (1970).
852. R. S. Razina and V. M. Al'bitskaya, *Zh. Org. Khim.*, **7**, 1637 (1971).
853. S. K. Pradhan and M. Girijavallabhan, *J. Chem. Soc., Chem. Commun.*, 591 (1975).
854. R. Huisgen, *Angew. Chem. (Intern. Ed.)*, **16**, 572 (1977).
855. G. Lamaty, R. Maleq, C. Selve, A. Sivade and J. Wylde, *J. Chem. Soc., Perkin II*, 1119 (1975).
856. R. A. Wohl, *Chimia*, **28**, 1 (1974).
857. H. E. Audier, J. F. Dupin and J. Jullien, *Bull. Soc. Chim. Fr.*, 2811 (1966).
858. H. E. Audier, J. F. Dupin and J. Jullien, *Bull. Soc. Chim. Fr.*, 3850 (1968).
859. H. E. Audier, J. F. Dupin and J. Jullien, *Bull. Soc. Chim. Fr.*, 3844 (1968).
860. G. Berti, B. Macchia and F. Macchia, *Tetrahedron*, **28**, 1299 (1972).
861. P. Crotti, B. Macchia and F. Macchia, *Tetrahedron*, **29**, 155 (1973).
862. A. Balsamo, P. Crotti, B. Macchia and F. Macchia, *Tetrahedron*, **29**, 199 (1973).
863. A. Balsamo, C. Battistini, P. Crotti, B. Macchia and F. Macchia, *Gazz. Chim. Ital.*, **106**, 77 (1976).
864. C. Battistini, A. Balsamo, G. Berti, P. Crotti, B. Macchia and F. Macchia, *J. Chem. Soc., Chem. Commun.*, 712 (1974).
865. C. Battistini, P. Crotti and F. Macchia, *Tetrahedron Letters*, 2091 (1975); *Gazz. Chim. Ital.*, **107**, 153 (1977).
866. G. Bellucci, G. Berti, M. Ferretti, G. Ingrosso and E. Mastrorilli, *J. Org. Chem.*, **43**, 422 (1978).

867. C. Battistini, G. Berti, P. Crotti, M. Ferretti and F. Macchia, *Tetrahedron*, **33**, 1629 (1977).
868. G. Berti, G. Catelani, M. Ferretti and L. Monti, *Tetrahedron*, **30**, 4013 (1974).
869. M. T. Langin and J. Huet, *Tetrahedron Letters*, 3115 (1974).
870. G. A. Morrison and J. B. Wilkinson, *Tetrehedron Letters*, 2713 (1975).
871. E. Glotter, P. Krinszky, M. Rejtoe and M. Weissenberg, *J. Chem. Soc., Perkin I*, 1442 (1976).
872. T. A. Campion, G. A. Morrison and J. B. Wilkinson, *J. Chem. Soc., Perkin I*, 2508 (1976).
873. C. Anselmi, G. Berti, B. Macchia, F. Macchia and L. Monti, *Tetrahedron Letters* 1209 (1972).
874. M. A. Khuddus and D. Swern, *J. Amer. Chem. Soc.*, **95**, 8393 (1973).
875. T. M. Santosusso and D. Swern, *J. Org. Chem.*, **40**, 2764 (1975).
876. J. W. Keller, N. G. Kundu and C. Heidelberger, *J. Org. Chem.*, **41**, 3487 (1976).
877. I. A. Beland and R. G. Harvey, *J. Amer. Chem. Soc.*, **98**, 4963 (1976).
878. D. L. Whalen, J. A. Montemarano, D. R. Thakker, H. Yagi and D. M. Jerina, *J. Amer. Chem. Soc.*, **99**, 5522 (1977).
879. S. K. Yang, D. W. McCourt, H. V. Gelboin, J. R. Miller and P. P. Roller, *J. Amer. Chem. Soc.*, **99**, 5124 (1977).
880. S. K. Yang, D. W. McCourt and H. V. Gelboin, *J. Amer. Chem. Soc.*, **99**, 5130 (1977).
881. V. F. Shvets and O. A. Tyukova, *Zh. Org. Khim.*, **7**, 1947 (1971).
882. A. Gagis, A. Fusco and J. T. Benedict, *J. Org. Chem.*, **37**, 3181 (1972).
883. B. L. Barili, G. Bellucci, G. Ingrosso and A. Vatteroni, *Gazz. Chim. Ital.*, **107**, 147 (1977).
884. J. G. Pritchard and I. A. Siddiqui, *J. Chem. Soc., Perkin II*, 452 (1973).
885. A. V. Willi, *Helv. Chim. Acta*, **56**, 2094 (1973).
886. M. D. Carr and C. D. Stevenson, *J. Chem. Soc., Perkin II*, 518 (1973).
887. R. Durand, P. Geneste, G. Lamaty and J. P. Roque, *Compt. Rend. (C)*, **277**, 1395 (1973).
888. V. F. Shvets, Yu. V. Lykov and A. R. Kugel, *Kinet. Katal.*, **16**, 639 (1975).
889. M. F. Sorokin, L. G. Shode and V. N. Stokozenko, *Zh. Org. Khim.*, **13**, 576 (1977).
890. V. F. Shvets and I. Al-Wahib, *Kinet. Katal.*, **16**, 785 (1975).
891. V. F. Shvets and I. Al-Wahib, *Zh. Fiz. Khim.*, **49**, 662 (1975).
892. N. N. Lebedev, V. F. Shvets, L. T. Kondrat'ev and L. L. Romaskina, *Kinet. Katal.*, **17**, 583 (1976).
893. N. N. Lebedev, V. F. Shvets, L. T. Kondrat'ev and L. L. Romaskina, *Kinet. Katal.*, **17**, 576 (1976).
894. N. N. Lebedev, V. F. Shvets, L. T. Kondrat'ev and L. L. Romaskina, *Kinet. Katal.*, **17**, 888 (1976).
895. G. S. Yoneda, M. T. Griffin and D. W. Carlyle, *J. Org. Chem.*, **40**, 375 (1975).
896. A. Kirrmann and R. Nouri-Bimorghi, *Bull. Soc. Chim. Fr.*, 3213 (1968).
897. L. A. Paquette, S. A. Lang, Jr., S. K. Porter and J. Clardy, *Tetrahedron Letters*, 3137 (1972).
898. L. Knothe and H. Prinzbach, *Tetrahedron Letters*, 1319 (1975).
899. H. Fumijoto, M. Katata, S. Yamabe and K. Fukui, *Bull. Chem. Soc. Japan*, **45**, 1320 (1972).
900. L. Birkhofer and H. Dickopp, *Chem. Ber.*, **102**, 14 (1969).
901. R. S. Razina, V. M. Al'bitskaya and V. V. Vasil'ev, *Zh. Org. Khim.*, **8**, 1816 (1972).
902. F. Asinger, B. Fell, J. Pfeifer and A. Saus, *J. Prakt. Chem.*, **71**, 314 (1972).
903. P. Bouchet and C. Coquelet, *Bull. Soc. Chim. Fr.*, 3153 (1973).
904. S. A. Kline and B. L. Van Duuren, *J. Heterocyclic Chem.*, **14**, 455 (1977).
905. I. Shahak, S. Manor and E. D. Bergmann, *J. Chem. Soc. (C)*, 2129 (1968).
906. H. Nakamura, H. Yamamoto and H. Nozaki, *Tetrahedron Letters*, 111 (1973).
907. J. L. Coke and R. S. Shue, *J. Org. Chem.*, **38**, 2210 (1973).
908. R. A. Izydore and R. G. Ghirardelli, *J. Org. Chem.*, **38**, 1790 (1973).
909. Gy. Schneider and B. Schönecker, *Acta Chim. Acad. Sci. Hung.*, **95**, 321 (1977).
910. W. Sucrow, M. Slopianka and D. Winkler, *Chem. Ber.*, **105**, 1621 (1972).

911.   P. F. Hudrlik, J. P. Arcoleo, R. H. Schwartz, R. N. Misra and R. J. Rona, *Tetrahedron Letters*, 591 (1977).
912.   C. Berse and R. Coulombe, *Can. J. Chem.*, **49**, 3051 (1971).
913.   E. Vioque, *J. Chromatogr.*, **39**, 235 (1969).
914.   G. H. Posner, D. Z. Rogers, C. M. Kinzig and G. M. Gurria, *Tetrahedron Letters*, 3597 (1975).
915.   G. H. Posner and D. Z. Rogers, *J. Amer. Chem. Soc.*, **99**, 8208 (1977).
916.   G. H. Posner and D. Z. Rogers, *J. Amer. chem. Soc.*, **99**, 8214 (1977).
917.   K. Ono, K. Okura and K. Murakami, *Chem. Letters*, 1261 (1977).
918.   S. N. Merchant, S. C. Sethi and H. R. Sonawane, *Indian J. Chem.*, **14B**, 460 (1976).
919.   A. Pasetti, F. Tarli and D. Sianesi, *Gazz. Chim. Ital.*, **98**, 290 (1968).
920.   E. Tobler, *Helv. Chim. Acta*, **52**, 408 (1969).
921.   P. Duhamel, L. Duhamel and J. Gralak, *Bull. Soc. Chim. Fr.*, 3641 (1970).
922.   H. G. Emblem, *J. Appl. Chem.*, **20**, 187 (1970).
923.   J. Sauleau, H. Bouget and J. Huet, *Compt. Rend. (C)*, **279**, 887 (1974).
924.   M. M. Tarnorutskii and N. V. Konyasheva, *Zh. Obshch. Khim.*, **45**, 155 (1975).
925.   S. S. Khripko, T. A. Alekseeva and A. A. Yasnikov, *Dokl. Akad. Nauk. Ukr. SSR (B)*, 55 (1977); *Chem. Abstr.*, **86**, 154840m (1977).
926.   V. B. Mochelin, Z. I. Smolina, A. N. Vul'fson, T. N. Dyumaeva and B. V. Unkovskii, *Zh. Org. Khim*, **7**, 825 (1971).
927.   G. Bernáth and M. Svoboda, *Tetrahedron*, **28**, 3475 (1972).
928.   K. Harada and Y. Nakajima, *Bull. Chem. Soc. Japan.*, **47**, 2911 (1974).
929.   N. S. Kozlov, K. A. Zhavnerko, L. S. Yakubovich and V. B. Prishchepenko, *Dokl. Akad. Nauk Belorussk. SSR*, **19**, 812 (1975).
930.   G. I. Polozov and I. G. Tisenko, *Izv. Akad. Nauk Belorussk. SSR, Ser. Khim. Nauk*, **2**, 59 (1975).
931.   U. Sulser and J. Widmer, *Helv. Chim. Acta*, **60**, 1676 (1977).
932.   D. R. Burfield, S. Gan and R. H. Smithers, *J. Chem. Soc., Perkin I*, 666 (1977).
933.   M. Inone, T. Sugita, Y. Kiso and K. Ichikawa, *Bull. Chem. Soc. Japan.*, **49**, 1063 (1976).
934.   T. Nakajima, Y. Nakamoto and S. Suga, *Bull. Chem. Soc. Japan.*, **48**, 960 (1975).
935.   V. P. Kukhar, L. A. Lazukina and A. V. Kirsanov, *Zh. Org. Khim.*, **9**, 304 (1973).
936.   N. S. Isaacs and D. Kirkpatrick, *Tetrahedron Letters*, 3869 (1972).
937.   Y. Echigo, Y. Watanabe and T. Mukaiyama, *Chem. Letters*, 1013 (1977).
938.   I. Tabushi, Y. Kuroda and Z. Yoshida, *Tetrahedron*, **32**, 997 (1976).
939.   P. E. Sonnet and J. E. Oliver, *J. Org. Chem.*, **41**, 3279 (1976).
940.   H. Nakai and M. Kurono, *Chem. Letters*, 995 (1977).
941.   E. J. Corey and C. U. Kim. *J. Amer Chem. Soc.*, **94**, 7586 (1972).
942.   F. De Reinach-Hirtzbach and T. Durst, *Tetrahedron Letters*, 3677 (1976).
943.   K. M. Foley and F. M. Vigo, *U.S. Patent*, No. 3,931,260; *Chem. Abstr.*, **84**, 164142h (1976).
944.   G. B. Sergeev, I. A. Leenson, M. M. Movsumzade, A. L. Shabanov and G. A. Sudakova, *Zh. Org. Khim.*, **12**, 506 (1976).
945.   M. M. Movsumzade, A. L. Shabanov, R. A. Babakhanov, P. A. Gurbanov and R. G. Movsumzade, *Zh. Org. Khim*, **9**, 1998 (1973).
946.   D. F. Lawson, *J. Org. Chem.*, **39**, 3357 (1974).
947.   V. I. Golikov, A. M. Aleksandrov, L. A. Alekseeva and L. M. Yagupol'skii, *Zh. Org. Khim.*, **10**, 297 (1974).
948.   V. I. Golikov, A. M. Aleksandrov, L. P. Glusko, V. G. Dryuk, L. A. Alekseeva, L. M. Yagupol'skii and M. S. Malinovskii, *Ukr. Khim. Zh.*, **41**, 495 (1975).
949.   H. Hoffmann, K. Merkel, F. Neumayr and J. Schossig, *German Patent*, No. 2,446,215 (1976); *Chem. Abstr.*, **85**, 45998c (1976).
950.   H. Gross and B. Costisella, *D.D.R. Patent*, No. 108,305 (1977); *Chem. Abstr.*, **83**, 10385g (1975).
951.   M. Braun and D. Seebach, *Chem. Ber.*, **109**, 669 (1976).
952.   M. Nagayama, O. Okumura, K. Yaguchi and A. Mori, *Bull. Soc. Chem. Japan*, **47**, 2473 (1974).

953. H. Gross and B. Costisella, *D.D.R. Patent*, No. 108,305 (1974); *Chem. Abstr.*, **83**, 10385g (1975).
954. J. Buddrus, *Angew. Chem.*, **84**, 1173 (1972).
955. S. Fumasoni, F. Pochetti and G. Roberti, *Ann. Chim. (Rome)* **63**, 873 (1973).
956. H. Koinuma, H. Kato and H. Hirai, *Chem. Letters*, 517 (1977).

CHAPTER **15**

Cyclic ethers

M. BARTÓK
Department of Organic Chemistry, József Attila University,
Szeged, Hungary

I. INTRODUCTION

The syntheses and reactions of the cyclic ethers (oxacycloalkanes) have been studied most extensively for the compounds with low numbers (3—6) of ring atoms. It is mainly these oxacycloalkanes that have acquired economic importance. Naturally, the oxiranes are of outstanding significance, and this has justified their review in a separate chapter[1].

The present chapter surveys cyclic ethers with 4—6 ring atoms, i.e. oxetanes, oxolanes and oxanes. The nature of this task and the limited space available preclude the treatment of the synthesis and reactions of compounds of these types also containing other functional groups. The most detailed reviews of the theme outlined above are those of Dittus[2–4] and Kröper[5]. Since the survey by Gritter[6], more recent reviews of certain aspects of the chemistry of cyclic ethers have also been published[7–9].

II. SYNTHESIS OF CYCLIC ETHERS

A. From Monofunctional Hydrocarbon Derivatives

As a result of wide-ranging investigations, a rational procedure has been developed for the synthesis of 2,5-dialkyloxolanes by means of the oxidative intramolecular cyclization of secondary alcohols[10] (equation 1). The yield is 35—95%, depending

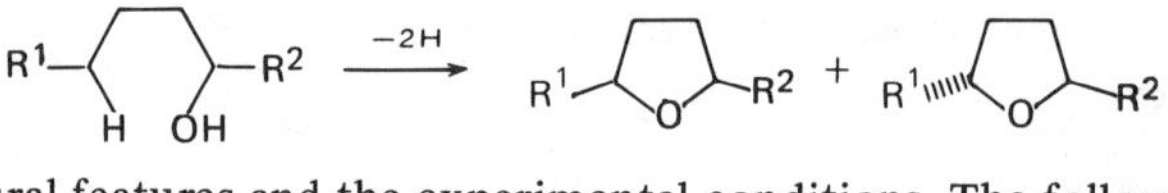

$$R^1 \underset{\overset{|}{H} \;\; \overset{|}{OH}}{\diagdown} R^2 \xrightarrow{-2H} R^1 \diagup\!\!\!\diagdown_O R^2 \;+\; R^1{}_{\prime\prime\prime\prime}\diagup\!\!\!\diagdown_O R^2 \tag{1}$$

on the structural features and the experimental conditions. The following have been used as reagents: $Pb(OAc)_4$; $Pb(OAc)_4 + I_2$; HgO or $Hg(OAc)_2 + I_2$ or Br_2; Ag_2O, AgOAc or $Ag_2CO_3 + I_2$ or Br_2. The procedures involving the halogens are known as hypohalite reactions.

Extensive studies have been carried out on the mechanism and stereochemistry of the cyclization[10,11], which were found to depend both on the configuration and conformation of the alcohol, and on the oxidizing agent employed. The mechanism of the $Pb(OAc)_4$ reaction is illustrated in equation (2), and its stereochemistry in equation (3).

The mechanism and stereochemistry of the hypohalite reaction (the course of which is similar to the previous one) are also treated in detail in the review by Mihailović[10], on the basis of his own results and those of Green and coworkers[11,12].

In spite of the fact that the reactions are not stereoselective, they may be used to advantage for the synthesis of optically active oxolanes: the configuration of the carbon atom bearing the OH group does not change in the course of the transformation, and thus, if the starting alcohol is an optically active one, optically

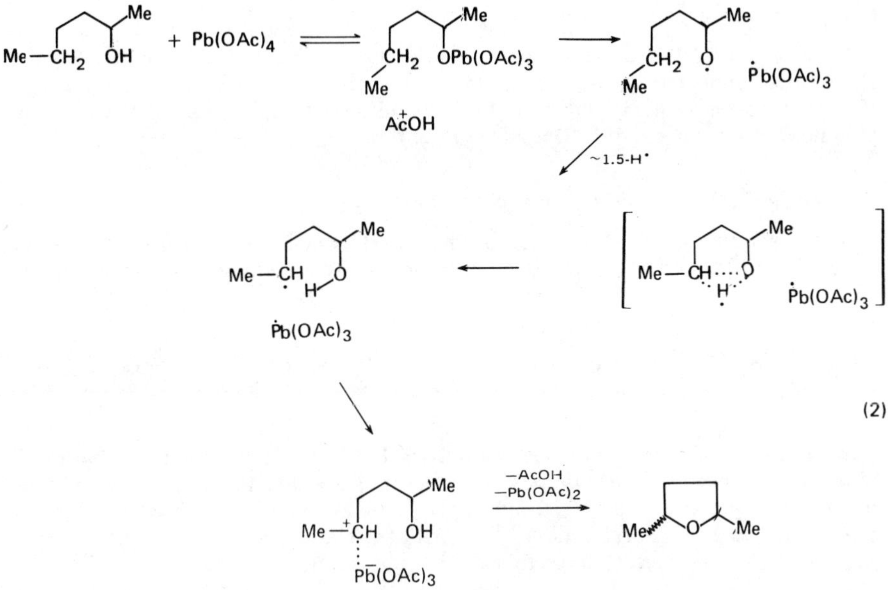

$$(2)$$

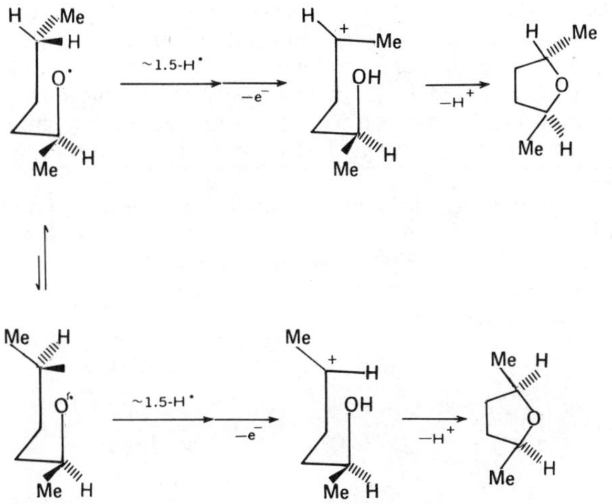

$$(3)$$

active *trans*-2,5-dialkyloxolane may be prepared from the diastereoisomer mixture (obtained in a ratio of nearly 1:1) after chromatographic separation.

B. From Difunctional Hydrocarbon Derivatives

The most general and most frequent procedures for the synthesis of oxacyclo-alkanes are the transformations under various experimental conditions of the 1,3-, 1,4- and 1,5-diols, and of difunctional compounds prepared from them, to oxetanes, oxolanes or oxanes.

1. Dehydration of diols

Using this method, oxolanes and oxanes can be prepared in very good yield. The results connected with the mechanism and stereochemistry of the dehydration of diols to cyclic ethers, and with the possibilities of application of the method, were surveyed in the chapter 'Dehydration of diols'[13].

2. Basic cyclization of difunctional compounds

The reaction scheme for this procedure is shown in equation (4). X is most frequently Cl, Br or OTs, while Y is H or Ac. The method may serve for the

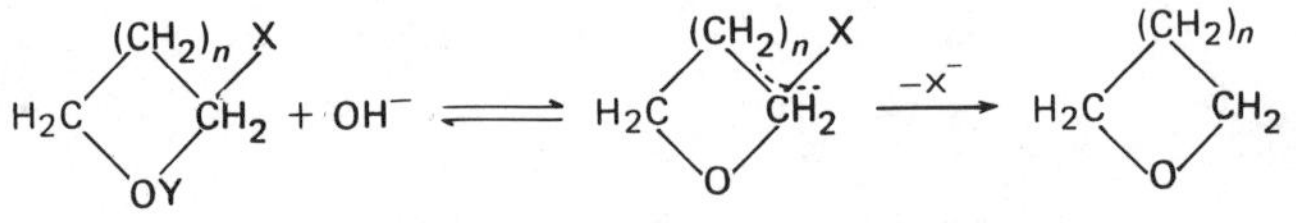

$$H_2C \underset{OY}{\overset{(CH_2)_n X}{\diagup}} CH_2 + OH^- \rightleftharpoons H_2C \underset{O}{\overset{(CH_2)_n X}{\diagup}} CH_2 \xrightarrow{-X^-} H_2C \underset{O}{\overset{(CH_2)_n}{\diagup}} CH_2 \qquad (4)$$

preparation of oxetanes, oxolanes and oxanes, but it is mainly used in the synthesis of oxetanes. The results of the past 10 years indicate that this procedure has been employed to prepare 2-aryl-[14], 2,2-dialkyl-[15], 3-alkyl- and 3-aryl-[16], 3,3-dialkyl-[17], 2,3-dialkyl-[18,19], 2-aryl-3-alkyl-[18], 2,4-dialkyl-[19] and 2,2,3,3-tetraalkyl-oxetanes[20], C≡C-substituted oxetanes[21-23], and various condensed polycyclic[24,25] and steroid[26,27] oxetanes. A number of publications deal with the preparation of the starting 1,3-chlorohydrins[28] and 1,3-chloroacetates[29,30], and also with the study of the mechanisms of the diol + acetyl chloride reactions[31,32]. Asymmetric induction occurs in the Grignard-type addition reaction of β-chlorobutyraldehyde[28].

The earlier finding that, in accordance with the method outlined in equation (4), oxetanes can be prepared in good yield only from compounds containing X in a primary position has been confirmed by additional experimental data[20,33,34], and has been convincingly justified by reaction kinetic and other examinations[19,20,35-37].

Investigations relating to the mechanism of the reaction, which have extended to the transition states of the molecules, confirm the reaction route of equation (4)[19,35-42]. Studies on the stereochemical course of the process[18,19,24-27,30], according to which the cyclization is stereospecific, similarly support the above mechanism (equations 5−7).

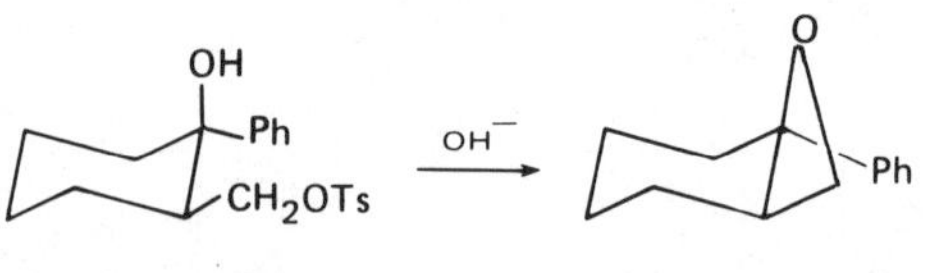

(Ref. 25)   (5)

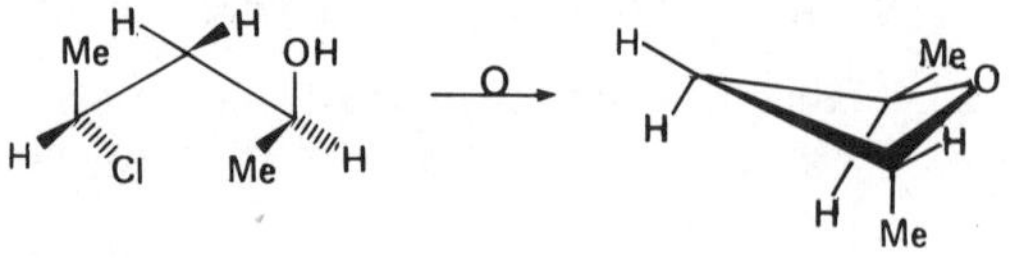

(Ref. 19)   (6)

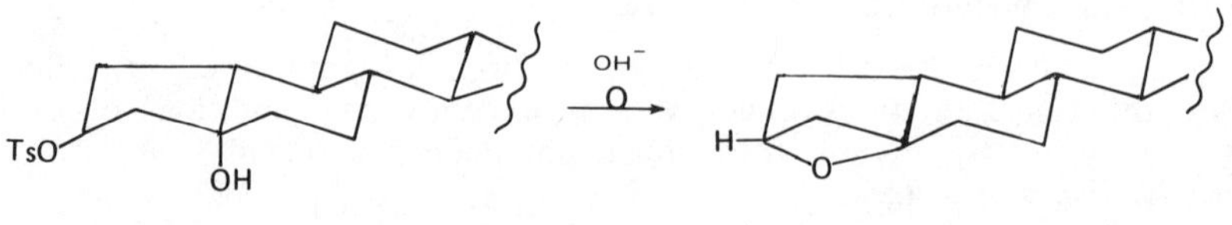

(Ref. 26)

(7)

The basic cyclization of the quaternary salts of 1,3-amino alcohols[43,44] and 1,4-amino alcohols[45] can be employed for the preparation of oxacycloalkanes only in the latter case.

Since new procedures have been elaborated for preparation of the starting compound, the method of equation (8) has been proposed for the synthesis of base-sensitive oxetanes[46].

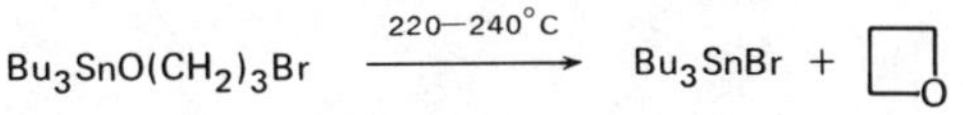

$$Bu_3SnO(CH_2)_3Br \xrightarrow{220-240°C} Bu_3SnBr + \text{[oxetane]} \qquad (8)$$

Oxetanes may be prepared too by the reaction of β-tosyloxycarbonyl compounds with organomagnesium or organolithium compounds (similarly by an S_Ni mechanism)[47] (e.g. equation 9).

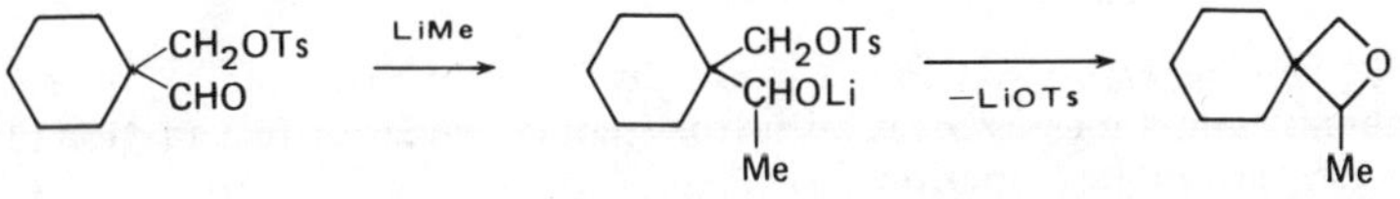

$$\qquad (9)$$

The basic cyclization of 1,4-diol dimesylates also occurs via an S_Ni mechanism[48] (equation 10). Since both reactions are accompanied by configuration changes, cis-oxolanes may be prepared from erythro-diols, and trans-oxolanes from threo-diols.

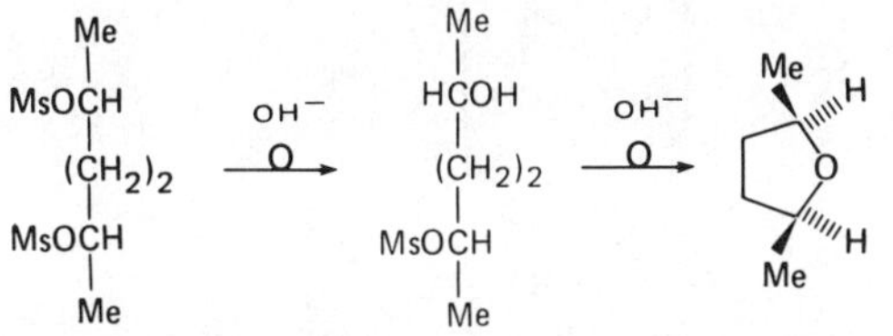

$$\qquad (10)$$

The presence of the corresponding oxonium salt intermediate has been proved experimentally in the cyclization of γ- and δ-methoxyalkyl halides in the presence of Lewis acids (e.g. equation 11)[49].

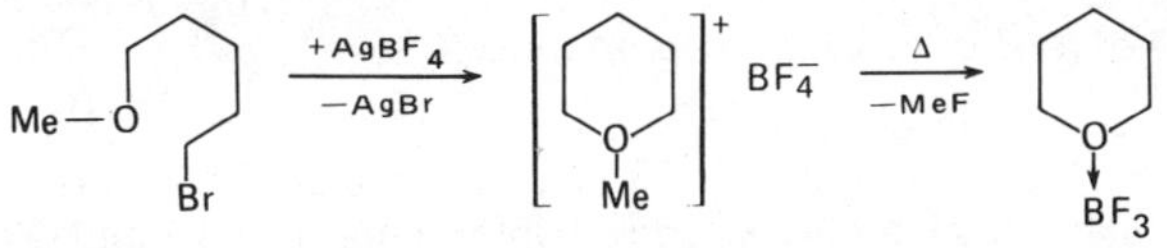

$$\qquad (11)$$

3. Transformation of unsaturated alcohols

Oxolanes and oxanes containing functional groups may be prepared in good yield from unsaturated alcohols under very varied experimental conditions and with various reagents (equation 12). The most recent literature data connected with the procedures are to be found in the review by Mihailović[10].

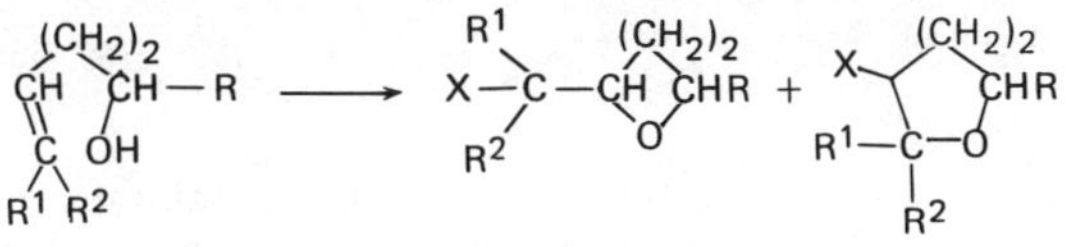

$$\qquad (12)$$

X = H, OH, OAc, Br, I, NO

Routes to oxolanes and oxanes not containing functional groups are shown in equations (13a) and (13b).

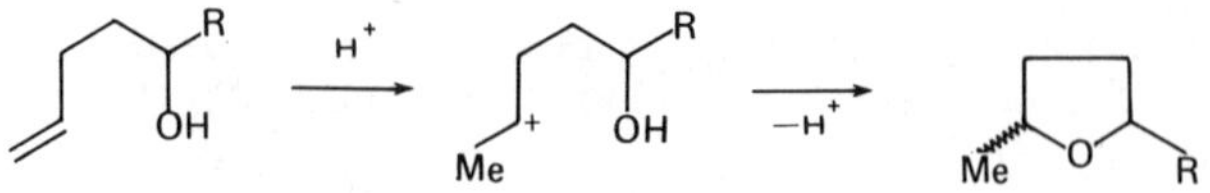

$$(13a)$$

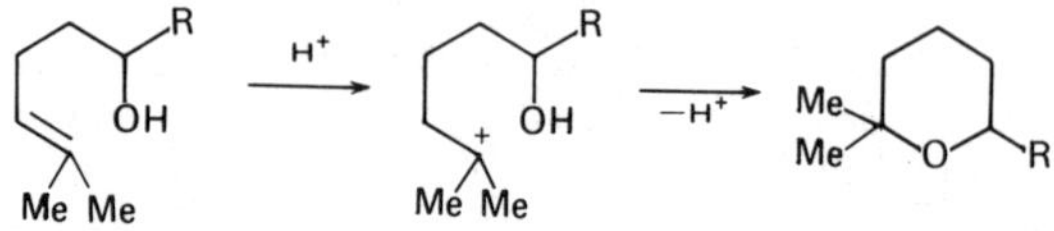

$$(13b)$$

PhSeCl can be employed in the synthesis of oxacycloalkanes[50] (equation 14). 2-Allylphenol undergoes cyclization in the manner outlined in equation (15), with neighbouring-group participation[51].

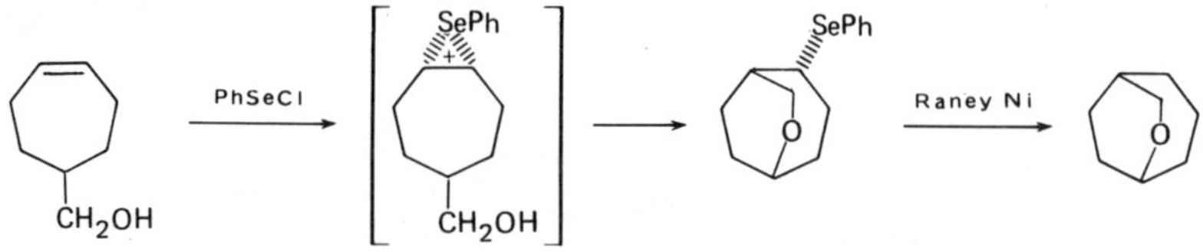

$$(14)$$

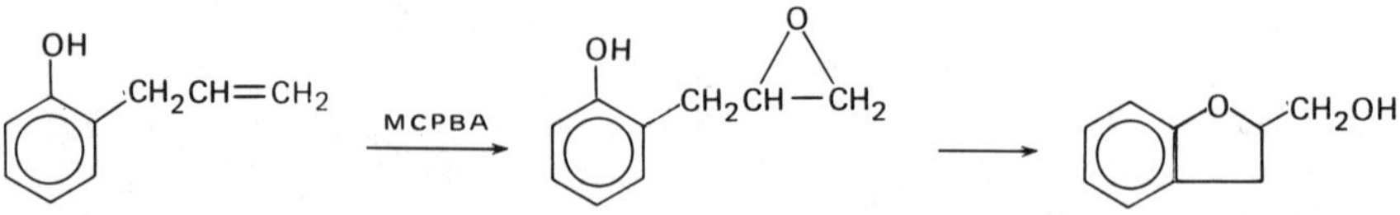

$$(15)$$

4. Cyclization of hydroxycarbonyl compounds

Although the intramolecular cyclizations of 1,4- and 1,5-hydroxycarbonyl compounds[52-55] to 2-hydroxy-oxolanes and -oxanes are reversible, subsequent dehydration makes these processes irreversible (equations 16 and 17). By catalytic

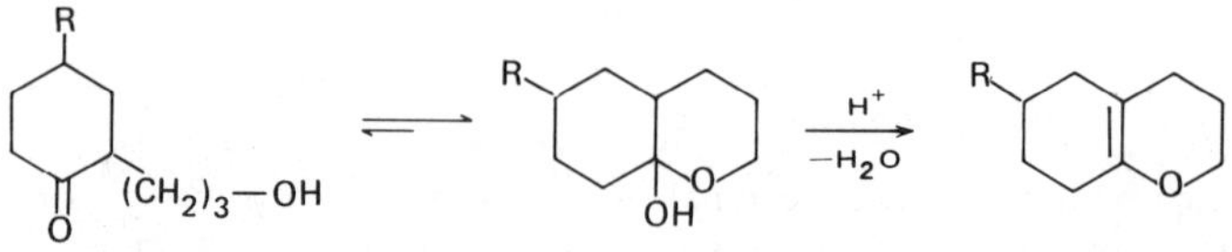

$$(Ref. 54) \quad (16)$$

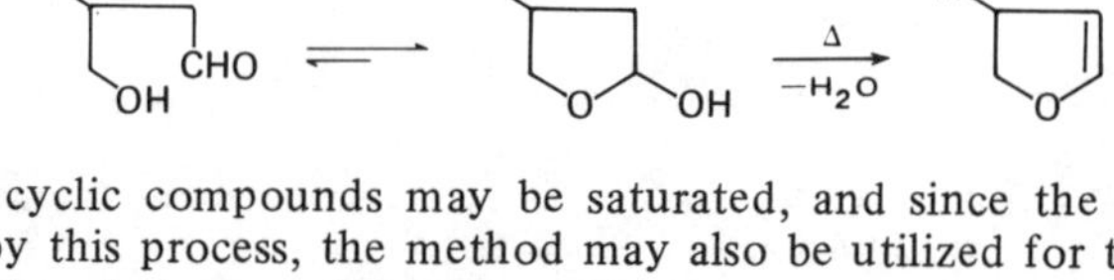

$$(Ref. 55) \quad (17)$$

reduction the cyclic compounds may be saturated, and since the chiral centre is not affected by this process, the method may also be utilized for the preparation of optically active oxolanes and oxanes.

C. From Heterocyclic Compounds

1. Formation from oxiranes

Oxiranes containing various functional groups can be transformed to oxetanes, oxolanes and oxanes.

By means of thermal rearrangement via alkoxytin intermediates, β-hydroxy-oxiranes may be converted to oxetane or oxolane derivatives, depending on the substituents on the carbon atoms of the oxirane ring[56] (equation 18). In the

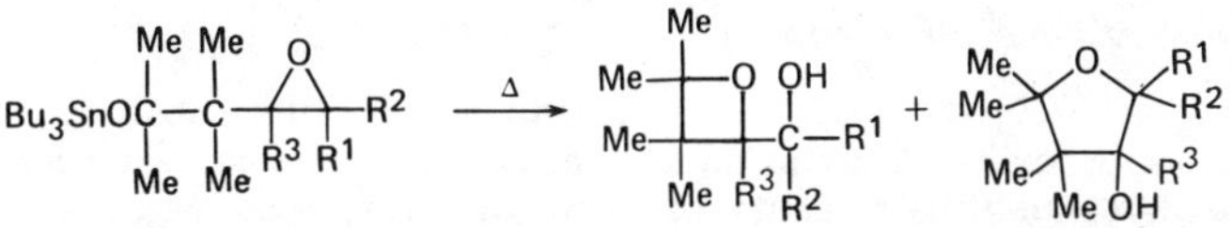

$$\text{(18)}$$

presence of bases, certain β-hydroxyoxiranes can be transformed to oxetanes directly in aqueous medium, by intramolecular cyclization[57,58] (e.g. equation 19).

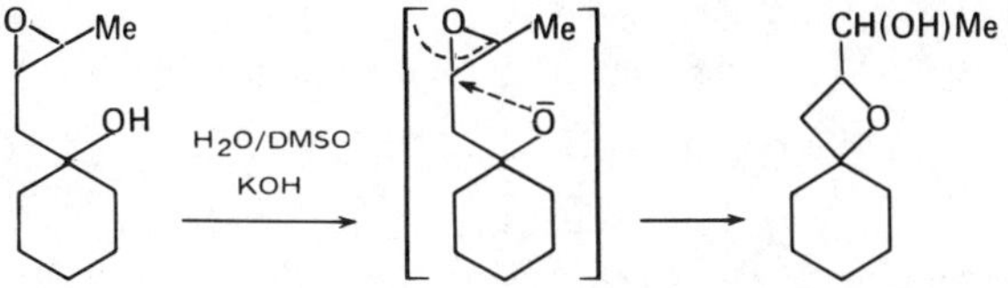

$$\text{(19)}$$

Oxolanes and oxanes may also be prepared from hydroxyoxiranes by either acid- or base-catalysed cyclization[59,60] (e.g. equation 20).

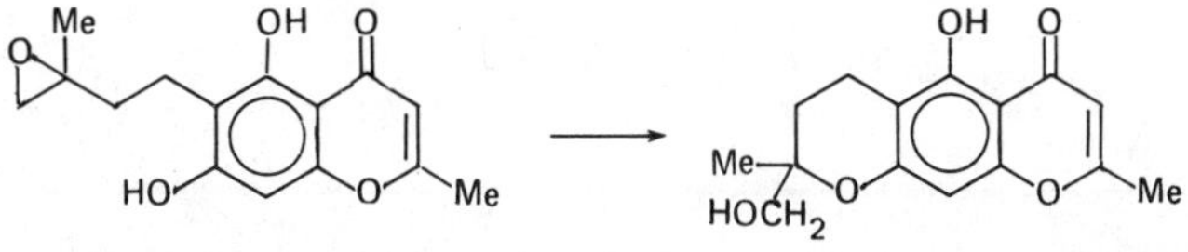

$$\text{(Ref. 59)} \quad \text{(20)}$$

β-Hydroxyoxiranes can be transformed to oxolanes by catalytic hydrogenolysis in the presence of acids, presumably via 1,4-diol intermediates[61] (equation 21).

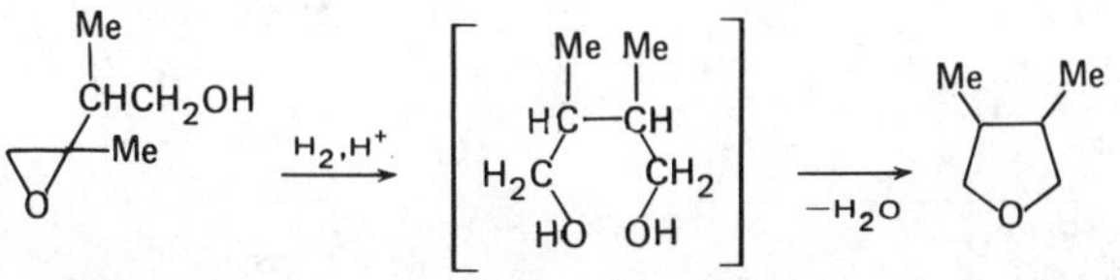

$$\text{(21)}$$

The vinyloxiranes undergo thermal rearrangement to dihydrofurans[62-65]. Equation (22) illustrates the mechanism of the much-examined rearrangement.

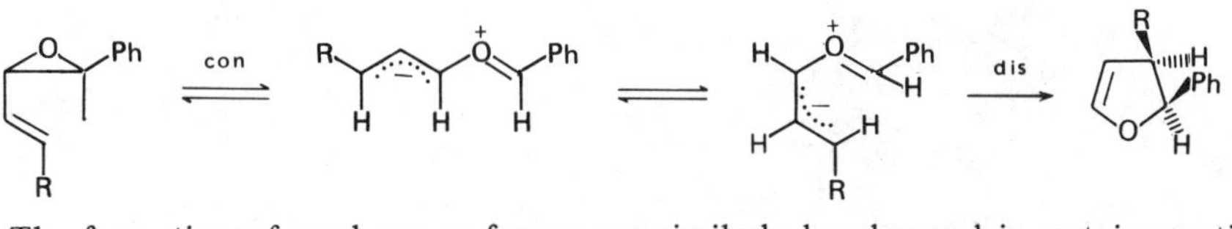

$$\text{(22)}$$

The formation of oxolanes or furans can similarly be observed in certain reactions of steroid oxiranes[66] or methoxyalleneoxiranes[67].

A comparatively simple method has been developed for the preparation of

2-aryl-3,6-dihydro-2H-pyrans, by means of the acid-catalysed rearrangement of cyclopropyloxiranes (equation 23)[68,69].

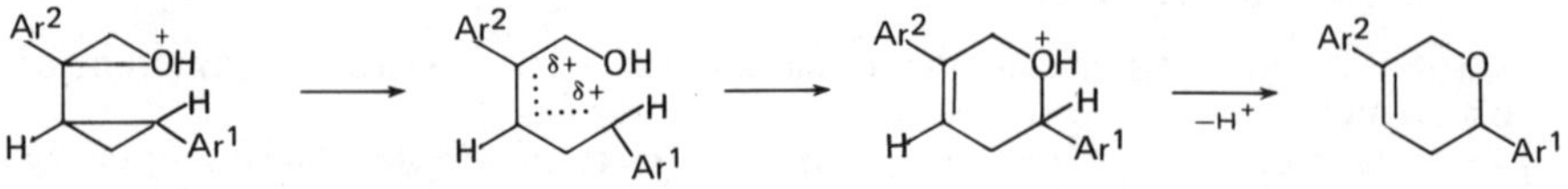

(23)

2. Reduction of oxacycloalkanones

Lactones can be converted to oxacycloalkanes with LiAlH$_4$, through the Grignard reaction or by catalytic reduction. Detailed studies have been carried out on various hydride-type reagents in the case of steroid lactones[70,71]. In the catalytic hydrogenation of maleic anhydride to oxolane, the effect of the composition of the bimetallic (Re—Ni) catalyst on the oxolane yield has been investigated[72].

Substituted oxolane-3-one can be utilized for the synthesis of 2,3-dihydrofurans[73], oxetanes[74] and oxolanes[74] (equation 24).

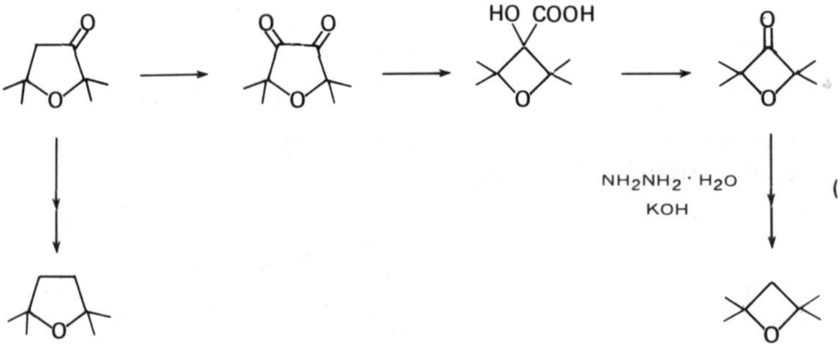

(Ref. 74) (24)

3. Reduction of dihydrofurans and furans

The reductions of furans have been reviewed by Armarego[8] and heterogeneous catalytic reductions (equation 25) by Bel'skii and Shostakovskii[7].

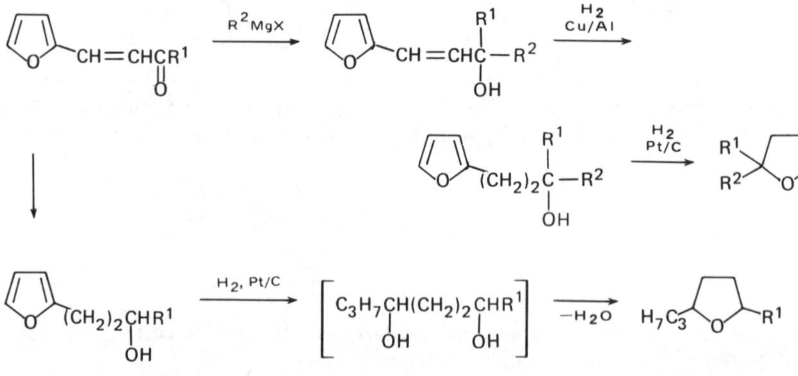

(25)

A new catalyst has been developed for the reduction of furan and alkylfurans[75].

The application of various zeolites as catalysts for the hydrogenation of alkylfurans has not proved satisfactory[76].

Optically active 2-methyloxolane can be prepared easily and in good yield as in equation (26)[77].

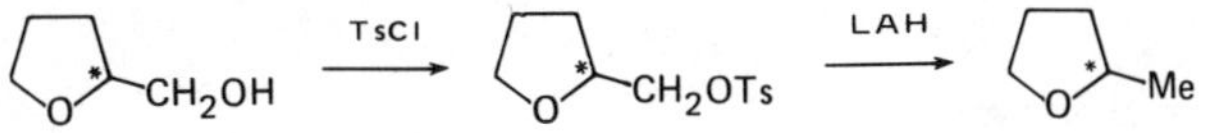

(26)

4. Preparation of oxanes from oxolanes

The procedures of Bel'skii[7,78] are also suitable for the preparation of oxanes (equations 27 and 28).

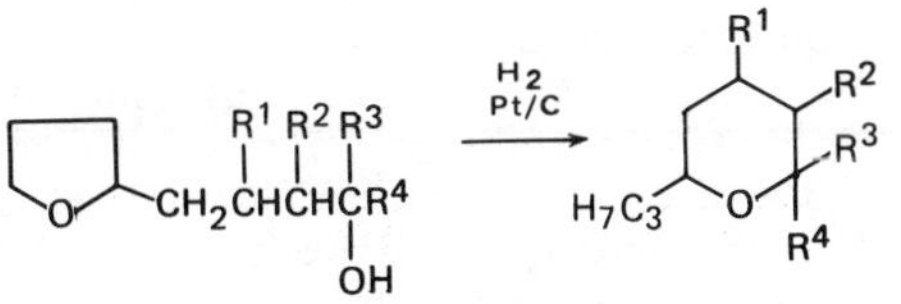

(27)

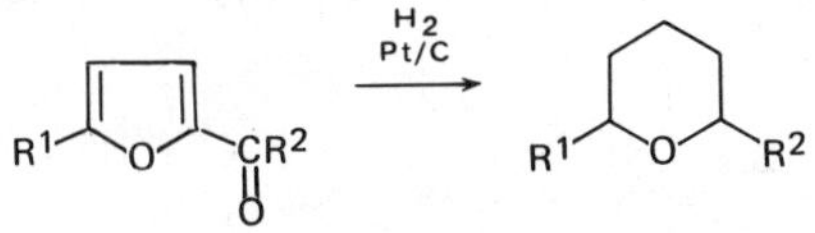

(28)

The mechanism of dehydration of 2-hydroxymethyloxolanes to yield dihydropyrans was studied[79,80]. The application of 2,3-dihydro-4H-pyrans as base-stable, acid-labile protective groups has been surveyed by Armarego[8].

5. Rearrangement of dioxacycloalkanes

A new procedure has been elaborated by Mousset and coworkers for the preparation of 3-acyloxolanes by means of the rearrangement of 5-vinyl-1,3-dioxolanes in the presence of electrophilic catalysts[81-84]. The stereoselective rearrangement is shown in equation (29). Alkyldimethyl-1,3-dioxanes undergo rearrangement to hydroxyoxanes in the presence of acids[85].

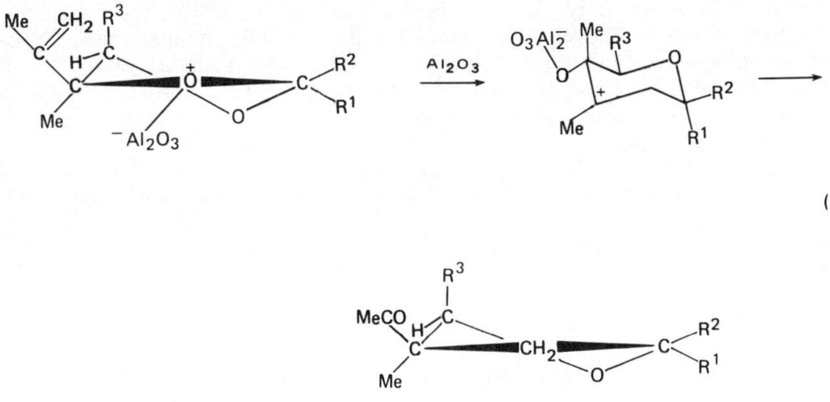

(29)

D. Via Cycloaddition Reactions

Cycloaddition can be employed for the preparation of oxetanes, oxolanes and oxanes. Many reviews of this topic[8,86-89] have appeared and we shall deal here mainly with the results published since 1974.

1. Synthesis of oxetanes

Oxetanes may be synthesized by the photocatalytic 1,2-cycloaddition of olefins and carbonyl compounds (Paterno–Büchi reaction). The carbonyl compounds used so far include aldehydes, ketones, diketones, quinones, carboxylic acid fluorides, urethanes, acyl nitriles, alkoxycarbonyl nitriles, thiocarbonyls and certain esters, while among the unsaturated compounds used are olefins, allenes, acetylenes, enones, ketene imines and ketene acetals[89].

The Paterno–Büchi reaction may occur either intermolecularly or intramolecularly. Meier[89] has tabulated the preparations of more than 200 oxetane derivatives. The yield varies from a few per cent to 80%. More recent papers deal with the regioselectivity, stereochemistry and mechanism of the cycloaddition.

Studies have been made of the cycloadditions of olefins and aldehydes[90,91], olefins and ketones[92-95], and reactants containing various functional groups[96-107] (e.g. equations 30–33). Reaction (30) is fairly regioselective ($3:4 = 9:1$). The

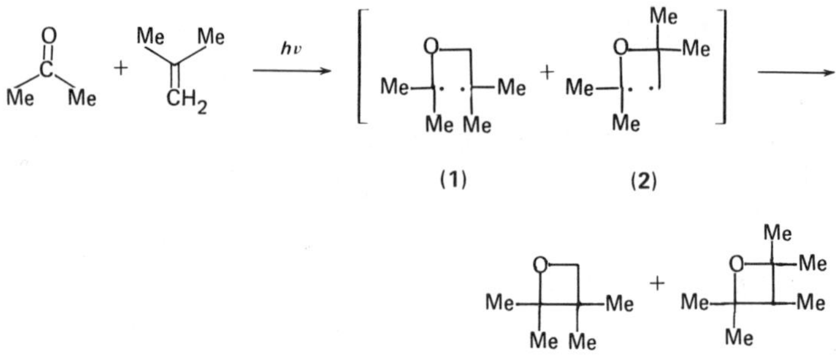

(1)   (2)

(3)   (4)

(Ref. 108) (30)

orientation of the cycloaddition is governed by the relative stabilities of the radicals 1 and 2. In general, a mixture of the (Z)- and (E)-isomers is formed in the addition[96]. However, only the cis-anellation (cis-fused) product, 5, is obtained in the course of the photocycloadditions of 1,4-dioxene and benzophenone (82%), or acetone (66%)[100] (equation 31). Particularly for rather complex molecules, the biradical formed during the photoreaction may have various structures and, depending on the relative stabilities of the individual radicals, many other products, including oxolanes, may be produced in addition to oxetane[109,110].

In the case of intramolecular photocycloaddition, the oxetanes formed may be 2,3- and/or 2,4-linked (equation 34). The course of the reaction may be strongly

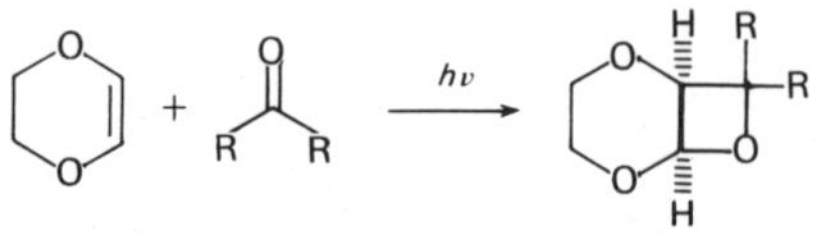

(Ref. 100) (31)

R = Me, Ph   (5)

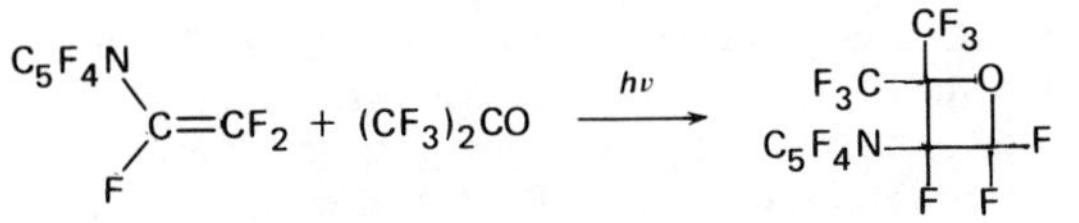

$$C_5F_4N \diagdown C=CF_2 + (CF_3)_2CO \xrightarrow{h\nu} \begin{array}{c} CF_3 \\ F_3C-\boxed{}-O \\ C_5F_4N-\boxed{}-F \\ F \quad F \end{array}$$ (Ref. 99) (32)

76%

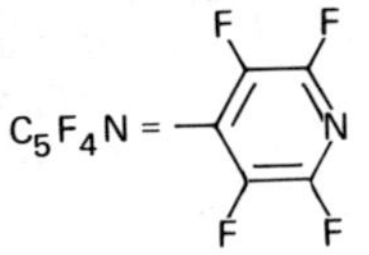

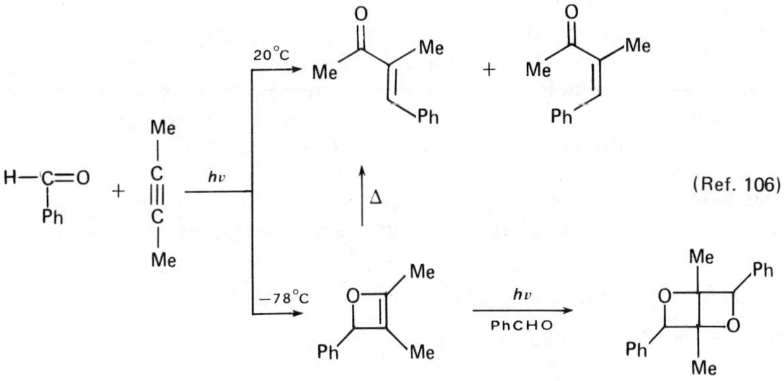

$$C_5F_4N = $$

(Ref. 106) (33)

$$H-\underset{Ph}{C}=O \ + \ \underset{Me}{\overset{Me}{\underset{|}{\overset{|}{\underset{C}{\overset{C}{\|}}}}}} \xrightarrow{h\nu}$$

20°C

$$\Delta$$

−78°C

$$\xrightarrow[PhCHO]{h\nu}$$

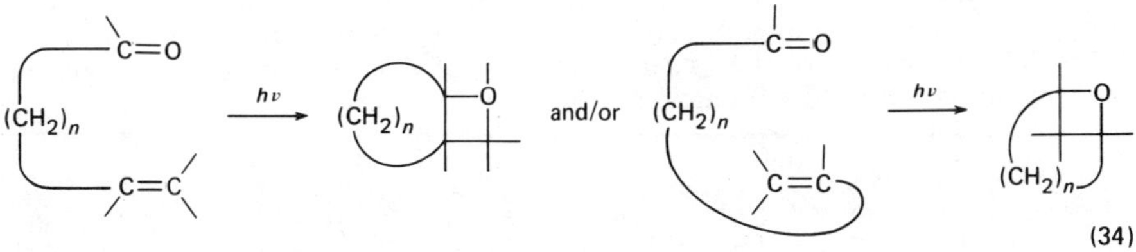

(34)

influenced by steric factors. Most reactions have been described for $n = 2$ and $n = 3$[111-113], but 2,2,3,4-tetramethyloxetane has also been prepared in good yield (70%) from a conjugated enone ($n = 0$)[114,115].

Many polycyclic oxetanes have been prepared from systems with rigid skeletons[116], and particularly by the photocycloaddition of 5-acylnorbornenes and their halogen and methoxy derivatives, in yields of 20–90%[117,118].

The Paterno–Büchi reaction is frequently used in more complex syntheses[119,120], and may, for example, yield intermediates in the syntheses of insect pheromones[121] or prostaglandin analogues[122].

Many hypotheses have been put forward for the mechanisms of the Paterno–Büchi reactions. A number of possibilities may be conceived for the radical formation itself, and for the reactions following this[92,93]. Moreover, if the triplet energy of the olefin is lower than that of the carbonyl, energy transfer may take place and olefin dimerization may become predominant.

As to the mechanism of the photochemical oxetane formation itself, no general theory exists that is valid for the overwhelming majority of the reactions. In principle, the reaction may be started by the excited (singlet or triplet) carbonyl, or by the olefin. In most cases, however, the initial step is the electrophilic attack of the excited carbonyl. In the first step of the excitation, singlet (^{1}n, π*) carbonyl is produced, which may pass over into a triplet (^{3}n, π*) state in a transition not involving radiation (intersystem crossing). Both states may be reactive (perhaps comparably so)[117], but in general one or other plays a predominant role. Transitions between the triplet and single states are possible by means of vibrational and spin-orbit couplings and other interactions[123]. A triplet state is often assumed in the reactions of aromatic ketones[94,95], while aldehydes and aliphatic ketones primarily react with a singlet carbonyl state[91,97,107,112,123].

As a result of the attack of the excited carbonyl, an excited transition complex (exciplex) is produced, which is converted to a 1,4-biradical, although the oxetane may also be formed from the exciplex via concerted development of two new σ-bonds[97]. The stereospecificity of the reaction in the singlet case is ensured by the higher rotational energy compared to that of the triplet state[124], and by the fact that (Z)–(E) isomerization at the radical site does not occur in general in a singlet biradical[125]. On the other hand, the regiospecificity is controlled by the relative stabilities of the radicals produced[108]. (A triplet 1,4-biradical may also be stabilized by cyclopropyl conjugation[126].) The biradicals may then be stabilized by ring-closure.

Meier gave a general scheme[89] for the possible reaction pathways of olefins and carbonyl compounds, though the transformations are not always reversible[95].

2. Synthesis of oxolanes and oxanes

The 2,3- and 2,5-cycloaddition reactions of furan and its derivatives, which can be used in many cases for the synthesis of condensed polycyclic oxolanes, have been reviewed by Armarego[8] (including the most recent literature data). Here, therefore, attention is merely drawn to the procedures outlined in equations (35)–(37), which show the general methods of synthesis of certain types of oxolanes by means of hydrogenation of the furans formed.

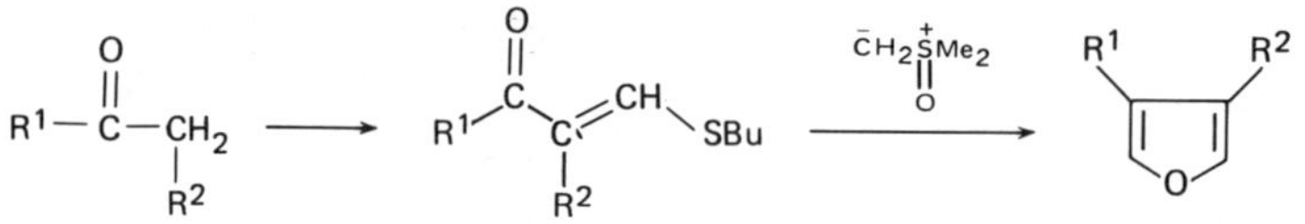

(Ref. 127)   (35)

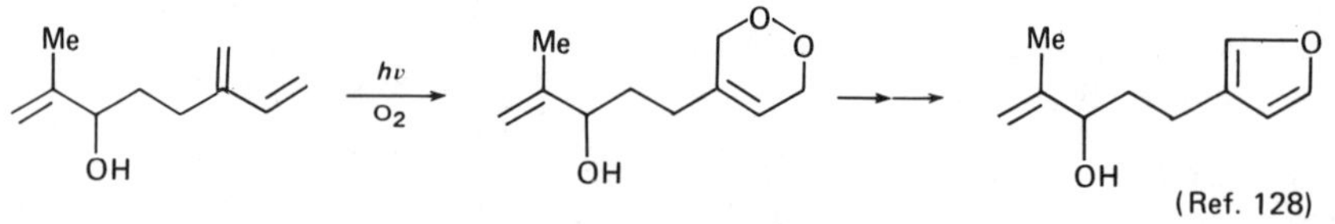

(Ref. 128)   (36)

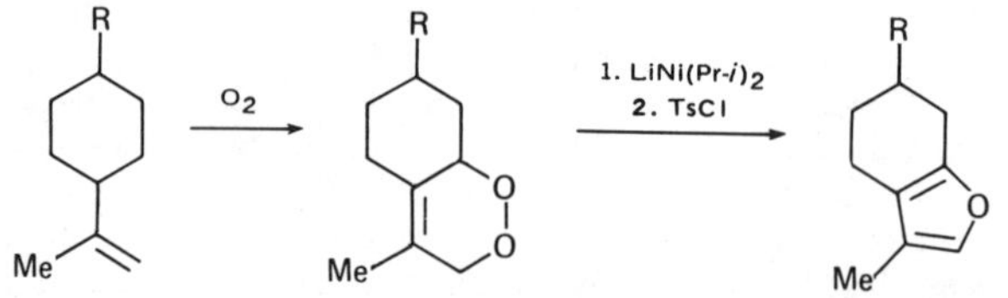

(Ref. 129)   (37)

Armarego[8] similarly gives a detailed account of the various procedures (among others by [2 + 2] π-cycloaddition from acrolein and olefins) for the synthesis of 2,3-dihydro-4H-pyrans and their cycloaddition transformations.

III. REACTIONS OF CYCLIC ETHERS

A. Deoxygenation

Cyclic ethers undergo deoxygenation on reaction with atomic carbon, to give the products outlined in equation (38)[130]. The mechanism of the deoxygenation is shown in equation (39) for the case of oxetane[130].

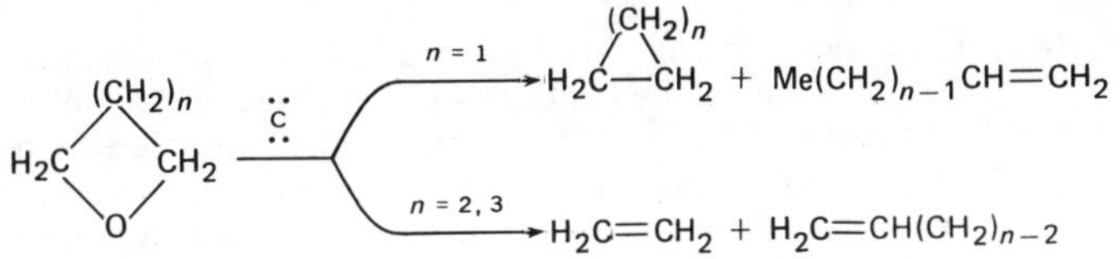

$$H_2C \overset{(CH_2)_n}{\underset{O}{\diamond}} CH_2 \quad \overset{\cdot\cdot}{C} \quad \begin{cases} \xrightarrow{n=1} H_2C{-}CH_2 \overset{(CH_2)_n}{\diagup} CH_2 + Me(CH_2)_{n-1}CH{=}CH_2 \\ \\ \xrightarrow{n=2,3} H_2C{=}CH_2 + H_2C{=}CH(CH_2)_{n-2} \end{cases} \tag{38}$$

$$\diamond\!\!-\!\!O \quad \overset{\cdot\cdot}{\underset{}{C}} \quad \left[\underset{C^-}{\overset{+}{\diamond}O} \right] \longrightarrow \left[\underset{\overset{\parallel}{C^-}}{\overset{+}{\triangle}O} \right] \overset{}{\underset{-CO}{\longrightarrow}} \left[\triangle \right] \longrightarrow \triangle + \diagup\!\!\diagdown \tag{39}$$

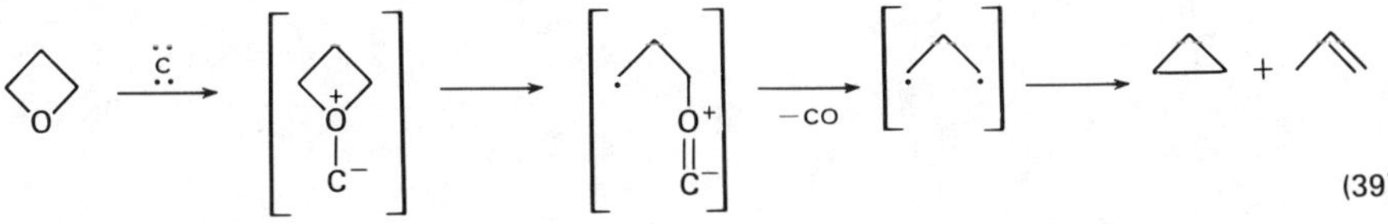

B. Dehydrogenation

Experimental observations are available only as regards the dehydrogenation of the dihydrofurans and the oxolanes[131-137]. The driving force of the dehydrogenation process is the striving towards aromatization, which is not possible for the oxetanes and the oxanes.

On a Pd/C catalyst, oxolane and the 2-alkyloxolanes are dehydrogenated to the corresponding furans (yield ~80%)[131]. If oxolane and 2,5-dihydrofuran are reacted with hydrogen acceptors transfer–hydrogenation reactions take place[132-134]. Oxolane does not disproportionate on Al_2O_3[135]. 2,3-Dihydrobenzofuran and its derivatives are dehydrogenated to the corresponding benzofurans via an ionic mechanism[136,137] (equation 40).

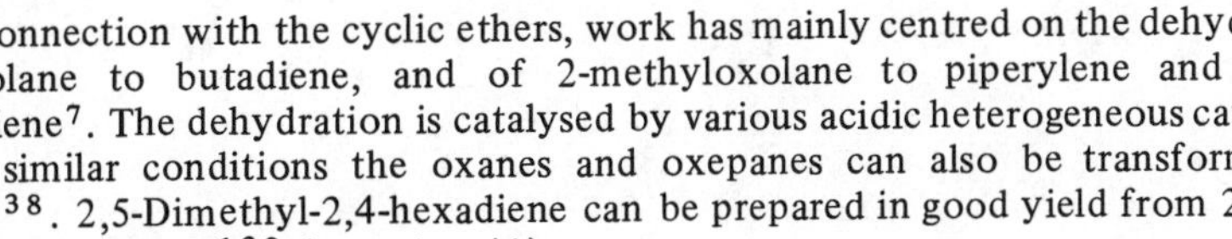

$$\tag{40}$$

C. Dehydration

In connection with the cyclic ethers, work has mainly centred on the dehydration of oxolane to butadiene, and of 2-methyloxolane to piperylene and cyclopentadiene[7]. The dehydration is catalysed by various acidic heterogeneous catalysts. Under similar conditions the oxanes and oxepanes can also be transformed to dienes[138]. 2,5-Dimethyl-2,4-hexadiene can be prepared in good yield from 2,2,5,5-tetramethyloxolane[139] (equation 41).

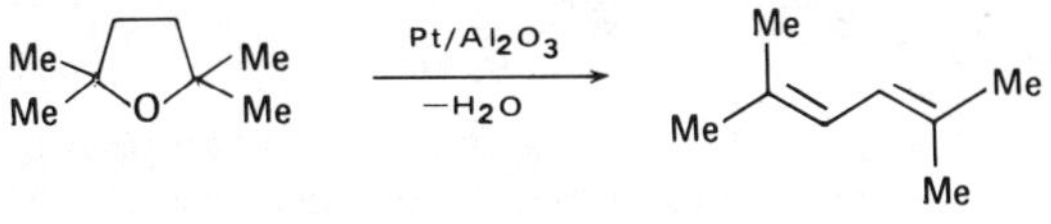

$$\underset{Me}{\overset{Me}{\diagdown}}\!\!\diamond\!\!\overset{Me}{\underset{Me}{\diagup}} \quad \xrightarrow[-H_2O]{Pt/Al_2O_3} \quad \underset{Me}{\overset{Me}{\diagdown}}\!\!\diagup\!\!\diagdown\!\!\overset{}{\underset{Me}{\diagup}}\!\!Me \tag{41}$$

D. Rearrangements

Two reviews have recently appeared on the rearrangements of cyclic ethers[7,8]. Because of the strained ring, the oxetanes (and the oxiranes) exhibit the highest reactivity of the cyclic ethers in rearrangement reactions.

1. Rearrangement of oxetanes

Comparatively few examinations have been made of the acid isomerizations of oxetanes[26,140-145]. By means of acid catalysis the oxetanes are mainly isomerized to unsaturated alcohols[26,143,145]. The isomerization depicted in equation (42)[145] proceeds with high selectivity on a g.l.c. column of acidic character to yield a β,γ-unsaturated alcohol. Another example is presented in equation (43)[26].

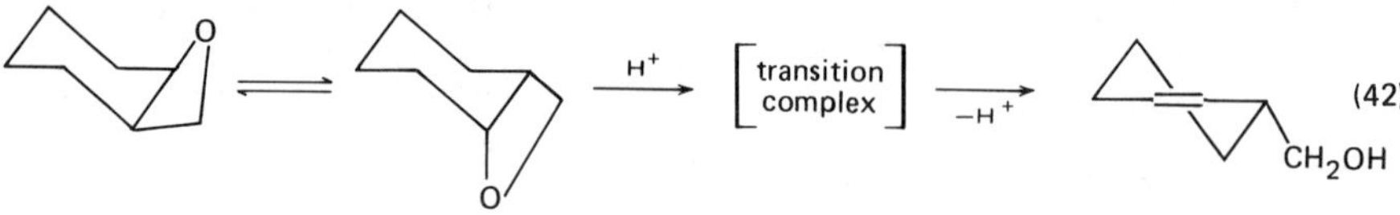

(42)

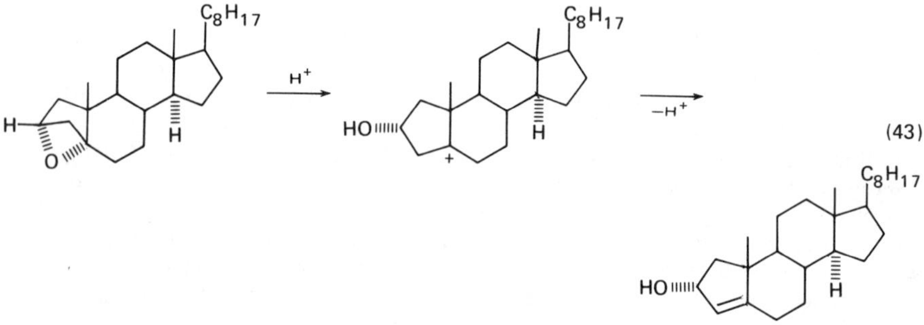

(43)

Equation (44) shows an acid-catalysed rearrangement of oxetane to oxolane[146]. On the action of neutral Al_2O_3, α-isopropylideneoxetanes are converted to the corresponding cyclobutanones in the course of rearrangement[147] (equation 45).

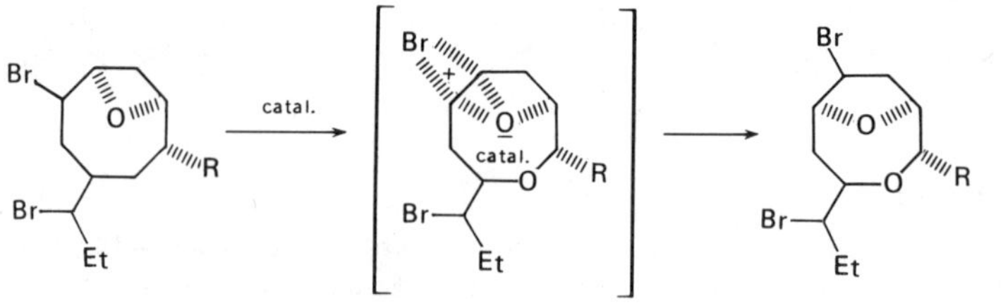

(44)

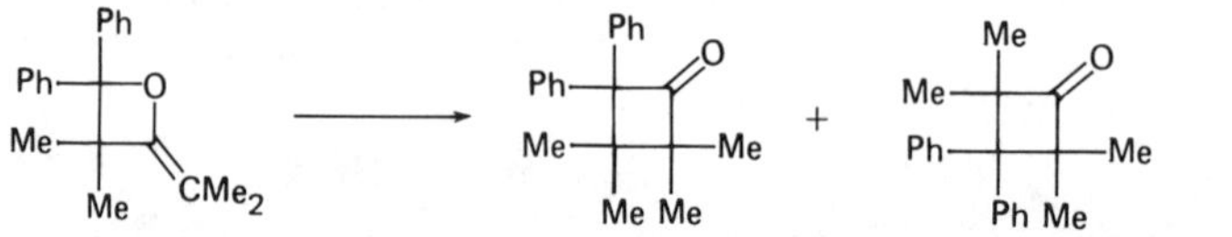

(45)

On Al_2O_3 and $Ca_3(PO_4)_2$ catalysts, isomeric carbonyl compounds are also formed in addition to the corresponding unsaturated alcohols in the rearrangement reactions[144]. The synthesis of 3-substituted furans is made possible by the rearrangement reaction shown in equation (46)[148]. In the presence of hydrogen

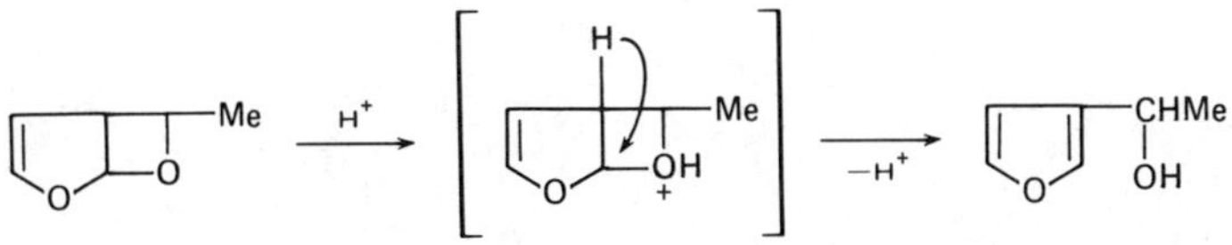

$$(46)$$

on supported metal catalysts, oxetanes undergo rearrangement to carbonyl compounds[140,149–153b] (equation 47). The mechanism of the reaction is very

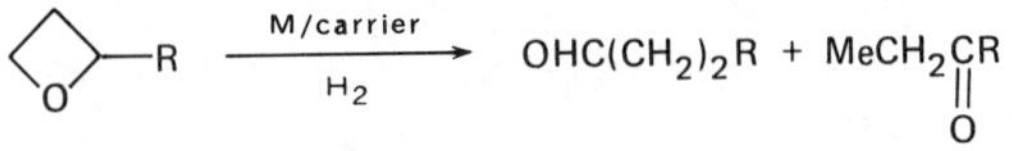

$$OHC(CH_2)_2R + MeCH_2\overset{\text{O}}{\underset{||}{C}}R \qquad (47)$$

complex, and depends to a great extent on the reaction conditions. From the examinations to date it is concluded[153b] that the formation of aldehydes can be explained by the participation of the electrophilic centres of the catalyst, while the presence of chemisorbed hydrogen is necessary for the formation of ketones.

On platinum metals, 2,2,4,4-tetramethyloxetane is rearranged to the corresponding ketone via a 1,3-bond shift mechanism[154] (equation 48).

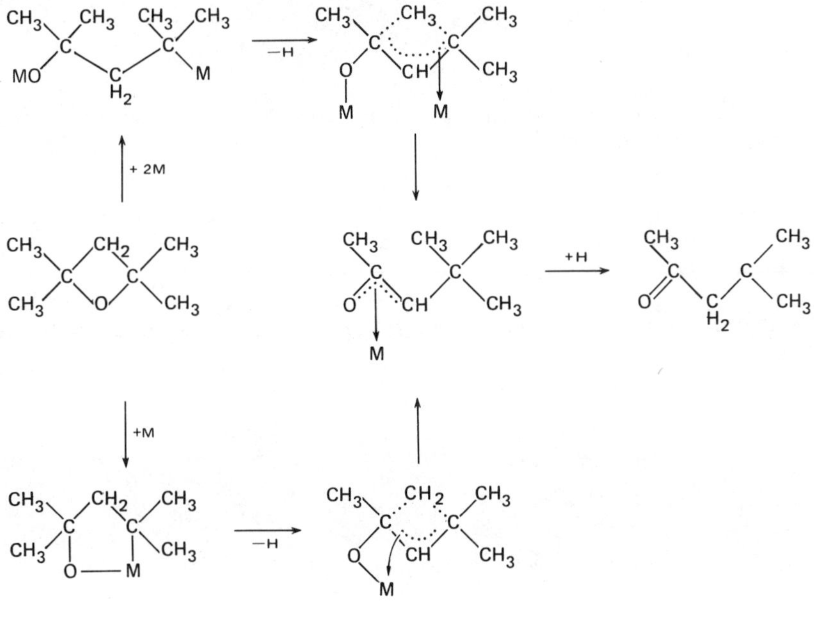

$$(48)$$

M = Pt, Pd, Rh

2. Rearrangement of oxolanes and oxanes

Oxolanes and oxanes are converted to ketones with very high regioselectivity on platinum metals[7]. In mechanistic studies[153b,155,156] it has been established that the presence of hydrogen is indispensable for the process to occur[157], while in all probability the reaction takes place according to a hydroisomerization mechanism[158] (equation 49). Some new results have also been reported on the rearrange-

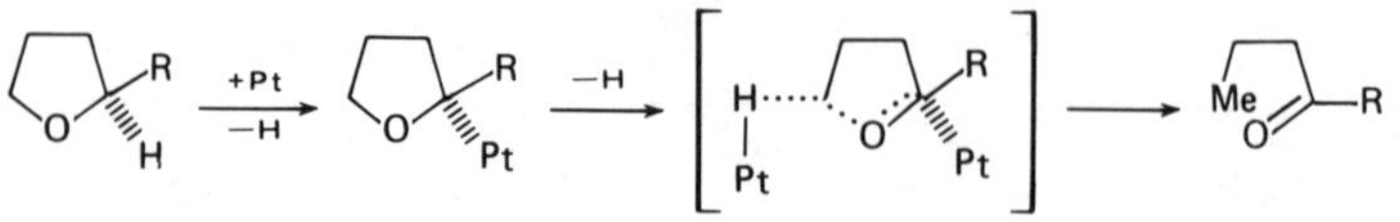

$$(49)$$

ment reactions of dihydrofurans and certain furan compounds. Studies have been made of the thermodynamics of isomerizations according to equation (50)[159].

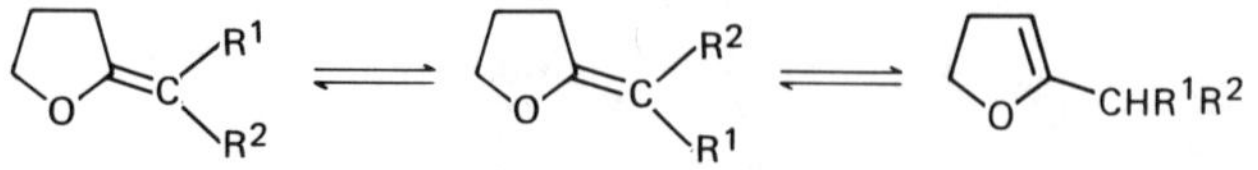

$$(50)$$

Examples of thermal isomerizations are the interconversions of **6** and **7**[160] (equation 51).

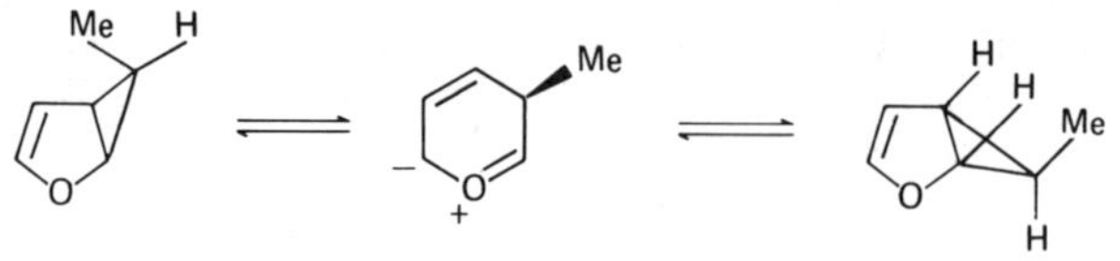

$$(51)$$

(6)   **(7)**

The acid-catalysed rearrangements of 2-furylcarbinols are electrocyclic reactions occurring with controtation[161] (equation 52). The process is stereospecific, only

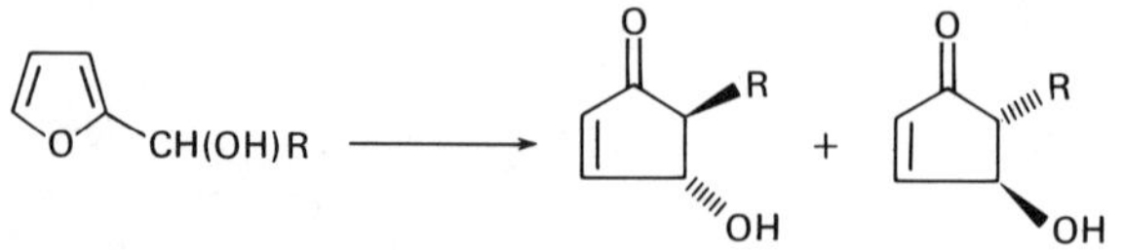

$$(52)$$

one of the enantiomer pairs being formed. Interesting rearrangements are to be seen in equations (53), (54)[162] and (55)[163].

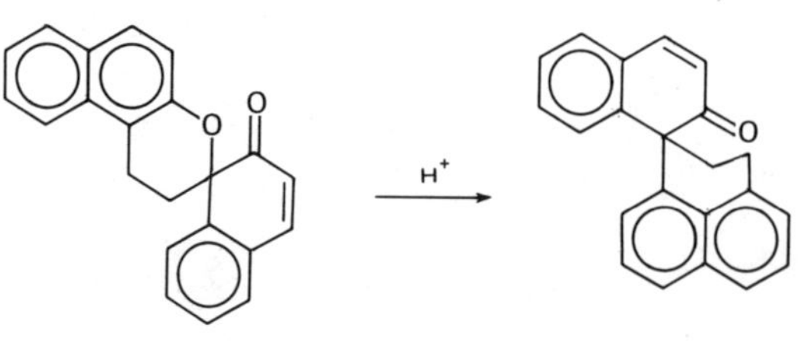

$$(53)$$

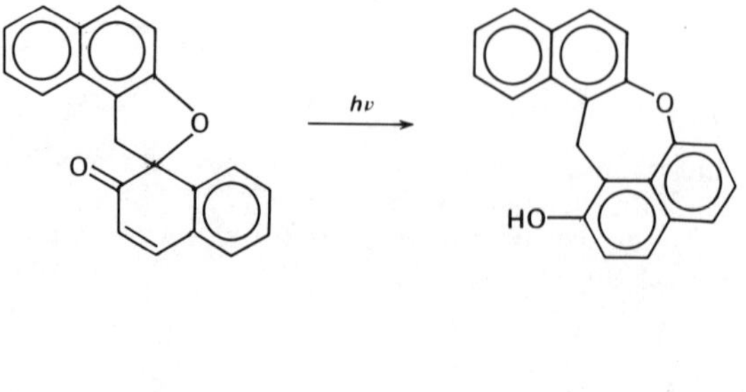

$$(54)$$

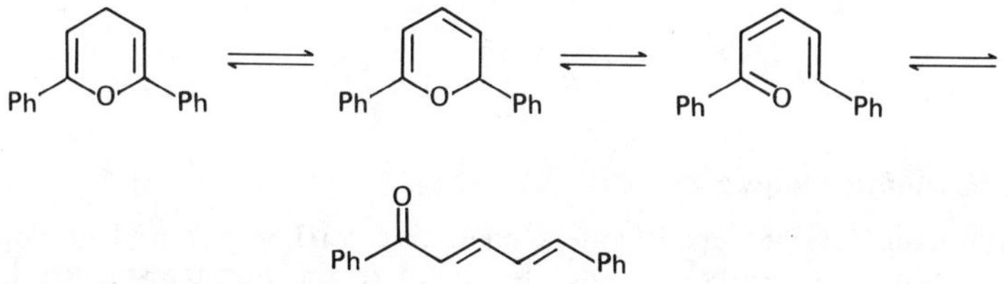

(55)

E. Oxidation

Much interest has been manifested recently in the reaction oxolane → γ-butyro-lactone. This process is of industrial importance; it can be carried out in the presence of catalysts[164,165], or electrochemically[166]. A procedure has been developed for the joint preparation of 2-hydroxyoxolane and γ-butyrolactone[164]. Investigations have been carried out on the kinetics and mechanism of the oxidation of oxolane with peroxydisulphate[167].

F. Reduction and Hydrogenolysis

Only a single review has appeared on the reduction and hydrogenolysis of 4-, 5- and 6-membered cyclic ethers[7]; this deals mainly with the hydrogenolysis of oxolanes and the reduction of furans and dihydrofurans. Since the reactivity decreases with the increase of the number of ring atoms, and only the oxetane ring can be opened with metal hydrides, the C–O bonds of oxolanes and oxanes can be cleaved by catalytic hydrogenolysis only.

1. Reduction with complex metal hydrides

With minor corrections, the regularities discovered for the oxiranes hold for the regioselectivity and mechanism of the reduction of oxetanes with $LiAlH_4$[1]. The regioselectivity is influenced by electronic and steric effects, and also by the nature of the reagent[168,169] (e.g. equation 56). The kinetics of the $LiAlH_4$ reduction of

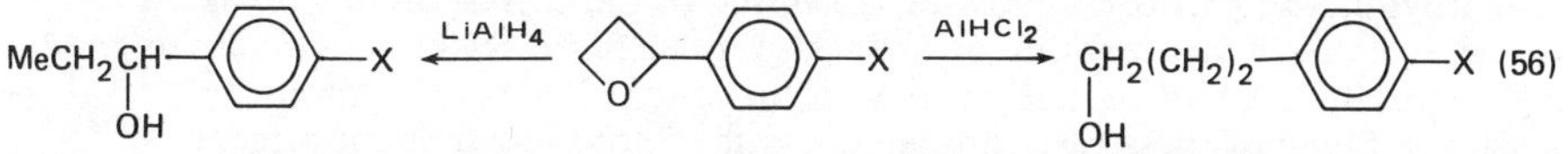

2-aryloxetanes can be well explained by an S_N2-type mechanism[170]. Studies have also been made of the reductions of certain 2-alkoxyoxetanes[171], polycyclic oxetanes[172] and spirooxetanes containing carbethoxy substituents[58,141] (equations 57–59).

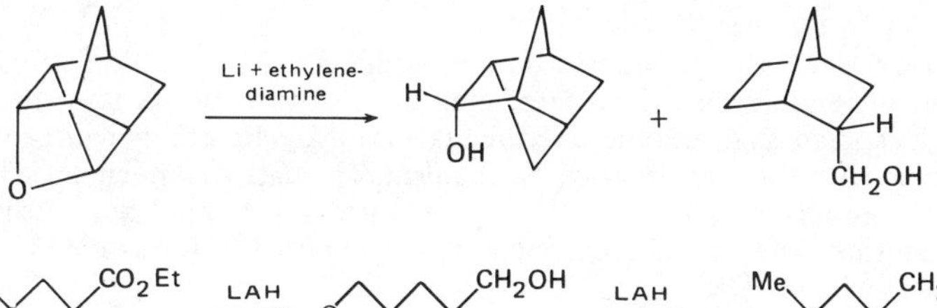

(57)

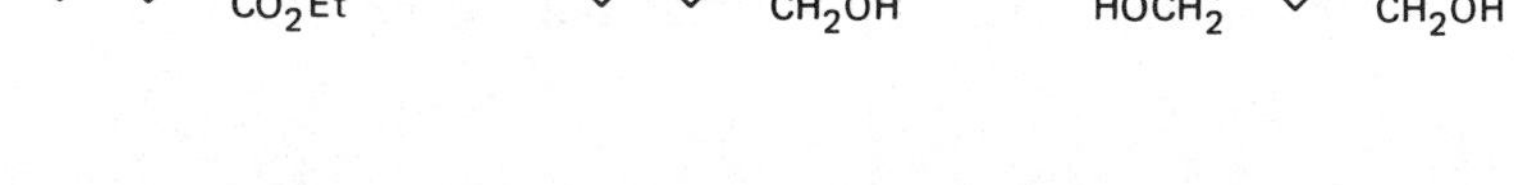

(58)

$$\text{(structure)} \quad \text{—CH}_2\text{OH} \xrightarrow{\text{LAH}} \text{(structure)} \text{—CH}_2\text{OH} \tag{59}$$

2. Catalytic hydrogenolysis

The hydrogenolysis of cyclic ethers on Group VIII metals and on copper has long been known. Recently, in order to elucidate the mechanism, use has been made of the pulse-microreactor technique[153a,b], selective catalyst poisoning[153b], isotope exchange[173], IR techniques[155], calculations of a thermodynamic and thermochemical nature[156] and other investigations relating to the end-products and intermediates[174].

The catalytic hydrogenolysis of oxetanes on various metal catalysts has been employed in syntheses and also in structure confirmations[22,172,175]. The isomerization of 2,2,4,4-tetramethyloxetane on platinum metals is accompanied by hydrogenolysis[154] (equation 60).

$$\text{Me} \underset{\text{Me}}{\overset{\text{Me}}{\diagdown}} \underset{\text{O}}{\diagup} \overset{\text{Me}}{\diagup} \xrightarrow[\text{H}_2]{\text{Pt, Pd or Rh}} \text{Me} \underset{\text{O}}{\overset{\text{Me}}{\diagdown}} \underset{\text{Me}}{\overset{\text{Me}}{\diagup}} + \text{Me} \underset{\text{Me}}{\overset{\text{Me}}{\diagdown}} \underset{\text{OH}}{\overset{\text{Me}}{\diagup}} \tag{60}$$

The variation in the regioselectivity of the hydrogenolysis of the oxacycloalkanes under pressure has been interpreted by its dependence on the number of ring atoms and on the catalyst (Raney Cu and Raney Ni)[176].

7-Hydroxyketones may be prepared by hydrogenolysis of the oxane ring[177] (equation 61):

$$\text{(structure)} \text{CHR}^1\text{CCH}_2\text{R}^2 \xrightarrow[\text{EtOH}]{\text{Raney Ni/M}^{2+}} \text{HO(CH}_2)_5\text{CHR}^1\text{CCH}_2\text{R}^2 \tag{61}$$

By hydrogenation on a Pt/C catalyst and subsequent hydrogenolysis, 2-alkyl-2-methyl-2,5-dihydrofurans may be converted to the corresponding isoalkanes[53].

By selective hydrogenation of furfurol, various furan skeleton compounds can be synthetized; the hydrogenation may occur with[178-180] or without[181-186] ring-cleavage. Some of these reactions are of synthetic or industrial importance.

G. Polymerization

The polymerization and copolymerization of cyclic ethers is important from an industrial aspect; this is best demonstrated by the large number of reviews that have appeared in the past decade[187-205].

As with oxiranes, the polymerization may take place by a cationic or an anionic mechanism, depending on the initiator employed. The view has recently begun to become widespread that anionic polymerization of cyclic ethers can proceed only in accordance with the coordination mechanism. The cationic mechanism[206,207] is illustrated in equations (62)–(65). The propagation steps may have either S_N1 or S_N2 mechanisms. The coordination anionic mechanism[208] is outlined in equations (66) and (67) with $Al(OR)_3$ as initiator.

$$\text{BF}_3 + \text{H}_2\text{O} \longrightarrow \text{H}^+[\text{BF}_3\text{OH}]^- = \text{H}^+\text{A}^- \tag{62}$$

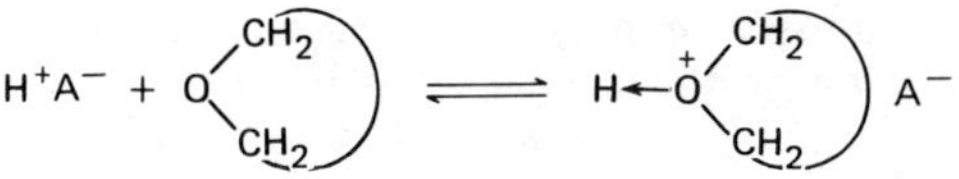

(63)

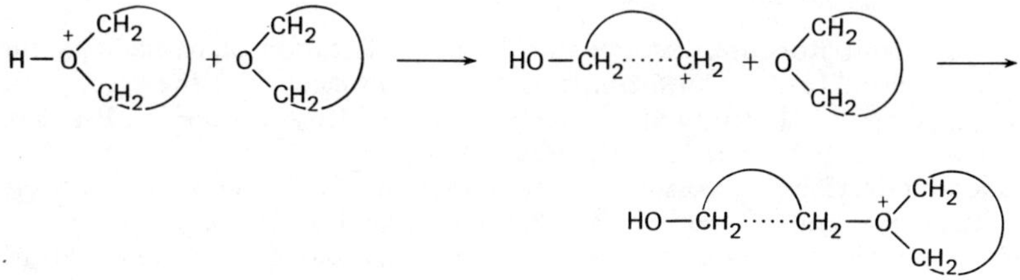

(64)

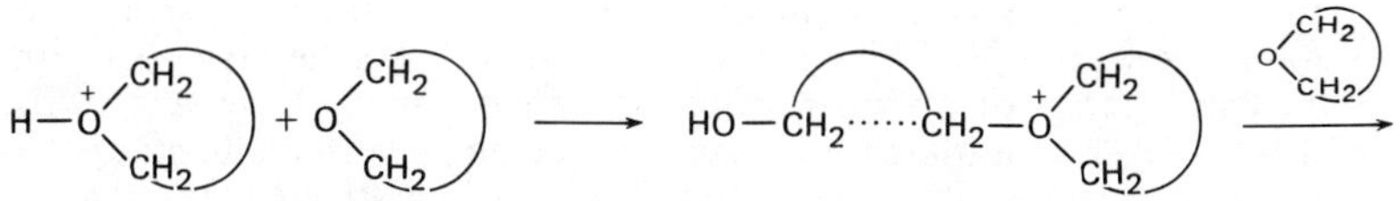

etc. (65)

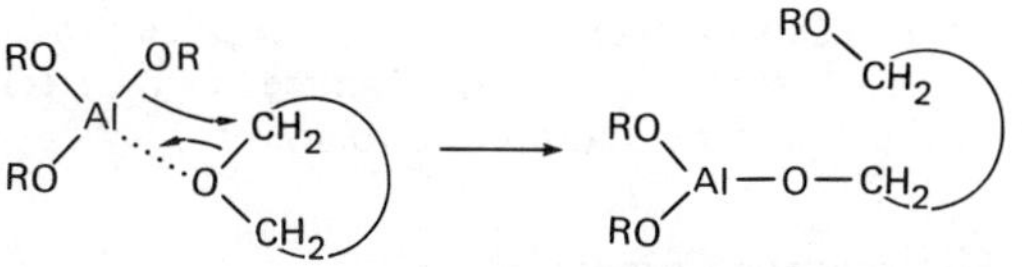

(66)

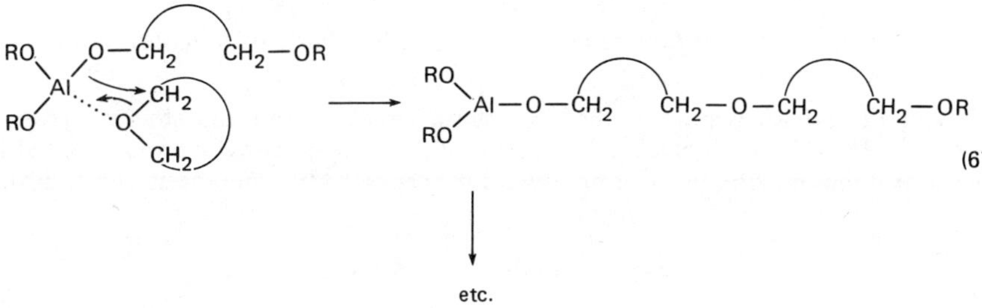

(67)

etc.

1. Polymerization of oxetane

According to recent investigations, the following initiators can be used for polymerization of oxetanes via the cationic mechanism: triethyloxonium salts[209], hexafluorophosphate salts[210] (e.g. Et_3O^+ PF_6^-, Ph_3C^+ PF_6^-) and ethyl trifluoro-methanesulphonate[211]. It is assumed[211,212] that both the oxonium ion produced in the initiation step, and the ester formed from it, are present in equilibrium (equation 68). With triethyloxonium salt initiators, oligomerization occurs in

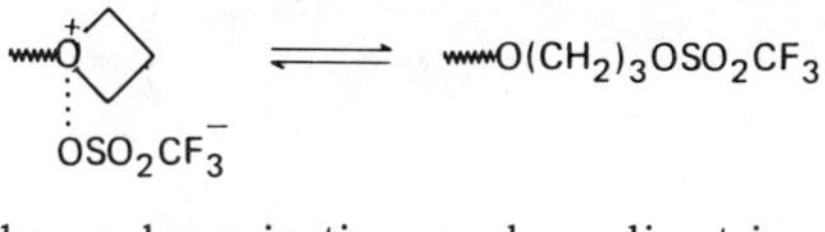

(68)

competition with the polymerization and cyclic trimers and tetramers are formed[209].

Cyclic ethers often undergo copolymerization on the action of CO_2. If triethyl-aluminium is used as initiator, the mechanism is anionic[213].

2. Polymerization of oxolane

The following initiators are employed in the polymerization of oxolane by the cationic mechanism: ethyl 2,4,6-trinitrobenzenesulphonate[214], the propylene oxide–BF_3 system[215], chlorosulphonic acid[216] and the trityl cation[217]. Esters of superacids have recently been frequently used as initiators[218-222].

The copolymerization of oxolane and methyloxirane has been comprehensively studied by Blanchard and coworkers[223-226]. An examination has been made of the effects of the polymerization of changes in the reaction parameters (temperature, catalyst, cocatalyst, solvent, oxolane–methyloxirane ratio, quantity of water in the reaction mixture). Dicarboxylic acid anhydrides may also be used as partners for oxolane in copolymerization[227]. Like other cyclic ethers, oxolane may also form oligomers[228]. The kinetics of polymerization of oxolane at high pressure in the presence of $Et_3O^+ BF_4^-$ as initiator have been subjected to systematic study[229]. The cationic polymerization of oxepane has also been investigated[230]. Modern methods (e.g. ^{13}C-NMR[212,231]) are being ever more frequently utilized for the study of the polymerization of cyclic ethers. By measurement of the ^{13}C-isotope effect, the pathway of formation of active centres can be followed throughout the course of the cationic polymerization of oxolane[232].

H. Formation of Heterocyclic Compounds

1. Ring-transformation of oxetanes to five- and six-membered heterocyclic compounds

With t-butyl isocyanide in the presence of boron trifluoride etherate, oxetane is converted to iminooxolane[233] (equation 69). With carbonyl compounds, substituted oxetanes may be transformed to 1,3-dioxanes[234a,b] (equations 70a and b).

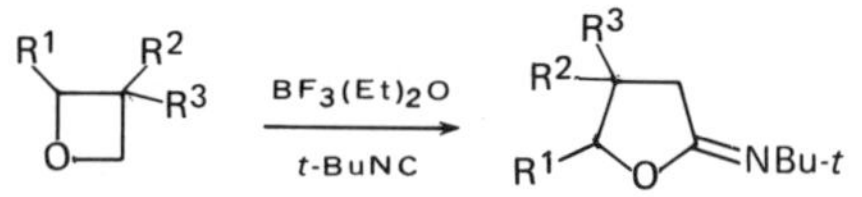

$$(69)$$

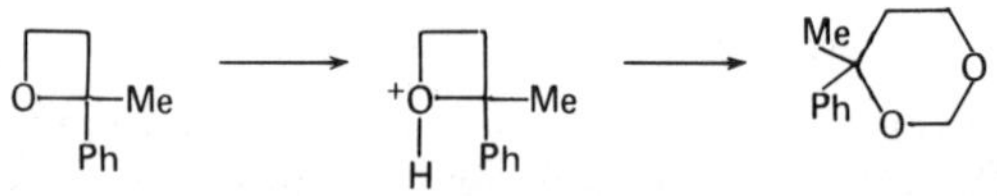

$$(70a)$$

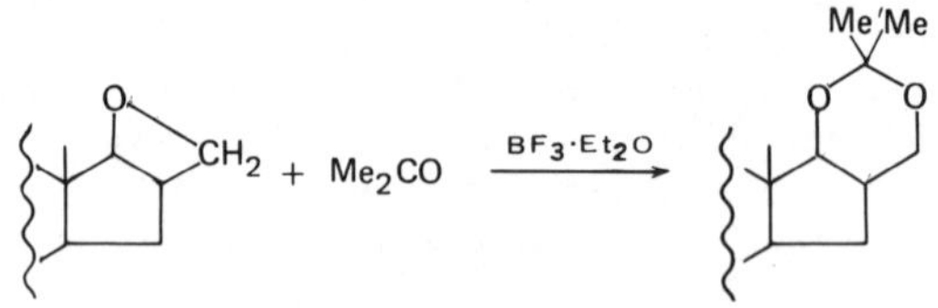

$$(70b)$$

2. Ring-transformation of oxolanes, furans and oxanes

The Yur'ev reaction[234c] is suitable for the preparation of five-membered heterocyclic compounds containing one heteroatom, and their perhydrogenated analogues, from furans and oxolanes. The publications and patents of the past decade have mainly described the application of new catalysts, and the use of new compound types. The importance of the Yur'ev reaction in the chemical industry is demonstrated by the numerous patents[235-240]. The literature provides information on the use of the following catalysts: Cr_2O_3[238], $CuO \cdot Cr_2O_3$[238], MoS_3[238], $CoCl_2/Al_2O_3$[240], HF/Al_2O_3[240], potassium phosphotungstate/Al_2O_3[237] and various zeolites[241-251]. Catalyst systems of complex composition are also used (e.g. metal/support + halo acid + sulphonated styrene—divinylbenzene copolymer[236], etc.). Some examples are given in equations (71)—(73).

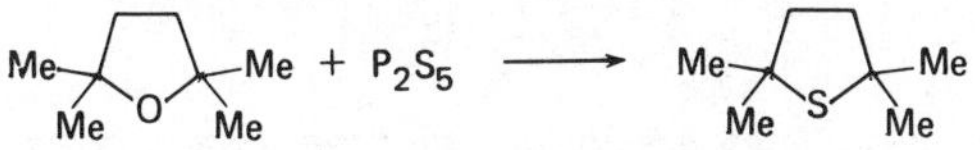

(Ref. 252) (71)

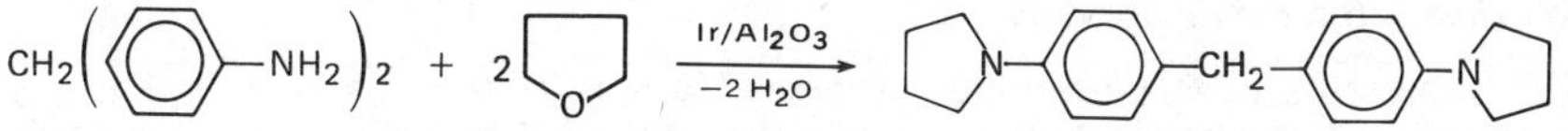

(Ref. 235) (72)

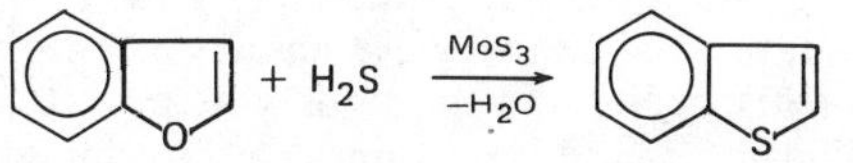

(Ref. 238) (73)

Synthetic zeolites* are effective catalysts of heteroatom exchange. On zeolites of moderate acidity (BaY), the transformation of furan to pyrrole with NH_3 proceeds with a selectivity of $\sim 100\%$[241,242]. With the use of an HL zeolite, oxolane can be converted to pyrrolidine with NH_3 with a selectivity of $\sim 90\%$[244].

It has been established that the active centres are the Brönsted sites formed in the zeolite lattice. The mechanism of the reaction is presented in equation (74)[244].

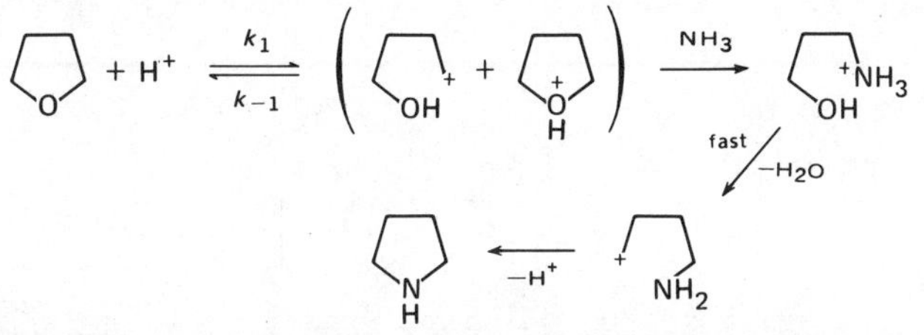

(74)

1-Propylpyrrolidine can be obtained from oxolane with propylamine on an AlY zeolite catalyst[248]. The transformation of γ-butyrolactone to 2-pyrrolidone is catalised with the greatest selectivity by the CuY zeolite[242,245]. The reaction of γ-butyrolactone and propylamine to give 1-propyl-2-pyrrolidone takes place with the highest yield in the presence of CaY, and with the best selectivity in the presence of CuY[247]. The product depends on the structure of the amine. The yield is lower with NH_3 than with primary amines. The reason for this is to be found in the different basicities, but it is very important that the steric effect too be taken into account.

The preparation of thiophen from furan with H_2S proceeds on Li^+ and Na^+ ion-exchange zeolites[246]. The activities of these catalysts increase with the decrease of the Si/Al ratio, and with the increase of the polarizing power of the cation. Alkali metal ion-exchange zeolites similarly catalyse the transformation of oxolane to

---

*X and Y zeolites are sodium aluminosilicates of faujasite type with different SiO_2/Al_2O_3 ratios; zeolite is potassium aluminosilicate.

thiolane[249]. It has been found that the X zeolites are more active than the corresponding Y zeolites. On CsY zeolite, γ-butyrolactone reacts with H_2S to give γ-thiobutyrolactone in a yield of 99%[250]. The catalytic activity is enhanced in the presence of pyridine, but disappears on the action of HCl; hence, basic sites play a very important role in the ring-transformation. The earlier results on the application of the zeolites in the Yur'ev reaction are reviewed by Venuto and Landis[253].

Oxane can be converted with NH_3 to piperidine on synthetic zeolite catalysts[244,251]. The hydrogen-form L zeolites display a higher selectivity than the Y zeolites; dealumination of the L zeolites enhances the catalytic activity and the selectivity[251].

3. Transformation of cyclic ethers containing functional groups to other heterocyclic compounds

This subsection deals with various types of furan-skeleton compounds that can be synthesized from furfurol, and outlines the methods for their transformation to other oxygen- and nitrogen-containing heterocycles. These new methods, using various supported metal catalysts, were developed by Bel'skii and coworkers[7]. Two methods for the preparation of oxanes have already been discussed in Section II. C. 3. Equations (75–(80) depict the methods whereby it is possible to prepare pyrroles and pyrrolidines[7,254–258], pyrrolines[257], pyridines[7], azepans[259], pyrazines[7] and 1,4-dioxanes[260,261]. All starting compounds may be obtained in good yields by classical syntheses from furfurol. The Yur'ev reaction has been utilized to develop a procedure for the formation of pyrrole from furfurol without isolation of

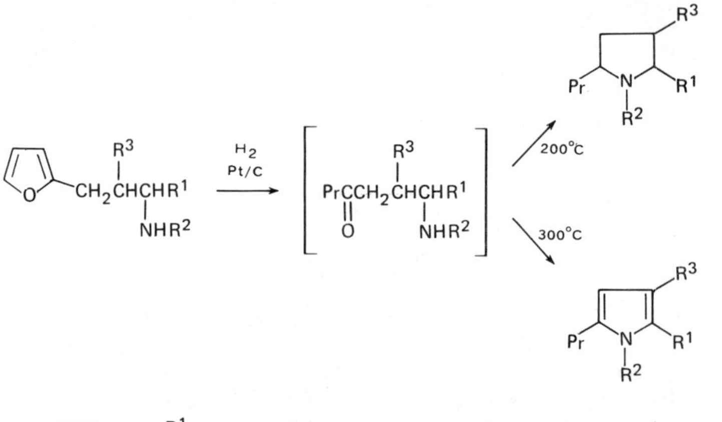

(75)

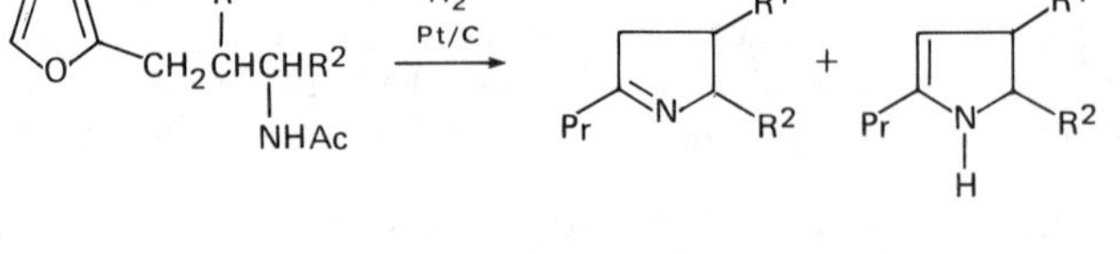

(76)

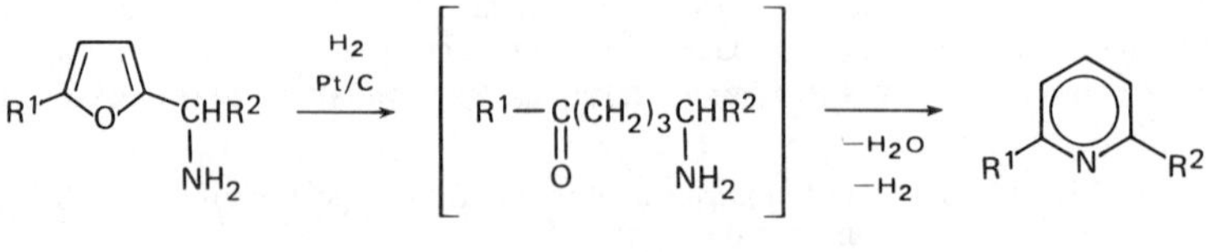

(77)

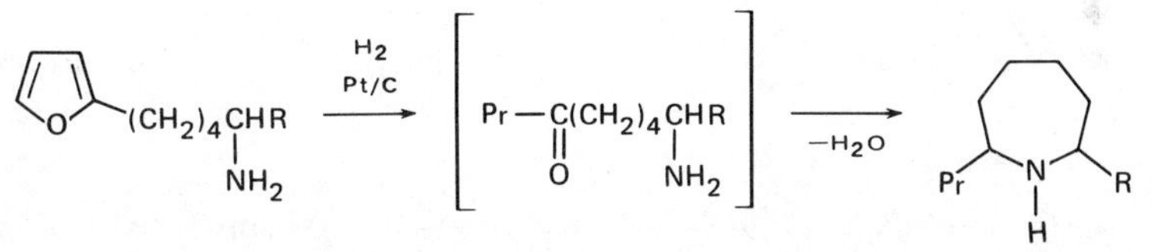

$$\text{(78)}$$

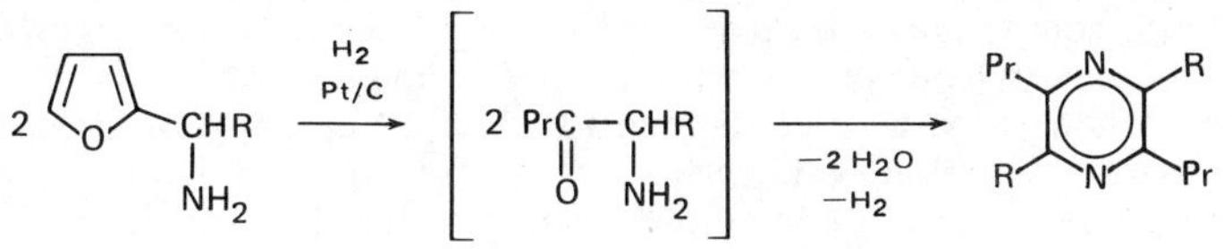

$$\text{(79)}$$

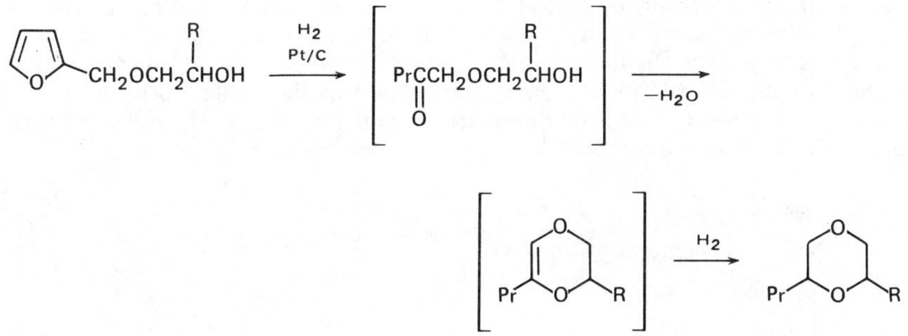

$$\text{(80)}$$

furan[262]. Finally, equation (81) illustrates a ring-expansion reaction in which two oxolane molecules take part[263].

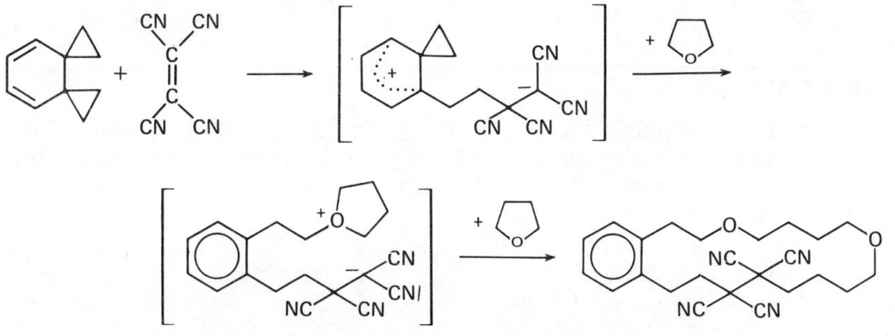

$$\text{(81)}$$

I. Reaction with Organometallic Compounds

Compared to oxiranes[1], the ring-opening of cyclic ethers occurs less readily, since the reactivity decreases with increase in the number of ring atoms. Three reviews on these reactions have appeared in recent years[264-266].

1. Reaction of oxetanes

This reaction is generally used for incorporation of the 3-hydroxypropyl group, with the involvement of either an organolithium[267-269] or a Grignard compound[171,270-273] (e.g. equations 82 and 83). In certain cases the reactions of

$$\text{(Ref. 269) (82)}$$

$$\text{(Ref. 273) (83)}$$

organolithium compounds are carried out in the presence of cuprous salts[274]. The reaction of 2-methyleneoxetane with phenyllithium results in methyl phenetyl ketone[275]. Whereas oxiranes containing a carbonyl function react regioselectively (via their oxirane function) with certain organometallic compounds[1], oxetan-3-one reacts with a Grignard compound either via its oxo function, or via both functional groups[276].

With trimethylchlorosilane, 2-alkyloxetanes yield the corresponding 1,3-chloro-hydrinsilyl ether isomers[277]. On the action of triethylaluminium, pentanol is formed to only a very slight extent[278]. 3-Ethyl-3-hydroxymethyloxetane reacts according to equation (84) with phenylmercurihydroxide, while 3,3-bis(hydroxy-methyl)oxetane gives 3,3-bis(phenylmercurioxymethyl)oxetane[279]. An interesting reaction is shown in equation (85)[280].

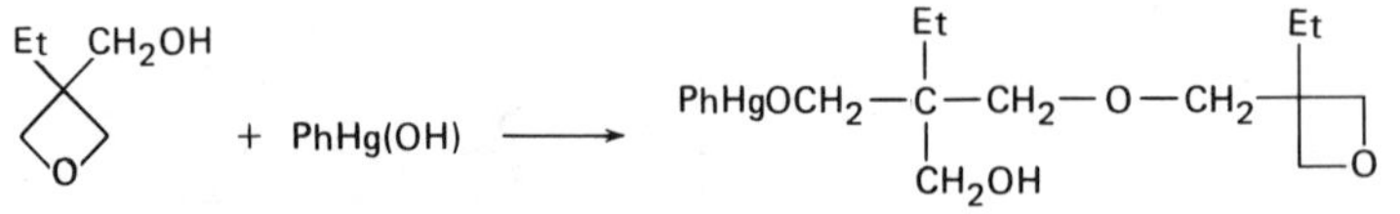

$$(84)$$

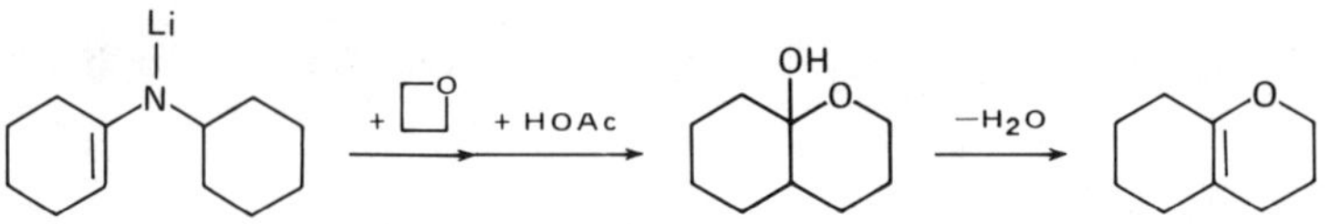

$$(85)$$

2. Reaction of oxolanes

On the action of alkyllithiums (e.g. *n*-BuLi), the oxolanes decompose to alkene and aldehyde enolate[281-284] after the splitting-off of an α-hydrogen. Alkyllithium and cuprous salt, or lithium dialkylcuprate, causes the ring of the 2-alkyloxolanes to open[274] (equation 86).

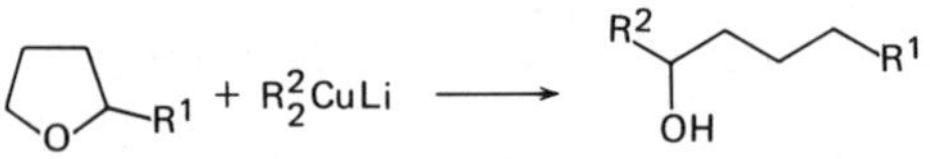

$$(86)$$

In the presence of tungsten hexachloride, oxolane undergoes α-phenylation with phenyllithium[285]. On the action of tri- and di-phenylmethyllithium, the corresponding butanol derivatives are obtained[286,287]. Lithium trialkylsilane converts oxolane to 4-trialkylsilanebutanol[288]. Trimethyliodosilane[289,290] and dimethyl-dichlorosilane[291] yield the corresponding 4-tri- and di-alkylsilyloxybutyl halides. In the presence of metals, trimethyliodosilane reacts with oxolane to give 1,8-bis-trimethylsilyloxyoctane[292].

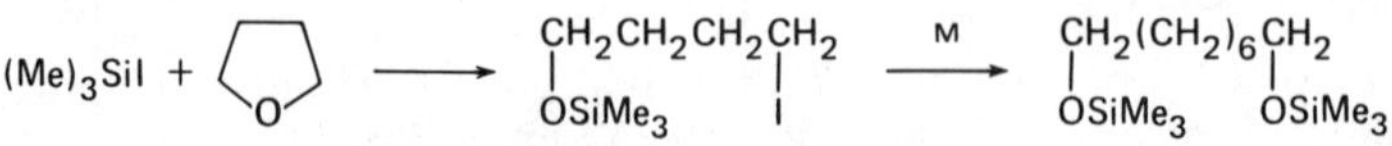

$$(87)$$

With a Grignard compound, 2-dialkylaminooxolane forms a 1,4-amino alcohol[293].

3-Oxolanone hydrazone can be opened with alkyllithium to give allene alcohol[294].
2-Alkoxyoxolane, which also contains an oxirane function, reacts regioselectively
with lithium dialkylcuprate via the oxirane function [295] (equation 88).

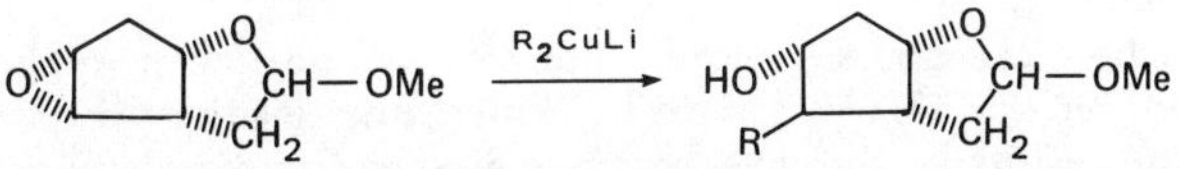

$$\text{(88)}$$

Oxolane forms various complexes and adducts with transition metal halides[296],
rare-earth metal salts[297] and metal complexes[298,299].
2-Hydroxymethyloxolane interacts via the hydroxy function with diphenylzinc
and phenylmercurihydroxide[279].

3. Reaction of oxanes

The six-membered oxacycloalkanes display a considerably lower reactivity towards organometallic compounds. On the action of n-BuLi, only a minimal amount
of 1-nonanol is obtained from oxane[274]. The 3-hydrazone derivative gives an allene
alcohol on reaction with n-BuLi[294] (equation 89).

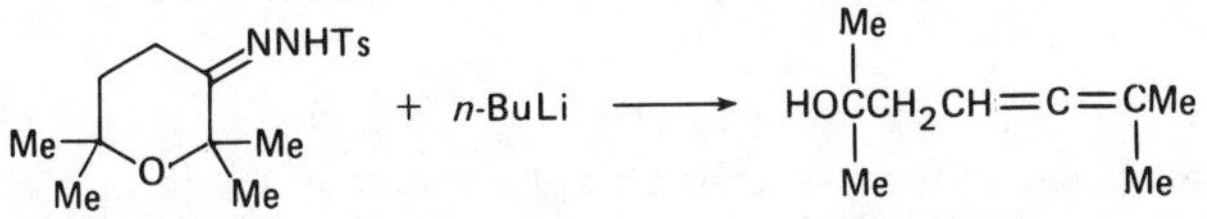

$$\text{HOCCH}_2\text{CH}=\text{C}=\text{CMe} \qquad \text{(89)}$$

With trimethyliodosilane, oxane may be opened to 1,5-iodohydrintrimethylsilyl
ether, while in the presence of metals (Li, Na, K, Mg) 1,10-decanediolbissilyl ether
may be obtained[292,300]. 2-Aminoalkyloxanes react with Grignard compounds to
give 1,5-amino alcohols[301].
New experimental data have been reported on the exchange of the 2-chloro
atom in 2-chlorooxanes[302] and 2,3-dichlorooxanes[303-308] for alkyl or aryl groups.
Equation (90) shows the double reactivity of 2-vinyloxyoxane[309]. With a
Grignard compound, 2-ethynyloxane gives an allene alcohol[308] (equation 91).

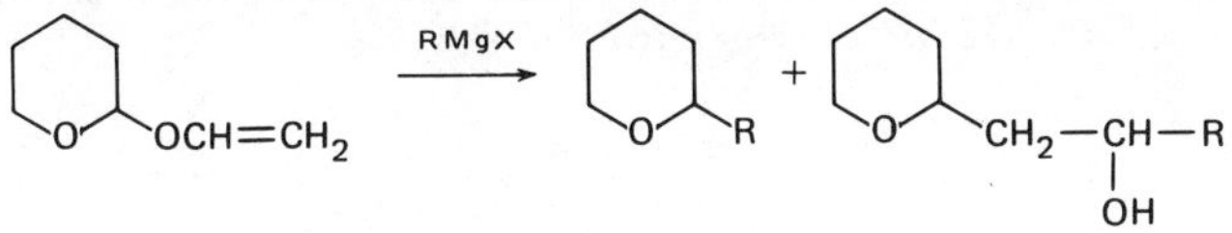

$$\text{(90)}$$

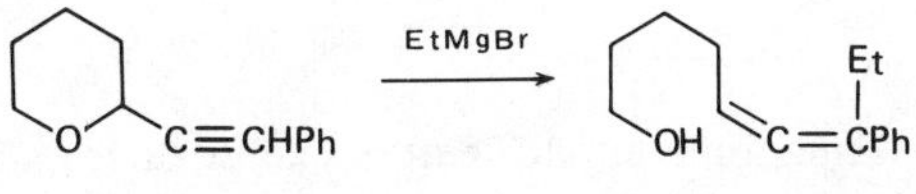

$$\text{(91)}$$

J. Free-radical Chemistry

Reactions of cyclic ethers that take place via a free-radical mechanism may be
induced thermally, with a free-radical initiator, photochemically in the presence or
the absence of an appropriate sensitizer, and by radiolysis.
In the pyrolysis of oxetanes, fission of the four-membered ring into two parts
proceeds with high selectivity. This reaction can be studied readily and permits
the understanding of the mechanism of the radical processes. These investigations
have extended to oxetane[310,311] and also to 2-alkyl- and 2-aryl-[312-314], 3-alkyl-
and 3-aryl-[315,316], 2,2-di-[316], 3,3-di-[317-319], 2,3-di-[314,320,321] and 2,4-di-
substituted[322] oxetanes, and to polysubstituted and functional derivatives of

oxetane[90,121,316,323-326]. The decomposition of oxetanes has also been studied in the presence of rhodium complexes[327,328]. The publications referred to above include investigations of the kinetics, the regioselectivity and the stereoselectivity of the transformation.

The stereochemical course of the thermolysis has been reported in many papers[90,314,316,320,321,323,328,329]. While not leading to totally uniform conclusions, the results of the investigations may be summarized briefly as follows. The gas-phase thermolysis of oxetanes to olefins and carbonyl compounds is a homogeneous, unimolecular process occurring via a biradical intermediate. The transformation is not completely stereoselective; cis—trans isomerization too may be observed during thermolysis.

The tendency of cyclic ethers to undergo radical reactions is due to the comparative weakness of the C—H bonds in the α-position. ESR studies have revealed the formation of the radicals 8 or 9 and 10 in the radiolysis of oxolane and 2-methyloxolane, respectively[330]. α-Radicals are also formed in the case of six-membered cyclic ethers[331,332]. The chemical evidence indicates that the tendencies

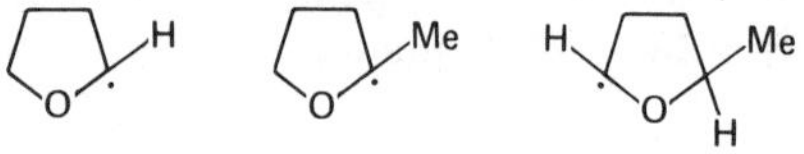

(8)   (9)   (10)

of oxolane and oxane to form radicals are approximately 10 times higher than those of oxetane and oxiranes, which corresponds with the fact that the C—H bond is stronger than the C—O bond in the latter.

Radical alkylations of cyclic ethers with olefins[333-336] are initiated by the radicals formed on the thermal decomposition of di-t-butyl peroxide. The reaction is suitable for the preparation of 2-alkyloxacycloalkanes from oxolane and oxane by utilization of the appropriate terminal olefin. The yield increases together with the molecular weight of the olefin, and in favourable cases attains 70—80%. The alkylation is a chain-reaction; the chain-propagating steps in the case of oxolane[334] are shown in equations (92) and (93). Chain-termination may be either disproportionation or combination of the radicals[337,338].

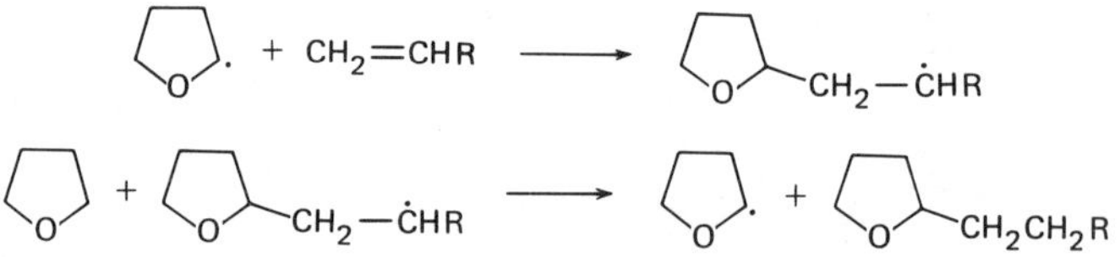

$$CH_2=CHR \qquad \longrightarrow \qquad CH_2-\dot{C}HR \qquad (92)$$

$$CH_2-\dot{C}HR \qquad \longrightarrow \qquad CH_2CH_2R \qquad (93)$$

Since both alkenes and ethers are difficult to excite, their photochemical reaction is achieved only in the presence of a sensitizer (e.g. acetone). Triplet-state acetone splits off an α-H atom from the ether, and the reaction proceeds by the same route as the radical-induced one[338]. Cyclic acetals too display an analogous reaction[337].

Similar reaction are also observed in the case of cumulated dienes[340,341]. Depending on the conditions, the reaction of oxacycloalkyl radicals with acetylenes produces either alkylation[342] or ring-opening[343]. By direct photochemical reaction with oxolane, a suitably excitable unsaturated compound such as 11, for

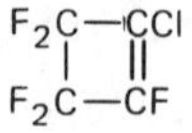

$$\begin{array}{c} F_2C-CCl \\ |\quad\;\; || \\ F_2C-CF \end{array}$$

(11)

example, gives the corresponding 2-oxolane derivative[344]. Oxolane similarly undergoes direct photochemical addition to maleic anhydride[345,346] and diethyl maleate[338,346]. The reaction may also be induced by radicals[346]. Oxolane may participate in a photoaddition reaction with 1,3-dimethyluracil (equation 94)[347],

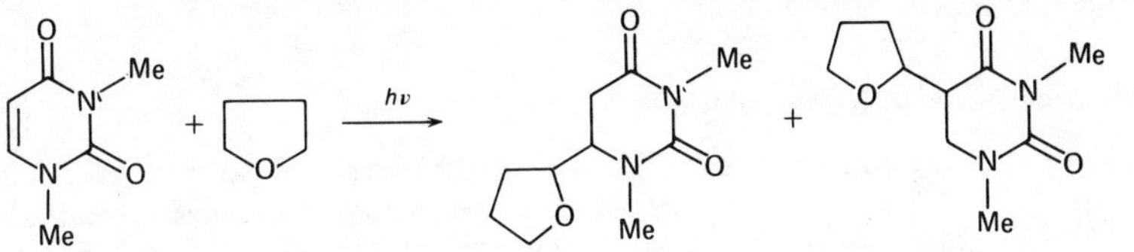

(94)

adenine, guanine and caffeine[337]. Excited purine and pyrimidine bases split off hydrogen from $C_{(2)}$ of oxolane, and the radical formed reacts as indicated.

In some reactions of cyclic ethers, ring-contraction occurs[348-350]. On the action of light, tetramethyloxetanone is converted to acetone and dimethylketene in an apolar solvent, and to tetramethyloxirane in a polar solvent[348]. 13 is formed selectively from 12 in a photochemical reaction[350] (equation 95). 2,3-Dihydropyran undergoes addition to benzene with very high stereoselectivity[351] (equation 96).

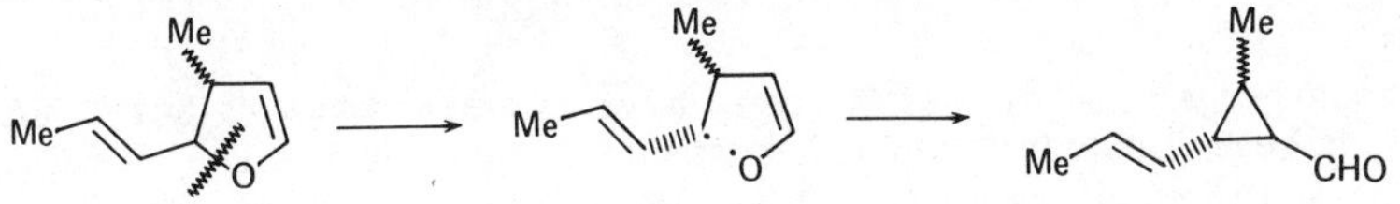

(95)

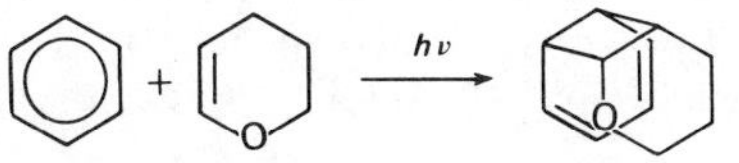

(96)

Nitrenes[352,353] and carbenes[354,355] are capable of insertion into the C—H bond. Studies have been made of the reactions of various cyclic ethers and carbethoxynitrene[353]. The mechanism of equation (97) has been proposed for the

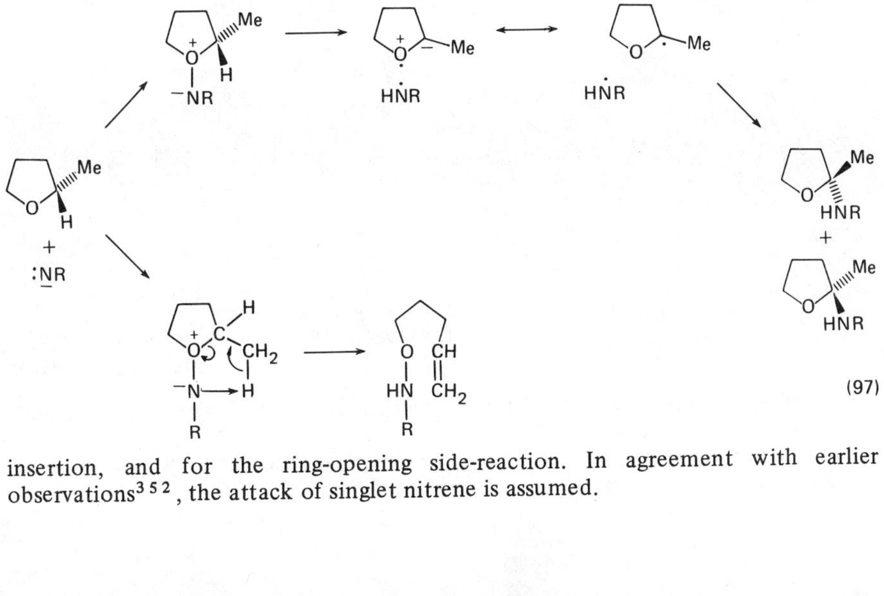

(97)

insertion, and for the ring-opening side-reaction. In agreement with earlier observations[352], the attack of singlet nitrene is assumed.

Dichlorocarbene is likewise inserted into the α-C—H bond. α-Dichloromethyl-oxacycloalkane can be prepared in good yield (80%) via this reaction[354].

Numerous publications have appeared on the fragmentation occurring during the mass-spectroscopic determination of oxetanes[356-358] and cyclic ethers with larger rings[359].

K. Ring-opening with Nucleophilic Reagents

Most of the experimental data in the literature relate to the acid-catalysed hydrolysis[24,26,141,360,361] of cyclic ethers (mainly oxetanes), their alcohol-ysis[25,362-366] and their transformations with hydrogen halides[362-370], carboxylic acids[25,371,372] and their derivatives[25,373-376]. These reactions are depicted in equation (98).

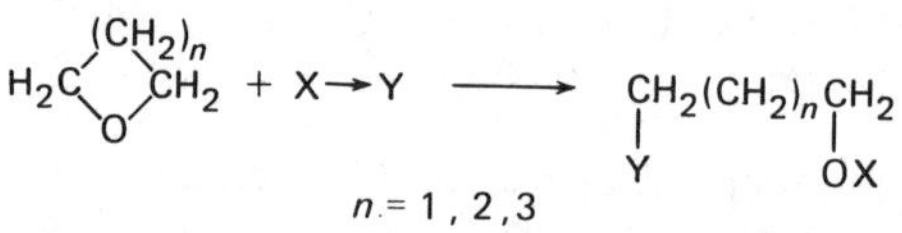

$$\underset{\substack{\text{(CH}_2)_n \\ \text{H}_2\text{C} \quad \text{CH}_2 \\ \text{O}}}{} + X{\rightarrow}Y \longrightarrow \underset{\substack{\text{CH}_2(\text{CH}_2)_n\text{CH}_2 \\ | \quad\quad | \\ Y \quad\quad \text{OX}}}{} \tag{98}$$

$$n = 1, 2, 3$$

X,Y = H_2O, hydrogen halides, ROH, RCOOH, RCOZ, etc.

Some investigations have been directed towards preparative uses, but the majority deal with regioselectivity, stereochemistry and mechanism. The overwhelming majority of the reactions take place via an S_N2 mechanism. However, some observations (mainly on oxetanes) can only be interpreted by an S_N1 mechanism. The mechanism of the reaction is greatly influenced by the number and type of the ring-atoms, the nature of the reagent and the experimental conditions. Some examples in support of this are presented in equations (99)–(102). In the acid-catalysed ring-opening of cyclic ethers, the first step is the formation of an oxonium

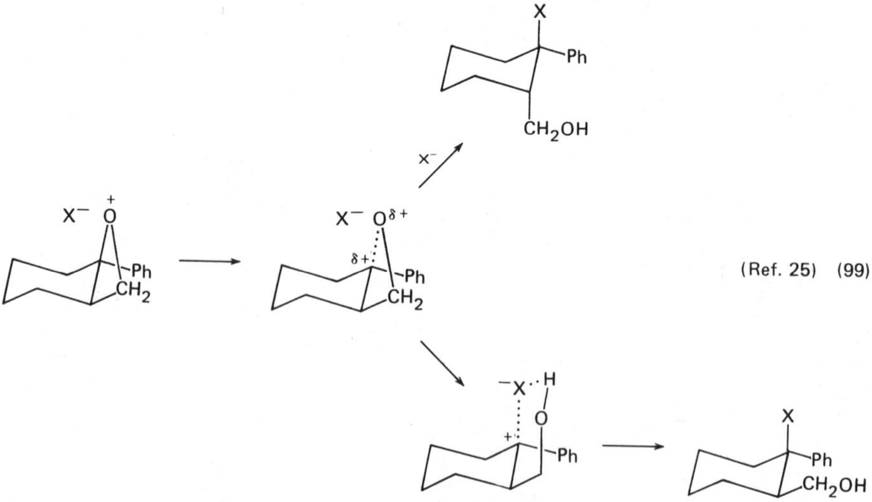

(Ref. 25) (99)

salt, which is a reversible process. Numerous stable oxonium salts have been isolated, e.g. in the case of cis-2,5-dimethyloxolane[377].

With steroid oxetanes, acid-catalysed cis ring-opening has been observed to occur with surprisingly high stereoselectivity[375] (equation 100).

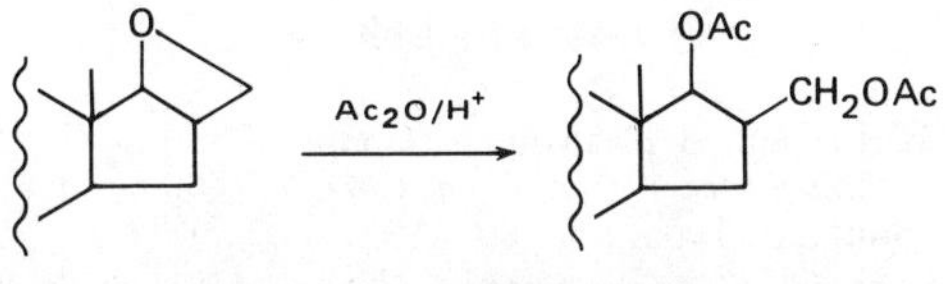

$$\text{(100)}$$

The ring-openings of *cis*- and *trans*-2,5-dimethyloxolanes take place by an S_N2 mechanism[372] (equations 101 and 102).

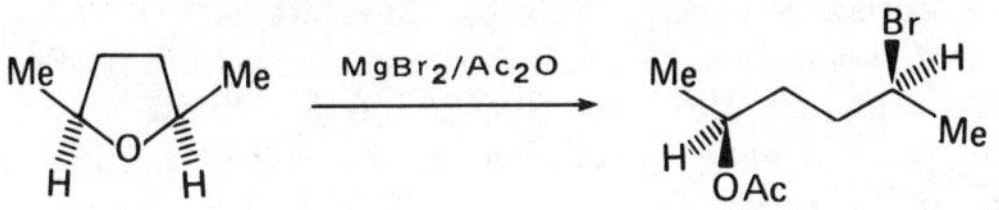

$$\text{(101)}$$

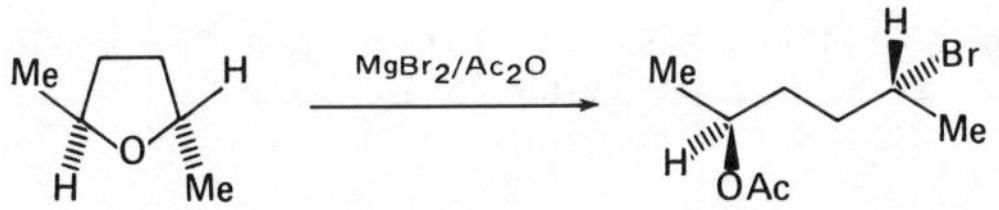

$$\text{(102)}$$

Other ring-opening reactions, mainly of oxetanes, occur, e.g. with phosphorus halides[378-380] or carbonic acid derivatives[373,381]. Some other unusual ring-openings of oxolanes take place with alkyl halides in the presence of mercuric salts[382,383], tetrafluorobenzene[384], alkyl chlorosulphonate[385] and phosgene[373] (equations 103–106):

(Ref. 386) (103)

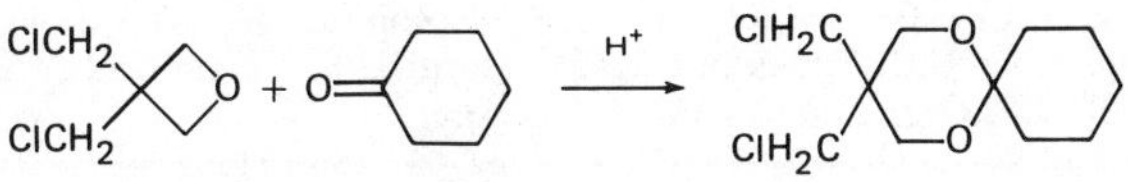

(Ref. 387) (104)

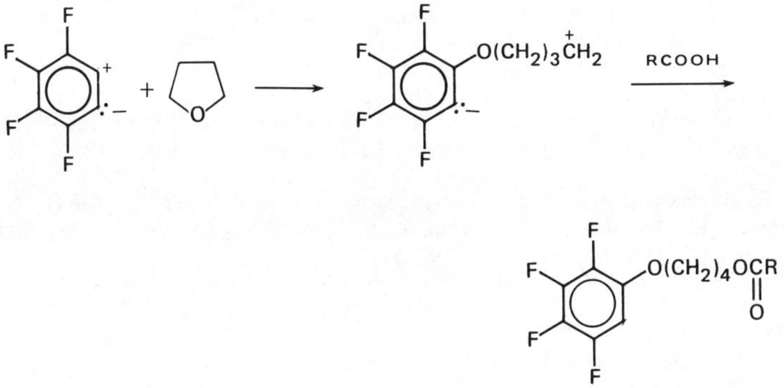

(Ref. 384) (105)

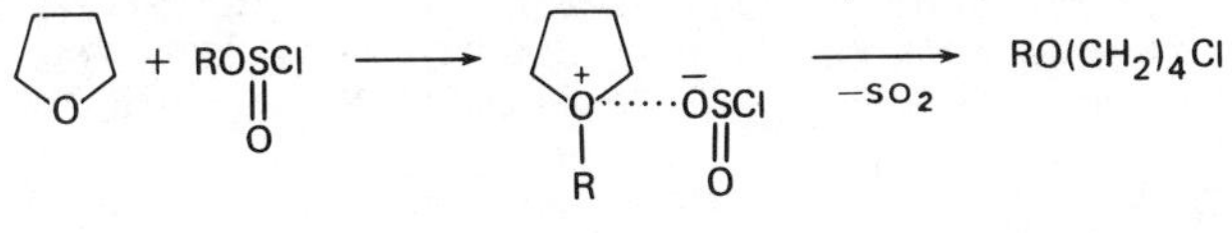

(Ref. 385) (106)

IV. REFERENCES

1. M. Bartók and K. L. Láng in This volume, Chap. 14.
2. G. Dittus in *Methoden der Organischen Chemie (Houben-Weyl)* Vol. VI/3, Georg Thieme Verlas, Stuttgart, 1965. pp. 489–517.
3. G. Dittus in *Methoden der Organischen Chemie (Houben-Weyl)* Vol. VI/4, Georg Thieme Verlag, Stuttgart, 1966, pp. 12–99.
4. G. Dittus and B. Zech in *Methoden der Organischen Chemie (Houben-Weyl)*, Vol. VI/4, Georg Thieme Verlag, Stuttgart, 1966, pp. 286–305.
5. H. Kröper in *Methoden der Organischen Chemie (Houben-Weyl)*, Vol. VI/3, Georg Thieme Verlag, Stuttgart, 1965, pp. 517–563, 648–673.
6. R. J. Gritter in *The Chemistry of the Ether Linkage* (Ed. S. Patai), John Wiley and Sons, London, 1967, pp. 411–443.
7. I. F. Bel'skii and V. M. Shostakovskii, *Kataliz v Khimii Furana,* Nauka, Moscow, 1972, p. 230.
8. W. L. F. Armarego in *Stereochemistry of Heterocyclic Compounds*, Part 2, John Wiley and Sons, New York, 1977, pp. 36–68, 78–100.
9. H. C. Van der Plas, *Ring Transformations of Heterocycles*, Vol. 1, Academic Press, London, 1973.
10. M. Lj. Mihailović in *Lectures in Heterocyclic Chemistry*, Vol. 3, 1976, pp. S-111–S-121.
11. M. M. Green, J. M. Moldowan and J. G. McGrew, *J. Chem. Soc., Chem. Commun.,* 451 (1973).
12. M. M. Green, J. M. Moldowan and J. G. McGrew, *J. Org. Chem.*, **39**, 2166 (1974).
13. M. Bartók and Á. Molnár in this volume, Chap. 16.
14. C. Schaal, *Compt. Rend. (C)*, **265**, 1264 (1967).
15. M. Bartók, *Acta Chim. Acad. Sci. Hung.*, **55**, 365 (1968).
16. M. Bartók, B. Kozma and N. I. Shuikin, *Izv. Akad. Nauk SSSR, Ser. Khim.*, 1241 (1966).
17. N. I. Shuikin, M. Bartók and B. Kozma, *Izv. Akad. Nauk SSSR, Ser. Khim.*, 153 (1967).
18. A. Balsamo, G. Ceccarelli, P. Crotti and F. Macchia, *J. Org. Chem.*, **40**, 473 (1975).
19. A. V. Bogatskii, Yu. Yu. Samitov, M. Bartók, S. A. Petrash, A. I. Gren' and G. B. Bartók, *Zh. Org. Khim.*, **12**, 215 (1976).
20. W. Fischer and C. A. Grob. *Helv. Chim. Acta*, **61**. 2336 (1978).
21. T. A. Favorskaya and Yu. M. Portnyagin, *Zh. Obshch. Khim.*, **34**, 699 (1964).
22. T. A. Favorskaya and Yu. M. Portnyagin, *Zh. Obshch. Khim.*, **35**, 440 (1965).
23. Yu. M. Portnyagin and N. E. Pak, *Zh. Org. Khim.*, **9**, 456 (1973).
24. Yu. M. Portnyagin and V. V. Sova, *Zh. Org. Khim.*, **4**, 1576 (1968).
25. A. Balsamo, P. Crotti, M. Ferretti and F. Macchia, *J. Org. Chem.*, **40**, 2870 (1975).
26. R. Heckendorn, *Helv. Chim. Acta*, **51**, 1068 (1968).
27. Gy. Schneider and I. Weisz-Vincze, *J. Chem. Soc., Chem. Commun.*, 1030 (1968).
28. A. V. Bogatskii, S. A. Petrash and M. Bartók, *Dokl. Akad. Nauk Ukr. SSR, Ser. B*, 793 (1976).
29. M. Bartók, B. Kozma and A. G. Schöbel, *Acta Phys. Chem. Szeged*, **11**, 35 (1965).
30. A. V. Bogatskii, G. A. Filip, S. A. Petrash, L. S. Semerdzhi, Yu. Yu. Samitov and G. V. P'yankova, *Zh. Org. Khim.*, **7**, 577 (1971).
31. F. Notheisz, M. Bartók and V. Remport, *Acta Phys. Chem. Szeged*, **18**, 89 (1972).
32. F. Notheisz, M. Bartók and V. Remport, *Acta Phys. Chem. Szeged*, **18**, 197 (1972).
33. T. A. Favorskaya and Yu. M. Portnyagin, *Zh. Obshch. Khim.*, **35**, 435 (1965).
34. N. I. Shuikin, M. Bartók and B. Kozma, *Izv. Akad. Nauk SSSR, Ser. Khim.*, 878 (1966).
35. M. Bartók, G. B. Bartók and K. Kovács, *Acta Chim. Acad. Sci. Hung.*, **66**, 115 (1970).
36. M. Bartók and G. B. Bartók, *Acta Chim. Acad. Sci. Hung.*, **72**, 423 (1972).
37. M. Bartók and G. B. Bartók, *Acta Chim. Acad. Sci. Hung.*, **72**, 433 (1972).
38. W. H. Richardson, C. M. Golino, R. H. Wachs and M. B. Yelvington, *J. Org. Chem.*, **36**, 943 (1971).
39. C. G. Swain, D. Á. Kuhn and R. L. Schowen, *J. Amer. Chem. Soc.*, **87**, 1553 (1965).
40. T. H. Cromartie and C. G. Swain, *J. Amer. Chem. Soc.*, **97**, 232 (1975)

41. M. Bartók, K. L. Láng and G. B. Bartók, *Acta Chim. Acad. Sci. Hung.*, **70**, 133 (1971).
42. M. Bartók, G. B. Bartók and K. Kovács, *Acta Chim. Acad. Sci. Hung.*, **72**, 297 (1972).
43. M. Bartók, Á. Molnár and K. Kovács, *Acta Chim. Acad. Sci. Hung.*, **58**, 337 (1968).
44. Á. Molnár, M. Bartók and K. Kovács, *Acta Chim. Acad. Sci. Hung.*, **59**, 133 (1969).
45. P. S. Portoghese and D. A. Williams, *J. Heterocyclic Chem.*, **6**, 307 (1969).
46. J. Biggs, *Tetrahedron Letters*, 4285 (1975).
47. H. Kaminski, 'Reaktionen von β-ständig elektronegativ substituierten Carbonyl-verbindungen mit Magnesium- und Lithium-organylen', *Ph.D. Dissertation*, Technical University of Berlin, 1969.
48. A. R. Jones, *J. Chem. Soc., Chem. Commun.*, 1042 (1971).
49. A. Kirrmann and L. Wartiski, *Bull. Soc. Chim. Fr.*, 3825 (1966).
50. K. C. Nicolaou and Z. Lysenko, *Tetrahedron Letters*, 1257 (1977).
51. B. Capon and J. W. Thomson, *J. Chem. Soc., Perkin II*, 917 (1977).
52. J. Huet, *Compt. Rend.*, **258**, 4570 (1964).
53. N. I. Shuikin, R. A. Karakhanov, I. I. Ibrakhimov and N. L. Komissarova, *Izv. Akad. Nauk SSSR, Ser. Khim.*, 122 (1966).
54. I. J. Borowitz and G. J. Williams, *J. Org. Chem.*, **33**, 2013 (1968).
55. C. Botteghi, G. Consiglio, G. Ceccarelli and A. Stefani, *J. Org. Chem.*, **37**, 1835 (1972); **38**, 2361 (1973).
56. J. P. Bats, J. Moulines and J. C. Pommier, *Tetrahedron Letters*, 2249 (1976).
57. A. Murai, M. Ono and T. Masamune, *J. Chem. Soc., Chem. Commun.*, 864 (1976).
58. A. Murai, M. Ono and T. Masamune, *Bull. Chem. Soc. Japan*, **50**, 1226 (1977).
59. M. F. Grundon and H. M. Okely, *J. Chem. Soc., Perkin I*, 150 (1975).
60. M. C. Sacquet, B. Graffe and P. Maitte, *Tetrahedron Letters*, 4453 (1972).
61. Sh. Suzuki, *U. S. Patent No.* 3,956,318 (1976); *Chem Abstr.*, **85**, 46364e (1976).
62. J. C. Paladini and J. Chuche, *Tetrahedron Letters*, 4383 (1971); *Bull. Soc. Chim. Fr.*, 197 (1974).
63. J. C. Pommelet, N. Manisse and J. Chuche, *Tetrahedron*, **28**, 3929 (1972).
64. V. Vukov and R. J. Crawford, *Can. J. Chem.*, **53**, 1367 (1975).
65. W. Eberbach and B. Burchardt, *Chem. Ber.*, **111**, 3665 (1978).
66. G. Berti, S. Catalano, A. Marsili, I. Morelli and V. Scartoni, *Tetrahedron Letters*, 401 (1976).
67. P. H. M. Schreurs, J. Meijer, P. Vermeer and L. B. Brandsma, *Tetrahedron Letters*, 2387 (1976).
68. J. A. Donnelly, J. G. Hoey, S. O'Brien and J. O'Grady, *J. Chem. Soc., Perkin I*, 2030 (1973).
69. J. A. Donnelly, S. O'Brien and J. O'Grady, *J. Chem. Soc., Perkin I*, 1674 (1974).
70. J. R. Dias and G. R. Pettit, *J. Org. Chem.*, **36**, 3485 (1971).
71. A. M. Maione and M. G. Quaglia, *Chem. Ind. (Lond.)*, 230 (1977).
72. J. Kanetaka, *Nippon Kagaku Kaishi*, 1195 (1974); *Chem. Abstr.*, **83**, 137406u (1975).
73. M. A. Gianturco, P. Friedel and V. Falanagan, *Tetrahedron Letters*, 1847 (1965).
74. I. Szabó, K. Kovács and M. Bartók, *Acta Chim. Acad. Sci. Hung.*, **51**, 411 (1967).
75. Z. Dudzik and M. Gasiorek, *Przemysl Chem.*, **54**, 637 (1975); *Chem. Abstr.*, **84**, 89903b (1976).
76. R. A. Karakhanov, V. I. Garanin, V. V. Kharlamov, M. A. Kapustin, B. B. Blinov and Kh. M. Minachev, *Izv. Akad. Nauk SSSR, Ser. Khim.*, 445 (1975).
77. D. C. Iffland and J. E. Davis, *J. Org. Chem.*, **42**, 4150 (1977).
78. I. F. Bel'skii and I. E. Grushko, *Khim. Geterotsikl. Soed.*, 6 (1969).
79. G. Descotes, B. Giround-Abel and J.-C. Martin, *Bull. Soc. Chim. Fr.*, 2466 (1967).
80. G. Descotes and G. Tedeschi, *Bull. Soc. Chim. Fr.*, 1378 (1969).
81. P. Martinet and G. Mousset, *Bull. Soc. Chim. Fr.*, 4093 (1971).
82. D. Chambenois and G. Mousset, *Compt. Rend. (C)*, **274**, 715 (1972).
83. D. Chambenois and G. Mousset, *Bull. Soc. Chim. Fr.*, 2969 (1974).
84. C. Malardeau and G. Mousset, *Bull. Soc. Chim. Fr.*, 988 (1977).
85. A. A. Gevorkyan and G. G. Tokmadzhyan, *Arm. Khim. Zh.*, **30**, 165 (1977).
86. O. L. Chapman and G. Lenz in *Organic Photochemistry*, Vol. 1, Marcel Dekker, New York, 1967, p. 238.

87. L. L. Müller and J. Hamer, *1,2-Cycloaddition Reactions*, Interscience, New York, 1967.
88. D. R. Arnold, *Advan. Photochem.*, **6**, 301 (1967).
89. H. Meier in *Methoden der Organischen Chemie (Houben-Weyl)* Vol. IV/5b, *Photochemie II*, Georg Thieme Verlag, Stuttgart, 1975, pp. 838–876.
90. G. Jones and S. C. Staires, *Tetrahedron Letters*, 2099 (1974).
91. C. W. Funke and H. Cerfontain, *J. Chem. Soc., Perkin II*, 1902 (1976).
92. H. A. J. Carless, *J. Chem. Soc., Chem. Commun.*, 316 (1973).
93. H. A. J. Carless, *J. Chem. Soc., Perkin II*, 834 (1974).
94. H. A. J. Carless, *Tetrahedron Letters*, 3173 (1973).
95. R. A. Caldwell, G. W. Sovocool and R. P. Gajewski, *J. Amer. Chem. Soc.*, **95** 2549 (1973).
96. K. Shima, T. Kawamura and K. Tanabe, *Bull. Chem. Soc. Japan*, **47**, 2347 (1974).
97. J. A. Barltrop and H. A. J. Carless, *J. Amer. Chem. Soc.*, **94**, 1951 (1972).
98. H. D. Scharf and J. Mattay, *Tetrahedron Letters*, 3509 (1976).
99. R. D. Chambers, J. Hutchinson and P. D. Philpot, *J. Fluorine Chem.*, **9**, 15 (1977).
100. N. R. Lazear and J. H. Schauble, *J. Org. Chem.*, **39**, 2069 (1974).
101. T. S. Cantrell, *J. Org. Chem.*, **39**, 2242 (1974).
102. R. J. C. Koster, D. G. Streefkerk, J. Ondshoorn van Veen and H. J. T. Bos, *Rec. Trav. Chim.*, **93**, 157 (1974).
103. R. J. C. Koster and H. J. T. Bos, *Rec. Trav. Chim.*, **94**, 79 (1975).
104. Z. I. Yoshida, M. Kimura and S. Yoneda, *Tetrahedron Letters*, 2519 (1974).
105. S. Farid and S. E. Shealer, *J. Chem. Soc., Chem. Commun.*, 296 (1973).
106. L. E. Friedrich and J. D. Bower, *J. Amer. Chem. Soc.*, **95**, 6869 (1973).
107. H. A. J. Carless and A. K. Maitra, *Tetrahedron Letters*, 1411 (1977).
108. T. Kubota, K. Shima, S. Toki and H. Sakurai, *J. Chem. Soc., Chem. Commun.*, 1462 (1969).
109. A. A. Gorman, R. L. Leyland, M. A. J. Rodgers and P. G. Smith, *Tetrahedron Letters*, 5085 (1973).
110. T. Kubota, K. Shima and H. Sakurai, *Chem. Letters*, 393 (1972).
111. J. Kossanyi, B. Guiard and B. Furth, *Bull, Soc. Chim. Fr.*, 305 (1974).
112. B. Guiard, B. Furth and J. Kossanyi, *Bull. Soc. Chim. Fr.*, 1553 (1976).
113. B. Furth, G. Daccord and J. Kossanyi, *Tetrahedron Letters*, 4259 (1975).
114. L. B. Friedrich and G. B. Schuster, *J. Amer. Chem. Soc.*, **94**, 1193 (1972).
115. R. C. Cookson and N. R. Rogers, *J. Chem. Soc., Perkin I*, 1037 (1974).
116. J. C. Dalton and F. H. Chan, *Tetrahedron Letters*, 3351 (1974).
117. R. R. Sauers, A. D. Rousseau and B. Byrne, *J. Amer. Chem. Soc.*, **97**, 4947 (1975).
118. R. R. Sauers, R. Bierenbaum, R. J. Johnson, J. A. Thich, J. Potenza and H. J. Schugar, *J. Org. Chem.*, **41**, 2943 (1976).
119. D. Bickham and M. Winnik, *Tetrahedron Letters*, 3857 (1974).
120. R. R. Sauers and T. M. Henderson, *J. Org. Chem.*, **39**, 1850 (1974).
121. G. Jones II, M. A. Acquardo and M. A. Carmody, *J. Chem. Soc., Chem. Commun.*, 206 (1975).
122. D. R. Morton and R. A. Morge, *J. Org. Chem.*, **43**, 2093 (1978).
123. N. C. Yang, M. Kimura and W. Eisenhardt, *J. Amer. Chem. Soc.*, **95**, 5058 (1973).
124. L. M. Stephenson, Th. A. Gibson, *J. Amer. Chem. Soc.*, **94**, 4599 (1972).
125. J. Saltiel, D. E. Townsend and A. Sykes, *J. Amer. Chem. Soc.*, **95**, 5968 (1973).
126. N. Shimizu, M. Ishikawa, K. Ishikura and S. Nishida, *J. Amer. Chem. Soc.*, **96**, 6456 (1974).
127. M. E. Garst and T. A. Spencer, *J. Amer. Chem. Soc.*, **95**, 250 (1973).
128. M. Matsumodo and K. Kondo, *Tetrahedron Letters*, 391 (1976).
129. B. Harirchian and P. D. Magnus, *Synth. Commun.*, **7**, 119 (1977).
130. P. S. Skell, K. Y. Klabunde, J. H. Klonka, J. S. Roberts and D. L. William-Smith, *J. Amer. Chem. Soc.*, **95**, 1547 (1973).
131. A. L. Tumolo, *U. S. Patent*, No. 3,857.859; *Chem. Abstr.*, **82**, 125265q (1975).
132. T. Nishiguchi, A. Kurooka and K. Fukuzumi, *J. Org. Chem.*, **39**, 2403 (1974).
133. T. Nishiguchi, H. Sakakibara and K. Fukuzumi, *Chem. Letters*, 649 (1976).

134. T. Tatsumi and K. Kizawa, *Chem. Letters*, 191 (1977).
135. G. I. Levi, A. A. Silakova and V. E. Vasserberg, *Izv. Akad. Nauk SSSR, Ser. Khim.*, 2050 (1977).
136. E. A. Karakhanov, M. V. Vagabov, A. V. Starkovskii and E. A. Viktorova, *Kinet. Katal.* **16**, 1198 (1975).
137. E. A. Karakhanov, E. A. Demianova and E. A. Viktorova, *Dokl. Akad. Nauk SSSR.*, **233**, 369 (1977).
138. L. Kh. Freidlin and V. Z. Sharf, *Neftekhimiya*, **5**, 558 (1965).
139. R. M. Thompson, *U. S. Patent*, No. 3,692,743; *Chem. Abstr.*, **78**, 5200k (1973).
140. M. Bartók and K. Kovács, *Acta Chim. Acad. Sci. Hung.*, **55**, 49 (1968).
141. W. P. Cochrane, P. L. Pauson and T. S. Stevens, *J. Chem. Soc. (C)*, 2346 (1969).
142. S. Farid and H. Scholz, *J. Org. Chem.*, **37**, 481 (1972).
143. Yu. M. Portnyagin and T. M. Pavel, *Zh. Org. Khim.*, **9**, 890 (1973).
144. L. Kh. Friedlin, V. Z. Sharf, M. Bartók and A. A. Nazaryan, *Izv. Akad. Nauk SSSR, Ser. Khim.*, 310 (1970).
145. M. Bartók and K. Felföldi, *Acta Chim. Acad, Sci. Hung.*, **85**, 339 (1975).
146. A. Fukuzawa, E. Kurosawa and T. Irie, *J. Org. Chem.*, **37**, 680 (1972).
147. H. Gotthardt and G. S. Hammond, *Chem. Ber.*, **107**, 3922 (1974).
148. A. Zamojski and T. Koźluk, *J. Org. Chem.*, **42**, 1089 (1977).
149. M. Bartók and B. Kozma, *Acta. Chim. Acad. Sci. Hung.*, **55**, 61 (1968).
150. M. Bartók and K. Kovács, *Acta Chim. Acad. Sci. Hung.*, **56**, 369 (1968).
151. M. Bartók and B. Kozma, *Acta Chim. Acad. Sci. Hung.*, **56**, 385 (1968).
152. M. Bartók, K. Kovács and N. I. Shuikin, *Acta Chim. Acad. Sci. Hung.*, **56**, 393 (1968).
153a. M. Bartók, I. Török and I. Szabó, *Acta Chim, Acad, Sci, Hung.*, **76**, 417 (1973).
153b. M. Bartók, *Acta Chim. Acad, Sci. Hung.*, **88**, 395 (1976).
154. M. Bartók, *J. Chem. Soc., Chem. Commun.*, 139 (1979).
155. J. Apjok, L. I. Lafer, M. Bartók and V. I. Yakerson, *Izv. Akad. Nauk SSSR, Ser. Khim.*, 24 (1977).
156. F. Notheisz and M. Bartók, *Acta Chim. Acad. Sci. Hung.*, **95**, 335 (1977).
157. M. Bartók, *Acta Phys. Chem. Szeged*, **21**, 79 (1975).
158. M. Bartók, *React. Kinet. Catal. Letters*, **3**, 115 (1975).
159. E. Taskinen, *Acta Chem. Scand. (B)*, **29**, 245 (1975).
160. J. Wolfhugel, A. Maujlan and J. Chuche, *Tetrahedron Letters*, 1635 (1973).
161. G. Piancatelli, A. Scettri and S. Barbadaro, *Tetrahedron Letters*, 3555 (1976).
162. A. W. S. Dick, F. M. Dean, D. A. Matkin and M. L. Robinson, *J. Chem. Soc., Perkin I*, 2204 (1977).
163. E. T. Østensen and M. M. Mishrikey, *Acta Chem. Scand. (B)*, **30**, 635 (1976).
164. D. P. Kreile, V. A. Slavinskaya, D. E. Apse, A. K. Strautinya, M. T. Brakmane and D. Ya. Eglite, *U.S.S.R. Patent*, No.484,214 (1976); *Chem. Abstr.*, **84**, 3085u (1976).
165. B. P. Krasnov and V. N. Alimov, *Khim. Prom.*, 74 (1976).
166. A. I. Kirsanova and M. G. Smirnova, *Izv. Ser. Kavk. Nauchn. Tsentra Vyssh. Shk., Ser. Tekh. Nauk.*, **5**, 108 (1977).
167. R. Curci, G. Delano, F. DiFuria, J. O. Edwards and A. R. Gallopo, *J. Org. Chem.*, **39**, 3020 (1974).
168. C. Schaal and J. Seyden-Penne, *Compt. Rend. (C)*, **266**, 217 (1968).
169. J. Seyden-Penne, *Bull. Soc. Chim. Fr.*, 3653 (1969).
170. H. Ruotsalainen, V. Palosaari and P. O. I. Virtanen, *Suom. Kemistilehti (B)*, **45**, 40 (1972).
171. S. H. Schroeter, *J. Org. Chem.*, **34**, 1188 (1969).
172. R. R. Sauers, W. Schinski, M. M. Mason, E. O'Hara and B. Byrne, *J. Org. Chem.*, **38**, 642 (1973).
173. J. C. Duchet and D. Cornet, *J. Catal.*, **44**, 57 (1976).
174. M. Bartók and I. Török, *Acta Chim, Acad. Sci. Hung.*, **88**, 35 (1976).
175. T. Irie, M. Izawa and E. Kurosawa, *Tetrahedron*, **26**, 851 (1970).
176. M. Bartók and R. A. Karakhanov, *Acta Phys. Chem. Szeged*, **20**, 453 (1974).
177. M. Lagrenée, *Compt. Rend. (C)*, **283**, 605 (1976).
178. A. Ya. Karmil'chik, V. V. Stonkus, E. Kh. Korchagova, Zh. G. Baikova,

N. I. Kalinovskaya, M. V. Shimanskaya and S. A. Giller, *Khim. Geterotsikl. Soed.*, 43 (1976).
179. T. M. Beloslyudova, L. A. Il'ina and O. K. Nikolaeva, *Zh. Prikl. Khim.*, **48**, 770 (1975).
180. T. M. Beloslyudova and L. A. Il'ina, *Zh. Prikl. Khim.*, **50**, 197 (1977).
181. M. Gasiorek and Z. Dudzik, *Przemysl. Chem.*, **55**, 542 (1976).
182. D. V. Sokol'skii, M. S. Erzhanova *et al.*, *Khimiya i Khim. Tekhnol.*, 113 (1974).
183. M. S. Erzhanova, T. Beisekov and E. V. Elemesov, *Khimiya i Khim. Tekhnol.*, 183 (1974).
184. M. S. Erzhanova, D. B. Daurenbekov and D. V. Sokol'skii, *Zh. Prikl. Khim.*, **51**, 346 (1978).
185. A. A. Ponomarev, *Khim. Geterotsikl. Soed.*, 163 (1966).
186. M. Bartók, K. L. Láng and L. G. Bogatskaya, *Dokl. Akad. Nauk SSSR*, **234**, 590 (1977).
187. P. Giusti, F. Andruzzi, P. Cerrai and G. Puce, *Corsi Semin. Chim.*, 141 (1968).
188. E. Z. Utyanskaya, *Vysokomol. Soedin. (A)*, **13**, 523 (1971).
189. G. Pruckmayr, *High Polym.*, **26**, 1 (1972).
190. N. S. Enikolopyan, *J. Macromol. Sci., Chem.*, **6**, 1053 (1972).
191. P. Dreyfuss, *Chem. Technol.*, **3**, 356 (1973).
192. I. Benedek and D. Feldman, *Stud. Cercet. Chim.*, **21**, 839 (1973).
193. P. Dreyfuss, *J. Macromol. Sci., Chem.*, **7**, 1361 (1973).
194. T. Saegusa and S. Kobayashi, *ACS Symp. Ser.*, **6**, 150 (1974).
195. P. Dreyfuss and M. P. Dreyfuss, *Compr. Chem. Kinet.*, **15**, 259 (1976).
196. P. Teyssie, T. Onhadi and J. P. Bioul, *Int. Rev. Sci., Phys. Chem. Ser. 2*, **8**, 191 (1975).
197. K. S. Kazanskii, *Itogi Nauki Tekh., Khim. Tekhnol. Vysokomol. Soedin.*, **9**, 5 (1977).
198. E. J. Goethals, *Pure Appl. Chem.*, **48**, 335 (1976).
199. V. A. Ponomarenko, *Vysokomol. Soedin. (A)*, **19**, 1670 (1977).
200. M. Sepulcher, N. Spassky, and P. Sigwalt, *Israel J. Chem.*, **15**, 33 (1977).
201. N. S. Enikolopyan, V. V. Ivanov and G. V. Korovina, *Vysokomol. Soedin. (A)*, **19**, 1924 (1977).
202. P. Dreyfuss and M. P. Dreyfuss in *Ring-opening Polymerization*, Vol. 2 (Eds. K. C. Frisch and S. L. Reegen), Marcel Dekker, New York, 1969, pp. 111–158.
203. Y. Ishii and S. Sakai in *Ring-opening Polymerization*, Vol. 2 (Eds. K. C. Frisch and S. L. Reegen), Marcel Dekker, New York, 1969, pp. 13–109.
204. T. Saegusa in *Jugo Hanno Ron, Kaikan Jugo (Ring-opening Polymerization) (I)*, Vol. 6, Kagakudojin, Kyoto, 1971.
205. T. Saegusa and S. Kobayashi in *Progress in Polymer Science Japan*, Vol. 6 (Eds. S. Onogi and K. Uno), Kodansha Ltd., Tokyo and John Wiley and Sons, New York, 1973, pp. 107–151.
206. H. Perst in *Oxonium Ions in Organic Chemistry*, Verlag Chemie, Weinheim, 1971, pp. 78, 141.
207. F. Frater, R. Mateva and R. Ludger, *Izv. Khim.*, **9**, 132 (1976).
208. J. Furukawa and T. Saegusa, *Polymerization of Aldehydes and Oxides*, John Wiley and Sons, New York, 1963.
209. P. Dreyfuss and M. P. Dreyfuss, *Polymer J.*, **8**, 81 (1976).
210. P. E. Black and D. J. Worsfold, *Can. J. Chem.*, **54**, 3325 (1976).
211. S. Kobayashi, H. Danda and T. Saegusa, *Bull. Chem. Soc. Japan*, **47**, 2699 (1974).
212. G. Pruckmayr and T. K. Wu, *Macromolecules*, **8**, 954 (1975).
213. H. Koinuma and H. Hirai, *Makromol. Chem.*, **178**, 241 (1977).
214. S. Kobayashi, H. Danda and T. Saegusa, *Bull. Chem. Soc. Japan*, **47**, 2706 (1974).
215. N. G. Taganov and G. N. Komratov, *Vysokomol. Soedin. (B)*, **19**, 510 (1977).
216. Y. Tanaka, *Kobunshi Roubunshu*, **34**, 491 (1977); *Chem. Abstr.*, **87**, 85339m (1977).
217. E. Yu. Bekhli, M. V. Fomina and S. G. Entelis, *Zh. Fiz. Khim.*, **50**, 816 (1976).
218. S. Kobayashi, H. Danda and T. Saegusa, *Bull. Chem. Soc. Japan*, **46**, 3214 (1973).
219. S. Kobayashi, T. Saegusa and Y. Tanaka, *Macromolecules*, **7**, 415 (1974).
220. S. Kobayashi, T. Saegusa and Y. Tanaka, *Bull. Chem. Soc. Japan*, **46**, 3220 (1973).

221. S. Smith and A. J. Hubin, *J. Macromol. Sci., Chem.*, **A7**, 1399 (1973).
222. G. Pruckmayr and T. K. Wu, *Macromolecules*, **6**, 33 (1973).
223. S. L. Malhotra and L. P. Blanchard, *J. Macromol. Sci., Chem.*, **A9**, 1485 (1975).
224. L. P. Blanchard, C. Raufast, H. H. Kiet and S. L. Malhotra, *J. Macromol Sci., Chem.*, **A9**, 1219 (1975).
225. L. P. Blanchard, G. G. Gabra and S. L. Malhotra, *J. Polym. Sci., Polym. Chem. Ed.*, **13**, 1619 (1975).
226. L. P. Blanchard, J. Singh and M. D. Baijal, *Can. J. Chem.*, **44**, 2679 (1966).
227. K. H. W. Reichert, *Chimia*, **29**, 453 (1975).
228. J. M. Kenna, T. K. Wu and G. Pruckmayr, *Macromolecules*, **10**, 877 (1977).
229. M. Okamoto, M. Sasaki and J. Osugi, *Rev. Phys. Chem. Japan*, **47**, 1 (1977).
230. U. Seitz, R. Höne and K. H. W. Reichert, *Makromol. Chem.*, **176**, 1689 (1975).
231. K. Brzezinska, W. Chwialkowska, P. Kubisa, K. Matyjaszewski and S. Penczek, *Makromol. Chem.*, **178**, 2491 (1977).
232. E. L. Berman, A. M. Sakharov, E. M. Galimov, A. P. Klimov, G. V. Isagulyants and V. A. Ponomarenko, *Dokl. Akad. Nauk SSSR*, **234**, 850 (1977).
233. T. Saegusa, N. Takaishi and Y. Ito, *Synthesis*, **2**, 475 (1970).
234a. O. Meresz, *Tetrahedron Letters*, 2797 (1972).
234b. Gy. Schneider, I. Weisz-Vincze, A. Vass and K. Kovács, *J. Chem. Soc., Chem. Commun.*, 713 (1972).
234c. Yu. K. Yur'ev, *Zh. Obshch. Khim.*, **6**, 972, 1669 (1936).
235. P. A. Pinke and S. N. Massie, *U. S. Patent*, No. 3,853,887; *Chem. Abstr.*, **82**, 97930g (1975).
236. S. N. Massie, *U. S. Patent*, No. 3,900,479; *Chem. Abstr.*, **84**, 17149n (1976).
237. B. Buchholz, B. Deger and R. H. Goshoru, *German Patent*, No. 1,228,273 (1967); *Chem. Abstr.*, **66**, 18665 (1967).
238. H. M. Foster, *U. S. Patent*, No. 3,381,018; *Chem. Abstr.*, **69**, 51981c (1968).
239. R. P. Gerhard, *German Patent*, No. 2,159,859; *Chem. Abstr.*, **77**, 126023t (1973).
240. A. A. Avots, V. S. Ajzbalts, I. Ya. Lazdyshyn, M. K. Sile and V. K. Ulaste, *U.S.S.R. Patent*, No. 467,069; *Chem. Abstr.*, **83**, 114198q (1975).
241. J. W. Ward, *J. Catal.*, **10**, 34 (1968).
242. K. Hatada, M. Shimada, K. Fujita, Y. Ono and T. Keii, *Chem. Letters*, 439 (1974).
243. K. Fujita, K. Hatada, Y. Ono and T. Keii, *J. Catal.*, **35**, 325 (1974).
244. Y. Ono, K. Hatada, K. Fujita, A. Halgeri and T. Keii, *J. Catal.*, **41**, 322 (1976).
245. K. Hatada, M. Shimada, Y. Ono and T. Keii, *J. Catal.*, **37**, 166 (1975).
246. K. V. Topichieva and A. A. Kubasov, *Sovrem. Probl. Fiz. Khim.*, **8**, 326 (1975).
247. K. Hatada and Y. Ono, *Bull. Chem. Soc. Japan*, **50**, 2517 (1977).
248. K. Hatada, K. Fujita and Y. Ono, *Bull. Chem. Soc. Japan*, **51**, 2419 (1978).
249. Y. Ono, T. Mori and K. Hatada, *Zeolite Symposium, Szeged*, Sept. 1978.
250. K. Hatada, Y. Takeyama and Y. Ono, *Bull. Chem. Soc. Japan*, **51**, 448 (1978).
251. Y. Ono, A. Halgeri, M. Kaneko and K. Hatada, *ACS Symposium Series* (Ed. J. R. Katzer), No. 40, 596 (1977).
252. V. V. Smirnov and S. B. Zotov, *Khim, Geterotsikl. Soed.*, 173 (1968).
253. P. B. Venuto and P. S. Landis, *Adv. Catal.*, **18**, 259 (1968).
254. I. F. Bel'skii, N. I. Shuikin and G. E. Skobtsova, *Izv. Akad. Nauk SSSR, Ser. Khim.*, 1118 (1964).
255. I. F. Bel'skii, N. I. Shuikin and G. E. Skobtsova in *Problemy Organicheskogo Sinteza*, Nauka, Moskva, 1965, p. 186.
256. G. E. Abgaforova, N. I. Shuikin and I. F. Bel'skii, *Izv Akad. Nauk SSSR, Ser. Khim.*, 734 (1965).
257. I. F. Bel'skii, N. I. Shuikin and G. E. Abgaforova, *Izv. Akad. Nauk SSSR, Ser. Khim.*, 160 (1965).
258. N. I. Shuikin, I. F. Bel'skii and G. E. Skobtsova, *Izv. Akad. Nauk SSSR, Ser. Khim.*, 1120 (1964).
259. I. F. Bel'skii and L. Ya. Barkovskaya, *Zh. Org. Khim*, **3**, 385 (1967).
260. I. F. Bel'skii, S. N. Khar'kov and N. I. Shuikin, *Dokl. Akad. Nauk SSSR*, **165**, 821 (1965).

261. N. I. Shuikin, I. F. Bel'skii and S. N. Khar'kov, *Izv. Akad. Nauk SSSR, Ser. Khim.*, 2109 (1967).
262. L. Mészáros, M. Bartók and A. G. Schöbel, *Acta Phys. Chem. Szeged*, **13**, 121 (1967).
263. D. Kaufmann, A. De Meijere, B. Hingerty and W. Saenger, *Angew. Chem.*, **87**, 842 (1975).
264. U. Schöllkopf in *Methoden der Organischen Chemie (Houben-Weyl)*, Vol. XIII/1, Georg Thieme Verlag, Stuttgart, 1970, p. 220.
265. K. Nützel in *Methoden der Organischen Chemie (Houben-Weyl)*, Vol. XIII/2a, Georg Thieme Verlag, Stuttgart, 1973, p. 349.
266. B. A. Trofimov and S. E. Korostova, *Usp. Khim.*, **44**, 78 (1975).
267. L. L. Darko and J. G. Cannon, *J. Org. Chem.*, **32**, 2352 (1967).
268. N. J. Foulger and B. J. Wakefield, *J. Chem. Soc., Perkin I*, 871 (1974).
269. D. Seebach and E. J. Corey, *J. Org. Chem.*, **40**, 231 (1975).
270. E. N. Marvell, D. Sturmer and R. S. Knutsen, *J. Org. Chem.*, **33**, 2991 (1968).
271. L. R. Kray and M. G. Reinecke, *J. Org. Chem.*, **32**, 225 (1967).
272. H. J. Bestmann, O. Wostrowsky and W. Stransky, *Chem., Ber.*, **109**, 337 (1976).
273. D. C. Chung and R. J. Kostelnik, *Makromol. Chem.*, **178**, 691 (1977).
274. J. Millon and G. Linstrumelle, *Tetrahedron Letters*, 1095 (1976).
275. P. F. Hudrlik and A. M. Hudrlik, *Tetrahedron Letters*, 1361 (1971).
276. A. Donnelly, J. G. Hoey and R. O'Donnel, *J. Chem. Soc., Perkin I*, 1218 (1974).
277. I. M. Gverdtsiteli, R. Yu. Papava and E. S. Gelashvili, *Zh. Obshch, Khim.*, **36**, 112 (1966).
278. D. B. Miller, *J. Organometal. Chem.*, **14**, 253 (1968).
279. P. E. Trockmorton and W. J. McKillip, *U. S. Patent*, No. 3,903,112 (1975); *Chem. Abstr.*, **83**, 206421k (1975).
280. P. F. Hudrlik and C. N. Wan, *J. Org. Chem.*, **40**, 2963 (1975).
281. S. C. Honeycutt, *J. Organometal. Chem.*, **29**, 1 (1971).
282. A. Maercher and J. Troesch, *J. Organometal. Chem.*, **102**, C 1 (1971).
283. R. B. Bates, L. M. Kroposki and D. E. Potter, *J. Org. Chem.*, **37**, 560 (1972).
284. P. Tomboulian, D. Amick, S. Beare, K. Dumke, D. Hart, R. Hites, A. Metzger and R. Nowak, *J. Org. Chem.*, **38**, 322 (1973).
285. J. Levisalles, H. Rudler and D. Villemin, *J. Organometal. Chem.*, **122**, C 15 (1976).
286. J. G. Carpenter, A. G. Evans and N. H. Rees, *J. Chem, Soc., Perkin II*, 1598 (1972).
287. T. Fujita, K. Suga and S. Watanabe, *Synthesis*, 630 (1972).
288. E. G. Evans, M. U. Jones and N. H. Rees, *J. Chem. Soc. (B)*, 894 (1969).
289. M. G. Voronkov, V. E. Puzanova, S. F. Pavlov and E. I. Dubinskaya, *Izv. Akad. Nauk SSSR, Ser. Khim.*, 448 (1975).
290. M. E. Yung and M. A. Lyster, *J. Org. Chem.*, **42**, 3761 (1977).
291. K. A. Andrianov, L. M. Volkova and O. G. Blokhina, *Zh. Obshch. Khim.*, **45**, 2206 (1975).
292. M. G. Voronkov, V. G. Komarov, A. I. Albanov. I. M. Korotaeva and E. I. Dubinskaya, *Izv. Akad. Nauk SSSR, Ser. Khim.*, 1415 (1978).
293. D. Conturier, *Ann. Chim.*, **7**, 19 (1972).
294. A. M. Foster and W. C. Agosta, *J. Org. Chem.*, **37**, 61 (1972).
295. E. J. Corey, K. C. Nicolaou and D. J. Beames, *Tetrahedron Letters*, 2439 (1974).
296. N. R. Chaudhuri and S. Mitra, *Bull. Chem. Soc. Japan*, **49**, 1035 (1976).
297. K. Rossmanith and C. Auer-Welsbach, *Monatsh. Chem.*, **96**, 606 (1965).
298. T. Kogane, H. Yukawa and R. Hirota, *Nippon Kagaku Kaishi*, 1115 (1975); *Chem. Abstr.*, **83**, 88140t (1975).
299. T. W. Leung and M. J. Hintz, *Inorg. Chem.*, **16**, 2606 (1977).
300. M. G. Voronkov. E. I. Dubinskaya, V. G. Komarov and S. F. Pavlov, *Zh. Obshch. Khim.*, **46**, 1908 (1976).
301. A. Gaumeton, *Ann. Chim.*, **8**, 457 (1963).
302. N. I. Shuikin, I. F. Bel'skii, R. A. Karakhanov. B. Kozma and M. Bartók, *Acta Phys. Chem. Szeged*, **9**, 37 (1963).
303. V. A. Arbuzov, E. N. Klimovitskii, L. K. Yuldasheva and A. B. Remizov, *Izv. Akad. Nauk SSSR, Ser. Khim.*, 377 (1974).
304. I. Ježo, *Chem. Zvesti*, **29**, 714 (1975); *Chem. Abstr.*, **85**, 32588u (1976).

305. R. Paul, O. Riobe and M. Maymy, *Org. Synth.*, **55**, 62 (1976).
306. T. E. Stone and G. D. Daves, *J. Org. Chem.*, **42**, 2151 (1977).
307. J. D. Woodyard and D. H. Corbin, *J. Heterocycl. Chem.*, **13**, 647 (1976).
308. D. J. Nelson and W. J. Miller, *J. Chem. Soc., Chem. Commun.*, 444 (1973).
309. J. D'Angelo, *Bull. Soc. Chim. Fr.*, 181 (1969).
310. N. I. Shuikin and M. Bartók, *Izv. Akad. Nauk SSSR., Ser. Khim.*, 129 (1968).
311. K. A. Holbrook and R. A. Scott, *J. Chem. Soc., Faraday I*, **71**, 1849 (1975).
312. E. A. Cavell, R. E. Parker and A. W. Scaplehorn, *J. Chem. Soc. (C)*, 389 (1966).
313. M. Bartók, *Acta Chim. Acad. Sci. Hung.*, **51**, 403 (1967).
314. M. J. Clarke and K. A. Holbrook, *J. Chem. Soc., Faraday I*, **73**, 890 (1977).
315. M. Bartók, B. Kozma and N. I. Shuikin, *Izv. Akad. Nauk SSSR, Ser. Khim.*, 132 (1968).
316. G. Jones, II and H. H. Kleinman, *Tetrahedron Letters*, 2103 (1974).
317. G. F. Cohoe and W. D. Walters, *J. Phys. Chem.*, **71**, 2326 (1967).
318. M. Bartók and B. Kozma, *Acta Chim. Acad. Sci. Hung.*, **52**, 83 (1967).
319. A. D. Clements, H. M. Frey and J. G. Frey, *J. Chem. Soc., Faraday I*, **71**, 2485 (1975).
320. K. A. Holbrook and R. A. Scott, *J. Chem, Soc., Faraday I*, **70**, 43 (1974).
321. N. Shimizu and Sh. Nishida, *J. Chem. Soc., Chem. Commun.*, 734 (1974).
322. M. Bartók and K. Kovács, *Acta Phys. Chem. Szeged*, **13**, 67 (1967).
323. H. A. J. Carless, *Tetrahedron Letters*, 3425 (1974).
324. G. Jones, II, S. B. Schwartz and M. T. Marton, *J. Chem. Soc., Chem. Commun.*, 374 (1973).
325. A. M. Hudrlik and C. N. Wan, *J. Org. Chem.*, **40**, 1116 (1975).
326. A. G. Hortmann and A. Bhattacharjya, *J. Amer. Chem. Soc.*, **98**, 7081 (1976).
327. G. Adams, C. Bibby and R. Grigg, *J. Chem. Soc., Chem. Commun.*, 491 (1972).
328. H. A. J. Carless, *J. Chem. Soc., Chem. Commun.*, 982 (1974).
329. H. Kwart, S. F. Sarner and J. Slutsky, *J. Amer. Chem. Soc.*, **95**, 5234 (1973).
330. S. Murabayashi, M. Shiotani and J. Sohma, *Chem. Phys. Letters*, **48**, 80 (1977).
331. C. Gaze and B. C. Gilbert, *J. Chem. Soc., Perkin II*, 754 (1977).
332. A. L. Blyumenfel'd and V. I. Trofimov, *Khim. Vys. Energ.*, **11**, 178 (1977).
333. N. I. Shuikin and B. L. Lebedev, *Z. Chem.*, **6**, 459 (1966).
334. N. I. Shuikin and B. L. Lebedev, *Usp. Khim.*, **35**, 1047 (1966).
335. N. I. Shuikin, B. L. Lebedev and I. P. Yakovlev, *Izv. Akad. Nauk SSSR, Ser. Khim.*, 644 (1967).
336. N. I. Shuikin and B. L. Lebedev, *Izv. Akad. Nauk SSSR, Ser. Khim.*, 639 (1967).
337. D. Leonov and D. Elad, *J. Org. Chem.*, **39**, 1470 (1974).
338. I. Rosenthal and D. Elad, *Tetrahedron*, **23**, 3193 (1967).
339. No reference.
340. E. Montaudon, J. Thepenier and R. Lalande, *Compt. Rend. (C)*, **280**, 1223 (1975).
341. E. Montaudon, J. Thepenier and R. Lalande, *Compt. Rend. (C)*, **284**, 581 (1977).
342. E. Montaudon and R. Lalande, *Bull. Soc. Chim. Fr.*, 2635 (1974).
343. W. T. Dixon, J. Foxall, G. H. Williams, D. J. Edge, B. C. Gilbert, H. Kazarians-Moghaddam and R. O. C. Norman, *J. Chem. Soc., Perkin II*, 827 (1977).
344. T. Ueda, H. Muramatsu and K. Inukai, *Nippon Kagaku Kaishi*, 100 (1975); *Chem. Abstr.*, **84**, 16495k (1976).
345. A. Ledwith and M. Sambhi, *J. Chem. Soc. (B)*, 670 (1966).
346. F. Krejči and J. Pichler, *Scripta Fac. Sci. Nat. Univ. Purkynianae Brun.*, **4**, 71 (1974); *Chem. Abstr.*, **84**, 150547g (1976).
347. M. D. Shetlar, *J. Chem. Soc., Chem. Commun.*, 653 (1975).
348. P. J. Wagner, C. A. Stout, S. Searles, Jr. and G. S. Hammond, *J. Amer. Chem. Soc.*, **88**, 1242 (1966).
349. R. T. K. Baker and J. A. Kerr, *J. Chem. Soc., Chem. Commun.*, 821 (1966).
350. J. Wiemann, N. Thoai and F. Weisbuch, *Bull. Soc. Chim. Fr.*, 575 (1966).
351. A. Gilbert and G. Taylor, *Tetrahedron Letters*, 469 (1977).
352. H. Nozaki, S. Fujita, H. Takaya and R. Noyori, *Tetrahedron*, **23**, 45 (1967).
353. N. Torimoto, T. Shingaki and T. Nagai, *Bull. Chem. Soc. Japan*, **49**, 2572 (1976).
354. M. Birchall, R. N. Haszeldine and P. Tissington, *J. Chem. Soc., Perkin I*, 1638 (1975).

355. S.-H. Goh, K.-C. Chan, T.-S. Kam and H.-L. Chong, *Australian J. Chem.*, **28**, 381 (1975).
356. P. O. I. Virtanen, A. Karjalainen and H. Ruotsalainen, *Suomen Kemistilehti*, **B43**, 219 (1970).
357. J. P. Brun, M. Ricard, M. Corval and C. Schaal, *Org. Mass Spectrom.*, **12**, 348 (1977).
358. G. Jones and L. P. McDonell, *J. Org. Chem.*, **43**, 2184 (1978).
359. R. Smakman and Th. J. De Boer, *Org. Mass Spectrom.*, **1**, 403 (1968).
360. P. O. I. Virtanen, *Suomen Kemistilehti*, **B39**, 58 (1966).
361. P. O. I. Virtanen, *Suomen Kemistilehti*, **B40**, 193 (1967).
362. P. O. I. Virtanen, *Suomen Kemistilehti*, **B40**, 185 (1967).
363. P. O. I. Virtanen and K. Manninen, *Suomen Kemistilehti*, **B40**, 341 (1967).
364. P. O. I. Virtanen and H. Ruotsalainen, *Suomen Kemistilehti*, **B42**, 69 (1969).
365. P. O. I. Virtanen, H. Malo and H. Ruotsalainen, *Suomen Kemistilehti*, **B43**, 512 (1970).
366. H. Ruotsalinen, J. Kaakkurivaara and P. O. I. Virtanen, *Suomen Kemistilehti*, **B45**, 35 (1972).
367. P. O. I. Virtanen, *Suomen Kemistilehti*, **B39**, 64 (1966).
368. A. Kankaanperä and S. Kleemola, *Acta Chem. Scand.*, **23**, 3607 (1969).
369. Ch. A. Smith and J. B. Grutzner, *J. Org. Chem.*, **41**, 367 (1976).
370. M. Aresta, C. F. Nobile and D. Petruzzelli, *Inorg. Chem.*, **16**, 1817 (1977).
371. W. W. Schmitt-Fumian and W. Hellein, *Makromol. Chem.*, **177**, 1613 (1976).
372. D. J. Goldsmith, E. Kennedy and R. G. Campbell, *J. Org. Chem.*, **40**, 3571 (1975).
373. K. O. Christe and A. E. Pavlath, *J. Org. Chem.*, **30**, 1639 (1965).
374. S. Sakai, H. Tanaka and Y. Ishii, *Kogyo Kagaku Zasshi*, **69**, 1388 (1966); *Chem. Abstr.*, **66**, 104866r (1967).
375. Gy. Schneider and I. Weisz-Vincze, *Kémiai Közlemények*, **31**, 383 (1969); *Chem. Abstr.*, **72**, 32109v (1970).
376. K. Baum and Ch. D. Beard, *J. Org. Chem.*, **40**, 81 (1975).
377. G. A. Olah and P. J. Szilagyi, *J. Org. Chem.*, **36**, 1121 (1971).
378. T. I. Zhukova, V. A. Vysotskii, K. E. V'yunov, M. P. Grinblam, G. L. Epshtein and E. G. Sochilin, *Zh. Org. Khim.*, **11**, 1989 (1975).
379. B. A. Arbuzov, O. N. Nuretdinova, L. Z. Nikonova and E. I. Gol'dfarb, *Izv. Akad. Nauk SSSR, Ser. Khim.*, 627 (1973).
380. B. A. Arbuzov, L. Z. Nikonova, O. N. Nuretdinova and N. P. Anoshina, *Izv. Akad. Nauk SSSR, Ser. Khim.*, 473 (1975).
381. C. Schaal, *Compt. Rend.(C)*, **264**, 1309 (1967).
382. N. Watanabe, S. Uemura and M. Okano, *Bull. Chem. Soc. Japan*, **48**, 3205 (1975).
383. N. Watanabe, S. Uemura and M. Okano, *Bull. Chem. Soc. Japan*, **49**, 2500 (1976).
384. S. Hayashi and N. Ishikawa, *Bull. Chem. Soc. Japan*, **48**, 1467 (1975).
385. Y. Hara and M. Matsuda, *J. Org. Chem.*, **40**, 2786 (1975).
386. H. Böhme, P. Wagner, *Chem. Ber.*, **102**, 2651 (1969).
387. D. L. Rakhmankulov, R. Kh. Nurieva, E. A. Kantor and P. S. Belov, *Zh. Prikl. Khim.*, **50**, 212 (1977).

CHAPTER **16**

Dehydration of diols

M. BARTÓK and Á. MOLNÁR

Department of Organic Chemistry, József Attila University, Szeged, Hungary

I. DEHYDRATION OF 1,2-DIOLS

The transformations of 1,2-diols accompanied by elimination of water can be summarized in three reactions:

(*i*) The classical process of pinacol rearrangement, first described by Fittig[1], and later studied by Butlerov[2]. Pinacol (1) is treated with cold concentrated sulphuric acid and thereby converted to methyl *t*-butyl ketone (pinacone) (2) (equation 1).

$$(CH_3)_2C - C(CH_3)_2 \xrightarrow[-H_2O]{H_2SO_4} (CH_3)_3CCOCH_3 \qquad (1)$$
$$\underset{OH \quad\ OH}{}$$

$$(1) \hspace{5cm} (2)$$

(*ii*) The formation of epoxides, which is observed mainly from tetrasubstituted and certain hindered trisubstituted diols.

(*iii*) The formation of unsaturated compounds, primarily dienes.

The pinacol rearrangement has been studied in great detail, particularly for the secondary—tertiary and ditertiary 1,2-diols. Transformations of other 1,2-diols, and the use of nonacidic reaction conditions differing from the classical ones, have received less attention.

A. Dehydration in the Solution Phase by the Action of Acids

1. Pinacol rearrangement

Rearrangement of pinacol with water elimination can be achieved in the presence of mineral acids. Sulphuric acid is used mainly, but use is frequently made of perchloric acid, aromatic sulphonic acids, organic acids (formic acid, oxalic acid) and acetic acid together with iodine, acetyl chloride or acetic anhydride.

Because of the very large volume of literature data we cannot give a full review of all the publications and shall restrict ourselves therefore to a brief survey of the still continuing research that has led to the currently accepted interpretation of the pinacol rearrangement. Using the earlier reviews[3-5] as a starting point, we shall mainly discuss the results of the past 15 years. In 1963, Bunton and Carr[6] came to the conclusion, still generally accepted, that there is no unique mechanism for the pinacol—pinacone rearrangement: depending on the structure of the diol and the reaction conditions, one or other mechanism predominates, or the rearrangement may occur via simultaneous processes.

In principle there are four fundamental routes for the pinacol rearrangement: via a carbonium cation, by a concerted mechanism, via an epoxide intermediate and via vinyl dehydration. Most reactions can be interpreted by means of the first two of these reaction paths, while the data so far obtained suggest that the final possibility may be excluded.

a. Rearrangement via a carbonium cation. This route for the rearrangement has been discussed in detail by Collins[3,7-9]. The reactions of the labelled diol (3) were studied in the presence of five different catalysts (concentrated H_2SO_4, formic acid, dilute H_2SO_4, oxalic acid, dioxan—H_2O—HCl). It was found that the process shown in equation (2) does not play a role in the rearrangement, while the three remaining possibilities (equations 3—5) depend to a large extent on the reaction conditions. The carbonium cation is formed reversibly (neighbouring-group par-

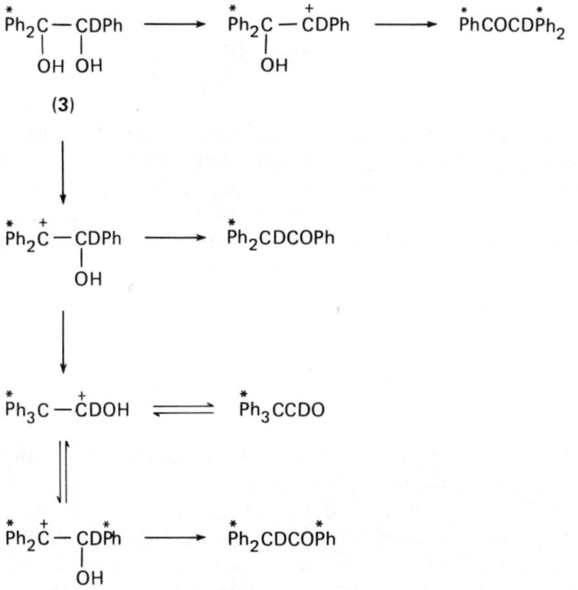

$$\overset{*}{Ph_2}C-CDPh \longrightarrow \overset{*}{Ph_2}C-\overset{+}{C}DPh \longrightarrow \overset{*}{Ph}COC\overset{*}{D}Ph_2 \qquad (2)$$
$$\underset{OH\ \ OH}{|\ \ \ |} \qquad\qquad \underset{OH}{|}$$

(3)

$$\overset{*}{Ph_2}\overset{+}{C}-CDPh \longrightarrow \overset{*}{Ph_2}CDCOPh \qquad (3)$$
$$\underset{OH}{|}$$

$$\overset{*}{Ph_3}C-\overset{+}{C}DOH \rightleftharpoons \overset{*}{Ph_3}CCDO \qquad (4)$$

$$\overset{*}{Ph_2}\overset{+}{C}-CD\overset{*}{P}h \longrightarrow \overset{*}{Ph_2}CDCO\overset{*}{P}h \qquad (5)$$
$$\underset{OH}{|}$$

ticipation cannot be observed); this is followed by irreversible hydrogen or aryl migration (equation 6).

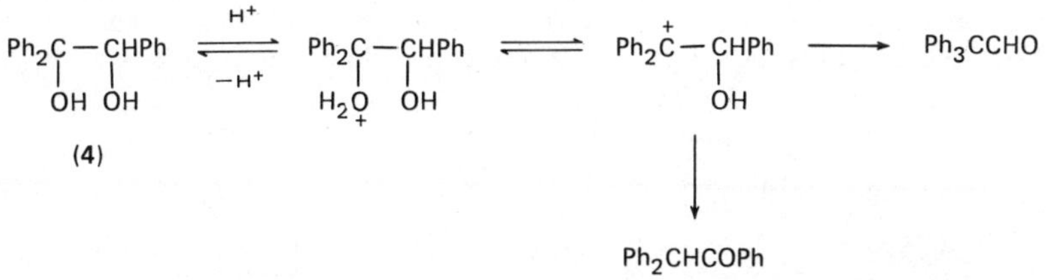

$$Ph_2C-CHPh \underset{-H^+}{\overset{H^+}{\rightleftharpoons}} Ph_2C-CHPh \rightleftharpoons Ph_2\overset{+}{C}-CHPh \longrightarrow Ph_3CCHO \quad (6)$$
$$\underset{OH\ \ OH}{|\ \ \ |} \qquad\qquad \underset{\underset{+}{H_2O}\ \ OH}{|\ \ \ \ |} \qquad\qquad \underset{OH}{|}$$

(4)

$$Ph_2CHCOPh$$

At the same time, the presence of the carbonium cation was proved by rate measurements in D_2O[10-12] and by ^{18}O-studies. Since the results reviewed earlier[3], many authors have confirmed the above observations under very varied experimental conditions and with diols of different structures[6,13-21]. Many of the most recent investigations[22-27] support the carbonium cation mechanism. Nevertheless, the results of a number of research groups are in agreement with the existence of a nonclassical bridged carbonium ion[28-37].

b. *Concerted mechanism.* From a study of the rearrangements of the diols **1**, **5** and **6** in 50% H_2SO_4, Stiles and Mayer[38] found that bond formation by the

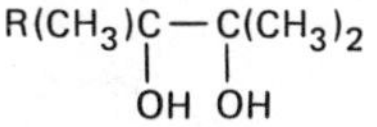

$$R(CH_3)C-C(CH_3)_2$$
$$\underset{OH\ \ OH}{|\ \ \ |}$$

(1) R = Me

(5) R = Et

(6) R = t-Bu

migrating group occurs in a slow step, which excludes the carbonium ion mechanism. They recommend a concerted reaction pathway (equation 7), in which water is

$$\underset{\substack{\smallfrown \\ \text{OH} \ \overset{+}{\text{OH}}_2}}{\overset{\displaystyle R \diagdown}{\underset{}{\text{RC}-\text{CR}_2}}} \longrightarrow \underset{\substack{\text{OH} \\ +}}{\overset{\displaystyle \text{RC}-\text{CR}_3}{\overset{\|}{}}} \qquad (7)$$

eliminated from the protonated pinacol with the anchimeric assistance of the migrating group. Since oxygen exchange takes place between the pinacol and the solvent during the rearrangement[12], formation of the carbonium hydrate (7) is probable in the first step. This formation is responsible for both the rearrangement and the oxygen exchange (equation 8). The rearrangement involves a backside displacement

$$\underset{\substack{\text{OH} \ \overset{+}{\text{OH}}_2}}{\text{R}_2\text{C}-\text{CR}_2} \rightleftharpoons \left[\underset{\substack{\text{OH} \ \text{OH}_2}}{\text{R}_2\text{C}-\text{C}\overset{R}{\underset{R}{\diagup}}}\right]^+ \overset{\text{oxygen exchange}}{\underset{\displaystyle R_3\text{CCOR}}{\diagup\diagdown}} \qquad (8)$$

$$(7)$$

of water by the neighbouring group R. Because of the geometry and hybridization of the carbon atom, this process should occur much more easily than the one shown in equation (7). The rates measured in the cases of the different substituents suggest that for **1** both the carbonium ion route and the concerted mechanism are involved in the rearrangement, while for **5** and **6** the latter pathway predominates. The role of the concerted mechanism is supported by the results of other investigations[39-41].

   c. *Role of the epoxide intermediate.* It has long been known[42-44] that epoxides are formed during the loss of water from certain tetrasubstituted and, hindered trisubstituted 1,2-diols (equation 9). More recently, in a kinetic investigation of the

$$\underset{\substack{\text{OH} \quad \text{OH}}}{\text{PhCH}-\text{C}(\alpha\text{-C}_{10}\text{H}_7)\text{Ph}} \xrightarrow{\text{dil. H}_2\text{SO}_4} \underset{\substack{\diagdown\diagup \\ \text{O}}}{\text{PhCH}-\text{C}(\alpha\text{-C}_{10}\text{H}_7)\text{Ph}}$$

$$\Big\downarrow \text{conc. H}_2\text{SO}_4 \qquad\qquad\qquad (9)$$

$$\text{PhCOC}(\alpha\text{-C}_{10}\text{H}_7)\text{Ph}$$

reaction of tetraphenylethylene glycol (8) in perchloric acid—acetic anhydride, it was found[45] that, besides the direct rearrangement, the transformation also takes place via the epoxide (equation 10) (~80% at 75°C). The ratio of the two processes

$$\underset{\substack{\text{OH} \ \text{OH} \\ (8)}}{\text{Ph}_2\text{C}-\text{CPh}_2} \xrightarrow{\hspace{3cm}} \begin{array}{l} \text{Ph}_3\text{CCOPh} \\ (9) \end{array} \qquad (10)$$

$$\underset{\substack{\diagdown\diagup \\ \text{O}}}{\text{Ph}_2\text{C}-\text{CPh}_2}$$

is determined by the stereostructure of the diol: the *trans*-diol has a favourable conformation for epoxide formation, whereas the *cis*-diol, present in lower amount, leads directly to the ketone (9).

   In studies of the various methods of rearrangement of *cis*- and *trans*-1,2-dimethylcyclopentanediol[6], pinacol[32,33] and other tetraaryl glycols[46,47] it has been similarly proved that the corresponding epoxides may act as intermediates under certain conditions.

In their study of the rearrangement of the tetraaryl glycols 10, Pocker and Ronald[46,47] described the transformation with the proved kinetic scheme of equation 11.

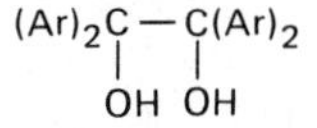

$$
\begin{array}{cc}
(Ar)_2C - C(Ar)_2 \\
\quad | \quad \ | \\
\quad OH \ \ OH
\end{array}
\qquad\qquad
\text{Glycol} \xrightarrow{\ k_1\ } R^+
\qquad
\begin{array}{c}
\text{Epoxide} \\
k_3 \nearrow \!\!\!\! / \!\!\!\! \nwarrow k_4 \\[4pt]
k_2 \searrow \\
\text{Ketone}
\end{array}
\qquad (11)
$$

(10)

Ar = Ph, $p\text{-}CH_3C_6H_4$, $p\text{-}CH_3OC_6H_4$

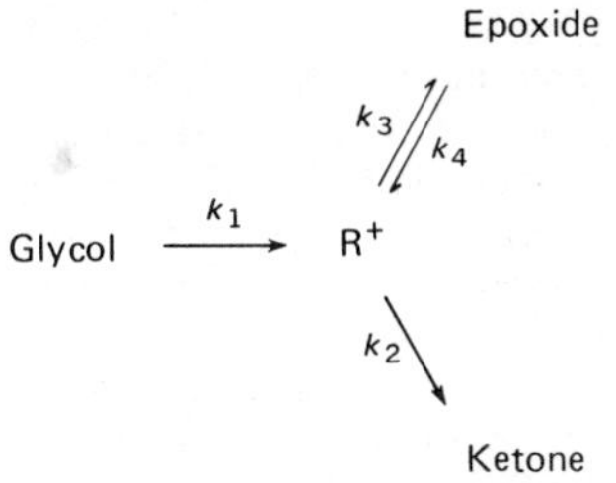

Compared to the previously discussed mechanisms, rearrangement via the epoxide is less general; this path is a rarely-occurring, special case of the rearrangement.

d. Vinyl dehydration. This possibility (equation 12) was first suggested in the

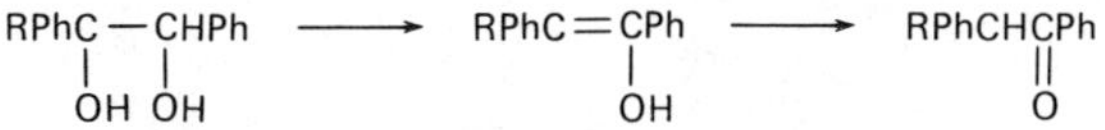

$$
\begin{array}{ccc}
RPhC - CHPh & \longrightarrow & RPhC = CPh & \longrightarrow & RPhCHCPh \\
\ \ | \quad\ | & & \qquad | & & \qquad\quad \| \\
\ \ OH\ OH & & \qquad OH & & \qquad\quad O
\end{array}
\qquad (12)
$$

1920s[43,48,49]. As a theoretical possibility, this mechanism was still mentioned by Kleinfelter and Schleyer in 1961[50].

The first evidence against vinyl dehydration was due to Mislow and Siegel[51]. In aqueous H_2SO_4 the dextrorotatory 1-phenyl-1-*o*-tolylethylene glycol (11) is converted to the optically active 12 (equation 13), which is incompatible with the formation of an enol intermediate.

$$
\begin{array}{cc}
Ph(o\text{-}CH_3C_6H_4)C - CH_2 & \longrightarrow & Ph(o\text{-}CH_3C_6H_4)CHCHO \\
\qquad\quad | \quad\ \ | \\
\qquad\quad OH\ \ OH
\end{array}
\qquad (13)
$$

(+)-(11) (+)-(12)

On the basis of other facts in the relevant literature reports[9,11,28-30], it may be stated that vinyl dehydration does not play a role in the rearrangement of the 1,2-diols in the cases examined.

e. Stereochemistry of the rearrangement. According to an older observation relating to open-chain 1,2-diols[52,53], under identical reaction conditions (in CH_3COOH/I_2 or in CH_3COCl) the *meso* and racemic forms of 1,2-diphenyl-1,2-di-α-naphthylethylene glycol give different products: the higher melting isomer is converted to the phenyl ketone, and the other isomer to the α-naphthyl ketone. Studies with the geometrical isomers of other 1,2-diols[45,54] clearly demonstrated the role of stereochemical factors in the pinacol rearrangement for the open-chain 1,2-diols too. The transformations of various ditertiary diols (13) with similar structures have been investigated, in an attempt to obtain a quantitative correlation

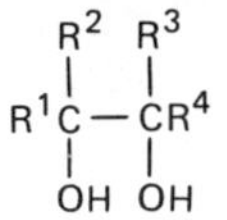

$$
\begin{array}{cc}
R^2\ \ R^3 \\
\ | \quad\ | \\
R^1C - CR^4 \\
\ | \quad\ | \\
OH\ \ OH
\end{array}
\qquad
\begin{array}{l}
R^1,\ R^2,\ R^3 = Me,\ Et \\[4pt]
R^4 = t\text{-}Bu, t\text{-}Am, t\text{-}Hex,\ Et_3C
\end{array}
$$

(13)

for the structure–migrating group interactions[23,24]. The number of carbon atoms in the substituents on the tertiary carbon atoms has an inverse effect on the rate of migration: if the number of carbon atoms on the carbon atom bearing the migrating group is increased, the reaction rate too increases. This may be explained by a relief in the steric strain caused by a change in the hydridization of this carbon atom: $sp^3 \rightarrow sp^2$. If the number of carbon atoms in the substituents on the carbonium ion is increased, the rate of rearrangement is influenced in the inverse way by the hybridization in the opposite direction. The experimental data also showed that another effect too is manifested: this originates from the conformational conditions of the molecule, is independent of the migrating group, and acts against migration. The effect may arise from the substituents of the two tertiary carbon atoms interacting in such a way as to destabilize the conformers favouring migration. Further studies are required, however, for the effect to be given in a quantitative form.

A very large number of publications deal with the stereochemistry of the pinacol rearrangements of alicyclic 1,2-diols with different ring sizes[6,20,25,34,55-66].

In the case of the cyclohexanediols **14** and **15**, even the first investigations[59,66] drew attention to the very characteristic transformations (equations 14 and 15).

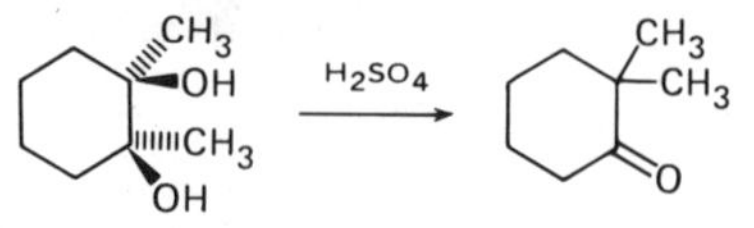

$$(14)$$

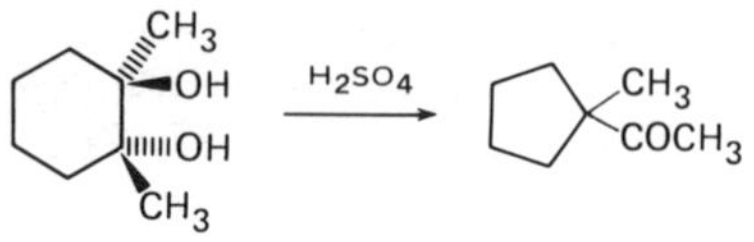

$$(15)$$

Later, different results were obtained[20,60], for it was found that transformation of the isomers led to the formation of the same product mixture[20] (equation 16).

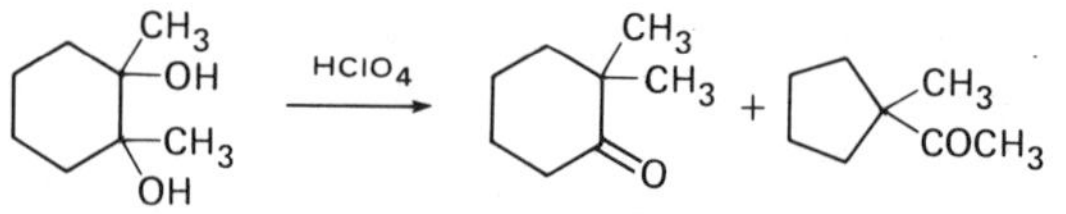

$$(16)$$

(14) or **(15)**   2–9%   91–98%

The stereochemistry of the process was studied in connection with the transformations of the four isomeric 1-phenyl-4-t-butyl-1,2-cyclohexanediols (**16–19**) in the BF_3-Et_2O complex[25]. The reaction always begins with the splitting of the benzyl C–O bond, which leads to the formation of the open carbonium ions **20** and **21**. The original configuration of $C_{(1)}$ no longer plays a role in the further reactions of these ions, the subsequent reactions being determined by the position

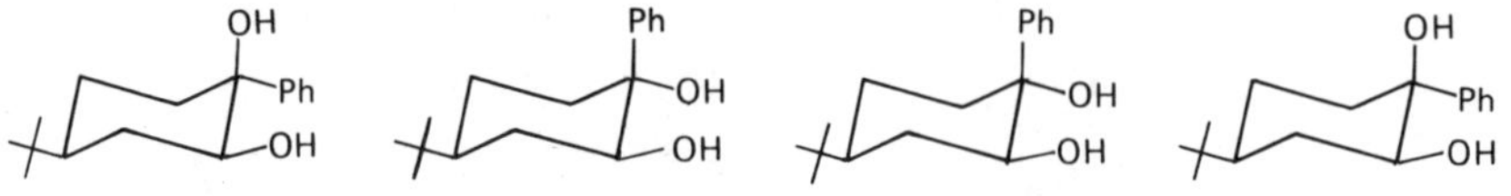

(16)   **(17)**   **(18)**   **(19)**

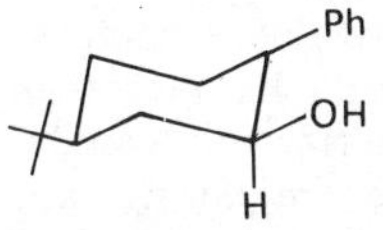

(20) 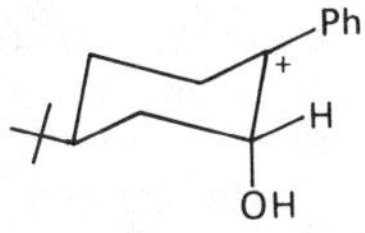 (21)

of the $C_{(2)}$ hydroxy group. In the carbonium ion (20) formed from 16 and 17, the axial hydrogen is readily able to migrate as a hydride anion, and thus the ketone 22 is obtained as product. In the ion 21 which may form from the diols 18 and 19, hydride anion migration is less favoured in the case of the equatorial hydrogen on $C_{(2)}$; here, therefore, in addition to ketone 23, a significant amount of aldehyde 24 is formed.

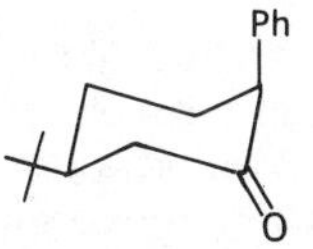

(22) 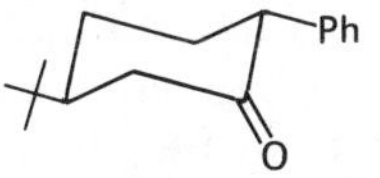 (23) 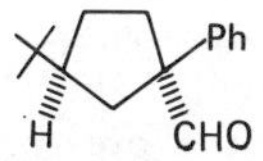 (24)

The study of some isomeric 1,2-cyclobutanediols revealed[34] that the isomers undergo transformation in the same way: in agreement with the stereochemical regularities, a single product, formed by ring contraction, is obtained (equation 17).

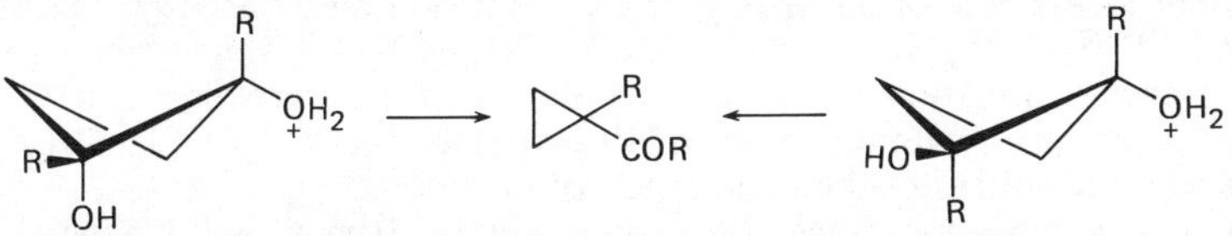

$$(17)$$

An investigation of the stereochemistry of the transformations of the diols 25 and 27 in 25% H_2SO_4 showed[13,14] that the diene 26 is formed as the main product from 25, while the spiroketone 28 is obtained from 27 via pinacol rearrangement (equations 18 and 19). The transformation can be well interpreted in terms of the

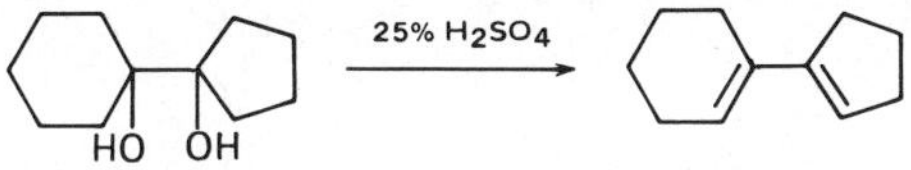

$$25\% \text{ } H_2SO_4 \qquad (18)$$

(25) (26)

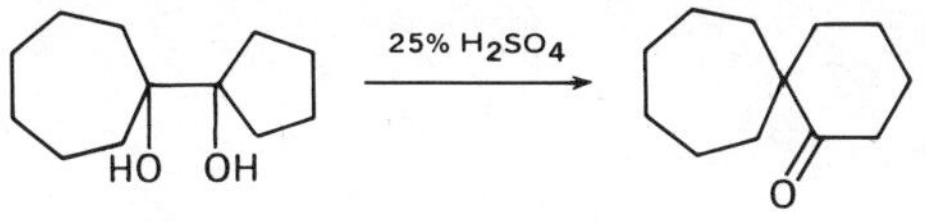

$$25\% \text{ } H_2SO_4 \qquad (19)$$

(27) (28)

role of the stereochemical factors. Otherwise, for compounds containing rings of the same size, spiroketone formation occurs primarily in the case of small (C_4 or C_5) rings[13,14,18,67-72], whereas larger rings are characterized by diene formation[13,14,18,73,74]. This is because, for small rings, ring expansion is a possibility for the relief of the ring strain; this factor is not of importance with larger rings, where diene formation will accordingly predominate[14].

M. Bartók and Á. Molnár

The data relating to the stereochemistry of the rearrangements of cyclic diols are in part contradictory, and numerous factors hamper their interpretation[75]. In many cases, for instance, *cis–trans* isomerization can be observed during the transformations; the products formed are frequently not stable, and undergo interconversion; and different results may be obtained by the application of different reaction conditions. These factors lead to uncertainty in the conclusions drawn.

2. Formation of unsaturated compounds

In certain cases the pinacol rearrangement is accompanied by diene formation. This has been observed with ditertiary glycols, sometimes on the use of reaction conditions differing from the classical ones.

By applying various reagents (HBr, HI, Cl_3COOH, aniline hydrobromide, $FeCl_3$), Kyriakides[76] prepared 2,3-dimethyl-1,3-butadiene from pinacol, in some cases with good yields. The HBr method has also been employed as a preparative procedure[77]. In the presence of phthalic anhydride[78], dienes are formed as well as ketones from 2,3-butanediol and pinacol. Diene formation has been studied in detail in the course of the dehydration of 3,4-dimethyl-3,4-hexanediol on various catalysts[79]. Catalysts of complex composition, containing WO_3, have also been used in the conversion of various 1,2-diols to dienes[80,81].

Dienes can be obtained in good yield from bicyclic, ditertiary diols (e.g. **25** and **27**), mainly in the case of larger (C_6 or C_7) rings. Various concentrations of dilute H_2SO_4[13,14,18,73,82], oxalic acid[18,74] or $(CH_3CO)_2O$[70] are suitable agents for the preferential formation of the diene rather than the spiroketone. With certain compounds, nonproton-donor catalysts [$HCl–(CH_3CO)_2O$, $CH_3COCl–(CH_3CO)_2O$] are similarly well suited for the preparation of dienes[83].

With acid catalysis, *p*-cymene (**30**) may be isolated from (−)-*cis*-pinane-*trans*-2,3-diol (**29**) (equation 20)[27], and isomeric dienes are also formed from **31**[84]. Similarly,

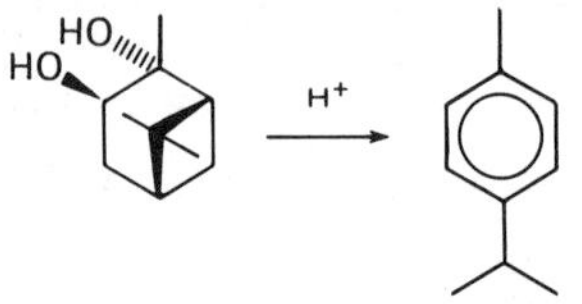

$$(20)$$

(29)   (30)

the transformation of **32** may also lead to aromatic products[22]. The ratio oxo compound/aromatic product depends on R and on the reaction conditions, but it is

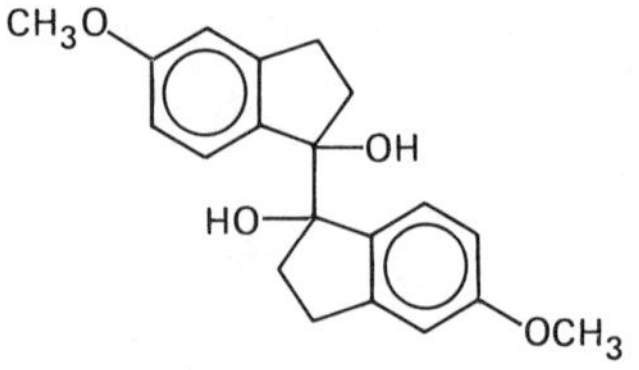

(31) 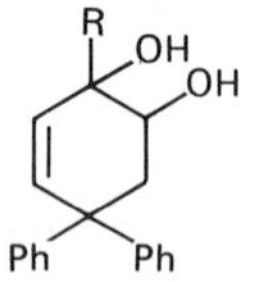   (32)

R = Me, Et, *n*-Pr, Ph

independent of the configuration of the diol. Transformation as in equation (21), leading to diene formation, proceeds at 100°C in the presence of 50% H_2SO_4[85].

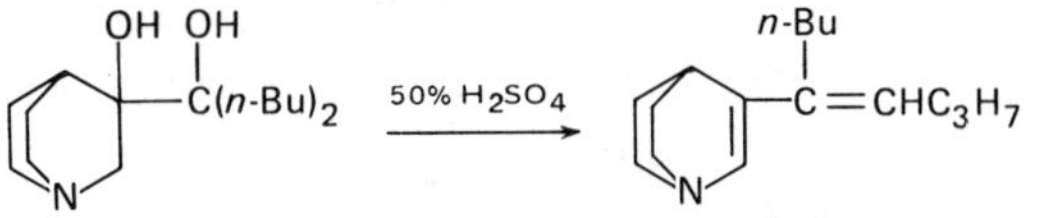

$$(21)$$

B. Transformations on Alumina

In the presence of alumina, the 1,2-diols are transformed to either a carbonyl compound or a diene. Pinacol yields 2,3-dimethyl-1,3-butadiene[77,86], while other publications[79,87-90] have dealt with the possibilities of ketone and diene formation. The diols **33** are converted to carbonyl compounds of type **34** (equation 22)[91], while 1,2-pentanediol yields pentanal[92].

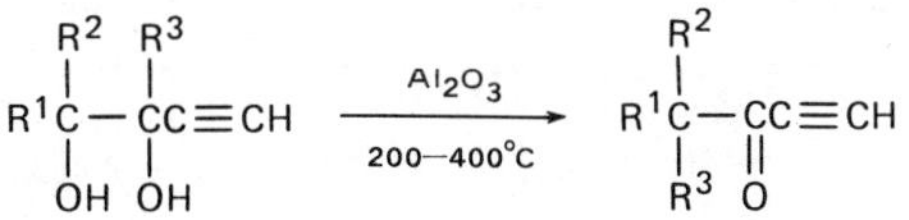

$$(22)$$

(33)   **(34)**

$$R^1, R^2, R^3 = C_1-C_6$$

A series of studies on *cis*- and *trans*-1,2-cyclohexanediol (**35** and **36**)[62,65,88] showed that mainly cyclohexanone (**37**) is formed from the *cis* isomer, and primarily cyclopentaneformaldehyde (**38**) from the *trans* compound, together with the 1,3-cyclohexadiene (**39**) generally accompanying the transformation (equation 23).

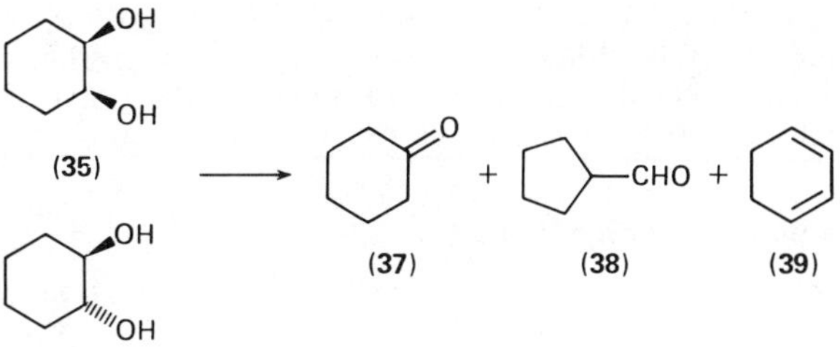

$$(23)$$

The isomeric 1-methyl-1,2-cyclohexanediols[63] gave a similar result. The transformations can be interpreted by analogy to the processes occurring in acidic medium. For the carbonyl product formation from the diols **40**[93], a carbonium cation mechanism was proposed, as shown in equation (24).

The stereochemistry of the heterogeneous catalytic reaction was studied with

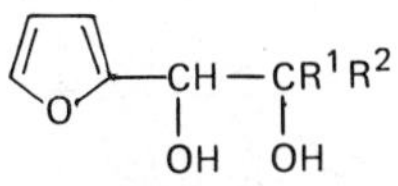

(40)

$$R^1 = H, Me$$

$$R^2 = Et, n\text{-}Pr, i\text{-}Pr,$$

vinyl, isopropenyl

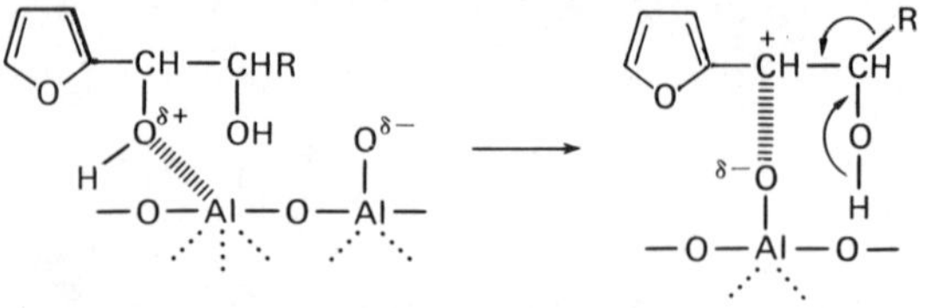

(24)

the *meso* and racemic forms of 2,3-butanediol (41) and 1,2-diphenylethylene glycol (42) in the vapour phase on alumina (equation 25)[94].

$$\underset{\substack{|\quad|\\ OH\quad OH}}{RCH-CHR} \xrightarrow[235-250°C]{Al_2O_3} \underset{\substack{||\\ O}}{RCH_2CR} + R_2CHCHO \qquad (25)$$

(41) R = Me	(*dl*-41)	92%	6%
	(*meso*-41)	84%	11%
(42) R = Ph	(*dl*-42)	77%	23%
	(*meso*-42)	38%	62%

In the case of the methyl- and phenyl-substituted *meso*-diols, the same conformation leads to the formation of both aldehyde and ketone. The formation of the aldehyde in the case of *meso*-42 is explained by the large eclipsing interaction of the phenyl substituents, and by the high migratory aptitude of the phenyl group. In the case of the *meso*-butanediol, the eclipsing effect is much less, while at the same time the migratory aptitude of the hydrogen exceeds that of the methyl group, and hence the ketone may be obtained as main product. For the *dl* isomers, the ketone and aldehyde are formed from different conformations, but there is no essential eclipsing effect in either case. Overall, therefore, the ketone is the main product from the *dl*-diols, in accordance with the fact that the methyl and phenyl groups are to be found in *anti* positions with regard to each other.

C. Dehydration on the Action of Metals

Metal, (especially copper) catalysts, catalyse the conversion of 1,2-diols to carbonyl compounds[95-104]. In the transformation of vinyl-substituted diols on supported copper catalysts, the vinyl group plays an essential role in the development of the carbonyl group[97,98], as its presence enables the diol to bind on the active centres of the catalyst in accordance with equation (26), with the formation of the π-allyl system.

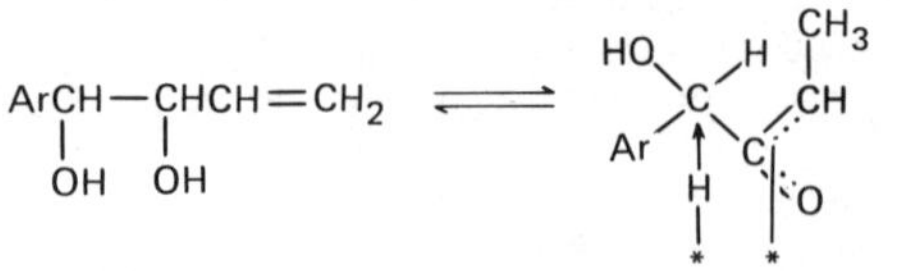

(26)

The product ratios obtained[103] with 2-ethyl-2,3-pentanediol (43) on copper catalysts with various properties and on Pt/C (equation 27) provide a possible

$$\underset{\substack{|\quad|\\ OH\ OH}}{(C_2H_5)_2C-CHCH_3} \longrightarrow \underset{\substack{||\\ O}}{(C_2H_5)_2CHCCH_3} + \underset{\substack{||\\ O}}{C_2H_5CCHCH_3(C_2H_5)} \qquad (27)$$

(43)

explanation of the reaction path of the rearrangement. Depending on the temperature and the catalyst the products are formed either directly by pinacol rearrangement, or via the α-hydroxyketone (44) which is formed by dehydrogenation and which then undergoes rearrangement to the isomeric 45 (equation 28), followed by dehydration and hydrogenation.

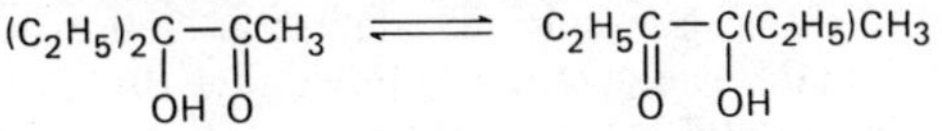

$$(C_2H_5)_2\underset{\underset{OH}{|}}{C}-\underset{\underset{O}{||}}{C}CH_3 \;\rightleftharpoons\; C_2H_5\underset{\underset{O}{||}}{C}-\underset{\underset{OH}{|}}{C}(C_2H_5)CH_3 \qquad (28)$$

(44)   (45)

On Pd/C catalysts[99,100] and in the presence of palladium-containing two-component alloys[101], 1,2-cyclohexanediol gives cyclohexanone as the main product, together with phenol, resulting from aromatization[99,100]. In the presence of Cu/Al and Cu catalysts[104], cis- and trans-1,2-cyclohexanediol yield (in addition to other products not formed by dehydration) cyclohexanone, and smaller amounts of formylcyclopentane, cyclohexanol and 2-cyclohexen-1-one. No essential differences were observed in the distributions of the products from these isomers.

D. Dehydration Under Other Conditions

On SiO_2, Na_2HPO_4 or a mixture of the two as catalysts, 1,2-propanediol gives propionaldehyde in good yield[105]. On phosphate catalysts[89,106,107], the transformation leads mainly to diene formation. On $Ca_3(PO_4)_2$[89] and $AlPO_4$[108], pinacol yields pinacone and 2,3-dimethyl-1,3-butadiene, whereas tetraphenylethylene glycol gives only triphenylmethyl phenyl ketone[108]. 2-Methyl-1,3-butadiene forms from 2-methyl-2,3-butanediol with Li_3PO_4 or Li_2HPO_4 catalysts on various supports[106].

The transformations of pinacol (1) leading to the diene 47 are as outlined in equation (29)[89]. Aluminium silicate, boron phosphate and silica gel primarily

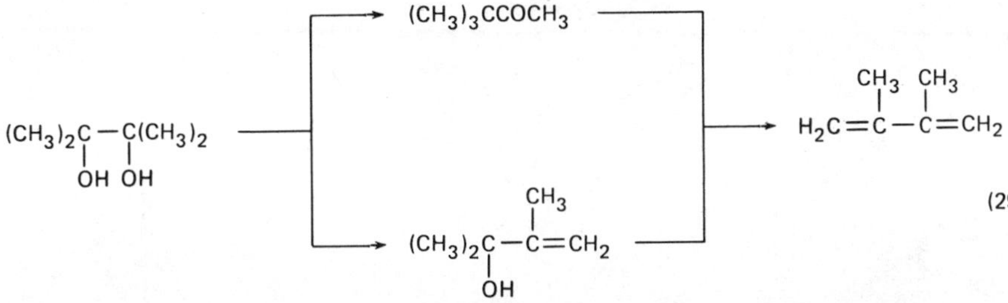

(1)   (46)   (47)

catalyse the rearrangement. On $Ca_3(PO_4)_2$ and Al_2O_3 the main product is the diene, formed via the unsaturated alcohol 46 on the former, and via both routes on the latter, catalyst.

Transformations yielding the carbonyl compound and the diene can also be observed with the use of dimethylsulphoxide[82].

With Friedel–Crafts catalysts (e.g. $AlBr_3$, BBr_3, $TiBr_4$, etc.)[109], 47 is formed in good yield from pinacol, while 1,3-butadiene may be obtained from 2,3-butanediol in the presence of ThO_2[110,111] or under thermal conditions[112].

In the thermal dehydration of ethylene glycol at 700–1000°C, microwave spectroscopy revealed vinyl alcohol, together with a little acetaldehyde and ethylene oxide[113].

The transformation of the derivatives of certain 1,2-diols to carbonyl compounds has also been achieved under photocatalytic conditions[114-116].

E. Intermolecular Water Elimination

In the presence of aluminium silicate catalyst at 200—400°C, ethylene glycol (48) is converted to diethylene glycol[117], but under the given experimental conditions this is not stable and takes part in further reactions (equation 30).

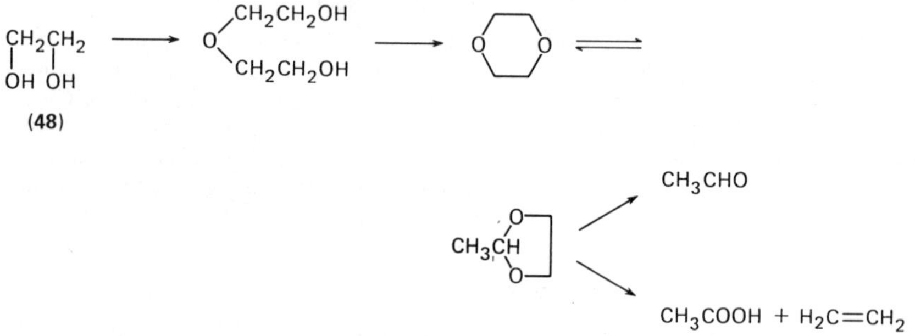

$$\text{(30)}$$

On the action of ion-exchange resins, the corresponding 1,4-dioxans (2,5-dimethyl- and 2,3,5,6-tetramethyl-1,4-dioxan, respectively) are formed from 1,2-propanediol and from 2,3-butanediol[118].

II. DEHYDRATION OF 1,3-DIOLS

A survey of the literature data relating to the transformations of the 1,3-diols indicates that the dehydration routes shown in equation (31) are characteristic.

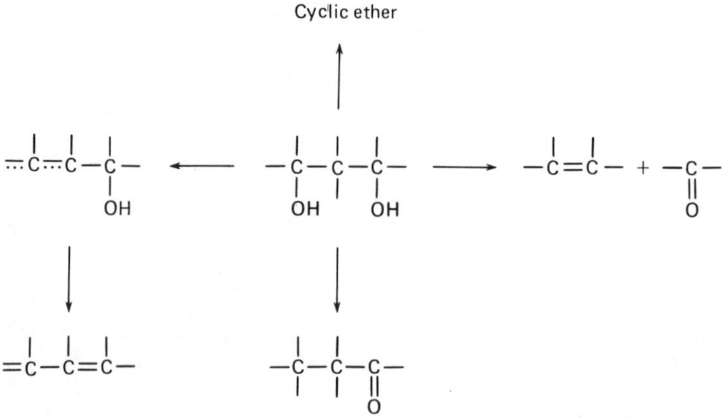

$$\text{(31)}$$

To illustrate these main transformations, some (mainly recent) literature data are listed in Table 1. The tendency in the case of unsubstituted or slightly substituted compounds (diprimary, primary—secondary, disecondary diols) is mainly the formation of carbonyl compounds; with the increase of the number of substituents, the formation of unsaturated alcohols and dienes or dehydration accompanied by fragmentation assume ever greater importance. However, the direction of the reaction depends on the catalyst too.

A. Formation of Carbonyl Compounds with the Same Number of Carbon Atoms as the Starting Diol

This process occurs for diprimary and primary — secondary diols on the action of H_2SO_4 or metal catalysts (primarily various copper catalysts). However, the selectivity of the reaction decreases as the number of substituents rises.

1. Dehydration on the action of sulphuric acid

The transformations of simple aliphatic 1,3-diols were studied at the beginning of this century[119-125] and the formation of carbonyl compounds was described. In the course of the dehydration of 2,2-disubstituted 1,3-propanediols and 1,3-butanediols, the reaction leading to the corresponding aldehyde is accompanied by the migration of the substituents on $C_{(2)}$[126-130]. A carbonium cation mechanism has been assumed (equation 32)[129]. In the case of the 2,2-disubstituted 1,3-butanediols, tetrahydrofuran derivatives are formed in parallel with the carbonyl compounds (see Section II.D.2).

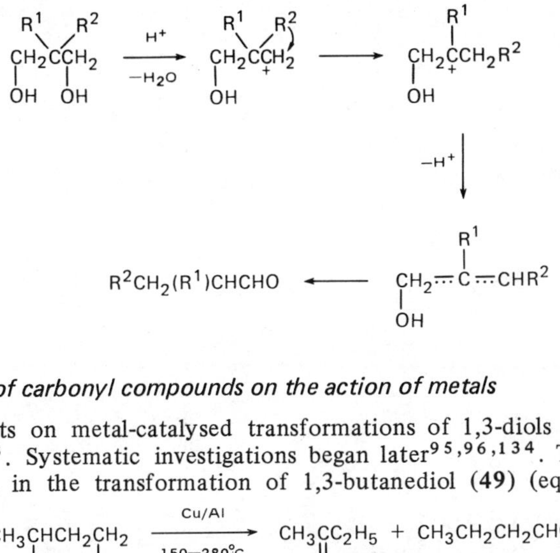

$$\begin{array}{c}
\underset{\substack{| \quad | \\ \text{OH} \quad \text{OH}}}{\overset{\substack{R^1 \quad R^2 \\ \diagdown \diagup}}{CH_2CCH_2}} \xrightarrow[-H_2O]{H^+} \underset{\substack{| \\ \text{OH}}}{\overset{\substack{R^1 \quad R^2 \\ \diagdown \diagup}}{CH_2CCH_2^+}} \longrightarrow \underset{\substack{| \\ \text{OH}}}{\overset{\substack{R^1 \\ |}}{CH_2CCH_2R^2}}
\end{array}$$

$$\downarrow{-H^+}$$ (32)

$$R^2CH_2(R^1)CHCHO \longleftarrow \underset{\substack{| \\ \text{OH}}}{\overset{\substack{R^1 \\ |}}{CH_2\!=\!C\!=\!CHR^2}}$$

2. Formation of carbonyl compounds on the action of metals

Initial results on metal-catalysed transformations of 1,3-diols are reported in patents[131-133]. Systematic investigations began later[95,96,134]. The role of the catalyst Cu/Al in the transformation of 1,3-butanediol (49) (equation 33) was

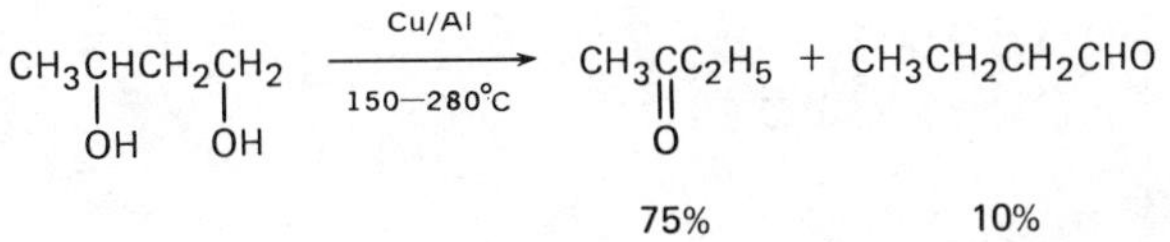

$$\underset{\substack{| \quad | \\ \text{OH} \quad \text{OH}}}{CH_3CHCH_2CH_2} \xrightarrow[150-280^\circ C]{Cu/Al} \underset{\substack{|| \\ O}}{CH_3CC_2H_5} + CH_3CH_2CH_2CHO \qquad (33)$$

$$ 75\% \qquad\qquad 10\%$$

(49)

studied[135], and results were later reported on the transformations of 1,3-propanediol[136] and 1,3-butanediol[137] on various Raney-type catalysts, and on the effects of different supported copper catalysts[138]. It subsequently became possible to generalize the observations on the basis of extensive studies with different types of open-chain[103,139-141] and alicyclic[104,142] diols. It was found that on various copper catalysts certain types of 1,3-diols are converted to carbonyl compounds, and the ditertiary diols to an unsaturated alcohol and a diene[140,143] (see Section II.B.2).

The comprehensive study of dehydration to the carbonyl compound (among others by the use of deuterium-labelled compounds) provides a possibility for the

TABLE 1. Selectivity of dehydration transformations of 1,3-diols

Compound	Catalyst	Selectivity (%)[a]				References
		Oxo compounds	Unsaturated alcohols + dienes	Fragmentation	Cyclic ethers	
HOCH₂CH₂CH₂OH	H₂SO₄	?				122
	Ca₃(PO₄)₂, 275°C	20	48	32		179
	Al₂O₃, 275°C	34	42	24		
	Cu/Al, Pd/Al, Ni/Al, Co/Al, Zn/Al	max. 75		min. 25		136
HOCH₂CH(i-Pr)CH₂OH	FSO₃H/SbF₅/SO₂	100				16
HOCH₂CH(n-Bu)CH₂OH	Cu/Al	60	20	40		140,143
HOCH₂C(CH₃)₂CH₂OH	Ca₃(PO₄)₂			80		179
HOCH₂CR¹R²CH₂OH (R¹, R² = alkyl, benzyl, phenyl)	20%H₂SO₄	98				127
	10 N HCl	70–100				128
	30% H₂SO₄	100				129,130
HOCH₂C(C₂H₅)₂CH₂OH	Ca₃(PO₄)₂, 300°C		18	71	11	179
HOCH₂CH(4-pyridyl)CH₂OH	HCl/HCHO				90	215
CH₃CH(OH)CH₂CH₂OH	Benzenesulphonic acid		47			78
	p-TsOH	26	> 23			167
	(CH₃)₂SO	31	54			82
	Ca₃(PO₄)₂, 300°C	40	39	21		179
	γ-Al₂O₃	50		50	9	137
	Ni, Cu	85				95,96
	Cu/Al					135
	Pd/Al, Co/Al, Ni/Al, Zn/Al, Cu/Zn/Al	max. 55		5–65		137
RCH(OH)CH₂CH₂OH (R = alkyl)	PdCl₂/CuCl₂	33				150
	RhCl₃/PPh₃	33–89				145
PhCH(OH)CH₂CH₂OH	Cu/Al	60		30		140,143
1-Hydroxymethylcyclohexanol-2	Cu/Al	90		10		142,143
PhCH(OH)CHPhCH₂OH	28% H₂SO₄			71		205

Substrate	Catalyst	Selectivity (%)				Ref.
$RCH(OH)C(CH_3)_2CH_2OH$ (R = alkyl)	20% H_2SO_4	19–69		1–8	18–80	208
	Triethylphosphate	52			32	223
	Wofatit KPS-200 ion exchange resin	27			34	225
$PhCH(OH)C(CH_3)_2CH_2OH$	20% H_2SO_4		8	100		208
	H_2SO_4					160
$(CH_3)_2C(OH)CH_2CH_2OH$	$NaHSO_4/Na_2SO_4$		85			175
	$Ca_3(PO_4)_2$		85			140,143
	Cu/Al	15	100			168
$CH_3CH(OH)CH_2CH(OH)CH_3$	p-TsOH		70	75		16
	$Ca_3(PO_4)_2$, 300°C		55	22		143
	$FSO_3H/SbF_5/SO_2$		100	45		145
	Cu/Al	85		15		206
	$RhCl_3/PPh_3$	97				224
$PhCH(OH)CH_2CH(OH)Ph$	$KHSO_4$		14	35	56	95,96
$i\text{-}PrCH(OH)CH_2CH(OH)CH_3$	p-TsOH			1		104
1,3-Cyclohexanediol	Cu, Ni	95				169
	Cu/Al, Cu	100				
5,5-Dimethyl-1,3-cyclohexanediol	Phthalic anhydride		84			185
$(CH_3)_2C(OH)CH_2CH(OH)CH_3$	H_2SO_4		50			82
	$(CH_3)_2SO$		74			181
	$Ca_3(PO_4)_2$, 275°C		8	92		184
	Al_2O_3, 190°C		77	23		183
	$CuSO_4$		98.5	1		146
	$FeCl_3 \cdot 6H_2O$		92			140
$Ph_2C(OH)CH(CH_3)CH(OH)Ph$	$RhCl_3/PPh_3$	85		75		207
	Cu/Al	20				146
$(CH_3)_2C(OH)CH_2(CH_3)_2COH$	H_2SO_4			65–89		140,143
	$RhCl_3/PPh_3$					199
	Cu/Al					16
$(CH_3)_2CC(OH)CH_2C(CH_3)_2$ (OH, OH)	H_2SO_4		100	100		
	$FSO_3H/SbF_5/SO_2$		100	100		
$R_2C(OH)CH_2CR_2OH$ (R = alkyl, phenyl)	$KHSO_4, H_2SO_4$			15–97		204

[a] Selectivity is based on 100 mole of the compound reacted.

elucidation of the mechanism[143,144]. As seen in the example of 1,3-butanediol-[3-^{2}H] (50), the process consists of three steps (equation 34).

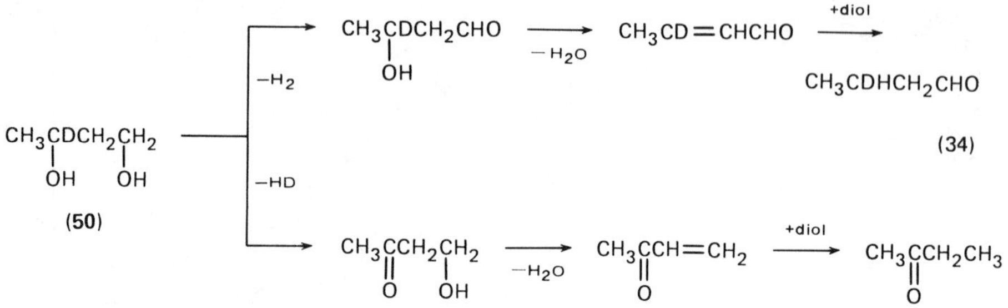

$$
\begin{array}{c}
CH_3CDCH_2CH_2 \\
\;\;\;|\quad\;\; | \\
OH\;\;\; OH \\
(50)
\end{array}
\begin{cases}
\xrightarrow{-H_2} \quad
\begin{array}{c}
CH_3CDCH_2CHO \\
\;\;\;| \\
OH
\end{array}
\xrightarrow{-H_2O} \quad CH_3CD{=}CHCHO \xrightarrow{+diol} CH_3CDHCH_2CHO \\[2mm]
\xrightarrow{-HD} \quad
\begin{array}{c}
CH_3CCH_2CH_2 \\
\;\;\;||\quad\;\; | \\
O\;\;\; OH
\end{array}
\xrightarrow{-H_2O} \;
\begin{array}{c}
CH_3CCH{=}CH_2 \\
\;\;\;|| \\
O
\end{array}
\xrightarrow{+diol}
\begin{array}{c}
CH_3CCH_2CH_3 \\
\;\;\;|| \\
O
\end{array}
\end{cases}
\tag{34}
$$

3. Formation of carbonyl compounds on the action of other agents

The formation of carbonyl compounds similarly proceeds in the presence of homogeneous catalysts, such as $RhCl_3/PPh_3$[145,146]. Apart from primary — secondary diols, disecondary and even secondary — tertiary diols[146] are transformed to the corresponding ketones (equation 35). On the action of $RhCl_3$ and chiral phosphines[147], it is also possible to achieve the enantioselective dehydration of 1,3-butanediol.

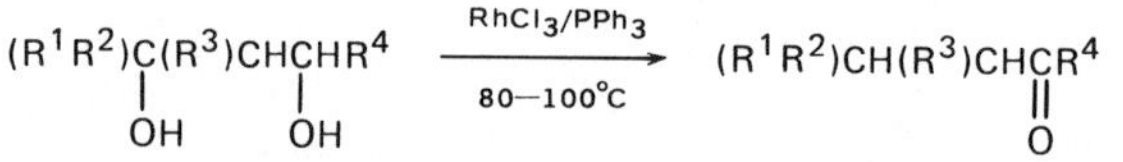

$$
(R^1R^2)C(R^3)CHCHR^4 \xrightarrow[\text{80—100°C}]{RhCl_3/PPh_3} (R^1R^2)CH(R^3)CHCR^4
\tag{35}
$$

with OH and OH on the left and O on the right carbonyl.

With other agents too, carbonyl compounds[16] and unsaturated carbonyl compounds[82,148-150] may be formed.

B. Formation of Unsaturated Alcohols and Dienes

The formation of these compounds via 1,2-elimination is a characteristic reaction of polysubstituted diols on the action of various organic and inorganic acids and organic acid anhydrides. The diols may be induced to undergo similar processes by other reactants too [$(CH_3)_2SO$, bromine, iodine, Al_2O_3]. Investigations in connection with the preparation of 1,3-butadiene and isoprene are of importance from an industrial point of view.

1. Dehydration on the action of acids

Sulphuric acid[151-161] and hydrogen bromide[75,162,163] are used most frequently with 1,3-diols. References may also be found to molybdic acid[164], HCl/CH_3COOH[165], organic acids (oxalic[157,166], benzenesulphonic[78], p-toluenesulphonic[167,168]), and acid anhydrides (phthalic[78,169], acetic[170,171]). Diene is likewise formed in $FSO_3H/SbF_5/SO_2$[16]. Similar agents may be employed for the dehydration of alicyclic compounds[169,172,173].

It is worthwhile to emphasize the results connected with 3-methyl-1,3-butanediol (51)[152-156,160,161,174-176]. Kinetic studies[157-159,177] have led to the scheme shown in equation (36). The fastest process in the system is the isomerization of dimethylallyl alcohol (52) to dimethylvinyl carbinol (53) and isoprene (54) and the side-products are formed from the equilibrium mixture of these two compounds.

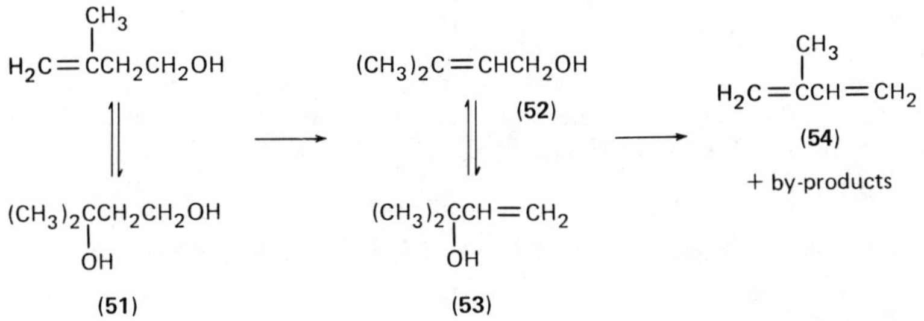

CH3
H2C=CCH2CH2OH
(CH3)2C=CHCH2OH
(52)
CH3
H2C=CCH=CH2
(54)
+ by-products
(CH3)2CCH2CH2OH
OH
(51)
(CH3)2CCH=CH2
OH
(53)

(36)

The dehydration of some 1,3-diols (1,3-propanediol[178,179], 1,3-butanediol[179], 2-methyl-1,3-butanediol[180], 2,2-diethyl-1,3-propanediol[179], 2-*n*-butyl-1,3-propanediol[179], 2,4-pentanediol[181] and 2-methyl-2,4-pentanediol[181]) have been studied on Al_2O_3 and $Ca_3(PO_4)_2$ catalysts, using heterogeneous catalysts with an acidic character. Equation (37) shows the reactions observed, and shows three dehydration

—CCH2CH—
O OH
—CHCH2CH—
OH OH
−H2O
—CH2CH2C—
O
−H2O
−H2O
—CCH3 + HC—
O O
—CH=CHCH—
OH
—CH2CH2CH—
OH
O
−H2O
−H2O
—CHCH3 + CH2—
OH OH
—CH=CHC—
O
—CH=CHCH=

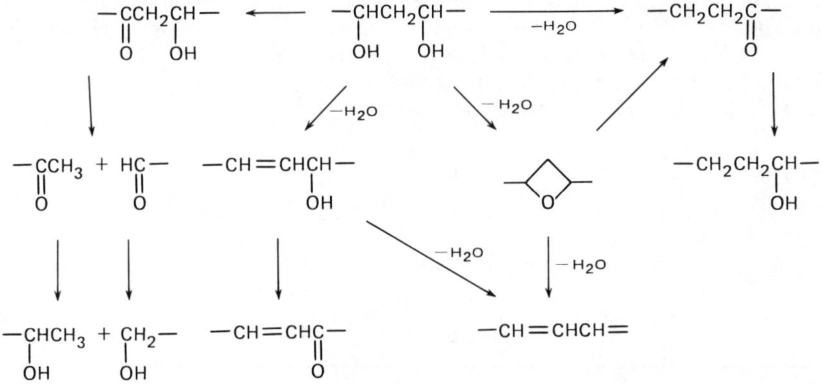

(37)

routes, and various types of fragmentation and hydrogen-transfer processes. The selectivity and the transformation mechanism[182] depend on both the reaction conditions (temperature, space-velocity) and the substrate. Dehydration accompanied by diene formation proceeds with high stereoselectivity[181].

2. *Dehydration on the action of other agents*

Unsaturated compounds can also be obtained on the action of various metal salts ($FeCl_3$, $CuSO_4$, Na_2SO_4)[170,183,184]. Dienes are obtained in good yield from 2-methyl-2,4-pentanediol on the action of iodine[183-185]. Dienes and an unsaturated alcohol are similarly formed in the presence of $(CH_3)_2SO$ (equation 38)[82].

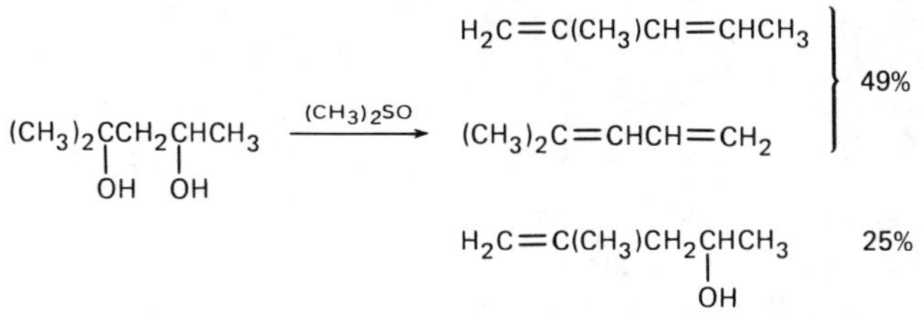

H2C=C(CH3)CH=CHCH3
49%
(CH3)2CCH2CHCH3
OH OH
(CH3)2SO
(CH3)2C=CHCH=CH2
H2C=C(CH3)CH2CHCH3
OH
25%

(38)

Bromine[186] and Al_2O_3[187] convert 1,3-cyclohexanediol to an unsaturated alcohol, while zeolite of NaX-type[188] gives dienes.

In the case of ditertiary diols, which undergo fragmentation on exposure to acids (see Section II.C), unsaturated alcohols and dienes are formed (equation 39) on the action of metal catalysts (Cu/Al, Cu, Pt/C)[140,143] or $RhCl_3$/PPh_3[146].

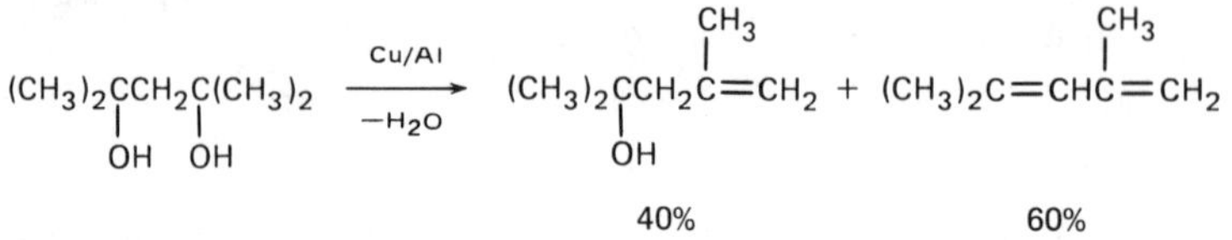

$$(CH_3)_2\underset{\underset{OH}{|}}{C}CH_2\underset{\underset{OH}{|}}{C}(CH_3)_2 \xrightarrow[-H_2O]{Cu/Al} (CH_3)_2\underset{\underset{OH}{|}}{C}CH_2\overset{\overset{CH_3}{|}}{C}=CH_2 + (CH_3)_2C=CH\overset{\overset{CH_3}{|}}{C}=CH_2 \quad (39)$$

$$40\% \qquad\qquad\qquad 60\%$$

3. Preparation of butadiene

The preparation of butadiene is dealt with in a great number of publications, mainly patents. These describe reaction conditions and catalysts which tend to favour diene formation, such as supported and support-free acidic and neutral phosphates[175,189-192], heterogeneous catalysts containing phosphoric acid[193,194], and other complex heterogeneous catalysts[195-197]. Butadiene may be obtained in a yield better than 90% with a catalyst of involved composition (carborundum − Al − Mg containing SiO_2 and WO_3)[81].

C. Dehydration Accompanied by Fragmentation

With ditertiary 1,3-diols on the action of H_2SO_4[198-201], $KHSO_4$[202-204] and FSO_3H/SbF_5/SO_2[16], water is eliminated, fragmentation occurs, and a carbonyl compound and an olefin are formed (equation 40). The process can also be observed

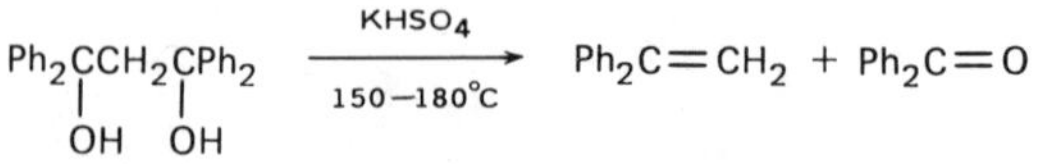

$$Ph_2\underset{\underset{OH}{|}}{C}CH_2\underset{\underset{OH}{|}}{C}Ph_2 \xrightarrow[150-180°C]{KHSO_4} Ph_2C=CH_2 + Ph_2C=O \quad (40)$$

for diols of lower order if the molecule contains a phenyl substituent[205-208], or if a substituent is found on the carbon atom enclosed between the carbon atoms bearing the hydroxy groups[200,205,207,209]. Similar findings hold for the cyclic diols too.

Studies with open-chain diols[200,203,204,206,207] indicate that the transformations take place via a concerted mechanism[200,206,207].

For asymmetrically-substituted compounds (e.g. 55), the possibility exists for two reactions, leading to different products (equation 41), and the direction depends

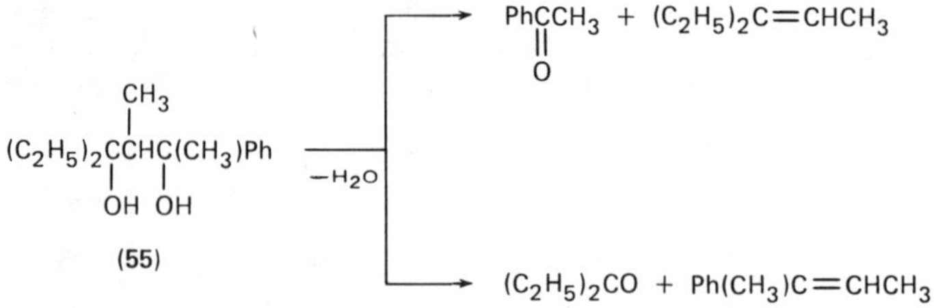

$$(C_2H_5)_2\underset{\underset{OH}{|}}{C}\overset{\overset{CH_3}{|}}{C}H\underset{\underset{OH}{|}}{C}(CH_3)Ph \xrightarrow{-H_2O} \begin{cases} Ph\underset{\underset{O}{||}}{C}CH_3 + (C_2H_5)_2C=CHCH_3 \\[2em] (C_2H_5)_2CO + Ph(CH_3)C=CHCH_3 \end{cases} \quad (41)$$

$$(55)$$

on the stereostructure of the diol[200,210]. In the case of the α- and β-isomers, 56 and 57, respectively, the conformations favour *trans* elimination.

An interesting example of the correlation between the structure and the reaction

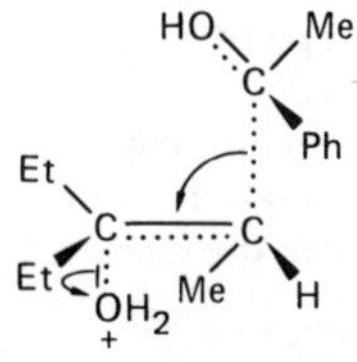

(56) 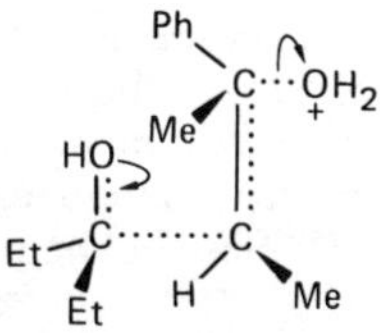 (57)

direction[208] is the difference between the transformations of 2,2-dimethyl-1,3-butanediol and 1-phenyl-2,2-dimethyl-1,3-propanediol: fragmentation occurs only for the latter compound.

With cyclic 1,3-diols of type 58[210], ring-splitting is the predominant process for the *cis*-diol, whereas ring-splitting and fragmentation of the side-chain are comparable for the *trans* derivative. With the *cis* compound, *trans* elimination occurs in the case of the formation of the ring-splitting product, (59), while with the *trans* compound both splitting products, 60 and 61, may be obtained by *trans* elimination.

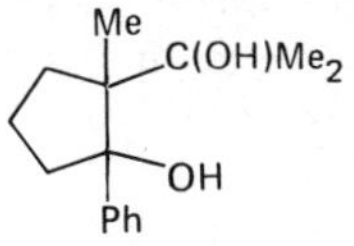

(58) 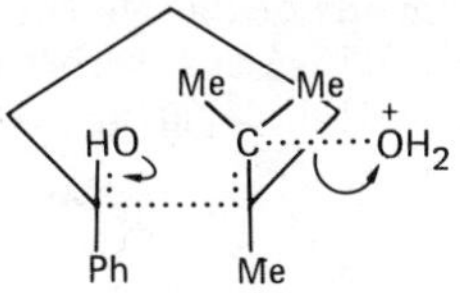 (59)

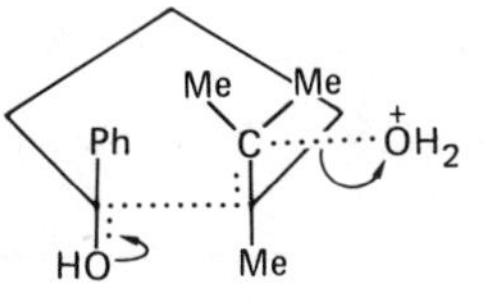

(60) 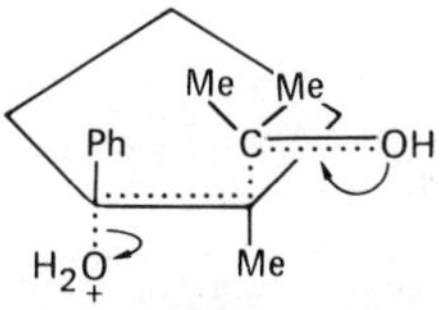 (61)

A similar interesting ring-splitting has been described for 2,2,4,4-tetramethyl-1,3-cyclobutanediol (62)[211] (equation 42). The *trans*-diol is converted to the un-

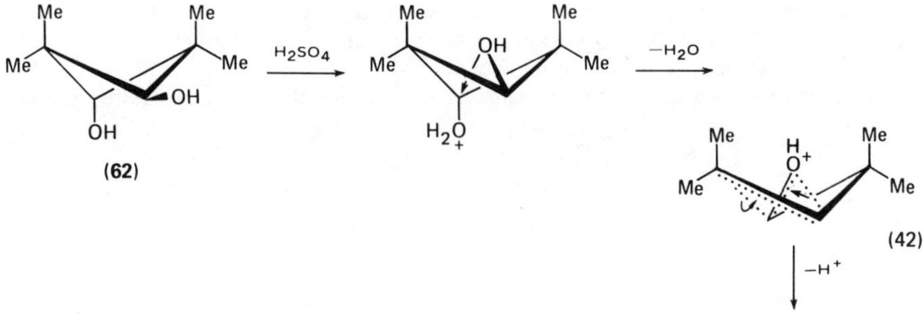

(62)

(42)

$$(CH_3)_2C=CH(CH_3)_2CCHO$$

(63)

saturated aldehyde 63 via the indicated intermediate, while the *cis* isomer can be recovered unchanged from the reaction mixture.

Similar processes accompanied by ring-opening or fragmentation of the side-chain can be observed for other cyclic 1,3-diols[212].

Stereochemically interesting processes have been observed for the transformation of 2,2,5,5-tetramethyl-1,3-cyclohexanediol (64) in the presence of $KHSO_4$ (equation 43)[213,214]. Besides the product formed by rearrangement (65), a ketone

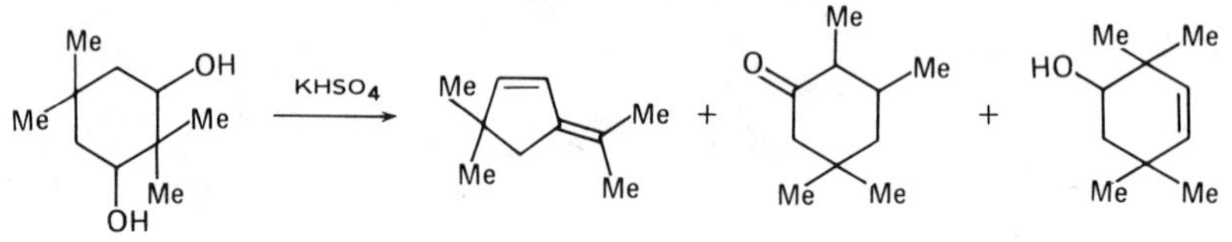

	(64)	(65)	(66)	(67)
cis		88%		12%
trans		53%	30%	17%

66 and an unsaturated alcohol 67 were detected. The formation of 65 is justified by the steric arrangements of the system. Additionally, in the case of the *trans*-diol, the formation of 66 or 67 is possible, because the methyl group and hydrogen are in *trans* axial positions to the departing axial hydroxy group (equation 44). For the

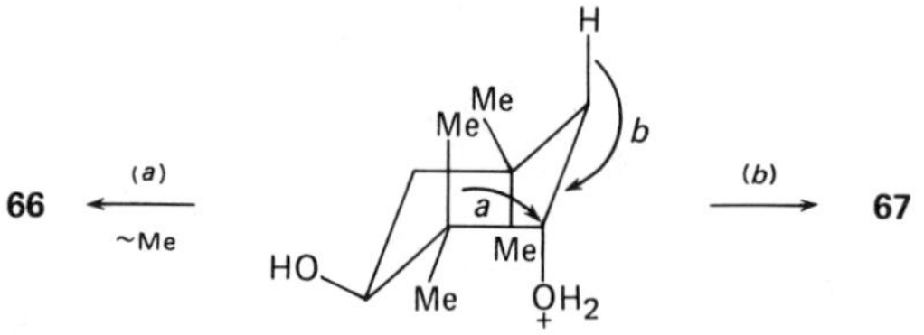

cis-diol, these latter transformations are not favoured as the equatorial hydroxy groups do not give rise to the above steric situation.

D. Formation of Cyclic Ethers

1. Formation of oxetanes

Oxetanes cannot usually be prepared from 1,3-diols by direct water elimination. Nevertheless, the process can be carried out with excellent yield from 68 in the presence of HCl and HCHO (equation 45)[215], and oxetanes were also obtained from some acetylene-1,3-diols (equation 46)[216].

In addition to the above, the presence of 2-methyloxetane and 3,3-diethyloxetane

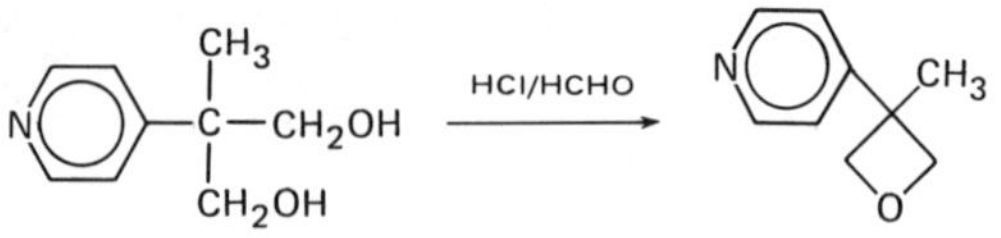

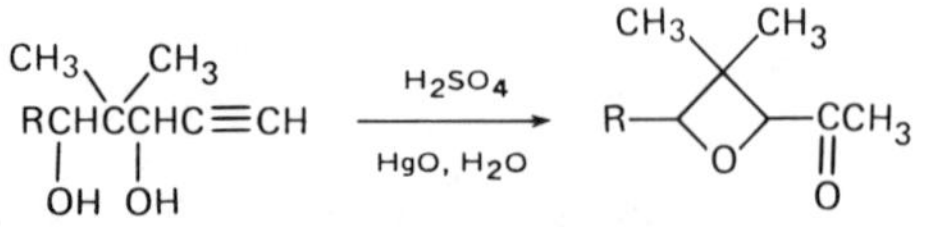

R = H, CH_3

was demonstrated in the dehydrations of 1,3-butanediol and 2,2-diethyl-1,3-propanediol on $Ca_3(PO_4)_2$ and Al_2O_3[179]. The other literature data[217-220] proved that the procedure is not reproducible.

2. Formation of oxolanes

Various 1,3-diols are dehydrated to yield oxolanes on the action of H_2SO_4[121,208,221], H_3PO_4[222], Et_3PO_4[222,223], p-toluenesulphonic acid[224] and ion-exchange resin[225]. In certain cases, isomeric aldehydes are formed together with the oxolanes (equation 47). Both products indicated result from rearrangement

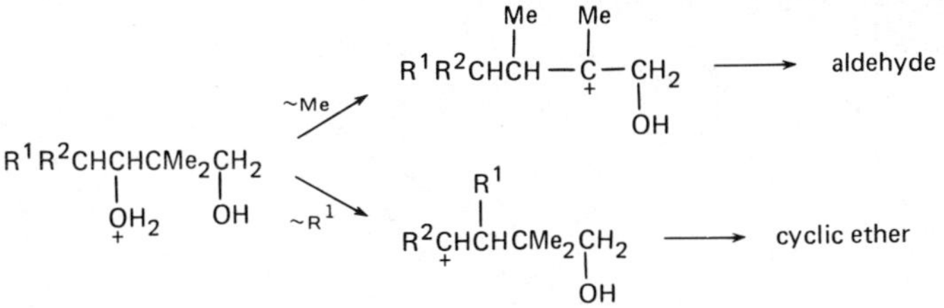

$$
\begin{array}{c}
& & \overset{Me}{\underset{|}{}} \quad \overset{Me}{\underset{|}{}} \\
& R^1R^2CHCH - \overset{}{\underset{+}{C}} - CH_2 \longrightarrow \text{aldehyde} \\
& \overset{\sim Me}{\nearrow} \qquad \overset{|}{OH} \\
R^1R^2CHCHCMe_2CH_2 & & \\
\overset{|}{\underset{+}{OH_2}} \quad \overset{|}{OH} & \overset{R^1}{\underset{|}{}} \\
\overset{\sim R^1}{\searrow} \quad R^2CHCHCMe_2CH_2 \longrightarrow \text{cyclic ether} \\
& \overset{+}{\underset{|}{}} \qquad \overset{|}{OH}
\end{array}
$$

(47)

(alkyl migration). The driving force of the process is the possibility of formation of more stable carbonium ions.

III. DEHYDRATION OF HIGHER DIOL HOMOLOGUES

The main reactions of this group of diols are summarized in equation (48). The most characteristic process is the transformation to the cyclic ether. Table 2 shows

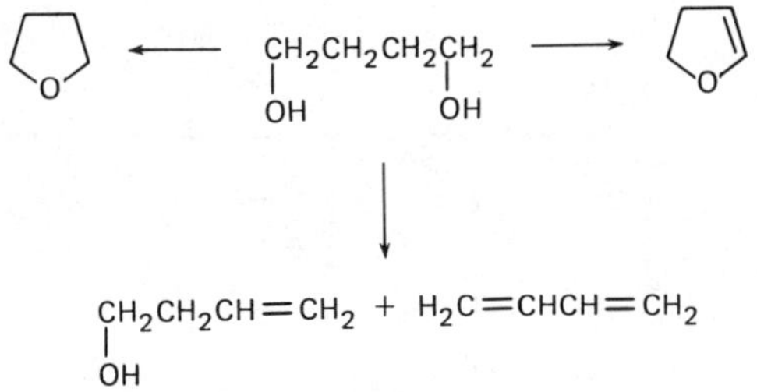

$$
\text{(oxolane)} \longleftarrow \underset{OH}{\overset{|}{CH_2}}CH_2CH_2\underset{OH}{\overset{|}{CH_2}} \longrightarrow \text{(furan)}
$$

$$
\downarrow
$$

(48)

$$
\underset{OH}{\overset{|}{CH_2}}CH_2CH=CH_2 + H_2C=CHCH=CH_2
$$

that the process is almost independent of the structure and degree of substitution of the starting diol.

The formation of unsaturated cyclic ethers can be observed on metal catalysts. Unsaturated alcohols and dienes are formed from primary — tertiary and ditertiary diols on the action of various acids, and the same processes are also induced by various oxides and by $Ca_3(PO_4)_2$ at high temperature, independently of the structures of the diols. Other special transformations may be observed also, but these only occur for individual diols.

A. Preparation of Oxacycloalkanes

1. Cyclodehydration on the action of various agents

The most general means of carrying our cyclodehydration is to perform the transformation in the presence of mineral acids (H_2SO_4[120,226-249], HCl[249-255],

TABLE 2. Selectivity of dehydration transformations of higher diol homologues

Compound	Catalyst	Selectivity (%)[a]			References
		Cyclic ethers	Unsaturated cyclic ethers	Unsaturated alcohols and dienes	
$HOCH_2CH_2CH_2OH$	80% H_3PO_4	100			249
	p-TsOH	90			
	Cr oxide	100			
	Na phosphate +			95–98	291
	Bu phosphate + graphite				
	$Ca_3(PO_4)_2$, 320°C	99			296
	426°C	27		47	249
	Morden bentonite	100			150
	Ion-exchange resin	98			303
	$PdCl_2/CuCl_2$	95			305
	$PdCl_2/CuCl_2/Cu(NO_3)_2$	98			330
	Ni/kieselgur	56			318
	Cu/Al	100			299
	Cu or Co/kieselgur		81		281
$HOCH_2CH=CHCH_2OH$	Pyridine · HCl	73			300
$HOCH_2CHMeCHMeCH_2OH$	Aluminium silicate	78			
1,2-Bis(hydroxymethyl)cyclohexane	H_2SO_4	94			
$HOCH_2(R)CH(CH_2)_2CH_2OH$ (R = alkyl)	Aluminium silicate	69–76			
$HOCH_2(CH_2)_4CH_2OH$	57% H_2SO_4	100			325
$CH_3CH(OH)CH_2CH_2CH_2OH$	Cu/Al, Pd/Al	80–90			304
$ArCH(OH)CH_2CH_2CH_2OH$ (Ar = Ph, p-$CH_3C_6H_4$)	50% H_3PO_4	70–75			259
$ArCH(OH)CHRCHRCH_2OH$ (R = H, CH_3; Ar = p-$CH_3C_6H_4$, Ph, p-$CH_3OC_6H_4$, 2-furyl)	p-TsOH	70–95			275
$RCHCH_2CH_2CH_2OH$ | OH (R = alkyl, Ph, thienyl, cyclohexyl)	60% H_2SO_4	57–72			242

Diol	Catalyst / Conditions				Ref.
$CH_3CH(OH)(CH_2)_3CH_2OH$	Cu				305
$CH_3CHRCH(CH_2)_2CH_2OH$, $\mid OH$ (R = alkyl)	Aluminium silicate	72–74	25		301
$CH_3(C_2H_5)C(OH)(CH_2)_2CH_2OH$	$(CH_3)_2SO$	67			310
$Ph(CH_3)C(OH)(CH_2)_2CH_2OH$	H_2SO_4	22–80			335
$(CH_3)_2C(CH_2)_3CH_2$, $\mid OH$ $\mid OH$	H_3PO_4	80			260
$(CH_3)_2C(CH_2)_3CHCH_2$, $\mid OH$ $\mid CH_3$ $\mid OH$	$Ca_3(PO_4)_2$			90	292
$CH_3CHCH_2CH_2CHCH_3$, $\mid OH$ $\mid OH$	Al_2O_3, 250°C	70		10	290
	400°C			80	
	$Ca_3(PO_4)_2$, 325°C	63.5		20	
	400°C	30		70	
	Aluminium silicate 200°C	70		15	
	350°C	5		50	
	Al_2O_3	35		25	248
	$PdCl_2/CuCl_2$	67			150
	$PdCl_2/CuCl_2/Cu(NO_3)_2$	85–95			
	$PdCl_2/Cu(NO_3)_2$	97			
	$PdCl_2/NaCl$	94			
	$(CH_3)_2SO$	68			309
	Cu/Al, Pd/Al	80–90			304
	Cu/Al	96			305
	Cu	29	55		
	Cu/SiO_2	92	19		
	Cu/Al				305
$CH_3CHCH_2CH_2CH_2CHCH_3$, $\mid OH$	Cu	3	61		
	Cu/SiO_2	4	73		
	p-TsOH	70–95			
$ArCH(OH)(CH_2)_2CH(OH)CH_3$ (Ar = Ph, 2-furyl)	p-TsOH	74		8	275
$PhCH(CH_2)_2CHPh$, $\mid OH$ $\mid OH$	p-TsOH/benzene	20		>50	311
	85% H_3PO_4			90	

TABLE 2 (continued)

Compound	Catalyst	Selectivity (%)[a]			References
		Cyclic ethers	Unsaturated cyclic ethers	Unsaturated alcohols and dienes	
PhCH(OH)(CH$_2$)$_3$CH(OH)Ph	p-TsOH	45		25	311
(CH$_3$)$_2$CCH$_2$CH$_2$CHCH$_3$ (OH, OH)	p-TsOH	60			224
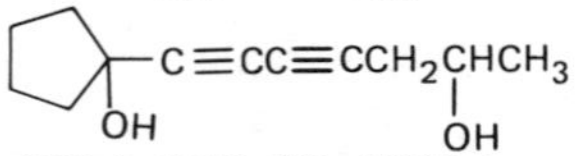 C≡CC≡CCH$_2$CHCH$_3$ (OH); cyclopentane–OH	KHSO$_4$			62	346
(CH$_3$)$_2$CCH$_2$CH$_2$C(CH$_3$)$_2$ (OH, OH)	H$_2$SO$_4$			42	238
	H$_3$PO$_4$	47		26	264
	Al$_2$O$_3$			77	349
	PdCl$_2$/CuCl$_2$	78			150
	Cu/Al, Pd/Al	60–70		30–40	304
	Cu/Al, Cu, Pt/C	65–83		13–17	103
	20% H$_2$SO$_4$	80			247
Ph(CH$_3$)CCH$_2$CH$_2$C(CH$_3$)Ph (OH, OH)	HCl/benzene	59		23	
	Formic acid	60			

[a]Selectivity is based on 100 mole of the compound reacted.

H_3PO_4[248,249,256-265]), or in certain cases acidic salts[266-270]. Organic acids too (benzenesulphonic[271], *p*-toluenesulphonic[224,249,272-276], formic[247,250,277], acetic[278,279], oxalic[249,280]) may be used in the same way. In most cases the cyclic ether formation is selective, and sometimes quantitative. These methods are primarily employed with aliphatic diols.

It is interesting to compare the transformations of 1,2-bis(hydroxymethyl)cyclohexane (**69**) and 1,3-bis(hydroxymethyl)cyclohexane (**70**). On the action of H_2SO_4 or H_3PO_4, the former gives a cyclic ether in excellent yield (equation 49)[268,281], while **70** yields scarcely any **71** (equation 50)[282]. The transformation may likewise occur for cyclic compounds with smaller rings[267,269].

$$
\begin{array}{ccc}
\text{(69)} & \xrightarrow{\;\text{H}^+\;} & \text{90–94\%}
\end{array}
\qquad (49)
$$

$$
\begin{array}{ccc}
\text{(70)} & \xrightarrow{\;\text{H}^+\;} & \text{(71)}
\end{array}
\qquad (50)
$$

Various metal salts (e.g. $MgCl_2$, $CaCl_2$, $ZnCl_2$, $AlCl_3$, $CuSO_4$) similarly catalyse the formation of tetrahydrofuran[235,254,270,283]. Cyclic ethers are also formed from 1,4- and 1,5-diols on the action of $RhCl_3/PPh_3$[145]. The use of $PdCl_2$ together with other salts [$CuCl_2$, $Cu(NO_3)_2$, $NaCl$] leads to the formation of five- and six-membered cyclic ethers in various yields[150]. The yield is good in the case of the ditertiary 1,4-diol **72** (equation 51).

$$
(CH_3)_2\underset{\underset{OH}{|}}{C}CH_2CH_2\underset{\underset{OH}{|}}{C}(CH_3)_2 \xrightarrow{PdCl_2/CuCl_2}
\qquad (51)
$$

(**72**)   70%

Oxides[249,270,280,283-286], and particularly alumina[92,245,249,270,280, 283-285,287-290], are also frequently used for the preparation of cyclic ethers. Acidic and neutral phosphates of various mono- and tri-valent metals can similarly be employed[178,266,270,283,284,290-292]. For example, oxolane is formed quantitatively from 1,4-butanediol in the vapour phase on the action of chromium oxide[249,270], alumina[270,285] and calcium phosphate[178,270,291]. Where $Ca_3(PO_4)_2$ is used as catalyst, however, the transformation is selective only at 250–320°C; at higher temperatures the cyclic ether formation becomes less important than the formation of unsaturated alcohols and dienes (see Section III.C.2).

The corresponding oxacycloalkanes can be obtained from the alicyclic compounds, in yields depending on the structure of the diol[287-289]. From *trans*-1,4-cyclohexanediol the main product is 1,4-epoxycyclohexane[293-295], whereas the *cis* isomer gives primarily 2-cyclohexen-1-ol (see Section III.C.3).

The effect of aluminium silicates has mainly been studied with simple diols. Nearly quantitative yields of oxolane are reported[285,296,297]. With diprimary, primary — secondary and disecondary 1,4- and 1,5-diols too, good yields can be attained[290,298-301]. In the case of zeolites, it has been established[297] that the

HNaX form and the decationized X form, at temperatures of 240–260°C, are optimal for oxolane formation.

Dehydrations of both 1,4- and 1,5-diols on supported Ni[302,303], Cu and Pd[139,304,305] and Pt[141] catalysts gave generally high yields. Because of the occurrence of other reactions, little 1,4-epoxycyclohexane is formed from the isomeric 1,4-cyclohexanediols, but the yield is always higher from the *trans* than from the *cis* compound[104,306]. The epimerization demonstrated by Pines and Kobylinski[306] (which has also been observed on Cu and Cu/Al catalysts[307]) strongly suggests that in the case of the *cis* compound too the 1,4-epoxycyclohexane is formed from the *trans*-diol, produced by epimerization.

With the diprimary diols, ion-exchange resins lead to cyclic ethers in excellent yields[118,249,308].

In some cases Me_2SO has also been employed to induce ring-closure. For the open-chain diprimary diols the reaction proceeds with a diol : Me_2SO molar ratio of 2:1, the yield of the cyclic ether decreasing with the distance between the hydroxy groups (oxolane: 70%, oxane: 47%, oxepane: 24%)[82]. The reagent is often used in a very great excess (diol : Me_2SO = 1:12)[248,309-311] without an appreciable change in the yield.

Ring-closure can also be achieved with diols containing a heteroatom. The dehydration has been carried out in the presence of $KHSO_4$[312-314], aluminium silicate[102] and ion-exchange resin[118] (equation 52). Cyclic ethers are likewise

$$HOCH_2CH_2XCH_2CH_2OH \longrightarrow \qquad\qquad (52)$$
$$X = S, N, O$$

formed from diols containing other groups too (trihydroxy compounds[249,315,316], epoxytriols[317], unsaturated diols[249,254,255,264,284,318]).

α, ω-Diols, with carbon chains consisting of six or more atoms, yield α-substituted oxolane and oxane derivatives, on the action of H_2SO_4 and H_3PO_4, independently of the number of carbon atoms (equation 53)[319-324]. In the reaction of 1,6-

$$CH_2(CH_2)_6CH_2 \xrightarrow{H_2SO_4} \qquad + \qquad\qquad (53)$$
$$\underset{OH}{|}\qquad\underset{OH}{|}$$

hexanediol, a small amount (1.5%) of oxepane too has been detected[325]. Similarly, all three cyclic ethers are also formed in the presence of Al_2O_3 and $Ca_3(PO_4)_2$[178,326].

The formation of cyclic ethers with rings of unexpected size may be promoted by the special electronic structure of the starting compound (as a consequence of electron shifts resulting from the presence of unsaturated bonds)[327,328].

2. Mechanism of oxacycloalkane formation

Mihailović and coworkers[248,329] have made a detailed study of the ring-closure of 2,5-hexanediol (73) under various conditions (H_3PO_4, H_2SO_4, Me_2SO, Al_2O_3) and have found that the process is stereoselective in every case: the *meso*-diol is selectively converted to *trans*-2,5-dimethyloxolane (equation 54) and the (±)-diol to *cis*-2,5-dimethyloxolane. It follows from this that the ring-closure takes place by intramolecular substitution of S_N2 type, inversion occurring on the chiral carbon atom bearing the departing protonated hydroxy group.

In the presence of Me_2SO, one of the hydroxy groups interacts with the reagent

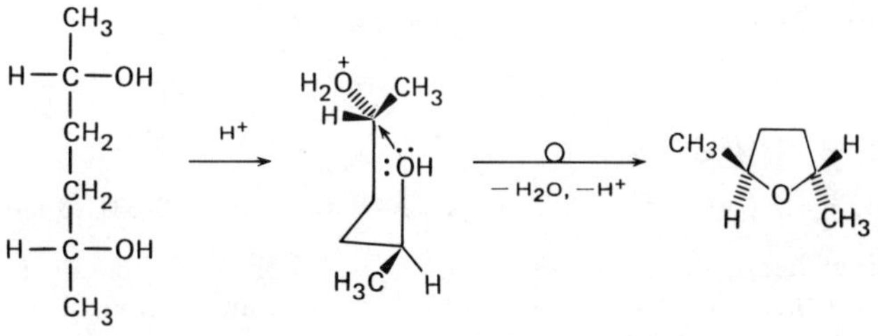

(equation 54 — structures)

(meso-73)

(74), thereby increasing the polarization of the C—O bond and facilitating cleavage of the bond. The results of Mihailović disprove the conception of Gillis and Beck[309], in whose view the cyclic transition state (75) is produced with the

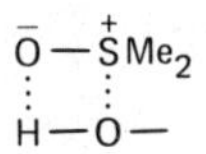

(74) 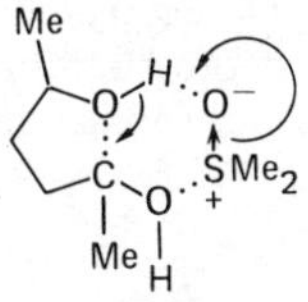   (75)

participation of both hydroxy groups of the diol, 75 then being convertible to the cyclic ether without inversion. The intramolecular $S_N 2$ mechanism is supported by other investigations[244,275].

Primary — tertiary diols[238,239,260] yield isomeric unsaturated alcohols by elimination of the tertiary hydroxy group on the action of acids; they then undergo isomerization to give the cyclic ether (equation 55).

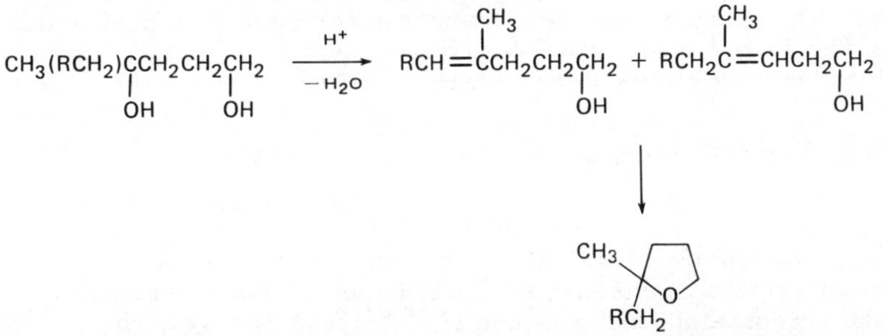

$$CH_3(RCH_2)CCH_2CH_2CH_2 \xrightarrow[-H_2O]{H^+} RCH=CCH_2CH_2CH_2 + RCH_2C=CHCH_2CH_2$$

(55)

On the reaction of (−)-4-methyl-1,4-hexanediol (76) in the presence of p-TsOH or Me$_2$SO, a racemic cyclic ether (77) is obtained[310]; this is a consequence of the fact that a carbonium ion is formed on the departure of the tertiary hydroxy group, a possibility thus arising for the cessation of the chirality of $C_{(4)}$ (equation 56). The

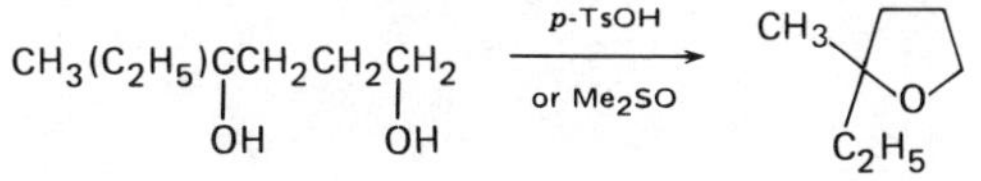

$$CH_3(C_2H_5)CCH_2CH_2CH_2 \xrightarrow[or \; Me_2SO]{p\text{-}TsOH}$$

(56)

(−)-(76)   racemic (77)

ring-closure proceeds via a carbonium cation in a similar way for diols not containing a tertiary hydroxy group (78, 79, 80), when the molecular structure promotes the formation of the carbonium ion and stabilizes it[275,311].

PhCH(CH$_2$)$_n$CHPh $\qquad$ ArCH(R^1)CH(R^2)CHCHR^3

$\quad$|$\qquad$|$\qquad\qquad\qquad$|$\qquad$|$\qquad$|

$\quad$OH$\qquad$OH$\qquad\qquad\qquad$OH$\qquad\quad$OH

$\qquad n = 2$ **(78)** $\qquad\qquad\qquad\qquad\qquad$ **(80)**

$\qquad n = 3$ **(79)** $\qquad$ Ar = Ph, p-CH$_3$C$_6$H$_4$, p-CH$_3$OC$_6$H$_4$, 2-furyl

On the other hand, the transformation of $(-)$-**76** on Al$_2$O$_3$ catalyst is to a slight extent stereospecific. This can be interpreted[310] by assuming that the C$_{(1)}$ hydroxy group first undergoes selective adsorption, followed by nucleophilic substitution of the C$_{(4)}$ hydroxy group. The low degree of stereoselectivity can be ascribed to the fact that the reaction takes place by another mechanism in addition to the above. Stereospecific dehydration has also been reported in the reactions of various di-secondary diols under similar conditions[245,248,329].

In the study of the Al$_2$O$_3$-catalysed transformation of *trans*-1,4-cyclohexanediol (**81**), it was found[294] that the ring-formation involves an S$_N$2 reaction even under these experimental conditions (equation 57).

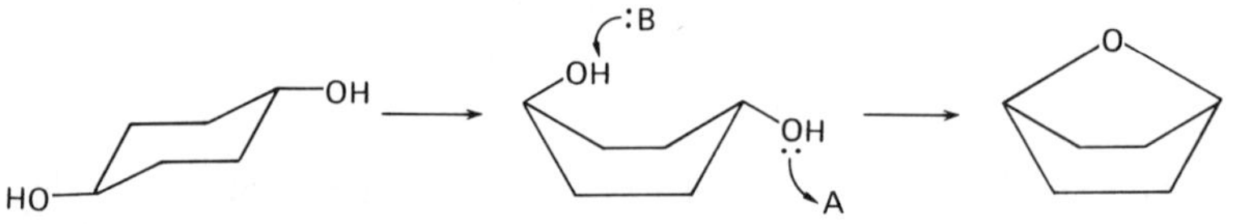

$\qquad$ **(81)**

Cyclic ethers are formed in an essentially similar manner on metal catalysts of Raney type (Cu/Al, Pd/Al). Cyclodehydration is promoted by the aluminium oxide hydroxides formed during the preparation and remaining on the surface of the catalyst. The intramolecular ring-closure has been proved in studies with 1,4-pentanediol-[4-^{2}H], and the results have been supported by data from measurements on the active centres of the catalysts[143,144,305].

B. Preparation of Unsaturated Cyclic Ethers

Certain metal catalysts may be used to prepare unsaturated oxacycloalkanes. Most of the data refer to supported[305,330-332] and support-free Cu[305] catalysts, but supported Co[330,333] and Ag[331] catalysts may also be employed. The transformation may be interpreted as a dehydration of the hemiacetal formed by dehydrogenation of the diol (equation 58)[305]. The process is explained in a similar way in investigations relating to nonmetallic catalysts[334].

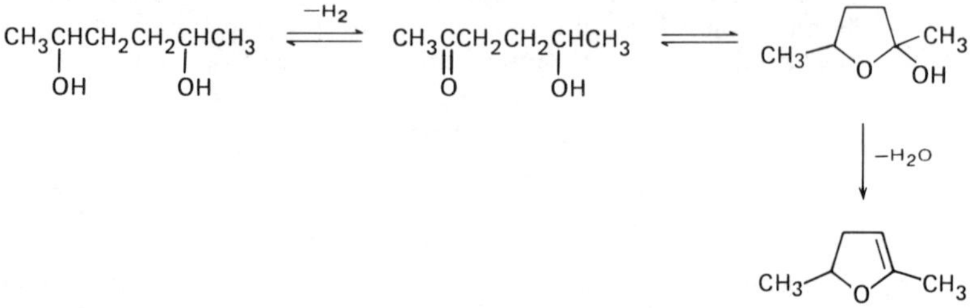

C. Preparation of Unsaturated Alcohols and Dienes

1. Dehydration on the action of acids

The dehydration can be induced with primary — tertiary and ditertiary diols, and is frequently accompanied by the formation of cyclic ethers. The proportions of the two reactions depend on the structure of the diol and on the reaction conditions. For example, with 5-phenyl-1,4-pentanediol[335], only a cyclic ether is formed on the action of H_2SO_4, whereas in the presence of other acids (H_3PO_4, formic acid, acetic acid/NaOAc) an unsaturated alcohol also appears in the product. On the action of p-TsOH/benzene and H_3PO_4, 1,4-diphenyl-1,4-butanediol (78) gives a diene, while under different reaction conditions (H_2SO_4, p-TsOH, Me_2SO) the main product is a cyclic ether[311].

Likewise, H_3PO_4/Al_2O_3 primarily catalyses diene formation from 2,5-dimethyl-2,5-hexanediol (82) (equation 59)[336], while with other ditertiary diols on different

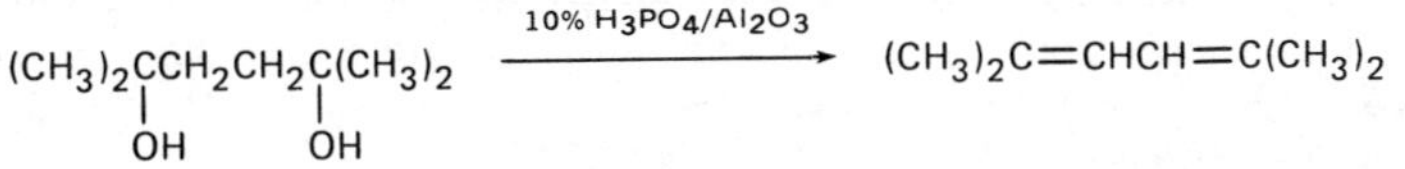

$$(CH_3)_2CCH_2CH_2C(CH_3)_2 \quad \xrightarrow{\text{10\% } H_3PO_4/Al_2O_3} \quad (CH_3)_2C=CHCH=C(CH_3)_2 \qquad (59)$$
$$\underset{OH}{|} \qquad \underset{OH}{|}$$

(82)   83%

catalysts the two main processes run in parallel[232,247,264,337], or the cyclic ether may become the main product[95,261]. For example, the diene 84 is formed in excellent yield from 83 (equation 60), whereas the diene cannot be prepared from the corresponding tetramethyl derivative[267].

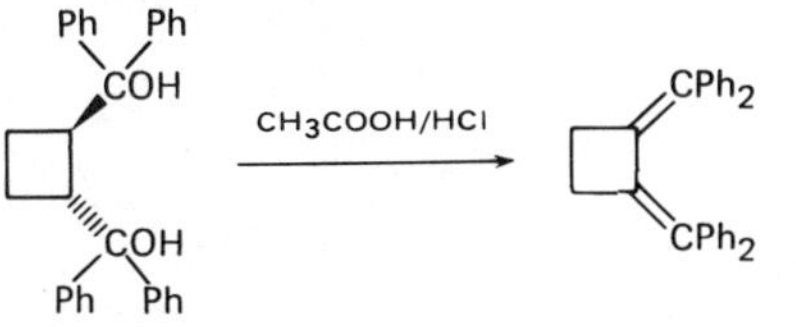

(83)   (84)

70%

Diene and unsaturated alcohol are formed from the isomeric 1,4-cyclohexanediols on the action of $KHSO_4$[295,338], $MgSO_4$[338], H_2SO_4[172,339,340] or oxalic acid[339].

1,4-, 1,5- and 1,6-diols have been studied in $FSO_3H/SbF_5/SO_2$[16]. The diprimary compounds do not react at all. 2,5-Hexanediol gives protonated 2,5-dimethyltetrahydrofuran, while 82 is converted to a diene.

Many authors have studied diols containing unsaturated bonds[341-346], a cyclic substituent often being present[344-347]. Here again the reaction direction is influenced by the structure of the diol. In the presence of 20% H_2SO_4 and $HgSO_4$, for instance, 85 gives the dihydropyran derivative 86 (equation 61)[345]; while the

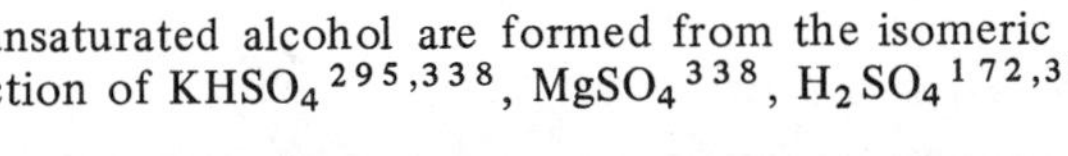

(85)   (86)

44%

dialkyl-substituted compound with similar structure (87) yields the unsaturated alcohol under the same reaction conditions (equation 62).

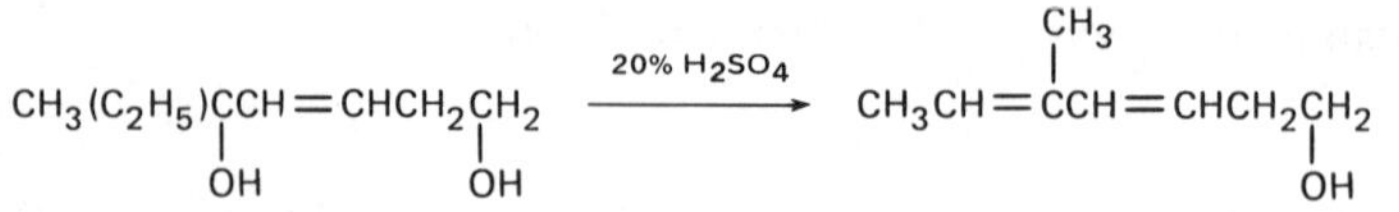

$$\underset{\underset{(87)}{\underset{|}{OH}}}{CH_3(C_2H_5)\overset{}{C}CH}=CHCH_2\underset{\underset{}{\underset{|}{OH}}}{CH_2} \quad \xrightarrow{20\%\ H_2SO_4} \quad CH_3CH=\overset{\overset{CH_3}{|}}{C}CH=CHCH_2\underset{\underset{49\%}{\underset{|}{OH}}}{CH_2} \quad (62)$$

2. Dehydration on phosphate catalysts

With $Ca_3(PO_4)_2$, detailed investigations present illustrative examples of how the reaction conditions can affect the pathways and hence the product composition. At low temperature, cyclic ether formation is dominant (see Section III.A.1). With the increase of the temperature, the product also includes unsaturated alcohols, and then dienes, and if the temperature is further elevated these become the main products[178,290-292,348].

Dienes may be formed from both the cyclic ether and the unsaturated alcohol, although the process occurs primarily via the unsaturated alcohol. The reaction scheme for general electrophilic catalysts is presented using the example of 1,4-butanediol (88) (equation 63)[178,182,290-292,348].

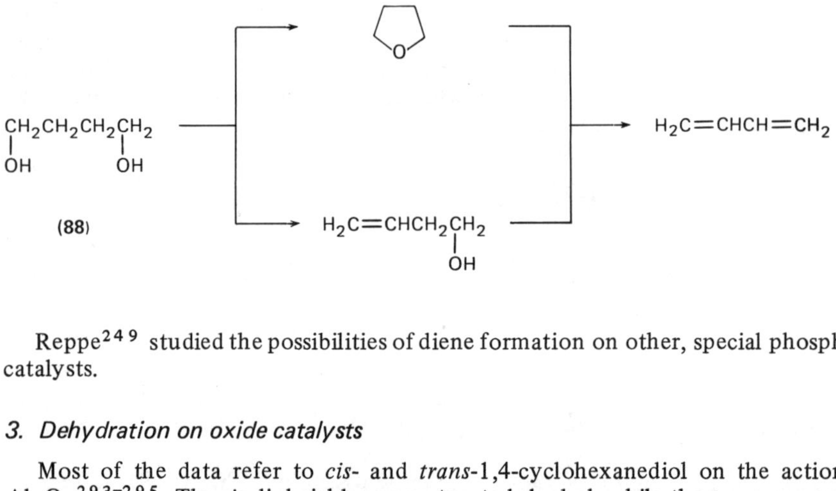

$$\underset{\underset{(88)}{\underset{|}{OH}}}{\underset{|}{OH}}{CH_2CH_2CH_2CH_2} \longrightarrow \qquad \longrightarrow H_2C=CHCH=CH_2 \quad (63)$$

$$H_2C=CHCH_2\underset{\underset{}{\underset{|}{OH}}}{CH_2}$$

Reppe[249] studied the possibilities of diene formation on other, special phosphate catalysts.

3. Dehydration on oxide catalysts

Most of the data refer to cis- and trans-1,4-cyclohexanediol on the action of Al_2O_3[293-295]. The cis-diol yields an unsaturated alcohol, while the trans compound possesses a favourable conformation for ring-closure[294] (see Section III.A.2). Detailed studies have been carried out[295] to clarify the dependence of the formation of the three possible products (1,4-epoxycyclohexane, 3-cyclohexen-1-ol, 1,3-cyclohexadiene) on the structure of the starting diol and on the reaction

In the case of open-chain diols, studies have been made with Al_2O_3[248,310,349,350], and also with other oxide catalysts[81,266,350], the latter primarily from the aspect of diene formation.

4. Dehydration on metal catalysts

Observations with Cu/Al, Cu and Pt/C catalysts[103] indicate that 2,6-dimethyl-2,6-heptanediol (89) is converted mainly to dienes (equation 64), in contrast with 2,5-dimethyl-2,5-hexanediol (82), yielding mainly the cyclic ether. This phenomenon can be explained by the rapid further reaction of the 90 formed.

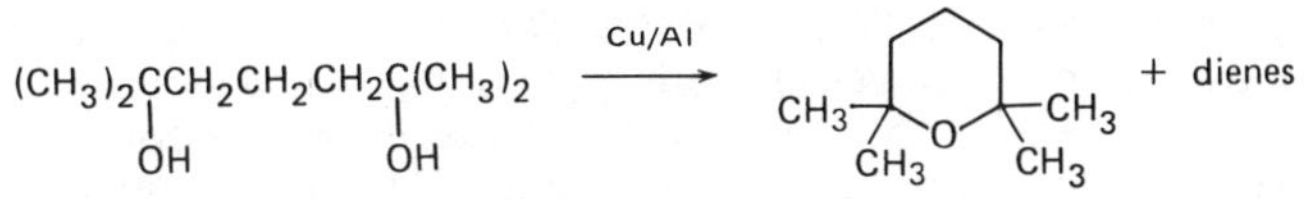

$$(CH_3)_2CCH_2CH_2CH_2C(CH_3)_2 \xrightarrow{Cu/Al}$$

with OH groups on both quaternary carbons

(89)

$$\text{(90)} \quad + \text{ dienes} \qquad (64)$$

28%   72%

The isomeric 1,4-cyclohexanediols have been investigated on Ni/SiO_2[306] as well as on Cu/Al and Cu[104] catalysts. On Ni/SiO_2 the transformation was carried out in the presence of hydrogen, and hence the unsaturated compounds could not be detected. On Cu/Al, 3-cyclohexenone and dienes are formed in addition to other products.

D. Other Transformations

Numerous observations[178,231,249,351-357] show that variously substituted 2-butene-1,4-diols are dehydrated to unsaturated oxo compounds in the presence of acids, $Ca_3(PO_4)_2$, Al_2O_3 and ThO_2 (in the case of the parent compound, 2-butene-1,4-diol, formation of 2,5-dihydrofuran is also found). By study of the isomers, it has been established[353,354] that the *trans* compound (**91**) is converted to crotonaldehyde (**93**), while both products are formed from the *cis*-diol (**92**) (equations 65 and 66). On the basis of the stereostructure, the ring-closure process should predominate for **92**, but here too crotonaldehyde is formed because of the *cis–trans* isomerization.

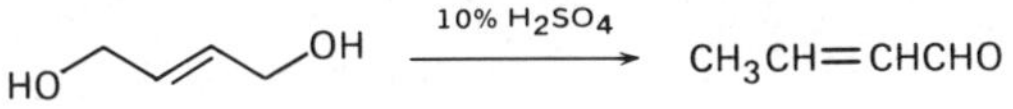

$$HO\text{—CH}_2\text{—CH}=\text{CH—CH}_2\text{—OH} \xrightarrow{10\% \text{ H}_2\text{SO}_4} CH_3CH{=}CHCHO \qquad (65)$$

(91)   **(93)**

80%

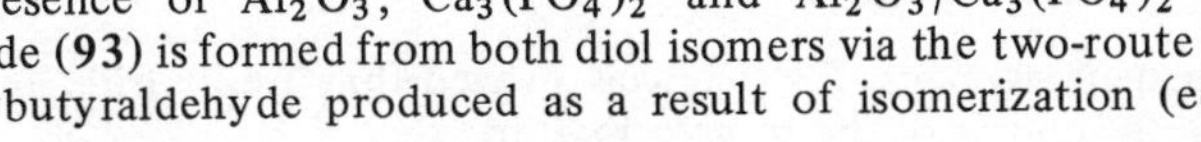

$$HO\text{—CH}_2\text{—CH}=\text{CH—CH}_2\text{—OH} \xrightarrow{10\% \text{ H}_2\text{SO}_4} CH_3CH{=}CHCHO + \quad \text{(furan ring)} \qquad (66)$$

(92)   **(93)**

65%   35%

In the transformations of the isomers of 1,1,4,4-tetraphenyl-2-butene-1,4-diol in acetic acid, on the other hand, only unidirectional processes can be observed[358].

In the presence of Al_2O_3, $Ca_3(PO_4)_2$ and $Al_2O_3/Ca_3(PO_4)_2$[178,355,356], crotonaldehyde (**93**) is formed from both diol isomers via the two-route dehydration of 4-hydroxybutyraldehyde produced as a result of isomerization (equation 67).

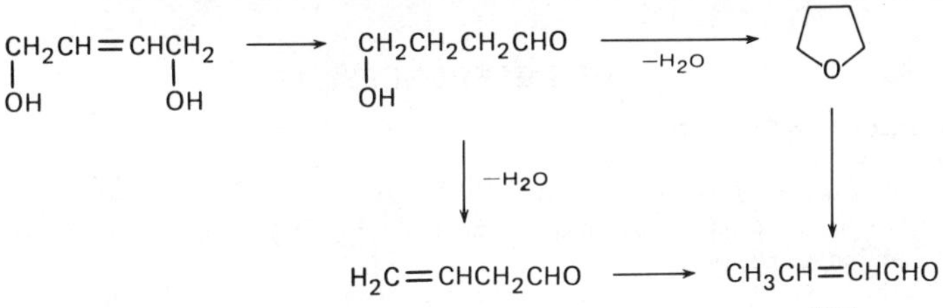

$$CH_2CH{=}CHCH_2 \longrightarrow CH_2CH_2CH_2CHO \xrightarrow{-H_2O} \quad \text{(furan ring)}$$
with OH on terminal carbons   with OH

$$\Big\downarrow -H_2O \qquad\qquad\qquad\qquad\qquad \Big\downarrow \qquad (67)$$

$$H_2C{=}CHCH_2CHO \longrightarrow CH_3CH{=}CHCHO$$

(93)

The two processes (isomerization of the diol, and dehydration) occur on different active centres.

3-Hexene-2,5-diol (94) in H_3PO_4 gives two isomeric ketones (equation 68)[357]. In an investigation of the mechanism with D_3PO_4/D_2O, it was proved that the product ratio is governed by the protonation of the dienol 95 formed by dehydration and subsequent rearrangement (equation 69). The stabilization is due almost exclusively to α-protonation.

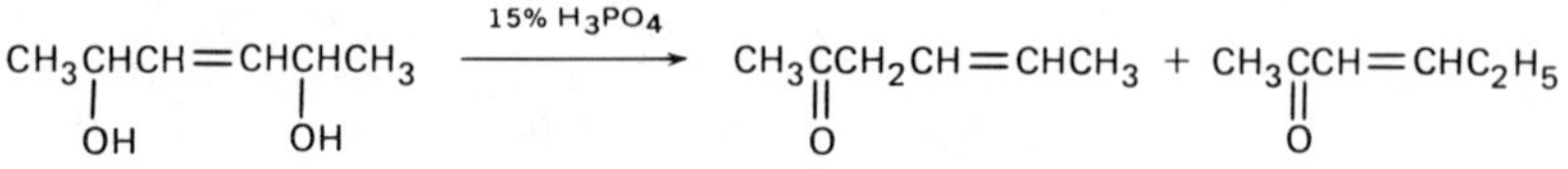

$$CH_3CHCH\!=\!CHCHCH_3 \xrightarrow{15\% \; H_3PO_4} CH_3CCH_2CH\!=\!CHCH_3 \;+\; CH_3CCH\!=\!CHC_2H_5 \quad (68)$$

positions: OH OH on (94); II O under first product; II O under second product

(94)   90–95%   5–10%

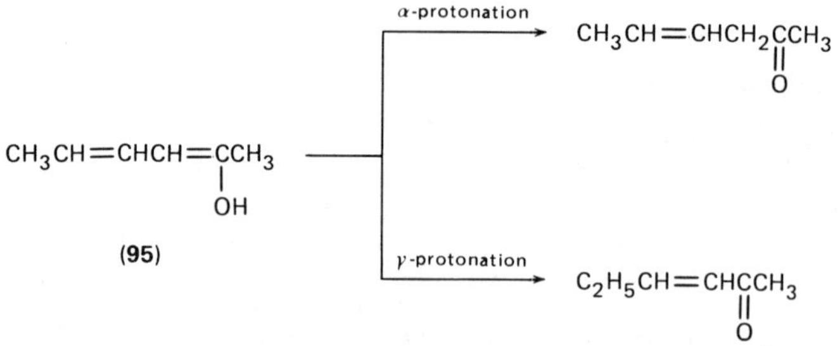

α-protonation

$$CH_3CH\!=\!CHCH_2CCH_3$$
(with II O)

$$CH_3CH\!=\!CHCH\!=\!CCH_3$$
|
OH

(95)   (69)

γ-protonation

$$C_2H_5CH\!=\!CHCCH_3$$
(with II O)

The diol 96 in acetic acid[359], $SOCl_2$[359] and $KHSO_4$[269] forms an unsaturated ketone by ring-opening and phenyl migration (equation 70).

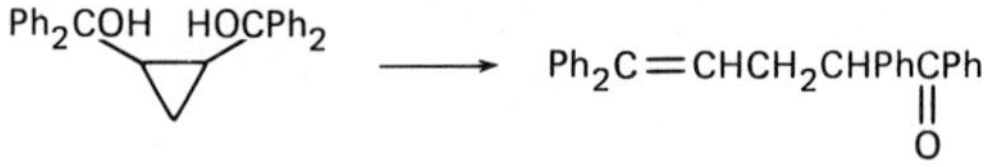

$$Ph_2C\!=\!CHCH_2CHPhCPh \quad (70)$$
(with II O under CPh, Ph₂COH HOCPh₂ cyclopropane starting material)

(96)

In the reaction of 1-methyl-1,6-cyclohexanediol (97), Prelog and Küng[360] isolated the ketone 98 (equation 71). By means of the reaction of the compound

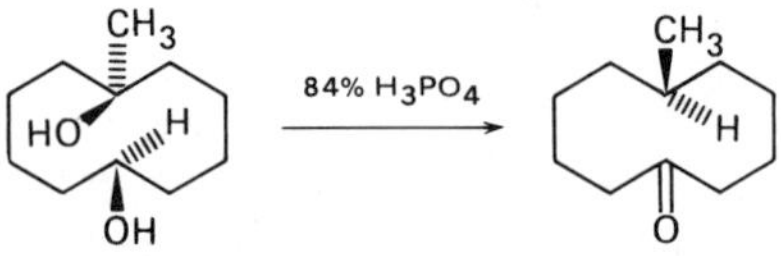

$$\xrightarrow{84\% \; H_3PO_4} \quad (71)$$

(97)   (98)

labelled with deuterium on $C_{(6)}$, it was proved that 1,6-hydride anion migration takes place in the course of the transformation.

IV. REFERENCES

1. R. Fittig, *Liebigs Ann. Chem.*, **110**, 17 (1859); **114**, 54 (1860).
2. A. Butlerov, *Liebigs Ann. Chem.*, **170**, 151 (1873); **174**, 125 (1874).
3. C. J. Collins, *Quart. Rev. (Lond.)*, **14**, 357 (1960).
4. Y. Pocker in *Molecular Rearrangements* (Ed. P. de Mayo), Interscience P., New York–London, 1963, pp. 1–25.
5. C. J. Collins and J. F. Eastham in *The Chemistry of the Carbonyl Group* (Ed. S. Patai), John Wiley and Sons, London, 1966, Chap. 15, pp. 762–767.
6. C. A. Bunton and M. D. Carr, *J. Chem. Soc.*, 5861 (1963).
7. C. J. Collins, *J. Amer. Chem. Soc.*, **77**, 5517 (1955).

8. P. M. Benjamin and C. J. Collins, *J. Amer. Chem. Soc.*, **78**, 4329 (1956).
9. C. J. Collins, W. T. Rainey, W. B. Smith and I. A. Kaye, *J. Amer. Chem. Soc.*, **81**, 460 (1959).
10. C. A. Bunton, T. Hadwick, D. R. Llewellyn and Y. Pocker, *Chem. Ind. (Lond.)*, 547 (1956).
11. J. B. Ley and C. A. Vernon, *J. Chem. Soc.*, 2987 (1957).
12. C. A. Bunton, T. Hadwick, D. R. Llewellyn and Y. Pocker *J. Chem. Soc.*, 403 (1958).
13. R. D. Sands and D. G. Botteron, *J. Org. Chem.*, **28**, 2690 (1963).
14. R. D. Sands, *Tetrahedron*, **21**, 887 (1965).
15. A. W. Bushell and P. Wilder, *J. Amer. Chem. Soc.*, **89**, 5721 (1967).
16. G. A. Olah and J. Sommer, *J. Amer. Chem. Soc.*, **90**, 927 (1968).
17. K. Matsumoto, *Bull. Chem. Soc. Japan*, **41**, 1356 (1968).
18. H. Christol, A. P. Krapcho and F. Pietrasanta, *Bull. Soc. Chim. Fr.*, 4059 (1969).
19. J. W. Huffman and L. E. Browder, *J. Org. Chem.*, **27**, 3208 (1962).
20. C. A. Bunton and M. D. Carr, *J. Chem. Soc.*, 5854 (1963).
21. D. C. Kleinfelter and T. E. Dye, *J. Amer. Chem. Soc.*, **88**, 3174 (1966).
22. W. Oppolzer, T. Sarkar and K. M. Mahalanabis, *Helv. Chim. Acta*, **59**, 2012 (1976).
23. J. E. Dubois and P. Bauer, *J. Amer. Chem. Soc.*, **98**, 6993 (1976).
24. P. Bauer and J. E. Dubois, *J. Amer. Chem. Soc.*, **98**, 6999 (1976).
25. P. L. Barili, G. Berti, B. Macchia, F. Macchia and L. Monti, *J. Chem. Soc. (C)*, 1168 (1970).
26. T. Shono, K. Fujita, S. Kumai, T. Watanabe and I. Nishiguchi, *Tetrahedron Letters*, 3249 (1972).
27. J. De Pascual Teresa, I. S. Bellido and J. F. S. Barrueco, *An. Quim.*, **72**, 560 (1976).
28. W. B. Smith, R. E. Bowman and T. J. Kmet, *J. Amer. Chem. Soc.*, **81**, 997 (1959).
29. W. B. Smith, T. J. Kmet and P. S. Rao, *J. Amer. Chem. Soc.*, **83**, 2190 (1961).
30. C. J. Collins, Z. K. Cheema, R. G. Werth and B. M. Benjamin, *J. Amer. Chem. Soc.*, **86**, 4913 (1964).
31. B. M. Benjamin and C. J. Collins, *J. Amer. Chem. Soc.*, **88**, 1556 (1966).
32. S. Wold, *Acta Chem. Scand.*, **23**, 1266 (1969).
33. S. Wold, *Acta Chem. Scand.*, **23**, 2978 (1969).
34. J.-P. Barnier and J.-M. Conia, *Bull. Soc. Chim. Fr.*, 285 (1976).
35. E. R. Alexander and D. C. Dittmer, *J. Amer. Chem. Soc.*, **73**, 1665 (1951).
36. J. F. Duncan and K. R. Lynn, *Australian J. Chem.*, **10**, 1 (1957).
37. J. F. Duncan and K. R. Lynn, *Australian J. Chem.*, **10**, 7 (1957).
38. M. Stiles and R. P. Mayer, *J. Amer. Chem. Soc.*, **81**, 1497 (1959).
39. J. G. Traynham and P. M. Greene, *J. Amer. Chem. Soc.*, **86**, 2657 (1964).
40. D. G. Botteron and G. Wood, *J. Org. Chem.*, **30**, 3871 (1965).
41. T. Moriyoshi and K. Tamura, *Rev. Phys. Chem. Japan*, **40**, 48 (1970).
42. M. Tiffeneau and J. Lévy, *Bull. Soc. Chim. Fr.*, **49**, 1738 (1931).
43. A. McKenzie and R. Roger, *J. Chem. Soc.*, **125**, 844 (1924).
44. A. McKenzie and W. S. Dennler, *J. Chem. Soc.*, **125**, 2105 (1924).
45. H. J. Gebhardt and K. H. Adams, *J. Amer. Chem. Soc.*, **76**, 3925 (1954).
46. Y. Pocker and B. P. Ronald, *J. Org. Chem.*, **35**, 3362 (1970).
47. Y. Pocker and B. P. Ronald, *J. Amer. Chem. Soc.*, **92**, 3385 (1970).
48. M. Tiffeneau and J. Lévy, *Bull. Soc. Chim. Fr.*, **33**, 759 (1923).
49. A. McKenzie, R. Roger and W. B. McKay, *J. Chem. Soc.*, 2597 (1932).
50. D. C. Kleinfelter and P. R. Schleyer, *J. Amer. Chem. Soc.*, **83**, 2329 (1961).
51. K. Mislow and M. Siegel, *J. Amer. Chem. Soc.*, **74**, 1060 (1952).
52. E. Bergmann and W. Schuchardt, *Liebigs Ann. Chem.*, **487**, 234 (1931).
53. W. E. Bachmann and R. V. Shankland, *J. Amer. Chem. Soc.*, **51**, 306 (1929).
54. L. W. Kendrick, B. M. Benjamin and C. J. Collins, *J. Amer. Chem. Soc.*, **80**, 4057 (1958).
55. G. Wittig, M. Leo and W. Wiemer, *Chem. Ber.*, **64**, 2405 (1931).
56. W. E. Bachmann and E. J.-H. Chu, *J. Amer. Chem. Soc.*, **57**, 1095 (1935).
57. W. E. Bachmann, *J. Amer. Chem. Soc.*, **54**, 1969 (1932).
58. W. E. Bachmann and E. J.-H. Chu, *J. Amer. Chem. Soc.*, **58**, 1118 (1936).
59. P. D. Bartlett and I. Pöckel, *J. Amer. Chem. Soc.*, **59**, 820 (1937).

60. H. Meerwein, *Liebigs Ann. Chem.*, **542**, 123 (1939).
61. P. D. Bartlett and R. F. Brown, *J. Amer. Chem. Soc.*, **62**, 2927 (1940).
62. S. Fujita and K. Nomura, *J. Chem. Soc. Japan*, **63**, 510 (1942); *Chem. Abstr.*, **41**, 3063c (1947).
63. S. Fujita, *J. Chem. Soc. Japan*, **72**, 539 (1951); *Chem. Abstr.*, **46**, 6601c (1952).
64. R. F. Brown, J. B. Nordmann and M. Madoff, *J. Amer. Chem. Soc.*, **74**, 432 (1952).
65. P. Richter and V. Ruzicka, *Chem. Prumysl*, **8**, 116 (1958).
66. S. Nametkin and N. Delektorsky, *Chem. Ber.*, **57**, 583 (1924).
67. M. Qudrat-i-Khuda and A. K. Ray, *J. Indian Chem. Soc.*, **16**, 525 (1939).
68. C. R. Walter, *J. Amer. Chem. Soc.*, **74**, 5185 (1952).
69. D. J. Cram and H. Steinberg, *J. Amer. Chem. Soc.*, **76**, 2753 (1954).
70. R. Ya. Levina, V. R. Skvarchenko and O. Yu. Oklobystin, *Zh. Obshch. Khim.*, **25**, 1466 (1955).
71. P. A. Naro and J. A. Dixon, *J. Amer. Chem. Soc.*, **81**, 1681 (1959).
72. E. Vogel, *Chem. Ber.*, **85**, 25 (1952).
73. N. V. Elagina and B. A. Kazanskii, *Dokl. Akad. Nauk SSSR*, **124**, 1053 (1959).
74. D. S. Greidinger and D. Ginsburg, *J. Org. Chem.*, **22**, 1406 (1957).
75. B. P. Mundy and R. D. Otzenberger, *J. Chem. Educ.*, **48**, 431 (1971).
76. L. P. Kyriakides, *J. Amer. Chem. Soc.*, **36**, 987 (1914).
77. L. W. Newton and E. R. Coburn, *Org. Synth.*, **22**, 39 (1942).
78. H. Waldmann and F. Petru, *Chem. Ber.*, **83**, 287 (1950).
79. W. Reeve and D. M. Reichel, *J. Org. Chem.*, **37**, 68 (1972).
80. W. J. Hale and H. Miller, *U.S. Patent*, No. 2,400,409; *Chem. Abstr.*, **40**, 4774[2] (1946).
81. W. J. Hale, *U.S. Patent*, No. 2,441,966; *Chem. Abstr.*, **42**, 7785c (1948).
82. V. J. Traynelis, W. L. Hergenrother, H. T. Hanson and J. A. Valicenti, *J. Org. Chem.*, **29**, 123 (1964).
83. J. F. Lane and L. Spialter, *J. Amer. Chem. Soc.*, **73**, 4411 (1951).
84. G. Majerus, E. Yax and G. Ourisson, *Bull. Soc. Chim. Fr.*, 4147 (1967).
85. A. D. Yanina, E. E. Mikhlina and M. V. Rubstov, *Zh. Org. Khim.*, **2**, 1707 (1966).
86. L. F. Fieser and A. M. Seligman, *J. Amer. Chem. Soc.*, **56**, 2690 (1934).
87. T. Kuwata, *J. Chem. Soc. Japan*, **62**, 1028, 1035, 1042 (1941); *Chem. Abstr.*, **41**, 3051d (1947).
88. H. Adkins and S. H. Watkins, *J. Amer. Chem. Soc.*, **73**, 2184 (1951).
89. L. Kh. Freidlin and V. Z. Sharf, *Izv. Akad. Nauk SSSR, Ser. Khim.*, 698 (1962).
90. S. Matida, *J. Chem. Soc. Japan*, **62**, 293 (1941); *Chem. Abstr.*, **37**, 4363[6] (1943).
91. J. P. Russel, *U.S. Patent*, No. 3, 235,602; *Chem. Abstr.*, **64**, 14095c (1966).
92. E. Beati and G. Mattei, *Ann. chim. applicata*, **30**, 21 (1940); *Chem. Abstr.*, **34**, 6930[6] (1940).
93. G. Dana and J. Wiemann, *Bull. Soc. Chim. Fr.*, 3994 (1970).
94. S. V. Kannan and C. N. Pillai, *Indian J. Chem.*, **7**, 1164 (1969).
95. A. Halasz, *Ann. Chim. (Paris)*, **14**, 318 (1940).
96. A. Halasz, *J. Chem. Educ.*, **33**, 624 (1956).
97. Y. L. Pascal, *Ann. Chim. (Paris)*, **3**, 67 (1968).
98. Y. L. Pascal, *Ann. Chim. (Paris)*, **3**, 245 (1968).
99. Ya. M. Paushkin, E. M. Buslova and S. A. Nizova, *Kinetika i Kataliz*, **10**, 918 (1969).
100. E. M. Buslova, S. A. Nizova and Ya. M. Paushkin, *Neftekhimiya*, **9**, 227 (1969); *Chem. Abstr.*, **71**, 38408 (1969).
101. M. E. Sarilova, A. P. Mishchenko, V. M. Gryaznov and V. S. Smirnov, *Izv. Akad. Nauk SSSR, Ser. Khim.*, 430 (1977).
102. T. Ishiguro, E. Kitamura and M. Matsumura, *J. Pharm. Soc. Japan*, **74**, 1162 (1954); *Chem. Abstr.*, **49**, 14767g (1955).
103. M. Bartók and Á. Molnár, unpublished results.
104. Á. Molnár and M. Bartók, *React. Kinet. Catal. Lett.*, **4**, 315 (1976).
105. J. L. Gear, *U.S. Patent*, No. 2,501,042; *Chem. Abstr.*, **44**, 5379c (1950).
106. W. G. Bowmann, *U.S. Patent*, No. 3,849,512; *Chem. Abstr.*, **82**, 74188g (1975).
107. N. N. Zelenetskii, G. A. Mazurova, N. D. Shcherbakova, E. I. Kirsankina and A. V. Gurevich, *Maslo-Zhir. Prom*, **36**, 39 (1970); *Chem. Abstr.*, **74**, 22831e (1971).

108. S. Esteban, J. M. Marinas, S. Perez-Ossorio and A. Alberola, *Annales de Quimica*, **70**, 944 (1974).
109. D. W. Young and C. E. Britton, *U.S. Patent*, No. 2,461,362; *Chem. Abstr.*, **43**, 3834c (1949).
110. M. E. Winfield, *J. Council Sci. Ind. Res.*, **18**, 412 (1945); *Chem. Abstr.*, **40**, 3719[1] (1946).
111. M. E. Winfield, *Australian J. Sci. Res.*, **3A**, 290 (1950); *Chem. Abstr.*, **45**, 1953f (1951).
112. A. M. Bourns and R. V. V. Nicholls, *Can. J. Res.*, **25B**, 80 (1947); *Chem. Abstr.*, **41**, 3051a (1947).
113. S. Saito, *Chem. Phys. Letters*, **42**, 399 (1976).
114. B. T. Golding, T. J. Kemp, E. Nocchi and W. P. Watson, *Angew. Chem. (Intern. Ed. Engl.)*, **14**, 813 (1975).
115. B. T. Golding, C. S. Sell and P. J. Sellars, *J. Chem. Soc. Chem. Commun*, 773 (1976).
116. B. T. Golding, T. J. Kemp, C. S. Sell, P. J. Sellars and W. P. Watson, *J. Chem. Soc., Perkin II*, 839 (1978).
117. R. D. Obolentsev and N. N. Gryazev, *Dokl. Akad. Nauk SSSR*, **73**, 319 (1950); *Chem. Abstr.*, **44**, 9916d (1950).
118. E. Swistak, *Compt. Rend.*, **240**, 1544 (1955).
119. F. X. Schmalzhofer, *Monatsh. Chem.*, **21**, 671 (1900).
120. T. Hackhofer, *Monatsh. Chem.*, **22**, 95 (1901).
121. A. Lieben, *Monatsh. Chem.*, **23**, 60 (1902).
122. F. Bauer, *Monatsh. Chem.*, **25**, 1 (1904).
123. M. Rix, *Monatsh. Chem.*, **25**, 267 (1904).
124. V. Kadiera, *Monatsh. Chem.*, **25**, 332 (1904).
125. J. Munk, *Monatsh. Chem.*, **26**, 663 (1905).
126. A. Fischer and B. Winter, *Monatsh. Chem..*, **21**, 301 (1900).
127. T. Yvernault and M. Mazet, *Bull. Soc. Chim. Fr.*, 2755 (1967); 3352 (1968).
128. M. Mazet, *Bull. Soc. Chim. Fr.*, 4309 (1969).
129. T. Yvernault and M. Mazet, *Bull. Soc. Chim. Fr.*, 638 (1969).
130. T. Yvernault, and M. Mazet, *Bull. Soc. Chim. Fr.*, 2652 (1971).
131. W. Reppe, *German Patent British Patent*, No. 318,124; *Chem. Zentr.*, **I**, 1218 (1930).
132. W. Reppe, *German Patent*, No. 528,360; *Chem. Zentr.*, **II**, 1488 (1931).
133. O. Schmidt, *German Patent*, No. 524,101; *Chem. Zentr.*, **II**, 767 (1931).
134. V. I. Ivanskii and B. N. Dolgov, *Zh. Prikl. Khim.*, **36**, 2256 (1963).
135. M. Bartók and B. Kozma, *Acta Phys. Chem. Szeged*, **9**, 116 (1963).
136. M. Bartók and L. Zalotai, *Acta Phys. Chem. Szeged*, **14**, 39 (1968).
137. L. Zalotai and M. Bartók, *Acta Phys. Chem. Szeged*, **14**, 47 (1968).
138. M. Bartók and B. Prágai, *Acta Phys. Chem. Szeged*, **18**, 85 (1972).
139. M. Bartók, Á. Molnár and F. Notheisz, *Acta Phys. Chem. Szeged.*, **18**, 85 (1972).
140. M. Bartók and Á. Molnár, *Acta Chim. Acad. Sci. Hung.*, **76**, 409 (1973).
141. M. Bartók and Á. Molnár, *Acta Chim. Acad. Sci. Hung.*, **78**, 305 (1973).
142. Á. Molnár and M. Bartók, *React. Kinet. Catal. Letters*, **3**, 421 (1975).
143. M. Bartók and Á. Molnár, *Kém. Közl.*, **45**, 335 (1976).
144. Á. Molnár and M. Bartók, *Acta Chim. Acad. Sci. Hung.*, **89**, 393 (1976).
145. K. Kaneda, M. Wayaku, T. Imanaka and S. Teranishi, *Chem. Letters*, 231 (1976).
146. K. Felföldi, Á. Molnár and M. Bartók in *Proceedings of Symposium on Rhodium in Homogeneous Catalysis*, Veszprém, Hungary, September, 1978, pp. 38–43.
147. O. Katsutoshi, H. Kyoshiro and Y. Kohji, *Inorg. Nucl. Chem. Letters*, **13**, 637 (1977).
148. A. Nef., *Liebigs. Ann. Chem.*, **335**, 206 (1904).
149. E. Arundale and H. O. Mottern, *U.S. Patent*, No. 2,620,357; *Chem. Abstr.*, **47**, 8089g (1953).
150. T. E. Nalesnik and N. L. Holy, *J. Org. Chem.*, **42**, 372 (1977).
151. A. Franke and M. Kohn, *Monatsh. Chem.*, **28**, 997 (1907).
152. V. Z. Sharf, L. Kh. Freidlin, E. N. German, G. K. Oparina and V. I. Kheifets, *Neftekhimiya*, **5**, 368 (1965); *Chem. Abstr.*, **63**, 8176f (1965).
153. K. M. Trenke, M. S. Nemtsov and S. K. Ogorodnikov, *U.S.S.R. Patent*, No. 181,090; *Chem. Abstr.*, **65**, 8760g (1966).
154. K. M. Trenke, M. S. Nemtsov and M. M. Kiseleva, *Zh. Org. Khim.*, **3**, 1365 (1967).

155. K. M. Trenke, M. S. Nemtsov and M. M. Kiseleva, *Zh. Org. Khim.*, **5**, 247 (1969).
156. V. Z. Sharf, L. Kh. Freidlin, V. I. Kheifets, V. V. Yakubenok and E. A. Shefer, *Neftekhimiya*, **13**, 832 (1973); *Chem. Abstr.*, **80**, 81644q (1974).
157. Yu. M. Blazhin, S. K. Ogorodnikov, N. S. Gurfein, G. S. Idlis and S. V. Kazakova, *Zh. Org. Khim.*, **11**, 238 (1975).
158. E. Z. Utyanskaya, *Kinetika i Kataliz*, **17**, 1396 (1976).
159. E. Z. Utyanskaya, *Kinetika i Kataliz*, **17**, 1405 (1976).
160. M. S. Nemtsov, M. M. Kiseleva, L. V. Fedulova and T. P. Surnova, *Zh. Prikl. Khim.*, **49**, 430 (1976).
161. M. S. Nemtsov, M. M. Kiseleva, L. V. Fedulova, M. I. Riskin, T. P. Surnova and L. D. Karelina, *Zh. Prikl. Khim.*, **49**, 435 (1976).
162. R. G. R. Bacon and E. H. Farmer, *J. Chem. Soc.*, 1065 (1937).
163. G. B. Bachman and C. G. Goebel, *J. Amer. Chem. Soc.*, **64**, 787 (1942).
164. P. Mastagli and C. de Fournas, *Compt. Rend.*, **250**, 3336 (1960).
165. O.. N. Chupakhin, Z. V. Pushkareva, Z. Yu. Kokoshko and V. G. Kitaeva, *Zh. Obshch. Khim.*, **34**, 3783 (1964).
166. Yu. M. Blazhin, S. K. Ogorodnikov, V. E. Kogan, L. N. Volkova, G. S. Idlis and A. I. Morozova, *U.S.S.R. Patent*, No. 432,121; *Chem. Abstr.*, **81**, 77459m (1974).
167. S. F. Birch and D. T. McAllan, *J. Chem. Soc.*, 2556 (1951).
168. P. Maroni, Y. Maroni-Barnaud and L. Cazaux, *Compt. Rend.*, **257**, 1715 (1963).
169. V. A. Mironov, A. D. Fedorovits and A. A. Ahrem, *Izv. Akad. Nauk SSSR, Ser. Khim.*, 1288 (1973).
170. A. St. Pfau and Pl. Plattner, *Helv. Chim. Acta*, **15**, 1250 (1932).
171. R. T. Arnold, *Helv. Chim. Acta*, **32**, 134 (1949).
172. J. B. Senderers, *Compt. Rend.*, **180**, 790 (1925); *Chem. Abstr.*, **19**, 1857[7,8] (1925).
173. L. Savidan and F. Chanon, *Compt. Rend. (C)* **264**, 716 (1967); *Chem. Abstr.*, **67**, 32112n (1967).
174. A. Laforgue, *Compt. Rend.*, **227**, 352 (1948).
175. V. Z. Sharf, L. Kh. Freidlin, G. K. Oparina, V. I. Kheifets, M. K. Bychkova, G. M. Kopylevich and V. V. Yakubenok, *Izv. Akad. Nauk SSSR, Ser. Khim.*, 1663 (1965).
176. M. S. Nemtsov, M. M. Kiseleva, L. V. Fedulova, M. I. Ryskin and S. S. Botkina, *Zh. Prikl. Khim.*, **49**, 617 (1976).
177. Yu. M. Blazhin, S. K. Ogorodnikov, L. N. Volkova, N. S. Gurfeyn, G. S. Idlis and T. M. Shapovalova, *Zh. Prikl. Khim.*, **47**, 2746 (1974).
178. L. Kh. Freidlin and V. Z. Sharf, *Dokl. Akad. Nauk SSSR*, **136**, 1108 (1961).
179. L. Kh. Freidlin, V. Z. Sharf, M. Bartók and A. A. Nazarjan, *Izv. Akad. Nauk SSSR, Ser. Khim*, 310 (1970).
180. L. Kh. Freidlin, V. Z. Sharf, G. I. Samokhvalov, M. A. Mironol'skaya, I. M. Privalova and M. Ts. Yanotovskii, *Neftekhimiya*, **3**, 104, (1963).
181. V. Z. Sharf, L. Kh. Freidlin and A. A. Nazarjan, *Izv. Akad. Nauk SSSR, Ser. Khim.*, 597 (1970).
182. M. Bartók, *Kém. Közl.*, **48**, 155 (1976).
183. V. N. Ipatieff and H. Pines, *J. Amer. Chem. Soc.*, **67**, 1200 (1945).
184. S. A. Ballard, R. T. Holm and P. H. Williams, *J. Amer. Chem. Soc.*, **72** 5734 (1950).
185. P. Maroni, Y. Maroni-Barnaud and J. Priéto, *Compt. Rend.*, **254**, 2170 (1962).
186. S. Sabetay and J. Bleger, *Bull. Soc. Chim. Fr.*, **47**, 463 (1930).
187. M. F. Clarke and L. N. Owen, *J. Chem. Soc.*, 2103 (1950).
188. I. B. Rapoport, L. B. Itsikson, E. M. Kheifets and G. V. Sidyakova *Neftekhimiya*, **5**, 738 (1965).
189. L. P. Kyriakides, *J. Amer. Chem. Soc.*, **36**, 980 (1914).
190. W. Reppe and U. Hoffmann, *German Patent*, No. 578,994; *Chem. Abstr.*, **28**, 777[1] (1934).
191. A. E. Lorch, *U.S. Patent*, No. 2,386,324; *Chem. Abstr.*, **40**, 1167[3] (1946).
192. Mitsui Chemical Industrial Company, *Japanese Patent*, No. 155,296; *Chem. Abstr.*, **44**, 3002g (1950).
193. *British Patent*, No. 326,185; *Chem. Abstr.*, **24**, 4051 (1930).
194. H. M. Guinot and A. Valet, *French Patent*, No. 942,088; *Chem. Abstr.*, **45**, 632g (1951).

195. H. Nagai, *J. Soc. Chem. Ind. Japan*, **44**, 64, 65 (1941); *Chem. Abstr.*, **35**, 3960³–3960⁵ (1941); **45**, 71, 95, 186, 188, 224, 226, 227 (1942); *Chem. Abstr.*, **46**, 412c–413a (1952).
196. H. Nagai, *Rept. Tokyo Ind. Testing Lab.*, **37**, 129, 143, 152, 162, 169, 177, 186 (1942); *Chem. Abstr.*, **43**, 5359i–5361b (1949).
197. H. Nagai, *J. Soc. Rubber Ind. Japan*, **15**, 350, 358 (1942); *Chem. Abstr.*, **43**, 2015f–2016 (1949).
198. A. Kalischew, *J. Russ. Phys. Chem. Soc.*, **46**, 427 (1914); *Chem. Zentr.*, **85**, II, 1261 (1914).
199. F. C. Whitmore and E. E. Stahly, *J. Amer. Chem. Soc.*, **67**, 2185 (1945).
200. H. E. Zimmerman and J. English, *J. Amer. Chem. Soc.*, **76**, 2294 (1954).
201. T. A. Favorskaya, Yu. M. Portnyagin and T. Y. Hsü, *Zh. Obshch. Khim.*, **29**, 2522 (1959).
202. A. Slawjanow, *J. Russ. Phys. Chem. Soc.*, **39**, 140 (1907); *Chem. Zentr.*, **78**, II, 134 (1907).
203. J. English, C. A. Russel and F. V. Brutcher, *J. Amer. Chem. Soc.*, **72**, 1653 (1950).
204. J. English and F. V. Brutcher, *J. Amer. Chem. Soc.*, **74**, 4279 (1952).
205. B. Freudenberg, *Chem. Ber.*, **85**, 78 (1952).
206. H. E. Zimmerman and J. English, *J. Amer. Chem. Soc.*, **76**, 2285 (1954).
207. H. E. Zimmerman and J. English, *J. Amer. Chem. Soc.*, **76**, 2291 (1954).
208. M. Mazet and M. Desmaison-Brut, *Bull. Soc. Chim. Fr.*, 2656 (1971).
209. F. Schubert, *Monatsh. Chem.*, **24**, 251 (1903).
210. T. E. Maggio and J. English, *J. Amer. Chem. Soc.*, **83**, 968 (1961).
211. R. H. Hasek, R. D. Clark and J. H. Chaudet, *J. Org. Chem.*, **26**, 3130 (1961).
212. F. V. Brutcher and H. J. Cenci, *J. Org. Chem.*, **21**, 1543 (1956).
213. A. W. Allan, R. P. A. Sneeden and J. M. Wilson, *J. Chem. Soc.*, 2186 (1959).
214. V. P. Hirsjärvi, *Suomen Kemistilehti*, **36B**, 51 (1963).
215. R. Lukes and V. Galik, *Chem. Listy*, **49**, 1832 (1955); *Coll. Czech. Chem. Commun.*, **21**, 620 (1956).
216. T. A. Favorskaya and Yu. M. Portnyagin, *Zh. Obshch. Khim.*, **34**, 1065 (1964).
217. H. Rupe and O. Klemm, *Helv. Chim. Acta*, **21**, 1538 (1938).
218. T. A. Geissman and L. Morris, *J. Amer. Chem. Soc.*, **66**, 716 (1944).
219. H. Rupe and K. Schäfer, *Helv. Chim. Acta*, **11**, 463 (1928).
220. T. A. Geissman and V. Tulagin, *J. Amer. Chem. Soc.*, **63**, 3352 (1941).
221. A. Franke, *Monatsh. Chem.*, **17**, 85 (1896).
222. M. A. Perry and R. E. De Busk, *U.S. Patent*, No. 2,870,214; *Chem. Abstr.*, **53**, 11229b (1959).
223. M. A. Perry, F. C. Canter, R. E. De Busk and A. G. Robinson, *J. Amer. Chem. Soc.*, **80**, 3618 (1958).
224. P. Maroni, Y. Maroni-Barnaud and L. Cazaux, *Compt. Rend.*, **257**, 1867 (1963).
225. H. Blatz, L. Schröder, S. Poredda and H. W. Zimny, *J. Prakt. Chem.*, **29**, 250 (1965).
226. G. Mossler, *Monatsh. Chem.*, **24**, 595 (1903).
227. H. Rupe and P. Schlochoff, *Chem. Ber.*, **38**, 1498 (1905).
228. R. G. Fargher and W. H. Perkin, *J. Chem. Soc.*, **105**, 1360 (1914).
229. A. Franke and F. Lieben, *Monatsh. Chem.*, **43**, 225 (1922).
230. J. Doeuvre, *Bull. Soc. Chim. Fr.*, **45**, 356 (1929).
231. A. F. Shepard and J. R. Johnson, *J. Amer. Chem. Soc.*, **54**, 4385 (1932).
232. R. F. Naylor, *J. Chem. Soc.*, 1106 (1947).
233. G. Wittig and O. Bub, *Liebigs Ann. Chem.*, **566**, 113, 127 (1950).
234. A. G. Brook, H. L. Cohen and G. F. Wright, *J. Org. Chem.*, **18**, 447 (1953).
235. R. Roger and D. M. Shepherd, *J. Chem. Soc.*, 812 (1954).
236. H. Dornow and W. Bartsch, *Chem. Ber.*, **87**, 633 (1954).
237. M. F. Ansell, W. J. Hickinbottom and A. A. Hyatt, *J. Chem. Soc.*, 1781 (1955).
238. T. A. Favorskaya and N. P. Ryzhova, *Zh. Obshch. Khim.*, **26**, 423 (1956).
239. T. A. Favorskaya, O. V. Sergievskaya and N. P. Ryzhova, *Zh. Obshch. Khim.*, **27**, 937 (1957).
240. T. A. Favorskaya, *Zh. Obshch. Khim.*, **31**, 86 (1961).

241. J. Schneiders, *German Patent*, No. 1,043,342; *Chem. Abstr.*, **55**, 2686a (1961).
242. V. G. Bukharov and T. E. Pozdnyakova, *Izv. Akad. Nauk SSSR, Ser. Khim.*, 135 (1961).
243. B. Waegell and G. Ourisson, *Bull. Soc. Chim. Fr.* 503 (1963).
244. B. G. Hudson and R. Barker, *J. Org. Chem.*, **32**, 3650 (1967).
245. H. Kessler, *Tetrahedron Letters*, 1461 (1968).
246. K. Yamakawa and M. Moroe, *Tetrahedron*, **24**, 3615 (1968).
247. I. L. Kotlyarevskii, M. S. Shvartsberg and Z. P. Trotsenko, *Zh. Obshch. Khim.*, **30**, 440 (1960).
248. M. Lj. Mihailović, S. Gojković and Ž. Čveković, *J. Chem. Soc. Perkin I*, 2460 (1972).
249. W. Reppe, *Liebigs Ann. Chem.*, **596**, 80 (1955).
250. M. C. Kloetzel, *J. Amer. Chem. Soc.*, **62**, 3405 (1940).
251. N. R. Easton, C. A. Lukach, V. B. Fish and P. N. Craig, *J. Amer. Chem. Soc.*, **75**, 4731 (1953).
252. A. T. Blomquist, E. S. Wheeler and Y. Chu, *J. Amer. Chem. Soc.*, **77**, 6307 (1955).
253. F. F. Blicke, P. E. Wright and W. A. Gould, *J. Org. Chem.*, **26**, 2114 (1961).
254. O. H. Huchler, S. Winderl, H. Mueller and H. Hoffmann, *German Patent*, No. 2,503,750; *Chem. Abstr.*, **85**, 142983b (1976).
255. J. R. Johnson and O. H. Johnson, *J. Amer. Chem. Soc.*, **62**, 2615 (1940).
256. I. G. Farbenindustrie A.-G., *British Patent*, No. 505,904; *Chem. Abstr.*, **33**, 9328[3] (1939).
257. W. Reppe and H. G. Trieschmann, *U.S. Patent*, No. 2,251,835; *Chem. Abstr.*, **35**, 7421[8] (1941).
258. W. Reppe, *U.S. Patent*, No. 2,251,292; *Chem. Abstr.*, **35**, 6982[9] (1941).
259. A. Pernot and A. Willemart, *Bull. Soc. Chim. Fr.*, 321 (1953).
260. C. Crisan, *Ann. Chim. (Paris)*, [13] **1**, 436, 462 (1956).
261. I. L. Kotlyarevskii, L. B. Fischer, A. S. Zanina, M. P. Terpugova, A. N. Volkov and M. S. Shvartsberg, *Izv. Vysshikh Ucheb. Zavedenii Khim. i Khim. Technol.*, **2**, 608 (1959).
262. J. Colonge and H. Robert, *Bull. Soc. Chim. Fr.*, 736 (1960).
263. J. Colonge and P. Lasfargus, *Bull. Soc. Chim. Fr.* 177 (1962).
264. A. S. Zanina, C. I. Shergina and I. L. Kotlyarevskii, *Zh. Prikl. Khim.*, **36**, 203 (1963).
265. W. Reppe and H. G. Trieschmann, *U.S. Patent*, No. 2,251,835; *Chem. Abstr.*, **35**, 7421[8] (1941).
266. I. G. Farbenindustrie A.-G., *French Patent*, No. 843,305; *Chem. Abstr.*, **35**, 1068[8] (1941).
267. K. B. Alberman and F. B. Kipping, *J. Chem. Soc.*, 779 (1951).
268. R. Ratouis and A. Willemart, *Compt. Rend.*, **233**, 1124 (1951).
269. T. Shono, A. Oku, T. Morikawa, M. Kimura and R. Oda, *Bull. Chem. Soc. Japan*, **38**, 940 (1965).
270. *British Patent*, No. 506,674; *Chem. Zentr.*, **II**, 3346 (1939); *Chem. Abstr.*, **33**, 9328[2] (1939).
271. C. Ferrero and H. Schinz, *Helv. Chim. Acta*, **39**, 2109 (1956).
272. R. K. Hill and S. Barcza, *J. Org. Chem.*, **27**, 317 (1962).
273. C. Kh. Begidov, I. A. D'yakonov and I. K. Korobitsyna, *Zh. Obshch. Khim.*, **33**, 2421 (1963).
274. D. G. Farnum and M. Burr, *J. Org. Chem.*, **28**, 1387 (1963).
275. G. Dana and J. P. Girault, *Bull. Soc., Chim. Fr.*, 1650 (1972).
276. E. H. Farmer, C. D. Lawrence and W. D. Scott, *J. Chem. Soc.*, 510 (1930).
277. T. A. Favorskaya and N. V. Shcherbinskaya, *Zh. Obshch. Khim.*, **23**, 2009 (1953).
278. W. J. Wasserman and M. C. Kloetzel, *J. Amer. Chem. Soc.*, **75**, 3036 (1953).
279. R. E. Lutz and C. L. Dickerson, *J. Org. Chem.*, **27**, 2040 (1962).
280. W. Reppe, O. Hecht and A. Steinhofer, *German Patent*, No. 700,036; *Chem. Abstr.*, **35**, 6982[9] (1941).
281. S. F. Birch, R. A. Dean and E. V. Whitehead, *J. Org. Chem.*, **19**, 1449 (1954).
282. G. A. Haggis and L. N. Owen, *J. Chem. Soc..*, 399 (1953).
283. I. G. Farbenindustrie A.-G., *British Patent*, No. 508, 548; *Chem. Abstr.*, **34**, 779[7] (1940).
284. I. G. Farbenindustrie A.-G., *British Patent*, No. 510,949; *Chem. Abstr.*, **34**, 5466[6] (1940).

285. General Electric Company, *Dutch Patent*, No. 74 16,316; *Chem. Abstr.*, **86**, 139823q (1977).
286. H. R. Arnold and J. E. Carnahan, *U.S. Patent*, No. 2,591,493; *Chem. Abstr.*, **47**, 1179d (1953).
287. G. A. Haggis and L. N. Owen, *J. Chem. Soc.*, 389 (1953).
288. E. L. Wittbecker, H. K. Hall and T. W. Campbell, *J. Amer. Chem. Soc.*, **82**, 1218 (1960).
289. Yu. K. Yur'ev, G. Ya. Kondrat'eva and E. P. Smyslova, *Zh. Obshch. Khim.*, **22**, 694 (1952).
290. L. Kh. Freidlin, V. Z. Sharf and M. A. Abidov, *Neftekhimiya*, **4**, 308 (1964).
291. L. Kh. Freidlin and V. Z. Sharf, *Izv. Akad. Nauk SSSR, Ser. Khim.*, 1700 (1960); *Zh. Prikl. Khim.*, **35**, 212 (1962).
292. L. Kh. Freidlin and V. Z. Sharf, *Izv. Akad. Nauk SSSR, Ser. Khim.*, 2055 (1960).
293. R. C. Olberg, H. Pines and V. N. Ipatieff, *J. Amer. Chem. Soc.*, **66**, 1096 (1944).
294. H. Pines and J. Manassen, *Advan. Catalysis*, **16**, 49 (1966).
295. M. T. Rincon and R. M. Perez Gutierrez, *Rev. Soc. Quim. Mex.*, **19**, 130 (1975); *Chem. Abstr.*, **84**, 163780c (1976).
296. A. M. Bourns and R. V. V. Nicholls, *Can. J. Res.* **26B**, 81 (1948); *Chem. Abstr.*, **42**, 4928 i (1948).
297. Kh. I. Areshidze and G. O. Chivadze, *Khim. Geterotsikl. Soedin.*, 195 (1969).
298. Yu. K. Yur'ev, G. Ya. Kondrat'eva and N. K. Sadovaya, *Zh. Obshch. Khim.*, **23**, 844 (1953).
299. Yu. K. Yur'ev and G. Ya. Kondrat'eva, *Zh. Obshch. Khim.*, **24**, 1645 (1959).
300. Yu. K. Yur'ev and O. M. Revenko, *Vest. Mosk. Univ., Ser. II., Khim.*, **17**, 68 (1962); *Chem. Abstr.*, **58**, 4500h (1963).
301. Yu. K. Yur'ev, Yu. A. Pentin, O. M. Revenko and E. I. Lebedeva, *Neftekhimiya*, **2**, 137 (1962); *Chem. Abstr.*, **59**, 557f (1963).
302. V. I. Ivanskii and B. N. Dolgov, *Kinetika i Kataliz*, **4**, 165 (1963).
303. H. Pines and P. Steingaszner, *J. Catal.*, **10**, 60 (1968).
304. N. I. Shuykin, M. Bartók, R. A. Karakhanov and V. M. Shostakovskii, *Acta Phys. Chem. Szeged*, **9**, 124 (1963).
305. M. Bartók and Á. Molnár, *Acta Chim. Acad. Sci. Hung*, **100**, 203 (1979).
306. H. Pines and T. P. Kobylinski, *J. Catal.*, **17**, 394 (1970).
307. Á. Molnár and M. Bartók, *React. Kinet. Catal. Lett.*, **4**, 425 (1976).
308. *German Patent*, No. 850,750; *Chem. Zentr.*, 3797 (1953).
309. B. T. Gillis and P. E. Beck, *J. Org. Chem.*, **28**, 1388 (1963).
310. J. Jacobus, *J. Org. Chem.*, **38**, 402 (1973).
311. H. Neudeck and K. Schlögl, *Monatsh. Chem.*, **106**, 229 (1975).
312. E. Fromm and B. Ungar, *Chem. Ber.*, **56**, 2286 (1923).
313. Yu. K. Yur'ev and K. Yu. Novitskii, *Dokl. Akad. Nauk SSSR*, **67**, 863 (1949); *Chem. Abstr.*, **44**, 1904g (1950).
314. G. J. Laemmle, *U.S. Patent*, No. 2,777, 846; *Chem. Abstr.*, **51**, 8810c (1957).
315. J. Colonge and G. Clerc, *Bull. Soc. Chim. Fr.*, 834 (1955).
316. R. Lukeš, O. Štrouf and M. Ferles, *Chem. Listy*, **50**, 1624 (1956).
317. V. I. Nikitin and M. M. Tulyaganov, *Zh. Obshch. Khim.*, **32**, 1433 (1962).
318. J. Egyed, P. Demerseman and R. Royer, *Bull. Soc. Chim. Fr.*, 3014 (1973).
319. A. Franke and O. Liebermann, *Monatsh. Chem.*, **43**, 589 (1922).
320. A. Franke, *Monatsh. Chem.*, **53/54**, 577 (1929).
331. T. C. Snapp and A. E. Blood, *U.S. Patent*, No 3,766179; *Chem. Abstr.*, **80**, 14939b
322. A. Franke, A. Kroupa and T. Panzer, *Monatsh. Chem.*, **60**, 106 (1932).
323. Yu. K. Yur'ev, V. I. Gusev, V. A. Tronova and P. P. Yurilin, *Zh. Obshch. Khim.*, **11**, 344 (1941); *Chem. Abstr.*, **35**, 5893[6] (1941).
324. A. Franke and F. Lieben, *Monatsh. Chem.*, **35**, 1431 (1914).
325. A. Franke and A. Kroupa, *Monatsh. Chem.*, **69**, 167 (1936).
326. C. Schuster and H. Lattermann, *German Patent*, No. 840,844; *Chem. Abstr.*, **52**, 16388g (1958).
327. G. Ohloff, K. H. Schulte-Elte and B. Willhalm, *Helv. Chim. Acta*, **47**, 602 (1964).
328. G. Ohloff, K. H. Schulte-Elte and B. Willhalm, *Helv. Chim. Acta*, **49**, 2135 (1966).

329. M. Lj. Mihailović, *Lectures in Heterocyclic Chemistry*, **3**, S-111 (1976).
330. P. Dimroth and H. Pasedach, *Agnew. Chem.*, **72**, 865 (1960).
331. T. C. Snapp and A. E. Blood, *U.S. Patent*, No. 3,766,179; *Chem. Abstr.*, **80**, 14939b (1974).
332. R. K. Summerbell, D. M. Jerina and R. J. Grula, *J. Org. Chem.*, **27**, 4433 (1962).
333. Badische Anilin- und Soda-Fabrik, A.-G., *German Patent*, No. 1,064,957; *Chem. Abstr.*, **56**, 455i (1962).
334. N. Clauson-Kaas, *Acta Chem. Scand.*, **15**, 1177 (1961).
335. T. A. Favorskaya and O. V. Sergievskaya, *Zh. Obshch. Khim.*, **25**, 1509 (1955).
336. J. A. S. Hammond, *U.S. Patent*, No. 2,715,649; *Chem. Abstr.*, **50**, 7840b (1956).
337. E. E. Connolly, *J. Chem. Soc.*, 338 (1944).
338. N. D. Zelinskiĭ and A. N. Titowa, *Chem. Ber.*, **64**, 1399 (1931).
339. L. N. Owen and P. A. Robins, *J. Chem. Soc.*, 320 (1949).
340. G. J. Gogek, R. Y. Moir and C. B. Purves, *Can. J. Chem.*, **29**, 946 (1951).
341. T. A. Favorskaya and O. V. Sergievskaya, *Zh. Obshch. Khim.*, **28**, 3232 (1958).
342. V. M. Vlasov, T. A. Favorskaya, A. S. Lozhenitsyna and T. S. Kuznetsova, *Izv. Akad. Nauk. SSSR, Ser. Khim.*, 764 (1966).
343. T. A. Favorskaya, A. S. Medvedeva, G. G. Chichkareva, N. D. Abdullaev and V. M. Vlasov, *Zh. Org. Khim.*, **4**, 1743 (1968).
344. A. S. Medvedeva, T. A. Favorskaya, V. M. Vlasov and L. P. Safranova, *Zh. Obshch. Khim.*, **38**, 43 (1968).
345. A. S. Medvedeva, T. A. Favorskaya, M. M. Demina, L. P. Safranova and V. M. Vlasov, *Zh. Org. Khim.*, **5**, 447 (1969).
346. M. F. Shostakovskii, T. A. Favorskaya, A. S. Medvedeva and M. M. Demina, *Zh. Org. Khim.*, **6**, 435 (1970).
347. T. A. Favorskaya, V. M. Vlasov, A. S. Lozhenitsyna and G. G. Chichkareva, *Zh. Obshch. Khim.*, **36**, 1892 (1966).
348. L. Kh. Freidlin, V. Z. Sharf and N. S. Andreev, *Izv. Akad. Nauk SSSR, Ser. Khim.*, 373 (1961).
349. Yu. K. Yur'ev and G. Ya. Kondrat'eva, *Zh. Obshch. Khim.*, **26**, 275 (1956).
350. I. N. Nazarov and M. V. Mavrov, *Zh. Obshch. Khim.*, **28**, 3061 (1958).
351. Yu. K. Yur'ev, I. K. Korobitsyna and E. K. Brige, *Zh. Obshch. Khim.*, **20**, 744 (1950).
352. C. Prévost, *Bull. Soc. Chim. Fr.*, **11**, 218 (1944).
353. A. Valette, *Compt. Rend.*, **223**, 907 (1946).
354. A. Valette, *Ann. Chim. (Paris)*, [12] **3**, 644 (1948); *Chem. Abstr.*, **43**, 2577i (1949).
355. V. Z. Sharf, L. Kh. Freidlin and A. A. Nazaryan, *Neftekhimiya*, **8**, 258 (1968); *Chem. Abstr.*, **69**, 76081x (1968).
356. L. Kh. Freidlin, V. Z. Sharf and A. A. Nazaryan, *Neftekhimiya*, **6**, 608 (1966); *Chem. Abstr.*, **65**, 18460d (1966).
357. H. Morrison and S. R. Kurowsky, *Chem. Commun.*, 1098 (1967).
358. R. E. Lutz, R. G. Bass and D. W. Boykin, *J. Org. Chem.*, **29**, 3660 (1964).
359. R. A. Darby and R. E. Lutz, *J. Org. Chem.*, **22**, 1353 (1957).
360. V. Prelog and W. Küng, *Helv. Chim. Acta*, **39**, 1394 (1956).

CHAPTER **17**

Enol ethers—structure, synthesis and reactions

PETER FISCHER
Institut für Organische Chemie, Biochemie und Isotopenforschung, Universität Stuttgart, Stuttgart, Bundesrepublik Deutschland

I. INTRODUCTION

The terms *enol ether* and *vinyl ether* are both generally used to designate *O*-alkyl derivatives of the enolized form of carbonyl compounds, specifically of aldehydes and ketones (equation 1). The proposed further differentiation into *enol* ethers[1], as

$$\text{\Large $\underset{}{\overset{}{>}}\!CH\!-\!C\!\!<\!\!\overset{O}{}}\quad \rightleftharpoons \quad \text{\Large $>\!C\!=\!C\!<\!\!\overset{OH}{}$}\quad \text{---}\!\!\rightarrow\quad \text{\Large $>\!C\!=\!C\!<\!\!\overset{OR}{}$} \tag{1}$$

$$\qquad\qquad\qquad\qquad \textit{en-ol}\qquad\qquad\qquad \textit{enol ether}$$

derived from parent compounds which are enolized extensively (for instance 1,3-diketones etc.), and *vinyl* ethers — derivatives of normal aldehydes and ketones where this is not the case — does not seem practical except for classifying the individual synthetic procedures[2]. However, there is a dual way of approaching the chemistry of the enol ethers: their *prima facie* structure allows them to be characterized either simply as α,β-unsaturated ethers (1) or, on the other hand, as $+M$-substituted, i.e. activated alkenes (2). Since organic chemistry utilizes enol

$$\text{\Large $>\!\!\overset{|}{C}{}^{\beta}\!=\!\overset{|}{C}{}^{\alpha}\!-\!O\!-\!\overset{|}{C}{}^{\alpha'}\!-$}\qquad\qquad\qquad \text{\Large $>\!\overset{}{C}{}^{2}\!-\!\overset{\bar{O}-\overset{|}{C}{}^{\alpha}-}{\underset{}{C}{}^{1}}\!\!<$}$$

$$\qquad\qquad (1)\qquad\qquad\qquad\qquad\qquad\qquad (2)$$

ethers as functional derivatives for the more facile chemical modification of the parent C=O compound, we shall consider almost exclusively the second aspect, as Effenberger has done in his review on the subject[3]. A note is still necessary on the naming of the enol ethers: they used to be designated according to the generic principle, alkyl alkenyl ether, until, with the latest collective index, *Chemical Abstracts* introduced systematic nomenclature for the enol ethers. However, we shall retain the ether nomenclature, where convenience and lucidity demand it; a concordance of systematic and established names is presented in Table 1 for some of the more common members.

Four basic types of enol ether reactions are outlined in Scheme 1; three of these (halogenation, hydrolysis and polymerization) had already been found by Wislicenus who first synthesized ethyl vinyl ether in 1878:[4]

(1) Polymerization in the presence of Lewis acids.
(2) Reaction with protonic species HX, leading either to restitution of the parent carbonyl compound (hydrolysis) or to derivatives such as acetals (addition of ROH).
(3) Electrophilic attack by reagents E—X; thus, addition and/or substitution products may be formed, the latter either directly via a σ-complex mechanism or in the course of an addition—elimination process.
(4) Cycloaddition, with the regiochemistry determined by the polarization of the enol ether π-system.

With the exception of truly concerted cycloadditions, the initial step in each case is the attack of an electrophile (Lewis acid, H^+, E^+) at the β-carbon of the enol ethers. Their chemistry is thus characterized by a close analogy to the chemistry of enamines which in the past 25 years have gained increasing preparative importance[5,6]. In both classes of compounds, excess π-electron density facilitates an electrophilic attack at the β-carbon, the higher relative nucleophilic potential of the enamines being due to the greater weight of the ammonium as compared with the oxonium resonance structure, **3b** vs. **4b**. This higher reactivity, i.e. the better

TABLE 1. Established *alkyl alkenyl ether* designation and systematic name for some of the more common enol ethers

Common name	Formula	Systematic nomenclature
Methyl vinyl ether	$CH_3-O-CH=CH_2$	Methoxyethene
Ethyl vinyl ether	$C_2H_5-O-CH=CH_2$	Ethoxyethene
Propyl vinyl ether	$C_3H_7-O-CH=CH_2$	1-(Ethenyloxy)propane
Isopropyl vinyl ether	$(CH_3)_2CH-O-CH=CH_2$	2-(Ethenyloxy)propane
Butyl vinyl ether	$C_4H_9-O-CH=CH_2$	1-(Ethenyloxy)butane
Isobutyl vinyl ether	$(CH_3)_2CHCH_2-O-CH=CH_2$	1-(Ethenyloxy)-2-methylpropane
t-Butyl vinyl ether	$(CH_3)_3C-O-CH=CH_2$	2-(Ethenyloxy)-2-methylpropane
Benzyl vinyl ether	$C_6H_5CH_2-O-CH=CH_2$	[(Ethenyloxy)methyl] benzene
Divinyl ether	$H_2C=CH-O-CH=CH_2$	1,1'-Oxybisethene
Propenyl ethyl ether	$C_2H_5-O-CH=CHCH_3$	1-Ethoxy-1-propene (E/Z)[a]
Propenyl isopropyl ether	$(CH_3)_2CH-O-CH=CHCH_3$	1-(1-Methylethoxy)-1-propene (E/Z)[a]
Isopropenyl ethyl ether	$C_2H_5-O-C(CH_3)=CH_2$	2-Ethoxy-1-propene (E/Z)[a]
Butenyl ethyl ether	$C_2H_5-O-CHC_2H_5$	1-Ethoxy-1-butene (E/Z)[a]
Isobutenyl ethyl ether	$C_2H_5-O-CH=C(CH_3)_2$	1–Ethoxy-2-methyl-1-propene (E/Z)[a]

[a]The *E/Z* designation replaces the usual *cis/trans* nomenclature.

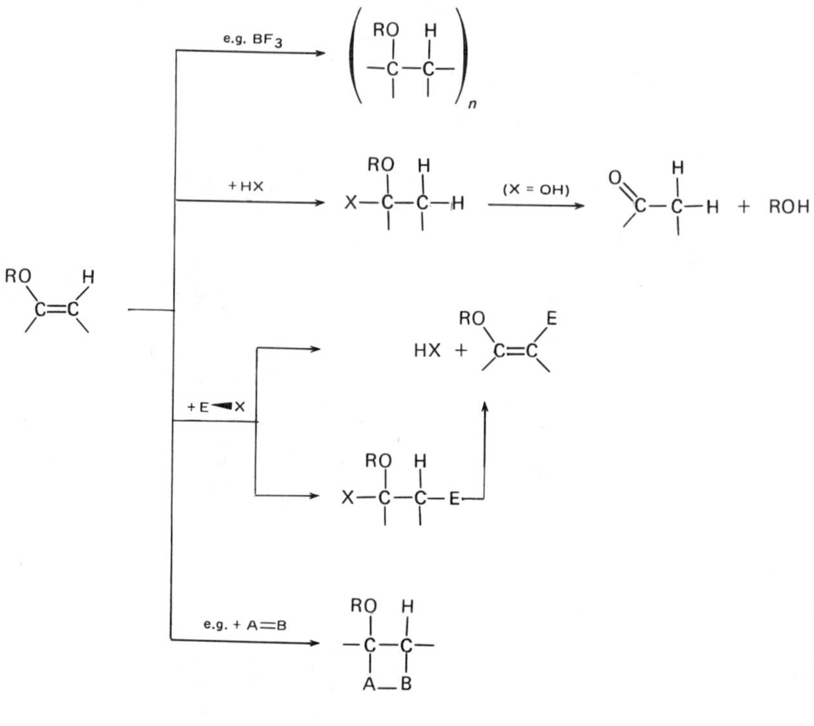

SCHEME 1.

(3a)   (3b) 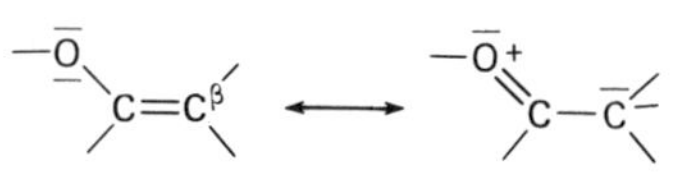   (4a)   (4b)

availability of the highest occupied MO for an electrophile, is tantamount, though, to a much lower oxidation potential. Since most electrophiles are at the same time oxidants, enamines are far more susceptible to radical side-reactions, e.g. in halogenation, than enol ethers. Actually, both classes of functional derivatives of carbonyl compounds complement each other rather well. C-Acylation with phosgene, oxalylchloride, or sulphonyl isocyanates, for instance, proceeds smoothly with enol ethers, while with enamines stable N-acyl products are formed which, as highly deactivated olefins, no longer undergo β-C reaction. On the other hand, it is sometimes rather difficult to find reagents with sufficient electrophilic potential to react with the enol ethers without at the same time inducing cationic polymerization (Friedel–Crafts-type activation is of course self-prohibitive).

In derivatizing the parent carbonyl compound, one is free as a rule to choose the ethereal component; the influence of a specific OR moiety on the reaction behaviour of the double bond is therefore an important aspect of enol ether chemistry. The dependence of enamine reactivity upon the nature of the nitrogen substituents is a well-established fact[7,8]. Towards an uncharged π-system in the ground state, the donor potential of the NR_2 groups decreases in the order, $N(C_2H_5)_2 \gtrsim$ pyrrolidino $> N(CH_3)_2 >$ piperidino $>$ morpholino[8,9]. This gradation is especially manifest from the C^β chemical shifts of the N-vinyl dialkylamines[9] (even though extreme care has to be taken if ground-state properties such as ^{1}H- or ^{13}C-NMR data are used for interpretation or prognostication of relative reactivities[8,9]). In *cis*-enamines, steric interaction forces the NR_2 group out of the olefinic plane, sacrificing $N(2p_z)/C=C(\pi)$ overlap (**5**); in Z-1-dialkylamino-1-propenes, the charge transfer from the amino moiety to the π-system is thus reduced to half its value in the corresponding vinyl- and *trans*-propenyl-amines[9]. For *cis* enol ethers, 180° rotation about the C^1-X bond relieves the steric strain and at the same time restores optimum C^1-O overlap conditions (**6**). This double rotational minimum for highest resonance interaction is one of the most significant features of enol ethers.

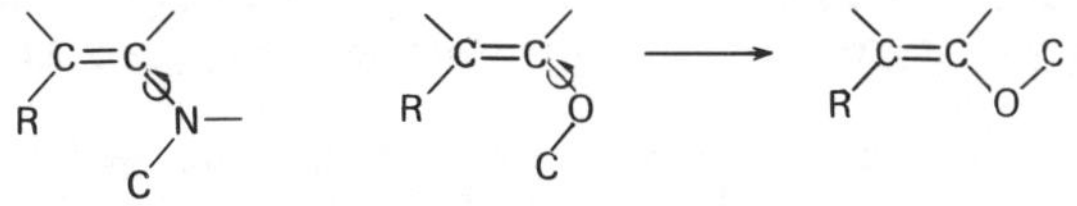

(5)   s-cis   (6)   s-trans

II. PHYSICAL PROPERTIES

A. Conformation

Methyl vinyl ether (**7**, Scheme 2) has been shown by infrared[10] and microwave[11] spectroscopy as well as by electron diffraction[12] to be most stable in a *cisoid* (*syn, s-cis*) form, with a planar heavy atom skeleton C=C—O—C. However, there is unequivocal evidence for the presence of a second conformer[10,12]; from the temperature dependence of the relative intensity of distinctive IR bands, it was shown to be less stable by 4.8 kJ mol^{-1} in the gas phase[10]. This second conformer was suggested to be a *gauche* form with a nonplanar skeleton[10], a result seemingly confirmed by electron diffraction (torsional angle $\phi = 80-110°$)[12]. When, however,

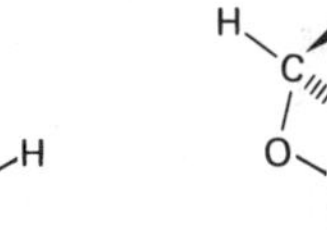

(7)

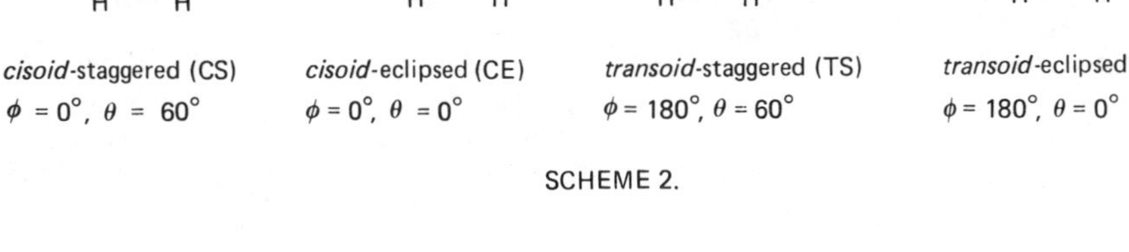

cisoid-staggered (CS)	*cisoid*-eclipsed (CE)	*transoid*-staggered (TS)	*transoid*-eclipsed (TE)
$\phi = 0°, \theta = 60°$	$\phi = 0°, \theta = 0°$	$\phi = 180°, \theta = 60°$	$\phi = 180°, \theta = 0°$

SCHEME 2.

ab initio calculations indicated the second conformer to be the planar *s-trans* form[13], the electron diffraction data were reevaluated[14] by including additional spectroscopic information. On this basis, a torsional angle $\phi \geqslant 150°$ was derived for the minor conformer.

In a detailed *ab initio* calculation of methyl vinyl ether by Epiotis and co-workers[15], the relative orientation of the methyl rotor (θ, see **7**) was also taken into account. Once again, on both the STO-3G and the 4-31G level (minimal and extended basis set), the *cisoid* conformation (CS) constitutes the minimum potential for rotation of the vinyl relative to the CH_3O moiety. A second minimum is obtained for $\phi = 180°$ (TS), 4.2 (STO-3G) or 10.5 kJ mol^{-1} (4-31G) higher than that for the CS orientation. The barrier of rotation (CS → TS) is calculated at about 20 kJ mol^{-1}, with a torsional angle $\phi \sim 70°$ in the transition state. The activation energy for the reverse process, TS → CS, has been determined at 15.5 kJ mol^{-1} by ultrasonic absorption[16]; since one has to add the 2.8 kJ mol^{-1} enthalpy difference in solution, the validity of the *ab initio* calculations appears experimentally well substantiated.

The authors[15] also present a descriptive rationale for understanding the conformational preference of methyl vinyl ether, utilizing Epiotis' concept of non-bonded attraction[17]. For this qualitative MO approach, a π-type CH_3-MO is included, incorporating the 1s AOs of the two methyl hydrogens $H^{a,b}$ in staggered position. (The procedure goes back to an idea of Hehre and Pople[18], and has, in a more general context, been pointed out also by Lister and Palmieri[19].) Since of

$H^{a,b}C^{\alpha}$ O $C^1=C^2$ Five-centre, 6-electron
(π) system

course finite overlap between the $H^{a,b}$ (1s) and $C^2(2p_z)$ orbitals is practical only in the CS orientation, the positive (π) bond order between these two nonbonded centres can exert a stabilizing influence only in the *cisoid* conformation. As a qualitative estimate of interaction energies for both the CS and TS form shows, it is this nonbonded stabilization which accounts for the predominance of the sterically more crowded form. The orbital symmetry approach likewise predicts relative π-bond orders and π-overlap populations in good agreement with the *ab initio* calculations.

The nonbonded attraction argument, as outlined above for methyl vinyl ether, may also be directly applied to the problem of conformational control of the relative stabilities of *geometric* (*E,Z*) isomers[15]. In a fastidious study of the mercuric acetate-catalysed *cis/trans* equilibration of various alkenyl alkyl ethers, Okuyama and collaborators[20] have determined relative thermodynamic stabilities for two homologous series of enol ether *E/Z* pairs (Table 2). In the case of the propenyl ethers (Nos. 1–5, Table 2), when R^2 is a bulky group (isopropyl or *t*-butyl), it is the *Z*-isomer which surprisingly proves to be more stable; for the primary alkyl substituents [$R^2 = CH_3$, C_2H_5, $CH_2CH(CH_3)_2$], on the other hand, the expected order holds (*E* > *Z*).

For an $H^{\alpha}_{a,b}(1s)-C^2(2p_z)$ attractive nonbonded interaction – which provides the additional stabilization for the *cisoid* conformer of methoxyethene – to be operative in other enol ethers also, two α-hydrogen atoms in a *cisoid* staggered position are clearly prerequisite (**8**). This structural condition can be met only in

TABLE 2. Relative E/Z isomer stabilities for alkenyl alkyl ethers (alkoxyalkenes), R^1—CH=CH—OR^2 [a] (equilibrium constants K and enthalpy and entropy changes for $cis \rightarrow trans$ isomerization, in bulk, mercuric acetate-catalysed[20])

No.	R^1	R^2	$K_{cis/trans}$ (25°C)	$\Delta H°$ (kJ mol^{-1})[b]	$\Delta S°$ (J K^{-1} mol^{-1})
1	CH$_3$	CH$_3$	0.968	3.82	6.2
2		C$_2$H$_5$	1.385	1.56	7.9
3		CH$_2$CH(CH$_3$)$_2$	1.431	1.80	9.0
4		CH(CH$_3$)$_2$	2.721	−2.38	0.4
5		C(CH$_3$)$_3$	3.378	−2.86	0.6
2	CH$_3$	C$_2$H$_5$	1.385	1.56	7.9
6	C$_2$H$_5$		0.874	2.37	6.8
7	C$_5$H$_{11}$		0.880	3.85	11.8
8	CH$_2$CH(CH$_3$)$_2$		0.901	2.74	8.3
9	CH(CH$_3$)$_2$		0.583	3.28	6.5
10	C(CH$_3$)$_3$		0.126	7.00	6.2
11	CH$_2$=CH	C$_2$H$_5$	0.450	3.87	6.2
12	C$_6$H$_5$		0.728	1.60	2.7
13	Cl		4.522	−2.77	3.3

[a] For thermodynamic data (K, $\Delta H°$, $\Delta S°$) of 2- and 3-alkoxy-2-alkenes and even higher substituted enol ethers, see the work of Taskinen and coworkers[21-25].
[b] Error limit in the last digit ±0.01–0.02.

E-methoxy- and -ethoxy-1-alkenes, but not in the corresponding isopropoxy and t-butoxy derivatives; in their E-form, these enol ethers are restricted to the *transoid* conformation, and thus lack nonbonded stabilization. For Z-propenyl ethers, *cisoid* orientation of the alkoxy group OR^2 is *a priori* impossible. However, via $C^3 H^{a,b}_3$ $(1s)-O(2p_z)$ interaction (9), a five centre, 6π-electron nonbonded stabilization,

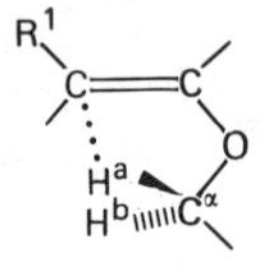

(8) 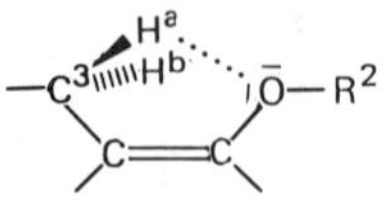   (9)

analogous to that for the E-isomers, may likewise be achieved for the Z-compounds. Though less effective than in 8, this nonbonded attraction (9) quite obviously suffices to swing the balance in favour of the Z-isomer for enol ethers with s- and t-alkoxy groups (see Table 2).

Additional experimental substantiation for this striking argument[15] has come forth recently[21]. Taskinen and his group have in a series of papers reported on the thermodynamics of vinyl ethers, determined from isomerization equilibria such as $10 \rightleftarrows 11 \rightleftarrows 12$ in an inert medium (hexane or cyclohexane, I_2-catalysed)[21]. From

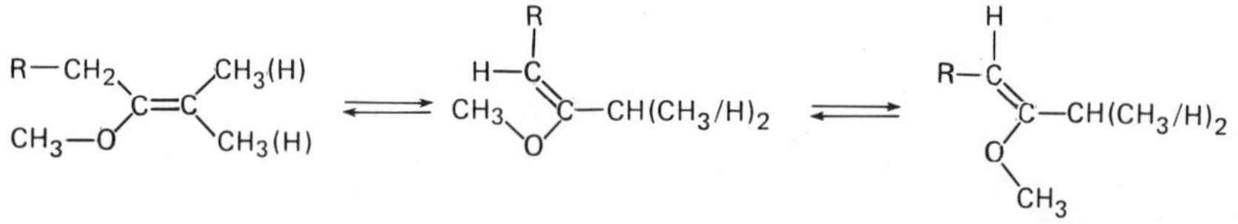

(10)   (11-*E*)   (12-*Z*)

the respective thermodynamic data for the isomerization of various substituted enol ethers[22,23], Taskinen and Anttila have evaluated interaction energies, $S[R^1 \leftrightarrow R^2]$, between two Z-substituents across the C=C double bond of enol ethers (Table 3)[21]. As the negative $S[O \leftrightarrow R^2]$ values reveal, *cis* interaction between CH_3O and alkyl groups is indeed stabilizing. This stabilizing effect decreases sharply, though, from CH_3 to $CH(CH_3)_2$; for $[CH_3O \leftrightarrow C(CH_3)_3]$, Z-interaction is destabilizing already.

TABLE 3. Steric interaction energies for two Z-substituents R^1, R^2 across the C=C bond of enol ethers[21]

R^1	R^2	$S[R^1 \leftrightarrow R^2]$ (kJ mol^{-1})
$C(CH_3)_3$	CH_3	18.2 ± 1.0
$CH(CH_3)_2$	C_6H_5	11 ± 2
	C_2H_5	6.1 ± 0.6
	$CH(CH_3)_2$	6.0 ± 0.6
OCH_3	$C(CH_3)_3$	2.9 ± 0.5
	$CH(CH_3)_2$	-0.7 ± 0.5
	C_2H_5	-1.5 ± 0.5
	C_6H_5	-2.1 ± 0.6
	CH_3	-2.9 ± 0.2

B. Spectral Properties

In photoelectron (PE) spectroscopy, unsaturated ethers are characterized by two low ionization potentials (IP), originated from π-type MOs[26]. The uppermost occupied orbital, as shown by the vibrational fine structure of the first PE band[26,27], is highly populated in the C=C bond, with partial charge transfer from the heteroatom[26,28] ($\pi_{C=C}$); the second MO corresponds mainly to the oxygen lone pair (n_O). By resonance interaction, π_{CC} and n_O, which *per se* have rather similar energies, are split 2–3 eV[29] (the effective mesomeric stabilization for, for example, 3,4-dihydropyran[30] is 1.2 eV). The separation between the first two ionization potentials $IP_{1,2}$ of enol ethers thus provides a sensitive probe for $C=C(\pi)/O(2p_z)$ collinearity[31].

For the *cisoid* conformers of *n*-alkoxyethenes and pyrans, $\Delta IP_{1,2}$ is generally 2.5–3.0 eV[26–31]. At elevated temperatures (510 K), bands of a second conformer emerge in the PE spectrum of methyl vinyl ether[31]; since $\Delta IP_{1,2}$ is even larger for this minor form, it likewise must have planar, i.e. *s-trans* conformation. Large ΔIP values argue a highly resonance-stabilized conformation also for the dominant form of isopropyl vinyl ether and of 2-methoxy-2-butene (14); for sterical reasons, this once again must be the *s-trans* orientation. The lesser conformer of 13 and 14, observed at 510 K, is characterized by a $\Delta IP_{1,2} < 0.5$ eV[31], clearly indicative of *gauche* orientation.

(13)   (14)

Even though conformational isomerism of vinyl ethers was first discovered from vibrational evidence[32,33], IR spectroscopy has proven a rather fickle tool for more detailed structural elucidation. Trofimov and collaborators[34] have ruled out a planar, resonance-stabilized conformation for alkoxyethenes with bulkier OR groups from the analysis of two bands each in the $\nu_{C=C}$, $\nu_{C=O}$ and $\omega_{=CH_2}$ region. They have completely neglected, however, the possibility of *two* planar conformations (CS, TS), considering only a 'planar' and 'nonplanar' form (without C=C/O resonance). In fact, a closer inspection of their published vibrational data reveals that the critical IR absorptions show coalescence rather than true alternate behaviour with increasing bulkiness of OR. For the sterically crowded *Z*-propenyl ethers, IR spectra clearly indicate the presence of only one, probably *gauche*, conformer[35].

In a recent extensive vibrational study of *n*-alkyl vinyl ethers in the gaseous, liquid and solid state[36], the enthalpy differences between major (*cisoid*) and minor conformers were determined from relative Raman intensities in good agreement with the results cited above[10,12,31]. However, the band assignment in this work[36] relies mainly on the — meanwhile revised[14] — electron diffraction results ($\phi = 80–110°$)[12]. Furthermore, a frequency decrease from 586 to 504 cm^{-1} is calculated for the C=C—O bending mode between the *cisoid* and *transoid* forms ($\phi = 0°/180°$); since the actual absorption comes at 526 cm^{-1}, the second conformer is definitely assigned the *skew* orientation ($\phi \sim 120°$). Owen and co-workers[37], in a painstaking comparative analysis of *E/Z*-methyl and -ethyl propenyl ether, likewise found evidence for nonplanarity; using mainly the observed band contours, they favour but slight deviation from the (planar) *s-trans* form. Ford, Katritzky and Topsom[38] also interpret their IR data in terms of a more or less coplanar second conformer for the *n*-alkyl vinyl ethers.

Both ^{13}C- and ^{1}H-NMR respond with a large upfield shift of C-2 and the β-vinyl protons to the increased C-2 π-charge density in the vinyl ethers (15), but detailed

$$(15)$$

analysis once more presents a rather confusing picture. In the first ^{1}H-NMR investigations on vinyl ethers[39,40], the chemical shift difference between cis and trans C-2 protons (15), which depends strongly on the nature of the alkoxy group, was taken as indicative of the relative contribution of the oxonium resonance structure[40]. In fact, however, only the cis proton moves downfield, from δ 4.23 (OCH₃) to 4.76 p.p.m. [OC(CH₃)₃], while $\delta_{H-trans}$ remains largely unaffected (the same behaviour was found for vinyl amines[9,41]). Actually, the authors[40] were interpreting the (anisotropic) shift differences[8] between the s-cis and s-trans form and not the graduation in resonance interaction: variations in π-charge density should affect both protons identically. We ourselves found[42] that the α-OR protons (OCH₂—, OCH₃) of propenyl and butenyl methyl and ethyl ether appear consistently 0.1 p.p.m. better shielded in the trans- than in the respective cis-ethers. In the CS conformation[15] (7), the two cisoid α-protons (H^{a,b}) come to lie well within the shielding region of the C=C anisotropy field[42]; the identical E/Z shift difference for methyl and ethyl ethers are a good argument for both trans compounds adopting the same (CS) conformation.

The groups of Hatada[43] and of Trofimov[44] also report a linear correlation between δ(C-2) and Taft's E_s constants for vinyl ethers with various OR groups. Their conclusion that with increasing bulkiness of R the gauche conformer becomes more and more favoured over the s-cis and s-trans forms is not valid, though, as a downfield shift of comparable magnitude is found for the structurally analogous alkenes[45] (with the ethereal O replaced by CH₂). Rojas and Crandall[46] have systematically investigated a series of alkenyl methyl ethers by ^{13}C-NMR: they report both the C-2 and the OCH₃ resonances at consistently higher field for the trans compounds, indicating the well-known cisoid γ-interaction [C² ↔ OCH₃] (Table 4). The pronounced downfield shift of C-2 in the cis compounds is probably due largely to the spatial interaction [O ↔ C³] and not to steric inhibition of resonance; it is practically independent of the size of both alkyl and alkoxy groups[46]. In the propenyl amine series, on the other hand, where sterical hindrance indeed causes torsion of the NR₂ group[41], thus effectively reducing N(2p$_z$)/C=C resonance, we have found large downfield shifts for C-2 between trans- and cis-enamine (e.g. 18 p.p.m. between E- and Z-1-diethylamino-1-propene)[9].

TABLE 4.

		δ (p.p.m.)			
		cis	trans	Δδ	Ref.
CH₃—C²H=CH—OCH₃	C-2	100.2	96.0	4.2	46
	OCH₃	58.5	54.9	3.6	
n-C₄H₉—C²H=CH—OCH₃	C-2	106.6	101.8	4.8	46
	OCH₃	58.4	54.6	3.8	
C₂H₅—C²H=CH—OC₂H₅				4.0	41

Steiger and coworkers[47] have calculated $^1H/^1H$ and $^1H/^{13}C$ coupling constants for vinyl compounds, and discussed the CNDO/2-derived values in terms of configuration and conformation about the double bond.

The fragmentation of alkyl vinyl ethers in electron impact mass spectrometry (EI-MS) is triggered by H-migration[48]; it proceeds by multiple H-transfer, via 2-methyl-substituted cyclic ether cations[48,49], the most prominent fragment being ionized vinyl alcohol, $CH_2=CH—OH]^{\overset{+}{\cdot}}$ (m/e 44)[48,50]. In ion cyclotron MS, unsaturated compounds undergo [2 + 2]cycloaddition with the molecular ion of methyl vinyl ether[51]. The cycloadducts are then cleaved orthogonally to the original cycloaddition orientation (equation 2), with the major radical cation **16** indicating the position of the double bond in the substrate.

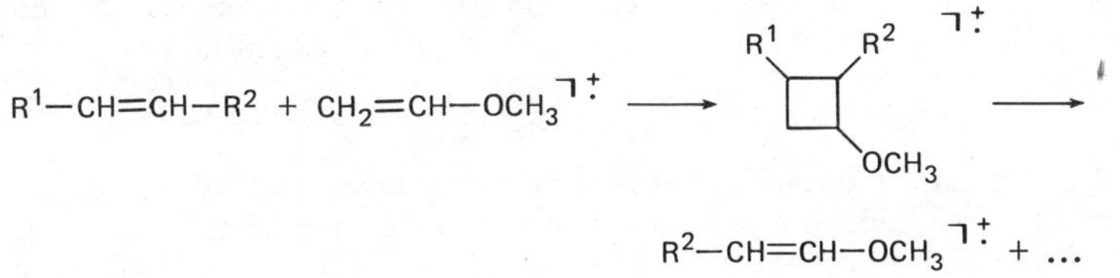

$$R^1—CH=CH—R^2 + CH_2=CH—OCH_3]^{\overset{+}{\cdot}} \longrightarrow$$

$$\text{(2)}$$

$$R^2—CH=CH—OCH_3]^{\overset{+}{\cdot}} + \dots$$

$$\text{(16)}$$

For a series of alkyl and aryl vinyl ethers, dipole moments were correlated with electronic and steric substituent constants[52], and also with relative basicities[53] (determined from ν_{O-H} shifts due to enol ether/phenol hydrogen bonding). From the temperature dependence of the dipole moment of methyl vinyl ether, an attempt was made to estimate μ for the different ethoxyethene conformations[54].

C. Summary: Conformation and Reactivity

The evidence of the reported physical investigations, probing for the molecular ground state of the enol ethers, may be summed up as follows:

For *trans*(*E*)-alkenyl ethers with primary alkoxy substituents, the *cisoid* conformation is always predominant; the second conformer of methyl vinyl ether — at least in the gas state — is either the *s-trans* form or a conformation with ϕ close to 180°.

The corresponding *cis*(*Z*)-alkenyl ethers, as well as vinyl and *E*-alkenyl ethers with bulkier OR groups, adopt the *s-trans* conformation; here, the less stable conformer has *gauche* orientation.

For sterically highly hindered enol ethers (with bulky substitution in geminal and/or *Z*-position at C-2), co-planar orientation is no longer feasible.

However, the electronic stabilization by $O(2p_z)/C=C(\pi)$ resonance in the neutral molecule is limited to interaction with unfilled antibonding MOs. Only in the more or less charged transition state of an electrophilic attack on enol ethers or of cycloaddition reactions, the full mesomeric potential of the +*M*-substituents (OR or NR$_2$) is challenged, and resonance stabilization may easily overcome steric barriers which are prohibitive in the ground state.

In contrast to the *prima facie* controversial interpretation of C=C/OR interaction in the ground state, the evidence on how the nature of the alkoxy group influences the *relative reactivity* of the enol ethers is unequivocal. For the hydrolysis, in charge-transfer complex spectra, towards electrophiles, and in cycloadditions, the inductive hierarchy is strictly observed: $OC(CH_3)_3 > OCH(CH_3)_2 >$

$OC_2H_5 > OCH_3$[55]. The reactivity of alkoxyethene monomers in cationic polymerization likewise follows this order, correlating with Taft's σ_I- or σ^*-constants[56,57].

III. PREPARATION

The various synthetic routes to enol ethers have been comprehensively summarized in a new volume of Houben—Weyl[1,2]. In the approved manner of this handbook, both scope and limitations are outlined for each procedure, and full experimental details given for one exemplary case. We shall therefore confine ourselves to a brief sketch of the most important synthetic pathways, emphasizing mainly recent developments.

The *vinylation of alcohols* by acetylene (equation 3) can be achieved under alkali catalysis (Favorskii[58] and Reppe[59]). For various substituted phenols, Zn, Cd

$$H-C{\equiv}C-H \ + \ ROH \xrightarrow[\text{180—200°C/20—50 bar}]{\text{KOH}} RO-CH{=}CH_2 \qquad (3)$$

and Hg(II) acetate and like catalysts have also been employed successfully[60]. Substantially lower temperatures are required in the case of activated alcohols[61]. With methyl- and *t*-butyl-acetylene, nucleophilic addition of aliphatic alcohols ROH $[R = CH_3 \ldots C(CH_3)_3]$ usually affords α-substituted ethenyl ethers,

$RO-\overset{|}{C}{=}CH_2$[62]; in the case of severe steric crowding, however, *cis*-propenyl ethers are obtained. An alternative, convenient laboratory procedure starts from the diphosphonium salt 17. Alcoholysis of one of the Ph_3P groups yields the intermediate 18 from which the vinyl ether is obtained by alkaline hydrolysis (equation 4)[63]. By using $NaOD/D_2O$ in the last step, β,β-dideuterated ethenyl ethers may be prepared.

$$[Ph_3P^+{-}CH{=}CH{-}^+PPh_3]\ 2\ Br^- \xrightarrow[\text{NEt}_3]{\text{ROH}} [RO{-}CH{=}CH{-}^+PPh_3]\ Br^- \xrightarrow[\text{H}_2\text{O}]{\text{NaOH}}$$
$$\qquad\qquad (17) \qquad\qquad\qquad\qquad\qquad\qquad (18)$$
$$RO{-}CH{=}CH_2 \quad (4)$$

Transvinylation (equation 5) is catalysed by Hg(II) salts of weak acids; the process is reversible[64]. Therefore, if the donating enol ether does not boil higher than the alcohol to be vinylated, or if 19 cannot be distilled off, ethyl vinyl ether

$$R^1O{-}CH{=}CH_2 \ + \ R^2OH \ \underset{\text{Hg(II)}}{\rightleftarrows} \ R^1OH \ + \ R^2O{-}CH{=}CH_2 \qquad (5)$$
$$\qquad\qquad\qquad\qquad\qquad\qquad (19)$$

has to be used in large excess, and the catalyst destroyed before work-up. Vinyl interchange under Pd(II) catalysis proceeds stereospecifically[65], with inversion of the configuration about the C=C double bond; thus, from *E*-propenyl ethyl ether and propanol, *Z*-propenyl propyl ether is formed. The drawback of the method — acetal formation above $-25°C$ — has been overcome with special bidentate Pd(II) complexes[66]. If optically active alcohols are converted to vinyl ethers by Hg(II)-catalysed transvinylation, and then recovered by acid hydrolysis (see below), their optical rotation is retained unimpaired[67] — unequivocal evidence that the vinylic (and not the alkylic) C—O bond is broken in vinyl interchange.

By far the most important laboratory synthesis for enol ethers is the elimination of alcohol from acetals[2] (acid-catalysed: $KHSO_4$, *p*-toluenesulphonic acid, $Ca_3(PO_4)_2$[68] etc.). For high preparative yields, careful separation of the alcohol formed is mandatory[69] since the overall sequence, $>C{=}O \rightleftarrows$ acetal/ketal $\rightleftarrows$ vinyl ether, is fully reversible, and the enol ether equilibrium concentration is only $\sim$50 p.p.m.[70]. (For acetaldehyde and its mono- and di-chloro derivative, the thermodynamics of this sequence have been carefully studied by [14]C- and [3]H-labelling[71].) If one or more isomeric enol ethers can be formed, thermodynamic equilibration of the product mixture may be achieved by traces of acid or, specifically, with iodine[72]. Acetals of acid-labile substrates can be decomposed thermally; especially for steroids, a number of special modifications has been devised[2] (e.g. reaction with 2,2-dimethoxypropane, which is not supposed to proceed via transacetalization). By the method of acid-catalysed pyrolysis ($\sim 150^\circ C/\leqslant 0.1$ Torr)[73], several nitroalkyl vinyl ethers could be prepared in excellent yield[74].

If the acet(ket)alization is carried out with orthoformates[75], the acetals/ketals, especially of cyclanones[69], need not be isolated; with *Amberlyst-15®* and ethyl orthoformate, the procedure can be run in one step ($0^\circ C$, N_2 atmosphere), the enol ethers being formed either directly, or by work-up distillation with a trace of *p*-toluenesulphonic acid[76]. The enols or enolate salts of 1,3-diketo compounds can be alkylated directly at one oxo function (in dipolar aprotic solvents, employing highly reactive alkylating agents with low $S_N 2$ potential and *hard* leaving groups)[77].

The *Horner–Wittig reaction* (equation 6) of triaryl(oxymethylidene)phosphoranes (**20**) with carbonyl compounds provides a versatile access to variously substituted enol ethers[78]; the yields are generally better for R = aryl than for the

$$Ph_3P{=}CH{-}OR^1 + R^2{-}CO{-}R^3 \longrightarrow \underset{R^3}{\overset{R^2}{\diagdown}}C{=}CH{-}OR^1 + Ph_3P{=}O \qquad (6)$$

(**20**)

alkoxymethylidene derivatives. A modified procedure (equation 7)[79], using phosphine oxides (**21**), is far superior to the process via the ylides in scope, yield, use of stable crystalline reagents and ease of product separation. Since the two diastereomeric adducts **22** can be separated chromatographically, sterically pure *E*- and *Z*-isomers of the vinyl ethers may thus be conveniently prepared[79].

$$\underset{(21)}{\overset{O}{\overset{\|}{Ph_2P}}{-}\underset{OCH_3}{\overset{R^1}{\diagup}}CH} + [(CH_3)_2CH]_2NLi \longrightarrow \overset{O}{\overset{\|}{Ph_2P}}{-}\underset{OCH_3}{\overset{R^1}{\diagup}}\overset{=}{C}\; Li^+ \xrightarrow[2.\, H_2O]{1.\, + R^2{-}CO{-}R^3}$$

$$\underset{CH_3{-}O\; R^3}{\overset{O\;\; R^1\; OH}{\overset{\|\;\; |\;\; |}{Ph_2P{-}C{-}C{-}R^2}}} \xrightarrow{NaH/THF} \underset{CH_3O}{\overset{R^1}{\diagup}}C{=}C\underset{R^3}{\overset{R^2}{\diagdown}}$$

(**22**)

$$(7)$$

Symmetrical divinyl ethers have become easily available from the reaction of bis(phosphonium) salts, $Ph_3 P^+{-}CH{=}CH{-}^+PPh_3$, alkoxides and carbonyl compounds[80]. From (alkoxymethane)phosphonic esters with $-M$-substituents in the α-position (**23**), various enol ethers with -acyl functions can be prepared[81]. The $C^1{-}OR$ element of the enol ether need not be supplied from the phosphorane

$$\overset{O}{\overset{\|}{(EtO)_2P}}{-}CH\overset{OR}{\underset{A}{\diagup}}$$

A = COOR, CONH$_2$, COR, Ph

(**23**)

component: examples for this 'reversal of polarity' are the reactions of triphenyl-(alkylidene)phosphoranes with ethyl fluoroacetates[82] or with (alkyl/arylmethoxy-carbene)pentacarbonyltungsten, $(OC)_5 W:CROCH_3$[83].

Rearrangement of allyl alkyl ethers with alkoxides in DMSO leads, stereo-specifically, to the corresponding cis-propenyl ethers[84]; the analogous procedure has been employed for the synthesis of cis-1-dialkylamino-1-propenes[85,41]. In carbohydrate chemistry, this reaction is utilized as the first step in cleaving off allylic protecting groups, followed by hydrolysis of the propenyl ethers[86,87]. Alkoxy-substituted arenes (benzenes, naphthalenes etc.) are transformed to cyclo-hexenyl enol ethers (1-alkoxy-1,4-cyclohexadienes) by either Birch or electrolytic reduction[2].

Further *special procedures* include: dehydrohalogenation of halo ethers and acetals[88,89]; decomposition of β-alkoxy-tosylhydrazones (NaOR, 160°C), yielding, via β-alkoxycarbenes, preferentially cis-enol ethers[90,91]; reaction of methoxyallene with organocopper(I) compounds[92]; CuBr-catalysed reaction of Grignard compounds with α,β-unsaturated acetals (equation 8)[93]; β-alkylation of β-bromovinyl

$$R^1MgX + R^2CH=C(R^3)-CH(OEt_2) \xrightarrow{CuBr} R^1R^2CH-CR^3=CH-OEt \qquad (8)$$

$$R^1 = CH_3 \ldots C(CH_3)_3$$

$$R^2, R^3 = H, CH_3$$

$$X = Cl, Br$$

ethyl ether with RMgBr, in the presence of catalytic amounts of nickel phosphine complexes[94]. Dehydrative decarboxylation of threo-3-hydroxycarbonic acids (24), which are formed with high stereoselectivity[95] from dilithiated carbonic acids and ketones[96] or aldehydes, provides another stereoselective access to enol ethers; reaction of 24 with tosylchloride leads, via the β-lactone, to the E-form, while reaction with the azodicarboxylate/Ph_3P adduct leads to the Z-form[95].

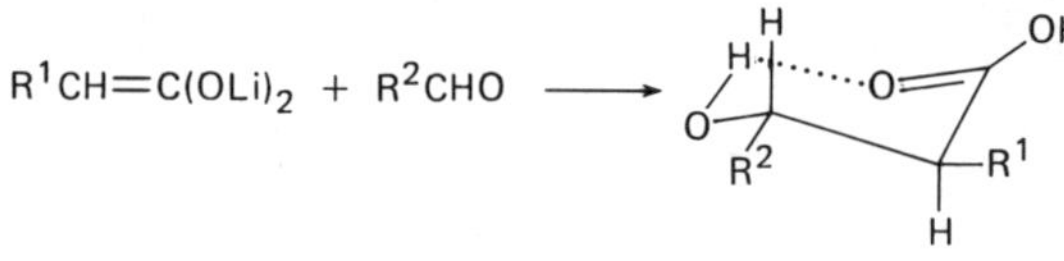

$$R^1CH=C(OLi)_2 + R^2CHO \longrightarrow \qquad (9)$$

(24)

IV. ELECTROPHILIC REACTIONS

In this section, reactions of the enol ethers with electrophilic reagents, $E \mathbin{\leftarrow} X$ or E^+X^-, shall be discussed, regardless of whether addition or substitution products are formed. Cycloadditions, on the other hand, will be dealt with separately.

A. Hydrolysis[97,98]

It is now well established that for the hydrolysis of simple vinyl ethers, proton transfer from the catalysing acid to the substrate is rate-determining (equation 10). Subsequently, the cationic intermediate (25) is rapidly hydrated to the hemiacetal/ketal (26) which in a last, fast step decomposes to the parent carbonyl compound and alcohol. Addition of H_2O to 25 has proven decidedly faster than retrodepro-tonation in all cases investigated so far[99,100], with but one special exception[101].

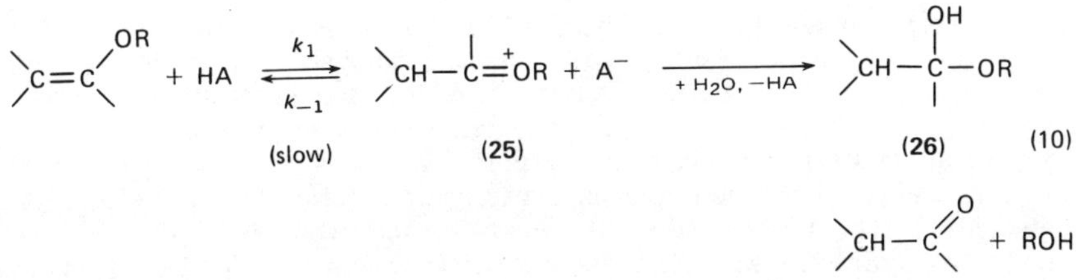

$$\underset{\substack{(\text{slow})}}{\overset{k_1}{\underset{k_{-1}}{\rightleftharpoons}}} \quad \text{(25)} \quad \xrightarrow{+ H_2O, -HA} \quad \text{(26)} \quad \text{(10)}$$

Even for the most reactive member, α-cyclopropylvinyl methyl ether (see Table 5), this mechanism still holds[100], although the margin for the limiting condition, $k_{-1}[A^-] < k_2[H_2O]$, cannot be very large; enamine protonation, for example, is rapidly reversible.

There is a linear relationship between the two sets of log k values for acid-catalysed hydrolysis of a series of vinyl ethers and of the corresponding formaldehyde acetals, $CH_2(OR)_2$[102]; this definitely excludes a nucleophilic function of

TABLE 5. Rates of H_3O^+-catalysed hydrolysis of various enol ethers in aqueous solution (25°C)

Enol ether	$k_{H_3O^+}$ (M^{-1} s^{-1})	Reference
$CH_2=CHOC_2H_5$	1.87	102[a]
$CH_2=CHOC_4H_9$-n	2.00	102[a]
$CH_2=CHOCH_2CH(CH_3)_2$	2.25	102[a]
$CH_2=CHOCH(CH_3)_2$	4.45	102[a]
$CH_2=CHOCH_2CH_2Cl$	0.165	102[a]
$CH_2=CHOC_2H_5$	2.13 ± 0.01	113[b]
$\underset{}{C_6H_5}\overset{OCH_3}{\underset{CH_3}{CH=C}}$	1.66 ± 0.02	108[b]
$CH_2=C\overset{OC_2H_5}{\underset{CH_3}{}}$	(5.79 ± 0.11) 10^2	103[b]
$CH_2=CHOC_6H_5$	(3.28 ± 0.02)10^{-3}	103[b]
$CH_2=C\overset{OC_6H_5}{\underset{CH_3}{}}$	5.98 ± 0.04	103[b]
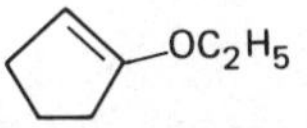 (cyclopentenyl)$-OC_2H_5$	(4.54 ± 0.17)10^2	103[b]
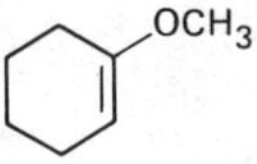 (cyclohexenyl)$-OCH_3$	(4.23 ± 0.04) 10^1	103[b]
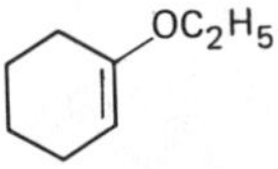 (cyclohexenyl)$-OC_2H_5$	(8.00 ± 0.12)10^1	103[b]
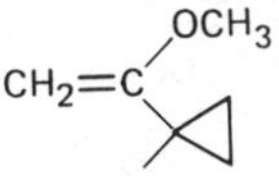 $CH_2=C\overset{OCH_3}{\underset{\triangleleft}{}}$	(7.49)10^3	100[b]

[a] Determined with HCl-catalysis in H_2O.
[b] Determined in aqueous $HClO_4$ solution.

the conjugate base of the catalyst, A^-, in the transition state of vinyl ether hydrolysis. The reaction is subject to general acid catalysis[102,103] for which H_3PO_4 has proven an unusually active catalyst[104]. A Brønsted factor, $\alpha = 0.63$, was determined[103] for the hydrolysis of cyclopentenyl and isopropenyl ethers with carboxylic acid catalysis. This can be interpreted in terms of a significant degree of proton transfer to the enol ether in the transition state[103]. A salt effect was not detected[105]. The unexpected small primary isotope effect, $k_H/k_D = 3.3 - 3.5$, for vinyl ether hydrolysis with HF/H_2O and DF/D_2O was attributed to strong hydrogenic bending vibrations in the transition state[99] (which are absent, of course, in the diatomic H/D donor).

(27)

All this evidence goes to show that the proton transfer is characterized by a rather late transition state, resembling the cationic species; the enol ether 27, for instance, incorporates D mainly in the axial position in deuteriolysis[106]. Consequently, the individual rates of hydrolysis (see Table 5) can be correlated with the stabilities of the intermediate carbenium ions (25), relative to that of the free vinyl ethers. (This is also important for understanding the mechanism of the reaction with electrophiles and of the stereospecific polymerization of enol ethers in homogeneous media[107].) The large rate increase upon α-alkyl substitution (10^2-10^4) thus becomes easily understandable. The slower hydrolysis of β-styryl ethers C_6H_5—CH=C(CH$_3$)—OR (equivalent to an increase in $\Delta G^{\neq}$ of ~12 kJ mol^{-1}) is attributed to additional (resonance) stabilization of the ground state[108]; β-alkyl substituents likewise retard the rate of hydrolysis. The higher reactivity of cis-1-alkenyl ethers, on the other hand, which generally are hydrolysed four times faster than the corresponding trans isomers[107] — irrespective of the relative cis/trans ground-state stability[109] — therefore cannot be due solely to their lesser thermodynamic stability[20]. Within the ethenyl ether series, CH_2=CH—OR, dependence of reactivity on the nature of OR follows the inductive order[110] [0.05 M HCl in acetone/water (80 : 20), 25°C]:

R	CH_3	C_2H_5	$CH_2CH(CH_3)_2$	$CH(CH_3)_2$	$C(CH_3)_3$	CH_2CH_2Cl
Relative rate of hydrolysis[110]	1.0	2.0	1.6	7.3	16.6	0.18

The relative rates are strongly dependent on medium polarity and the acid catalyst[110]; only two sets of vinyl ether hydrolysis data, each obtained for pure H_3O^+ catalysis under identical conditions, are therefore presented in Table 5.

Butadienyl ethers (28) are protonated exclusively at the terminal carbon, C-4[111]; for 29, hydrolysis proceeds via both the normal pathway (rate-limiting C-3 protonation) and protonation at the carbonyl group[112].

C^{4}H$_2$=CH—CH=CH—OR

(28)

(29)

The reaction of vinyl ethers with protic agents other than H_2O[114] (alcohols, mercaptans, acids etc.) follows the same mechanistic course as hydrolysis, with rate-limiting H$^+$-transfer to the olefinic C-2[113]; true electrophilic addition is therefore always in the Markownikoff direction. Within structurally related series of X—H compounds, reactivities towards alkoxyalkenes have been correlated with a variety of σ-constants (see for example Reference 115).

B. Halogenation

The addition of Cl_2 and Br_2 to vinyl ethers has been studied extensively by Shostakovskii and coworkers[116]. The reaction is highly exothermic, often leading to substantial amounts of by-products; by HHal elimination, for instance, and subsequent addition of a second Hal_2 molecule, trihalo ethers are formed (equation 11)[117]. If carried out at $-20°C$ in the dark, however, the reaction of

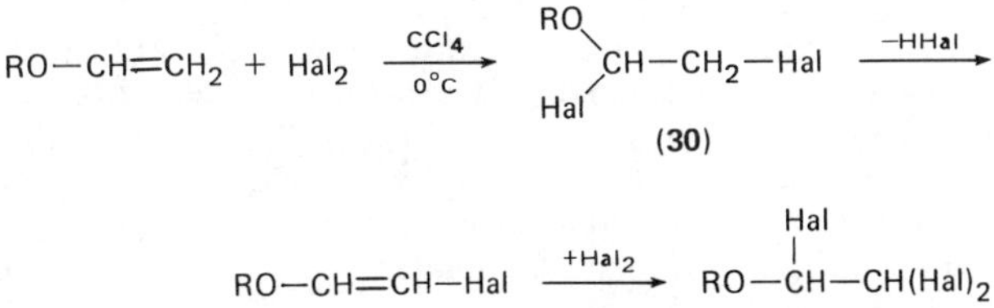

$$RO-CH=CH_2 + Hal_2 \xrightarrow[0°C]{CCl_4} \underset{Hal}{\overset{RO}{>}}CH-CH_2-Hal \xrightarrow{-HHal}$$

$$(30)$$

$$(11)$$

$$RO-CH=CH-Hal \xrightarrow{+Hal_2} RO-\overset{\overset{\displaystyle Hal}{|}}{CH}-CH(Hal)_2$$

Cl_2, Br_2 and ICl even with the more reactive aliphatic enol ethers can be held at the stage of the primary addition compounds (30)[118]. Direct iodination gives only polymers[119]. Fluorination of enol ethers has gained importance in the steroid field; with $FClO_3$ in pyridine, fluorine can be introduced into steroids with excellent yields under mild conditions[120].

The stereochemistry of the reaction with electrophilic halogen is controlled by several factors. Addition of Cl_2 to the dihydropyran 31 in pentane gives stereoselectively the cis-dichloro derivative (80% 32a), while in CH_2Cl_2 the stereochemistry is inverted (66% 32b)[121]; this solvent dependence has been confirmed repeatedly[122]. (HCl addition to 33, on the other hand, is exclusively syn.)

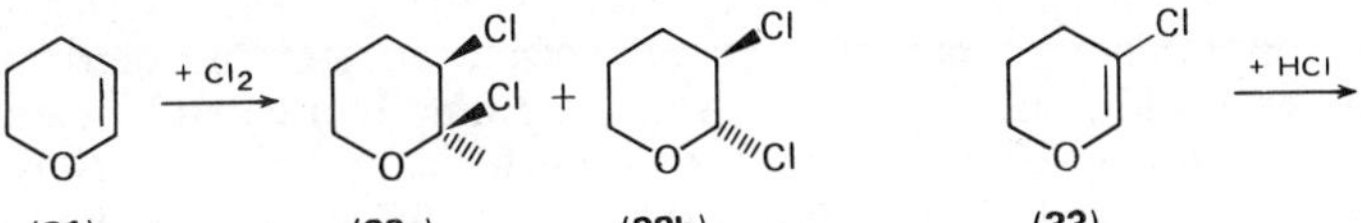

$$(12)$$

(31)   (32a)   (32b)   (33)

Primarily, a 'syn' ion pair is supposed to be formed (34) which in nonpolar solvents rapidly collapses to the cis-dichloro product[121]. The trans reaction can be triggered in three different ways: (1) dissociation of the Cl^-, (2) attack of a protic solvent molecule at C-1 from the backside or (3), for acyclic substrates, rotation of the $RO^+=C^1$ moiety about the C^1-C^2 bond $(34 \to 35)$[121]; this results in trans addition from the collapse of the 'anti' $Cl \cdots Cl$ ion pair (35).

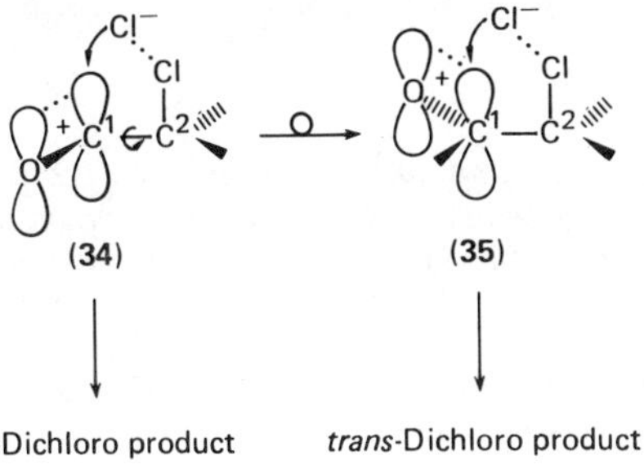

$$(13)$$

(34)   (35)

cis-Dichloro product   trans-Dichloro product

The percentage of *anti* addition increases in the order $Cl_2 < Br_2 < ICl$ and likewise from *p*-methoxy- to *p*-chlorophenyl enol ethers[118] (i.e. with decreasing availability of the ether oxygen lone pair); apparently, halonium stabilization competes more and more with the RO resonance interaction which is the decisive factor in chlorination[121]. This argument has been confirmed by kinetic investigations of iodination and bromination in water: they demonstrate that much less charge is localized at C-1 in the transition state of electrophilic I_2 attack than in protonation[123] (see above); this must be due to iodine participation. For the reaction with Br_2, such halonium stabilization is much less effective[124]. The bromination of acetone in methanol, by the way, proceeds almost exclusively via the enol ether present in the equilibrium, $CH_3-C(OH)=CH_2 \rightleftarrows (CH_3)_2CO \rightleftarrows (CH_3)_2C(OR)_2 \rightleftarrows CH_3-C(OR)=CH_2$[125].

With *N*-bromophthalimide in alcohol or carboxylic acids, cyclic and acyclic enol ethers are transformed into α-bromoacetals in excellent yield[126]; the reaction is definitely ionic and not radical. From the reaction in CCl_4, the addition product of Br^+ and phthalimide can be isolated (65%)[127]; *N*-chloro-, -bromo- and -iodo-succinimide have also been employed successfully[128]. With *t*-butyl hypochlorite in ROH (equation 14), *trans* addition predominates (85%)[129]; in benzyl alcohol or

$$\text{(diagram)} + (CH_3)_3COCl/Br \xrightarrow{ROH} \text{(diagram)} + Br/Cl \text{(diagram)} \tag{14}$$

carboxylic acids, and likewise with hypobromite, the percentage of *anti* reaction is even higher. Chlorination of aliphatic enol ethers and dihydropyrans with iodoso-benzenedichloride, $C_6H_5ICl_2$, is >95% *trans*[130]; it has been described as a radical chain reaction, with short chain-length.

Halogenation of intermediates with an enol ether partial structure has gained increasing importance in carbohydrate chemistry. Reaction of Cl_2 with D-glucal triacetate (**36**) in non-polar solvents gives exclusively *cis* and in polar medium predominantly *trans* addition[131]. **36** has also been bromofluorinated in good yield with AgF/Br_2 in CH_3CN (equation 15)[132]; although the reaction is mainly *trans* (**37**, **38**), 20% *cis* product (**39**) is still formed. This addition likewise works with AgF/I_2 or with *N*-bromo(iodo) succinimide and HF[132].

$$\underset{(36)}{\text{(diagram)}} + AgF + Br_2 \xrightarrow{-AgBr} \underset{(37)}{\text{(diagram)}} + \underset{(38)}{\text{(diagram)}} + \underset{(39)}{\text{(diagram)}}$$

$$\tag{15}$$

If the halogenation of the enol ethers is not used solely for the specific introduction of an α-halogen into the parent carbonyl compound, the halo ethers are usually transformed further by HHal elimination and/or nucleophilic substitution. Among these follow-up reactions, a specific synthesis for mixed ketene acetals should be mentioned[133]: bromination of $EtO-CH=CH_2$ with Br_2 (in Et_2O at $-30°C$), followed by substitution of the α-Br with RO^-, and then by dehydrohalogenation, yields the mixed ketene acetal.

C. Reactions with Electrophilic O, S, N and P

Enol ethers are fairly stable against O_2 and react only with stronger oxidants (O_3, peracids etc.). The epoxides formed from peracids and enol ethers are usually hydrolysed immediately in the acidic reaction medium[134]. If the epoxidation is carried out in alcohol (equation 16), α-hydroxyacetals can be isolated in excellent

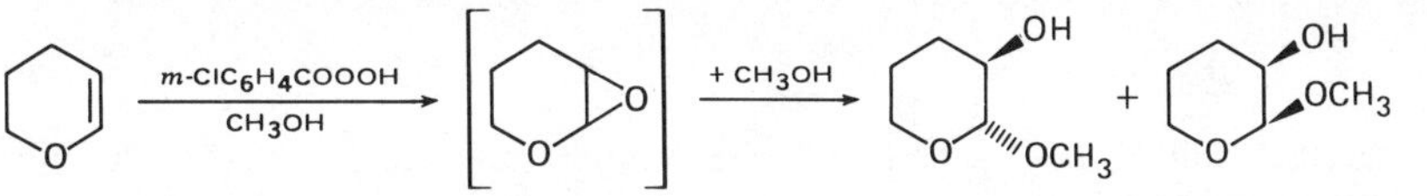

$$(16)$$

$$(9:1)$$

yield, with the addition of ROH preferentially *trans*[135]. The procedure works equally well with 1-methoxycyclohexene, affording 1,1-dimethoxy-2-hydroxycyclohexane, and allows the facile synthesis of mixed α-hydroxyacetals if the enol ether bears an OR function different from that of the epoxidation medium[135]. With enol esters[136] and with some special enol ethers (equation 17)[137], the epoxides can be isolated.

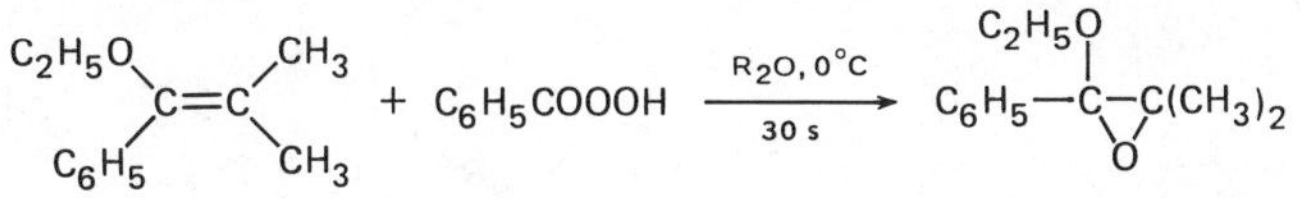

$$(17)$$

90%

With ground-state (^3P) oxygen atoms (generated by Hg-sensitized photodecomposition of nitrous oxide), methyl vinyl ether is transformed into the oxirane **40** with 45% yield[138] (total yield of oxygenation products 86%, equation 18). **40**

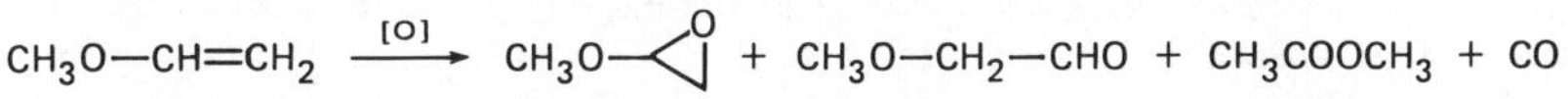

$$CH_3O\!-\!CH\!=\!CH_2 \xrightarrow{[O]} CH_3O\!-\!\underset{\substack{ \\ \textbf{(40)}}}{\overset{O}{\triangleleft}} + CH_3O\!-\!CH_2\!-\!CHO + CH_3COOCH_3 + CO \quad (18)$$

| 45% | 26% | 2% | 13% |

is stable in $CDCl_3$ solution at 25°C for several hours, but attempts at isolation or purification failed. With the exception of 2,3-dihydrofuran, 2-alkoxyoxiranes could be obtained from various enol ethers in 40% yield[138] (though not yet on a larger preparative scale).

The ozonization of enol ethers (equation 19) is of analytical value since it allows the definite cleavage of an α-C—C bond in the parent carbonyl compound[139]; from enol ethers of cyclic ketones, ω-formylcarboxylic acids thus become readily available[140].

$$R^2CH\!=\!CR^1\!-\!OR^3 \xrightarrow[\text{2. } H_2/Pd]{\text{1. } O_3} R^1COOR^3 + R^2CHO \quad (19)$$

Anodic oxidation of 1-alkenyl alkyl ethers[141,142] in methanol (equation 20) yields 50% 1,4-dialkoxy-1,4-dimethoxybutanes (**41**)[142] (acetals of 1,4-dicarbonyl compounds); analogous β,β'-dimerization of 1-alkoxycycloalkenes affords, after hydrolysis, 2,2'-bis(cycloalkanones) with 30—50% current yield[142].

Two O-functions (e.g. $OCOCH_3$) are usually incorporated into enol ethers upon

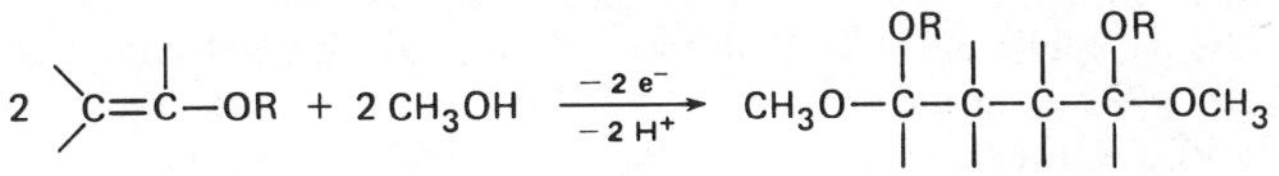

$$(20)$$

$$(41)$$

oxidation with Pb(IV)[143] or Tl(III) acetate[144], with benzoyl peroxide[145], and with HO· radicals[146]. Reaction of Co(III) derivatives (cobalamines, cobaloximes) with vinyl ethers gives, very probably via the π-bonded complexes **42**, the corresponding σ-bonded α-Co acetals (equation 21)[147].

$$X-Co(III) + CH_2=CH-OR \longrightarrow \underset{\underset{\overset{|}{Co(III)}}{\cdot\overset{+}{\cdot}\cdot}}{H_2C\cdots CH-OR} + X^- \xrightarrow[-HX]{+ROH} Co(III)-CH_2-CH(OR)_2$$

$$(42)$$

$$(21)$$

Thiols RSH add to enol ethers at lower temperatures (e.g. $-20°C$ in SO_2) to yield the respective mixed O,S-acetals[148]; reaction at elevated temperature with either azoisobutyronitrile[149,150] or UV irradiation[151], on the other hand, gives the anti-Markownikoff adducts (1-alkoxy-2-alkylthio-) in high yield. With sulphenyl chlorides, both addition and substitution products are formed[152-154] (equation 22), depending on the reaction conditions and the nature of the substituents. The addition is exclusively *trans*, with the RS moiety always at C-2, owing probably to the intermediacy of a thiirenium structure[154].

$$R^1O-CH=CH-R^2 + R^3SCl \longrightarrow \left[\underset{\underset{\overset{|}{R^3}}{\overset{\cdot\cdot}{S}}}{R^1O-CH-CH-R^2} \right] Cl^- \begin{array}{c} \nearrow R^1O-\underset{\overset{|}{Cl}}{CH}-\underset{\overset{|}{SR^3}}{CH}-R^2 \\ \\ \searrow_{-HCl} R^1O-CH=CR^2-SR^3 \end{array}$$

$$(22)$$

The 1 : 2 adducts of SCl_2 with vinyl ethers (**43**) are stable in solution but cannot be isolated[155]; hydrolytic work-up yields both the expected dialdehydes **44** and the oxathianes **45** in comparable amounts (equation 23). The primary addition products of enol ethers with dichlorodisulphane, S_2Cl_2, are even less stable; the dithianes can be isolated, though, after nucleophilic Cl/OR exchange[156] or alkaline hydrolysis[157].

$$2 R^1CH=CH-OR^2 + SCl_2 \xrightarrow[0°C]{R_2O} S(CHR^1-CH(OR^2)Cl)_2 \xrightarrow[H_2O]{CaCO_3}$$

$$(43)$$

$$R^1 = H, C_2H_5$$

$$S(CHR^1CHO)_2 +$$

$$(23)$$

$$(44) \qquad\qquad (45)$$

Thionyl chloride, too, reacts with two molecules of ethenyl ethers (equation 24)[158]. The bis(β-alkoxy-β-chloroethyl)sulphoxides **46** can be transformed to the dienamines **47**; tertiary amines give double HCl elimination, partially accompanied by rearrangement[158]. Only one-sided 1:1-addition is observed with the higher 1-alkenyl ethers. 1-Alkoxycyclohexenes react with SO_2 (equation 25), reversibly forming a 1,3-dipole (**48**) not stabilized by conjugation[159]; **48** can also be reached directly from the acetal in SO_2. [2 + 3]Cycloaddition of **48** to another cyclohexenyl ether molecule, followed by ROH elimination, yields the tricyclic **49**[159]. β-Sulphonylation of vinyl ethers is also possible with the pyridine–SO_3 adduct[160].

Nitrosyl halides smoothly add to enol ethers with the expected regiochemistry $(ON^{\delta+}-Cl^{\delta-})$, but the (probably dimeric) α-halo-β-nitroso ethers so formed (e.g.

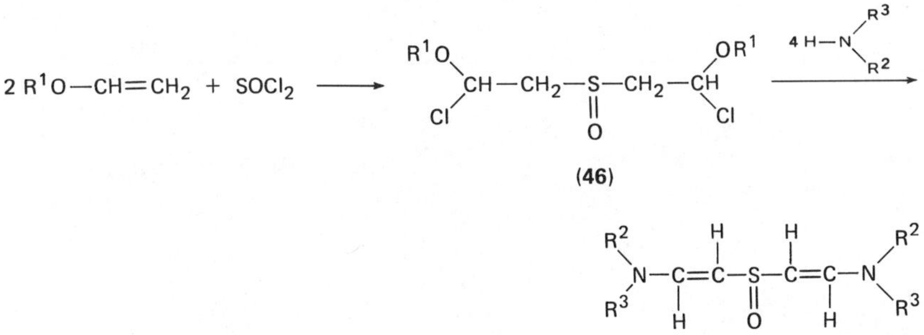

$$2\,R^1O{-}CH{=}CH_2 + SOCl_2 \longrightarrow$$

(46)

(24)

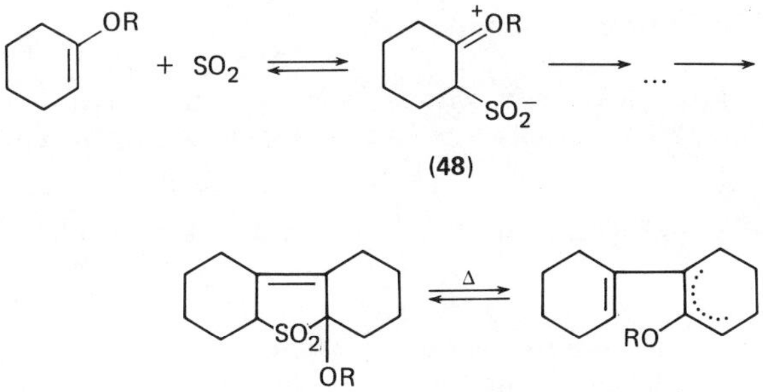

(47)

(25)

85%

(49)

50) are very labile[161]. Alcoholysis in basic medium yields the corresponding nitroso acetals which are generated directly from enol ethers and alkyl nitrites[161]. Nitrosation in the presence of alcohol, or work-up without a HCl scavenger, affords oximes[162]; thus, cyclohexenoneoximes (51) are obtained from 1-alkoxycyclohexenes (equation 26)[162]. If the nitrosation leads to tertiary nitrosyl compounds

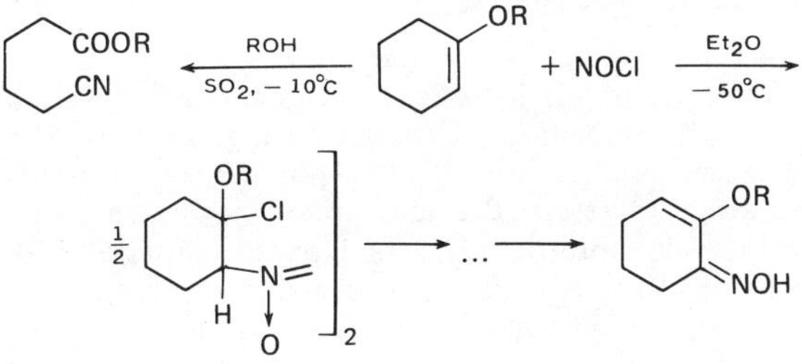

(26)

(50)   (51)

(52) where tautomerization to the oxime is impossible, the original enol ether C=C bond is broken upon alcoholysis (equation 27)[163]. Under proper reaction conditions, 'nitrosolytic' C—C cleavage can also be achieved for less substituted enol ethers (see equation 26)[162]. This reaction has been put to elegant use in makrolide synthesis[164]. Nitrosation of 53 in the presence of stoichiometric quantities of ROH and H_2O (equation 28) results in cleavage of the central C=C bond, yielding the dioximes 54 and, upon hydrolysis, the ketolactones 55.

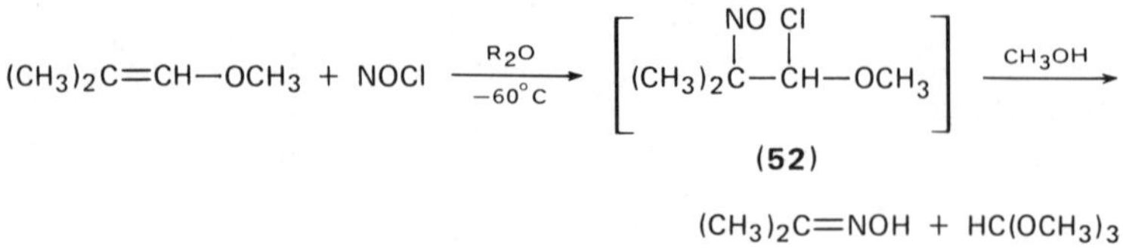

$$(CH_3)_2C=CH-OCH_3 + NOCl \xrightarrow[-60°C]{R_2O} \left[(CH_3)_2\overset{\overset{\displaystyle NO}{|}}{C}-\overset{\overset{\displaystyle Cl}{|}}{CH}-OCH_3 \right] \xrightarrow{CH_3OH}$$

$$\text{(52)}$$

$$\tag{27}$$

$$(CH_3)_2C=NOH + HC(OCH_3)_3$$

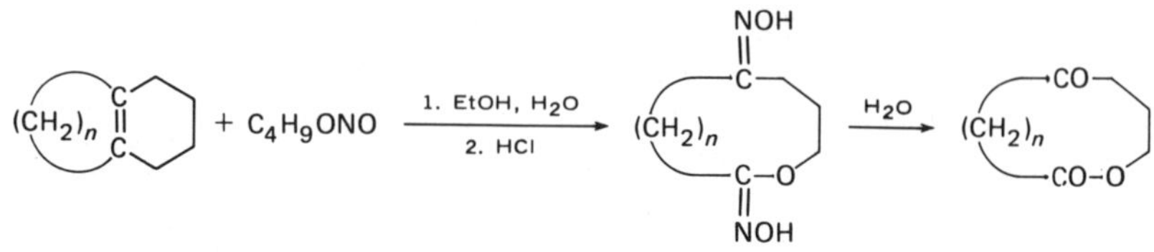

$$\tag{28}$$

(53)   (54)   (55)

Diazonium salts couple readily with α- and β-substituted vinyl ethers in the β-position (equation 29), but only the hydrolysed glyoxalhydrazones **56** can be

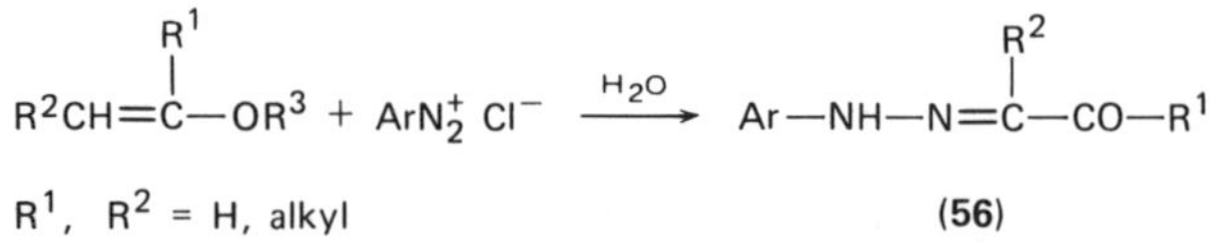

$$R^2CH=\overset{\overset{\displaystyle R^1}{|}}{C}-OR^3 + ArN_2^+ Cl^- \xrightarrow{H_2O} Ar-NH-N=\overset{\overset{\displaystyle R^2}{|}}{C}-CO-R^1 \tag{29}$$

$$R^1, R^2 = H, \text{ alkyl} \qquad\qquad \text{(56)}$$

isolated[165,166]. The analogous reaction of α-ethoxystyrene with azo esters gives – apart from Diels–Alder cycloaddition – the β-hydrazino-substituted styrene[167] (probably via a dipolar intermediate).

In the presence of, for example, azoisobutyronitrile, H—PO(OR)$_2$ and other P(III) derivatives are smoothly added to vinyl ethers (of course with anti-Markownikoff orientation)[168]; phosphine itself gives mono-, bis- and tris-(β-alkoxyalkyl)phosphines[169]. With PCl$_5$ and tetrahalophosphoranes, β-*substitution* products are formed via an ionic mechanism[170].

D. Reactions with Carbon Electrophiles

While β-alkylation of enamines is a facile process with a variety of alkylating agents RX[5,7,171], the nucleophilicity of the C-2 in enol ethers is not sufficient for uncatalysed reactions[172]. Activation of the alkylating agents with Friedel–Crafts catalysts as a rule is self-prohibitive with the polymerization-prone enol ethers. Vol'pin and collaborators[173] report the addition of tropylium bromide to alkenyl ethers which leads to cycloheptatrienyl acetaldehydes (equation 30); the reaction conditions have to be carefully adjusted since usually the action of tropylium salts results in polymerization of vinyl ethers[174].

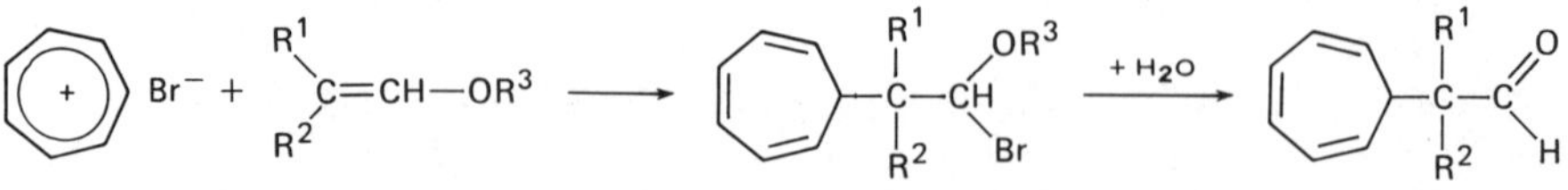

$$\tag{30}$$

Alkoxonium ions, $>C^+-OR \leftrightarrow >C=^+OR$, represent the necessary compromise between sufficient activation of the electrophilic carbon centre and suppression of enol ether polymerization, and the polar C—C linkage of aldehydes or ketones and their derivatives with enol ethers has found widespread application[175,176]. The BF_3-catalysed reaction of enol ethers with acetals[177], for instance, constitutes a valuable alternative to the classic aldol condensation, the main preparative advantage lying in the unequivocal course of the reaction[178] since enol ether and acetal can each act only as electrophilic and nucleophilic (methylene and carbonyl) component, respectively[179]. Depending on the nature of the reactants and reaction conditions, either β-alkoxyacetals or α,β-unsaturated aldehydes are formed (equation 31); at the acetal stage, addition of a second vinyl ether molecule is possible.

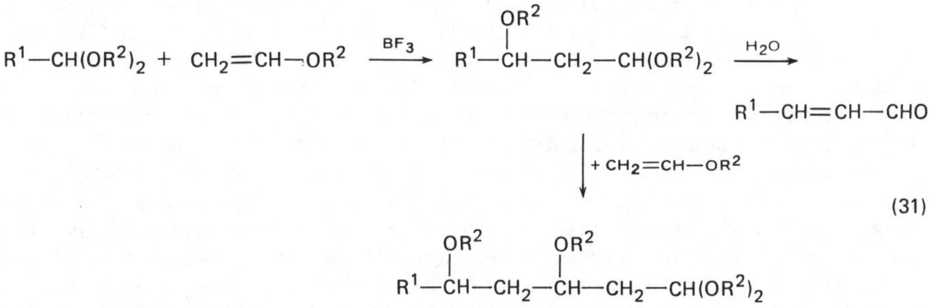

$$R^1\text{—CH}(OR^2)_2 + CH_2\text{=}CH\text{—}OR^2 \xrightarrow{BF_3} R^1\text{—}\overset{\displaystyle OR^2}{\overset{|}{CH}}\text{—}CH_2\text{—}CH(OR^2)_2 \xrightarrow{H_2O}$$

$$R^1\text{—}CH\text{=}CH\text{—}CHO$$

$$\downarrow {+\, CH_2\text{=}CH\text{—}OR^2}$$

(31)

$$R^1\text{—}\overset{\displaystyle OR^2}{\overset{|}{CH}}\text{—}CH_2\text{—}\overset{\displaystyle OR^2}{\overset{|}{CH}}\text{—}CH_2\text{—}CH(OR^2)_2$$

(57)

Hoaglin and Hirsh have proposed a carbenium ion mechanism (equation 32) for the overall reaction[178], analogous to that for the acid-catalysed aldol reaction. The first step in this sequence is the dissociation of the primary $\text{>O} \rightarrow BF_3$ complexation product to the alkoxonium species **58** (probably in the form of an ion pair). Electrophilic attack of **58** upon a vinyl ether molecule leads to a new alkoxonium ion (**59**) which then either adds an alkoxy moiety, forming the β-alkoxyacetal **60**, or another $CH_2\text{=}CH\text{—}OR^2$ to the 1 : 2-adducts **57**. The partitioning between these two pathways is governed, of course, through the relative electrophilicity of **58** and **59**. Yet even with the least reactive saturated aliphatic acetals, the reaction can be held at the 1 : 1-addition stage (**60**) with at least 80% yield if a large (5:1 or

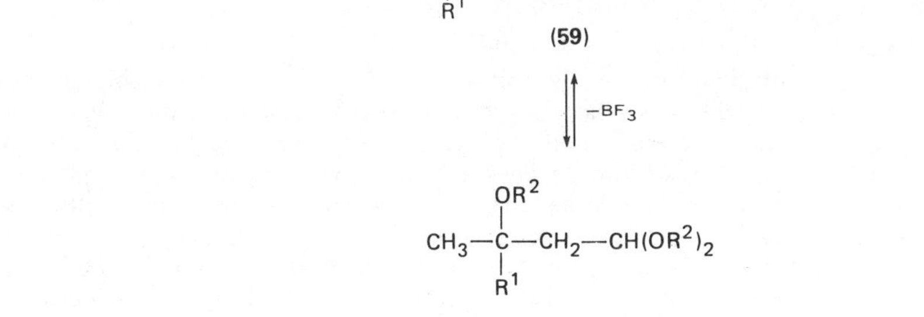

$$CH_3\text{—}\overset{\displaystyle R^1}{\overset{|}{C}}(OR^2)_2 + Et_2O\cdot BF_3 \xrightarrow{-\,Et_2O} CH_3\text{—}\overset{\displaystyle R^1}{\overset{|}{C}}\text{=}^+OR^2\ R^2OBF_3^- \xrightarrow{+\,CH_2\text{=}CHOR^2}$$

(58)

$$CH_3\text{—}\overset{\displaystyle OR^2}{\underset{\displaystyle R^1}{\overset{|}{\underset{|}{C}}}}\text{—}CH_2\text{—}CH\text{=}^+OR^2\ R^2OBF_3^- \underset{\displaystyle \xrightleftharpoons{}}{\overset{\displaystyle -BF_3}{}} \textbf{(57)}$$

(59)   (32)

$$\Big\updownarrow {-BF_3}$$

$$CH_3\text{—}\overset{\displaystyle OR^2}{\underset{\displaystyle R^1}{\overset{|}{\underset{|}{C}}}}\text{—}CH_2\text{—}CH(OR^2)_2$$

(60)

more) excess of acetal is used[178]. From ketals (R^1 = alkyl), on the other hand, practically no 1:1-product is obtained since in this case **59** is so much more reactive than **58**.

Mainly 1:1-products are formed even from the equimolar reaction of aromatic aldehyde acetals[180]; with α,β-unsaturated acetals, which show the highest reactivity towards enol ethers, the aspect is still more propitious. Because of its well-defined (1:1) stoichiometry and definite regiochemistry, the condensation of unsaturated aldehyde acetals with vinyl ethers could thus be successfully employed in the synthesis of polyene aldehydes[181] (with $ZnCl_2$ catalysis) and of carotinoids[182]. If 1-alkoxy-1,3-dienes are used as the enol ether component, the electrophilic alkoxonium centre of the acetal adds exclusively at C-4[183]. Alkoxydienes and 1-substituted enol ethers (ketone derivatives) which have a much higher polymerization tendency than the enol ethers of saturated aldehydes[180], can be coupled only with the more reactive (aromatic and unsaturated) acetals since the Lewis-acid catalysts, used in the acetal condensation, at the same time promote polymerization.

Dioxolanes and other cyclic acetals have also been employed in enol ether condensations[184]; with the much less reactive thioacetals, the reaction is limited to phenyl vinyl and divinyl ethers[185] which do not polymerize so easily. The enhanced electrophilicity of the carbenium ions generated from α-halo ethers[186] and Schiff bases in HOAc[187], on the other hand, makes for especially smooth addition to enol ethers. Mechanistically, the dimerization of vinyl ethers with BF_3 in the presence of Hg(II) salts (equation 33)[188] must also be classified among the condensation reactions with activated acetals.

$$CH_2{=}CHOR + Hg(OAc)_2 \rightleftharpoons AcOHgCH_2\overset{\underset{\displaystyle OR}{|}}{C}HOAc \xrightarrow[BF_3]{+CH_2=CHOR} AcOHgCH_2\overset{\underset{\displaystyle OAc}{|}}{C}HCH_2CH(OR)_2$$

$$\text{(61)}$$

$$\downarrow {+CH_2=CHOR}$$

$$\text{(33)}$$

$$61 + CH_2{=}CHCH_2CH(OR)_2$$

Hoaglin and Hirsh also report the BF_3-catalysed *direct* condensation of aliphatic aldehydes with enol ethers[189] leading, via 1,3-dioxanes, to α,β-unsaturated aldehydes (equation 34). Their findings have been confirmed by a Japanese group[190];

$$R^1CH{=}CHOEt + 2R^2CHO \xrightarrow{Et_2O{\cdot}BF_3} \text{(62)} \xrightarrow{H^+/H_2O} R^2CH{=}\overset{\underset{\displaystyle R^1}{|}}{C}CHO \qquad (34)$$

$$\text{(62)}$$

if the catalyst is neutralized before hydrolysis, the dioxanes **62** can be isolated and cleaved independently. These authors[190] have also extended the vinyl ether condensation to acetone and to methyl ethyl ketone. The rather poor yields are due to the lesser carbonyl activity of the ketones and the concomitant increase in side-reactions; among these, *trans*-enoletherification between vinyl ether and ketone is most important[191] (thus, the regiospecificity of the reaction is lost). But even if ketones are subjected to BF_3-catalysed condensation with their own enol ethers (to avoid the product mixture due to *trans*-enoletherification), the yields of definite 1:1-products are unsatisfactory (20–50%)[192].

Excellent yields are reported for cross aldol condensation via enol ethers with titanium catalysts (equation 35)[193]; essential for the success of the reaction is that both components are present in equimolar quantities, and that $TiCl_4$ *and* $Ti(OR)_4$ are applied together.

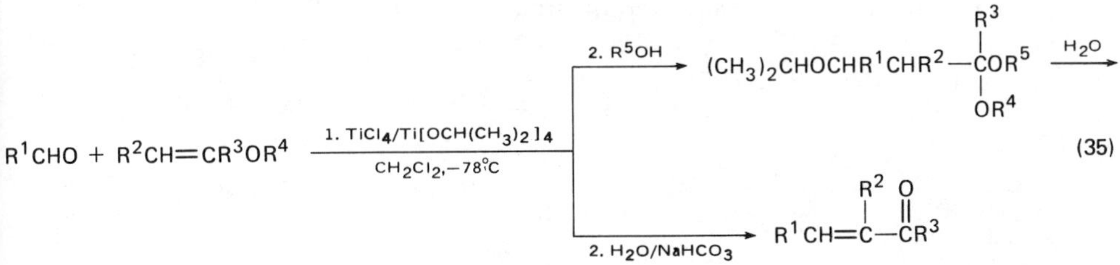

$$R^1CHO + R^2CH{=}CR^3OR^4 \xrightarrow[CH_2Cl_2, -78°C]{1. \ TiCl_4/Ti[OCH(CH_3)_2]_4}$$

$$\xrightarrow{2. \ R^5OH} (CH_3)_2CHOCHR^1CHR^2-\underset{\underset{OR^4}{|}}{\overset{\overset{R^3}{|}}{C}}OR^5 \xrightarrow{H_2O}$$

$$\xrightarrow{2. \ H_2O/NaHCO_3} R^1CH{=}\overset{\overset{R^2}{|}}{C}-\overset{\overset{O}{\|}}{C}R^3 \tag{35}$$

The *formylation* of enol ethers with orthoformates[176,194] (equation 36) follows the same mechanistic course as the acetal condensations; the malonaldehyde derivatives thus formed constitute valuable building blocks for heterocyclic syntheses[195]. Among the Vilsmeier–Haack reagents, **63** (derived from DMF and phosgene) has the least Lewis-acid properties, and so has been employed most successfully for the formylation of vinyl ethers[196,197] (equation 37).

$$HC(OR^1)_3 + R^2R^3C{=}CHOR^1 \xrightarrow{H^+} (R^1O)_2CH-\underset{\underset{R^3}{|}}{\overset{\overset{R^2}{|}}{C}}-CH(OR^1)_2 \tag{36}$$

$$R^1CH{=}CHOR^2 + [H-\overset{\overset{Cl}{|}}{C}{=}^+N(CH_3)_2]\,Cl^- \longrightarrow [(CH_3)_2N\overset{\overset{Cl}{|}}{C}H-\overset{\overset{R^1}{|}}{C}H-CH{=}^+OR^2]\,Cl^-$$

$$(63)$$

$$\Big\downarrow {H_2O/OH^-} \tag{37}$$

$$(CH_3)_2NCH{=}CR^1CHO$$

The decreasing reactivity of the higher ortho esters bars enol ether acylation beyond the orthoacetate stage, and has not been used much even there[198]. Tetramethoxymethane (methyl orthocarbonate), on the other hand, can be added smoothly to vinyl ethers under $SnCl_4$ catalysis[199].

The *acylation* of vinyl ethers requires strong activation of the acylating agents. Employment of Friedel–Crafts catalysts is naturally limited to enol ethers with negligible polymerization tendency, a prerequisite met fully by steroid enol ethers (equation 38)[200]. Electronegative substituents likewise raise the carbonyl activity

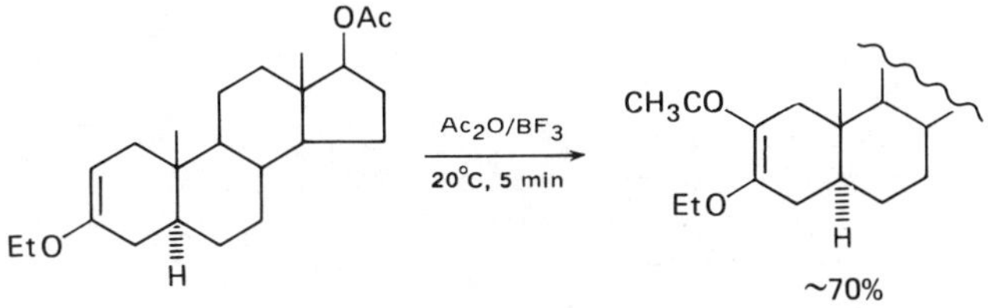

$$\tag{38}$$

~70%

of the acyl component; with trifluoro(chloro)acetic anhydride, or the mixed trihaloacetic acetic anhydrides, vinyl ethers are β-trihaloacyl-substituted in quantitative yield[201]. Effenberger and Maier have demonstrated strikingly how the course of the reaction depends on the electrophilic potential of the acyl function[202]: while acetyl chloride does not react at all with ethyl vinyl ether, and chloroacetyl chloride causes polymerization, dichloroacetyl chloride gives the addition, trichloroacetyl chloride the substitution product (equation 39).

$$\underset{\substack{\\ Cl}}{\underset{\displaystyle Cl_2CHCCH_2CH}{\overset{\displaystyle \overset{O}{\|}\quad OEt}{}}} \quad \xleftarrow[\text{0°C, 48 h}]{Cl_2CHCOCl} \quad CH_2{=}CHOEt \quad \xrightarrow[\text{0°C, 48 h}]{Cl_3CCOCl} \quad \overset{\overset{\displaystyle O}{\|}}{Cl_3CCCH}{=}CHOEt \qquad (39)$$

Oxalyl chloride, with similarly enhanced electrophilicity, readily adds two moles of enol ether at room temperature[203]; the resulting double α-halo ethers can be dehydrohalogenated facilely with tertiary amines. In the case of 3,4-dihydro-2H-pyran, the addition of $(COCl)_2$ is exclusively *cis*[204]. Substitution of vinyl ethers with phosgene at 0°C yields β-alkoxyacryl chlorides (equation 40)[205]

$$ROCH{=}CH_2 + COCl_2 \xrightarrow{-HCl} ROCH{=}CHCOCl \qquad (40)$$

which represent valuable reagents in heterocyclic syntheses[206]; β-CO—NCO substitution is found in the reaction of vinyl ethers with isocyanatocarbonyl chloride[207]. β-Carboxamidation of enol ethers with isocyanates[208], though likewise a substitution reaction, proceeds via cycloaddition, and will be dealt with in Section V.B.

In the presence of radical initiators or with UV irradiation, tetrahalomethanes can be added to the enol ether double bond (equation 41)[209]; if mixed tetrahalomethanes are used, the halogen which is easiest cleaved off radically (Br·) is found in the α-position of the halo ether 64[210]. These primary adducts (64) are thermally

$$R^1CH{=}CHOR^2 + CHal_4 \xrightarrow{\text{rad.}} \underset{\text{(64)}}{Hal_3C\overset{\displaystyle R^1}{\underset{\displaystyle}{C}}\overset{\displaystyle Hal}{\underset{\displaystyle}{C}}HOR^2} \xrightarrow{-HHal} Hal_2C{=}\overset{\displaystyle R^1}{\underset{\displaystyle}{C}}{-}\overset{\displaystyle Hal}{\underset{\displaystyle}{C}}HOR^2 \qquad (41)$$

extremely labile, and give off HHal on distillation[209]. From the reaction of glycol divinyl ethers with CCl_4 and azoisobutyronitrile, up to 50% bis(trichloroallyl) ethers, $(-CH_2OCHClCH{=}CCl_2)_2$, could be isolated[211]. The reaction has been utilized for the introduction of a CH_3 group into the 6-position of steroids[212] (equation 42). Tetranitromethane has likewise been added radically to enol ethers, forming isoxazolidines[212a].

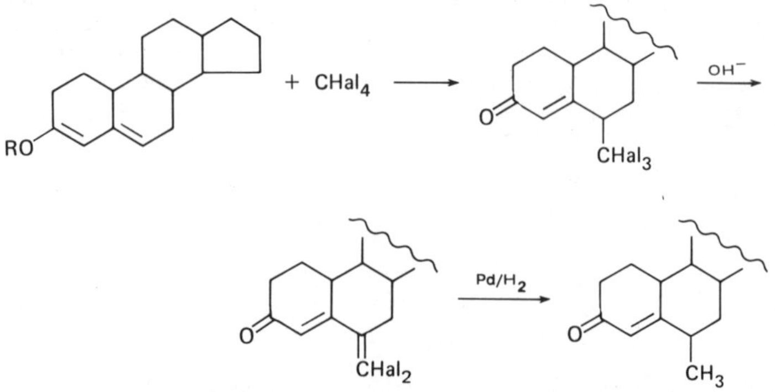

$$(42)$$

V. CYCLOADDITIONS

Between enamines and nonhetero-substituted alkenes, enol ethers hold a midway position in overall reactivity as electron-rich olefins, as well as in the polarization of the C=C π-bond. In thermal [2 + 2] cycloadditions (for definitions, see Reference 213), that asymmetry in the π-electron system is of paramount importance since the principle of orbital symmetry conservation[214] forbids a concerted course, with parallel approach of the two π-systems, for this reaction[214,215]. (The orthogonal $(_\pi 2_s + _\pi 2_a)$ mode[216,217], which allows synchronous bond closure, is definitely operative only with ketenes[217], and perhaps with heterocumulenes.) Consequently, [2 + 2] cycloadditions with alkenes proceed via a (singlet) biradicalic intermediates[218] while the highly polarized π-system of enamines tends towards a polar, two-step mechanism[219,220], with a concomitant shift in the product spectrum from addition to substitution derivatives. For electrophilic additions to enol ethers (e.g. acylation), a similar predominance of substitution over addition with increasing reactivity, i.e. higher polarization of the attacking electrophile, has been noted above.

The moderate activation of alkoxy-substituted alkenes designates them as mechanistic borderline cases. Also, the free choice of the OR moiety, with definite gradation in electron release, and the possibility of selective synthesis of geometric isomers (or, alternatively, the ease of their separation) allows the construction of substrates specifically adapted to individual mechanistic problems. Enol ethers have thus become favourite subjects for studying the mechanism of [2 + 2] cyclo-addition; in particular, the query 'concerted or not concerted' has instigated some highly sophisticated work.

A. [2 + 2] Cycloadditions with Tetracyanoethylene

The cycloaddition reaction of enol ethers with tetracyanoethylene (TCNE) can now be considered as definitely cleared up in almost every mechanistic detail[221,222]. Even though reaction mechanisms are more or less based on circumstantial evidence, 'the network of mechanistic criteria and experimental findings'[221] which Huisgen and his coworkers have accumulated in this case, must be regarded as extremely tight, and their ratiocination as very compelling indeed: The cycloaddition is not stereospecific with respect to the electron-rich double bond, and proceeds via a zwitterionic intermediate (66 in Scheme 3).

Vinyl ethers, and even the phenylogous p-alkoxystyrenes, are sufficiently electron-rich to form cyclobutanes with TCNE at room temperature[223] (TCNE is characterized by a highly electron-deficient C=C bond with low-lying MOs). The reaction of either E- or Z-butenyl ethyl ether with TCNE, for instance, is completed within a few seconds and yields, quantitatively, two cyclobutane derivatives (Scheme 3): in the major product, the configuration of the alkenyl ether is retained, in the minor one, inverted (^{1}H-NMR evidence). This stereochemical leakage increases with solvent polarity (Table 6) since rotation about the C-1/C-2 bond in the zwitterionic intermediate 66 becomes more and more favoured relative to ring-closure by better solvation and reduced Coulomb attraction. But even in acetonitrile, ring-closure is still five times faster than this rotation for both cis- and trans-66[224]. In contrast, rotation is much faster than cyclobutane formation for the biradical from tetrafluoroethene and (Z)-2-butene[225]; the [2 + 2] cyclo-addition of benzyne to (E)- and (Z)-1-propenyl ethyl ether, supposedly proceeding via a biradical, likewise shows substantial nonstereospecific portions[226]. The addition of fumaro- and maleo-nitrile to tetramethoxyethene, on the other hand, though very probably still proceeding via zwitterionic intermediates, gives sterically

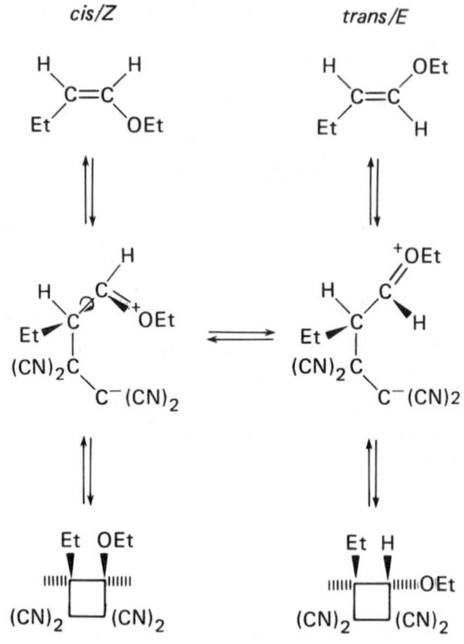

SCHEME 3.

pure E- and Z-dicyanocyclobutanes, respectively[227]. Thus, TCNE/enol ether cyclo-additions appear to rank at the lower end of the stereoselectivity scale among [2 + 2] cycloadditions via zwitterions. The nevertheless fairly high stereochemical fidelity (Table 6), compared with the biradicals, can easily be rationalized in terms of Coulomb attraction of the charge centres (see below); however, 'through-bond coupling' seems to contribute significantly to the height of the rotational barrier around the C-1/C-2 bond[221,228].

If TCNE is reacted in CH_3CN with 1.1 equivalents of (Z)-1-butenyl ethyl ether of $\geqslant 99.5\%$ configurational purity, the 0.1 equivalents of enol ether recovered turn out to be 18% $Z \to E$-isomerized[224]. For this, the simplest mechanism is formation of Z-66, rotation to E-66, and dissociation into the starting materials (see Scheme 3). The zwitterion thus enters into three competitive processes: ring-closure, rotation about the former enol ether double bond, and redissociation[221]. But the mechanistic picture is still more complex. If a $CHCl_3$ solution of the

TABLE 6. % Cyclobutane (67) with inverted configuration (see text), starting from (Z)- and (E)-1-butenyl ethyl ether

Solvent	(Z)	(E)
Benzene	2	2
CH_2Cl_2	7	3
Ethyl acetate	10	5
Acetonitrile	18	16

TCNE/α-methoxystyrene cycloadduct is heated to 50°C, the red–violet colouring of the CT complex between TCNE and enol ether develops reversibly. The TCNE present in the equilibrium, although not measurable directly, can be intercepted with the more reactive ethyl vinyl ether, and so be transferred quantitatively from the 1-methoxy-1-phenyl- to the 1-ethoxy-2,2,3,3-tetracyanocyclobutane. (In a similar situation, we have found ⩾30% cycloreversion for ketene acetal/isocyanate cycloadducts at 65°C; in this case, both constituents are easily identified by ^{1}H-NMR[229].) In view of these results, it is not surprising that the stereoisomeric cyclobutanes (**67**), which are stable in nonpolar solvents, slowly isomerize in CH_3CN solution[224].

The zwitterion thus turns out to be the pivot around which the whole cycloaddition scene revolves[221], yet so far its intermediacy has been *inferred* from kinetic and mechanistic evidence only. If the cycloadduct from TCNE and ethyl vinyl ether (**69**) is incubated with $CH_3C{\equiv}N$, $(CH_3)_2C{=}O$, or $C_6H_5CH{=}NCH_3$, though, the 1,4-dipole of the zwitterion **68** is *intercepted*, and **69** converted quantitatively into six-membered ring-products (equation 43)[230]. Since addition of these dipolarophiles to **68** is rather slow, only 4–6% of **70** can be isolated directly from the cycloaddition in acetonitrile or acetone.

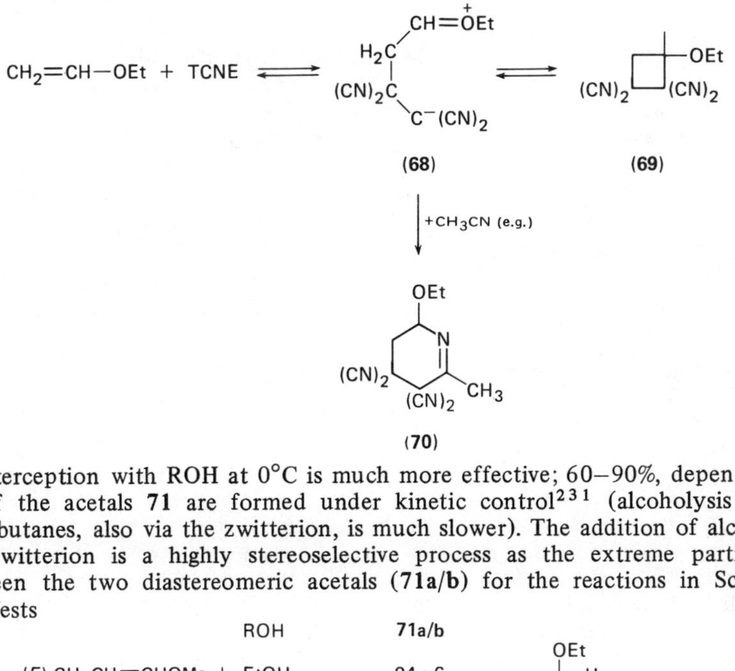
CH2=CH—OEt + TCNE
CH=OEt
H2C
(CN)2C
C−(CN)2
OEt
(CN)2
(CN)2
(68)
(69)
+CH3CN (e.g.)
OEt
N
(CN)2
(CN)2
CH3
(70)

(43)

Interception with ROH at 0°C is much more effective; 60–90%, depending on R, of the acetals **71** are formed under kinetic control[231] (alcoholysis of the cyclobutanes, also via the zwitterion, is much slower). The addition of alcohol to the zwitterion is a highly stereoselective process as the extreme partitioning between the two diastereomeric acetals (**71a/b**) for the reactions in Scheme 4 manifests

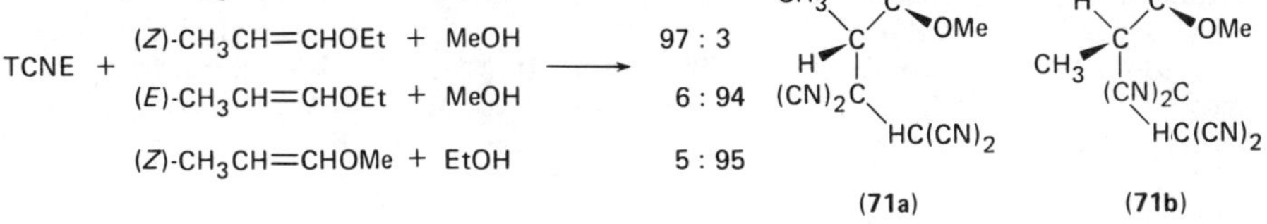
ROH
71a/b
TCNE +
(E)-CH3CH=CHOMe + EtOH 94 : 6
(Z)-CH3CH=CHOEt + MeOH 97 : 3
(E)-CH3CH=CHOEt + MeOH 6 : 94
(Z)-CH3CH=CHOMe + EtOH 5 : 95
OEt
H
CH3
OMe
H
(CN)2C
HC(CN)2
(71a)
(71b)

SCHEME 4.

In the *gauche* or *cis* conformation of the zwitterion, tacitly assumed in Scheme 3, the $(CN)_2C|^-$ group offers 'built-in solvation'[232] from the inner side to the carboxonium pole; in fact, **66** represents an intramolecular ion pair with substantial charge transfer. Nucleophilic attack of ROH should thus be from the outside. This could be verified by intercepting the 1,4-dipole from (*Z*)-1-propenyl methyl ether and TCNE with (*S*)-2-butanol. One of the two diastereomeric acetals, formed in comparable amounts, was isolated by crystallization from (*S*)-2-butanol, and demonstrated by X-ray analysis to have *RRS*-structure (**72**)[233], that is, indeed the result of outside attack of ROH.

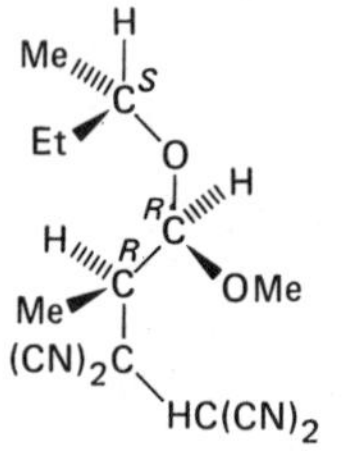

(**72**)

The rate of TCNE/enol ether cycloaddition strongly depends on the polarity of the reaction medium[230]; the immense acceleration, $\sim 10^4$ from cyclohexane to acetonitrile, is unique among cycloadditions. A plot of log k vs. E_T for the reaction of four different enol ethers in ten solvents displays very good linearity over practically the whole polarity range[234,235]. Since TCNE cycloadditions, for example equation (44), are accompanied by a considerable increase in substrate

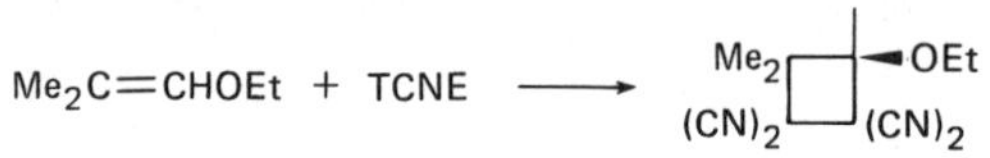

$$Me_2C=CHOEt + TCNE \longrightarrow \underset{(CN)_2\ \ (CN)_2}{\overset{Me_2\ \ OEt}{\boxed{}}} \tag{44}$$

$\mu(D)$ 1.28   0.0   6.05

polarity, these rate enhancements do not represent *prima facie* evidence of (di)polar intermediates. From the experimental solvent dependency, dipole moments of 10–14 D were calculated for the transition state; these values, representing about 2/3 of the fully developed charge in the zwitterion, are definitely larger than expected for a concerted pathway[221]. The large negative value for the volume of activation $\Delta V_{exp}^{\neq}$ (−36 ml/mol, constant for a series of enol ethers)[235] and the solvent dependence of $\Delta V^{\neq}$ [236] can be explained only in terms of a two-step process via zwitterionic intermediates. The CT complex between TCNE and the enol ethers is a dead-end (side) equilibrium[235]; it is *not* traversed in the course of the cycloaddition as usually formulated[237].

Acrylo- and fumaro-nitrile do not react with enol ethers, owing to insufficient stabilization of the zwitterion by only one CN group. Between 1,1-di-, tri- and tetra-cyanoethene, on the other hand, no great difference in cycloaddition reactivity is found[238] (Table 7); in fact, TCNE reacts slowest. In Diels–Alder cycloadditions, these cyanoethenes exhibit a gradation of 10^7-10^9 in relative reactivity[239] (Table 7); the comparison once again demonstrates the fundamental disparity between these established concerted processes and the [2 + 2] cycloaddition of TCNE.

As expected for the zwitterionic mechanism, the TCNE cycloaddition rate is enhanced tremendously by a second α-substituent in the vinyl ether (R, Ar, OR);

TABLE 7. Relative rates for [2 + 2] cycloadditions of polycyanoethenes[238,239]

| | [2 + 2] (benzene, 25°C) | Diels—Alder (dioxane, 20°C) | |
	Isobutenyl methyl ether	Cyclopentadiene	Dimethyl-anthracene
Acrylonitrile	0	0.52	0.45
Fumaronitrile	0	4.1×10^1	7.0×10^1
1,1-Dicyanoethene	16.0	2.3×10^4	6.4×10^4
Tricyanoethene	1.2	2.4×10^5	3.0×10^6
Tetracyanoethene	1.0^a	2.2×10^7	6.5×10^9

$^a k_2 = 3.97 \times 10^{-5} \, M^{-1} \, s^{-1}$; this value has to be divided by a statistical factor of 2 for the relative rate.

TABLE 8. Experimental rate constants k_2 [$10^{-3} \, M^{-1} \, s^{-1}$] for TCNE cycloaddition to enol ethers (in ethyl acetate, 25°C)[240]

R	CH_2=CHOR	(Z)-C_2H_5CH=CHOR	(E)-C_2H_5CH=CHOR
C_6H_5	0.0043	—	—
CH_3	—	5.5	4.2
C_2H_5	18	15	17
n-C_4H_9	20	—	—
$CH(CH_3)_2$	—	28	57
c-C_6H_{11}	112	—	—
$C(CH_3)_3$	255	80	140

α-methoxystyrene, for instance, reacts 10^5 times faster than β-methoxystyrene[240]. In contrast, the acceleration by β-substituents is moderate, ~50-fold for CH_3, but rapidly dropping again with increasing bulkiness (Table 8)[240]. Between (Z)- and (E)-1-alkenyl ethers, there is but little difference in TCNE cycloaddition reactivity (Table 8) — in striking contrast to ketene cycloadditions[216,241] (see below). The higher relative reactivity of the (Z)-1-butenyl methyl and ethyl ethers is due to the additional ground-state stabilization of the corresponding E-compounds by non-bonded attraction in the s-cis conformation.

These mechanistic findings for the TCNE addition are also pertinent for the reaction of enol ethers with other highly electron-deficient cyano- or (alkoxycarbonyl)-ethenes[242]. Furthermore, since the zwitterion is structurally analogous to the species produced in the initial step of the cationic enol ether polymerization, relative reactivities towards TCNE can be directly correlated with relative polymerization rates for vinyl ethers with various alkoxy moieties and different α- and/or β-substituents (e.g $(CH_3)_3C$—O > $(CH_3)_2CH$—O > C_2H_5—O ~ n- or i-alkyl)[243,244].

B. Other [$_\pi 2 + _\pi 2$] Cycloadditions

The addition of diphenylketene to enol ethers (discovered as early as in 1920[245]) leads exclusively to 3-alkoxycyclobutanones (75). By now, the concerted nature of this cycloaddition, following the [$_\pi 2_a + _\pi 2_s$] mechanism of Woodward and Hoffman (equations 45 and 46)[215], can be considered as safely estab-

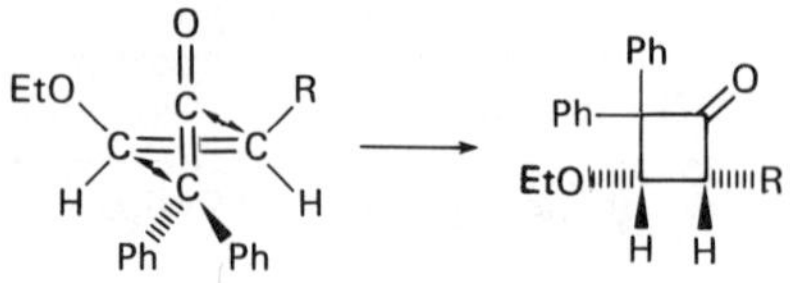

(45)

(73)   (Z)-(75)

(46)

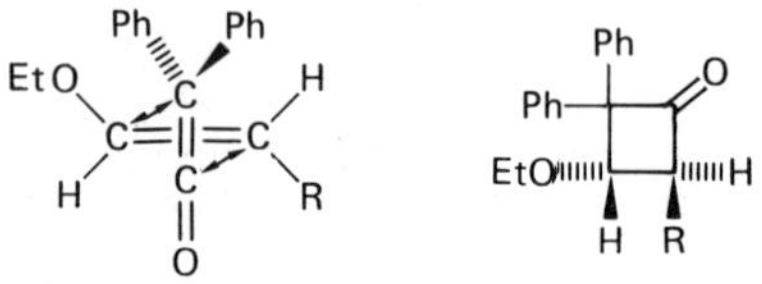

(74)   (E)-(75)

lished[216,217,246]. The decisive factor in favour of the orthogonal approach is the additional stabilization, provided through the interaction of the *unoccupied C=O* orbital in the ketene and the HOMO of the ketenophile[217]. This interaction is also responsible for the regiochemistry of the cycloaddition, i.e. for the addition of the ketene *C=C* bond to the enol ether[217]; bis(trifluoromethyl)ketene adds to enol ethers with the *C=O*[247], bis(trifluoromethyl)ketene imines with the *C=N* double bond[248].

The PMO treatment[217] predicts that successive replacement of the β-hydrogens in ethyl vinyl ether should accelerate the ketene addition (by raising the enol ether HOMO energy). (Z)-1-Propenyl ethyl ether indeed reacts slightly faster (Table 9), addition to the *E*-isomer, however, is retarded almost 100-fold[241]. This rate enhancement of $\sim 10^2$ for *cis*- over the respective *trans*-olefins appears to be a unique feature of ketene ($_\pi 2_a + _\pi 2_s$) cycloadditions[216,249,250], and must be due to the extremely stringent steric requirements for the antarafacial approach. Huisgen and Mayr[241] have advanced cogent arguments for diverse ketene orientation in the transition states of *Z*- and *E*-enol ether addition (since cyclobutane bonding cannot be far progressed in the transition state[241], the orientation complexes 73 and 74 represent appropriate models). The different steric interaction in

TABLE 9. Cycloaddition rate constants k_2 $(10^{-4}\ M^{-1}\ s^{-1})$ of diphenylketene to E/Z-isomeric 1-alkenyl ethers

	$(Z)/(E)$-$C_2H_5OCH{=}CHR^1$ (in benzonitrile, 40°C)[241]				
	R^1 = H	CH_3	C_2H_5	$(CH_3)_2CH$	$C(CH_3)_3$
$k_{cis}\ \rightarrow (Z)$-(75)	{45}	109	128	117	~ 3.7
$k_{trans} \rightarrow (E)$-(75)		1.29	1.20	0.742	0.054
k_{cis}/k_{trans}	–	84	107	158	~ 70
	$(Z)/(E)$-$R^2OCH{=}CHR^1$ (in CCl_4, 23°C)[249]				
k_{cis}/k_{trans} $R^2 = C_2H_5$	–	120	115	–	–
$R^2 = CH_3$	–	160	150	–	–

73 and **74** is self-evident. Increasing the bulkiness of the β-vinyl substituent from methyl to isopropyl (Table 9) leaves both the *cis* and *trans* rate and the k_{cis}/k_{trans} ratio nearly unchanged; in the case of the quasi-isotropic *t*-butyl rotor, however, where no special conformation is possible which would minimize steric interaction in the transition state, the rate drops sharply (Table 9), but once more the *cis/trans* ratio is hardly affected.

Detailed mechanistic and kinetic investigations have also been reported for dimethylketene[250] and other ketene derivatives. With unsymmetrical ketenes, the large substituent is turned to the outside in the orientation complex[251], and − in cyclobutanone formation with alkyl vinyl ethers − ends up predominantly (though by no means always exclusively) in the *E*-position to OR. The (*Z/E*) stereochemistry of the enol ethers which enter the concerted process as the suprafacial component of course always remains unimpaired.

The cycloaddition of *E*- and *Z*-enol ethers to *heterocumulenes* (e.g. isocyanates) likewise proceeds with very high stereoselectivity even in polar solvents such as CH_3CN[42,249] (Scheme 5). For the two azetidinones, (*Z*)-**76** and (*E*)-**76**, obtained from tosyl isocyanate and *cis*- and *trans*-enol ethers, respectively, stereoselectivity can be assessed at $\geqslant 95\%$ since the isomers are easily differentiated by ^{1}H-NMR[42,252]. Unlike the cyclobutanones **75**, the NCO adducts are thermally unstable: in solution, the sterically pure azetidinones are converted to an equilibrium *E/Z* mixture (60–75% *E*) and, finally, into the acrylamides **78**. The rate enhancement for the ethoxy over the methoxy derivatives is much more pronounced in epimerization − which must traverse the zwitterion **77** − than in cycloaddition; thence, and from the stereochemistry of the cycloaddition, a concerted $(_\pi2_a + _\pi2_s)$ mechanism was advanced also for the −N=C=O addition[249]. In view of the high stereochemical fidelity of the two-step TCNE addition and its overall kinetics, this view will probably have to be revised. The low k_{cis}/k_{trans} ratios for the tosyl

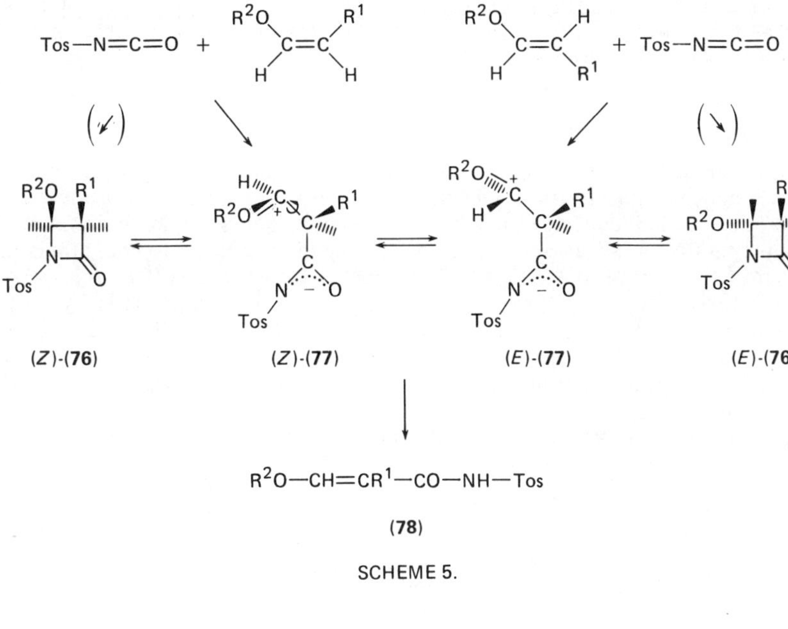

(Z)-(**76**)   (Z)-(**77**)   (E)-(**77**)   (E)-(**76**)

$$R^2O—CH=CR^1—CO—NH—Tos$$

(**78**)

SCHEME 5.

isocyanate cycloaddition (5–10 in CCl_4 and 3–4 in CD_3CN) likewise argue against the orthogonal π–π approach (see above).

Reaction of reactive N-acyl (CCl_3CO) isocyanates with enol ethers affords both [2 + 2]- and [2 + 4]-cycloaddition[253]; both products are unstable and isomerize to the respective β-substitution products. With N-thioaroyl isocyanates, only [2 + 4] products are found[254].

The efficiency of electron-rich olefins, e.g. vinyl ethers, in quenching singlet and triplet n,π* ketone fluorescence and/or phosphorescence correlates well with TCNE charge-transfer data and gas-phase ionization potentials[255]. Quenching involves an *exciplex* which partitions either to generate ground states, or to yield biradicals and thence oxetanes[255]. The [2 + 2] photocyloaddition of enol ethers to 2-cyclohexenone, which affords 7-alkoxybicyclo[4.2.0]-2-octanones in good yield, is likewise formulated via a π-complex with the excited ketone[256]. The same regiochemistry is observed for the photoaddition of t-butyl vinyl ether to 1,3-dimethyluracil[257]. Irradiation of adamantanethione in the presence of enol ethers yields alkoxyspirothietanes (equation 47)[258], but in extremely low quantum yield. From the n,π* triplet, only **79** is obtained, with the C=C stereochemistry scrambled as becomes a biradical; with the π,π*-excited thione (singlet), on the other hand, both **79** and **80** are formed. Addition in this case is no longer regio-, yet

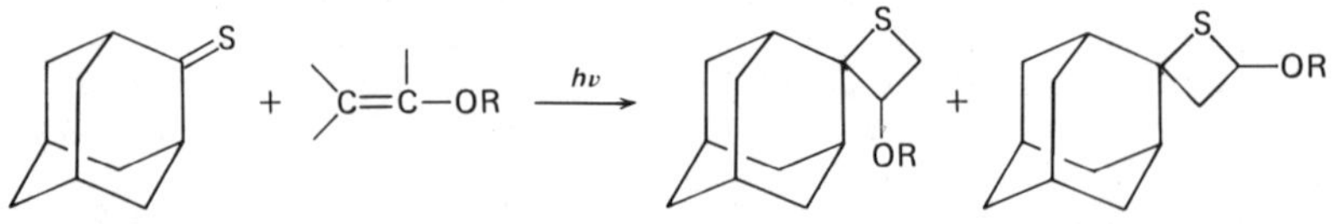

(47)

(79)   (80)

Excitation: 500 nm (n, π*)   100%   0%
            254 nm (π, π*)   67%   33%

fully *stereo-specific*[258]. In photoaddition to benzene, ethyl vinyl ether gives the largest amount of [2 + 2] addition of all olefins; in polar solvents, the [2 + 2]/[2 + 4] ratio is even higher[259].

C. [1 + 2] Cyloadditions (Carbene Reactions)

Singlet carbenes and nitrenes react with enol ethers in a straightforward manner: there is practically no insertion, and the cycloaddition is stereospecifically *cis*[260], i.e. in a more or less concerted fashion[261]. Dihalocarbenes (which have found the widest preparative application) as electrophilic agents add faster to enol ethers than to alkenes[262]; within the CH_2=CHOR series [R = CH_3 ... $C(CH_3)_3$], relative reactivity towards CCl_2 follows the well-known inductive order as in hydrolysis, polymerization etc.[263]. The bicyclic products formed from cyclic enol ethers can undergo thermal cyclopropane ring cleavage (equation 48 and 49); in the dihydrofuran adduct **81** this rearrangement is an extremely facile process[264], in **82** it requires 140°C[265].

Dichlorocarbene addition to enol ethers of cyclic ketones with subsequent ring

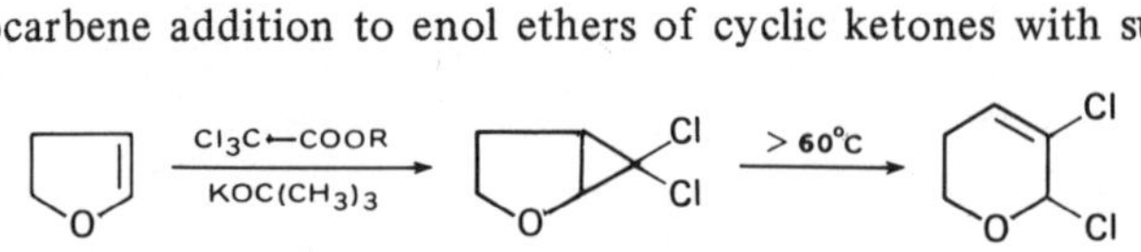

(48)

(81)

$$
\text{(82)} \qquad \xrightarrow[\text{−HCl}]{140°\text{C}} \qquad \text{(49)}
$$

enlargement has been utilized for an elegant muscone synthesis (equation 50)[266], and also for the preparation of steroids with a tropone structure of the A-ring[267].

$$
\xrightarrow{Cl_2} \qquad \xrightarrow[\text{2. } H_2O/H^+]{\text{1. } CH_3Li \text{ in THF/HMPT}} \qquad \xrightarrow{H_2} \text{D, L- muscone}
$$

$$n = 10, 12 \qquad (50)$$

1,1-Dibromo-2-alkoxycyclopropanes, formed in 50% yield by CBr_2 addition to vinyl ethers, offer a convenient access to alkoxyallenes or, alternatively, to propargyl aldehyde acetals (equation 51)[268]. Chlorocarbene likewise adds to vinyl

$$
CH_2{=}C{=}CHOR \xleftarrow{CH_3Li} \underset{Br_2}{\overset{-OR}{\triangleleft}} \xrightarrow{EtOH/EtONa} HC{\equiv}CCH(OEt)_2 \qquad (51)
$$

ethers in fair yield; the alkoxychlorocyclopropanes obtained are predominantly *cis* (*cis/trans* 20 : 1)[269]. *Cis*-Disubstituted cyclopropanes are formed preferentially, too, with alkoxycarbene while phenoxycarbene gives the *trans*-diethers[270,271]. Cyclopropane formation from simple vinyl ethers in moderate to good yields has been reported also for difluoro-[272], fluorobromo-[273] and phenylthio-carbene[274,275]. The addition of cyclohexylidene carbene to *t*-butyl vinyl ether, yielding cyclohexylidenecyclopropane[276], is noteworthy, too.

D. 1,3-Dipolar [2 + 3] Cycloadditions

Among 1,3-dipolar cycloadditions[277] to enol ethers, both mechanistic[278] and preparative studies have been focused on the reaction with aryl, acyl and sulphonyl azides (less activated azides do not react, and some less reactive enol ethers are inert even towards *p*-nitrophenyl azide[279]). The overall reaction (Scheme 6) offers an extreme width in its product spectrum, depending on the number and nature of the substituents in both reactants[278,280].

The primary 1,3-addition of the azide has been demonstrated to proceed stereospecifically *cis*[281]; the terminal azido nitrogen always attacks the π-bond at the electron-rich β-position, while the more nucleophilic N-1 bonds to C-1, in the α-position to OR. The addition rate is strongly accelerated with increasing solvent polarity[282] and is, for instance, 5×10^4 times faster with picryl than with phenyl azide[283]; however, a concerted reaction mechanism, though with partial charges in the transition state at N-1 (δ^-) and C-5 (δ^+) of the incipient triazoline structure (85)[282], is now generally accepted[278] (but not by Firestone, see below). The triazolines from *p*-nitrophenyl azide and β,β-unsubstituted vinyl ethers (83) ($R^3 = R^4 = H$) lose alcohol R^1OH at 130–150°C to form triazoles (86), e.g. 1-nitro-phenyltriazole from butyl vinyl ether (Scheme 6)[279]. The triazolines from vinyl ethers and phosphoryl azides (84, $R^5 = R_2P(=O)-$), on the other hand, undergo thermal 1,3-dipolar cycloreversion to diazo compounds[284] (more generally observed with enamine/azide cycloadducts[285]). The triazolines from 1-alkenyl and isobutenyl ethers and *p*-nitrophenyl azide are much more labile[280], owing probably

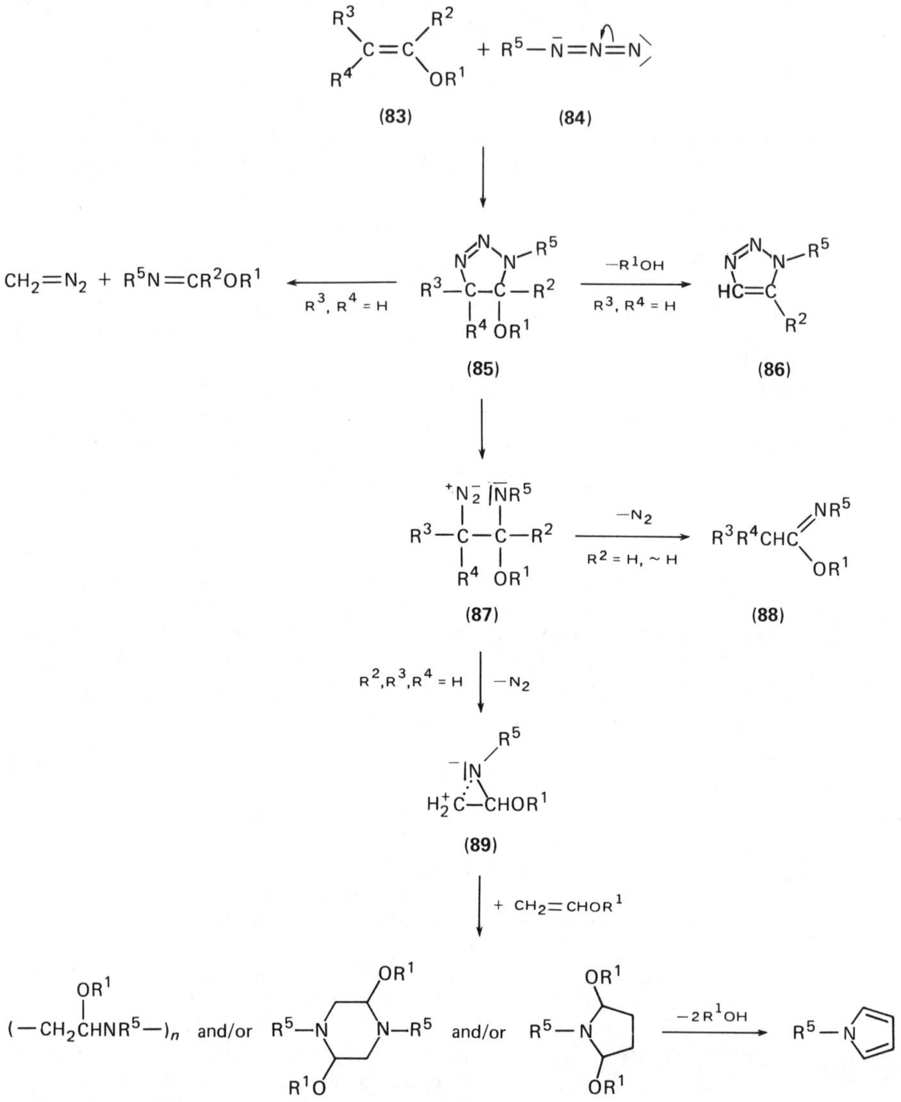

SCHEME 6.

to better stabilization of the incipient carbenium centre in **85**; the N_2 expulsion is accompanied by a 1,2-hydrogen shift, with formation of imino ethers (**88**). Since both cycloaddition to Z-alkenyl ethers and N—N bond scission in the respective *cis*-triazolines are much faster than for the corresponding *trans* compounds[281], only *trans*-4-alkyl-5-alkoxytriazolines (**85**) ($R^2 = R^4 = H$) are obtained besides imino ethers from E/Z mixtures of 1-propenyl and 1-butenyl ethers[280]; with tosyl azide[280] or trichloroacetyl azide[286] (where the negative charge in **87** is especially well stabilized), only imino ethers (**88**) are isolated (some in quantitative yield[280]). Tosyl azide reaction with unsubstituted vinyl ethers usually produces only polymeric oily material; under special conditions, however, either piperazines or pyrrols can be obtained, some in very good yields[287]. Whether nucleophilic attack of the second enol ether molecule — with either subsequent ring-closure to a 2,5-dialkoxytetrahydropyrrol or further $CH_2=CHOR^1$ addition, followed by polymerization — is to the zwitterion **87** or to **89**, cannot be decided[280]. However, acetolysis[279] and alcoholysis[288] of the triazolines **85**, in which the R^5NH group ends up at the β-carbon of the former enol ether, must by necessity proceed via intermediate aziridine structures.

Thermal decomposition of the N-aryltriazolines from cyclic enol ethers (equation 52)[279] or direct cycloaddition with tosyl azides[279,289] affords the

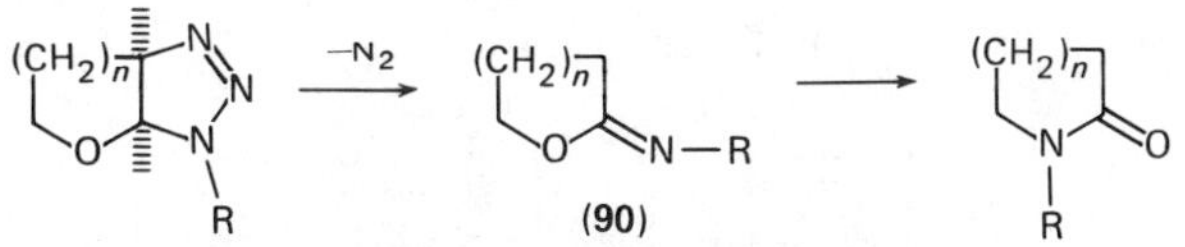

$$(52)$$

iminolactones **90** which can then undergo Chapman rearrangement[289]. If no α-hydrogen is present and ROH elimination not feasible, as in the derivatives of alkoxycycloalkenes (equation 53), imino ether formation occurs via Wagner–Meerwein rearrangement[290].

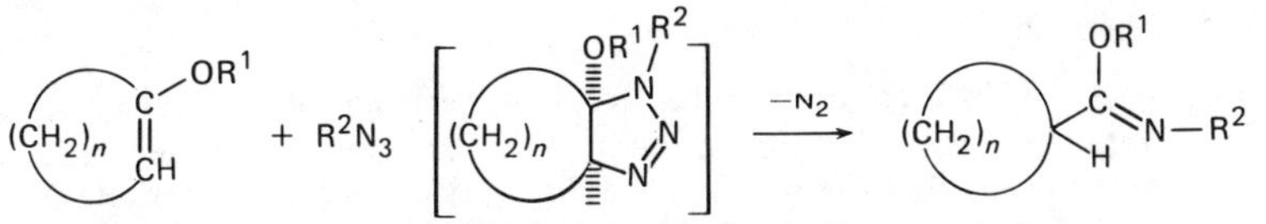

$$(53)$$

As Huisgen has repeatedly emphasized[278], the directionality in 1,3-dipolar cycloadditions still remains a fairly dark phenomenon. The addition of trichloroacetyl azide to methyl and ethyl vinyl ether, for instance, affords two oxazolines (**91a,b**) after N_2 elimination[291] which can obtain only from two cycloadducts with opposite regiochemistry. The nitrile ylide **92** combines with simple vinyl ethers to form 4-RO-substituted pyrrolines (equation 54); but with phenyl vinyl ether, 12% of the inverted addition product is found besides 88% **93**[292]. The slightly reduced polarity of the phenoxyalkene apparently suffices to overturn the usual addition direction.

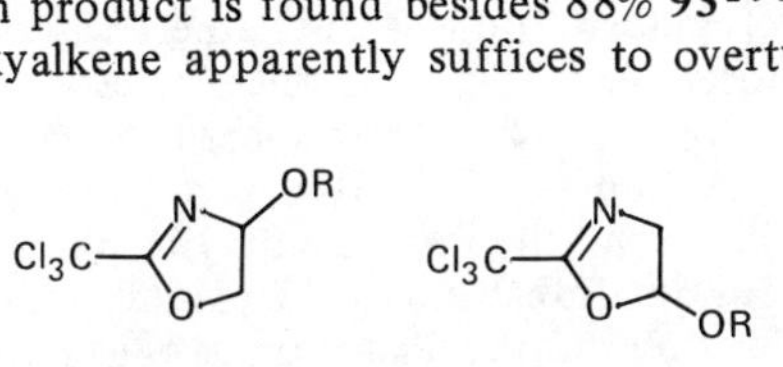

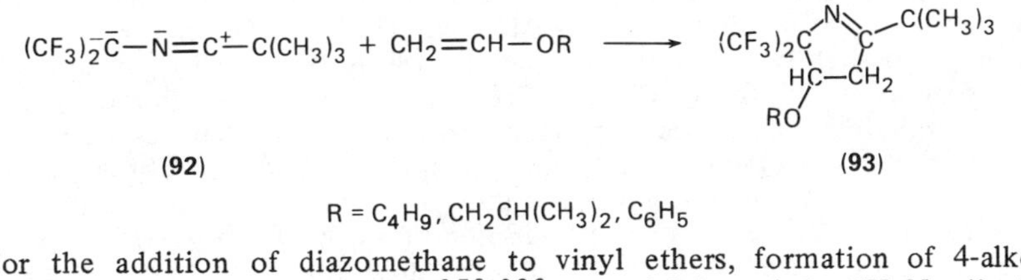

(**91a**)   (**91b**)

$$(CF_3)_2\bar{C}-\bar{N}{=}\overset{+}{C}-C(CH_3)_3 + CH_2{=}CH-OR \longrightarrow$$

$$(54)$$

(**92**)   (**93**)

$$R = C_4H_9, CH_2CH(CH_3)_2, C_6H_5$$

For the addition of diazomethane to vinyl ethers, formation of 4-alkoxy-pyrazolines (**94a**) has been reported[278,293], i.e. addition of the CH_2N_2 dipole in the same sense as to ethoxyacetylene. Firestone (who has fought for diradical intermediates in 1,3-dipolar cycloadditions from the beginning[294]) reports, however, the formation of 3-ethoxypyrazoline (**94b**) from a 38-day reaction of CH_2N_2 with ethyl vinyl ether in the dark[295] (the combined $^1H/^{13}C$-NMR evidence is irrefutable). This result could be accommodated by Firestone's biradical theory, but would invalidate the only theoretical 'silver lining' in the dark world of 1,3-dipolar cycloaddition directionality. This rationalization is based upon the frontier orbital concept of Fukui[296], and argues that the direction of 1,3-dipolar cycloaddition is governed by HOMO(1,3-dipole)/LUMO(dipolarophile) interaction[297,298]; for the CH_2N_2/enol ether reaction, addition is predicted between the terminal N atom of CH_2N_2 and the C-2 of the vinyl ether, as in **94a**.

(**94a**)   (**94b**)

Further heterocyclic syntheses via enol ethers include 1,3-cycloadditions of nitrile oxides, generated *in situ*[299], and of phenylsydnone[300].

E. [2 + 4] Cycloadditions

Normal Diels—Alder reactions, with electron-deficient dienophiles[301], are of course facilitated by alkoxy groups in the diene[302]; however, as a rule the entropic term contributes more than half to the free activation energy, so that steric effects frequently override the electronic influence[303] as is often found for truly concerted processes. 1-Alkoxy-1,3-cyclohexadienes, readily accessible by Birch reduction of alkoxyarenes and subsequent KNH_2 rearrangement[2], add twice to p-benzoquinone[304]. For Diels—Alder additions with inverse electron demand[301,305], enol ethers (like enamines) are ideally suited substrates; they smoothly react with cyclones, hexachloropentadiene and 1,2,4,5-tetrazines[306-309].

Desimoni and coworkers[310] have extensively studied the mechanism and stereochemistry of the [2 + 4] cycloaddition of vinyl ethers with α,β-unsaturated carbonyl compounds, namely with 4-benzal-5-pyrazolones (equation 55). The reaction is first order in each reactant, stereospecific with respect to the enol ether double bond, as demonstrated for the addition of (Z/E)-1-propenyl propyl ether to 4-benzal-1,3-diphenyl-5-pyrazolones[311], and the resulting dihydropyran is formed preferentially with the 4-aryl and 6-alkoxy substituents *cis* to each other (**96a**)[312]. The underlying additional stabilization of the *endo* transition state (here via $R\underline{O} \leftrightarrow C=O$ interaction) is analogous to that found for the regular Diels-Alder reaction[301]. Thus, all kinetic and stereochemical evidence indicates a concerted mechanism[310], with the rate of addition controlled solely by HOMO(vinyl ether)/LUMO-(benzalpyrazolone) interaction[297]. Variation of the p-benzal substituent in the diene component (**95**) from NO_2 to $N(CH_3)_2$ leads to a decrease in rate by a factor of $\sim 10^2$ while the *cis/trans* ratio remains practically unaffected; for both k_{cis} and k_{trans} perfect Hammett plots vs. σ_p^+ are obtained[310]. Variation of the OR moiety in the enol ether has much less effect; although the inductive order basically holds, the influence of steric effects on the addition rate, e.g. in t-butyl vinyl ether, is of the same order of magnitude[310]. It would be interesting now to test the kinetics of the benzalpyrazolone cycloaddition to various (Z/E)-1-alkenyl ethers.

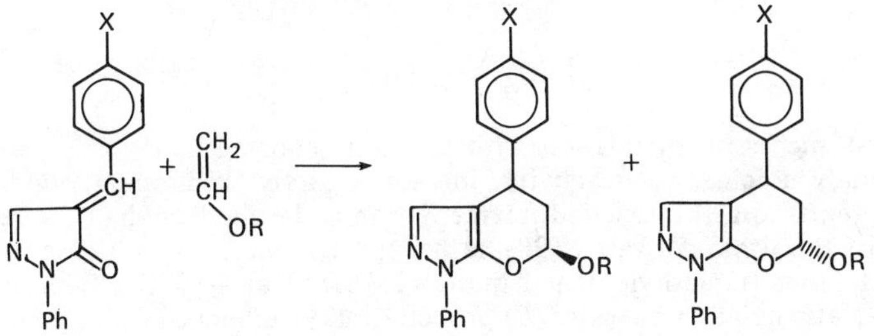

(95)   (96a)   (96b)

R = CH(CH₃)₂   a/b 2.25—2.97 for X = NO₂ ⟶ N(CH₃)₂

For the addition to 3,4-dihydro-2H-pyran, some *trans* addition to the pyran double bond is found; a small fraction of the reaction thus must proceed via a zwitterionic intermediate[313]. The cycloadditions of 2-alkylidenecycloalkanones with enol ethers require 170°C and show definite acid catalysis[314]; thence, an electrophilic attack on the vinyl ether, with polar intermediates, has been postulated. The BF_3 catalysis in the reaction of enol ethers with N-aryl Schiff bases[315] likewise argues a polar mechanism.

Diels—Alder additions of electron-rich olefins $C=C-\overline{X}$ to electron-rich dienes (with a +M-substituent, $-\overline{X}$, in the 1- or 2-position) are virtually unknown. In one example, the reaction of the diene **97** (with OCH_3 in a vinylogous 2-position) and ethyl vinyl ether (equation 56), two cycloadducts are formed with moderate

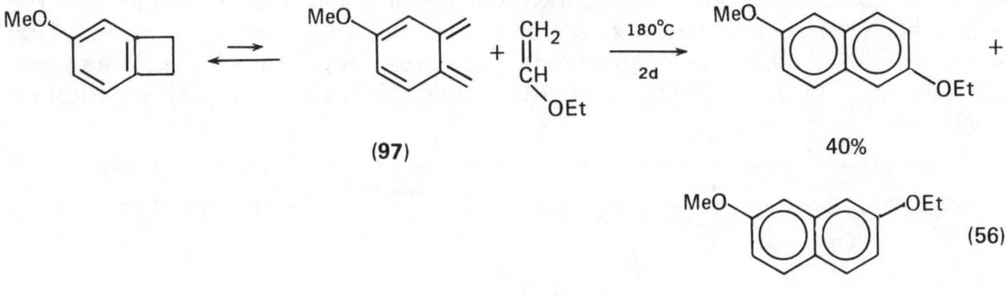

(97)   40%

(56)

60%

regioselectivity[316]; the major product, however, is the one expected on the basis of frontier orbital theory while the biradical formalism predicts the opposite polarization.

1,4-Dipolar cycloaddition of the dipolar species, generated from α-chloroaldonitrones and $AgBF_4$, with cyclic enol ethers of varying ring-size offers a further convenient route to medium-ring lactones[317]. In the presence of Lewis acids, 1,5-dipolar addition of 1,3-oxazolidines to cyclic enol ethers leads to 1,4-oxazepines in good yield[318].

VI. METALATION

Both α- and β-vinylic hydrogen atoms in enol ethers can be substituted with pentylsodium[319]. The α-sodium derivative can be trapped with CO_2; β-metalation, on the other hand, results in immediate cleavage into alcohol and acetylene, as in equation (57)[319].

$$\text{(furan-CH}_2\text{)} \xrightarrow{\text{C}_5\text{H}_{11}\text{Na}} \left[\text{(furan-CH}^-\text{Na}^+\text{)} \right] \longrightarrow \text{NaOCH}_2\text{CH}_2\text{CH}_2\text{C}\equiv\text{CH} \quad (57)$$

The first successful metalation with LiR was reported in 1972[320]. Because of their generally much lower reactivity, the most reactive lithio compounds must be employed (equation 58); to avoid fragmentation and effectively halt the reaction at the stage of the lithio derivative (98), rather special reaction conditions are needed. With *t*-BuLi in tetramethylethanediamine (TMEDA) at −30°C[320], or in THF at −65°C[321], ethenyl as well as (Z/E)-propenyl alkyl ethers can be lithiated in the α-position in essentially quantitative yield. Once formed, 1-methoxyvinyllithium (98b), for instance, is surprisingly stable up to 0°C[321].

$$\text{CH}_2{=}\text{CH}{-}\text{OEt} \xrightarrow[\text{TMEDA/}-30°\text{C}]{t\text{-BuLi}} \text{CH}_2{=}\text{C}\overset{\text{OEt}}{\underset{\text{Li}}{\big\langle}} \quad (58)$$

$$\textbf{(98a)}$$

$$[\text{CH}_2{=}\overset{=}{\text{C}}{-}\text{OR}] \longrightarrow [\text{CH}_3{-}\overset{=}{\text{C}}{=}\text{O}]$$

The usual enol ether polarity is inverted in 98, electrophilic substitution now being directed to C-1 ('*Umpolung*'); at the same time, 1-alkoxyvinyllithium represents a masked acetyl anion, i.e. a synthon which allows *nucleophilic* acetylation. It readily adds to aldehyde and ketone C=O functions (equation 59), even in the sterically demanding case of 17-ketosteroids[321], and causes *no* enolization in the carbonyl substrate. If the addition reaction is quenched with NH$_4$Cl at 0°C, the enol ether (99) is recovered and can be further modified electrophilically at C-2; work-up with H$_2$O/H$^+$ directly gives the α-hydroxylacyl product (100). Reaction of 98b with ethyl carboxylates results in double CH$_2$=C−OCH$_3$ substitution (101)[320].

$$\text{CH}_2{=}\text{CHOCH}_3 \xrightarrow[\text{THF/}-65°\text{C}]{t\text{-BuLi}} \text{CH}_2{=}\text{C(OCH}_3)\text{Li} \xrightarrow[\text{2. NH}_4\text{Cl/0°C}]{1.\ +\ \text{R}^1\text{R}^2\text{CO}} \text{R}^1\text{R}^2\text{C}{-}\overset{\text{OCH}_3}{\underset{\text{OH}}{\text{C}}}{=}\text{CH}_2$$

$$\textbf{(98b)} \qquad\qquad\qquad\qquad\qquad\qquad \textbf{(99)}$$

1. + R³COOEt   H₂O/H⁺   (59)

$$\text{1. + R}^3\text{COOEt} \qquad\qquad\qquad\qquad \xrightarrow{\text{H}_2\text{O/H}^+} \quad (59)$$
$$\text{2. + NH}_4\text{Cl}$$

$$\text{H}_3\text{CO}\overset{\text{R}^3\ \ \text{OH}}{\diagdown\underset{}{\text{C}}\diagup}\text{OCH}_3 \qquad\qquad\qquad \text{R}^1{-}\overset{\text{R}^2}{\underset{\text{OH}}{\text{C}}}{-}\text{COCH}_3$$

$$\textbf{(101)} \qquad\qquad\qquad\qquad\qquad\qquad \textbf{(100)}$$

α-Lithiation of cyclic enol ethers (dihydro-furans, -pyrans) likewise requires *t*-BuLi (*n*- or *s*-BuLi are not sufficient)[322]; the solvent THF is best kept at the minimum of 0.5−0.75 equivalents which are necessary for LiR dissociation. (Z)-1,2-Dimethoxyethene, in contrast, is smoothly monolithiated with *n*-BuLi (0°C in THF/TMEDA) and added to various carbonyl compounds, e.g. 17-ketosteroids[323].

With specially prepared Cu(I) salts, **98a** can be transformed into bis(α-methoxy-ethenyl)cuprate, $(R_2Cu)Li$; this reagent is highly selective, adding to α,β-unsaturated cyclohexenones exclusively in the 1,4-position[324,325] (**102**), though rather sensitive towards sterical crowding at the electrophilic site.

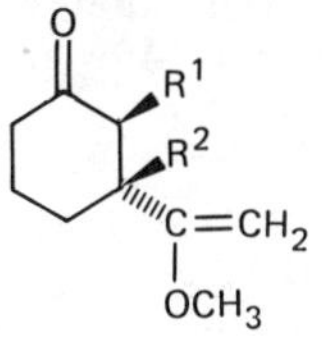

(102)

1-Alkoxy-2-propenyllithium (**103**) is readily accessible by α-metalation of allyl ethers with n- or s-BuLi($-65°C$ in THF)[326,327]; both alkylation and C=O addition take place, however, at the terminal C-3 [only after transformation into the corresponding zinc dialkyl (equation 60) can quantitative α-reaction be enforced[326]].

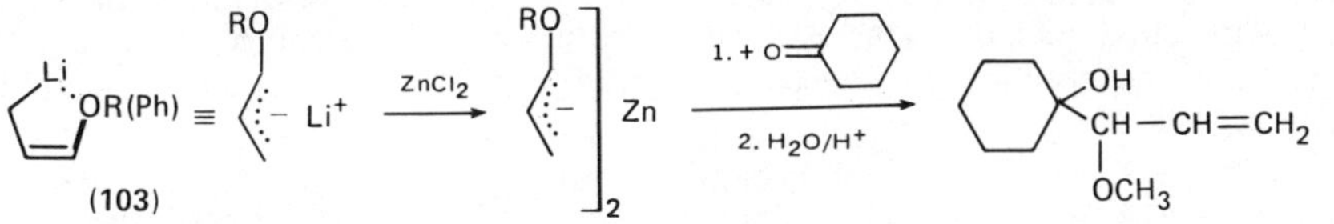

Allylic lithiation, and subsequent γ-alkylation, is likewise observed for (Z)-1-propenyl phenyl ether with n-BuLi/$(CH_3)_3COK$[327], due probably to cheletropic Li $\leftrightarrow$ OR interaction. If, however, the 2-tetrahydropyranyl moiety is employed as ethereal component, the respective vinyl, (Z)-1-alkenyl, and also isobutenyl ethers are metalated exclusively in the α-position with s-BuLi/t-BuOK ($-78°C$ in THF)[328], owing probably once more to cheletropic stabilization (**104**).

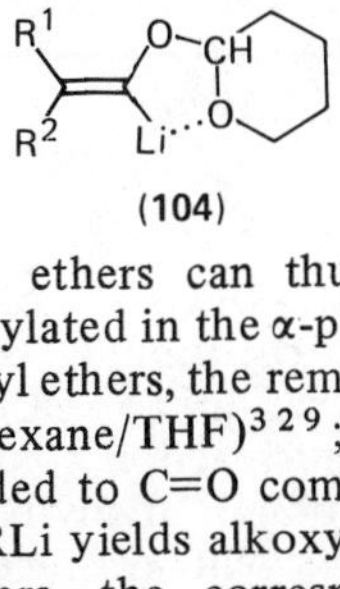

(104)

These 2-tetrahydropyranyl enol ethers can thus be readily alkylated, α- or β-hydroxyalkylated, and even formylated in the α-position[328].

In both (E)- and (Z)-2-halovinyl ethers, the remaining β-hydrogen can be lithiated with BuLi (**105**) ($-100°C$ in hexane/THF)[329]; **105** can either be trapped with CO_2, alkylated with RX, or added to C=O compounds (equation 61). HCl elimination with a second mole of RLi yields alkoxyethinyllithium, $Li-C≡C-OR$[329]. From the 2-stannyl vinyl ethers, the corresponding nonhalogenated β-lithio-

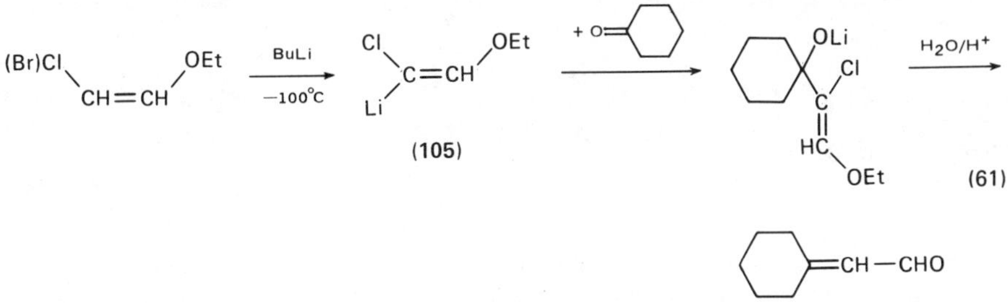

(105)   (61)

alkoxyethenes are accessible[330]. Reaction of 5-bromo-3,4-dihydro-2H-pyran, on the other hand, with t-BuLi at $-110°C$ yields the β-lithio derivative (106) via metal–halogen exchange[331]. 106 is significantly less stable than Li–CH=CH–OR, and shows alkyne cleavage already above $-90°C$. At $-110°C$, though, it can be added in high yield to ketones (equation 62); after transformation into the corresponding dialkyllithium cuprate, 106 also gives selective 1,4-addition to α,β-unsaturated ketones[331].

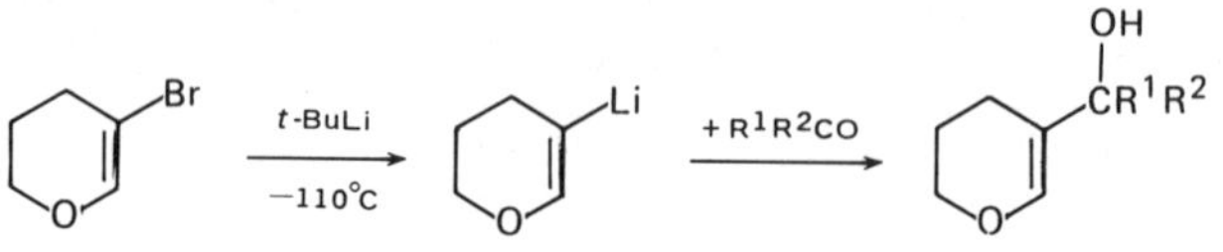

$$(62)$$

(106)

1-Ethoxyvinyllithium also reacts readily with trialkylboranes (equation 63)[332]. For sterically undemanding n-alkyl BR_3 groups, oxidation of the 'ate' complex (107) is faster than a second R transfer, and ketones (108) are obtained in good yield. With bulkier alkyl groups, or in the presence of acid, rearrangement is much faster, and the reaction is directed quantitatively towards the dialkyl methyl carbinols (109)[332].

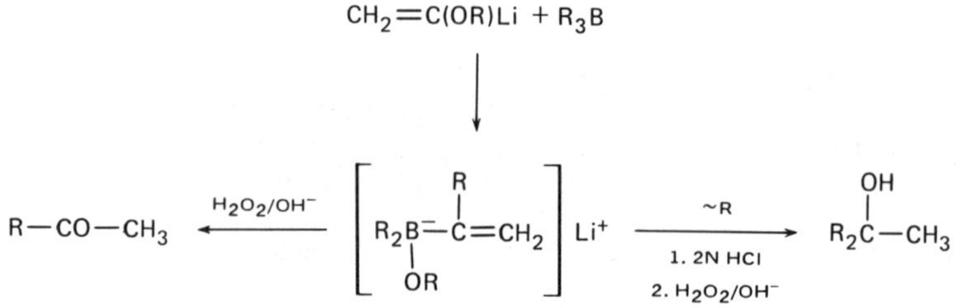

$$CH_2=C(OR)Li + R_3B$$

$$(63)$$

(108)   (107)   (109)

$$R = n\text{-alkyl} \ldots\ldots c\text{-}C_6H_{11}$$

Silanes can be added in good yield to enol ethers with $H_2[PtCl_6]$ or Pd/C catalysts[333]; in general, though, partitioning between addition and the usually prevailing cleavage of the vinyl ether linkage, =C—OR, by silane or borane reagents depends critically on catalysts and reaction conditions[334,335] (low temperature usually favouring addition). Reaction of triallylboranes with vinyl ethers, proceeding probably via a Claisen-type cyclic rearrangement (equation 64), affords a convenient synthesis of 1,4-dienes[336].

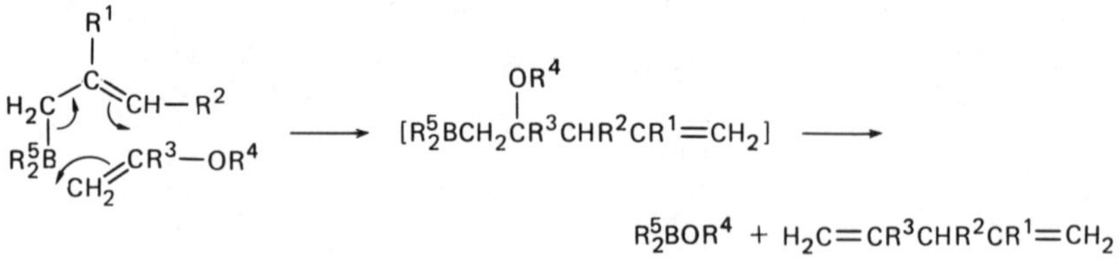

$$R_2^5BOR^4 + H_2C=CR^3CHR^2CR^1=CH_2 \quad (64)$$

With Grignard reagents, either the vinyl or the alkyl ether C—O bond is broken, depending largely on the size and nature of the vinyl ether C-1 substituent[337].

VII. SILYL ENOL ETHERS[338,339]

A. Preparation and Reactivity

The OR group of alkyl alkenyl ethers as a rule is introduced via nucleophilic reactions (Section III); silyl enol ethers, R_3Si—O—$\overset{\displaystyle |}{C}$=$C\overset{\displaystyle \diagup}{\diagdown}$, on the other hand, are without exception prepared by O-silylation of either the parent carbonyl compound or its enolate, and not by incorporation of a silyloxy moiety[338]. The standard procedure for O-silylation is refluxing the carbonyl substrate with chloro-trimethylsilane (Me_3SiCl) and triethylamine or diazabicyclo[2.2.2]octane in DMF[340]. Silylation is much faster and can be effected under far milder conditions with some new reagents such as trimethylsilyl trifluoromethanesulphonate[341], or alternatively, Me_3SiCl in the presence of $C_4F_9SO_3K$ (with NEt_3 in cyclo-hexane)[342], or trimethylsilyl ethyl acetate[343]. The latter reagent, in the presence of quaternary ammonium fluorides, also allows highly *stereo*selective ($\geqslant 99\%$) preparation of Z-enol ethers (equation 65)[344]. E-Enol silyl ethers are best prepared

$$\underset{}{\overset{O}{\underset{\diagdown\diagup\diagdown}{||}}} + Me_3SiCH_2COOC_2H_5 \xrightarrow[-78°C]{1\% \ Bu_4N^+F^-} \underset{99.5\% \ (Z)}{\overset{OSiMe_3}{\diagdown\diagup\diagdown}} \qquad (65)$$

with Me_3SiCl and lithium diisopropylamide[345] or lithium 2,2,6,6-tetramethyl-piperidide[343].

Enolate ions can be generated *regio*specifically with $(R_2Cu)Li$ from α,α'-di-bromo- or α,β-unsaturated ketones (equation 66) or by Li/NH_3 reduction of such

$$\underset{(110)}{\text{[structures for equation 66]}} \qquad (66)$$

alkenones, and then trapped by reaction with Me_3SiCl as silyloxyalkenes (110)[338]. Potassium hydride[346] or lithium hexamethyldisilazane[347] have been employed successfully for the metalation (with subsequent silylation) of sterically hindered, e.g. t-butyl, ketones.

1,4-Addition of hydrosilanes to α,β-unsaturated aldehydes and ketones is affec-ted with Pt, Ni, and especially Rh catalysts[338]; among these, $(Ph_3P)_3RhCl$ has been found the most effective[348]. Some special catalysts also allow the dehydro-genative silylation of saturated C=O compounds.

Trimethylsilyl vinyl ether is most stable in the s-trans conformation (owing to the larger size of the $SiMe_3$ group and, probably, to the lack of nonbonded attractive stabilization). Relative to Me_3SiOMe, the Si—O force constant in silyl vinyl ethers appears diminished by ~25%[349]; for silyl phenyl ether, $H_3SiOC_6H_5$, an unusually large Si—O distance has been determined[350]. Both findings indicate an especially high mesomeric potential of the silyoxy oxygen — as has indeed been verified by the great reactivity of silyl enol ethers[339]. Conversely, the SiR_3 moiety in silyl vinyl ethers is rather labile; it is often removed directly by the nucleophilic counterion, X^-, of the attacking electrophile, thus regenerating the parent carbonyl compound, now in the α-substituted form (equation 67). Ethenyloxytrimethyl-silane has consequently been employed as *silylating* agent for alcohols, thiols, amines and acids[351].

$$CH_2=\underset{\underset{R}{|}}{C}-OSiMe_3 \ + \ E-X \ \longrightarrow \ E-CH_2-CO-R \ + \ Me_3SiX \qquad (67)$$

B. Reactions with Heteroelectrophiles

Among the reactions of silyl enol ethers with protic reagents HX, that with liquid anhydrous HCN is noteworthy, affording α-silyloxynitriles (111) in ~50% yield (equation 68)[352]. With HN_3, the α-azido silyl ethers 112a, and with HN_3 and excess alcohol in the presence of $TiCl_4$, the α-azido alkyl ethers 112b are formed[353].

$$\underset{(111)}{\underset{\underset{R^2 \quad R^3}{|\qquad|}}{\overset{\overset{R^1 \quad OSiMe_3}{|\qquad\;\;|}}{CH-C-CN}}} \quad\xleftarrow{\text{HCN}}\quad \underset{(112a)}{\underset{\underset{R^2}{|}}{\overset{\overset{R^1}{|}}{C}}=\underset{\underset{R^3}{}}{\overset{\overset{OSiMe_3}{}}{C}}} \quad\xrightarrow[\text{2. }HN_3/R^4OH/TiCl_4]{\text{1. }HN_3}\quad \underset{\underset{R^2 \quad R^3}{|\qquad|}}{\overset{\overset{R^1 \quad OR}{|\qquad|}}{CH-C-N_3}} \qquad (68)$$

(111)   (112a) R = SiMe$_3$

(112b) R = R^4

Acid-catalysed addition of α,ω-diols or 2-mercaptoethanol provides a rapid and high-yield synthesis of O,O- and O,S-acetals (113), respectively[354]; with isopropenyl trimethylsilyl ether, $trans$-cyclohexane-1,2-diol has thus successfully been transformed into the corresponding acetonide for the first time.

$$\underset{R^2}{\overset{R^1}{>}}CH-CH\overset{X}{\underset{O}{<}}(CH_2)_n$$

(113)

X = O, S

Halogenation of silyl enol ethers (with molecular Cl_2, Br_2 or, alternatively, N-halosuccinimides) yields not the addition products but rather the desilylated α-halocarbonyl compounds[355], and is especially suited for the preparation of α-halo aldehydes[356]. However, bromination in the presence of triethylamine (in CH_2Cl_2, $-60°C$) smoothly affords 2-bromo-substituted 1-silyloxy-1-alkenes[357]. Peroxidation of ketone-derived silyl enol ethers with, for example, m-perbenzoic acid gives, via an intramolecular Si migration, the α-silyloxyketones 114 in 70–90% yield (equation 69)[358]; from 1-silyloxyalkenes, the α-hydroxyacetals 115 are obtained[359].

$$\underset{(115)}{\underset{\underset{R^2 \quad OCOR^4}{|\qquad|}}{\overset{\overset{OH \quad OSiMe_3}{|\quad\;\;|}}{R^1-C-CH}}} \quad\xleftarrow[R^3=H]{R^4COOH}\quad \underset{\underset{R^2}{|}}{\overset{\overset{R^1}{|}}{C}}=\underset{\underset{R^3}{}}{\overset{\overset{OSiMe_3}{}}{C}} \quad\xrightarrow{R^4COOH}\quad \underset{(114)}{\underset{\underset{R^2}{|}}{\overset{\overset{Me_3SiO \quad O}{|\qquad\;\;\|}}{R^1-C-C-R^3}}} \qquad (69)$$

(115)   (114)

$$R^4 = m\text{-}ClC_6H_4$$

Me_3Si migration (116) likewise occurs in the photosensitized addition of singlet oxygen (equation 70), the second peroxide (117) being formed via an ene-reaction pathway[360]. $Pb(OCOR)_4$ oxidation[361] and ozonolytic cleavage[362] of the C=C—OSi bond proceed as in the case of alkyl enol ethers; the silyl enol ether of camphor, however, is simply oxidized by ozone (again with a $SiMe_3$ shift) to α-silyloxycamphor.

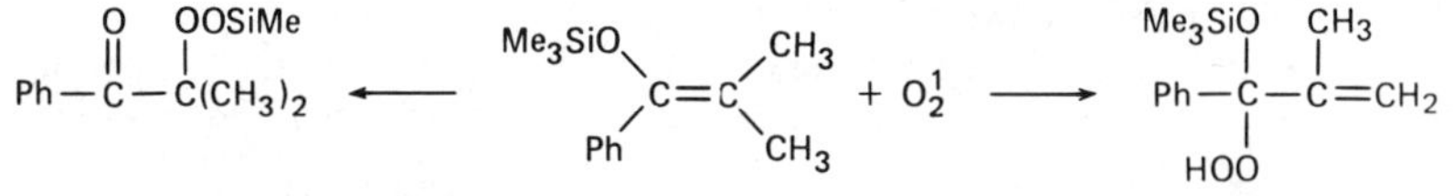

$$\underset{(116)}{\overset{\displaystyle O \quad OOSiMe}{Ph-\overset{\parallel}{C}-\underset{\mid}{C}(CH_3)_2}} \longleftarrow \underset{Ph}{\overset{Me_3SiO}{\underset{}{\diagdown}}}C=C\overset{CH_3}{\underset{CH_3}{\diagup}} + O_2^1 \longrightarrow \underset{(117)}{\overset{Me_3SiO \quad CH_3}{Ph-\underset{\mid}{\overset{\mid}{C}}-\overset{\mid}{C}=CH_2}} \quad (70)$$

Oxidation of (ketone-derived) silyl enol ethers with $AgNO_3$ in polar aprotic medium yields (β,β-coupled) 1,4-diketones, and may also be applied for cross-coupling reactions (equation 71)[363]; the high specificity is rationalized in terms of a silver enolate intermediate, generated regiospecifically.

$$2\ R^1R^2C{=}CR^3{-}OSiMe_3 \xrightarrow[\substack{60-100°C}]{AgNO_3/DMSO} R^3COCR^1R^2CR^1R^2COR^3 \quad (71)$$

Attack of sulphur or nitrogen electrophiles at the β-carbon of silyl enol ethers always proceeds with concomitant desilylation; for this, a smooth six-centre mechanism can be envisaged (equation 72). Thus, reaction of sulphenyl or sulphonyl

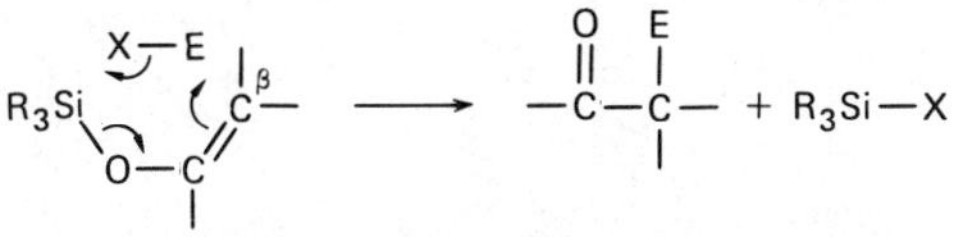

$$R_3Si \overset{X-E}{\underset{O-C}{\diagdown}} \overset{\mid}{\underset{\mid}{\overset{\beta}{C}-}} \longrightarrow \overset{O \quad E}{-\overset{\parallel}{C}-\underset{\mid}{\overset{\mid}{C}-}} + R_3Si-X \quad (72)$$

halides with silyl enol ethers gives β-keto sulphides[364] and sulphones[365], respectively, in good yield; sulphonation with sulphamoyl chlorides requires metal halide catalysis and is rather troublesome[366]. Nitrosation with NOCl produces, as in the case of the alkoxy analogues, the tautomeric α-oximino derivatives[367]; with nitryl chloride, NO_2Cl, the α-*halo* C=O derivatives are formed[368], due probably to the inverted regiochemistry of the electrophilic attack. β-Nitration of acyclic, cyclic and bicyclic silyl enol ethers can be effected in excellent yields with nitronium tetra-fluoroborate (in CH_3CN, $-25°C$)[369].

C. Reactions with Carbon Electrophiles; Metalation

Even with the highly reactive silyl enol ethers, C—C linkage requires strong activation of the carbon electrophile, $TiCl_4$ having proven the most versatile among the various Lewis acids. Thus, Reetz and Maier[370] have developed the first direct and general *t*-alkylation procedure by treating a mixture of silyl enol ether (118) and *t*-butyl chloride in CH_2Cl_2 at -45 to $-78°C$ with one equivalent $TiCl_4$ (equation 73). The reaction opens a facile route to compounds with two adjacent quaternary carbon centres (hexasubstituted ethanes) as in 119, proceeding even then with

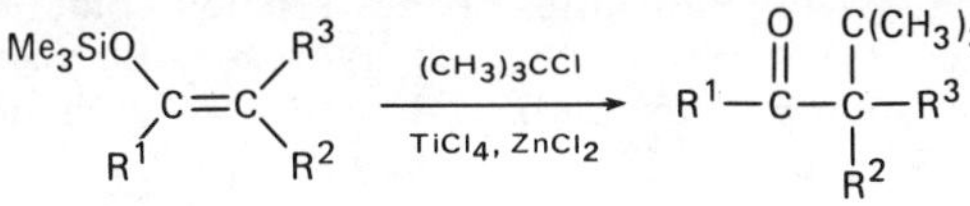

$$\underset{(118)}{\overset{Me_3SiO}{\underset{R^1}{\diagdown}}}C=C\overset{R^3}{\underset{R^2}{\diagup}} \xrightarrow[\substack{TiCl_4,\ ZnCl_2}]{(CH_3)_3CCl} \underset{(20-90\%)}{\overset{O \quad C(CH_3)_3}{R^1-\overset{\parallel}{C}-\underset{\mid}{\overset{\mid}{C}-R^3}}} \quad (73)$$

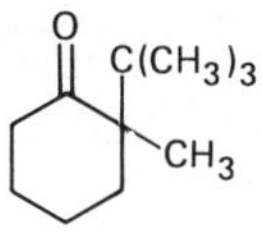

(119)

86%

$\geqslant 95\%$ regioselectivity. It works equally well with α-bromoadamantane as the alkylating agent[371], and is being extended to other alkyl halides[370]. Less heavily substituted silyloxyalkenes require $ZnCl_2$ (in catalytic amount) and give decidely lower yields.

As a rule, however, the directed enolates, regenerated from the silyl enol ethers with CH_3Li (e.g. **120**), are used as substrates for the uncatalysed alkylation with either alkyl or allyl halides[340,372,373]. Dialkylation and insufficient regio-selectivity remain problematic, even if the anionic substrates are set free under nonequilibrating conditions by a specific reaction (desilylation with CH_3Li, perhaps again via a six-centre process?). This can be overcome by generating the enolates, with either stoichiometric[374] or catalytic[375] amounts of $NR_4{}^+F^-$, in the form of their quaternary ammonium salts. A new procedure for the annelation of cyclo-hexanones utilizes the (Michael-type) addition of α-silylated vinyl ketones to cyclohexanone enolates (equation 74)[372,376].

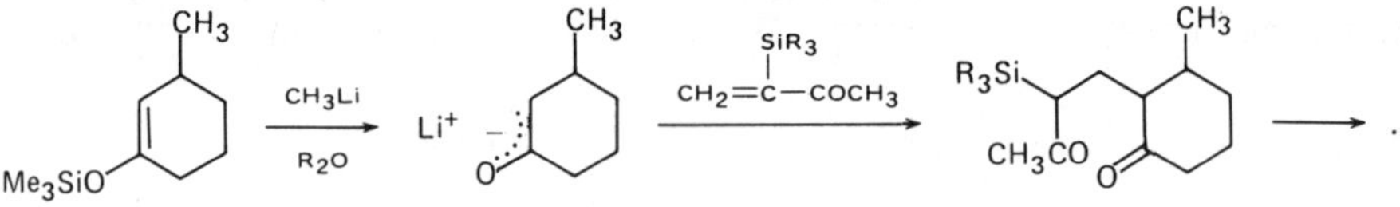

(120)   (74)

Reaction of a silyl enol ether-derived enolate with trifluoromethanesulphonic anhydride represents the most convenient route to primary vinyl triflates and thence vinylidene carbenes (equation 75)[377].

$$R^1R^2C{=}CHOSiMe_3 \xrightarrow[\text{2. }(CF_3SO_2)O]{\text{1. }CH_3Li} R^1R^2C{=}CHOSO_2CF_3 \longrightarrow R^1R^2C{=}C| \quad (75)$$

Enolate substitution with 'functionalized' C-electrophiles is limited to CH_2O[372]. If the carbonyl component is strongly activated by one equivalent of $TiCl_4$, how-ever, both aldehydes and ketones[378] as well as the respective acetals and ketals[379] undergo smooth condensation with the parent silyl enol ethers (equation 76). The

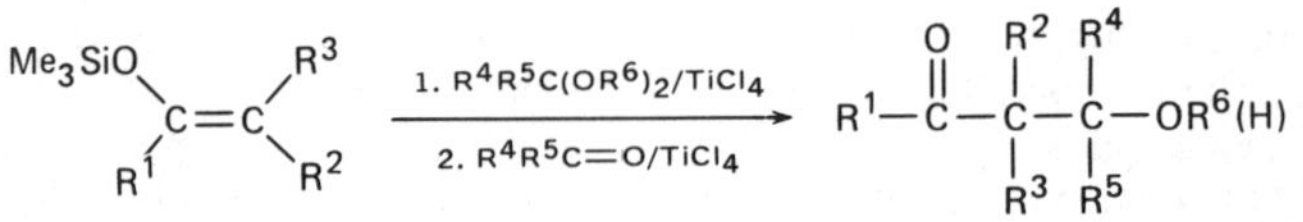

(121)

regioselectivity of these cross-aldol reactions is exceptional, differentiating even between two unlike C=O functionalities in the carbonyl component; at least one substituent (R^{1-3} in **121**), though, must be hydrogen. By using $TiCl_4$ in con-junction with Ti(IV) isopropoxide, the acetal condensation could be extended to 1-trimethylsilyloxy-1,3-butadiene[380]. In the presence of $TiCl_4$ or, better, of $TiCl_4$ *and* $Ti[OCH(CH_3)_2]_4$, the Michael reaction of α,β-unsaturated ketones, the respec-tive acetals and esters with silyl enol ethers affords 1,5-dicarbonyl compounds in good to excellent yield[381]; with the acetals of α,β-unsaturated aldehydes, Ti(IV) *t*-butoxide must be employed.

Despite the high nucleophilic potential of silyl enol ethers, their acylation requires di- or tri-haloacyl halides[382] and anhydrides[383], respectively; since the primary addition products immediately lose Me_3SiX, the α-acylated carbonyl derivatives are formed under nonacidic conditions. In the presence of $HgCl_2$[383] or N-(4-pyridyl)-pyrrolidine[338], silyl enol ethers are O-acylated even with nonactivated acyl halides. Acylation with oxalyl chloride provides the first general route to furandiones (equation 77)[384].

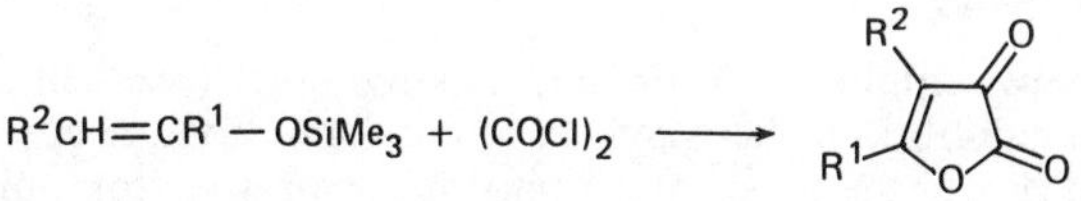

$$R^2CH= CR^1- OSiMe_3 + (COCl)_2 \longrightarrow \qquad (77)$$

D. Cycloaddition Reactions

The cycloaddition behaviour of silyl enol ethers fully parallels that of alkyl enol ethers. *trans*-1-Methoxy-3-trimethylsilyloxy-1,3-butadiene, for instance, has proven a valuable and highly reactive diene component in Diels—Alder additions[385], especially because of the ease with which the C=O function can be regenerated from the C=C—OSi functionality in the [2 + 4] cycloadduct. 1,3-Dipolar cycloaddition of arenesulphonyl azides offers a convenient route to N-sulphonyl cycloalkanecarboxamides, (equation 78)[386]. [2 + 2] Cycloadditions, yielding either cyclobutane derivatives or β-substitution products, likewise present no surprising aspects[338].

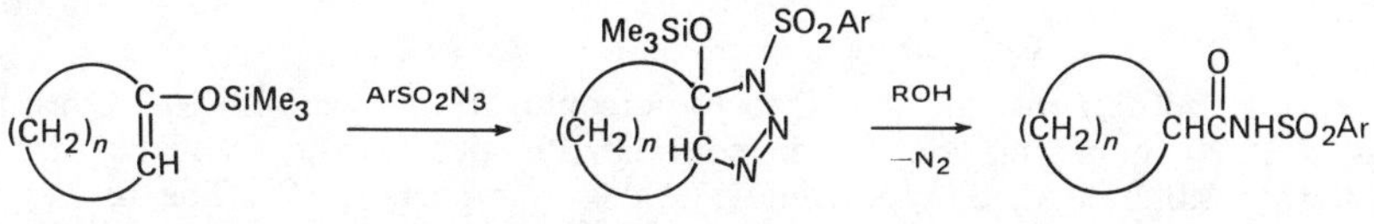

$$\text{(CH}_2)_n \quad \xrightarrow{ArSO_2N_3} \quad \text{(CH}_2)_n \quad \xrightarrow[-N_2]{ROH} \quad \text{(CH}_2)_n \quad CHCNHSO_2Ar \quad (78)$$

The Simmons—Smith cyclopropanation of silyloxyalkenes and subsequent transformation of the resultant silyloxycyclopropanes has been developed as a general synthetic procedure by Conia and his group[387]; cyclopropanols (equation 79), α-methyl carbonyl compounds (122), cyclobutanones (equation 80) and cyclopentanones have thus become readily available (average yields ≥90%). At the same time, equation (80) presents a general route to α-spirocyclobutanones[387].

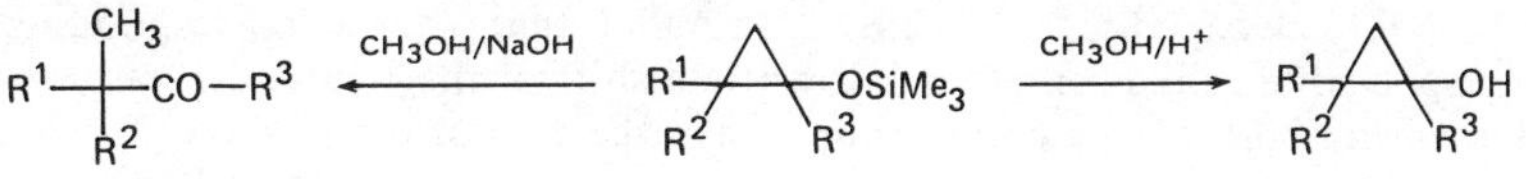

$$R^1\!-\!\overset{CH_3}{\underset{R^2}{\mid}}\!-\!CO\!-\!R^3 \quad \xleftarrow{CH_3OH/NaOH} \quad R^1\!\!-\!\!\overset{}{\underset{R^2}{\triangle}}\!\!-\!\!OSiMe_3 \quad \xrightarrow{CH_3OH/H^+} \quad R^1\!\!-\!\!\overset{}{\underset{R^2}{\triangle}}\!\!-\!\!OH \quad (79)$$

(122)

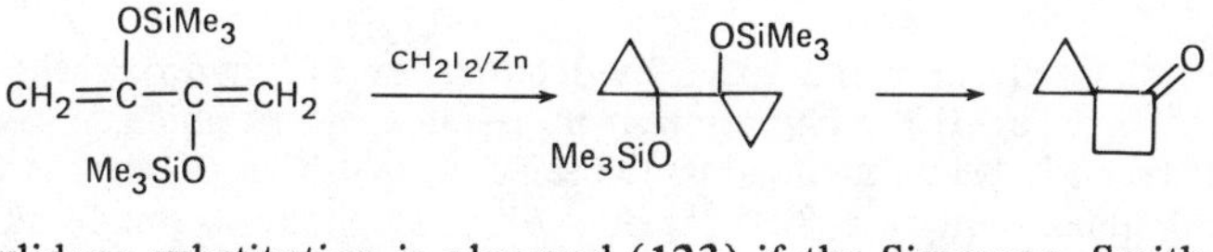

$$CH_2\!=\!\overset{OSiMe_3}{\underset{Me_3SiO}{\mid}}\!C\!-\!C\!=\!CH_2 \quad \xrightarrow{CH_2I_2/Zn} \quad \xrightarrow{} \quad (80)$$

β-Methylidene substitution is observed (123) if the Simmons—Smith reaction is carried out with one third the amount of solvent usually employed[388]. The 1-silyloxy-2,2-dihalocyclopropanes from CCl_2 or CBr_2 addition upon acidic hydrolysis undergo ring enlargement (124), with excellent overall yields[389].

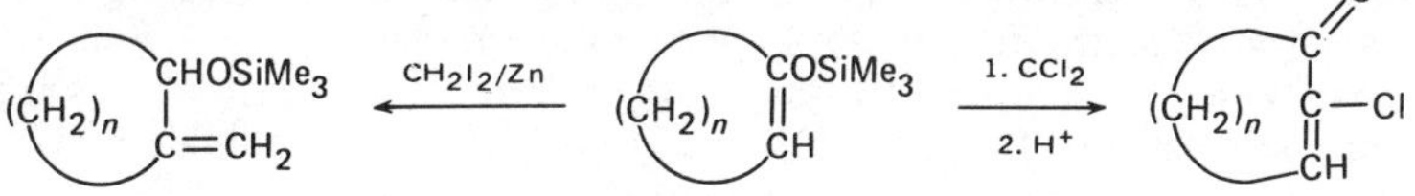

$$\text{(CH}_2)_n \quad \xleftarrow{CH_2I_2/Zn} \quad \text{(CH}_2)_n \quad \xrightarrow[2.\ H^+]{1.\ CCl_2} \quad \text{(CH}_2)_n \quad (81)$$

(123)   (124)

VIII. THIOENOL ETHERS

A. Physical Properties

For methyl vinyl sulphide, as for methyl vinyl ether (see Section II), a temperature-dependent equilibrium between two conformations, s-cis and gauche, has
been established. From the most recent photoelectron-spectroscopic data, measured
in the range 20–600°C[390], the energy difference between the two forms was
determined at 9.6 ± 0.8 kJ/mol, with an equilibrium concentration of 94% cis at
25–40°C, and of 81% at 200°C. These values are in good accord with earlier PE[27],
electron diffraction[391] and IR results[392], but differ sharply from the electron
diffraction data interpretation of Samdal and Seip[14] (33–38% cis at 200°C). There
is general agreement, though, bolstered by ab initio calculations[14], that the lesser
conformer of methyl vinyl sulphide has a gauche orientation ($\phi \sim 105°$)[14].

The first PE ionization potential for s-cis methyl vinyl sulphide (8.45 eV)[393,26]
is lower than for the s-cis conformer of the oxo analogue[31]; nevertheless, the $n_S-\pi$
resonance interaction of SCH_3 is much less pronounced than for OCH_3 as shown,
for instance, by the calculated gross atomic populations in the frontier orbital[26]:

$$CH_2=CH-S-CH_3 \qquad CH_2=CH-O-CH_3$$

$$0.63 \quad 0.20 \ 1.10 \qquad\qquad 1.03 \quad 0.53 \ 0.39$$

Calculations also demonstrate that C–S hyperconjugation lowers the σ_{C-S} orbital
energy in a 90° conformation[390] (in contrast, PE spectroscopy indicates that in
allyl methyl sulphide C–S hyperconjugation is unimportant[394]). The barrier of
rotation from gauche to s-cis (8 kJ/mol, as determined by ultrasonic relaxation[16])
is only half that for methyl vinyl ether (s-trans → s-cis), but rather similar for the
reverse process[19,395].

Since the smaller bond angle $=C-S-CH_3$ ($\sim95°$)[392] induces significant steric
strain in the cisoid orientation even for methyl vinyl sulphide, the homologous
alkylthioethenes [$R = C_2H_5 \ ... \ C(CH_3)_3$] probably assume s-trans conformation[38,396]. Within the methyl ... t-butyl vinyl sulphide series, both ${}^1H$-[397]
and ${}^{13}C$-NMR behaviour[44,398] closely parallel that of the corresponding enol
ethers, especially in the pronounced downfield shift of the β-vinyl carbon resonance
with increasing bulk of the alkyl group. As in the case of the alkoxyethenes, this is
most probably not due to steric inhibition of resonance (see Section II.B).

B. Preparation

Thioenol ethers are prepared either by dehydration of β-hydroxyethyl sulphides
with KOH[399], or by HX elimination from β-haloethyl sulphides[400]. The latter
reaction has recently been extended to the selective synthesis of, alternatively, 1- or
2-alkenyl sulphides (equation 82)[401]; at −78°C, sulphenyl bromide addition and
subsequent dehalogenation affords 125 and 126 in 85:15 ratio; at elevated temperature, the product ratio is reversed (5:95).

The alkoxide-catalysed rearrangement of allyl sulphides in ethanol yields propenyl sulphides at reflux temperature[402]; under these conditions, the corresponding allyl ethers are recovered unchanged. Wittig–Horner reaction of the ylides,
generated from (methylthio)methyl phosphine oxides[403] or from (methylthio)methanephosphonic esters[404], succeeds with alkyl and aryl ketones as well as
with aldehydes; usually, though, only the respective phenyl sulphides have been
prepared. The most general route to phenyl alkenyl sulphides so far is the elimination of thiophenol from thioketals with Cu(I) ions (equation 83)[405].

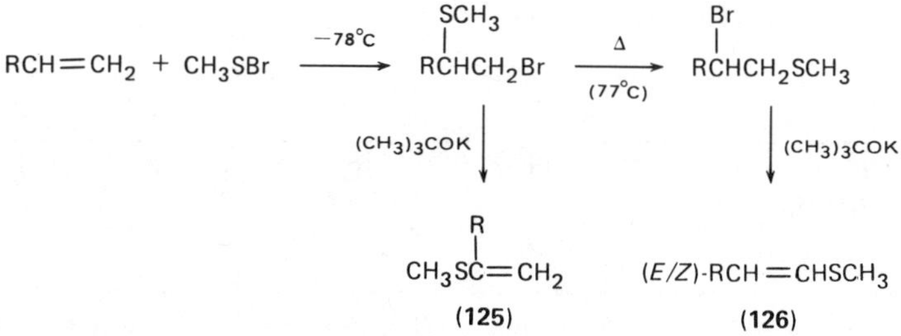

$$RCH{=}CH_2 + CH_3SBr \xrightarrow{-78°C} \underset{\underset{SCH_3}{|}}{RCHCH_2Br} \xrightarrow[(77°C)]{\Delta} \underset{\underset{Br}{|}}{RCHCH_2SCH_3}$$

$$\downarrow {(CH_3)_3COK} \qquad\qquad\qquad\qquad \downarrow {(CH_3)_3COK} \qquad (82)$$

$$\underset{\underset{R}{|}}{CH_3SC}{=}CH_2 \qquad (E/Z)\text{-}RCH{=}CHSCH_3$$

$$\textbf{(125)} \qquad\qquad\qquad \textbf{(126)}$$

$$R^1R^2CHC(SPh)_2R^3 + Cu^+ \xrightarrow[THF]{C_6H_6} R^1R^2C{=}CR^3SPh + CuSPh + H^+ \qquad (83)$$

Condensation of a silyl- *and* thienyl-substituted methyllithium (**127**) with ketones (equation 84) yields alkenyl phenyl sulphides (**128**) in good to excellent

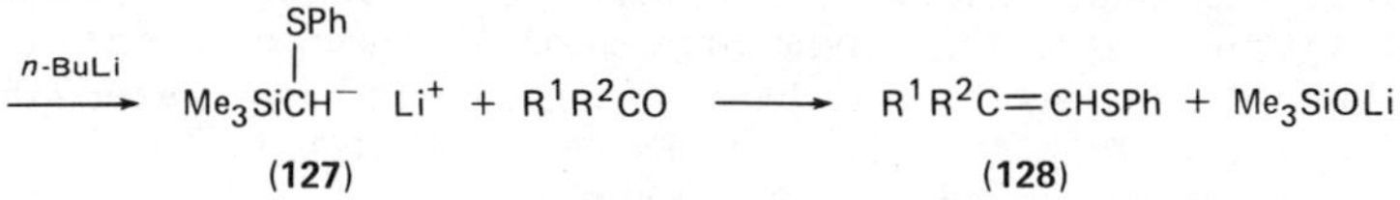

$$\xrightarrow{n\text{-BuLi}} \underset{\underset{SPh}{|}}{Me_3SiCH^-} Li^+ + R^1R^2CO \longrightarrow R^1R^2C{=}CHSPh + Me_3SiOLi \qquad (84)$$

$$\textbf{(127)} \qquad\qquad\qquad\qquad \textbf{(128)}$$

yield even in the case of sterically hindered and α,β-unsaturated substrates (e.g. pinacolone, cyclohexenone)[406]. While the oxygen analogues require t-BuLi to minimize nucleophilic attack at the oxygen (i.e. ether cleavage), n-BuLi is sufficient for thienyl carbanion generation (equation 84).

Terminally unsaturated thioenol ethers (**130**) are formed selectively if metalated 2-methyl-2-methylthiocarboxylic acids (**129**) are treated with N-chlorosuccinimide (NCS) (equation 85)[407]. A highly polar aprotic solvent like dimethoxyethane is prerequisite for the practically specific regiochemistry of the elimination (no trace of the alternative thioenol ether **131** is detectable). By treatment with anhydrous acid, **130** is rapidly converted to the thermodynamically more stable isomer **131**.

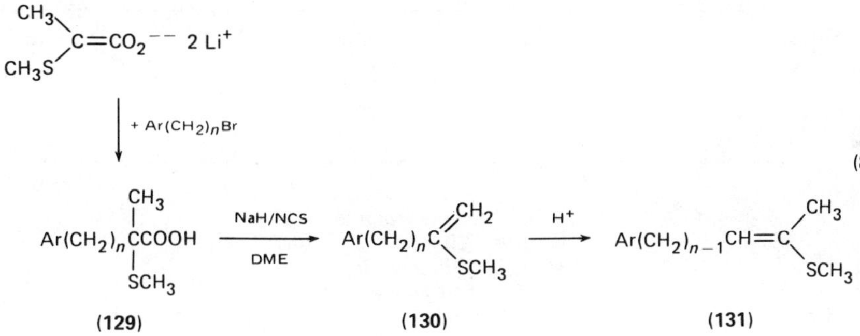

$$\underset{CH_3S}{\overset{CH_3}{\diagdown}}C{=}CO_2{}^{--}\ 2\,Li^+$$

$$\downarrow {+\ Ar(CH_2)_nBr}$$

$$(85)$$

$$\underset{\underset{SCH_3}{|}}{\overset{\overset{CH_3}{|}}{Ar(CH_2)_nCCOOH}} \xrightarrow[DME]{NaH/NCS} Ar(CH_2)_nC\underset{SCH_3}{\overset{CH_2}{\diagup}} \xrightarrow{H^+} Ar(CH_2)_{n-1}CH{=}C\underset{SCH_3}{\overset{CH_3}{\diagup}}$$

$$\textbf{(129)} \qquad\qquad\qquad \textbf{(130)} \qquad\qquad\qquad \textbf{(131)}$$

C. Reactivity

For the hydrolysis of alkenyl sulphides, too, a mechanism with rate-determining β-carbon protonation has been definitively established (Brønsted factor $\alpha = 0.7$)[398,109]. Thus, hydrolysis can be considered as a model reaction for electrophilic addition/substitution processes. Generally, it proceeds 100–1000 times slower[398] (Table 10) than with the structurally analogous enol ethers (see Section IV.A, Table 5). The gross substituent effects are the same as in the enol ether

TABLE 10. Rates of H_3O^+-catalysed hydrolysis[a] of various alkyl alkenyl sulphides, $k_{H_3O^+}$ $(10^{-3}$ M^{-1} $s^{-1})$

	R =	CH_3	C_2H_5	$CH(CH_3)_2$	$C(CH_3)_3$
CH_2=CHSR		11.7	10.4	8.98	4.17
CH_3CH=CHSR (Z)		–	0.49	0.79	0.54
(E)		–	0.28	0.45	0.34
CH_2=C(CH_3)SR		–	814	–	–

[a]Determined with aqueous HCl in 10% aqueous CH_3CN solution, ionic strength adjusted to $\mu = 0.50$ by addition of KCl, 25°C.

series[107], β-alkyl substituents retarding the rate by a factor of ~100, while α-CH_3 increases the reactivity about hundredfold (Table 10)[398]. The reversed order in the hierarchy of the *S*-alkyl substituents, which is in contrast to that found for CH_2=CHOR, has been rationalized in terms of decreasing hyperconjugative potential $[C(CH_3)_3 \ll CH_3]$[398]; hyperconjugation of course can operate only via vacant sulphur orbitals. As the interchange of relative reactivity between ethyl and isopropyl vinyl and propenyl sulphides, respectively, indicates, the balance between the various effects is rather delicate in the ground state.

The rates of cycloaddition of thioenol ethers with TCNE, in striking contrast, are much higher than for the corresponding enol ether reactions[408]. This must be due to a specific sulfur effect since the relative gradation between the various alkenyl substrates, as well as the gradation between the individual SR functions within each series (Table 11)[408], are practically identical with that found for the alkoxyalkenes

TABLE 11. Experimental rate constants, k_2 $(10^{-3}$ M^{-1} $s^{-1})$, for TCNE cycloaddition to alkyl alkenyl sulphides (in CH_2Cl_2, 25°C)[408]

	R =	CH_3	C_2H_5	$CH(CH_3)_2$	$C(CH_3)_3$
CH_2=CHSR		21.0	34.2	85.4	252.0
CH_3CH=CHSR (Z)		–	7.69	14.3	51.3
(E)		–	25.6	52.7	93.1
CH_2=C(CH_3)SR		–	2150	–	–
CH_2=C(CH_3)OR		–	19.9	–	–

TABLE 12. Experimental rate constants, k_2 $(10^{-5}$ M^{-1} $s^{-1})$, for TCNE cycloaddition to vinyl phenyl ethers and sulphides:

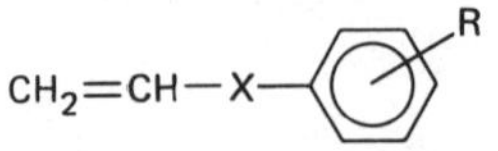

R	X = O	X = S
p-OCH_3	2.4	4470
p-CH_3	0.98	719
m-CH_3	0.65	215
H	0.35	102

(see Section V.A, Table 8). The higher cycloaddition reactivity of (E)- compared to (Z)-propenyl compounds is even more pronounced in the alkylthio series[408]. The same authors have also demonstrated that, in the cycloaddition of TCNE to vinyl phenyl ethers and vinyl phenyl sulphides, the effect of a m- or p-aryl substituent is transmitted far better through the S than the O linkage[408] (Table 12), an effect predicted by CNDO/2 calculations[409].

IX. REFERENCES

1. G. Hesse, 'Methoden zur Darstellung und Umwandlung von Enolen bzw. deren O-Derivaten', in *Methoden der Organischen Chemie* (Houben–Weyl), 4th ed., Vol. 6/1d, Georg Thieme Verlag, Stuttgart, 1978, pp. 1–216.
2. Reference 1, pp. 136 ff.
3. F. Effenberger, *Angew. Chem.*, **81**, 374 (1969).
4. J. Wislicenus, *Liebigs Ann. Chem.*, **192**, 106 (1878).
5. G. Stork, R. Terrell and J. Szmuszkovicz, *J. Amer. Chem. Soc.*, **76**, 2029 (1954).
6a. A. G. Cook (Ed.), *Enamines: Synthesis, Structure and Reactions*, Marcel Dekker, New York, 1969.
6b. S. F. Dyke, *Chemistry of Enamines*, Cambridge University Press, London, 1973.
7. G. H. Alt, 'Electrophilic substitutions and additions to enamines', in Reference 6a, pp. 115–168.
8. F. Effenberger, P. Fischer, W. W. Schoeller and W.-D. Stohrer, *Tetrahedron*, **34**, 2409 (1978).
9. P. Fischer and D. Müller, *Discussion Paper* presented at the Chemiedozententagung, Berlin, 1978; P. Fischer, *Habilitationsschrift*, University of Stuttgart, 1978.
10. N. L. Owen and N. Sheppard, *Trans. Faraday Soc.*, **60**, 634 (1964).
11. P. Cahill, L. P. Gold and N. L. Owen, *J. Chem. Phys.*, **48**, 1620 (1968).
12. N. L. Owen and H. M. Seip, *Chem. Phys. Letters*, **5**, 162 (1970).
13. B. Cadioli and U. Pincelli, *J. Chem. Soc., Faraday Trans.*, **2**, 991 (1972).
14. S. Samdal and H. M. Seip, *J. Mol. Struct.*, **28**, 193 (1975).
15. F. Bernardi, N. D. Epiotis, R. L. Yates and H. B. Schlegel, *J. Amer. Chem. Soc.*, **98**, 2385 (1976).
16. E. Wyn-Jones, K. R. Crook and W. J. Orville-Thomas, *Advan. Mol. Relaxation Processes*, **4**, 193 (1972).
17. N. D. Epiotis, S. Sarkanen, D. Bjorkquist, L. Bjorkquist and R. Yates, *J. Amer. Chem. Soc.*, **96**, 4075 (1974); N. D. Epiotis, *J. Amer. Chem. Soc.*, **95**, 3087 (1973).
18. D. Cremer, J. S. Binkley, J. A. Pople and W. J. Hehre, *J. Amer. Chem. Soc.*, **96**, 6900 (1974).
19. D. G. Lister and P. Palmieri, *J. Mol. Struct.*, **32**, 355 (1976).
20. T. Okuyama, T. Fueno and J. Furukawa, *Tetrahedron*, **25**, 5409 (1969).
21. E. Taskinen and M. Anttila, *Tetrahedron*, **33**, 2423 (1977).
22. E. Taskinen, *Tetrahedron*, **31**, 957 (1975).
23. E. Taskinen and K. Jokila, *Acta Chem. Scand.*, **B 29**, 249 (1975).
24. E. Taskinen, *Acta Chem. Scand.*, **B 29**, 245 (1975).
25. E. Taskinen and R. Virtanen, *J. Org. Chem.*, **42**, 1443 (1977).
26. A. A. Planckaert, J. Doucet and C. Sandorfy, *J. Chem. Phys.*, **60**, 4846 (1974).
27. H. Bock, G. Wagner, K. Wittel, J. Sauer and D. Seebach, *Chem. Ber.*, **107**, 1869 (1974).
28. G. W. Mines and H. W. Thompson, *Spectrochim. Acta*, **A 29**, 1377 (1973).
29. C. Batich, E. Heilbronner, C. B. Quinn and R. J. Wisemann, *Helv. Chim. Acta*, **59**, 512 (1976).
30. M. Bloch, F. Brogli, E. Heilbronner, T. B. Jones, H. Prinzbach and O. Schweikert, *Helv. Chim. Acta*, **61**, 1388 (1978).
31. H. Friege and M. Klessinger, *J. Chem. Res. S.*, 208 (1977).
32. A. Kirrmann and P. Chancel, *Bull. Soc. Chim. Fr.*, **21**, 1338 (1954).
33. Y. Mikawa, *Bull. Chem. Soc. Japan*, **29**, 110 (1956).

34. B. A. Trofimov, N. I. Shergina, A. S. Atavin, Z. I. Kositsyna, A. V. Gusarov and G. M. Gavrilova, *Izv. Akad. Nauk SSSR, Ser. Khim.*, 116 (1972); see also further literature cited here.
35. F. Marsault-Herail, G. S. Chiglien, J. P. Dorie and M. L. Martin, *Spectrochim. Acta*, **A** **29**, 151 (1973).
36. M. Sakakibara, F. Inagaki, I. Harada and T. Shimanouchi, *Bull. Chem. Soc. Japan*, **49**, 46 (1976).
37. S. W. Charles, F. C. Cullen and N. L. Owen, *J. Mol. Struct.*, **18**, 183 (1973).
38. G. P. Ford, A. R. Katritzky and R. D. Topsom, *J. Chem. Soc., Perkin Trans.* 2, 1378 (1975); see also former papers of Katritzky's and Topsom's groups cited here.
39. A. Ledwith and H. J. Woods, *J. Chem. Soc. (B)*, 753 (1966).
40. K. Hatada, M. Takeshita and H. Yuki, *Tetrahedron Letters*, 4621 (1968).
41. D. Müller, *Dissertation*, University of Stuttgart, 1977.
42. F. Effenberger, P. Fischer, G. Prossel and G. Kiefer, *Chem. Ber.*, **104**, 1987 (1971).
43. K. Hatada, K. Nagata and H. Yuki, *Bull. Chem. Soc. Japan*, **43**, 3195 (1970).
44. G. A. Kalabin, B. A. Trofimov, V. M. Bzhezovskii, D. F. Kushnarev, S. V. Amosova, N. K. Gusarova and M. L. Al'pert, *Izv. Akad. Nauk SSSR, Ser. Khim.*, 576 (1975).
45. D. E. Dorman, M. Jautelat and J. D. Roberts, *J. Org. Chem.*, **36**, 2757 (1971).
46. A. C. Rojas and J. K. Crandall, *J. Org. Chem.*, **40**, 2225 (1975).
47. T. Steiger, E. Gey and R. Radeglia, *Z. Phys. Chem. (Leipzig)*, **255**, 1102 (1974); **256**, 49 (1975); **257**, 172 (1976).
48. M. Katoh, D. A. Jaeger and C. Djerassi, *J. Amer. Chem. Soc.*, **94**, 3107 (1972).
49. P. Krenmayr, *Monatsh. Chem.*, **106**, 925 (1975).
50. C. C. van de Sande and F. W. McLafferty, *J. Amer. Chem. Soc.*, **97**, 4613 (1975), and following papers by McLafferty's group.
51. A. J. V. Ferrer-Correia, K. R. Jennings and D. K. S. Sharma, *J. Chem. Soc., Chem. Commun.*, 973 (1975).
52. B. A. Trofimov, V. B. Modonov, T. N. Bazhenova, N. A. Nedolya and V. V. Keiko, *Reakts. Sposobn. Org. Soedin.*, **11**, 747 (1975).
53. B. A. Trofimov, N. I. Shergina, S. E. Korostova, E. I. Kositsyna, O. N. Vylegzhanin, N. A. Nedolya and M. G. Voronkov, *Reakts. Sposobn. Org. Soedin.*, **8**, 1047 (1971).
54. O. N. Vylegzhanin, V. B. Modonov and B. A. Trofimov, *Tetrahedron Letters*, 2243 (1972).
55. A. Ledwith and H. J. Woods, *J. Chem. Soc. (B)*, 310 (1970).
56. T. Masuda, *J. Polym. Sci., Polym. Chem. Ed.*, **11**, 2713 (1973).
57. G. Heublein, G. Agatha, H. Dawczynski and B. Zaleska, *Z. Chem.*, **13**, 432 (1973).
58. A. Favorskii, *J. Prakt. Chem.*, **37**, 531 (1888); **44**, 208 (1891).
59. W. Reppe *et al.*, *Liebigs Ann. Chem.*, **601**, 81 (1956).
60. D. E. Stepanov, L. I. Belousova and N. G. Evsyutina, *Zh. Org. Khim.*, **8**, 788 (1972).
61. M. F. Shostakovskii, V. V. An, G. G. Skvortsova and L. M. An, *Ref. Zh. Khim.*, 9Zh244 (1971); *Chem. Abstr.*, **76**, 140367u (1972).
62. V. I. Laba, A. A. Kron and E. N. Prilezhaeva, *Izv. Akad. Nauk SSSR, Ser. Khim.*, 1546 (1976).
63. H. Christol, H.-J. Christau and M. Soleiman, *Synthesis*, 736 (1975).
64. W. H. Watanabe and L. E. Conlon, *J. Amer. Chem. Soc.*, **79**, 2828 (1957).
65. J. E. McKeon, P. Fitton and A. A. Griswold, *Tetrahedron*, **28**, 227 (1972).
66. J. E. McKeon and P. Fitton, *Tetrahedron*, **28**, 233 (1972).
67. E. Chiellini, *Gazz. Chim. Ital.*, **102**, 830 (1972).
68. M. G. Katsnelson and A. L. Uzlyaner-Neglo, *Neftepererab. Neftekhim.*, 39 (1976); *Chem. Abstr.*, **85**, 77569e (1976).
69. R. A. Wohl, *Synthesis*, 38 (1974).
70. J. Toullec and J. E. Dubois, *Tetrahedron Letters*, 1281 (1976).
71. A. Kankaanperä, P. Salomaa, P. Jühala, R. Aaltonen and M. Mattsén, *J. Amer. Chem. Soc.*, **95**, 3618 (1973).
72. S. J. Rhoads, J. K. Chattopadhyay and E. E. Waali, *J. Org. Chem.*, **35**, 3352 (1970).
73. M. G. Voronkov, *Zh. Obshch. Khim.*, **20**, 2060 (1950).
74. M. D. Coburn, *Synthesis*, 570 (1977).

75. H. Meerwein, 'Herstellung und Umwandlung von Acetalen' in *Methoden der Organischen Chemie* (Houben–Weyl), 4th ed., Vol. 6/3, Georg Thieme Verlag, Stuttgart, 1965, pp. 221–222.
76. S. A. Patwardhan and S. Dev, *Synthesis*, 348 (1974).
77. Reference 1, pp. 158 ff.
78. G. Wittig and M. Schlosser, *Chem. Ber.*, **94**, 1373 (1961).
79. C. Earnshaw, C. J. Wallis and S. Warren, *J. Chem. Soc., Chem. Commun.*, 314 (1977).
80. K. Dimroth, G. Pohl and H. Follmann, *Chem. Ber.*, **99**, 634, 642 (1966).
81. W. Grell and H. Machleidt, *Liebigs Ann. Chem.*, **699**, 53 (1966).
82. H. J. Bestmann, H. Dornauer and K. Rostock, *Chem. Ber.*, **103**, 2011 (1970).
83. C. P. Casey and T. J. Burkhardt, *J. Amer. Chem. Soc.*, **94**, 6543 (1972); C. P. Casey, S. H. Bertz and T. J. Burkhardt, *Tetrahedron Letters*, 1421 (1973).
84. C. C. Price and W. H. Snyder, *J. Amer. Chem. Soc.*, **83**, 1773 (1961).
85. J. Sauer and H. Prahl, *Tetrahedron Letters*, 2863 (1966); *Chem. Ber.*, **102**, 1917 (1969).
86. J. Gigg and R. Gigg, *J. Chem. Soc (C)*, 82 (1966).
87. B. Fraser-Reid, S. Y. K. Tam and B. Radatus, *Can. J. Chem.*, **53**, 2005 (1975).
88. H. Meerwein, 'Methoden zur Herstellung und Umwandlung von Äthern', in *Methoden der Organischen Chemie* (Houben–Weyl), 4th ed., Vol 6/3, Georg Thieme Verlag, Stuttgart, 1965, p. 102.
89. M. G. Voronkov, G. G. Balezina, S. F. Malysheva and S. M. Shostakovskii, *Zh. Prikl. Khim. (Leningrad)*, **48**, 1172 (1975).
90. W. Kirmse and M. Buschhoff, *Chem. Ber.*, **100**, 1491 (1967).
91. D. P. G. Hamon and K. M. Pullen, *J. Chem. Soc., Chem. Commun.*, 459 (1975).
92. K. Klein, H. Eijsinga, H. Westmijze, J. Meijer and P. Vermeer, *Tetrahedron Letters*, 947 (1976).
93. J. J. Normant, A. Commerçon, M. Bourgain and J. Villieras, *Tetrahedron Letters*, 3833 (1975).
94. K. Tamao, M. Zembayashi and M. Kumada, *Chem. Letters*, 1237 (1976).
95. J. Mulzer, G. Brüntrup and M. Zippel, *Discussion Paper*, presented at the Chemiedozententagung, Darmstadt, 1979.
96. G. Caron and J. Lessard, *Can. J. Chem.*, **51**, 981 (1973).
97. B. Capon, 'Reaction of aldehydes and ketones and their derivatives', in *Organic Reaction Mechanisms,* John Wiley and Sons, London, 1965–1977.
98. P. Salomaa, in *The Chemistry of the Carbonyl Group* (Ed. S. Patai), John Wiley and Sons, London, 1966, p. 177.
99. A. J. Kresge, H. J. Chen and Y. Chiang, *J. Amer. Chem. Soc.*, **99**, 802 (1977).
100. A. J. Kresge and W. K. Chwang, *J. Amer. Chem. Soc.*, **100**, 1249 (1978).
101. J. D. Cooper, V. P. Vitullo and D. L. Whalen, *J. Amer. Chem. Soc.*, **93**, 6294 (1971).
102. P. Salomaa, A. Kankaanperä and M. Lajunen, *Acta Chem. Scand.*, **20**, 1790 (1966).
103. A. J. Kresge, H. L. Chen, Y. Chiang, E. Murrill, M. A. Payne and D. S. Sagatys, *J. Amer. Chem. Soc.*, **93**, 413 (1971).
104. G. M. Loudon and D. E. Ryono, *J. Org. Chem.*, **40**, 3574 (1975).
105. G. M. Loudon and C. Berke, *J. Amer. Chem. Soc.*, **96**, 4508 (1974).
106. P. W. Hickmott and K. N. Woodward, *J. Chem. Soc., Chem. Commun.*, 275 (1974).
107. T. Okuyama, T. Fueno, H. Nakatsuji and J. Furukawa, *J. Amer. Chem. Soc.*, **89**, 5826 (1967).
108. A. J. Kresge and H. J. Chen, *J. Amer. Chem. Soc.*, **94**, 2818 (1972).
109. T. Okuyama and T. Fueno, *J. Org. Chem.*, **39**, 3156 (1974).
110. A. Ledwith and H. J. Woods, *J. Chem. Soc. (B)*, 753 (1966).
111. J. P. Gouesnard and M. Blain, *Bull. Soc. Chim. Fr.*, 338 (1974).
112. A. Kankaanperä and M. Mattsén, *Acta Chem. Scand.*, **A 29**, 419 (1975).
113. A. J. Kresge and Y. Chiang, *J. Chem. Soc. (B)*, 53 (1967).
114. Reference 88, pp. 185–189.
115. N. Nedolya and B. A. Trofimov, *Zh. Fiz. Khim.*, **51**, 398 (1977).
116. M. F. Shostakovskii, Y. B. Kagan and F. P. Sidel'kovskaya, *Zh. Obshch. Khim.*, **17**, 957 (1947); M. F. Shostakovskii, *Chemie (Prague)*, **10**, 273 (1958); *Chem. Abstr.*, **54**, 1250e (1960).

814   Peter Fischer

117. M. F. Shostakovskii and F. P. Sidel'kovskaya, *Zh. Obshch. Khim.*, **21**, 1610 (1951).
118. G. Dana, O. Convert and C. Perrin, *J. Org. Chem.*, **40**, 2133 (1975).
119. C. E. Schildknecht, A. O. Zoss and C. McKinley, *Ind. Eng. Chem.*, **39**, 180 (1947).
120. S. Nakanishi, K. Morita and E. V. Jensen, *J. Amer. Chem. Soc.*, **81**, 5259 (1959).
121. T. E. Stone and G. D. Daves, *J. Org. Chem.*, **42**, 2151 (1977).
122. R. U. Lemieux and B. Fraser-Reid, *Can. J. Chem.*, **43**, 1460 (1965).
123. B. Barbier, J. Toullec and J.-E. Dubois, *Tetrahedron Letters*, 3629 (1972).
124. J.-E. Dubois, J. Toullec and G. Barbier, *Tetrahedron Letters*, 4485 (1970).
125. J. Toullec and J.-E. Dubois, *J. Amer. Chem. Soc.*, **98**, 5518 (1976).
126. E. M. Gaydoŭ, *Tetrahedron Letters*, 4055 (1972).
127. J. R. Shelton and T. Kasuga, *J. Org. Chem.*, **28**, 2841 (1963).
128. G. Greenwood and H. M. R. Hoffmann, *J. Org. Chem.*, **37**, 611 (1972).
129. A. J. Duggan and S. S. Hall, *J. Org. Chem.*, **42**, 1057 (1977).
130. E. Vilsmaier and G. Adam, *Liebigs Ann. Chem.*, **757**, 181 (1972).
131. K. Igarashi, T. Honma and T. Imagawa, *J. Org. Chem.*, **35**, 610 (1970).
132. L. D. Hall and J. F. Manville, *Can. J. Chem.*, **47**, 361 (1969).
133. G. R. Cliff and D. J. Dunn, *Org. Prep. Proced. Int.*, **7**, 23 (1975).
134. M. Bergmann and A. Miekeley, *Ber. dtsch. chem. Ges.*, **62**, 2297 (1929)
135. A. A. Frimer, *Synthesis*, 578 (1977).
136. A. L. Draper, W. J. Heilman, W. E. Schaefer, H. J. Shine and J. N. Shoolery, *J. Org. Chem.*, **27**, 2727 (1962).
137. C. L. Stevens and J. Tazuma, *J. Amer. Chem. Soc.*, **76**, 715 (1954).
138. J. J. Havel and K. H Chan, *J. Org. Chem.*, **41**, 513 (1976).
139. H. P. Crocker and R. H. Hall, *J. Chem. Soc.*, 2052 (1955).
140. U. Schmidt and P. Grafen, *Liebigs Ann. Chem.*, **656**, 97 (1962).
141. B. Belleau and Y. K. Anyung, *Can. J. Chem.*, **47**, 2117 (1969).
142. D. Koch, H. Schaefer and E. Steckhan, *Chem. Ber.*, **107**, 3640 (1974).
143. Y. Yukawa and M. Sakai, *Bull. Soc. Chem. Japan*, **36**, 761 (1963).
144. S. Uemura, R. Kito and K. Ichikawa, *Nippon Kagaku Zasshi*, **87**, 986 (1966); *Chem. Abstr.*, **65**, 19962f (1966).
145. M. F. Shostakovskii, N. A. Gershtein and V. A. Neterman, *Dokl. Akad. Nauk SSSR*, **103**, 265 (1955).
146. D. J. Edge, B. C. Gilbert, R. O. C. Norman and R. P. West, *J. Chem. Soc. (B)*, 189 (1971).
147. R. B. Silverman and D. Dolphin, *J. Amer. Chem. Soc.*, **96**, 7094, 7096 (1974); **98**, 4626, 4633 (1976).
148. M. F. Shostakovskii, E. N. Prilezhaeva and E. S. Shapiro, *Izv. Akad. Nauk SSSR, Otdel. Khim. Nauk*, 284 (1951); 357 (1953); 292 (1954).
149. E. N. Prilezhaeva, N. P. Petukhova and M. F. Shostakovskii, *Dokl. Akad. Nauk SSSR*, **154**, 160 (1964).
150. G. G. Skvortsova, V. V. An, L. M. An and V. K. Voronov, *Khim. Geterosikl. Soedin.*, 1155 (1972).
151. K. Yamagishi, *Nippon Kagaku Zasshi*, **80**, 1361 (1959); *Chem. Abstr.*, **55**, 4405e (1961).
152. A. Senning and S. O. Lawesson, *Tetrahedron*, **19**, 695 (1963).
153. A. V. Kalabina, E. F. Kolmakova, T. I. Bychkova, Y. K. Maksyutin, E. A. Denisevich and G. I. Smolina, *Zh. Obshch. Khim.*, **35**, 979 (1965).
154. M. J. Baldwin and R. K. Brown, *Can. J. Chem.*, **46**, 1093 (1968).
155. M. Muehlstaedt, D. Martinetz and P. Schneider, *J. Prakt. Chem.*, **315**, 940 (1973).
156. E. Kobayashi and R. Sakata, *Yakugaku Zasshi*, **82**, 455 (1962); *Chem. Abstr.*, **58**, 4552h (1963).
157. M. Seefelder and H. Pasedach (BASF-A.G.), *German Patent*, No. 960095 (1957); *Chem. Abstr.*, **52**, 1629i (1958); M. Muehlstaedt and D. Martinetz, *Z. Chem.*, **18**, 297 (1974).
158. F. Effenberger and J. Daub, *Chem. Ber.*, **102**, 104 (1969).
159. M. M. Rogić and J. Vitrone, *J. Amer. Chem. Soc.*, **94**, 8642 (1972).
160. A. P. Terent'ev and N. P. Volynskii, *Zh. Obshch. Khim.*, **19**, 784 (1949).

161. S. Tchelitcheff (Société des Usines Chimiques Rhône-Poulenc), *U.S. Patent*, No. 2674625 (1954); *Chem. Abstr.*, **49**, 6301h (1955).
162. M. M. Rogić, J. Vitrone and M. D. Swerdloff, *J. Amer. Chem. Soc.*, **99**, 1156 (1977).
163. K. A. Ogloblin and D. M. Kunowskaya, *Zh. Org. Khim.*, **4**, 897 (1968).
164. J. R. Mahajan, G. A. L. Ferreira and H. C. Araŭjo, *J. Chem. Soc., Chem. Commun.*, 1078 (1972).
165. A. P. Terent'ev and V. A. Zagorvskii, *Zh. Obshch. Khim.*, **26**, 200 (1956).
166. M. Seefelder and H. Eilingsfeld, *Angew. Chem.*, **75**, 724 (1963).
167. R. Huisgen and H. Pohl, *Chem. Ber.*, **93**, 527 (1960).
168. T. Nishiwaki, *Tetrahedron*, **22**, 711 (1966).
169. M. M. Rauhut, H. A. Currier, A. M. Semsel and V. P. Wystrach, *J. Org. Chem.*, **26**, 5138 (1961).
170. V. S. Tsivunin, G. Kamai and D. B. Sultanova, *Zh. Obshch. Khim.*, **33**, 2149 (1963).
171. G. Stork, A. Brizzolara, H. Landesman, J. Szmuszkovicz and R. Terrell, *J. Amer. Chem. Soc.*, **85**, 207 (1963).
172. G. Stork and R. L. Danheiser, *J. Org. Chem.*, **38**, 1755 (1973).
173. M. E. Vol'pin, I. S. Akhrem and D. N. Kursanov, *Zh. Obshch. Khim.*, **30**, 159 (1960).
174. C. E. Bawn, C. Fitzsimmons and A. Ledwith, *Proc. Chem. Soc. (Lond.)*, 391 (1964).
175. L. S. Povarov, *Russ. Chem. Rev.*, 639 (1965).
176. O. Bayer, 'Methoden zur Herstellung und Umwandlung von Aldehyden', in *Methoden der Organischen Chemie* (Houben–Weyl), 4th ed., Vol. 7/1, Georg Thieme Verlag, Stuttgart, 1954, pp. 115–120.
177. M. Mueller-Cunradi and K. Pieroh (I. G. Farbenindustrie A.G.), *U.S. Patent*, No. 2165962 (1939); *Chem. Abstr.*, **33**, 8210² (1939).
178. R. I. Hoaglin and D. H. Hirsh, *J. Amer. Chem. Soc.*, **71**, 3468 (1949).
179. H. Normant and G. Martin, *Bull. Soc. Chim. Fr.*, 1646 (1963).
180. M. F. Shostakovskii and E. P. Gracheva, *Zh. Obshch. Khim.*, **26**, 1679 (1956).
181. I. N. Nazarov, I. I. Nazarova and I. V. Torgov, *Dokl. Akad. Nauk SSSR*, **122**, 82 (1958).
182. O. Isler, M. Montavon, R. Rüegg and P. Zeller, *Liebigs Ann. Chem.*, **603**, 129 (1957).
183. B. M. Mikhailov and L. S. Povarov, *Zh. Obshch. Khim.*, **29**, 2079 (1959).
184. B. M. Mikhailov and L. S. Povarov, *Izv. Akad. Nauk SSSR, Otdel. Khim. Nauk*, 1903 (1960).
185. B. M. Mikhailov and G. S. Ter-Sarkisyan, *Izv. Akad. Nauk SSSR, Otdel. Khim. Nauk*, 1888 (1960).
186. R. Y. Popova, T. V. Protopopova, V. G. Vinokurov and A. P. Skoldinov, *Zh. Obshch. Khim.*, **34**, 114 (1964).
187. L. S. Povarov and B. M. Mikhailov, *Izv. Akad. Nauk SSSR, Ser. Khim.*, 1910 (1964).
188. R. I. Hoaglin, D. G. Kubler and A. E. Montagna, *J. Amer. Chem. Soc.*, **80**, 5460 (1958).
189. R. I. Hoaglin and D. H. Hirsh (Union Carbide and Carbon Corporation), *U.S. Patent*, No. 2628257 (1953); *Chem. Abstr.*, **48**, 1423d (1954).
190. S. Satsumabayashi, K. Nakajo, R. Soneda and S. Motoki, *Bull. Chem. Soc. Japan*, **43**, 1586 (1970).
191. A. Guerrero, F. Camps, J. Albaigés and J. Rivera, *Tetrahedron Letters*, 2221 (1975).
192. J. Albaigés, F. Camps, J. Castells, J. Fernandez and A. Guerrero, *Synthesis*, 378 (1972).
193. E. Kitazawa, T. Imamura, K. Saigo and T. Mukaiyama, *Chem. Letters*, 569 (1975).
194. K. C. Brannock, *J. Org. Chem.*, **25**, 258 (1960).
195. H. Bredereck, R. Gompper and G. Morlock, *Chem. Ber.*, **90**, 942 (1957).
196. Z. Arnold, *Coll. Czech. Chem. Commun.*, **25**, 1308 (1960); Z. Arnold and J. Žemlička, *Coll. Czech. Chem. Commun.*, **24**, 786 (1959); Z. Arnold and F. Sorm, *Coll. Czech. Chem. Commun.*, **23**, 452 (1958).
197. G. Martin and M. Martin, *Bull. Soc. Chim. Fr.*, 637, 1646 (1963).
198. L. A. Yanovskaya, V. T. Kucherov and B. A. Rudenko, *Izv. Akad. Nauk SSSR, Otdel. Khim. Nauk*, 2182 (1962).

199. J. W. Copenhaver (General Aniline and Film Corporation), *U.S. Patent*, No. 2527533 (1950); *Chem. Abstr.*, **45**, 1622i (1951).
200. R. D. Youssefyeh, *J. Amer. Chem. Soc.*, **85**, 3901 (1963).
201. M. Hojo, R. Masuda, Y. Kokuryo, H. Shioda and S. Matsuo, *Chem. Letters*, 499 (1976).
202. R. Maier, *Diplomarbeit*, University of Stuttgart, 1964.
203. F. Effenberger, *Chem. Ber.*, **98**, 2260 (1965).
204. N. S. Zefirov, N. M. Shekhtman and R. A. Karakhanov, *Zh. Org. Khim.*, **3**, 1925 (1967).
205. R. E. Paul and S. Tchelitcheff (Société des Usines Chimiques Rhône-Poulenc), *U.S. Patent*, No. 2768174 (1956); *Chem. Abstr.*, **51**, 5818f (1957).
206. F. Effenberger and W. Hartmann, *Angew. Chem.*, **76**, 188 (1964).
207. V. I. Gorbatenko and N. V. Mel'nichenko, *Zh. Org. Khim.*, **11**, 2227 (1975).
208. B. A. Arbuzov, N. N. Zobova and I. I. Andronova, *Izv. Akad. Nauk SSSR, Ser. Khim.*, 1566 (1974); see also former work by Arbuzov and coworkers cited here.
209. M. Levas and E. Levas, *Compt. Rend.*, **230**, 1669 (1959); *Bull. Soc. Chim. Fr*, 1800 (1959).
210. P. Tarrant and E. C. Stump, *J. Org. Chem.*, **29**, 1198 (1964).
211. A. S. Atavin, G. M. Gavrilova and B. A. Trofimov, *Izv. Akad. Nauk SSSR, Ser. Khim.*, 2040 (1971).
212. H. van Kamp and S. J. Halkes, *Rec. Trav. Chim.*, **84**, 904 (1965).
212a. L. M. Andreeva, K. V. Altukhov and V. V. Perekalin, *Zh. Org. Khim.*, **5**, 220 (1969).
213. R. Huisgen, *Angew. Chem.*, **80**, 329 (1968).
214. R. B. Woodward and R. Hoffmann, *J. Amer. Chem. Soc.*, **87**, 2046 (1965).
215. R. B. Woodward and R. Hoffmann, *Angew. Chem.*, **81**, 797 (1969).
216. R. Huisgen, L. A. Feiler and G. Binsch, *Chem. Ber.*, **102**, 3460 (1969).
217. R. Sustmann, A. Ansmann and F. Vahrenholt, *J. Amer. Chem. Soc.*, **94**, 8099 (1972).
218. P. D. Bartlett, *Quart. Rev.*, **24**, 473 (1970).
219. A. G. Cook, 'Cycloaddition reactions of enamines', in Reference 6a, pp. 211 ff.
220. P. Otto, L. A. Feiler and R. Huisgen, *Angew. Chem.*, **80**, 759 (1968); R. Huisgen, L. A. Feiler and P. Otto, *Chem. Ber.*, **102**, 3444 (1969).
221. R. Huisgen, *Acc. Chem. Res.*, **10**, 117 (1977).
222. R. Huisgen, *Acc. Chem. Res.*, **10**, 199 (1977).
223. J. K. Williams, D. W. Wiley and B. C. McKusick, *J. Amer. Chem. Soc.*, **84**, 2210 (1962).
224. R. Huisgen and G. Steiner, *J. Amer. Chem. Soc.*, **95**, 5054, 5055 (1973).
225. P. D. Bartlett, K. Hummel, S. P. Elliott and R. A. Minns, *J. Amer. Chem. Soc.*, **94**, 2898 (1972).
226. H. H. Wasserman, A. J. Solodar and L. S. Keller, *Tetrahedron Letters*, 5597 (1968); I. Tabushi, R. Oda and K. Okazaki, *Tetrahedron Letters*, 3743 (1968).
227. R. W. Hoffmann, U. Bressel, J. Gehlhaůs and H. Haeuser, *Chem. Ber.*, **104**, 873 (1971).
228. R. Hoffmann, S. Swaminathan, B. G. Odell and R. Gleiter, *J. Amer. Chem. Soc.*, **92**, 7091 (1970); R. Gleiter, *Angew. Chem.*, **86**, 770 (1974); and unpublished results cited in Reference 222.
229. F. Effenberger, P. Fischer and E. Schneider, *Discussion Paper*, presented at the *V*th *International Congress of Heterocyclic Chemistry*, Ljubljana, 1975; P. Fischer and E. Schneider, unpublished results.
230. R. Schug and R. Huisgen, *J. Chem. Soc., Chem. Commun*, 60 (1975).
231. R. Huisgen, R. Schug and G. Steiner, *Angew. Chem.*, **86**, 47, 48 (1974).
232. J. F. Bunnett and R. J. Morath, *J. Amer. Chem. Soc.*, **77**, 5051, 5055 (1955).
233. I. Karle, J. Flippen, R. Huisgen and R. Schug, *J. Amer. Chem. Soc.*, **97**, 5285 (1975).
234. G. Steiner and R. Huisgen, *J. Amer. Chem. Soc.*, **95**, 5056 (1973).
235. J. v. Jouanne, H. Kelm and R. Huisgen, *J. Amer. Chem. Soc.*, **101**, 151 (1979).
236. F. K. Fleischmann and H. Kelm, *Tetrahedron Letters*, 3773 (1973).
237. T. Arimoto and J. Osugi, *Rev. Phys. Chem. Japan*, **44**, 25 (1974); *Chem. Letters*, 271 (1974).

238. R. Huisgen and R. Schug, *J. Amer. Chem. Soc.*, **98**, 7819 (1976).
239. J. Sauer, *Angew. Chem.*, **79**, 76 (1967).
240. R. Huisgen and G. Steiner, *Tetrahedron Letters*, 3763 (1973).
241. R. Huisgen and H. Mayr, *Tetrahedron Letters*, 2965 (1975).
242. H. K. Hall and P. Ykman, *J. Amer. Chem. Soc.*, **97**, 800 (1975).
243. R. F. Tarvin, S. Aoki and J. K. Stille, *Macromolecules*, **5**, 663 (1972).
244. J. K. Stille and D. C. Chung, *Macromolecules*, **8**, 114 (1975).
245. H. Staudinger and E. Suter, *Ber. dtsch. chem. Ges.*, **53**, 1092 (1920).
246. R. Montaigne and L. Ghosez, *Angew. Chem.*, **80**, 194 (1968).
247. D. C. England and C. G Krespan, *J. Org. Chem.*, **35**, 3312 (1970).
248. D. P. Del'tsova and N. P. Gambaryan, *Izv. Akad. Nauk SSSR, Ser. Khim*, 858 (1976).
249. F. Effenberger, G. Prossel and P. Fischer, *Chem. Ber.*, **104**, 2002 (1971).
250. N. S. Isaacs and P. Stanbury, *J. Chem. Soc., Perkin Trans. 2*, 166 (1973).
251. R. Huisgen and H. Mayr, *Tetrahedron Letters*, 2969 (1975); H. Mayr, *Angew. Chem.*, **87**, 491 (1975).
252. H. Mayr and R. Huisgen, *Tetrahedron Letters*, 1349 (1975).
253. B. A. Arbuzov and N. N. Zobova, *Synthesis*, 461 (1974) (esp. p. 467).
254. J. Goerdeler and H. Schenk, *Chem. Ber.*, **98**, 3831 (1965).
255. N. E. Schore and N. J. Turro, *J. Amer. Chem. Soc.*, **97**, 2482 (1975).
256. E. J. Corey, J. D. Bass, R. Le Mahieu and R. B. Mitra, *J. Amer. Chem. Soc.*, **86**, 5570 (1964).
257. J. S. Swenton, J. A. Hyatt, J. M. Lisy and J. Clardy, *J. Amer. Chem. Soc.*, **96**, 4885 (1974).
258. A. H. Lawrence, C. C. Liao, P. de Mayo and V. Ramamurthy, *J. Amer. Chem. Soc.*, **98**, 2219, 3572 (1976).
259. D. Bryce-Smith, A. Gilbert, B. Orger and H. Tyrrell, *J. Chem. Soc., Chem. Commun.*, 334 (1974).
260. L. Skattebøl, *J. Org. Chem.*, **31**, 1554 (1966).
261. P. S. Skell and A. Y. Garner, *J. Amer. Chem. Soc.*, **78**, 5430 (1956).
262. W. v. E. Doering and W. A. Henderson, *J. Amer. Chem. Soc.*, **80**, 5274 (1958).
263. A. Ledwith and H. J. Woods, *J. Chem. Soc. (B)*, 973 (1967).
264. J. C. Anderson, D. G. Lindsay and C. B. Reese, *Tetrahedron*, **20**, 2091 (1964).
265. E. E. Schweizer and W. E. Parham, *J. Amer. Chem. Soc.*, **82**, 4085 (1960).
266. T. Hiyama, T. Mishima, K. Kitatani and H. Nozaki, *Tetrahedron Letters*, 3297 (1974).
267. A. J. Birch, J. M. H. Graves and J. B. Siddall, *J. Chem. Soc.*, 4234 (1963).
268. V. S. Aksenov and V. A. Filimoshkina, *Izv. Sib. Otd. Akad. Nauk SSSR, Ser. Khim. Nauk*, 147 (1975); *Chem. Abstr.*, **83**, 147154a (1975).
269. S. M. Shostakovskii, L. N. Aksenova, V. I. Erofeev and V. S. Aksenov, *Ref. Zh., Khim.*, 4Zh120 (1973); *Chem. Abstr.*, **79**, 78166j (1973); **82**, 30989q (1975).
270. S. M. Shostakovskii, T. K. Voropaeva, V. N. Voropaev and M. G. Voronkov, *Dokl. Akad. Nauk SSSR*, **228**, 861 (1976).
271. D. F. Kushnarev, G. A. Kalabin, S. M. Shostakovskii and T. K. Voropaeva, *Izv. Akad. Nauk SSSR, Ser. Khim*, 787 (1976).
272. M. Kamel, W. Kimpenhaus and J. Buddrus, *Chem. Ber.*, **109**, 2351 (1976).
273. Y. V. Savinykh, V. S. Aksenov and T. A. Bogatyreva, *Izv. Sib. Otd. Akad. Nauk SSSR, Ser. Khim. Nauk*, 111 (1975); *Chem. Abstr.*, **84**, 89410p (1976).
274. G. Boche and D. R. Schneider, *Tetrahedron Letters*, 4247 (1975).
275. R. I. Polovnikova, E. N. Sukhomazova, S. M. Shostakovskii and V. I. Parygina, *Khim. Vysokomol. Soedin. Neftekhim.*, 59 (1973); *Chem. Abstr.*, **80**, 145507g (1974).
276. M. S. Newman and Zia-ud-Din, *J. Org. Chem.*, **38**, 547 (1973).
277. R. Huisgen, *Angew. Chem.*, **75**, 604 (1963).
278. R. Huisgen, *J. Org. Chem.*, **41**, 403 (1976).
279. R. Huisgen, L. Moebius and G. Szeimies, *Chem. Ber.*, **98**, 1138 (1965).
280. O. Gerlach, P. L. Reiter and F. Effenberger, *Liebigs Ann. Chem.*, 1895 (1975).
281. R. Huisgen and G. Szeimies, *Chem. Ber.*, **98**, 1153 (1965).
282. R. Huisgen, G. Szeimies and L. Moebius, *Chem. Ber.*, **100**, 2494 (1967).
283. A. S. Bailey and J. E. White, *J. Chem. Soc. (B)*, 819 (1966).

284. K. D. Berlin and M. A. R. Khayat, *Tetrahedron,* **22,** 975 (1966).
285. M. Regitz and G. Himbert, *Liebigs Ann. Chem.,* **734,** 70 (1970).
286. K. A. Ogloblin, V. P. Semenov and E. V. Vasil'eva, *Zh. Org. Khim.,* **8,** 1613 (1972).
287. M. A. R. Khayat and F. S. Al-Isa, *Tetrahedron Letters,* 1351 (1970).
288. V. P. Semenov, C. B. Filippova and K. A. Ogloblin, *Zh. Org. Khim.,* **11,** 298 (1975).
289. J. E. Franz, M. W. Dietrich, A. Henshall and C. Osuch, *J. Org. Chem.,* **31,** 2847 (1966).
290. R. A. Wohl, *Tetrahedron Letters,* 3111 (1973).
291. V. P. Semenov, I. V. Volkov and K. A. Ogloblin, *Zh. Org. Khim.,* **9,** 2119 (1973).
292. K. Burger, K. Einhellig, W. D. Roth and L. Hatzelmann, *Tetrahedron Letters,* 2701 (1974).
293. I. A. D'yakonov, *Zh. Obshch. Khim.,* **17,** 67 (1947).
294. R. A. Firestone, *J. Org. Chem.,* **33,** 2285 (1968); **37,** 2181 (1972).
295. R. A. Firestone, *J. Org. Chem.,* **41,** 2212 (1976).
296. K. Fukui, *Topics Curr. Chem.,* **15,** 1 (1970).
297. L. Salem, *J. Amer. Chem. Soc.,* **90,** 543 (1968); A. Devaquet and L. Salem, *J. Amer. Chem. Soc.,* **91,** 3793 (1969); R. Sustmann, *Tetrahedron Letters,* 2717 (1971).
298. J. Bastide and O. Henri-Rousseau, *Bull. Soc. Chim. Fr.,* 2294 (1973); 1037 (1974).
299. R. Paul and S. Tchelitcheff, *Bull. Soc. Chim. Fr.,* 2215 (1962).
300. V. F. Vasil'eva and V. G. Yashunskii, *Zh. Obshch. Khim.,* **34,** 2059 (1964).
301. J. Sauer, *Angew. Chem.,* **78,** 233 (1966); **79,** 76 (1967).
302. I. I. Guseinov and G. S. Vasil'ev, *Usp. Khim.,* **32,** 20 (1963).
303. D. Craig, J. J. Shipman and R. B. Fowler, *J. Amer. Chem. Soc.,* **83,** 2885 (1961).
304. A. J. Birch, D. N. Butler and J. B. Sidall, *J. Chem. Soc.,* 2932 (1964).
305. J. Sauer and H. Wiest, *Angew. Chem.,* **74,** 353 (1962).
306. V. S. Abramov and A. P. Pakhomova, *Zh. Obshch. Khim.,* **24,** 1198 (1954).
307. M. F. Shostakovskii, A. V. Bogdanova and T. M. Ushakova, *Izv. Akad. Nauk SSSR, Otdel. Khim. Nauk,* 709 (1961).
308. A. V. Kalabina, D. E. Stepanov, V. A. Kron and A. B. Chernov, *Izv. Sib. Akad. Nauk SSSR, Ser. Khim. Nauk,* 106 (1964).
309. J. Sauer, A. Mielert, D. Lang and D. Peter, *Chem. Ber.,* **98,** 1435 (1965).
310. G. Desimoni, A. Gamba, M. Monticelli, M. Nicola and G. Tacconi, *J. Amer. Chem. Soc.,* **98,** 2947 (1976).
311. G. Desimoni, A. Gamba, P. P. Righetti and G. Tacconi, *Gazz. Chim. Ital.,* **101,** 899 (1971).
312. G. Desimoni, L. Astolfi, M. Cambieri, A. Gamba and G. Tacconi, *Tetrahedron,* **29,** 2627 (1973).
313. G. Desimoni, G. Cellerina, G. Minoli and G. Tacconi, *Tetrahedron,* **28,** 4003 (1972).
314. M. Mizuta, A. Zuzuki and Y. Ishii, *Kogyo Kagaku Zasshi,* **69,** 77, 79 (1966); *Chem. Abstr.,* **65,** 5320a,b (1966).
315. V. I. Grigos, L. S. Povarov and B. M. Mikhailov, *Izv. Akad. Nauk SSSR, Ser. Khim,* 2163 (1965).
316. I. Fleming, F. L. Gianni and T. Mah, *Tetrahedron Letters,* 881 (1976).
317. E. Shalom, J.-L. Zenou and S. Shatzmiller, *J. Org. Chem.,* **42,** 4213 (1977).
318. H. Griengl and A. Bleikolm, *Tetrahedron Letters,* 2565 (1975).
319. R. Paul and S. Tchelitcheff, *Bull. Soc. Chim. Fr.,* 808 (1952); *Compt. Rend.,* **235,** 1226 (1952).
320. U. Schoellkopf and P. Haenssle, *Liebigs Ann. Chem.,* **763,** 208 (1972).
321. J. E. Baldwin, G. A. Hoefle and O. W. Lever, *J. Amer. Chem. Soc.,* **96,** 7125 (1974).
322. R. K. Boeckman and K. J. Bruza, *Tetrahedron Letters,* 4187 (1977).
323. H. A. F. Heinemann and W. Kreiser, *Discussion Paper,* presented at the Chemiedozententagung, Darmstadt, 1979.
324. R. K. Boeckman, K. J. Brŭza, J. E. Baldwin and O. W. Lever, *J. Chem. Soc., Chem. Commun.,* 519 (1975).
325. C. G. Chavdarian and C. H. Heathcock, *J. Amer. Chem. Soc.,* **97,** 3822 (1975).
326. D. A. Evans, G. C. Andrews and B. Buckwalter, *J. Amer. Chem. Soc.* **96,** 5560 (1974).
327. J. Hartmann, R. Muthukrishnan and M. Schlosser, *Helv. Chim. Acta,* **57,** 2261 (1974).

328. J. Hartmann, M. Staehle and M. Schlosser, *Synthesis*, 888 (1974).
329. J. Ficini and J.-C. Depezay, *Tetrahedron Letters*, 937 (1968).
330. J. Ficini, S. Falou, A.-M. Touzin and J. D'Angelo, *Tetrahedron Letters*, 3589 (1977).
331. J. Ficini, P. Kahn, S. Falou and A.-M. Touzin, *Tetrahedron Letters*, 67 (1979).
332. A. B. Levy and S. J. Schwartz, *Tetrahedron Letters*, 2201 (1976).
333. R. K. Freidlina, N. A. Kuz'mina and E. T. Chukovskaya, *Izv. Akad. Nauk SSSR, Ser. Khim.*, 176 (1966).
334. B. M. Mikhailov and A. N. Blokhina, *Izv. Akad. Nauk SSSR, Otdel. Khim. Nauk*, 5373 (1962).
335. P. Pino and G. P. Lorenzi, *J. Org. Chem.*, **31**, 329 (1966).
336. B. M. Mikhailov and Y. N. Bubnov, *Zh. Obshch. Khim.*, **41**, 2039 (1971).
337. C. M. Hill, R. Woodberry, D. E. Simmons, M. E. Hill and L. Haynes, *J. Amer. Chem. Soc.*, **80**, 4602 (1958).
338. J. K. Rasmussen, *Synthesis*, 91 (1977).
339. E. W. Colvin, *Chem. Soc. Rev.*, **7**, 15 (1978).
340. H. O. House, L. J. Czuba, M. Gall and H. D. Olmstead, *J. Org. Chem.*, **34**, 2324 (1969).
341. G. Simchen and W. Kober, *Synthesis*, 259 (1976).
342. H. Vorbrüggen and K. Krolikiewicz, *Synthesis*, 35 (1979).
343. E. Nakamura, T. Murofushi, M. Shimizu and I. Kuwajima, *J. Amer. Chem. Soc.*, **98**, 2346 (1976).
344. E. Nakamura, K. Hashimoto and I. Kuwajima, *Tetrahedron Letters*, 2079 (1978).
345. R. E. Ireland, R. H. Mueller and A. K. Willard, *J. Amer. Chem. Soc.*, **98**, 2868 (1976).
346. C. A. Brown, *J. Org. Chem.*, **39**, 1324 (1974).
347. M. Tanabe and D. F. Crowe, *J. Chem. Soc., Chem. Commun.*, 564 (1973).
348. I. Ojima, T. Kogure and Y. Nagai, *Tetrahedron Letters*, 5035 (1972); *Bull. Chem. Soc. Japan*, **45**, 3506 (1972).
349. A. N. Lazarev, I. S. Ignat'ev, L. L. Schukovskaya and R. I. Pal'chik, *Spectrochim. Acta*, **23A**, 2291 (1971).
350. G. Glidewell, D. W. H. Rankin, A. G. Robiette and G. M. Sheldrick, *Trans. Faraday Soc.*, **65**, 2621 (1969).
351. M. Donike and L. Jaenicke, *Angew. Chem.*, **81**, 995 (1969).
352. W. E. Parham and C. S Roosevelt, *Tetrahedron Letters*, 923 (1971).
353. R. Fibiger and A. Hassner, unpublished results cited in Reference 338.
354. G. L. Larson and A. Hernandez, *J. Org. Chem.*, **38**, 3935 (1973); *Synth. Commun.*, **4**, 61 (1974).
355. L. Blanco, P. Amice and J.-M. Conia, *Synthesis*, 194 (1976).
356. R. M. Reuss and A. Hassner, *J. Org. Chem.*, **39**, 1785 (1974).
357. M. Zembayashi, K. Tamao and M. Kumada, *Synthesis*, 422 (1977).
358. G. M. Rubottom, M. A. Vazqüez and D. R. Pelegrina, *Tetrahedron Letters*, 4319 (1974).
359. A. Hassner, R. H. Reuss and H. W. Pinnick, *J. Org. Chem.*, **40**, 3427 (1975).
360. G. M. Rubottom and M. I. L. Nieves, *Tetrahedron Letters*, 2423 (1972).
361. G. M. Rubottom, J. M. Gruber and K. Kincaid, *Synth. Commun.*, **6**, 59 (1976); G. M. Rubottom, J. M. Gruber and G. M. Mong, *J. Org. Chem.*, **41**, 1673 (1976).
362. R. D. Clark and C. H. Heathcock, *Tetrahedron Letters*, 1713 (1974); *J. Org. Chem.*, **41**, 1396 (1976).
363. Y. Ito, T. Konoike and T. Salegusa, *J. Amer. Chem. Soc.*, **97**, 649 (1975).
364. S. Murai, Y. Kuroki, K. Hasegawa and S. Tsutsumi, *J. Chem. Soc., Chem. Commun.*, 946 (1972).
365. Y. Kuroki, S. Murai, N. Sonida and S. Tsutsumi, *Organometal. Chem. Synth.*, **1**, 465 (1972).
366. J. K. Rasmüssen and A. Hassner, *Tetrahedron Letters*, 2783 (1973).
367. J. K. Rasmussen and A. Hassner, *J. Org. Chem.*, **39**, 2558 (1974).
368. J. K. Rasmussen and A. Hassner, unpublished results cited in Reference 338.
369. I. S. Shvarts, V. N. Yarovenko, M. M. Krayushkin, S. S. Novikov and V. V. Sevost'janova, *Izv. Akad. Nauk SSSR, Ser. Khim.*, 1674 (1976).
370. M. T. Reetz and W. F. Maier, *Angew. Chem.*, **90**, 50 (1978).

371. M. T. Reetz, W. F. Maier, J. Schwellnus and I. Chatziiosifidis, *Angew. Chem.*, **91**, 78 (1979); T. Sasaki, A. Usuki and M. Ohno, *Tetrahedron Letters*, 4925 (1978).
372. G. Stork. *Pure Appl. Chem.*, **43**, 553 (1975).
373. R. M. Coates, L. O. Sandefur and R. D. Smillie, *J. Amer. Chem. Soc.*, **97**, 1619 (1975).
374. I. Kuwajima and E. Nakamura, *J. Amer. Chem. Soc.*, **97**, 3257 (1975).
375. R. Noyori, K. Yokoyama, J. Sikata, I. Kuwajima, E. Nakamura and M. Shimazu, *J. Amer. Chem. Soc.*, **99**, 1265 (1977).
376. G. Stork and B. Ganem, *J. Amer. Chem. Soc.*, **95**, 6152 (1973).
377. P. J. Stang, M. G. Magnum, D. P. Fox and P. Haak, *J. Amer. Chem. Soc.*, **96**, 4562 (1974).
378. T. Mukaiyama, K. Banno and K. Narasaka, *J. Amer. Chem. Soc.*, **96**, 7503 (1974).
379. T. Mukaiyama and M. Hayashi, *Chem. Letters*, 15 (1974).
380. T. Mukaiyama and A. Ishida, *Chem. Letters*, 319, 1167 (1975).
381. K. Narasaka, K. Soai, Y. Aikawa and T. Mukaiyama, *Bull. Chem. Soc. Japan*, **49**, 779 (1976).
382. S. Murai, Y. Kuroki, K. Hasegawa and S. Tsutsumi, *J. Chem. Soc., Chem. Commun.*, 946 (1972).
383. E. P. Kramarova, Y. I. Baukov and I. F. Lutsenko, *Zh. Obshch. Khim.*, **43**, 1857 (1973); **45**, 478 (1975).
384. S. Murai, K. Hasegawa and N. Sonoda, *Angew. Chem.*, **87**, 668 (1975).
385. S. Danishefsky and T. Kitahara, *J. Amer. Chem. Soc.*, **96**, 7807 (1974); *J. Org. Chem.*, **40**, 538 (1975).
386. R. A. Wohl, *Helv. Chim. Acta*, **56**, 1826 (1973).
387. J.-M. Conia, *Pure Appl. Chem.*, **43**, 317 (1975); C. Girard, P. Amice, J. P. Barnier and J.-M. Conia, *Tetrahedron Letters*, 3329 (1974).
388. S. Murai, T. Aya, T. Renge, I. Aya and N. Sonoda, *J. Org. Chem.*, **39**, 858 (1974).
389. G. Stork and T. L. Macdonald, *J. Amer. Chem. Soc.*, **97**, 1264 (1975); P. Amice, L. Blanco and J.-M. Conia, *Synthesis*, 196 (1976).
390. C. Mueller, W. Schaefer, A. Schweig, N. Thon and H. Vermeer, *J. Amer. Chem. Soc.*, **98**, 5440 (1976).
391. J. L. Derissen and J. M. Bien, *J. Mol. Struct.*, **16**, 289 (1973).
392. J. Fabian, H. Kroeber and R. Mayer, *Spectrochim. Acta*, **A24**, 727 (1968).
393. H. Bock and G. Wagner, *Angew. Chem.*, **84**, 119 (1972).
394. W. Schaefer and A. Schweig, *J. Chem. Soc., Chem. Commun.*, 824 (1972).
395. R. E. Penn and F. A. Curl, *J. Mol. Spectry.*, **24**, 235 (1967).
396. A. R. Katritzky, R. F. Pinzelli and R. D. Topsom, *Tetrahedron*, **28**, 3441 (1972).
397. G. Ceccarelli and E. Chiellini, *Org. Magn. Res.*, **2**, 409 (1970).
398. T. Okuyama, M. Nakada and T. Fueno, *Tetrahedron*, **32**, 2249 (1976).
399. C. C. Price and R. G. Gillis, *J. Amer. Chem. Soc.*, **75**, 4750 (1953).
400. H. J. Boonstra, L. Brandsma, A. M. Wiegman and J. F. Arens, *Rec. Trav. Chim.*, **78**, 252 (1959).
401. B. Giese and S. Lachhein, *Chem. Ber.*, **112**, 2503 (1979).
402. D. S. Tarbell and W. E. Lovett, *J. Amer. Chem. Soc.*, **78**, 2259 (1956).
403. J. I. Grayson and S. Warren, *J. Chem. Soc., Perkin Trans. 1*, 2263 (1977).
404. I. Shahak and J. Almog, *Synthesis*, 145 (1970).
405. T. Cohen, G. Herman, J. R. Falck and A. J. Mura, *J. Org. Chem.*, **40**, 812 (1975).
406. F. A. Carey and A. S. Court, *J. Org. Chem.*, **37**, 939 (1972).
407. B. M. Trost, M. J. Crimmin and D. Butler, *J. Org. Chem.*, **43**, 4549 (1978).
408. T. Okuyama, M. Nakada, K. Toyoshima and T. Fueno, *J. Org. Chem.*, **43**, 4546 (1978).
409. O. Kajimoto, M. Kobayashi and T. Fueno, *Bull. Chem. Soc. Japan*, **46**, 2316 (1973).

CHAPTER **18**

Oxathiacyclanes: preparation, structure and reactions

K. PIHLAJA and P. PASANEN
Department of Chemistry, University of Turku, SF-20500 Turku 50, Finland

I. INTRODUCTION

The purpose of this chapter is to discuss the chemistry of different oxathia-cyclanes emphasizing their distinctive features in relation to their oxygen or sulphur counterparts. We have also included compounds containing O—S, S—S or S=O bonds. Sultones are excluded since they are mainly synthetic intermediates and can be prepared from sultines by oxidation or even directly without an attack on the hydroxyl or mercapto group.

The material in this chapter has not been extensively reviewed earlier, although it has been touched on lightly[1,2].

II. FOUR-MEMBERED RINGS

A. 1,2-Oxathietane

The geometry of 1,2-oxathietane (1), which is known only as its 2-oxide (2), has been optimized using the CNDO/B parametrization[3].

    (1) 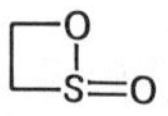   (2)

If capable of existence, 1 can be expected to exhibit equilibrium behaviour similar to that of oxetane[2-5] and thietane[6-8].

B. 2-Oxo-1,2-oxathietane

Durst and coworkers[9] found that β-hydroxysulphoxides (3) react with N-bromo-succinimide, N-chlorosuccinimide or SO_2Cl_2 to give initially 2-oxo-1,2-oxathietanes (4), which are probably formed via intramolecular cyclization of the initially formed β-hydroxychlorosulphonium chloride to an alkoxyoxosulphonium salt which fragments to 4 and t-butyl chloride (equation 1). They were, however, able to

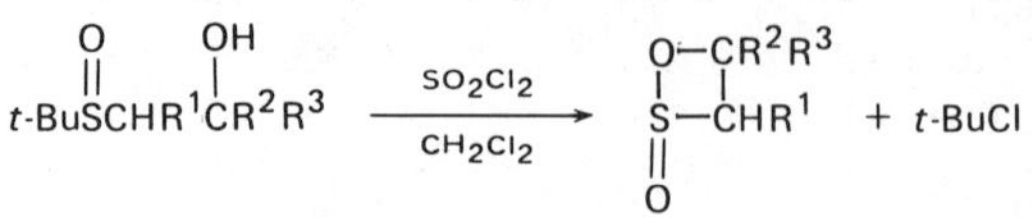

        (3)   (4)

characterize only the 4-phenyl derivative of **4** with its ^{1}H-NMR spectrum from a crude product since 2-oxo-1,2-oxathietanes exhibit only limited thermal stability.

Later on, a crystalline derivative (**6**) was isolated in a 45% yield from the reaction of **5** with SO_2Cl_2 at 203 K (equation 2)[10]. **6** decomposed quantitatively into

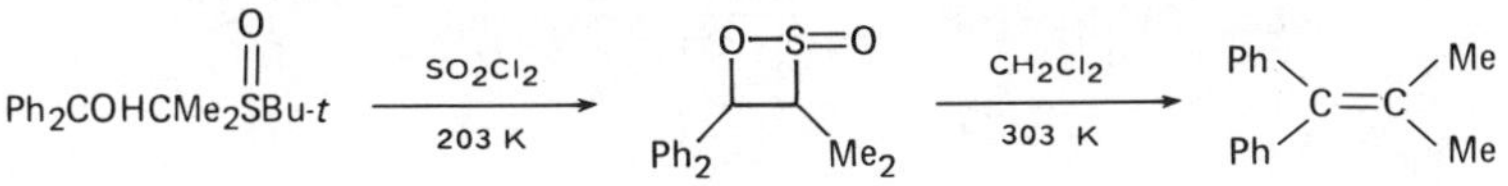

$$Ph_2COHCMe_2SBu\text{-}t \xrightarrow[203\ K]{SO_2Cl_2} \quad \xrightarrow[303\ K]{CH_2Cl_2} \quad (2)$$

(5)   (6)

1,1-diphenyl-2,2-dimethylethylene when warmed in CH_2Cl_2 at 303 K for 24 h. The authors[10] suggest that the conformation of **6** is nonplanar with the substituents on $C_{(3)}$ and $C_{(4)}$ as far apart as possible. On these grounds increasing substitution decreases the stability of the transition state for decomposition which in turn increases the relative stability of **6**.

The geometry of **2** has been optimized by CNDO/B parametrization[11] and the *exo*-oxygen is predicted to lie 62° out of the average plane of the ring[12]. The potential energy surface for the [2 + 2] retrocycloaddition of **2** has also been partially investigated[12].

III. FIVE-MEMBERED RINGS

A. 2-Oxo-1,2-oxathiolanes

1. *Preparation*

Since 2-oxo-1,2-oxathiolanes (**7**) are also cyclic sulphinate esters they can be synthesized by a reductive desulphurization of thiosulphonates[13,14] (equation 3).

$$\overset{S}{\underset{SO_2}{\bigcirc}} + P(NEt_2)_3 \longrightarrow \overset{O}{\underset{SO}{\bigcirc}} + S{=}P(NEt_2)_3 \qquad (3)$$

(7)

Treatment of *cis*- and *trans*-2,4-diphenylthietane-1,1-dioxides (**8** and **9**) with *t*-butoxymagnesium bromide gave *cis*-3,*cis*-5-diphenyl-*r*-2-oxo- (**10**) and *cis*-3,*trans*-5-diphenyl-*r*-2-oxo-1,2-oxathiolanes (**11**), respectively[15]. The mechanism of this reaction has been discussed and believed to resemble closely that of the Stevens rearrangement[16,17].

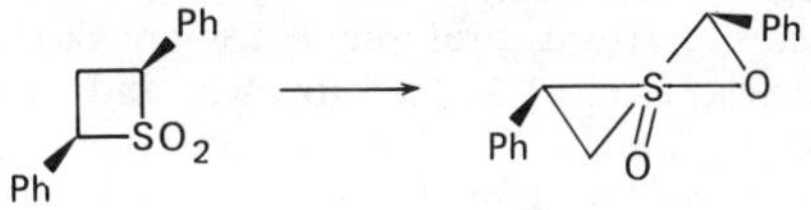

(8)   (10)

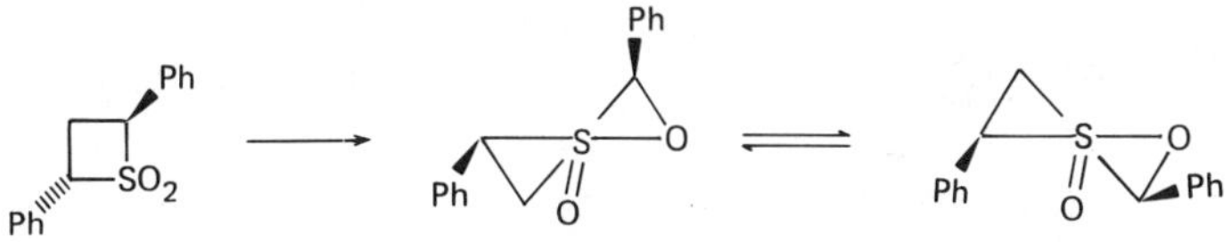

(9)   (11a)   (11b)

The best route to 2-oxo-1,2-oxathiolanes and other related cyclic sulphinate esters is, however, the cyclization of *t*-butyl hydroxyalkylsulphoxides with *N*-chlorosuccinimide or SO_2Cl_2 in CH_2Cl_2 [18,19] (equation 4).

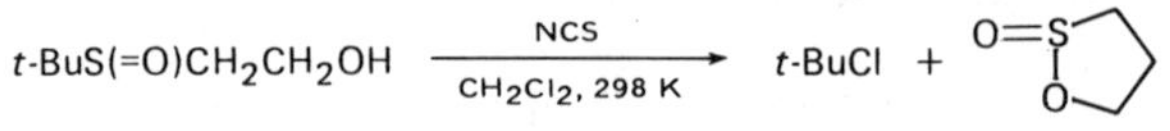

$$t\text{-BuS}(=O)CH_2CH_2OH \xrightarrow[\text{CH}_2\text{Cl}_2,\ 298\ K]{\text{NCS}} t\text{-BuCl} + \underset{(7)}{\overset{O=S}{\diagup}} \tag{4}$$

2. Structure

Dodson and colleagues[15] analysed the ^{1}H-NMR spectra of **10** and **11** and found that the observed parameters are best explained by assuming that the sultines exist in half-chair conformations where the bond connecting $C_{(4)}$ and $C_{(5)}$ bisects the plane including $C_{(3)}$, $S_{(2)}$ and $O_{(1)}$.

The *cis*-sultine (**10**) is practically anancomeric whereas the *trans* isomer behaves like a 9 : 1 mixture of **11a** and **11b**, although calculation of dihedral angles suggests that the conformation of **11** resembles an envelope (**11c**) more closely than the

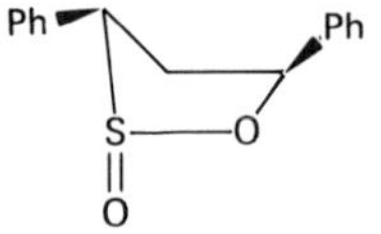

(11c)

half-chair forms (**11a** and **11b**). The latter conclusion is in good accord with the structural properties of 2-oxo-1,3,2-dioxathiolanes (see Section III.F.2) and the conclusion based on ^{1}H- and ^{13}C-NMR spectra[19] that *cis*- (**12**) and *trans*-4-phenyl-2-oxo-1,2-oxathiolanes (**13**) prefer $C_{(3)}$ envelope conformations. Despite

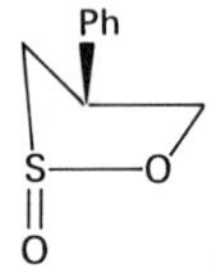

(13) 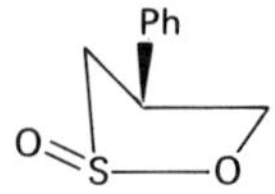   (12)

the preparation of various substituted 2-oxo-1,2-oxathiolanes[18,19] their detailed structural features are still largely unknown.

Exner and coworkers[20] tried to correlate the magnitude of the dipole moments with the postulate that the strong preference for an axial S=O configuration in 2-oxo-1,3,2-dioxathiolanes, 2-oxo-1,2-oxathiolanes, and in the corresponding six-membered rings[14] results from a dipolar interaction analogous to the anomeric effect[21]. Their attempts to estimate the dipole moment of **7** failed, however, although they were able to evaluate the dipole moment of **14** by assuming that 2-oxo-1,2-oxathiane exists predominantly in the S=O axial chair form (**14**).

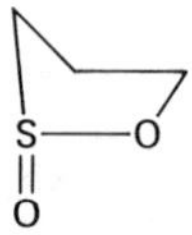

(7) 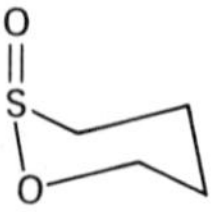   (14)

3. Reactions

Najam and Tillett[22] studied the alkaline hydrolysis of **7** and **14** and determined their enthalpies and entropies of activation. The close similarity in the rates of hydrolysis is surprising and also the order of magnitude (**7** > **14**) opposite to that observed for the hydrolysis of other cyclic esters of sulphur or for the hydrolysis of cyclic carbonates and lactones[23-26]. The authors[22] were not, however, able to make any definite conclusions as to the detailed mechanism of the decomposition except that 2-oxo-1,2-oxathiolane (**7**) does not undergo ^{18}O-exchange with the solvent during the hydrolytic reaction.

Preparation of chiral sulphoxides from **7** and **14** with various Grignard and/or organocopper lithium reagents has also been studied. The latter reagents were found to give better yields[27].

B. 1,3-Oxathiolanes

1. Introduction

1,3-Oxathiolane (**15**) and its substituted derivatives are the most widely studied five-membered oxathiacyclanes. This is due to several factors. Firstly, they can be easily prepared, and secondly (see Section III.B.2), they are interesting intermediates between their symmetric counterparts, 1,3-dioxolanes (**16**) and 1,3-dithiolanes (**17**), and hence offer a simple opportunity to make a thorough study of the kind of similarities and differences existing in **15–17**. Moreover, epimeric 1,3-oxathiolanes can be equilibrated to obtain energetic information from the structural properties and their ^{1}H-NMR spectra are normally reasonably well resolved at least at 220 MHz.

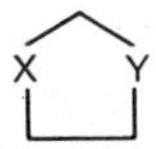

$$\begin{array}{ll}
& \text{(15) } X = O; \, Y = S \\
X \quad Y & \text{(16) } X = Y = O \\
& \text{(17) } X = Y = S
\end{array}$$

2. Preparation

In most cases 1,3-oxathiolanes have been synthesized conventionally (equation 5) by the p-toluenesulphonic acid–benzene (or CH_2Cl_2) azeotrope method[28-39]. Wilson and coworkers[38] obtained somewhat higher yields by using BF_3-Et_2O instead of p-TOS–benzene.

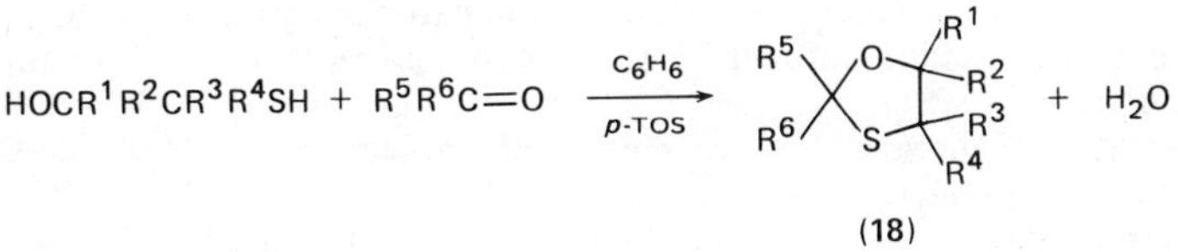

$$HOCR^1R^2CR^3R^4SH + R^5R^6C{=}O \xrightarrow[p\text{-}TOS]{C_6H_6} \quad \begin{array}{c} R^5 \\ R^6 \end{array}\!\!\!\! \begin{array}{c} O \\ S \end{array}\!\!\!\! \begin{array}{c} R^1 \\ R^2 \\ R^3 \\ R^4 \end{array} + H_2O \qquad (5)$$

$$(18)$$

The preparation of some 2-alkylimino- and 2-acylimino-1,3-oxathiolanes has also been reported[40-42].

3. Structure

Cooper and Norton[43] have determined the crystal structure of the 1,3-oxathiolane ring in cholestan-4-one-3-spiro(2,5-oxathiolane) and found that it has an envelope conformation where the methylene group next to the ring oxygen lays 51 pm out of the plane of the remainder of the oxathiolane ring ($C_{(5)}$-envelope,

15a). Pasto and coworkers[44] analysed the ^{1}H-NMR spectra of some 2-substituted derivatives and concluded from the chemical shift data that the $O_{(1)}$-envelope **(15b)** is most compatible with their NMR results. Their approach was, however, rather complicated and based on the postulation that the 2-t-butyl derivative is conformationally homogeneous, an assumption which is valid only if there is no other strongly interacting substituent[45]. Nevertheless they were able to estimate conformational energies for the 2-methyl and 2-ethyl groups fairly accurately[30,46] but greatly overestimated that for 2-isopropyl. Wilson and colleagues[47] concluded from the $^3J_{HH}$ coupling constants for a set of 2-substituted 1,3-oxathiolanes that the $C_{(5)}$-envelope **(15a)** is the preferred conformation, although they could not altogether exclude the existence of the $O_{(1)}$-envelope **(15b)** which they regarded as the next stable ring conformation. Later on Wilson[48] carried out conformational

(15a)   (15b)   (15c)

energy calculations on 1,3-oxathiolane **(15)** and 2-methyl-1,3-oxathiolane **(19)** and pointed out that the conformational energy minima for both compounds are quite

(19)

shallow and the lowest energy transition states for pseudorotation are of the order of 13 kJ mol^{-1}. In both cases the minimum exists at the $C_{(5)}$-envelope **(15a)** where in the case of **19** the methyl group is *anti* to $C_{(5)}$ with respect to the ring plane.

A systematic study of the ^{1}H-NMR spectra of several diastereomeric alkyl-substituted 1,3-oxathiolanes[28-31,45,49,50] in conjunction with the chemical equilibration of epimeric derivatives[29-31,45,46] has been proved to be very fruitful. Table 1 lists the results of chemical equilibration of several epimeric 1,3-oxathiolanes together with some comparable data for 1,3-dioxolanes **(16)**[51] and 1,3-dithiolanes **(17)**[52].

Together with the values of vicinal coupling constants ($^3J_{45}$), these results confirmed that the most favoured ring conformation is the $C_{(5)}$-envelope **(15a)**, although in some cases the $O_{(1)}$-envelope or the half-chair form where $C_{(4)}$ is above and $S_{(3)}$ below the plane defined by the remaining three atoms **(15c)** may appear to be favoured[31,53].

Conformational energies[31,53] increase in the order Me-4 $\ll$ Me-2 $<$ Et-2 $<$ Me-5 $< i$-Pr-2 $\ll t$-Bu-2 in such a way that $-\Delta G^{\ominus}$(4-Me) ~ 0 and $-\Delta G^{\ominus}$(2-t-Bu) = 8.6 kJ mol^{-1} whereas the rest of the values are between 4.6 and 5.7 kJ mol^{-1}.

It is interesting to note that all of the available evidence is in accordance with the observation that a sulphur atom tends to increase the puckering of a five-membered ring[52,54] whereas an oxygen atom appears to do the opposite[51]. Furthermore, the distortion due to the greatly different bond lengths[43] in **15** is responsible for the special features of this ring system, at least to the extent that 1,3-oxathiolane can almost better be compared with 1,3-oxathianes than with its symmetric counterparts, **16** and **17**.

Eliel and coworkers[51] pointed out that the steric requirements of the 1,3-dioxolane ring are very small and only the most bulky substituents may raise

TABLE 1. Thermodynamic parameters for the chemical equilibration of various 1,3-oxathiolanes. Some data for 1,3-dioxolanes and 1,3-dithiolanes have been included for comparison

Compound	$-\Delta H^{\ominus}$ (kJ mol⁻¹)	$-\Delta S^{\ominus}$ (J mol⁻¹ K⁻¹)	$-\Delta G^{\ominus}$ (kJ mol⁻¹)	Reference
2,5-Me₂[a]	4.7 ± 0.3	5.6 ± 1.0	3.0	29
2-Et-5-Me[a]	4.87 ± 0.07	7.6 ± 0.25	2.6	29
2-i-Pr-5-Me[a]	3.2 ± 0.1	3.1 ± 0.4	2.3	29
2-t-Bu-5-Me[a]	4.60 ± 0.05	2.9 ± 0.2	3.7	43
2-Et-2,5-Me₂[b]	1.11 ± 0.08	0.2 ± 0.25	1.05	29
2-t-Bu-2,5-Me₂[b]		—	4.0	29
2,4-Me₂[a]	-0.18 ± 0.02	-0.1 ± 0.06	-0.15	30
2-Et-2,4-Me₂[b]	-0.41 ± 0.01	-1.95 ± 0.02	0.17	30
2-i-Pr-2,4-Me₂[b]	-0.29 ± 0.04	-5.5 ± 0.1	1.1	30
2,4,5-Me₃[c]	4.57 ± 0.08	4.9 ± 0.2	3.1	31
2,4,5-Me₃[d]	5.2 ± 0.2	7.2 ± 0.65	3.0	31
2,4,4,5-Me₄[a]	4.44 ± 0.03	3.4 ± 0.1	3.4	31
2,4-Me₂-1,3-dithiolane[a]	-0.01 ± 0.06	-1.0 ± 0.2	-0.3	1,52
2-t-Bu-4-Me-1,3-diothiolane[a]		—	0.5	49
2-Et-2,4-Me₂-1,3-dithiolane[b]		—	0.7	49
2,4-Me₂-1,3-dioxolane[a]	0.8	1.1 ± 0.2	1.15	1,51
2,4,5-Me₃-1,3-dioxolane[c,e]	3.1 ± 0.2	1.2 ± 0.6	2.8	51

[a] K = cis/trans.
[b] K = (r-2-alkyl-t-2, t-5-Me₂)/(r-2-alkyl-t-2, c-5-Me₂) or K = (r-2-alkyl-t-2, c-4-Me₂)/(r-2-alkyl-t-2, t-4-Me₂).
[c] K = (r-2, t-4, t-5-Me₃)/(r-2, c-4, c-5-Me₃).
[d] K = (r-2, c-4, t-5-Me₃)/(r-2, t-4, c-5-Me₃).
[e] Recalculated from the equilibration data in Reference 51.

the barriers for the otherwise fairly free pseudorotation. This is understandable since the deviation of the ring atoms from the average plane of the molecule is fairly small. The importance of the ring atoms as structure-forming factors is seen when comparing 1,3-dithiolanes and 1,3-oxathiolanes with 1,3-dioxolanes. The former has relatively great preference towards the half-chair form where $C_{(2)}$ is at the isoclinal position (17c)[52,54]. Due to the long C—S bonds, isomeric 2,4-dimethyl-1,3-dithiolanes (Table 1) are almost equally stable. The same situation

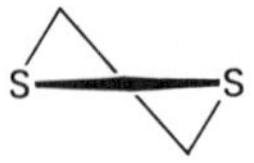

(17c)

also prevails in the case of 2,4-dialkyl-1,3-oxathiolanes. A comparison of the equilibria shown in Table 1 demonstrates the similarities and differences in 15, 16 and 17 quite well. Recent ^{13}C-NMR chemical-shift correlations[55] for alkyl-substituted 1,3-oxathiolanes lend further support to the above structural views.

The relative stabilities of the ethyl rotamers[56] of 4-ethyl-1,3-oxathiolane (20) and its 2- and 5-alkyl-substituted derivatives have been determined[50] using the Karplus equation and the values of J_{AX} and J_{BX} from the methylene protons of the ethyl group to $H_{(4)}$. In general 21a is 1.7 ± 0.2 kJ mol^{-1} more stable than 21b and 3.0 ± 0.4 kJ mol^{-1} more stable than 21c, although their relative amounts

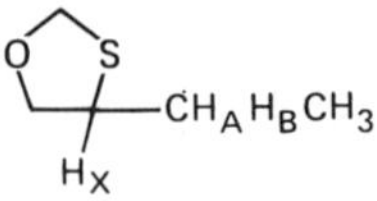

(20)

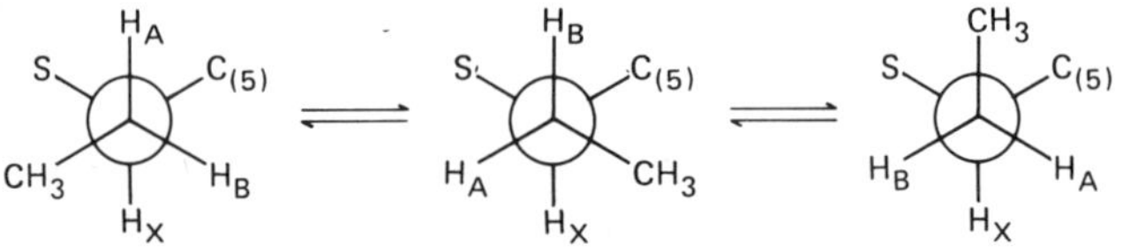

(21a)   (21b)   (21c)

depend also on the accessible ring conformations. Bushweller and colleagues[57,58] investigated the rate of the t-butyl rotation in 22–26 with the aid of the ^{1}H-DNMR spectra and determined their activation parameters. In going from 22 to 23 and from 24 to 25 the barrier increases by about 12 kJ mol^{-1} indicating that methyl is substantially more hindering to t-butyl rotation than hydrogen as expected. Replacement of O by S in 26 increases $\Delta G^{\neq}$ by about 10 (25) and 13 kJ mol^{-1} (23); in other words the second step enhances the barrier much less than the first as expected in the light of the structural differences (see above).

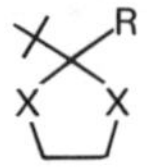

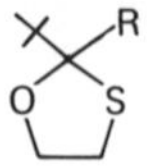

(22) X = S, R = H   (24) R = H
(23) X = S, R = Me   (25) R = Me
(26) X = O, R = Me

Aromatic solvent-induced shifts in the ^{1}H-NMR spectra (C_6H_6 or $C_6H_5CH_3$ vs. CCl_4) of the methylene protons in the 4,5-position of 2,2-dimethyl-1,3-dioxolane $(27)^{59}$, -dithiolane $(28)^{59}$ and -oxathiolane $(29)^{60}$ are close to each other $(0.3-0.4$ p.p.m.) and 29 has been shown to solvate with toluene similarly to 2,2-dimethyl-1,3-oxathiane (30).

(27) X = Y = O
(28) X = Y = S
(29) X = O, Y = S

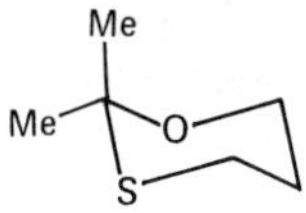

(30)

Optical rotatory dispersion, circular dichroism and IR data for a series of 3-spiro-1,3-dioxolane, -1,3-oxathiolane and -1,3-dithiolane derivatives of 4-oxosteroids have been discussed and an axial sulphur substituent α to the carbonyl found to greatly enhance the carbonyl Cotton effect[61].

Mass spectrometric fragmentation pathways of 1,3-oxathiolane (15) and its alkyl derivatives have been well documentated[62-64]. Types I and II are the main fragmentation routes, although the parent compound (15) decomposes also by type V[63] and 19 by type III[64]. The various modes of fragmentation of 1,3-dithiolanes and 1,3-oxathiolanes resemble each other closely but differ considerably from those of 1,3-dioxolanes[63,64] (Table 2). This is in agreement with the general observation that sulphur increases the relative stability of the parent and large fragment ions. The intensity of the parent-less-methyl ion of 2-substituted 1,3-oxathiolanes (Table 2) is less than that of the corresponding 1,3-dithiolanes or 1,3-dioxolanes which is probably due to a weaker resonance stabilization in the former.

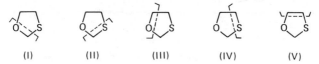

(I)   (II)   (III)   (IV)   (V)

4. Reactions

a. Acid-catalysed hydrolysis. De and Fedor[32] studied the acid-catalysed hydrolysis of 2-(substituted phenyl)-1,3-oxathiolanes (31) and concluded that protonation occurs predominantly on the oxygen atom which actually means that the ring

TABLE 2. The relative intensities of the $[M]^+$ and $[M - Me]^+$ ions of some 1,3-diheterocyclopentanes at 70 eV[64]

Compound	$[M]^+$ (%)	$[M - Me]^+$ (%)
2-Methyl-1,3-dioxolane	11	100
2-Methyl-1,3-oxathiolane	51	16
2-Methyl-1,3-dithiolane	100	91
2,2-Dimethyl-1,3-dioxolane	–	53
2,2-Dimethyl-1,3-oxathiolane	21	6
2,2-Dimethyl-1,3-dithiolane	49	53

cleavage should principally involve the acetal carbon—oxygen bond. Furthermore, these authors have proposed the A2 mechanism for the hydrolytic decomposition.

Fife and Jao[33] came to the conclusion that the ring rupture proceeds via the sulphur-protonated conjugate acid which would require breaking of the acetal carbon—sulphur bond in the critical transition state. Moreover, they proposed A1 mechanism for the hydrolysis reaction.

Pihlaja[65] has shown that the data for the hydrolytic decomposition of 15, 19 and 29 are, however, best consistent with an A1 mechanism in which the ring cleavage occurs at the acetal carbon—oxygen bond (equation 6). A peculiar feature

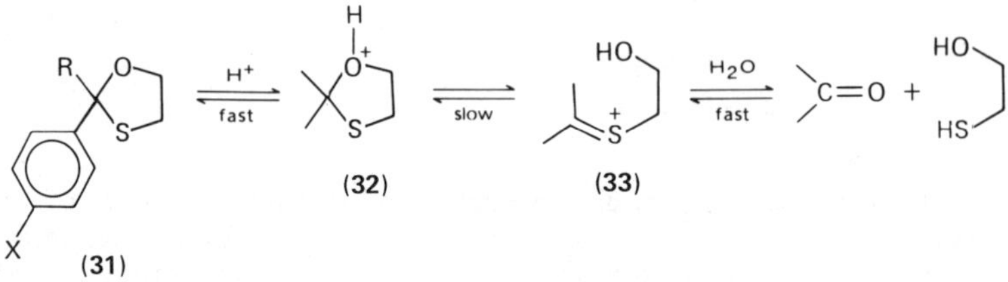

of the acid-catalysed hydrolysis of 1,3-oxathiolanes is the specific solvent deuterium isotope effect since the rate is much higher when the deuterium atom fraction approaches unity than one would expect. This is best understood by assuming that a carbonium—sulphonium ion intermediate is formed, in which the hybridized p- and d-orbitals of sulphur have a significant contribution[65]. Another explanation is that the reaction involves parallel routes[66]. There is, however, very little support for this view and all the available evidence seems to point to the mechanism involving 32 and 33[32,33,67,68].

Guinot and Lamaty[67,68] found that the protonation of 2,2-dimethyl-1,3-oxa-thiolane (29) in FSO_3H-SbF_5 led exclusively to the formation of the carbonium—sulfonium cation (34) which despite the extreme conditions[67] accords with the

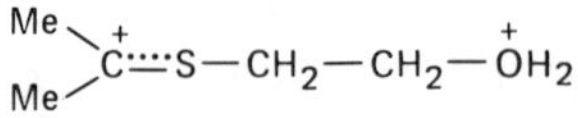

(34)

results for the hydrolytic decomposition[65]. They also concluded[68] from the magnitude of the kinetic deuterium isotope effects of 35 and 36 that the acid-catalysed hydrolysis proceeds through the C—O bond rupture since in the case of the C—S bond cleavage k_H/k_D should have been greater for 36 and not for 35 as observed:

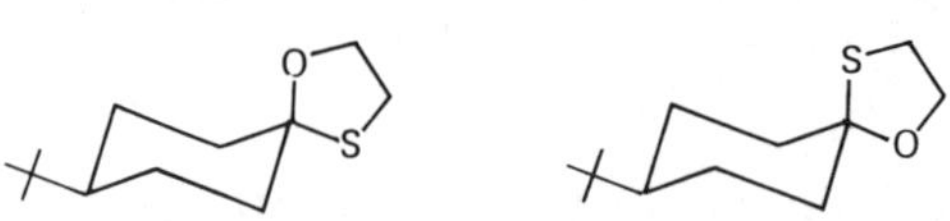

k_H/k_D	1.32	1.11
	(35)	(36)

The relative rates in the oxathiolane series 38 are very similar to those in the corresponding 1,3-dioxolane series (37)

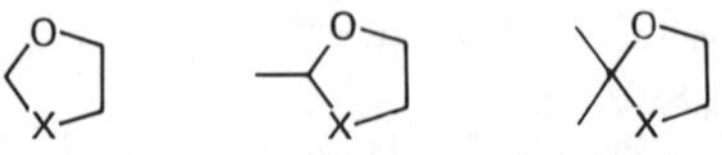

		k_{rel}			
(37)	X = O;	k_{rel}	1	4300	43,000
(38)	X = S;	k_{rel}	1	2250	114,000

whereas 1,3-dithiolanes are practically inert under the same conditions[65]. The rate increase due to the second methyl substituent is 57-fold in 38 but only 10-fold in 37[69]. In both cases, however, one of the groups bound by the bonds attached to the acetal carbon is forced to bend inward and the other outward in relation to the ring. This steric retardation is different in 2,2-dimethyl-1,3-dioxolane[65,69] and 2,2-dimethyl-1,3-oxathiolane[65] since in the latter the interaction between a bending 2-methyl and a 4-hydrogen is at least initially very small (Table 2) whereas that between the bending 2-methyl and a 5-hydrogen is even initially around 4 kJ mol^{-1}.

   b. *Photochemically initiated reactions.*   These reactions have been studied in $CFCl_3$ at 273 K[35]. In the presence of benzophenone 2-alkyl-1,3-oxathiolanes (39) are photolysed considerably more slowly than the corresponding 2-alkyl-1,3-dioxolanes (40) under similar conditions and furthermore the former react very selectively (equation 7). Only the S-2-chloroethyl thio ester (42) is formed with no

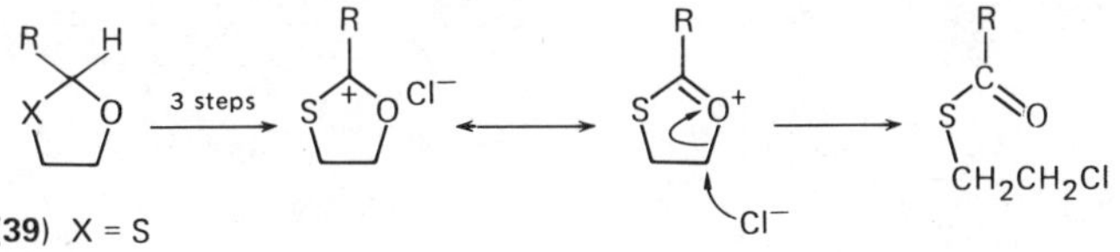

(39) X = S

(40) X = O   (41a)   (41b)   (42)

(7)

trace of the O-2-chloroethyl thio ester. Assuming a similar mechanism as for the photolyses of 2-substituted 1,3-dioxolanes the observed reaction products can be explained by the resonance structure 41b. The higher stability of 42 as compared with O-alkyl thio esters[35] may also contribute to the occurrence of the specific ring-opening.

   c. *Reduction.*   The reduction of 43 with $LiAlH_4$-$AlCl_3$ (equation 8) leads to the corresponding β-hydroxyethyl and γ-hydroxypropyl sulphides (44)[36] whereas the reduction with metal–liquid ammonia combinations (equation 9) gives rise to β- and γ-alkoxythiols (45)[34,70]. The hydrogenolysis by the 'mixed hydride'[71] in

(44)   (43)   (45)

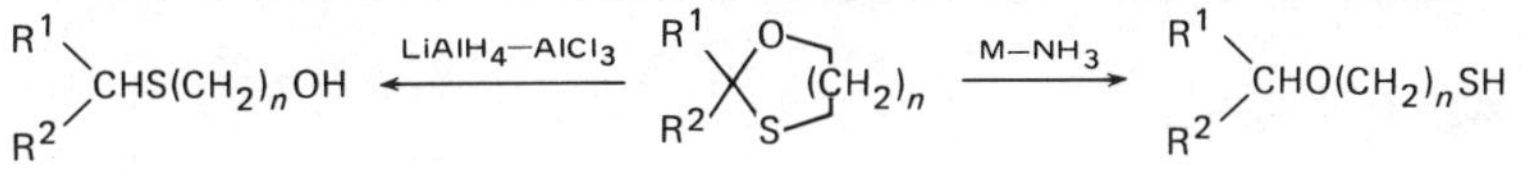

(8)   (9)

ether solution involves selective cleavage of the C—O bond but the M—NH$_3$ reduction occurs principally through a C—S bond rupture, although in some cases the yields remain low[34].

   d. *Miscellaneous reactions.*   Wilson and Huang[72] used halogenation of 1,3-oxathiolanes derived from benzophenone, diisopropyl ketone and cycloheptanone for regeneration of the ketone.

   Emerson and Wynberg[73] reported good to excellent yields of the corresponding aldehydes and ketones in the treatment of 1,3-oxathiolanes with a solution of sodium N-chloro-p-toluenesulphonamide in water, ethanol or methanol under mild conditions. This method is a useful addition to the older more tedious methods[72] in protecting carbonyl groups during synthesis.

C. 3-Oxo-1,3-oxathiolanes

   Very little is known about 3-oxo-1,3-oxathiolane (46) and its derivatives, although Hoge and Fischer determined the crystal structure (bond lengths in pm and torsion angles) of 2-p-nitrophenyl-3-oxo-1,3-oxathiolane (47)[74]. The purpose of this analysis was to solve the configuration of the single product obtained in the oxidation of 2-p-nitrophenyl-1,3-oxathiolane instead of two diastereoisomeric sulphoxides. The results showed that the oxygen was introduced in the *trans* position (47) and torsion angles show the ring to be in the half-chair form with $O_{(1)}$ and $C_{(5)}$ above and below the plane of the other three atoms of the ring. This observation is in accordance with the conclusions reached from the 300 MHz ^{1}H spectra which also suggest that **46** greatly favours the half-chair form where the oxo group is *anti* to the ring oxygen (**46a**). From $^3J_{4\,5}$ values it has been estimated that **46a** is about 4.6 kJ mol^{-1} more stable than **46b**[75]. The above conclusion is also in accordance with the observations of Harpp and Gleason[14].

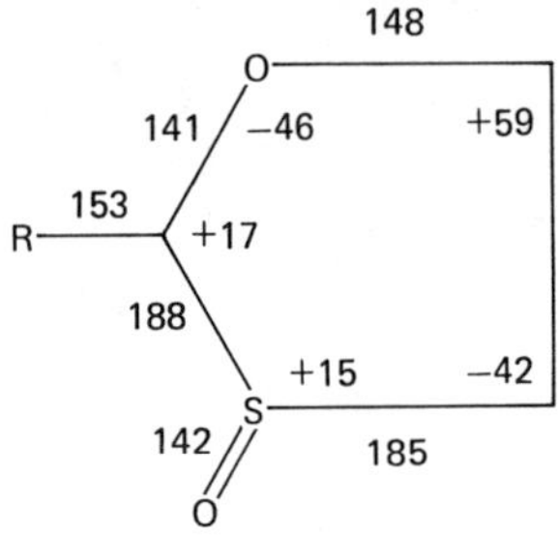

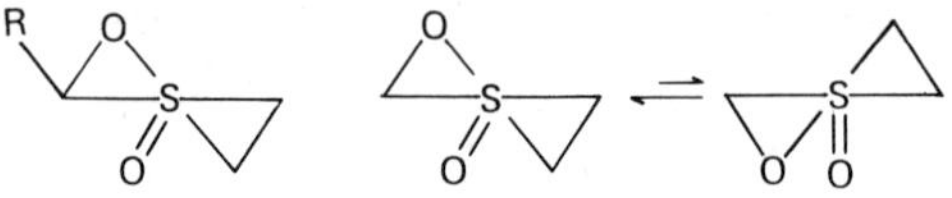

(47)   (46a)   (46b)

Schank and coworkers report a cyclofragmentation of 3-oxo-1,3-oxathiolanes (equation 10) to vinyl sulphenates (**48**)[76]. Kellogg[77] mentioned the oxidative forma-

$$R^1CHO + HSCHR^2COOH \longrightarrow R^1CH \overset{S-CHR^2}{\underset{O-C=O}{|}} + H_2O \qquad (11)$$

tion of substituted 3-oxo-1,3-oxathiolanes from the corresponding *trans*-2,4-disubstituted 5-diphenylmethylene-1,3-oxathiolanes but did not characterize them very well.

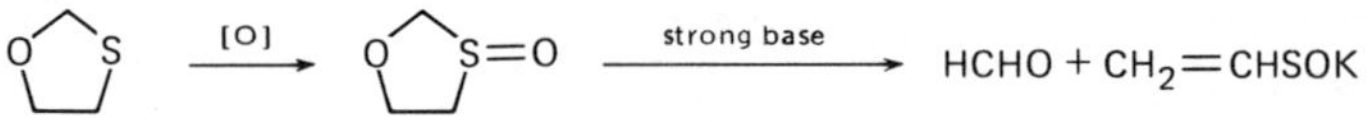

D. 5-Oxo-1,3-oxathiolanes

1. Preparation

The title compounds are both thioacetals and esters and can be prepared by various methods from aldehydes or ketones and α-mercaptocarboxylic acids (**49**)[78]. Satsumabayashi and colleagues[79] used three different modifications to obtain **50** (equation 11). In method A the reactions were carried out in refluxing benzene with azeotropic removal of the water eliminated. Method B produced **50** by stirring equimolar amounts of the reactants without any solvent or dehydrating agent, followed by direct distillation. Method C also required boiling benzene but with p-toluenesulphonic acid catalyst and without azeotropic removal of the water

(49)

(50)

$$R^1 = \text{Me, Et, Pr, Ph, } p\text{-NO}_2\text{C}_6\text{H}_4, p\text{-CH}_3\text{C}_6\text{H}_4$$

$$R^2 = \text{H, Me}$$

formed in the reaction. Yields are not high (14—56%), partly because of the formation of side-products (51) and/or intermediates (52).

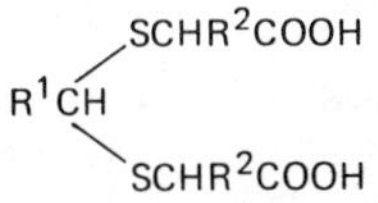

$$R^1CH \begin{array}{c} SCHR^2COOH \\ \\ SCHR^2COOH \end{array} \qquad R^1CH \begin{array}{c} SCHR^2COOH \\ \\ OH \end{array}$$

$$\text{(51)} \qquad\qquad \text{(52)}$$

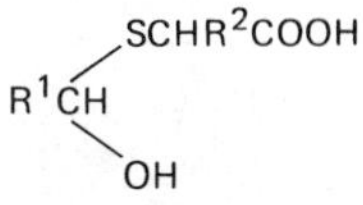

Pailer and coworkers[80] prepared several 2-substituted 4,4-diphenyl-5-oxo-1,3-oxathiolanes (53) by transacetalization and purified them as their hydrochlorides. Some authors[81-83] have used $BF_3.Et_2O$ as catalyst to enhance the yield of some 5-oxo-1,3-oxathiolanes.

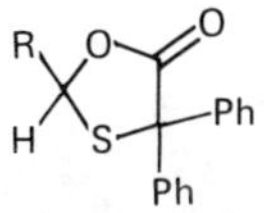

$$\text{(53)}$$

$$R = Me_2NCH_2, Et_2NCH_2, Me_2NCH_2CH_2, p\text{-}Me_2NC_6H_4, \text{ etc.}$$

2. Structure

Due to the lactone grouping $-C-C(=O)-O-C-$, the conformational situation in 54 is clearly different from that in 1,3-oxathiolane (15) which has been shown to favour the $C_{(5)}$ envelope form (Section III.B.3). The only significant conformation of 5-oxo-1,3-oxathiolane (54) is an envelope where $S_{(3)}$ is the flap atom[83]. Chemical

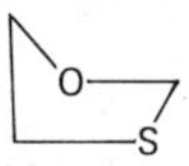

$$\text{(15)} \qquad\qquad\qquad \text{(54)}$$

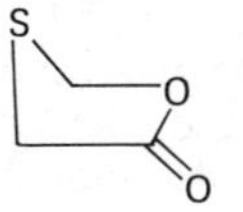

equilibration of epimeric 2,4-dimethyl and 2-t-butyl-4-methyl derivatives have shown that 2-Me, 2-t-Bu and 4-Me favour equatorial positions by 7.6, 9.8 and 1.2 kJ mol^{-1}, respectively[83]. The enhanced magnitude of the conformational energies is in accordance with the structural difference between 54 and 15.

^{1}H-NMR spectra have shown that trans-2,4-dialkyl-5-oxo,1,3-oxathiolanes (55) exhibit larger $^4J_{24}$ values than the cis forms (56) in good agreement with the relative magnitude of the $^4J_{2,5}$ values in correspondingly substituted 1,3-dioxolanes[84].

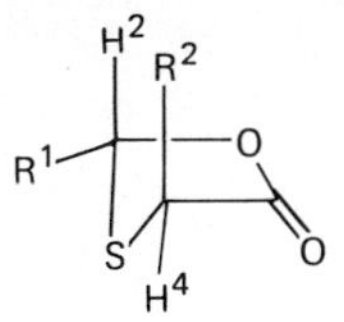

$$\text{(55)} \quad ^4J_{24} = 0.7 \text{ Hz} \qquad\qquad \text{(56)} \quad ^4J_{24} = 0.4 \text{ Hz}$$

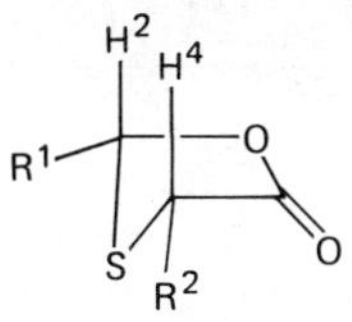

Accordingly, 2-phenyl-5-oxo-1,3-oxathiolane (57) is mainly in the equatorial envelope form and $^4J_{2a4e}$ is about 0.6 Hz and $^4J_{2a4a}$ about 0.4 Hz as reported by Brink, although he was not able to assign the relative orientation of the protons in 57[85]. The characteristic IR bands of several alkyl-substituted 5-oxo-1,3-oxathiolanes have also been reported[83].

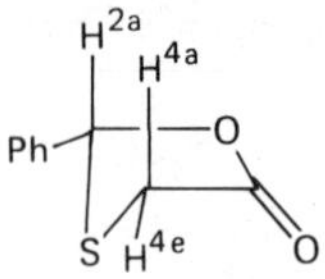

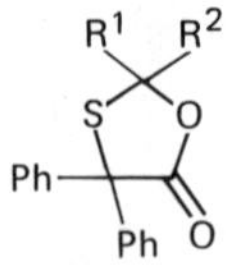

(57)   (58)

Møller and Pedersen[81] studied the electron impact mass spectra of some 2-mono-
and 2,2-dialkyl-substituted 4,4-diphenyl-5-oxo-1,3-oxathiolanes (58) and came to
the naïve conclusion that the ester function appreciably changes the balance between
the different fragmentation modes from that of 1,3-oxathiolane (15) and its alkyl
derivatives[62] for which types I and II predominate (Section III.B.3). The main
fragmentation mode of 2,2-dialkyl-substituted derivatives[81] (58: $R^1 = R^2$ = alkyl)
is III followed by types V and II whereas for 2-monoalkyl derivatives[80,81] (58:
R^1 = H, R^2 = alkyl) the most important mode is IV, followed by III, I, II and V.

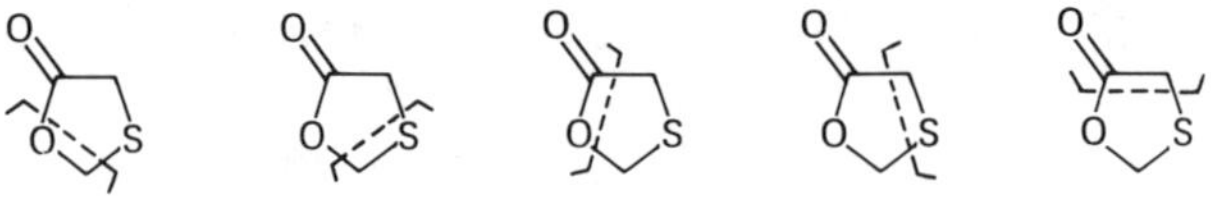

(I)   (II)   (III)   (IV)   (V)

3. Reactions

The reaction of 2-aryl-4,4-diphenyl-5-oxo-1,3-oxathiolane (58) with ethyl-
magnesium bromide[86] (equation 12) gives the acid 59 when R = Ph or p-CH$_3$OC$_6$H$_4$,
whereas no 59 is formed when R = p-NO$_2$C$_6$H$_4$. The acids are the result of an attack
on C$_{(2)}$ and a subsequent cleavage of the carbon—oxygen bond[86,87].

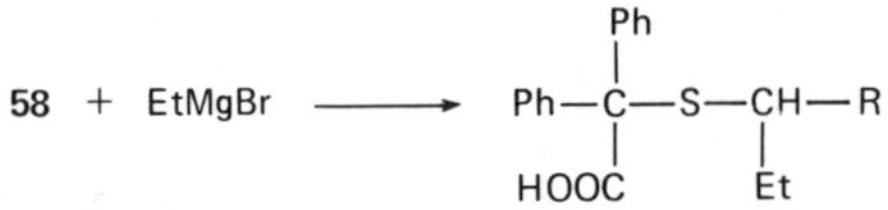

$$58 + EtMgBr \longrightarrow \underset{\underset{HOOC}{|}}{\overset{\overset{Ph}{|}}{Ph-C}}-S-\underset{\underset{Et}{|}}{CH}-R \qquad (12)$$

(59)

On treatment with concentrated sulphuric acid and dilution with water (equation
13), 58 gives isobenzothiophenes (60) as primary products[88,89]. Pyrolysis of 58
(equation 14) gives rise to 1,1,2-triarylethylenes (61) via thiirane intermediates (62)[90].

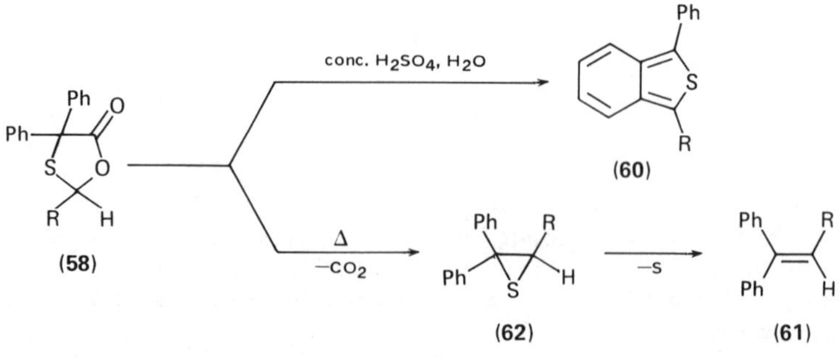

(13)

(14)

Oxidation of **58** with either peroxysebacic acid or H_2O_2 gives both diastereo-isomeric forms of **63**, the relative configurations of which are assigned on the basis of their ^{1}H-NMR spectra[91]. The assignment has been carried out by postulating the signal of H$_{(2)}$ of the *trans* form (**63a**) at higher field. In most cases this isomer was also the main product of oxidation[91] in agreement with the orientation of the S=O group in **63** (Section III.C)[74,75].

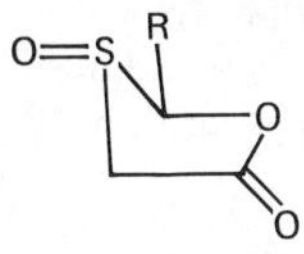

(63a) 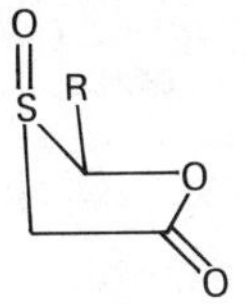   (63b)

Glue and colleagues[92] prepared some 2,2-dialkyl-substituted 3,5-dioxo-1,3-oxathiolanes (**64**) by smooth oxidation (H_2O_2, glacial AcOH, 298 K) and studied their reactions with acetic anhydride (equation 15). A highly stereoselective process (Pummerer rearrangement) gives the corresponding 4-acetoxy-2,2-dialkyl-5-oxo-1,3-oxathiolanes (**67**), in which the acetoxy group stereochemically retains the orientation of the S=O bond in **64**. The stereoselectivity has been explained by an intramolecular process, possibly proceeding via the acetoxysulphonium ion **65** and the ion pair **66** to **67**[92].

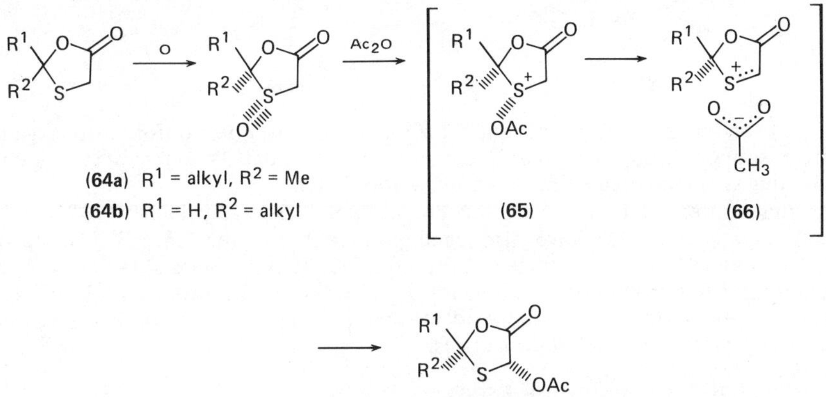

(64a) R^1 = alkyl, R^2 = Me
(64b) R^1 = H, R^2 = alkyl

(65)   (66)

(15)

(67a) R^1 = alkyl, R^2 = Me
(67b) R^1 = Me, R^2 = alkyl

E. 2-Oxo-1,3,2-dioxathiolanes

Of the various methods of preparation[93,94] of cyclic sulphites the best yields have been obtained by the condensation of 1,2-diols with thionyl chloride in the presence of pyridine[95] (equation 16). The ring geometry of **68** and its alkyl and phenyl derivatives have been extensively studied by electron diffraction[96,97], IR[98] ^{1}H-NMR[95,98,99], CD techniques[100] and ^{13}C-NMR[101]; these reports review the older literature fairly thoroughly. Although the electron diffraction study of 2-oxo-1,3,2-dioxathiolane (**68**) itself postulates an essentially planar structure for

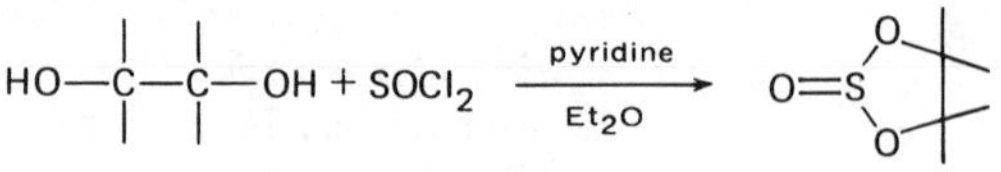

$$\text{HO—C—C—OH} + \text{SOCl}_2 \xrightarrow[\text{Et}_2\text{O}]{\text{pyridine}} \quad O=S \cdots \text{ring} \qquad (16)$$

(68)

the ring, a later report[97] shows that the experimental findings for cis-4, trans-5, r-2-oxo- **(69)** and trans-4, trans-5, r-2-oxo-1,3,2-dioxathiolanes **(70)** can be best explained by the existence of the twist-envelope forms in agreement with the CD[100], IR[98] and the most recent NMR work[98,101]. These twist-envelope forms are interconverted by rapid pseudorotatory **(68a** and **68b)** paths not involving

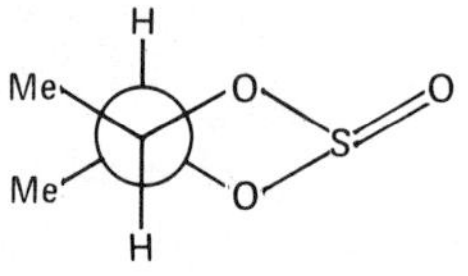

(69) 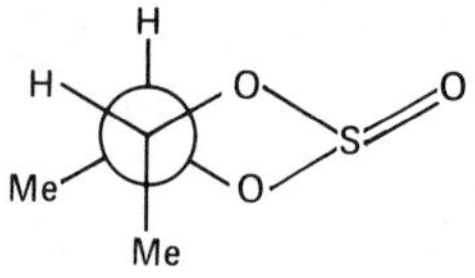   **(70)**

(68a)   ⇌   **(68b)**

inversion at sulphur. The existence of the twist-envelope conformations gains indirect support from the great preference of the axial S=O group in the corresponding six-membered sulphites (see Section IV.G).

In this context it is worth noting that 2-oxo-1,2,3-oxadithiolane **(71)** and its 5-methyl derivative **(72)** have also been prepared[102] (equation 17). Thompson and coworkers[103] have obtained 2-thioxo-1,3,2-dioxathiolanes **73—75** from the reaction of sulphur monochloride with 1,2-ethanediol, 1,2-propanediol and 2,3-butanediols (equation 18), and found by spectroscopic means that they resemble structurally **68** and its methyl derivatives.

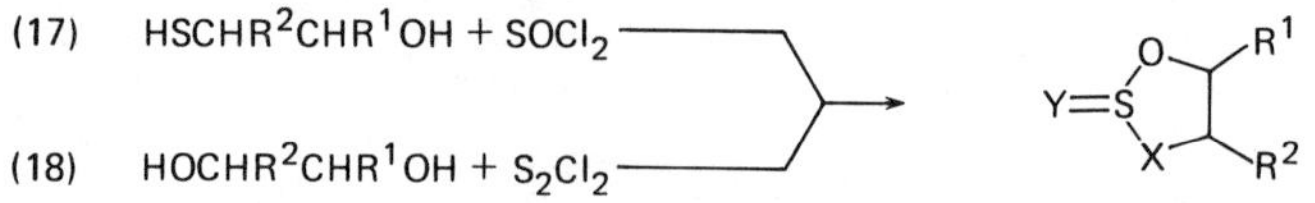

(17) $\text{HSCHR}^2\text{CHR}^1\text{OH} + \text{SOCl}_2$ ⟶

(18) $\text{HOCHR}^2\text{CHR}^1\text{OH} + \text{S}_2\text{Cl}_2$ ⟶

$$Y=S \overset{O}{\underset{X}{\bigg\langle}} \begin{matrix} R^1 \\ R^2 \end{matrix}$$

	X	Y	R^1	R^1
(71)	S	O	H	H
(72)	S	O	Me	H
(73)	O	S	H	H
(74)	O	S	H	Me
(75)	O	S	Me	Me

IV. SIX-MEMBERED RINGS

A. 2-Oxo-1,2-Oxathianes

1. Preparation

The 1,2-oxathiane system is unknown and apparently unstable, but the 2-oxides are fairly well described in the literature. As sulphinate esters, the title compounds, e.g. the nonsubstituted molecule (76), can be prepared via the reductive desulphurization coupled with rearrangement of the six-membered thiosulphonate (77) in the presence of tris(diethylamino)phosphine (equation 19)[13a,14]. The conformation

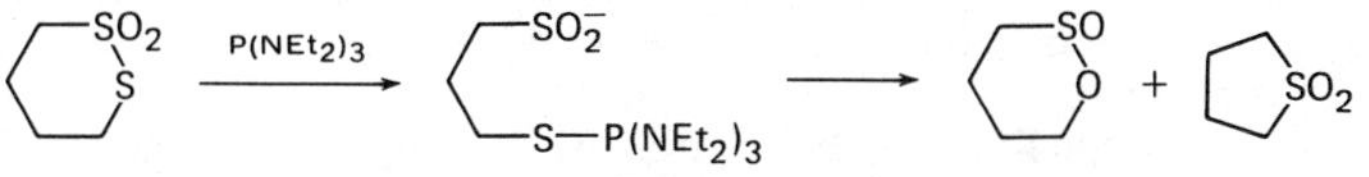

$$\underset{(77)}{\left\langle \overset{-SO_2}{\underset{S}{}} \right\rangle} \xrightarrow{P(NEt_2)_3} \underset{(79)}{\left\langle \overset{-SO_2^-}{\underset{-S-P(NEt_2)_3}{}} \right\rangle} \longrightarrow \underset{(76)}{\left\langle \overset{-SO}{\underset{O}{}} \right\rangle} + \underset{(78)}{\left\langle \overset{}{\underset{SO_2}{}} \right\rangle} \qquad (19)$$

of 76 (90%) and 78 (10%) was attributed to the ambivalent nature of the intermediate sulphinate anion (79) capable of cyclizing through either the sulphur or oxygen atom[13a].

Certain 4-chloro derivatives (80) are obtained by treatment of 3-butenols with SOCl$_2$ (equation 20)[104,105].

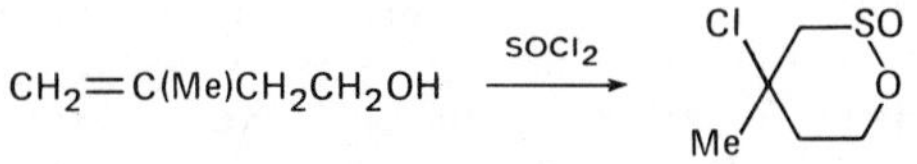

$$CH_2=C(Me)CH_2CH_2OH \xrightarrow{SOCl_2} \underset{(80)}{\underset{Me}{\overset{Cl}{\bigtimes}}\overset{-SO}{\underset{O}{}}} \qquad (20)$$

The most general route to cyclic sultines developed by Sharma and coworkers[19] utilizes cleavage of t-butyl (δ-hydroxyalkyl)sulfoxides (81) by SO$_2$Cl$_2$, and enables preparation of several specifically substituted derivatives such as 82 from relatively simple precursors in isolated yields of ca. 75% (equation 21). Although this method

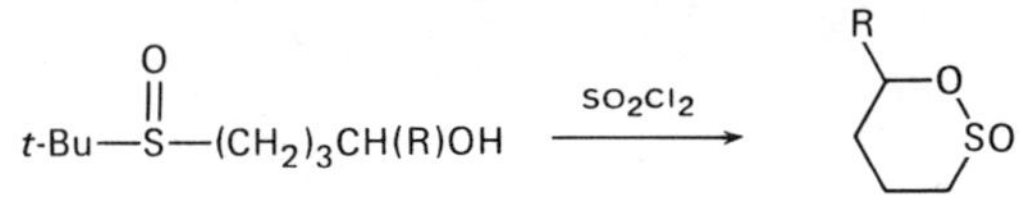

$$\underset{(81)}{t\text{-Bu}-\overset{\overset{O}{\parallel}}{S}-(CH_2)_3CH(R)OH} \xrightarrow{SO_2Cl_2} \underset{(82)}{\overset{R}{\bigtimes}\overset{-O}{\underset{SO}{}}} \qquad (21)$$

reduced the problem of preparing various alkylated 2-oxo-1,2-oxathianes mainly to the synthesis of properly substituted derivatives of 81, it was unsuccessful in a few cases, e.g. in the production of sultines with a phenyl or two methyl groups α to the sulphur atom[19]. Furthermore, the products obtained showed high diastereomeric purity, which was reasoned to follow from great stability differences between isomers and/or their facile epimerization under the reaction or isolation conditions[19].

2. Structure

The main interest in the structural study of 2-oxo-1,2-oxathianes is concerned with the steric disposition and different interactions of the S=O group, and hence, with the more general question of the conformational behaviour of molecules possessing polar groups or atoms.

The 100 MHz ^{1}H-NMR spectrum of 76, as temperature-independent from −90 to $+150°C$, is best interpreted in terms of a rigid chair conformation with a strong preference for the axial structure (76a) over the equatorial one (76b)[14]. The energy

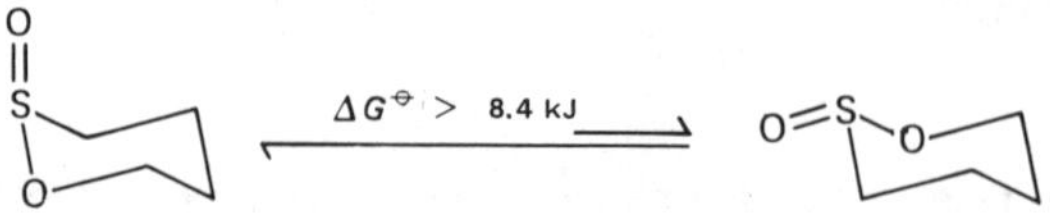

(76a)   (76b)

difference, ca. 8.4 kJ mol^{-1}, estimated indirectly from the ^{1}H-NMR spectrum, is well in line with those presented for the sulphoxide group in thiane oxides $(0.8-2\ \text{kJ})$[107], and in 2-oxo-1,3,2-dioxathianes $(12-20\ \text{kJ})$[108], if the reasoning based on dipole—dipole interactions between the exocyclic and ring hetero-atoms[14,21] is relevant.

This concept is qualitatively supported by the dipole moment measurements of Exner and coworkers[20], who found that the experimental value for 76 is consistent with the estimated one only if a chair form with the axial S=O group (76a) is assumed to predominate.

By ^{1}H-NMR and ^{13}C-NMR measurements as complementary tools Buchanan and his colleagues[106] came to the conclusion that even molecules like 83 with an axial substituent in their 6-position still exist in a chair—chair equilibrium which prefers the *syn*-axial alternative (83a). Possible reasons for this somewhat conflicting behaviour[94,109,110] are not discussed[106], but it finds some resemblance in the results of the combined ^{1}H-NMR and IR study by Dhami[104,105], who noted that *cis*-4-chloro-, *trans*-4-methyl, *r*-2-oxo-1,2-oxathiane (84) exists in a single chair form (84a) where both the Cl and S=O groupings are axially orientated (see also Sections IV.C. and IV.G).

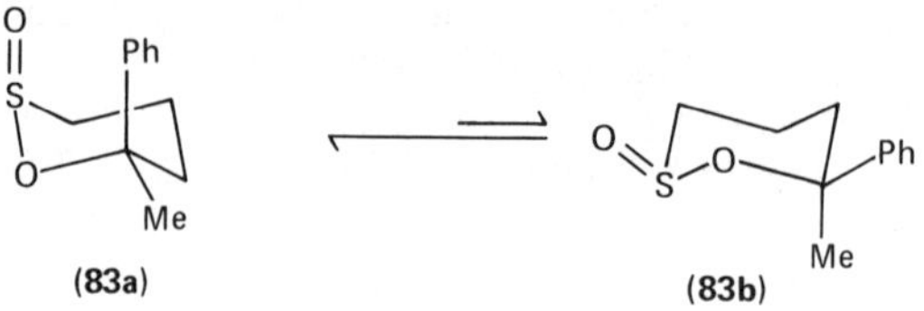

(83a)   (83b)

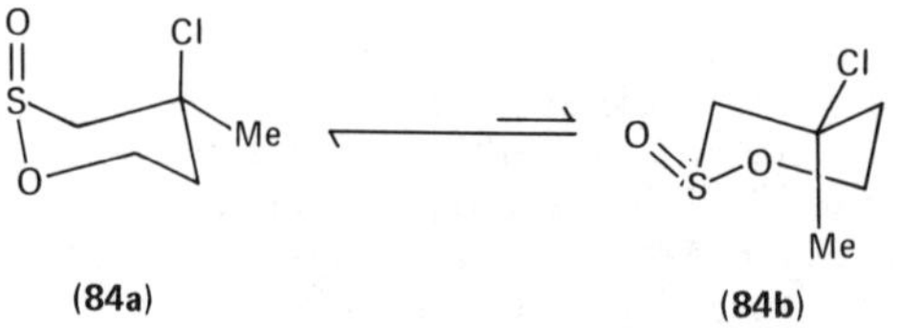

(84a)   (84b)

3. Reactions

The kinetics of the alkaline hydrolysis of 76 was studied by Najam and Tilett[22], who reported some anomalous features in the relative reactivity along the series from five- to six-membered and open-chain analogues (see Section III.A.3). The facile oxidation of 2-oxo-1,2-oxathianes has served as a proof of their structure[13a], and also as a means of preparing cyclic sultones[19] which are difficult to synthesize by direct methods.

B. 1,3-Oxathianes

1. General remarks

The 1,3-oxathiane system (85), as structurally intermediate between 1,3-dioxane (86) and 1,3-dithiane (87), offers an interesting opportunity to compare

(85) X = O, Y = S
(86) X = Y = O
(87) X = Y = S

conformational and other structural properties and to test the degree of additivity of such effects. A strict parallelism in the features of these three analogues is not expected, since structural parameters such as bond lengths and angles may undergo different compromizing alterations in each case to optimize the ring geometry.

2. Preparation

Most alkyl-substituted 1,3-oxathianes (88) can be synthesized by conventional acid-catalysed condensation of a suitable mercaptoalkanol and a carbonyl compound or its acetal in the case of sterically constrained molecules (equation 22)[111-119].

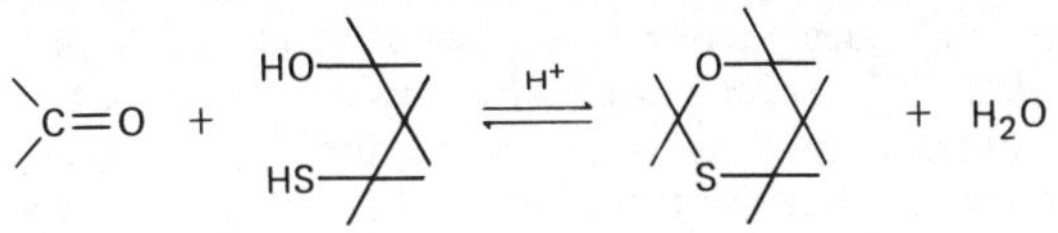

$$\text{C=O} + \begin{array}{c}\text{HO} \\ \text{HS}\end{array} \xrightleftharpoons{\text{H}^+} \begin{array}{c}\text{O} \\ \text{S}\end{array} + H_2O \qquad (22)$$

(88)

Certain derivatives are obtained in 90% yields by the ring-closure of mixed diesters (equation 23)[119,120]. Due to the stereospecificity of the whole reaction sequence from 1,3-diol (89) via the dimesylate (90) and the 2-mesyloxy-4-thiolacetoxypentane (91), optically active forms of 1,3-oxathianes (92) may be obtained if enantiomers of 89 are available in reasonable purity[120]. While the 1,3-dioxane (93) is not formed from 90, the disubstitution product (94) leads to 1,3-dithiane (95),

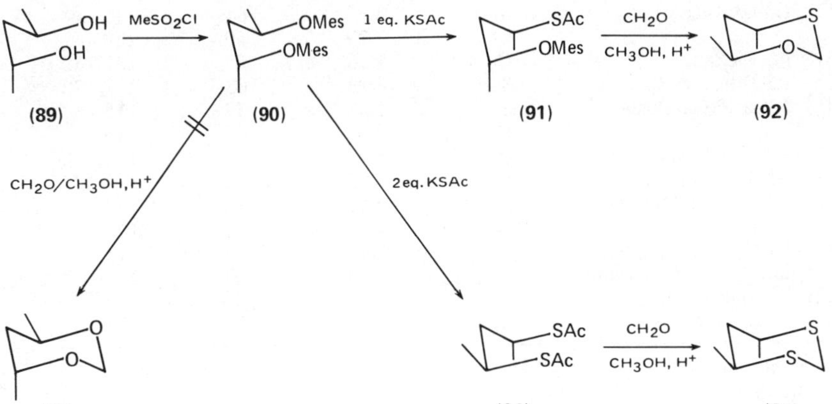

(89) $\xrightarrow{\text{MeSO}_2\text{Cl}}$ (90) $\xrightarrow{\text{1 eq. KSAc}}$ (91) $\xrightarrow[\text{CH}_3\text{OH, H}^+]{\text{CH}_2\text{O}}$ (92)

(89) —OH —OH   (90) —OMes —OMes   (91) —SAc —OMes   (92)

$\xrightarrow{\text{CH}_2\text{O/CH}_3\text{OH,H}^+}$ 2 eq. KSAc   (23)

(93)   (94) —SAc —SAc $\xrightarrow[\text{CH}_3\text{OH, H}^+]{\text{CH}_2\text{O}}$ (95)

with the same configuration as **92**, indicating that the mechanism of the ring-closure does not involve hydrolysis of the OMes group, but rather that of the SAc function, followed by internal $S_N 2$ displacement of the sulphonate grouping[120].

Some synthetic utility may be derived also from the reaction of **85** with *s*-butyl-lithium to give 1,3-oxathianyl-2-lithium, which on subsequent treatment with alkyl halides yields a variety of 2-alkylated 1,3-oxathianes[121]. Similarly, the acid-catalysed equilibration of **85** with 2-*R*-substituted 1,3-dioxanes leads to 2-*R*-1,3-oxathianes[122,123] in 79–95% yields.

3. Structure

There are no exact studies on the ring geometry of **85**, but ^{1}H-NMR data[111, 114-116,118,119,124-135] for variously substituted derivates show that its fundamental conformation is a chair form with some special features due to the coexistence of oxygen and sulphur atoms in the same ring.

Although Gelan and Anteunis[128] tried to construct two deformed models for the chair form of **85**, this has later been shown to be a misinterpretation of the dissymmetric character of the 1,3-oxathiane ring itself[115,130-132,136].

Table 3 presents a collection of Buys–Lambert *R*-values [$R = {}^3J_{trans}/{}^3J_{cis}$][130-133] and torsional angles [$\cos^2 \psi = 3/(2 + 4R)$][130-133] determined recently for some 1,3-oxathianes as well as those for certain 1,3-dioxanes and 1,3-dithianes. These values clearly demonstrate that oxygen-containing rings have an inherent tendency to flatten the $C_{(4)}-C_{(5)}-C_{(6)}$ moiety (in **86** and **97** $\psi_{4,5} = \psi_{5,6} = 55°$), while their sulphur analogues favour a somewhat puckered shape (in **87** and **98** $\psi_{4,5} = \psi_{5,6} = 63°$)[131-133]. Interestingly, **85** can still adopt a normally staggered arrangement, and contrary to a previous conclusion[126] the type of substitution does not seem to engender any profound effect ($\psi_{4,5} = \psi_{5,6} \sim 60°$ in **85, 96, 100** and **101**), with the exception of derivatives with severe steric crowding in their

TABLE 3. *R*-values and torsional angles (ψ) in some 1,3-oxathianes, 1,3-dioxanes and 1,3-dithianes

	Compound	Side	R	ψ	Reference
(85)	1,3-Oxathiane	S	2.97	61	130
		O	2.29	59	130
(86)	1,3-Dioxane	O	1.76	55	131,132
(87)	1,3-Dithiane	S	3.23	63	131,132
(96)	2-Me-1,3-Oxathiane	S	2.38	59	115
		O	2.44	60	115
(97)	2-R-1,3-Dioxane*a*	O	1.81	55	132, 133
(98)	2-R-1,3-Dithiane*b*	S	3.23	63	132,133
(99)	2,2-Me$_2$-1,3-Oxathiane	S	2.47	60	130
		O	1.94	56	130
(100)	4,4-Me$_2$-1,3-Oxathiane	O	2.40	59	115
(101)	6,6-Me$_2$-1,3-Oxathiane	S	2.50	60	115
(102)	2,2-*trans*-4,6-Me$_4$-1,3-Oxathiane	O,S	1.65*c*	<54*c*	115

*a*R = *p*-chlorophenyl[132] or *t*-butyl[133].
*b*R = phenyl[132].
*c*Average value for the $C_{(4)}-C_{(5)}-C_{(6)}$ moiety.

chair forms, e.g. 99 and 102[115]. In fact, the average torsional angle ($\sim 54°$) determined for 2,2-*trans*-4,6-tetramethyl-1,3-oxathiane (102) is characteristic of a 2,5-twist—boat (102b). In general, 1,3-oxathianes having various *syn*-axial 2,4- or 2,6-

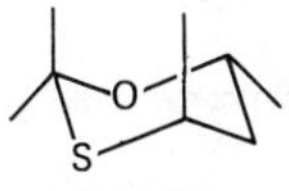

(102a) 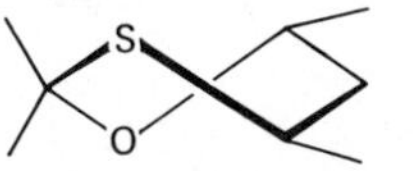   (102b)

methyl—methyl interactions in their chair conformations favour some of the three conceivable (1,4-, 3,6-, or 2,5-) twist forms[114,115,137,138]. The basic geometry of the 1,3-oxathiane ring is also consistent with aromatic solvent-induced ^{1}H-NMR shifts of the ring protons[60] (see Section III.B).

As to the quantitative evaluation of different stereochemical preferences, most information apart from ^{1}H-NMR data has been provided by chemical equilibration[114,116,138-140] of proper epimeric 1,3-oxathianes. For instance, the chair—twist energy parameters for 85 were estimated[138] and recently recalculated[115] from equilibrium data for *r*-2-*t*-butyl-2,*cis*-6- (103), and *r*-2-*t*-butyl-2,*trans*-6-dimethyl-1,3-oxathianes (104) by making some relevant assumptions about the

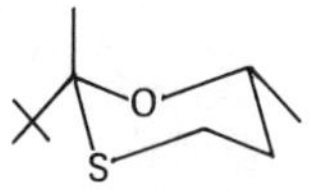

(103) 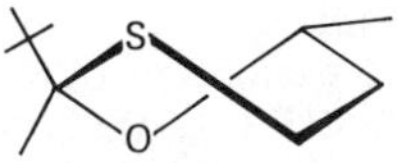   (104)

plausibility of contributing twist forms[115,137]. The values thus obtained for the chair—2,5-twist equilibrium (equation 24) are roughly intermediate to those evaluated for 86 and 87[1,129] (Table 4), and indicate a fair additivity of the opposite

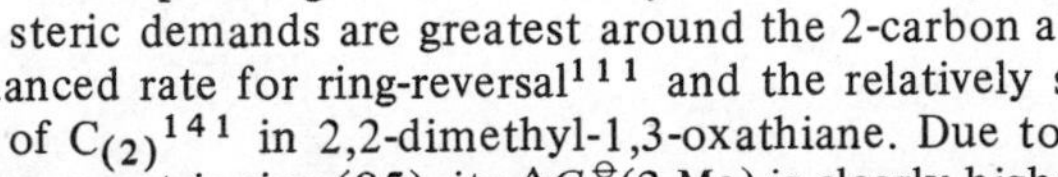

$$\text{(24)}$$

Chair   2,5-Twist

trends. Conformational energies of methyl groups at different positions of the 1,3-oxathiane ring derived from equilibration data[138-140] are presented in Table 5 together with the corresponding values for the symmetric analogues 86 and 87[1,129].

As expected, steric demands are greatest around the 2-carbon atom as suggested also by the enhanced rate for ring-reversal[111] and the relatively short spin-lattice relaxation time of $C_{(2)}$[141] in 2,2-dimethyl-1,3-oxathiane. Due to the constrained nature of the dissymmetric ring (85), its $\Delta G^{\ominus}$(2-Me) is clearly higher than the mean value of the same interactions in 86 and 87[111,129-139]. Positions 6 in 85 and 86 are energetically comparable whereas $\Delta G^{\ominus}$(4-Me) is enhanced in going from 87 to

TABLE 4. Chair—twist free energy, enthalpy and entropy differences for 1,3-oxathiane, 1,3-dioxane and 1,3-dithiane

Compound	$\Delta G^{\ominus}_{CT}$ (kJ mol^{-1})	$\Delta H^{\ominus}_{CT}$ (kJ mol^{-1})	$\Delta S^{\ominus}_{CT}$ (J mol^{-1} K^{-1})	Reference
(85) 1,3-Oxathiane	23.5	27.0	11.6	115,138
(86) 1,3-Dioxane	33.5	35.8	9.1	1,129
(87) 1,3-Dithiane	11.0	16.7	19.0	1,129

TABLE 5. Conformational preferences of the methyl groups at different positions in 1,3-oxathianes, 1,3-dioxanes and 1,3-dithianes

Parent compound	Conformational energy (kJ mol^{-1}) for different locations				References
	2a	4a	5a	6a	
(85) 1,3-Oxathiane	13.6	7.4	2.9–3.7	12.3	115,116,139,140
(86) 1,3-Dioxane	16.7	12.2	ca. 4	12.2	1,129
(87) 1,3-Dithiane	8.0	6.5	4.9	6.5	1,129

85 (6.5 vs. 7.4 kJ mol^{-1}) at the expense of some loss in torsional strain [111,115,129]. The relatively low ΔG° (5-Me) of 85 is, apart from the small steric requirements of the heteroatoms[111,129], a manifestation of the fundamental deformation of the ring, whereas the slightly higher estimate for 87 is attributed to the enhanced puckering of its $C_{(4)}-C_{(5)}-C_{(6)}$ region and/or to the large van der Waals' radius of sulphur as compared to oxygen[111,114,129].

From ^{1}H-NMR[126,134,135] and equilibration[126,135,138–140] studies it appears that the acceptance of the additivity principle is justified or at least of value in the semiquantitative evaluation of steric effects in simple 1,3-oxathianes. For instance, trans-4,6-dimethyl-1,3-oxathiane (105) involves the 4- (105a) and 6-axial (105b) conformations in a ratio of 87 : 13 as concluded from the vicinal [1]H–[1]H coupling constants[126,134], in fair agreement with the energy difference obtained directly from the respective interactions in 1,3-dioxanes and 1,3-dithianes (12.2−6.5 = 5.7 kJ mol^{-1}) (Table 5).

Later on, however, the above result was argued in a study based on a chemical-shift method[119], which led to controversial thermodynamic parameters. The results of a chemical equilibration of suitable anancomeric model compounds, r-2-cis-4-trans-6- (106) and r-2-trans-4-cis-6-trimethyl-1,3-oxathianes (107) at various temperatures[139,140] firmly confirmed the original estimates[126,134] and made the chemical-shift method questionable.

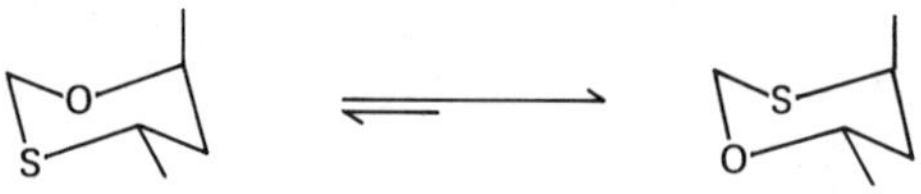

(105b)   (105a)

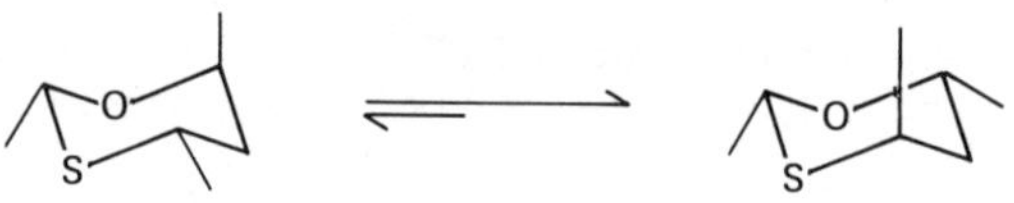

(106)   (107)

Additional structural knowledge about the title compounds comes from electron impact mass spectrometric studies[142–145]. The main features in the positive-ion mass spectra of 1,3-oxathianes[142] are the relatively high intensity of molecular and large fragment ions, the abundance of metastable transitions and the preferential charge retention on sulphur-containing fragments over the oxygen analogues, probably due to the ability of sulphur to stabilize the electron deficiency with the aid of its d-shell electrons. The course of fragmentation depends somewhat on the substitution pattern but only two principal modes of ring cleavage (I and II) are found[142], which is different from the behaviour of 1,3-dioxanes but comparable rather to that of 1,3-oxathiolanes (see Section III.B.3)[62-64].

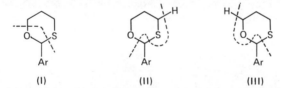

(I)   (II)

Measurement of the ionization and appearance potentials for a series of stereo-isomeric 1,3-oxathianes has yielded information about their conformational energies in the gas phase[143]. According to the principles derived originally by Pihlaja and Jalonen[146] it was found that in the formation of the M^+ or $[M-Me]^+$ ions the nonbonding interactions are mainly released, so that differences in the ground-state enthalpies of isomeric structures can be evaluated from equations (25) and (26)[143,146,147], where AP is the appearance potential of the primary fragment

$$IP([M]^+) - IP([M_1]^+) = \Delta H_f^{\ominus}(M_1) - \Delta H_f^{\ominus}(M) \qquad (25)$$

$$AP([M-R]^+) - AP([M_1-R]^+) = \Delta H_f^{\ominus}(M_1) - \Delta H_f^{\ominus}(M) \qquad (26)$$

ion, IP the ionization potential and $\Delta H_f^{\ominus}$ the standard enthalpy of formation of the compound in question. The most interesting point was the observation[143] that $\Delta H_{CT}^{\ominus}$ for the 1,3-oxathiane family in the gaseous state (25 kJ mol^{-1}) is not far from the result obtained by chemical equilibration (ca. 27 kJ mol^{-1})[115,138]. Also the values of other conformational energies from appearance and/or ionization potentials are in fair agreement with the liquid-phase values[115,139].

Bowie and Ho[144] studied negative-ion mass spectra of 2-aryl-1,3-oxathianes (2-aryl = o-, m- or p-nitrophenyl). The spectra were characterized by intense molecular anions and large fragment ions produced by simple (I–III) or complex modes of cleavage. With the aid of deuterated derivatives the authors[144] were able

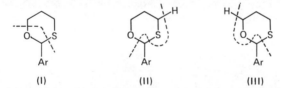

Ar   Ar   Ar
(I)   (II)   (III)

to show that the extent of hydrogen randomization between the 2-, 4- and 6-positions depends in a specific way upon the isomer in question, the behaviour of which parallels that noted for corresponding 1,3-dithianes[145] but is in marked contrast to isomeric 2-nitrophenyl-1,3-dioxanes[144] which display mutually very similar spectra and exhibit no hydrogen scrambling.

4. Reactions

Pihlaja and coworkers[148] determined the relative rates for the acid-catalysed hydrolysis of 1,3-oxathiane (85), 2-methyl-1,3-oxathiane (108) and 2,2-dimethyl-1,3-oxathiane (109) and found that the acceleration effect for 108 is exceptionally low in comparison with 1,3-oxathiolanes[65]. A possible explanation is the acidic character of the protons at position 2 in 85 and 108, but the exact mechanism for the hydrolytic decomposition of **85, 108** and related molecules is not clear and requires further study[149].

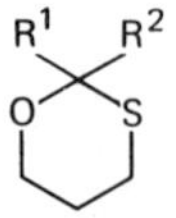

	(85) $R^1 = R^2 = H$
	(108) $R^1 = H, R^2 = Me$
	(109) $R^1 = R^2 = Me$

Eliel and his colleagues[150] described an asymmetric synthesis of (*S*)-(+)-atrolactic acid methyl ether **(110)** proceeding either from (*S*)-(−)-4,6,6-trimethyl-1,3-oxathiane **(111**, R = H) with about 100% optical yield, or from (*R*)-(+)-4,4,6-trimethyl-1,3-oxathiane **(112**, R = H) with ca. 92% optical yield. The reaction sequence

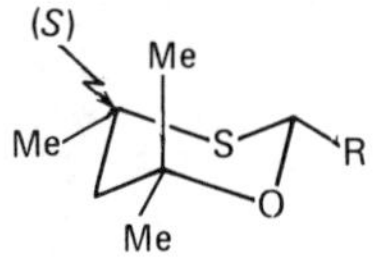

(111)	**(110)**	**(112)**

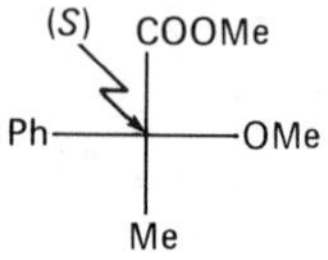

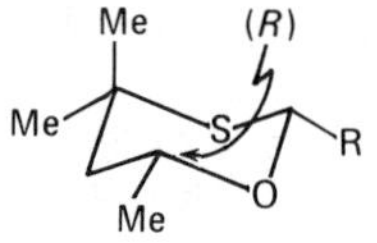

involves a stereoselective electrophilic attack on a biased 2-lithio-1,3-oxathiane leading exclusively to equatorial substitution, where the original chirality at $C_{(4)}$ or $C_{(6)}$ is transferred to $C_{(2)}$, and an asymmetric reaction of a Grignard reagent to yield **111** (R = (*S*)-C(OH)MePh) with an exocyclic asymmetric centre which after methylation, ring-cleavage and oxidation produces **110**[150].

For the reduction of cyclic monothio-acetals and -ketals, see Section III.B.4.

C. 3-Oxo-1,3-oxathianes

Only a few reports[151-153] have appeared on the properties of the title compounds, although they offer an easily preparable model to study the often unexpected interactions between polar functions. For instance, **113** was obtained in high yield (94%) by treatment of 1,3-oxathiane (**85**, see Section IV.B.2) with sodium metaperiodate in water—methanol solution[151].

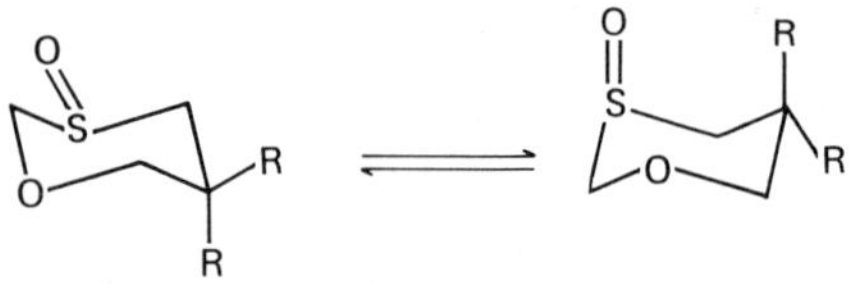

(a)	(b)

(113) R = H
(114) R = Me

Conformational preferences in **113**[152,153] and **114**[153] were examined by ^{1}H-NMR spectra. Interestingly, at ambient temperatures **113a** and **113b** are nearly equally populated, whereas at −95°C **113b** is reported to predominate in a ratio of 8 : 1[152] which is approximately in agreement with the result of 84 : 16 at −98°C $(-\Delta G^{\ominus} = 2.4$ kJ mol^{-1})[153]. In **114** with *gem*-dimethyl grouping at the 5-position, the proportion of **114b** is drastically lowered (**114b/114a** ~ 1 : 9 and $-\Delta G^{\ominus}$ ~ 3.0 kJ mol^{-1})[153]. Consequently, the disfavouring effect caused by *syn*-axial S=O and methyl groups would amount to 5.4 kJ mol^{-1}.

These results are pronouncedly different from those observed for thiane-1-oxides ($-\Delta G^{\circ}$ = 0.73 kJ mol^{-1} in favour of the axial S=O form)[154], and for 1,3-dithiane-3-oxides (equatorial preference of S=O at and below ambient temperatures)[152,153], emphasizing the difficulties in evaluating interactions between polar groups and lone-pair electrons.

D. 1,4-Oxathianes

1. Preparation

1,4-Oxathianes have been synthesized with a variety of different methods of which the most recent ones will be described in the following. Karabinos and Hazdra[155] obtained 1,4-oxathiane (116) and 1,4-dithiane (117) in a 7 : 1 ratio from the cyclization of thiodiethyleneglycol (115) upon treatment with PF$_5$ (equation 27).

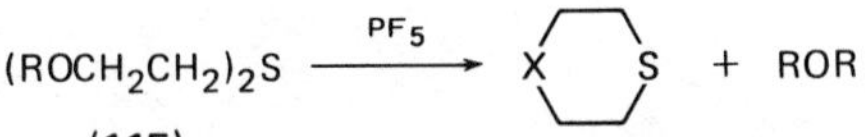

$$(ROCH_2CH_2)_2S \xrightarrow{PF_5} X \underset{}{\overset{}{\bigcirc}} S + ROR \tag{27}$$

(115)

(116) X = O

(117) X = S

Black prepared 2-methyl-1,4-oxathiane (118) from 1-(2-hydroxyethylthio)-2-propanol by dehydration with orthophosphoric acid and some 2-oxo-1,4-oxathianes (119) from the reactions of thioglycolic acid with oxiranes[156].

(118)

(119) R = H, Me, Ph

3-Chloro-1,4-oxathiane (120) has been prepared by chlorinating the parent compound with N-chlorosuccinimide[157] or with Cl$_2$ in CCl$_4$ at ca. 260 K[158]. In a reaction with RNa 120 gave different 3-substituted 1,4-oxathianes (equation 28)[158].

$$S \overset{}{\bigcirc} O + RNa \longrightarrow S \overset{}{\bigcirc} O + NaCl \tag{28}$$

(120)

Evans and Mason[159] used a modification of the Haubein method[160] to synthesize 2,2,3- (121) and 2,3,3-trichloro-1,4-oxathianes (122), both of which were shown with the aid of the hydrolysis products to be substituted on the same side. The structure of 121 was confirmed by desulphurization (equation 29) which

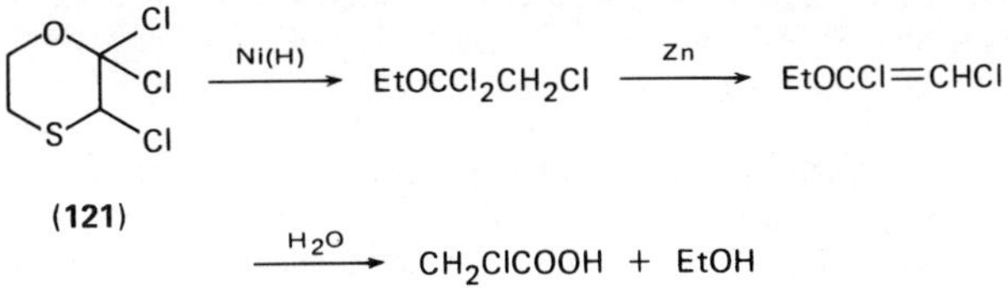

$$\text{(121)} \xrightarrow{Ni(H)} EtOCCl_2CH_2Cl \xrightarrow{Zn} EtOCCl=CHCl$$

$$\xrightarrow{H_2O} CH_2ClCOOH + EtOH \tag{29}$$

(121)

resulted in the formation of monochloroacetic acid and ethanol. Similarly, **122** gave dichloroacetaldehyde and ethanol (equation 30). The formation of **122** is con-

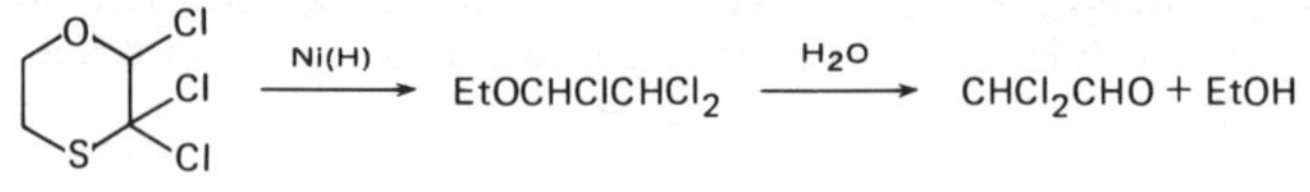

(122)

$$\text{(122)} \xrightarrow{\text{Ni(H)}} \text{EtOCHClCHCl}_2 \xrightarrow{\text{H}_2\text{O}} \text{CHCl}_2\text{CHO} + \text{EtOH} \qquad (30)$$

sistent with the chlorination and thermal dehydrohalogenation of diethyl ether[161] whereas for that of **121** at 353 K an entirely different mechanism must be postulated (equation 31)[159,162].

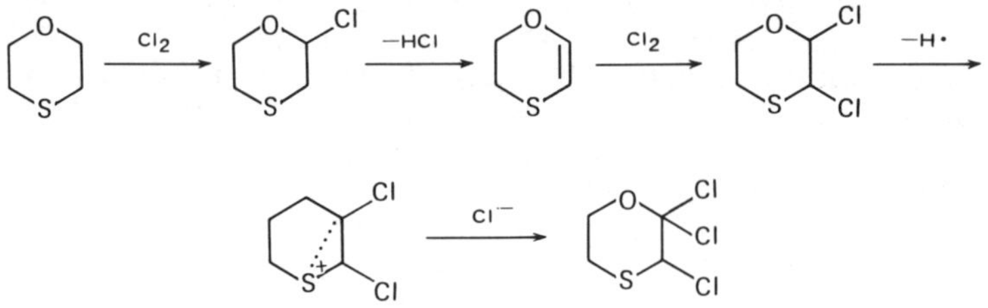

(31)

(121)

Hydrogenolysis of the acetal function of 2,8-dioxa-6-thiabicyclo[3.2.1]octanes also affords 1,4-oxathianes[163]. Reaction of $(\text{CH}_2\text{=CCH}_3\text{CH}_2)_2\text{O}$ with SCl_2 gives 3,3-dichloromethyl-3,3-dimethyl-1,4-oxathiane **(123)**[164] and hydrolysis of $(\text{ROCHClCH}_2)_2$ S yields 2,6-dialkoxy-1,4-oxathianes **(124)**[165].

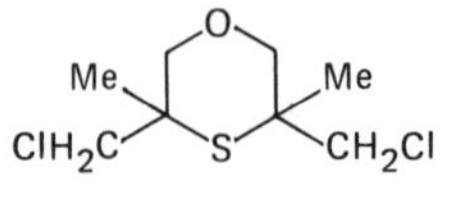

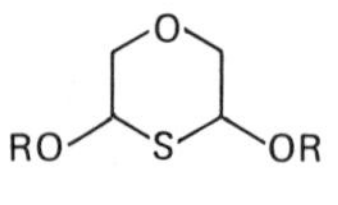

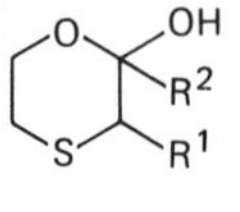

(123)   **(124)**   **(126)**

Blagoveschchenskii and colleagues[166] prepared several 2- and 3-substituted 1,4-oxathianes by treatment of 1,4-oxathiene and 3-chloro-1,4-oxathiane with RH $(\text{R} = \text{Me}_3\text{CO}, \text{PrS}, \text{Me}_3\text{CS}, \text{Me}_2\text{EtCS}, \text{PhCH}_2\text{S}, \text{PhS}, (\text{MeO})_2\text{P(S)S}$ and $(\text{EtO})_2\text{P(S)S})$, respectively. Acetamido-substituted 1,4-oxathianes **(125)** can be obtained through the base- or acid-catalysed intramolecular cyclization of S-hydroxyalkylated 2-acetamidopropenethiolates (equation 32)[167].

(125)

1,4-Oxathianes have also been prepared by mercuric salt ring-closure from diallyl sulphide[168] and by cyclization upon electrochemical fluorination of **115** (R = H)[169]. The reaction of 2-mercaptoethanol with $\text{R}^1\text{CHXCOR}^2$ (X = halogen) in C_6H_6 containing KOAc gives substituted 1,4-oxathianes **(126)** in good yields[170].

Some 2-oxo-1,4-oxathianes have been prepared by heating mercaptoacetic acid with oxirane or substituted oxiranes[171] as well as by intramolecular dehydration of $\text{HOCH}_2\text{CHRSCH}_2\text{CO}_2\text{H(R} = \text{H, Me)}$[172]. These methods have been used to synthesize the corresponding seven-membered compounds, 2-oxo-1,4-oxathiepanes (see Section V.A.).

2. Structure

The conformation and structural characteristics of 1,4-oxathiane and many of its derivatives[155,173,174] have been extensively studied. Jensen and Neese[175] determined the activation parameters for the ring-reversal process (chair to twist) of 1,4-oxathiane and found them to be $\Delta H^{\neq}$ 37 ± 3 kJ mol^{-1}, $\Delta S^{\neq}$ 2 ± 1 J mol^{-1} K^{-1} and $\Delta G^{\neq}$ 36 ± 1 kJ mol^{-1}. The free energies of activation measured for the ring-reversal of 2-oxo-1,4-oxathiane and its 6-methyl, 5-phenyl and 6-phenyl derivatives are 41, 73, 79 and 79 kJ mol^{-1}, respectively[176].

The microwave[177], IR and Raman[178], and electron diffraction[179,180] results for 1,4-oxathiane are all in accord with a chair form. The crystal structure of *trans*-2,3-dichloro-1,4-oxathiane (127) shows that the molecule has a chair conformation with the chlorine atoms in axial positions[181]. The overall geometry of 127 is half-way between the conformations of the corresponding *trans*-2,3-dichloro derivatives of 1,4-dioxane (128) and 1,4-dithiane (129)[182–185]. The torsional angle in 127 is 60° from the values of the vicinal ^{1}H-coupling constants using the Buys–Lambert approach[132,186] in good agreement with the diffraction results[181]. 128 is somewhat less puckered since its torsion angle is only 57°.

Crossley and coworkers tried to apply an improved microwave procedure to the detailed conformational study of 1,4-diheterocyclohexanes (116, 117 and 130) but with a relatively small amount of new information[187]. In a number of papers Zefirov and colleagues[188,189] have studied the conformational properties of 2-substituted 1,4-oxathianes and heteroanalogues of bicyclo[3.3.1]nonane[189,190]. The results of these investigations have been already reviewed[2].

(127)	X = O, Y = S	(116)	X = O, Y = S
(128)	X = O, Y = O	(117)	X = S, Y = S
(129)	X = S, Y = S	(130)	X = O, Y = O

Burdon and Parsons synthesized different highly fluorinated 1,4-oxathianes[191] and deduced their structures from the ^{19}F-NMR spectra by a chemical-shift parameter scheme[192]. The majority of the compounds exist in chair conformations, with a strong anomeric effect or its equivalent operating both α to oxygen and α to sulphur[192]. A comparison with a similar set of polyfluorinated 1,4-dioxanes has also been made. Phillips and Wray[193] evaluated an additive method of calculating $^2J_{HF}$ in polyfluoro-1,4-dioxanes and -oxathianes and stated that the approach may be useful in stereochemical and conformational studies of related molecules.

Szarek and colleagues[194] studied the ^{13}C-NMR spectra of a number of 1,4-oxathianes including 4-oxo and 4,4-dioxo derivatives and applied the results to carrying out a structural differentiation of the two nucleotides 131 and 132.

Condé-Caprace and Collin[195] discussed the various modes of fragmentation of 116 and 117 and found that they are qualitatively very similar but differ considerably from those of 1,4-dioxane (130)[196] (see also Sections III.B.3 and IV.B.3).

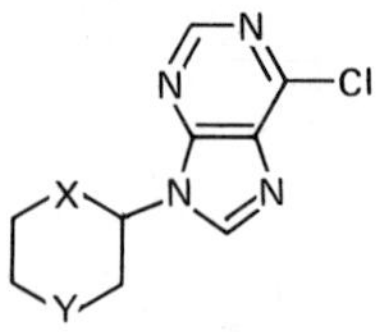

(131) X = O, Y = S
(132) X = S, Y = O

Obviously the influence of the sulphur atom predominates in the case of oxathiacyclanes[62-64,142,195].

Sweigart and Turner[197] studied the photoelectron spectra and lone-pair ionization potentials in some oxygen and sulphur heterocycles including 116 and interpreted the lone-pair interactions in terms of through-space and through-bond mechanisms; the latter is favoured in 116, 117 and 130, whereas 1,3-diheterocyclanes prefer the former.

3. Reactions and 4-oxo-1,4-oxathiane

Havinga and coworkers[198] studied the chlorination of 116 under different conditions and prepared 3-chloro- (133), trans-2,3-dichloro- (127), 3,3-dichloro- (134), 2,3,3-trichloro- (121), cis-3,5-dichloro- (135) and trans-2,3,3,5-tetrachloro-1,4-oxathianes (136). The substitution takes place preferentially at $C_{(3)}$ and, up to three chlorine atoms, in the same half of the ring. The use of peroxides favours substitution at $C_{(2)}$.

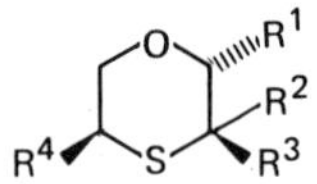

	R^1	R^2	R^3	R^4
(133)	H	H	Cl	H
(134)	H	Cl	Cl	H
(135)	H	H	Cl	Cl
(136)	Cl	Cl	Cl	Cl

The oxidation of 116 by H_2O_2 in several solvents and mixed-solvent systems and the influence of solvent on the mechanistically related t-butyl hydroperoxide oxidation have also been studied[199]. Foster and colleagues[200,201] treated 137 with $NaIO_4$ and obtained a 10 : 1 mixture of the axial (138) and equatorial (139)

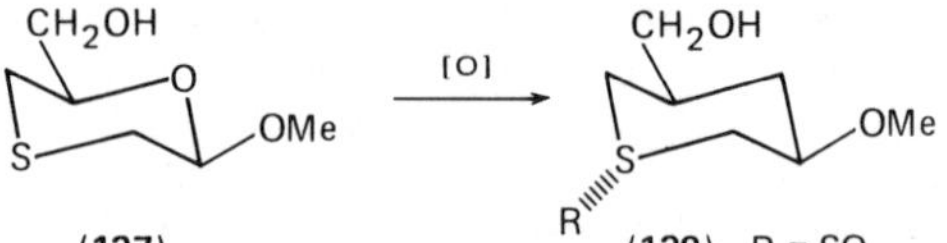

(137)

(138) R = SO_{ax}
(139) R = SO_{eq}

sulphoxides. With O_3 the sulphoxides were obtained in about equimolar yields. In general, the control of S-oxidation was best achieved by variation of the configuration at the anomeric centre, $C_{(2)}$; an axial substituent engenders equatorial S-oxygenation whereas an equatorial substituent leads to an axial S-oxide.

Foster and coworkers[202] also used ¹H-NMR spectra to assign sulphoxide configuration using the significantly different shielding effects of axial and equatorial S=O groups[203]. A crystal structure determination[204] established that the major sulphoxide obtained by NaIO$_4$ oxidation of *trans*-2-methoxy-6-hydroxymethyl-1,4-oxathiane (140) has the *trans*-4-methoxy, *cis*-6-hydroxymethyl, *r*-4-oxo configuration (141) in agreement with ^{1}H-NMR results[205].

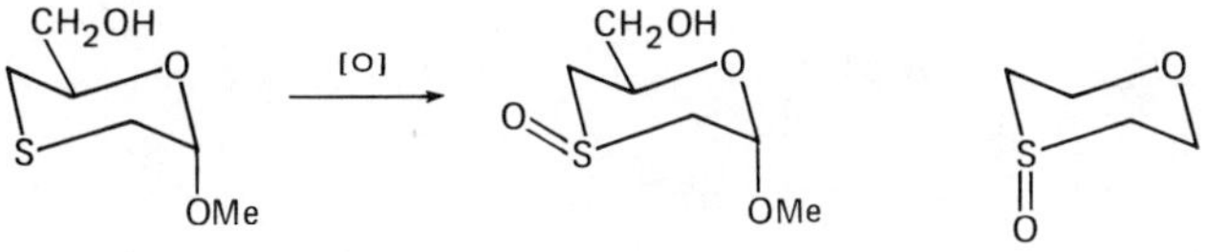

(140)   (141)   (142)

^{13}C-NMR data[203,206] for 4-oxo-1,4-oxathiane (142) indicate that the sulphinyl oxygen prefers the axial position by 2.8 kJ mol^{-1} at 205 K. This is at least in qualitative agreement with the report[207] which on the basis of IR and Raman spectra (solid and liquid samples) considers the oxide to have the C_s chair-axial conformation (142).

1,4-Oxathiane (116), like 1,4-dioxane (130), easily forms complexes with iodine[208], ZnMe$_2$[209] and many metal halogenides[210-213], but this topic will not be considered here.

E. 1,3,5-Oxadithianes and -Dioxathianes

The title triheterocyclohexanes are not well characterized. The reaction of saturated aldehydes with gaseous H$_2$S has been reported[214] to give 143—145. Some 4-alkoxy-4-alkyl derivatives of 144 can be obtained in 40—50% yield by treating

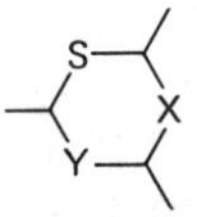

(143) X = O, Y = O
(144) X = S, Y = O
(145) X = S, Y = S

sodium oxydimethylenedithiosulphate with CH$_3$COOR in absolute propanol in the presence of HCl for 7—8 h[215]. Dipole moment and ^{1}H-NMR studies show that at least the most stable isomer of 143 exists in a chair conformation with three equatorial substituents[216].

Oxidation of 144 with perhydrol for 2.5 h at 333—338 K gives the 3,3,5,5-tetroxide in nearly quantitative yield[215]. The H—D exchange at $C_{(2)}$ of 145 accelerates when $S_{(5)}$ is converted to a sulphinyl or sulphonyl group. The remote participation of the sulphur atom is also seen in the slow H—D exchange at $C_{(6)}$ of 143[217].

F. 1,3,2-Dioxathianes

Very little attention has been paid to 1,3,2-dioxathianes, although their 2-oxides have been widely studied (see Section IV.G). Since the early attempts[103] [see also Section III.E] to prepare 1,3,2-dioxathiane (146) and some of its alkyl derivatives Wood and coworkers[218,219] have reported the synthesis of a number of methyl-substituted 1,3,2-dioxathianes. The barrier ($\Delta G^{\neq}$) to the ring-reversal of 146 is somewhat higher than that for cyclohexane or 1,3-dioxane but lower than that for 1,2,3-trithiane[219]. ^{13}C-chemical shifts for *trans*-4,6-dimethyl-1,3,2-dioxathiane (147) have also been reported[110]. The above results are still in some doubt since one of the present authors[220] has not been able to repeat the preparation of the materials.

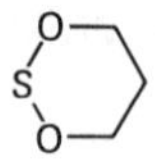

(146) 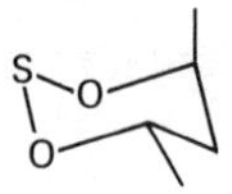   (147)

	δ (p.p.m.)
$C_{(4,6)}$	74.8
$C_{(5)}$	39.4
$\underline{C}H_{3(4,6)}$	19.7

G. 2-Oxo-1,3,2-dioxathianes

These compounds are cyclic sulphites and can be easily prepared from $SOCl_2$ and 1,3-diols[94,221-224]. Despite the fact that during the last 10−15 years some 40 papers have been published on the structure of **148** and its alkyl and halo derivatives a considerable extent of controversy has been left in the detailed explanation of the results. Very recently it was pointed out that both ^{1}H-NMR spectra and dipole moments can be interpreted consistently only if **148** and its alkyl derivatives greatly prefer chair forms (usually with an axial S=O group)[224]. In the same context the chair form of **146** is estimated to be ca. 31 kJ mol^{-1} more stable than the twist form[224]. Two other recent reports[225,226] confirm the conclusions made by Pihlaja and coworkers[224] as to the high preference of the chair form and withdraw, together with the latter and some other consistent publications[131,227-241], the significance of the discussion based on the existence of simple alkyl-substituted derivatives in the twist form[94,109,110,222,242-253]. In a forthcoming report[230] a correct assignment of the IR bands in the 1180−1250 cm^{-1} region also disproves the necessity of twist forms in contrast to opposite claims[247,252,253]. The only substituted 2-oxo-1,3,2-dioxathiane which has been proved to attain a twist conformation is trans-5-chloro, cis-4, trans-6-di-t-butyl, r-2-oxo-1,3,2-dioxathiane (**149**)[227]. A complete discussion as to the detailed structure and properties of **148** and its derivatives will be published in a separate review[230] and in some future reports[228-230].

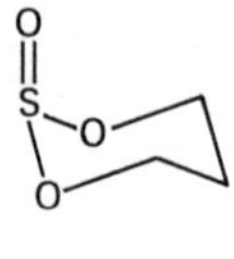

(148) 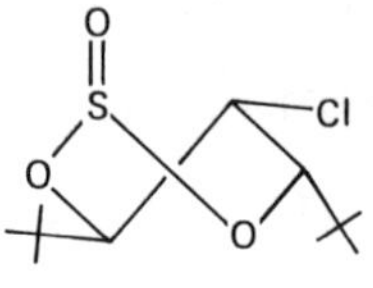   (149)

V. SEVEN-MEMBERED AND LARGER RINGS

A. 1,4-Oxathiepanes

Acetamido substituted 1,4-oxathiepane (**150**) can be prepared via acid- (or base-) catalysed intramolecular cyclization of the Z-isomer of S-hydroxyalkylated 2-acet-amidopropenethiol (**151**) (equation 33)[254]. The structure of **150** was stated to be confirmed by conventional means but no data were reported[254]. Attempts to synthesize eight-membered rings by lengthening the hydroxyalkyl chain failed[254].

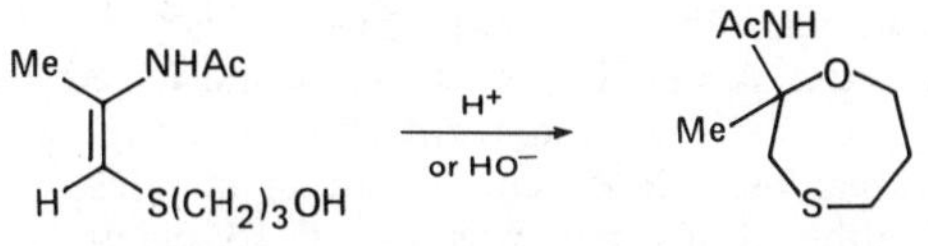

$$
\text{(33)}
$$

(151)   (150)

Preparations of 2-oxo-1,4-oxathiepane (152) (equation 34)[172], and 7-oxo-1,4-oxathiepane (153) (equation 35)[171] have also been reported. The latter synthesis utilizing the ring-cleavage of oxiranes by β-mercaptopropionic acid (154) leads to appreciable amounts of 155 as a by-product[171].

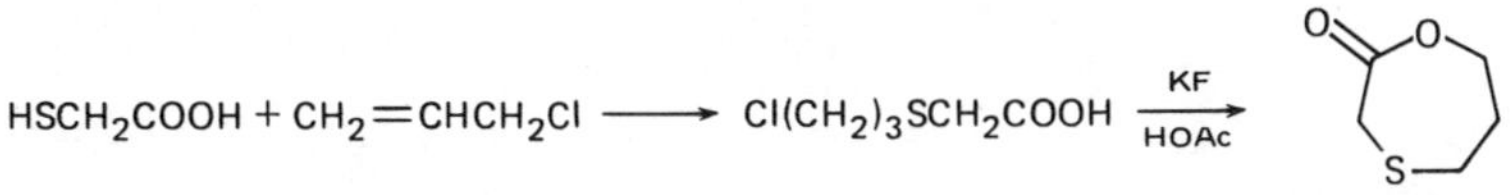

$$
HSCH_2COOH + CH_2{=}CHCH_2Cl \longrightarrow Cl(CH_2)_3SCH_2COOH \xrightarrow[\text{HOAc}]{\text{KF}}
$$

$$
\text{(34)}
$$

(152)

$$
HSCH_2CH_2COOH + \quad\longrightarrow\quad + HOCH_2CH_2SCH_2CH_2COOH \qquad \text{(35)}
$$

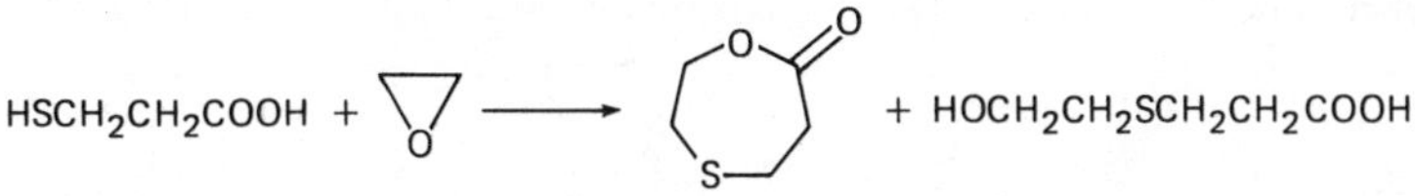

(154)   (153)   (155)

B. 1,4,5-Oxadithiepanes

1,4,5-Oxadithiepane (157) is obtained by treating 156 with Na_2S_4 at 358–363 K in aqueous solution in the presence of sodium alkylnaphthalenesulphonate, NaOH and $MgCl_2$ (equation 36)[255]. Cannizzaro reaction of 2,2′-dithiobis(2-methylpropa-

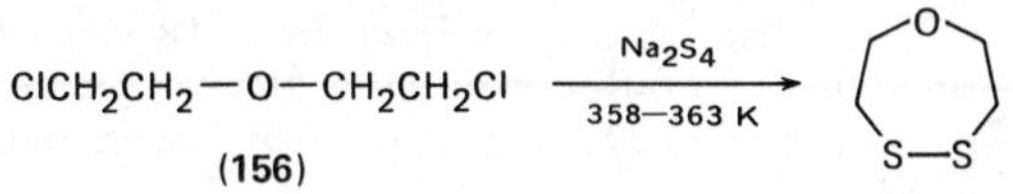

$$
ClCH_2CH_2{-}O{-}CH_2CH_2Cl \xrightarrow[\text{358–363 K}]{Na_2S_4}
$$

$$
\text{(36)}
$$

(156)   (157)

nal) in aqueous NaOH yields 158 which is readily cyclized in the presence of Ac_2O to 2-oxo-1,4,5-oxadithiepane (159) (equation 37)[256]. Both 157[255,257] and 159[256] are readily polymerized by alkoxide, alkylaluminium and metal hydride catalysis.

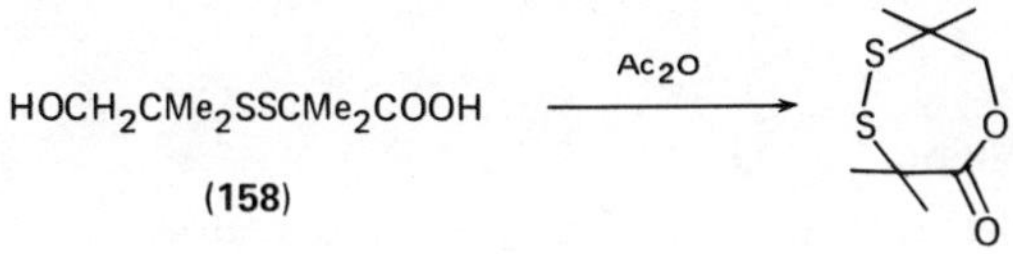

$$
HOCH_2CMe_2SSCMe_2COOH \xrightarrow{Ac_2O}
$$

$$
\text{(37)}
$$

(158)

(159)

Heats of polymerization for 157 in bulk, C_6H_6 and 1,4-dioxane solutions were measured by Dainton and coworkers[257] who suggested an anionic mechanism for iodine-catalysed reaction in which I^- is assumed to be the initiator.

C. 2-Oxo-1,3,2-dioxathiepanes

Seven-membered cyclic sulphites (160) can be similarly synthesized, though in lower yields than their six-membered homologues (see Section IV.G.), e.g. by treatment of the corresponding diol with $SOCl_2$[258-261]. Structural information about **160** is very limited. According to ^{1}H-NMR and IR measurements by Faucher and Guimaraes[262] the most favoured form at room temperature is a chair, but the detailed conformational behaviour remains an open question. Hydrolysis of **160** under acidic or alkaline conditions was found to occur by a bimolecular (A2) mechanism[259,263] which is also normal for lower homologues and acyclic sulphites[263].

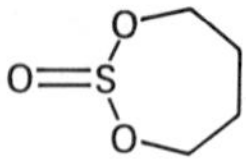

(160)

D. 1,3,6-Dioxathiocanes

1,3,6-Dioxathiocane (**161**) is a true acetal, and can be smoothly prepared via condensation of thiodiethylene glycol (**162**) and formaldehyde (equation 38)[264,265].

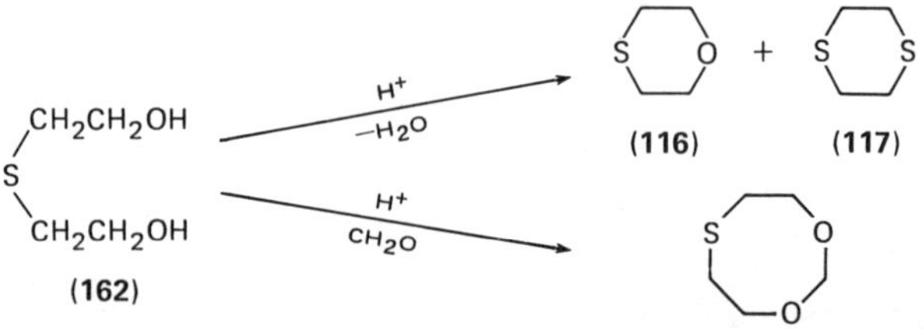

(38)

Direct cyclization of **162** gives 1,4-oxathiane (**116**), or after disproportionation, 1,4-dithiane (**117**)[264] (see also Section IV.D.1). In an IR study of **161** and related heterocycles Tarte and Laurent[266] discovered that the oxygen atom has little effect on the CH_2 deformation frequency whereas sulphur lowers that of adjacent CH_2 groups by ca. 40 cm^{-1}.

Mass spectrometric fragmentation of **161**[267] includes the loss of CH_2O and the formation of the 1,4-oxathiane molecular ion in the primary stage, and secondary transitions lead to the same fragment ions with similar relative abundances as observed for **116** which is a common mode for seven- and six-membered oxygen heterocycles[268].

E. Macrocyclic Rings

Several polyether sulphides containing 9–21 ring atoms have been prepared[269] by treatment of an oligoethylene glycol with a suitable dithiol or Na_2S in ethanol solution. The crystal structure of the molecules exhibits certain nonplanar regular arrangements of the ethylene 1,4-dithia, 1,4-oxathia and 1,4-dioxa fragments as evidenced by X-ray analysis. Also ^{1}H-NMR spectra recorded for some members

(163)

such as 1,4-dithia(12-crown-4) (163) are reported to be consistent with the assumed stereostructures[269].

VI. REFERENCES

1. K. Pihlaja, *Kemia-Kemi*, 1, 492 (1974).
2. W. L. F. Armarego, *Stereochemistry of Heterocyclic Compounds*, Part 2, John Wiley and Sons, New York, 1977, Chap. 4, pp. 314–325.
3. J. P. Snyder and L. Carlsen, *J. Amer. Chem. Soc.*, 99, 2931 (1977).
4. (a) T. Ueda and T. Shimanouchi, *J. Chem. Phys.*, 47, 4042 (1967).
   (b) J. Jokisaari and J. Kauppinen, *J. Chem. Phys.*, 59, 2260 (1973).
5. K. Pihlaja, J. Jokisaari, P. O. I. Virtanen, H. Ruotsalainen and M. Anteunis, *Org. Mag. Res.*, 7, 286 (1975).
6. D. O. Harris, H. W. Harrington, A. C. Luntz and W. D. Gwinn, *J. Chem. Phys.*, 44, 3467 (1966).
7. K. Karakida and K. Kuchitsu, *Bull. Chem. Soc. Japan*, 48, 1691 (1975).
8. T. R. Borgers and H. L. Strauss, *J. Chem. Phys.*, 45, 947 (1966).
9. F. Jung, N. Sharma and T. Durst, *J. Amer. Chem. Soc.*, 95, 3422 (1973).
10. T. Durst and B. P. Gimbarzensky, *Chem. Commun.*, 724 (1975).
11. L. Carlsen and J. P. Snyder, *Tetrahedron Letters*, 2045 (1977).
12. R. J. Boyd and M. A. Whitehead, *J. Chem. Soc., Dalton Trans.*, 73, 78, 81 (1972).
13. (a) D. N. Harpp and J. G. Gleason, *Tetrahedron Letters*, 1447 (1969).
    (b) D. N. Harpp. J. G. Gleason and D. K. Ash. *J. Org. Chem.*, 36, 322 (1971).
14. D. N. Harpp and J. G. Gleason, *J. Org. Chem.*, 36, 1314 (1971).
15. R. M. Dodson, P. D. Hammen and R. A. Davis, *J. Org. Chem.*, 36, 2693 (1971).
16. U. Schöllkopf, *Angew. Chem. (Intern. Ed. Engl.)*, 9, 763 (1970).
17. R. M. Dodson, P. D. Hammen and J. Yu Fan, *J. Org. Chem.*, 36, 2703 (1971).
18. N. K. Sharma, F. Jung and T. Durst, *Tetrahedron Letters*, 2863 (1973).
19. N. K. Sharma, F. de Reinach-Hirtzbach and T. Durst, *Can. J. Chem.*, 54, 3012 (1976).
20. O. Exner, D. N. Harpp and J. G. Gleason, *Can. J. Chem.*, 50, 548 (1972).
21. E. L. Eliel, N. L. Allinger, S. J. Angyal and G. A. Morrison, *Conformational Analysis*, Interscience, New York, 1965, p. 375.
22. A. A. Najam and J. G. Tillett, *J. Chem. Soc., Perkin 2*, 858 (1975).
23. S. Sarel, I. Levin and L. A. Pohoryles, *J. Chem. Soc.*, 3082 (1960).
24. J. G. Tillett and D. E. Wiggins, *J. Chem. Soc. (b)*, 1359 (1970).
25. H. K. Hall, M. K. Brandt and R. M. Mason, *J. Amer. Chem. Soc.*, 80, 6420 (1958).
26. H. K. Hall, *J. Org. Chem.*, 28, 2027 (1963).
27. D. N. Harpp, S. M. Vines, J. P. Montillier and T. H. Chan, *J. Org. Chem.*, 41, 3987 (1976).
28. K. Pihlaja, *Suomen Kemistilehti (B)*, 43, 143 (1970).
29. R. Keskinen, A. Nikkilä and K. Pihlaja, *Tetrahedron*, 28, 3943 (1972).
30. R. Keskinen, A. Nikkilä and F. G. Riddell, *J. Chem. Soc., Perkin 2*, 466 (1974).
31. R. Keskinen, A. Nikkilä and K. Pihlaja, *J. Chem. Soc., Perkin 2*, 343 (1977).
32. N. C. De and L. R. Fedor, *J. Amer. Chem. Soc.*, 90, 7266 (1968).
33. T. H. Fife and L. K. Jao, *J. Amer. Chem. Soc.*, 91, 4217 (1969).
34. E. L. Eliel and T. W. Doyle, *J. Org. Chem.*, 34, 2716 (1970).
35. J. W. Hartgerink, L. C. J. van der Laan, J. B. F. N. Engberts and Th. J. de Boer, *Tetrahedron*, 27, 4323 (1971).
36. E. L. Eliel, L. A. Pilato and V. G. Badding, *J. Amer. Chem. Soc.*, 84, 2377 (1962).
37. N. Indictor, J. W. Horondniak, H. Jaffe and D. Miller, *J. Chem. Eng. Data*, 14, 76 (1969).
38. G. E. Wilson, Jr., M. G. Huang and W. W. Schloman, Jr., *J. Org. Chem.*, 33, 2133 (1968).
39. R. Böhm and E. Hannig, *Pharmazie*, 26, 598 (1971).
40. D. Hoppe and R. Follmann, *Angew. Chem. (intern. Ed. Engl.)*, 16, 462 (1977).
41. R. Feinauer, M. Jacobi and K. Hamann, *Chem. Ber.*, 98, 1782 (1965).

854 K. Pihlaja and P. Pasanen

42. J. Burkhardt, R. Feinauer, E. Gulbins and K. Hamann, *Chem. Ber.,* **99**, 1912 (1966).
43. A. Cooper and P. A. Norton, *J. Org. Chem.,* **33**, 3535 (1968).
44. D. J. Pasto, F. M. Klein and T. W. Doyle, *J. Amer. Chem. Soc.,* **89**, 4368 (1967).
45. K. Pihlaja, R. Keskinen and A. Nikkilä, *Bull. Soc. Chim. Belg.,* **85**, 435 (1976).
46. K. Pihlaja and E. Taskinen, *Physical Methods in Heterocyclic Chemistry* (Ed. A. R. Katrizky), Academic Press, New York, 1974, pp. 210–214.
47. G. E. Wilson, Jr., M. G. Huang and F. A. Bovey, *J. Amer. Chem. Soc.,* **92**, 5907 (1970).
48. G. E. Wilson, Jr., *J. Amer. Chem. Soc.,* **96**, 2426 (1974).
49. K. Pihlaja and R. Keskinen, *Org. Mag. Res.,* **9**, 177 (1977).
50. K. Pihlaja, T. Nurmi and P. Pasanen, *Acta Chem. Scand. (B),* **31**, 895 (1977).
51. W. E. Willy, G. Binsch and E. L. Eliel, *J. Amer. Chem. Soc.,* **92**, 5394 (1970).
52. R. Keskinen, A. Nikkilä and K. Pihlaja, *J. Chem. Soc., Perkin 2,* 1376 (1973).
53. R. Keskinen, *Ph. Lc. Thesis,* University of Turku, Turku, Finland, 1976.
54. L. A. Sternson, D. A. Coviello and R. S. Egan, *J. Amer. Chem. Soc.,* **93**, 6529 (1971).
55. K. Pihlaja, R. Keskinen, A. Nikkilä and T. Nurmi, unpublished results.
56. F. G. Riddell and M. J. T. Robinson, *Tetrahedron,* **27**, 4163 (1971).
57. C. H. Bushweller, G. U. Rao, W. G. Anderson and P. E. Stevenson, *J. Amer. Chem. Soc.,* **94**, 4744 (1972).
58. P. E. Stevenson, G. Bhat, C. H. Bushweller and W. G. Anderson, *J. Amer. Chem. Soc.,* **96**, 1067 (1974).
59. N. E. Alexandrou and P. M. Hadjimihalakis, *Org. Mag. Res.,* **1**, 401 (1969).
60. K. Pihlaja and M. Ala-Tuori, *Acta Chem. Scand.,* **26**, 1904 (1972).
61. C. H. Robinson, L. Miewich, G. Snatzke, W. Klyne and S. R. Wallis, *J. Chem. Soc. (C),* 1245 (1968).
62. D. J. Pasto, *J. Heterocycl. Chem.,* **6**, 1975 (1969).
63. G. Condé-Caprace and J. E. Collin. *Org. Mass. Spectrom.,* **6**, 415 (1972).
64. K. Pihlaja and P. Kuosmanen, unpublished results.
65. K. Pihlaja, *J. Amer. Chem. Soc.,* **94**, 3330 (1972).
66. J. Albery in *Proton-transfer Reactions* (Ed. E. F. Caldin and V. Gold), Chapman and Hall, London, 1975, Chap. 9, pp. 292–294.
67. F. Guinot, G. Lamaty and H. Münsch, *Bull. Soc. Chim. Fr.,* 541 (1971).
68. F. Guinot and G. Lamaty, *Tetrahedron Letters,* 2569 (1972).
69. A. Kankaanperä, *Ann. Univ. Turku. Ser. AI,* No. 95 (1966).
70. E. D. Brown, S. M. Iqbal and L. N. Owen, *J. Chem. Soc. (C),* 415 (1966).
71. E. L. Eliel, *Rec. Chem. Progr.,* **22**, 129 (1961).
72. G. E. Wilson, Jr. and M.-G. Huang, *J. Org. Chem.,* **41**, 966 (1976).
73. D. W. Emerson and H. Wynberg, *Tetrahedron Letters,* 3445 (1971).
74. R. Hoge and K. F. Fischer, *Cryst. Struct. Commun.,* **4**, 505 (1975).
75. K. Pihlaja, A. Nikkilä and T. Nurmi, unpublished results.
76. K. Schank, R. Wilmes and G. Ferdinand, *Int. J. Sulfur Chem.,* **8**, 397 (1973).
77. R. M. Kellogg, *J. Org. Chem.,* **38**, 844 (1973).
78. A. Bistrzycki and B. Brenken, *Helv. Chim. Acta,* **3**, 447 (1920).
79. S. Satsumabayashi, S. Irioka, H. Kudo, K. Tsujimoto and S. Motoki, *Bull. Chem. Soc. Japan,* **45**, 913 (1972).
80. M. Pailer, W. Streicher and F. Takacs, *Monatsh. Chem.,* **99**, 891 (1968).
81. J. Møller and C. Th. Pedersen, *Acta Chem. Scand.,* **24**, 2489 (1970).
82. M. Farines, *Dissertation,* University of Perpignan, Perpignan, France (1973).
83. K. Pihlaja, A. Nikkilä, K. Neuvonen and R. Keskinen, *Acta Chem. Scand. (A),* **30**, 457 (1976).
84. Y. Asabe, S. Takitani and Y. Tsuzuki, *Bull. Chem. Soc., Japan,* **48**, 966 (1975).
85. M. Brink, *Org. Mag. Res.,* **4**, 195 (1972).
86. E. G. Frandsen and C. Th. Pedersen, *Acta Chem. Scand.,* **26**, 1301 (1972).
87. S. Cabiddu, A. Maccioni and A. Secci, *Gazz. Chim. Ital.,* **99**, 1095 (1969).
88. C. Th. Pedersen, *Acta Chem. Scand.,* **20**, 2314 (1966).
89. C. Dufraisse and D. Daniel, *Bull. Soc. Chim. Fra.,* 2063 (1937).
90. C. Th. Pedersen, *Acta Chem. Scand.,* **22**, 247 (1968).

91. C. Th. Pedersen, *Acta Chem. Scand.*, **23**, 489 (1969).
92. S. Glue, I. T. Kag and M. R. Kipps, *J. Chem. Soc. (D)*, 1158 (1970).
93. H. F. Van Woerden, *Chem. Rev.*, **63**, 557 (1963).
94. G. Wood, J. M. McIntosh and M. H. Miskow, *Can. J. Chem.*, **49**, 1202 (1971).
95. C. H. Green and P. G. Hellier, *J. Chem. Soc., Perkin 2*, 243 (1973).
96. B. A. Arbuzov, V. A. Naumov, N. M. Zaripov and L. D. Pronicheva, *Dokl. Akad. Nauk. SSSR*, **198**, 1333 (1970).
97. H. J. Geise and E. Van Laere, *Bull. Soc. Chim. Belg.*, **84**, 775 (1975).
98. C. H. Green and D. G. Hellier, *J. Chem. Soc., Perkin 2*, 243 (1973).
99. C. H. Green and D. G. Hellier, *J. Chem. Soc., Perkin 2*, 1966 (1973).
100. V. Usieli, A. Pilersdorf, S. Shor, J. Katzhendler and S. Sarel, *J. Org. Chem.*, **39**, 2073 (1974).
101. G. W. Buchanan and D. G. Hellier, *Can. J. Chem.*, **54**, 1428 (1976).
102. H. C. F. Su, G. Segebarth and K. C. Tsou, *J. Org. Chem.*, **26**, 4993 (1961).
103. Q. E. Thompson, M. M. Crutchfield and M. W. Dietrich, *J. Org. Chem.*, **30**, 2696 (1965).
104. K. S. Dhami, *Chem. Ind. (Lond.)*, 1004 (1968).
105. K. S. Dhami, *Indian J. Chem.*, **12**, 278 (1974).
106. G. W. Buchanan, N. K. Sharma, F. De Reinach-Hirtzbuch, and T. Durst, *Can. J. Chem.*, **55**, 44 (1977).
107. N. A. Allinger, J. A. Hirsch, M. A. Miller and L. J. Tyminski, *J. Amer. Chem. Soc.*, **91**, 337 (1969).
108. J. A. Deyrup and C. L. Moyer, *J. Org. Chem.*, **34**, 1975 (1969).
109. G. Wood, G. W. Buchanan and M. H. Miskow, *Can. J. Chem.*, **50**, 521 (1972).
110. G. W. Buchanan, J. B. Stothers and G. Wood, *Can. J. Chem.*, **51**, 3746 (1973).
111. H. Friebolin, H. G. Schmid, S. Kabuss and W. Faisst, *Org. Mag. Res.*, **1**, 67 (1969).
112. K. Pihlaja and P. Pasanen, *Acta Chem. Scand.*, **24**, 2257 (1970).
113. P. Pasanen and K. Pihlaja, *Acta Chem. Scand.*, **25**, 1908 (1971).
114. P. Pasanen, *Dissertation*, University of Turku, Turku, Finland, 1974.
115. K. Pihlaja, P. Pasanen and J. Wähäsilta, *Org. Mag. Res.*, **12**, 331 (1979).
116. P. A. White, *Dissertation*, University of New Hampshire, Durham, N.H., U.S.A., 1968.
117. A. V. Bogatskii, A. M. Turyanskaya, A. I. Gren, E. Baltrusch and A. Voigt, *Vopr. Stereokhim.*, **4**, 49 (1974).
118. Y. Allingham, T. A. Crabb and R. F. Newton, *Org. Mag. Res.*, **3**, 37 (1971).
119. J. Gelan and M. Anteunis, *Bull. Soc. Chim. Belg.*, **79**, 313 (1970).
120. D. Danneels, M. Anteunis, L. Van Acker and D. Tavernier, *Tetrahedron*, **31**, 327 (1975).
121. K. Fuji, M. Ueda and E. Fujita, *Chem. Commun.*, 814 (1977).
122. D. L. Rakhmankulov, E. A. Kantor and R. S. Musavirov, *Zh. Org. Khim.*, **13**, 897 (1977).
123. D. L. Rakhmankulov, E. A. Kantor and R. S. Musavirov, *Zh. Prikl. Khim. (Leningrad)*, **50**, 2130 (1977).
124. M. Anteunis, G. Swaelens and J. Gelan, *Tetrahedron*, **27**, 1917 (1971).
125. M. Anteunis, *Bull. Soc. Chim. Belg.*, **80**, 3 (1971).
126. P. Pasanen, *Suomen. Kemistelehti (B)*, **45**, 363 (1972).
127. J. Gelan and M. Anteunis, *Bull. Soc. Chim. Belg.*, **77**, 447 (1968).
128. J. Gelan, G. Swaelens and M. Anteunis, *Bull. Soc. Chim. Belg.*, **79**, 321 (1970).
129. K. Pihlaja, *J. Chem. Soc., Perkin 2*, 890 (1974).
130. N. De Wolf and H. R. Buys, *Tetrahedron Letters*, 551 (1970).
131. H. R. Buys, *Rec. Trav. Chim.*, **89**, 1244 (1970).
132. H. R. Buys, *Rec. Trav. Chim.*, **89**, 1253 (1970).
133. K. Bergesen, B. M. Carden and M. J. Cook, *J. Chem. Soc., Perkin 2*, 345 (1976).
134. J. Gelan and M. Anteunis, *Bull. Soc. Chim. Belg.*, **77**, 423 (1968).
135. P. Pasanen, *Finn. Chem. Letters*, 49, (1974).
136. F. G. Riddell in *SPR*, Vol. 1, Part 3: *Aliphatic, Alicyclic and Saturated Heterocyclic Chemistry*, The Chemical Society, London, 1973, p. 89.
137. K. Pihlaja, G. M. Kellie and F. G. Riddell, *J. Chem. Soc., Perkin 2*, 252 (1972).
138. K. Pihlaja and P. Pasanen, *J. Org. Chem.*, **39**, 1948 (1974).
139. P. Pasanen and K. Pihlaja, *Tetrahedron*, **28**, 2617 (1972).
140. P. Pasanen and K. Pihlaja, *Tetrahedron Letters*, 4515 (1971).

141. K. Pihlaja and P. Pasanen, *Suomen Kemistilehti (B)*, **46**, 273 (1973).
142. K. Pihlaja and P. Pasanen, *Org. Mass Spectrom.*, **5**, 763 (1971).
143. J. Jalonen, P. Pasanen and K. Pihlaja, *Org. Mass Spectrom.*, **7**, 949 (1973).
144. J. H. Bowie and A. C. Ho, *Australian J. Chem.*, **26**, 2009 (1973).
145. J. H. Bowie and P. Y. White, *Org. Mass Spectrom.*, **6**, 75 (1972).
146. K. Pihlaja and J. Jalonen, *Org. Mass Spectrom.*, **5**, 1363 (1971).
147. J. Jalonen and K. Pihlaja, *Org. Mass Spectrom.*, **7**, 1203 (1973).
148. K. Pihlaja, J. Jokila and U. Heinonen, *Finn. Chem. Letters*, 275 (1974).
149. K. Pihlaja and U. Heinonen, unpublished results.
150. E. L. Eliel, J. K. Koskimies and B. Lohri, *J. Amer. Chem. Soc.*, **100**, 1614 (1978).
151. R. M. Carlson and P. M. Helqwist, *J. Org. Chem.*, **33**, 2596 (1968).
152. K. Bergesen, M. J. Cook and A. P. Tonge, *Org. Mag. Res.*, **6**, 127 (1974).
153. L. Van Acker and M. Anteunis, *Tetrahedron Letters*, 225 (1974).
154. J. B. Lambert and R. G. Keske, *J. Org. Chem.*, **31**, 3429 (1966).
155. J. V. Karabinos and J. J. Hazdra, *J. Org. Chem.*, **27**, 4253 (1962).
156. D. K. Black, *J. Chem. (C)*, 1708 (1966).
157. D. L. Tuleen and R. H. Bennett, *J. Heterocycl. Chem.*, **6**, 115 (1969).
158. V. S. Blagoveschchenskii, I. V. Kazimirchik, A. A. Alekseeva and N. S. Zefirov, *Zh. Org. Khim.*, **8**, 1325 (1972).
159. M. L. Evans and C. T. Mason, *J. Org. Chem.*, **33**, 1643 (1968).
160. A. H. Haubein, *J. Amer. Chem. Soc.*, **81**, 144 (1959).
161. G. E. Hull and F. M. Ubertini, *J. Org. Chem.*, **15**, 715 (1950).
162. H. Kwart and E. R. Evans, *J. Org. Chem.*, **31**, 413 (1966).
163. J. Gelas and S. Veyssieres-Rambaud, *Carbohydr. Res.*, **37**, 303 (1974).
164. M. Muehlstaedt, P. Schneider and D. Martinetz, *J. Prakt. Chem.*, **315**, 929 (1973).
165. M. Muehlstaedt, D. Martinetz and P. Schneider, *J. Prakt. Chem.*, **315**, 940 (1973).
166. V. S. Blagoveschchenskii, I. V. Kazimirchik, O. P. Yakovleva, N. S. Zefirov and V. K. Denisenko, *Probl. S-kh. Nauki Mosk. Univ.*, 260 (1975).
167. S. Hoff, A. P. Blok and E. Zwanenburg, *Rec. Trav. Chim.*, **92**, 890 (1973).
168. R. K. Summerbell and E. S. Polkacki, *J. Org. Chem.*, **27**, 2074 (1962).
169. T. Abe, S. Nagase and H. Baba, *Bull. Chem. Soc. Japan*, **46**, 2524 (1973).
170. H. Schubert, P. Goehmann, H. J. Dietz and L. Schroeder, *Z. Chem.*, **16**, 147 (1976).
171. K. Jankowski, R. Coulombe and C. Berse, *Bull. Acad. Pol. Sci., Ser. Sci. Chim.*, **19**, 661 (1971).
172. D. I. Davies, L. Hughes, Y. D. Vankar and J. E. Baldwin, *J. Chem. Soc., Perkin 2*, 2476 (1977).
173. W. B. Smith and B. A. Shoulders, *J. Phys. Chem.*, **69**, 579 (1965).
174. J. Barrett and M. J. Hitch, *Spectrochim. Acta (A)*, **25**, 407 (1969).
175. F. R. Jensen and R. A. Neese, *J. Amer. Chem. Soc.*, **97**, 4922 (1975).
176. K. Jankowski and R. Coulombe, *Tetrahedron Letters*, 991 (1971).
177. R. W. Kitchin, T. K. Avirah, T. B. Malloy and R. L. Cook, *J. Mol. Struct.*, **24**, 337 (1975).
178. O. H. Ellestad, P. Klaboe and G. Hagen, *Spectrochim. Acta (A)*, **28**, 137 (1972).
179. G. Schultz, I. Hargittai and H. Laszlo, *J. Mol. Struct.*, **14**, 353 (1972).
180. G. Hagen, I. Hargittai and G. Schultz, *Acta Chim. Acad. Sci. Hunt.*, **86**, 219 (1975).
181. N. de Wolf, C. Romers and C. Altona, *Acta Cryst.*, **22**, 715 (1967).
182. C. Altona and C. Romers, *Acta Cryst.*, **16**, 1225 (1963).
183. C. Altona and C. Romers, *Rec. Trav. Chim.*, **82**, 1080 (1963).
184. H. T. Kalff and C. Romers, *Acta Cryst.*, **18**, 164 (1965).
185. H. T. Kalff and C. Romers, *Rec. Trav. Chim.*, **85**, 198 (1966).
186. L. Cazaux and J. Navech, *Org. Mag. Res.*, **7**, 26 (1975).
187. J. Crossley, A. Holt and S. Walker, *Tetrahedron*, **21**, 3141 (1965).
188. N. S. Zefirov, V. S. Blagoveschchenskii, I. V. Kazimirchik and N. S. Surova, *Tetrahedron*, **27**, 3111 (1971).
189. N. S. Zefirov and S. V. Rogozina, *Tetrahedron*, **30**, 2345 (1974).
190. R. Gleiter, M. Kobayashi, N. S. Zefirov and V. A. Palyulin, *Dokl. Akad. Nauk. SSSR*, **235**, 347 (1977).

191. J. Burdon and I. W. Parsons, *Tetrahedron*, **27**, 4533 (1971).
192. J. Burdon and I. W. Parsons, *Tetrahedron*,**27**, 4553 (1971).
193. L. Phillips and V. Wray, *J. Chem. Soc., Perkin 2*, 928 (1974).
194. W. A. Szarek, D. M. Vyas, A. M. Sepulchre, S. D. Gero and G. Lukacs, *Can. J. Chem.*, **52**, 2041 (1974).
195. G. Condé-Caprace and J. E. Collin, *Org. Mass Spectrom.*, **2**, 1277 (1969).
196. J. E. Collin and G. Condé, *Bull.. Acad. Roy, Belges Cl. Sci.*, **52**, 978 (1966).
197. D. A. Sweigart and D. W. Turner, *J. Amer. Chem. Soc.*, **94**, 5599 (1972).
198. N. de Wolf, P. W. Henniger and E. Havinga, *Rec. Trav. Chim.*, **86**, 1227 (1967).
199. M. A. P. Dankleff, R. Curci, J. O. Edwards and H.-Y. Pyun, *J. Amer. Chem. Soc.*, **90**, 3209 (1968).
200. K. W. Buck, A. B. Foster, A. R. Perry and J. M. Webber, *Chem. Commun.*, 433 (1965).
201. A. B. Foster, Q. H. Hasan, D. R. Hawkins and J. M. Webber, *Chem. Commun.*, 1084 (1968).
202. A. B. Foster, T. D. Inch, M. H. Qadir and J. M. Webber, *Chem. Commun.*, 1086 (1968).
203. M. J. Cook, *Kemia-Kemi*, **3**, 16 (1976).
204. K. W. Buck, T. A. Hamor and D. J. Watkin, *Chem. Commun.*, 759 (1966).
205. K. W. Buck, A. B. Foster, W. D. Pardoe, M. H. Qadir and J. M. Webber, *Chem. Commun.*, 759 (1966).
206. D. M. Frieze and S. A. Evans, *J. Org. Chem.*, **40**, 2690 (1975).
207. Y. Hase and Y. Kawano, *Spectry. Letters*, 151 (1978).
208. P. C. Dwivedi, *Indian J. Chem.*, **9**, 1408 (1971).
209. K. H. Thide, *Z. Anorg. Allgem. Chem.*, **322**, 71 (1963).
210. K. Feenan and G. W. A. Fowles, *Inorg. Chem.*, **4**, 310 (1965).
211. G. W. A. Fowles, R. A. Hoodless and R. A. Walton, *J. Chem. Soc.*, 5873 (1963).
212. K. Feenan and G. W. A. Fowles, *J. Chem. Soc.*, 2449 (1965).
213. R. A. Walton, *J. Chem. Soc. (A)*, 1852 (1967).
214. M. Boelens, L. M. Van der Linde, P. J. De Valois, H. M. Van Dort and H. J. Takken, *J. Agric. Food Chem.*, **22**, 1071 (1974).
215. M. G. Gadzhieva, I. G. Alizade and A. I. Abdullaev, *Uch. Zap.-Minist. Vyssh. Sredn. Spets. Obraz. Az. SSR, Ser. Khim. Nauk*, 50 (1975).
216. B. A. Arbuzov, E. N. Klimovitskii, L. K. Yuldasheva and L. V. Guzovskaya, *Izv. Akad. Nauk. SSSR, Ser. Khim.*, 2346 (1974).
217. M. Fukunaga and M. Oki, *Chem. Letters*, 1081 (1972).
218. G. Wood and R. M. Strivastava, *Tetrahedron Letters*, 2937 (1971).
219. G. Wood, R. M. Strivastava and B. Adlam, *Can. J. Chem.*, **51**, 1200 (1973).
220. K. Pihlaja and H. Nikander, unpublished observations.
221. P. C. Lauterbur, J. G. Pritchard and R. L. Vollmer, *J. Chem. Soc.*, 5307 (1963).
222. L. Cazaux and P. Maroni, *Bull. Soc. Chim. Fr.*, 773 (1972).
223. L. Cazaux, G. Chassaing and P. Maroni, *Tetrahedron Letters*, 2517 (1975).
224. H. Nikander, V. M. Mukkala, T. Nurmi and K. Pihlaja, *Org. Mag. Res.*, **8**, 375 (1976).
225. G. W. Buchanan, C. M. E. Cousineau and T. C. Mundell, *Tetrahedron Letters*, 2775 (1978).
226. G. W. Buchanan, C. M. E. Cousineau and T. C. Mundell, *Can. J. Chem.*, **56**, 2019 (1978).
227. P. A.-C. Carbonnelle, Y. Jeannin and F. Robert, *Acta Cryst. (B)*, **34**, 1631 (1978).
228. K. Pihlaja, H. Nikander and E. Rahkamaa, unpublished results.
229. K. Pihlaja, D. M. Jordan and H. Nikander, unpublished results.
230. K. Pihlaja and H. Nikander, to be published.
231. R. S. Edmundson, *Tetrahedron Letters*, 1649 (1965).
232. H. F. van Woerden, *Tetrahedron Letters*, 2407 (1966).
233. H. F. van Woerden, H. Erfontain, C. H. Green and R. J. Reijerkerk, *Tetrahedron Letters*, 6107 (1968).
234. H. F. van Woerden and A. T. de Vries-Miedema, *Tetrahedron Letters*, 1687 (1971).
235. F. J. Mustoe and J. L. Hencher, *Can. J. Chem.*, **50**, 3892 (1972).

236. P. Albriktsen, *Acta Chem. Scand.*, **25**, 478 (1971).
237. P. Albriktsen, *Acta Chem. Scand.*, **26**, 1783 (1972).
238. C. H. Green and D. G. Hellier, *J. Chem. Soc., Perkin 2*, 458 (1972).
239. L. Ḳ. Yuldasheva, R. P. Arshinova and Yu. Yu. Samitov, *Izv. Akad. Nauk SSSR, Ser. Khim.*, 2461 (1970).
240. L. K. Yuldasheva and R. P. Arshinova, *Vop. Stereokhim.*, **1**, 57 (1971).
241. A. B. Remizov, *Zh. Strukt. Khim.*, **12**, 1101 (1971).
242. G. Wood and M. H. Miskow, *Tetrahedron Letters*, 4433 (1966).
243. P. C. Hamblin, R. F. M. White, G. Eccleston and E. Wyn-Jones, *Can. J. Chem.*, **47**, 2731 (1969).
244. G. Wood and M. H. Miskow, *Tetrahedron Letters*, 1109 (1969).
245. W. Wucherpfennig, *Liebigs Ann. Chem.*, **737**, 144 (1970).
246. L. Cazaux and P. Maroni, *Tetrahedron Letters*, 3667 (1969).
247. L. Cazaux and P. Maroni, *Bull. Soc. Chim. Fr.*, 773, 780, 794 (1972).
248. L. Cazaux, G. Chassaing and P. Maroni, *Org. Mag. Res.*, **8**, 461 (1976).
249. P. Albriktsen, *Acta Chem. Scand.*, **26**, 3678 (1972).
250. P. Albriktsen, *Acta Chem. Scand. (A)*, **29**, 824 (1975).
251. P. Albriktsen, *Acta Chem. Scand. (A)*, **30**, 763 (1976).
252. P. Albriktsen and T. Thorstenson, *Acta Chem. Scand. (A)*, **31**, 83 (1977).
253. D. G. Hellier and F. J. Webb, *J. Chem. Soc., Perkin 2*, 612 (1977).
254. S. Hoff, A. P. Blok and E. Zwanenburg, *Rec. Trav. Chim.*, **92**, 890 (1973).
255. F. O. Davis, *Macromol Synth.*, **4**, 69 (1972).
256. K. Hayashi, *Macromolecules*, **3**, 5 (1970).
257. F. S. Dainton, J. A. Davies, P. P. Manning and S. A. Zahir, *Trans. Faraday Soc.*, **53**, 813 (1957).
258. H. H. Szmant and W. Emerson, *J. Amer. Chem. Soc.*, **78**, 454 (1956).
259. C. A. Bunton, P. B. D. de la Mare, P. M. Creaseley, D. R. Llewellyn, N. H. Pratt and J. G. Tillett, *J. Chem. Soc.*, 4751 (1958).
260. R. G. Gillis, *J. Org. Chem.*, **25**, 651 (1960).
261. J. Lichtenberger and J. Hincky, *Bull. Soc. Chim. Fr.*, 1495 (1961).
262. H. Faucher and A. Guimaraes, *Tetrahedron Letters*, 1743 (1977).
263. P. A. Bristow, M. Khowaja and J. G. Tillett, *J. Chem. Soc.*, 5779 (1965).
264. P. A. Laurent and P. Tarte, *Bull. Soc. Chim. Fr.*, 954 (1960).
265. D. Weichert, *Plaste Kautschuk*, **10**, 579 (1963).
266. P. Tarte and P. A. Laurent, *Bull. Soc. Chim. Belg.*, **70**, 43 (1961).
267. G. Condé-Caprace and J. E. Collin, *Org. Mass Spectrom.*, **6**, 341 (1972).
268. J. E. Collin and G. Condé-Caprace, *J. Mass. Spectrom. Ion Phys.*, **1**, 213 (1968).
269. J. S. Bradshaw, J. Y. Hui, B. L. Haymore, J. J. Christensen and R. M. Izatt, *J. Heterocycl. Chem.*, **10**, 1 (1973).

CHAPTER **19**

Allene oxides and related species

PETER J. STANG

Chemistry Department, The University of Utah, Salt Lake City, Utah 84112, U.S.A.

I. INTRODUCTION

Allene oxide (**1**) is a member of the family of strained small-ring compounds. The parent compound itself is part of the C_3H_4O energy surface that besides **1** includes cyclopropanone (**2**), oxyallyl (**3**) and **4**, all of which are valence tautomers. Allene oxides contain within their framework the structural elements of an enol ether, a double bond and an epoxide, elements that cause them to be of considerable intrinsic interest.

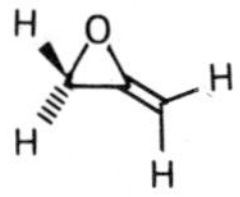

(1) 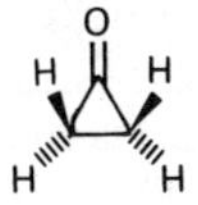   (2) 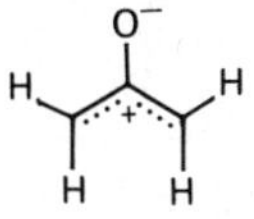   (3) 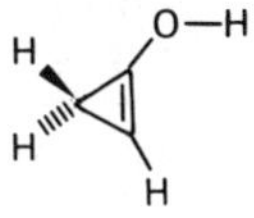   (4)

Despite their intrinsic interest as well as their close relationship to the well-known and extensively investigated epoxides, little was known about allene oxides until recently. This surprising lack of investigation is due to their considerable instability and high reactivity, particularly in comparison to normal epoxides. However, within the last dozen years, allene oxides have been the subject of both theoretical and experimental attention. This chapter will provide an account and summary of the available data through late 1978. Separate sections will deal with theoretical calculations, preparation and chemistry of allene oxides. A final section will briefly cover the little that is known about related species such as allene episulphides, oxaspiropentanes, etc.

II. THEORETICAL CALCULATIONS

A number of quantum-mechanical calculations dealing with the C_3H_4O energy surface have appeared. The majority of these calculations deal with the interconversions between cyclopropanone (2) and oxyallyl (3) but several also treat allene oxide (1). The results of these calculations are summarized in Table 1.

It is evident from the data in Table 1 that with the exception of EHMO, calculations predict 2 to be more stable than either 1 or 3. In fact, 1 is predicted to be some 6–21 kcal/mol less stable than 2. All calculations except EHMO also predict that singlet oxyallyl (3) resulting from the disrotatory ring-opening[8] of 2 is a high-energy species with some 36–232 kcal/mol above 2 and therefore higher in energy than even 1.

Besides disagreement on the relative stabilities of these species as determined by the various calculational methods, there is the question of the exact mechanism of interconversion or isomerization between 1, 2 and 3. Liberles and coworkers[4,5] consider 3 to be an intermediate (or at least a transition state) in the known (see below) isomerization of 1 to 2. Although substituted oxyallyls have been postulated as intermediates[9] and even as stable entities[10], the actual evidence for their existence is rather scant.

A novel pathway, shown in Figure 1 and Scheme 1, for the isomerization of allene oxide (1) to cyclopropanone (2) was proposed by Zandler and coworkers[6]

TABLE 1. Theoretical calculations of the C_3H_4O energy surface

Calculation[a]	Relative energies (kcal/mol)[b]			
	1	2	3	Reference
EHMO	−21	0.0	−23	1
MINDO/2	c	0.0	78	2
INDO	c	0.0	220	3
INDO	6	0.0	232	4
ab initio SCF	21	0.0	83	4
MINDO	c	0.0	36	5
CNDO/2	c	0.0	110	6
MINDO/3	c	0.0	66	7

[a]See original reference for definition and details.
[b]Relative to cyclopropanone (2): negative energy indicates greater stability than 2, positive energy indicates lower stability than 2.
[c]Not given.

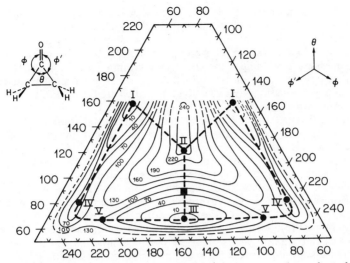

FIGURE 1. Contour diagram of the CNDO/2 energy surface for the allene oxide–oxallyl–cyclopropanone system. The contour spacing is 30 kcal/mol. Reprinted with permission from M. E. Zandler, C. E. Choc and C. K. Johnson, *J. Amer. Chem. Soc.*, **96**, 3317 (1974). Copyright by the American Chemical Society.

The Zandler pathway primarily involves bending motions via I → IV → V → III of Scheme 1 for the allene oxide–cyclopropanone isomerization. Such a pathway has only one half the energy barrier of the pathway via II. The lower barrier was suggested[6] to be the result of the lower energy requirements of bond bending compared to bond stretching. Stabilization due to delocalization in II is apparently insufficient to compensate for destabilization due to bond breakage[6]. On the

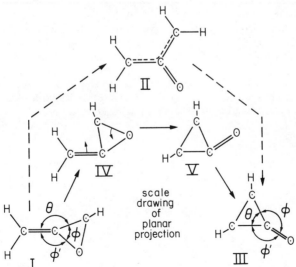

SCHEME 1. Reprinted with permission from M. E. Zandler, C. E. Choc and C. K. Johnson, *J. Amer. Chem. Soc.*, **96**, 3317 (1974). Copyright by the American Chemical Society.

CNDO/2 energy surface, the three-membered ring is preserved intact until bending allows another ring to form with minimal bond stretch (Figure 1). However, as the authors point out[6], the reliability of this novel isomerization mechanism is hard to assess since CNDO/2 is known to overestimate bond force constants causing excess resistance in bond stretching. Ring strain may be improperly estimated as well. It will be interesting to see if these results hold up under more sophisticated *ab initio* calculations or if the pathway via oxyallyl (II) is the true theoretically predicted one for the allene oxide–cyclopropanone interconversion. The Zandler mechanism versus the oxyallyl pathway may be subject to experimental verification. Rearrangement of an optically active allene oxide should result in an optically active cyclopropanone via the Zandler pathway, whereas it should give racemic cyclopropanone via the intermediacy (or transition state) of the planar symmetrical oxyallyl. No such experimental data are available to date.

A semiempirical calculation has also been done on the ring-opening of substituted cyclopropanones to the corresponding oxyallyls[5]. Unfortunately, the corresponding substituted allene oxides were not considered. This calculation shows that methyl-, methoxy- and fluorine-substituted cyclopropanones undergo ring-opening to oxyallyl more readily than the parent compound but the exact magnitudes of the energy differences between the appropriate isomeric cyclopropanone and oxyallyl are unreliable[5].

Recently, an estimate of the thermodynamic energy difference between allene oxide (1) and cyclopropanone (2) has been made[11] by means of the appropriate bond dissociation energies[12]. This estimate showed 2 to be 22 kcal/mol more stable than the isomeric 1. This 'thermodynamic' value of 22 kcal/mol is remarkably close to the 21 kcal/mol difference between 1 and 2 predicted by *ab initio* calculations[4] (see Table 1). Although this agreement is likely to be fortuitous, other indirect data from microwave[13] and photoelectron spectroscopy[14] studies on 2 also suggest 2 to be the most stable isomer on the C_3H_4O energy surface.

III. PREPARATION

Allene oxides have been proposed as intermediates, along with cyclopropanones, in the Favorskii reaction[15,16]. As yet, no allene oxides and only a few cyclopropanones have been trapped in the Favorskii reaction[17]. Indeed very few allene oxides at all have been isolated as stable compounds at room temperature.

There are two main approaches to the synthesis of allene oxides: peracid oxidation of allenes and exocyclic β-elimination of an epoxide. Each of these will be discussed in turn together with some miscellaneous methods.

A. Peracid Oxidation of Allenes

Analogously to normal epoxidation of olefins via peracids, peracid oxidation would seem the logical and simplest entry into allene oxides (equations 1 and 2).

$$\text{\\C=C/} + RCO_3H \xrightarrow{\text{solvent}} \text{\\C-C/} \qquad (1)$$

$$\text{\\C=C=C/} + RCO_3H \xrightarrow{\text{solvent}} \text{\\C-C=C/} \qquad (2)$$

As part of an extensive investigation of epoxidation reactions, Boeseken[18] investigated the reaction of peracetic acid with 1,1-dimethylallene and reported 3-acetoxy-

3-methyl-2-butanone but no allene oxide as part of the products. Early Russian work[19] reported dioxidation products, albeit with meagre evidence, in the peracid reaction of several substituted allenes. More recently, an extensive investigation of peracid oxidation of various allenes has been carried out by Crandall and coworkers[20–25]. As will be shown in Section IV there is little doubt that allene oxides are involved in many of these peracid oxidations of allenes. However, except in two instances they could not be isolated as stable compounds owing to their great propensity to react with nucleophiles, undergo further epoxidation to the allene dioxide and in some instances isomerize to the corresponding cyclopropanone.

The first stable allene oxide was prepared and isolated by Camp and Greene[26] by the reaction of *m*-chloroperbenzoic acid with excess 1,3-di-*t*-butylallene in hexane (equation 3). The allene oxide **5** is a colourless liquid, stable for long

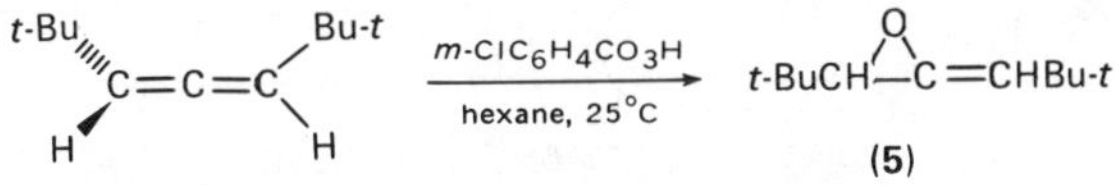
t-Bu
Bu-t
C=C=C
H
H
m-ClC6H4CO3H
hexane, 25°C
O
t-BuCH—C=CHBu-t

(5) (3)

periods at room temperature, with spectral properties fully consistent with its structure[26]. There are two geometric isomers possible for any 1,3-disubstituted allene oxide, **6a** and **6b**. The simplicity of the NMR spectrum of **5** (CCl_4), δ0.98 (s, 9H), 1.08 (s, 9H), 3.25 (s, 1H), 4.82 (s, 1H) suggests that it is a single species but of unknown geometry.

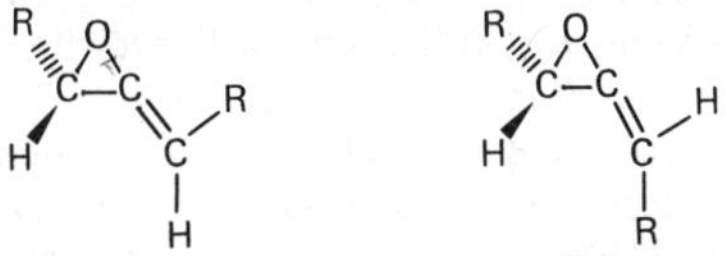

(6a) (6b)

Reaction of 1,1,3-tri-*t*-butylallene (**7**) with *m*-chloroperbenzoic acid gave the stable tri-*t*-butylallene oxide (**8**)[24] (equation 4).

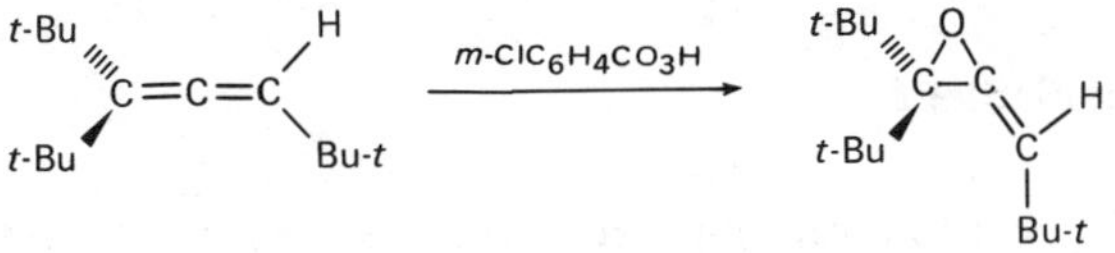
t-Bu
H
C=C=C
t-Bu
Bu-t
m-ClC6H4CO3H
t-Bu
O
C—C
H
t-Bu
C
Bu-t

(4)

(7) (8)

Allene oxides **5** and **8** undoubtedly owe their considerable stability to the bulky *t*-butyl substituents that provide steric stabilization by preventing the usual (see below) interaction with nucleophiles.

B. Exocyclic β-Elimination of Epoxides

An elegant synthesis of allene oxides has been developed by Chan and coworkers[27–30] via dehalosilylation[31] of epoxides (**9**) (equation 5). This approach

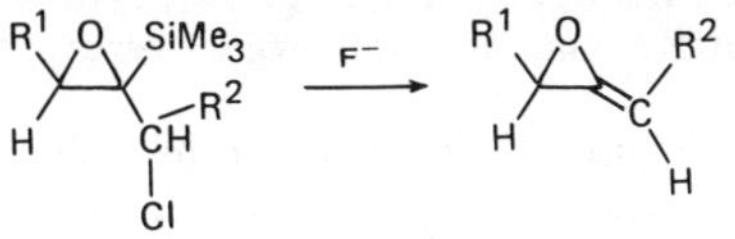
R1
O
SiMe3
R2
H
CH
Cl
F−
R1
O
R2
H
C
H

(5)

(9) (10)

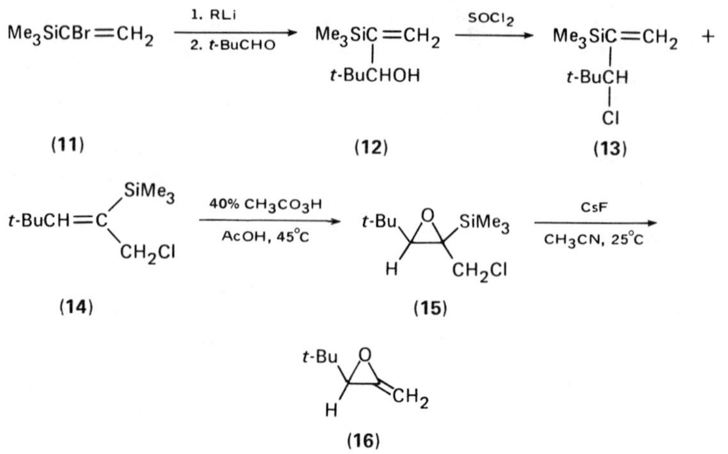

$$\text{Me}_3\text{SiCBr}=\text{CH}_2 \xrightarrow[\text{2. } t\text{-BuCHO}]{\text{1. RLi}} \underset{\substack{|\\ t\text{-BuCHOH}}}{\text{Me}_3\text{SiC}=\text{CH}_2} \xrightarrow{\text{SOCl}_2} \underset{\substack{|\\ t\text{-BuCH}\\ |\\ \text{Cl}}}{\text{Me}_3\text{SiC}=\text{CH}_2} +$$

(11)   (12)   (13)

$$t\text{-BuCH}=\underset{\text{CH}_2\text{Cl}}{\overset{\text{SiMe}_3}{C}} \xrightarrow[\text{AcOH, 45°C}]{40\% \text{ CH}_3\text{CO}_3\text{H}} \quad t\text{-Bu}\overset{O}{\triangle}\text{SiMe}_3 \xrightarrow[\text{CH}_3\text{CN, 25°C}]{\text{CsF}}$$

(14)   (15)

$$t\text{-Bu}\overset{O}{\triangle}\text{CH}_2$$

(16)

SCHEME 2.

has the advantage that the epoxide ring is preformed by standard techniques with a subsequent elimination under very mild conditions to generate the double bond and hence the final allene oxides (10). This approach has been successfully applied to the preparation and isolation of 1-t-butylallene oxide (16) as shown in Scheme 2. Reaction of vinylsilane (11) with alkyllithium followed by pivaldehyde gave alcohol (12) which gave a mixture of chlorides 13 and 14 upon treatment with SOCl_2[32]. Epoxidation of 14 gave epoxide 15 which upon fluoride-initiated dehalosilylation gave the product 16[30]. Allene oxide (16) was formed, in 55% yield from 15, as a colourless liquid which is stable in dilute solutions at room temperature for 1–2 h followed by polymerization[30]. Numerous other allene oxides were prepared *in situ* via this technique and reacted with various nucleophiles as will be discussed in Section IV.

C. Miscellaneous Methods

An interesting approach to allene oxides consists of the addition of an unsaturated carbene[33] (17) to a carbonyl group (equation 6). Such a reaction has been

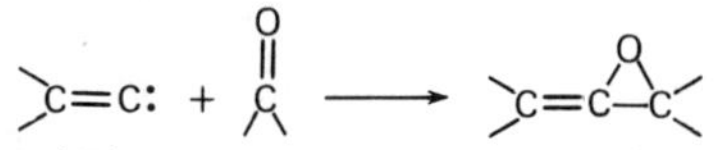

$$\underset{(17)}{\text{C}=\text{C:}} + \overset{O}{\underset{}{\overset{||}{C}}} \longrightarrow \text{C}=\text{C}\overset{O}{\triangle}\text{C} \tag{6}$$

investigated by Kuo and Nye[34] as well as Newman and Liang[35]. Kuo and Nye[34] investigated the addition of carbene (19), obtained via deamination of the precursor 18, to a variety of carbonyl groups resulting in diadducts (21) as shown in Scheme 3. The diadducts (21) were postulated to arise via the addition of a second carbonyl group to allene oxides (20) although no direct evidence was provided for the actual intermediacy of 20. Both aldehydes such as pivaldehyde and *p*-tolualdehyde as well as ketones such as acetone and acetophenone were employed as substrates in Scheme 3[34].

Completely different results were obtained by Newman and Liang[35]. Under phase-transfer conditions, the carbene precursor 22 gave adducts 24 and 25 with isobutyraldehyde and pivaldehyde, respectively, as shown in Scheme 4. These

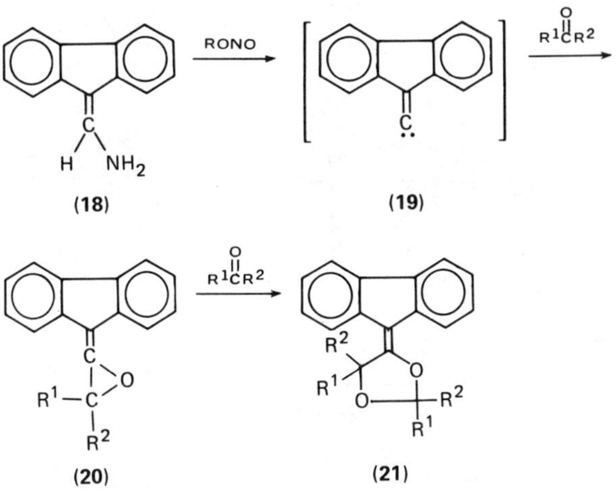

SCHEME 3.

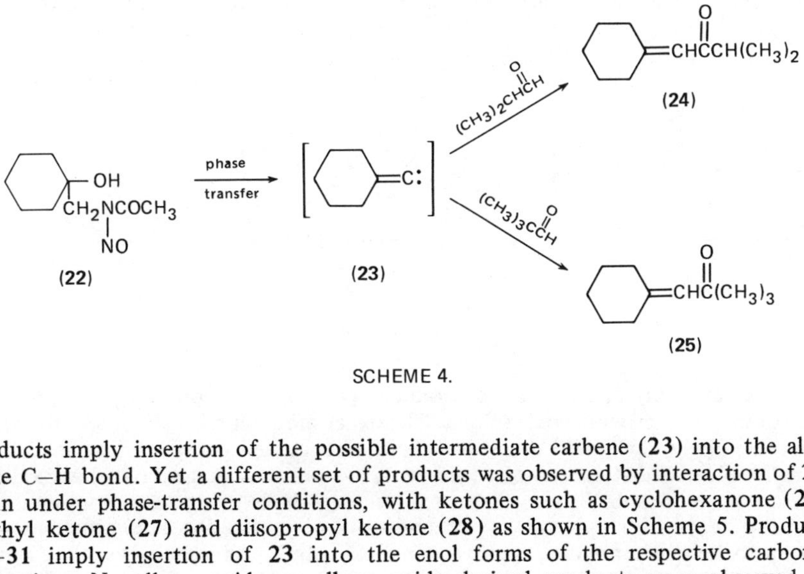

SCHEME 4.

products imply insertion of the possible intermediate carbene (23) into the aldehyde C–H bond. Yet a different set of products was observed by interaction of 22,
again under phase-transfer conditions, with ketones such as cyclohexanone (26),
diethyl ketone (27) and diisopropyl ketone (28) as shown in Scheme 5. Products
29–31 imply insertion of 23 into the enol forms of the respective carbonyl
derivatives. No allene oxide or allene oxide derived products were observed by
Newman and Liang[35]. The reasons for the discrepancy of the results of Kuo and
Nye and Newman and Liang is not clear. It may be the result of the differing modes
of carbene generation or the different reaction conditions. It is possible that
unsaturated carbenes may not be involved in the reactions of Newman and Liang[35].

There has been a claim made[36] that tetramethylallene oxide (33) was obtained
by the zinc–copper debromination of ketone 32 in dimethylformamide (equation
7). However, subsequent results have shown that the actual product was 4-iso-
propylidene-5,5-dimethyl-2-dimethylamino-1,3-dioxolane (34) rather than 33[37].

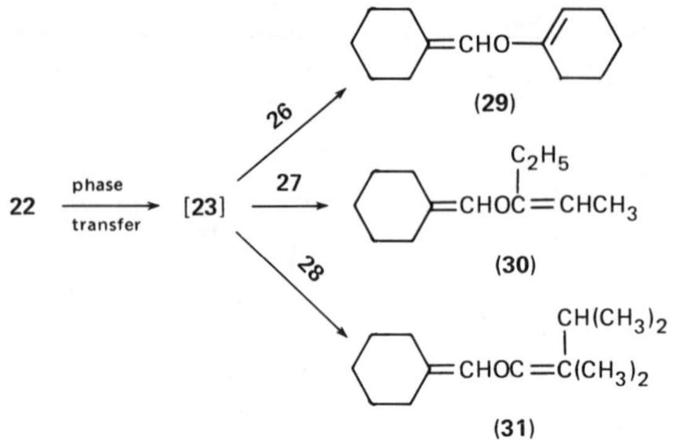

SCHEME 5.

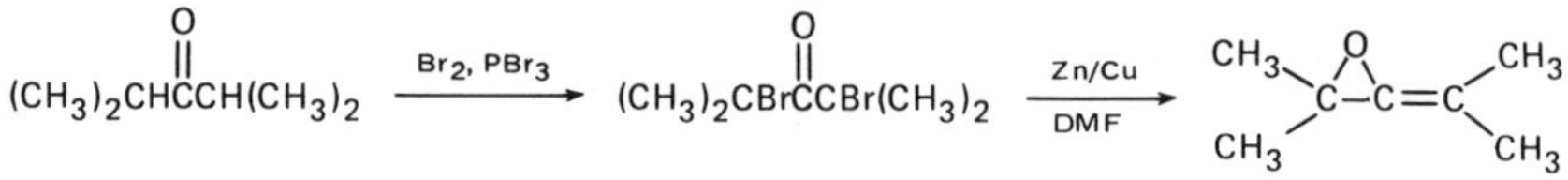

Formally, **34** may be looked upon as a 1,3-dipolar adduct between tetramethyl-oxyallyl (**3** : CH_3 instead of H) and dimethylformamide. Whether an allene oxide is involved in the above reactions is at present unclear.

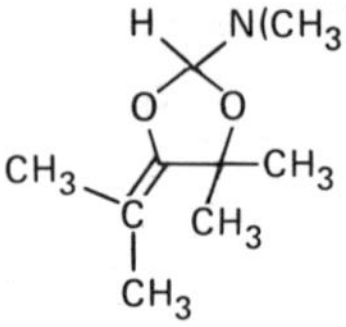

(34)

IV. REACTIONS

The reactions of allene oxides generally fall in three categories: (a) further oxidation and formation of spiro dioxides; (b) isomerization to cyclopropanones and (c) interaction with nucleophiles. The exact mode of reaction of specific allene oxides is highly dependent upon the substituents as well as the reaction conditions. Spiro dioxide formation can of course only occur under peracid or other oxidizing conditions. Monosubstituted allene oxides as well as the parent compound, **1**, generally react with nucleophiles or undergo polymerization rather than isomerization to the corresponding cyclopropanones[32]. Bulky substituents such as *t*-butyl that provide steric *hindrance* to interaction with nucleophiles allow isomerization to cyclopropanone. For aryl- or di-substituted allene oxides the rate of isomerization to cyclopropanone is generally faster than nucleophilic attack[32]. Each of these reactions will now be discussed in more detail.

A. Further Oxidation

In the presence of peracids used to form the allene oxides from the precursor allenes, the former generally undergo further oxidation. The initial intermediate is believed to be a 1,4-dioxaspiro[2,2]pentane (35) that itself undergoes further reaction (equation 8).

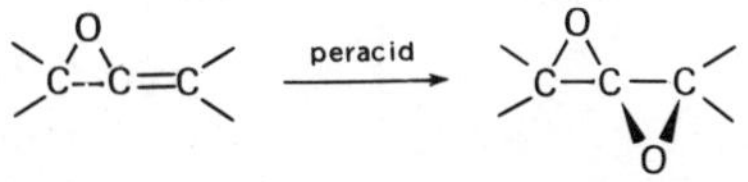

$$\underset{(35)}{\text{peracid}}$$

(8)

Epoxidation of tetramethylallene[21] with peracetic acid gave 52% of 2-acetoxy-2,4-dimethyl-3-pentanone (see below), 39% of 2-acetoxy-4-hydroxy-2,4-dimethyl-3-pentanone (37), 4% of 4-hydroxy-2,4-dimethylpent-1-en-3-one (38), 3% tetramethyl-3-oxetanone (39) and 2% of lactone (40) as shown in Scheme 6. These products were rationalized via the intermediacy of the spiro dioxide 36. Protonation followed by isomerization of 36 results in 37 and 38. Acid-catalysed or thermal isomerization of 36 results in 39 which upon Baeyer–Villiger oxidation gives lactone 40.

Similarly the spiro dioxide 43 has been invoked to account for the observed products in the peracid oxidation[25,38] of 1,2-cyclononadiene (41) as shown in equation (9). In the peracid oxidation of 2,5,5-trimethyl-2,3-hexadiene (44), the

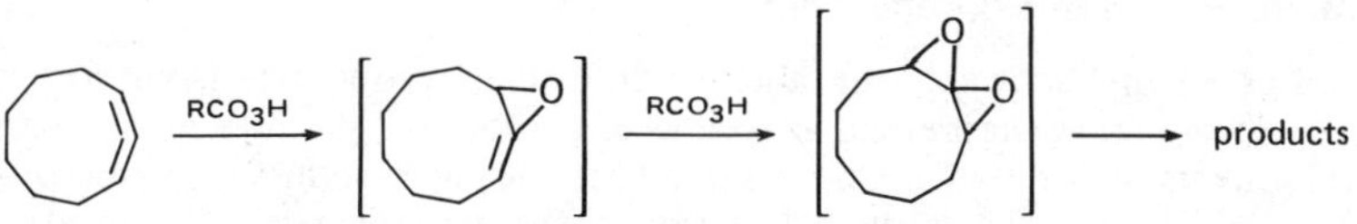

$$\underset{(41)}{}\xrightarrow{\text{RCO}_3\text{H}}\underset{(42)}{\left[\quad\right]}\xrightarrow{\text{RCO}_3\text{H}}\underset{(43)}{\left[\quad\right]}\longrightarrow \text{products}$$

(9)

spiro dioxide 45 could be isolated as a stable compound and spectrally characterized (equation 10)[22]. Reaction of 45 with HCl was shown[22] to give oxetanone (46) and an unsaturated ketone (47) (equation 11). The products 46 and 47 are

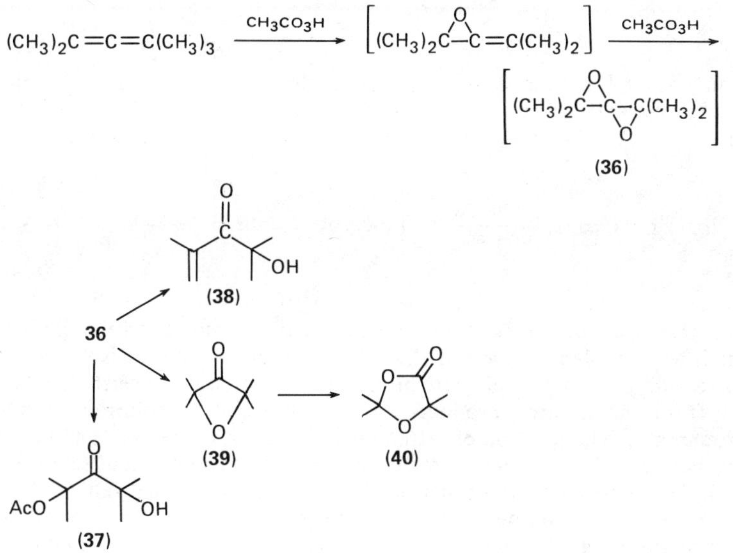

$$(CH_3)_2C\!=\!C\!=\!C(CH_3)_3 \xrightarrow{CH_3CO_3H} \left[(CH_3)_2C\!-\!C\!=\!C(CH_3)_2\right] \xrightarrow{CH_3CO_3H}$$

$$\left[(CH_3)_2C\!-\!C\!-\!C(CH_3)_2\right]$$

(36)

SCHEME 6.

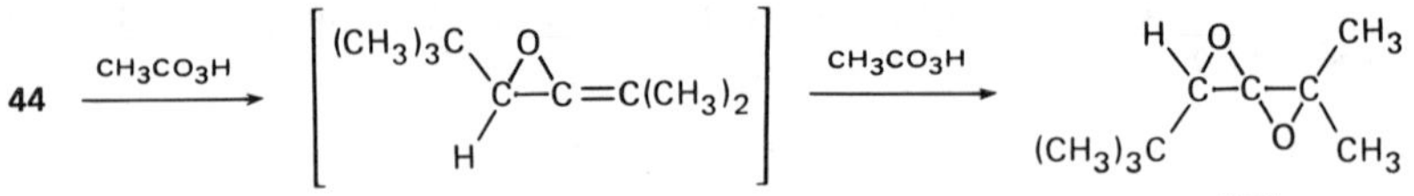

44 $\xrightarrow{\text{CH}_3\text{CO}_3\text{H}}$ $\left[\begin{array}{c} (\text{CH}_3)_3\text{C} \quad \text{O} \\ \quad \text{C—C}=\text{C}(\text{CH}_3)_2 \\ \text{H} \end{array} \right]$ $\xrightarrow{\text{CH}_3\text{CO}_3\text{H}}$ (structure **45**) (10)

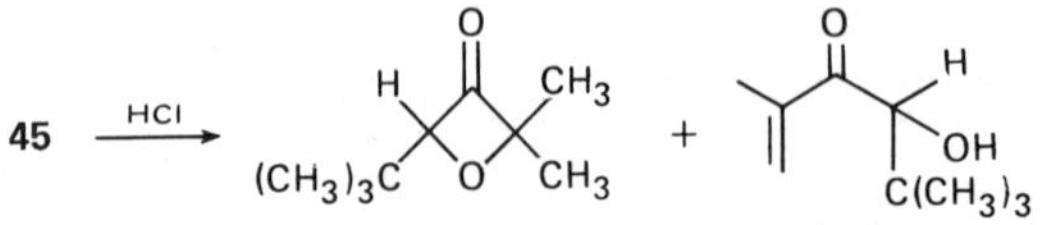

(**45**)

45 $\xrightarrow{\text{HCl}}$ (structure **46**) + (structure **47**) (11)

(**46**)   (**47**)

analogous to products observed in the peracid oxidation of tetramethylallene where the intermediate spiro dioxide could not be isolated. Hence, there is little doubt that spiro oxides or diepoxides are viable reaction intermediaries or products in the further reaction of the intermediate allene oxides resulting from the peracid treatment of certain allenes.

B. Isomerization to Cyclopropanones

As indicated in Section I, the allene oxide—cyclopropanone isomerization has attracted considerable theoretical as well as experimental[8,17] interest. It represents the interconversion of two highly strained small-ring systems[39]. Numerous such isomerizations have been observed. Perhaps the most clear-cut example is the isomerization of 1,3-di-t-butylallene oxide (**5**) to *trans*-2,3-di-t-butylcyclopropanone (**48**) with a $t_{1/2}$ of five hours at 100°C (equation 12). Similarly, peracid

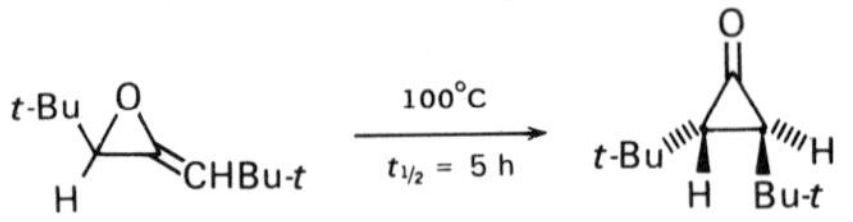

(structure **5**) $\xrightarrow[t_{1/2}\ =\ 5\ h]{100°\text{C}}$ (structure **48**) (12)

(**5**)   (**48**)

oxidation of 1,1-di-t-butylallene in methylene chloride gave 2,2-di-t-butylcyclopropanone (**50**) as the sole product presumably via the intermediacy of the isomeric allene oxide **49**, which could not be isolated (equation 13)[23].

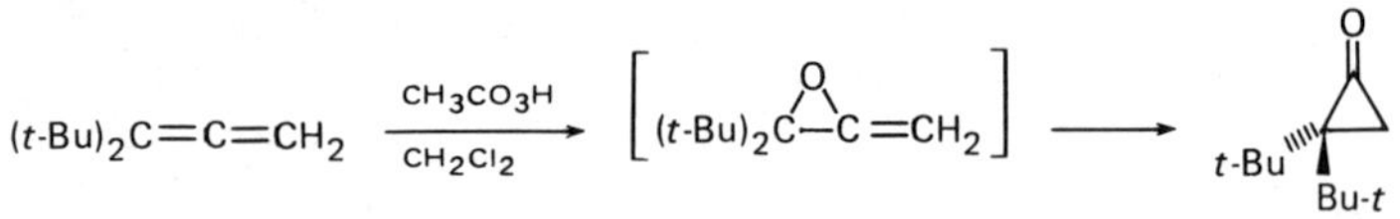

$(t\text{-Bu})_2\text{C}=\text{C}=\text{CH}_2$ $\xrightarrow[\text{CH}_2\text{Cl}_2]{\text{CH}_3\text{CO}_3\text{H}}$ $\left[(t\text{-Bu})_2\text{C}-\text{C}=\text{CH}_2 \right]$ $\longrightarrow$ (structure **50**) (13)

(**49**)   (**50**)

Other instances of allene oxide—cyclopropanone isomerization involve cases where neither the allene oxide nor the cyclopropanone could be isolated under the reaction conditions employed, but the isomerization could nevertheless be clearly inferred from the isolated products and the known[8,17] solution chemistry of cyclopropanones. The reaction of tetramethylallene with peracetic acid in methanol leads to, besides other products already mentioned in the previous section, 37% of tetramethylethylene oxide and other products that were rationalized via the isomerization of the allene oxide to tetramethylcyclopropanone and the subsequent reactions of the latter[25]. Cyclooctene epoxide obtained in the peracid oxidation of 1,2-cyclononadiene was similarly rationalized[25,38]. Cyclopropane-derived products

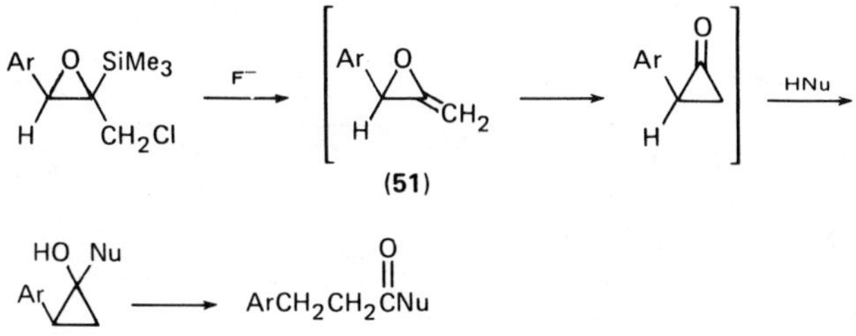

(51)

(52) Ar = Ph, p-CH$_3$C$_6$H$_4$

Nu = OMe, EtS, PhNH

SCHEME 7.

(52) were also observed[30] in the reaction of 1-arylallene oxides (51) with various nucleophiles as shown in Scheme 7. The 1-aryl substituents in the reactions of Scheme 7 are essential for isomerization to occur. In the case of 1-alkyl- or 3-aryl-allene oxides the allene oxide itself was intercepted by the nucleophiles (see below) prior to rearrangement to the corresponding cyclopropanone[30]. Similar results were obtained[30] in the reaction of spiroadamantylallene oxide (54) obtained via desilylation of 53 as shown in Scheme 8. In this case the rearranged cyclopropanone intermediate (55) could be trapped as the hemithioketal (56) by reaction with ethanethiol[30]. The rearrangement of 51 and 54 were interpreted[30] as

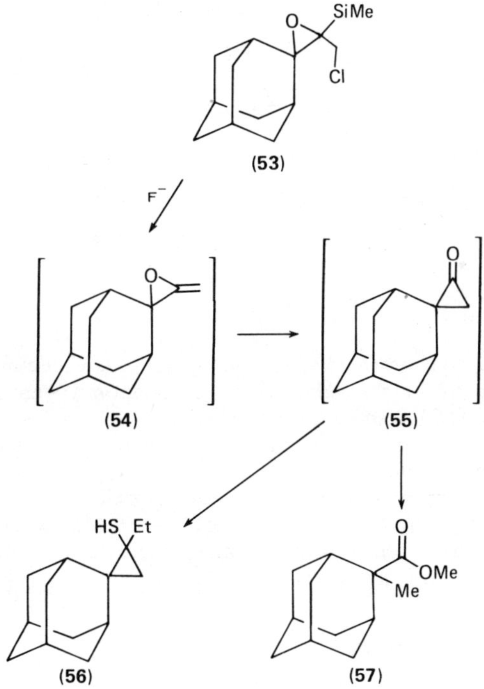

(53)

(54) — **(55)**

(56) **(57)**

SCHEME 8.

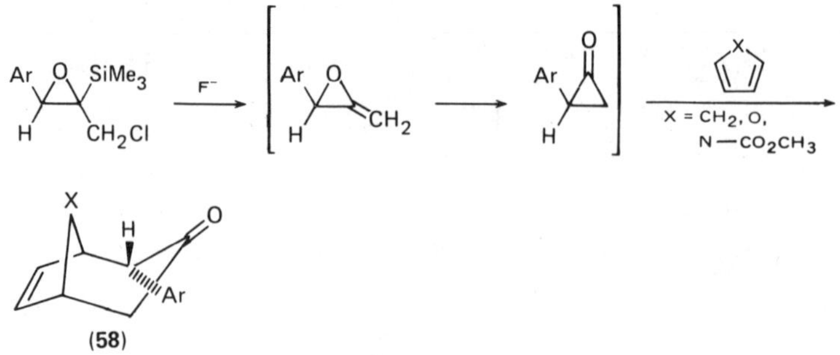

(58)

SCHEME 9.

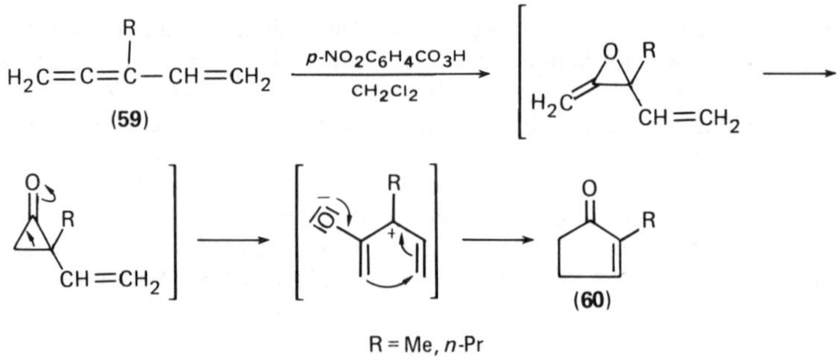

R = Me, n-Pr

SCHEME 10.

R = Me, Ph

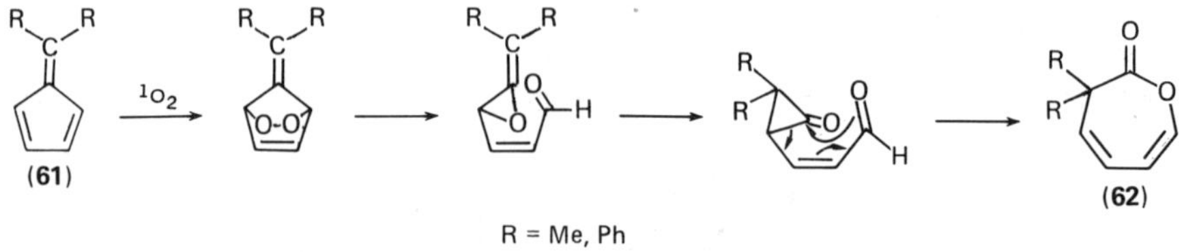

SCHEME 11.

evidence for an oxyallyl intermediate in the allene oxide–cyclopropanone isomerization. In the case of **51** the isomeric cyclopropanones were also trapped[40] as Diels–Alder adducts (**58**) as shown in Scheme 9.

Allene oxide–cyclopropanone isomerization was also invoked[41] to account for the observed cyclopentenone products (**60**) in the peracid oxidation of vinylallenes (**59**) as shown in Scheme 10. Similarly, the allene oxide–cyclopropanone isomerization was used to explain[42] the formation of 3,3-disubstituted-2-(3*H*)-oxepinones (**62**) in the dye-sensitized photooxygenation via singlet oxygen of 6,6-disubstituted fulvenes (**61**) as shown in Scheme 11.

C. Reaction with Nucleophiles

Allene oxides monosubstituted by a 1- or 3-alkyl group and other sterically unhindered allene oxides readily undergo nucleophilic substitution with a variety of nucleophiles. The major product (53%) in the peracetic acid oxidation of tetra-methylallene is the acetoxyketones (63) resulting from interaction of the inter-mediate allene oxide with $HOAc^{21}$ (equation 14). Similarly, a wide variety of

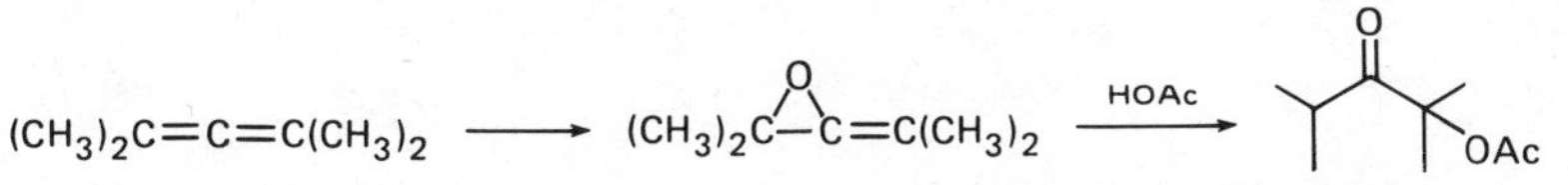

$$(CH_3)_2C=C=C(CH_3)_2 \longrightarrow (CH_3)_2C-\overset{O}{\underset{}{C}}=C(CH_3)_2 \xrightarrow{HOAc} \qquad (14)$$

(63)

1-monoalkyl-substituted allene oxides (64) gave^{32} ketone products (65) as shown in equation (15). The reactions of 64 with the nucleophiles (HNu) were found to be

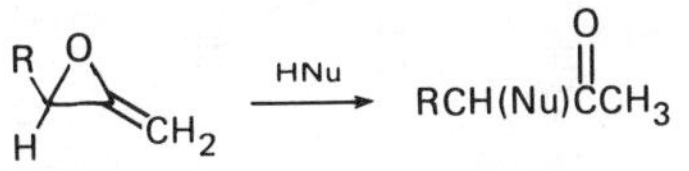

$$\underset{H}{\overset{R}{\diagdown}}\overset{O}{\diagup}_{CH_2} \xrightarrow{HNu} RCH(Nu)\overset{O}{\overset{\|}{C}}CH_3 \qquad (15)$$

(64) (65)

R = H, Me, *i*-Pr, *t*-Bu, *c*-C₆H₁₁, *n*-C₁₀H₂₁
Nu = OH, OMe, EtS, Cl, PhO, PhS, PhNH

regiospecific as demonstrated by the behaviour of isomeric allene oxides (66) shown in Scheme 12. Isomer 66a upon reaction with methanol gave exclusively the methoxyketone (67a) with none of 67b as product, whereas isomer 66b gave only the methoxyketone (67b) under identical conditions^{32}. This regiospecificity indi-cates exclusive nucleophilic attack upon the 'tetrahedral carbon' of the isomeric epoxides (66) and also rules out a common species such as an oxyallyl or cyclo-propanone as the intermediate in these reactions.

Allene oxides (64) have also been trapped by cyclopentadiene to give ketones (69) presumably via the intermediacy of zwitterions (68)^{40} (equation 16). This reaction further demonstrates the electrophilic nature of allene oxides.

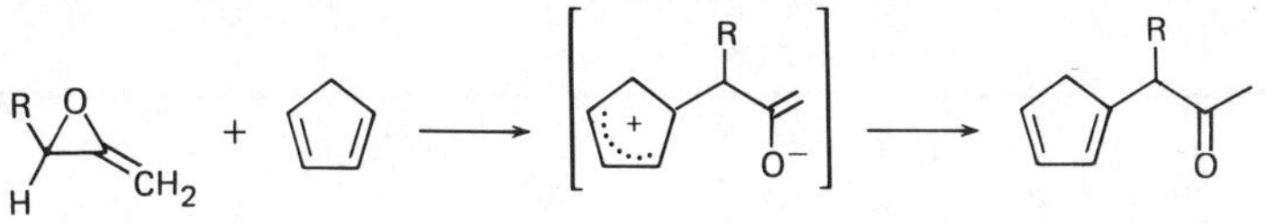

$$\underset{H}{\overset{R}{\diagdown}}\overset{O}{\diagup}_{CH_2} + \overset{}{\bigcirc} \longrightarrow \left[\overset{R}{\underset{O^-}{\diagup}} \right] \longrightarrow \overset{R}{\underset{O}{\diagup}} \qquad (16)$$

(64) (68) (69)

A summary of known allene oxides, their mode of generation and major reaction products are given in Table 2.

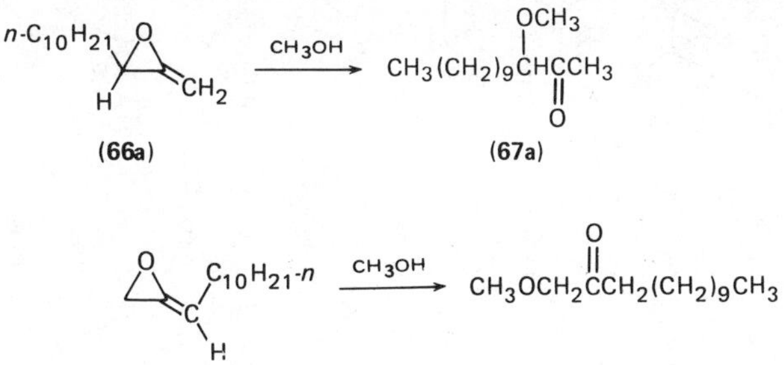

$$\underset{H}{\overset{n-C_{10}H_{21}}{\diagdown}}\overset{O}{\diagup}_{CH_2} \xrightarrow{CH_3OH} CH_3(CH_2)_9\overset{OCH_3}{\underset{}{C}H}\overset{O}{\underset{\|}{C}}CH_3$$

(66a) (67a)

$$\overset{O}{\diagdown}\underset{C}{\diagup}\overset{C_{10}H_{21}\text{-}n}{\underset{H}{}} \xrightarrow{CH_3OH} CH_3OCH_2\overset{O}{\overset{\|}{C}}CH_2(CH_2)_9CH_3$$

(66b) (67b)

SCHEME 12.

TABLE 2. Summary of the preparation and reactions of allene oxides

Allene oxide	Reaction conditions	Products	Reference
	, CsF, CH₃CN, PhOH, 25°C	PhOCH₂CCH₃ (with C=O)	32
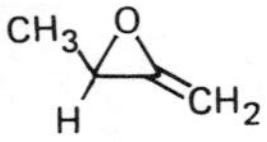	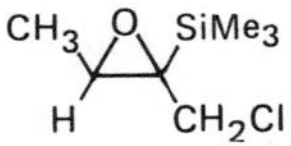 , CsF, CH₃CN, HNu, 25°C	CH₃CH(Nu)CCH₃ (with C=O); Nu = Cl, PhO, (i-Pr)₂N	32
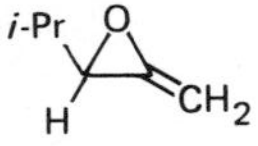	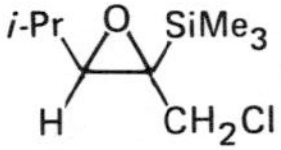 , CsF, CH₃CN, HNu, 25°C	i-PrCH(Nu)CCH₃ (with C=O); Nu = Cl, PhO, EtS, PhS	32
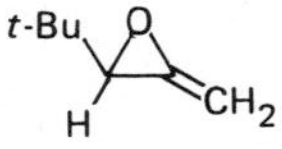	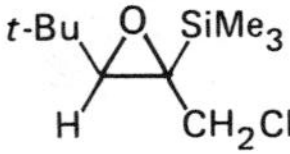 , CsF, CH₃CN, HNu, 25°C	t-BuCH(Nu)CCH₃ (with C=O); Nu = OH, EtS, CCl₃CO₂	32
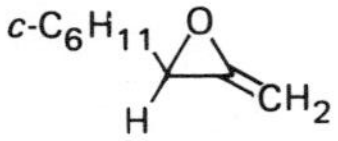	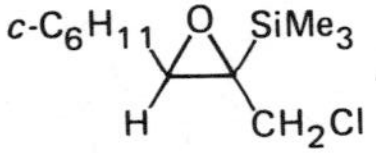 , CsF, CH₃CN, HNu, 25°C	c-C₆H₁₁CH(Nu)CCH₃ (with C=O); Nu = Cl	32
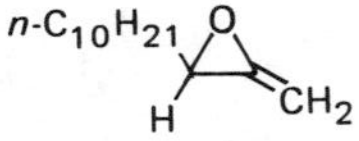	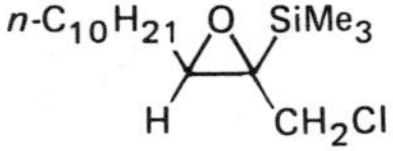 , CsF, CH₃CN, HNu, 25°C	n-C₁₀H₂₁CH(Nu)CCH₃ (with C=O); Nu = Cl, MeO	32
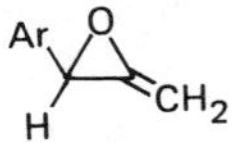 Ar = Ph, p-CH₃C₆H₄	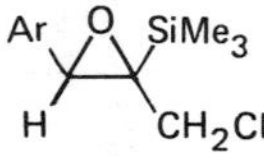 , CsF, CH₃CN, HNu, 25°C	ArCH₂CH₂CNu (with C=O); Nu = MeO, EtS, PhNH, PhO; Ar = Ph, p-CH₃C₆H₄	32
	, CsF, CH₃CN, HNu, 25°C	NuCH₂CCH₂(CH₂)₉CH₃ (with C=O); Nu = MeO, PhO	32

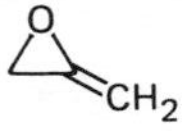

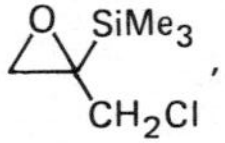

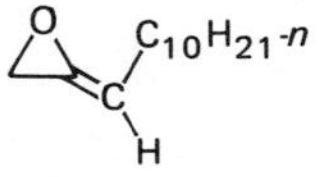

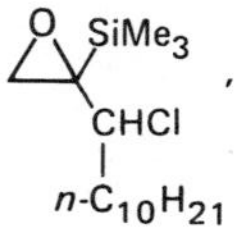

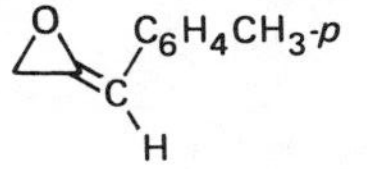

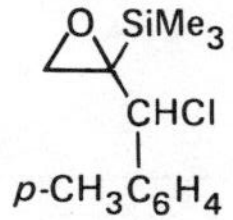

Starting material	Conditions	Products	Ref.
		$\underset{O}{\parallel}$ $NuCH_2CCH_2C_6H_4CH_3$; Nu = EtS, PhO	32

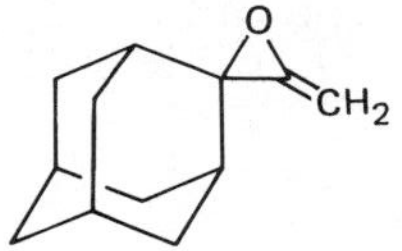

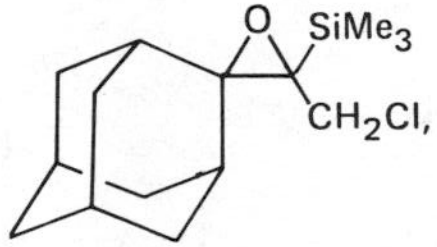

$\underset{H}{\overset{O}{\diagdown}}\overset{SiMe_3}{\underset{C}{\diagup}}\,CHCl\,—\,C_6H_4CH_3\text{-}p$ CsF, CH_3CN, HNu, 25°C $p\text{-}CH_3C_6H_4$

CH_2Cl, Et_4NF, CH_3CN, HNu, 25°C

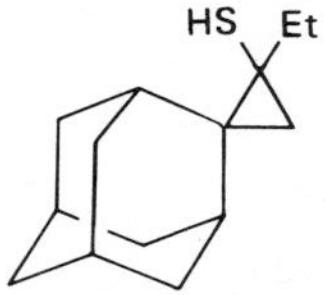

Nu = EtS   Nu = MeO   32

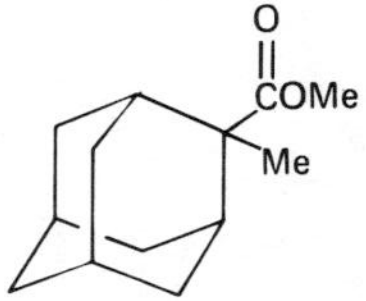

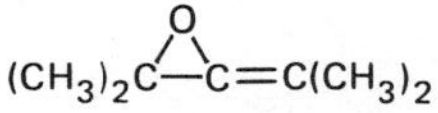

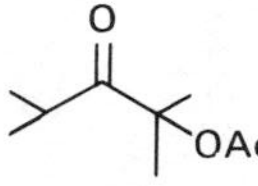

$(CH_3)_2C—C≡C(CH_3)_2$, 40% CH_3CO_3H, Na_2CO_3, CH_2Cl_2, 0°C

$\underset{O}{\overset{}{\diagdown}}$ + $\underset{OAc}{\overset{O}{}}$ 52% $\underset{OH}{\overset{O}{}}$ 39%   + others   21

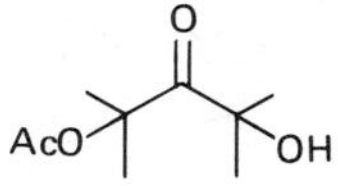

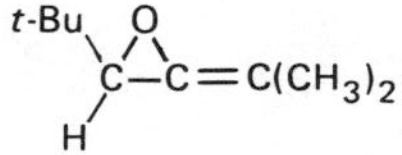

$t\text{-BuCH}=C=C(CH_3)_2$, CH_3CO_3H, Na_2CO_3, CH_2Cl_2, 0°C

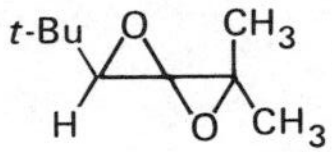

+ $t\text{-BuCHCC(CH}_3)_2OAc$ 22

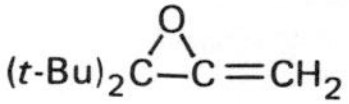

$(t\text{-Bu})_2C=C=CH_2$, CH_3CO_3H 23

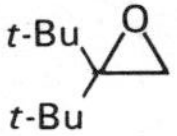

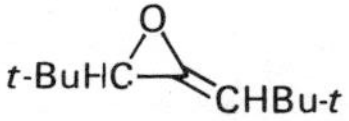

$t\text{-BuCH}=C=CHBu\text{-}t$, $m\text{-ClC}_6H_4CO_3H$, hexane, 25°C   at 25°C   at 100°C   26

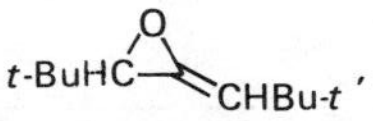

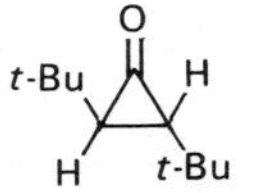

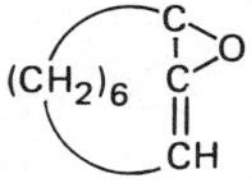

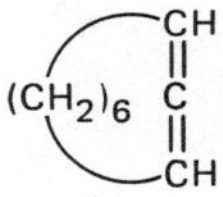

$(CH_2)_6$, RCO_3H, $NaNO_3$, CH_2Cl_2, 0°C   + others   25,38

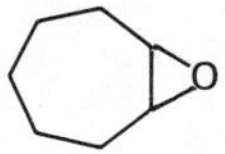

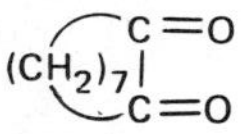

TABLE 2. (continued)

Allene oxide	Reaction conditions	Products	Reference
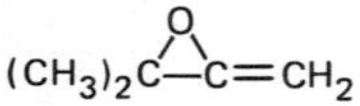 $(CH_3)_2C{-}C{=}CH_2$ (epoxide)	$(CH_3)_2C{=}C{=}CH_2$, CH_3CO_3H, $NaNO_3$, CH_2Cl_2, $0°C$	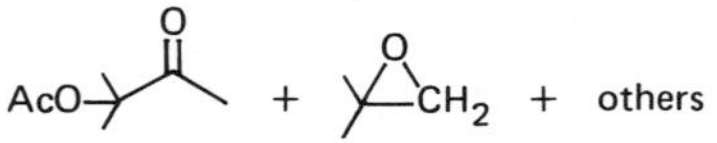 $AcO{-}$ (ketone) $+$ (epoxide)CH_2 $+$ others	25
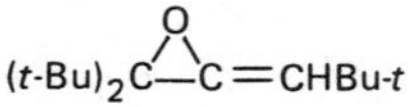 $(t\text{-}Bu)_2C{-}C{=}CHBu\text{-}t$ (epoxide)	$(t\text{-}Bu)_2C{=}C{=}CHBu\text{-}t$, $m\text{-}ClC_6H_4CO_3H$,	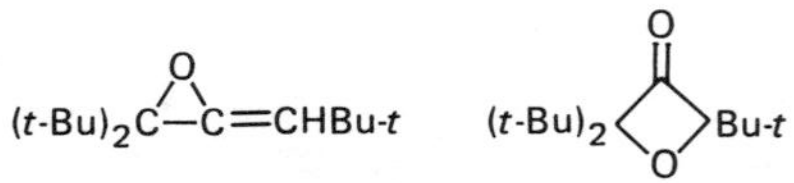 $(t\text{-}Bu)_2C{-}C{=}CHBu\text{-}t$ (epoxide) **10** ; fluorenylidene dioxolane (R^2, R^1) ; $(t\text{-}Bu)_2$ oxetanone **1**	24
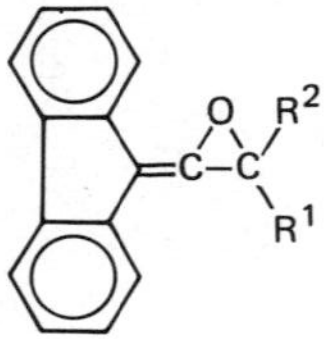 fluorenylidene allene oxide (R^2, R^1)	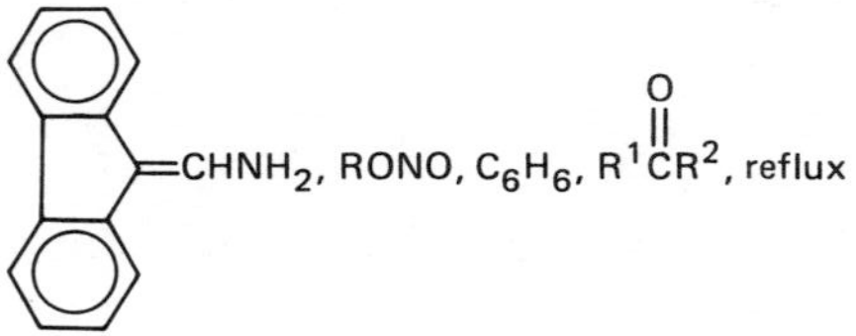 $=CHNH_2$, RONO, C_6H_6, R^1CR^2 ($C{=}O$), reflux	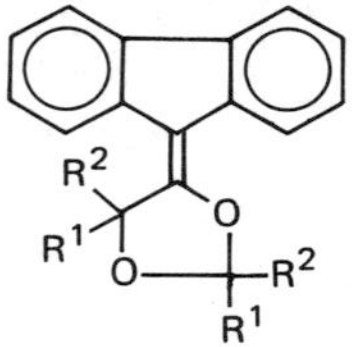	34
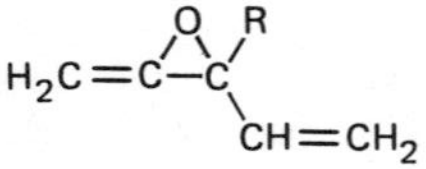 $H_2C{=}C{=}C(R)(C)CH{=}CH_2$ allene oxide	$H_2C{=}C{=}C(R){-}CH{=}CH_2$, $p\text{-}NO_2C_6H_4CO_3H$, CH_2Cl_2	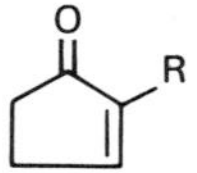 cyclopentenone (R)	41
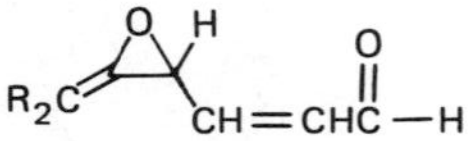 $R_2C{=}C{=}CH{-}CHC{-}H$ oxide, $CH{=}CHC({=}O){-}H$	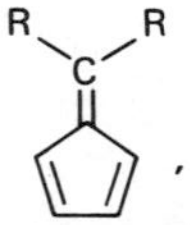 cyclopentadiene (R, R), $^1O_2$	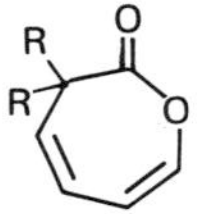 oxepinone (R, R)	42

V. RELATED SPECIES

To the best of our knowledge, from a limited literature search, cumulene oxides are not known. Since most cumulenes are relatively thermally unstable or unstable towards oxygen, their oxides presumably would be too unstable to isolate. In this section we will discuss two species related to allene oxides namely oxaspiropentanes (70) and allene episulphides (71).

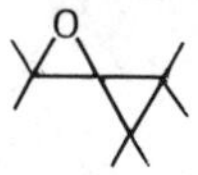

(70) 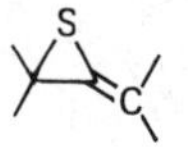   (71)

A. Oxaspiropentanes

To the extent that one normally considers the electron-rich bonds of a cyclopropane as being analogous to the π-system of a double bond, oxaspiropentanes (70) are related to allene oxides. Furthermore oxaspiropentanes like allene oxides are highly strained small-ring molecules. The strain energy of cyclopropane and its oxygen analogue ethylene oxide are within one kcal/mol the same at 28 kcal/mol[43]. Therefore the strain energy of oxaspiropentane is likely to be close to the 65 kcal/mol of strain energy in spiropentane[44].

Oxaspiropentanes with different alkyl and aryl substituents, as well as the parent compound, are readily available by the peracid oxidation of alkylidene cyclopropanes developed by Crandall and coworkers[45] (equation 17) and subsequently applied to a large number of systems[46-48].

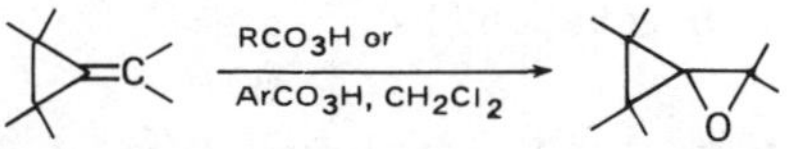

$$\text{(17)}$$

A second major route to oxaspiropentanes is the reaction of sulphur ylides with carbonyl compounds as developed by Trost and coworkers[49-51]. In particular, reaction of diphenylsulphonium cyclopropylylide with carbonyl compounds gave high yields of oxaspiropentanes (equation 18).

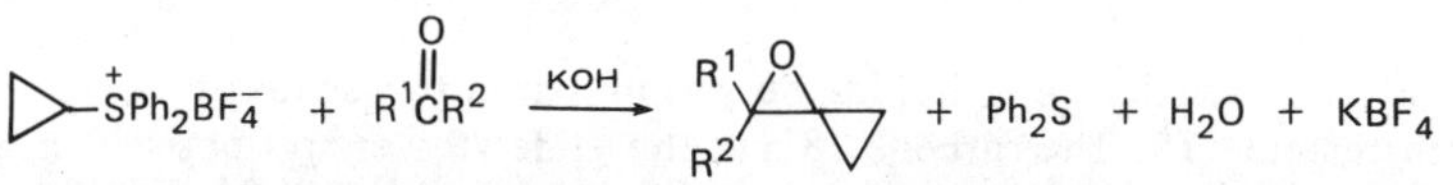

$$\text{(18)}$$

There are two major modes of reaction of oxaspiropentanes: Lewis or Brønsted acid catalysed rearrangement to cyclobutanones[45-49] and base-catalysed rearrangement to vinyl cyclopropanols[51] isolated as the silyl ether as shown in Scheme 13.

The oxaspiropentane—cyclobutanone rearrangement has been invoked to explain

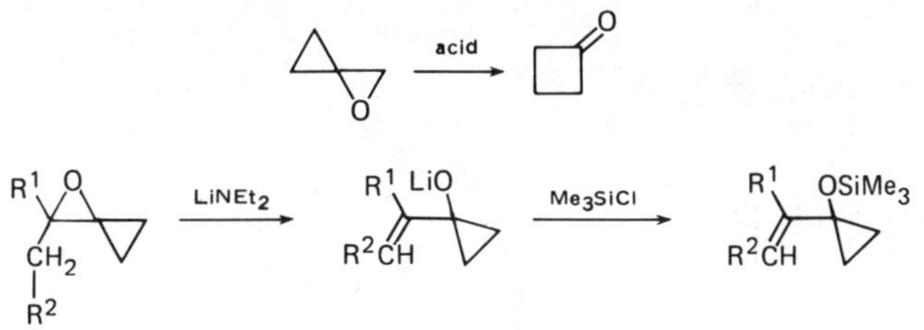

SCHEME 13.

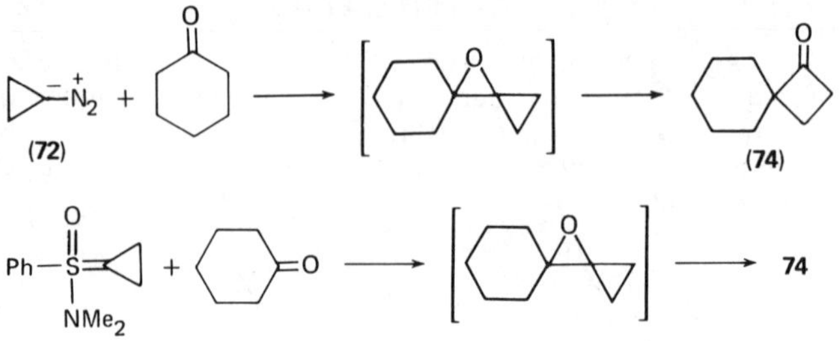

SCHEME 14.

the formation of the spiroketone 74 in the reaction of the diazocyclopropane 72[52] as well as the ylide 73[53] to cyclohexanone as shown in Scheme 14.

Finally, oxaspiropentanes, virtually unknown ten years ago, have proven to be versatile synthetic intermediates[50,51].

B. Allene Episulphides

Allene episulfides (71) are the sulphur analogues of allene oxides. To date, only three examples of allene episulphides are known. The first synthesis by Middleton[54] involved the reaction of bistrifluoromethyl thioketene with bistrifluoromethyl diazomethane to give thiadiazoline (75) as a stable compound. Heating of 75 at reflux for 24 hours gave the tetratrifluoromethylallene episulphide 76 as a stable colourless liquid as shown in equation (19).

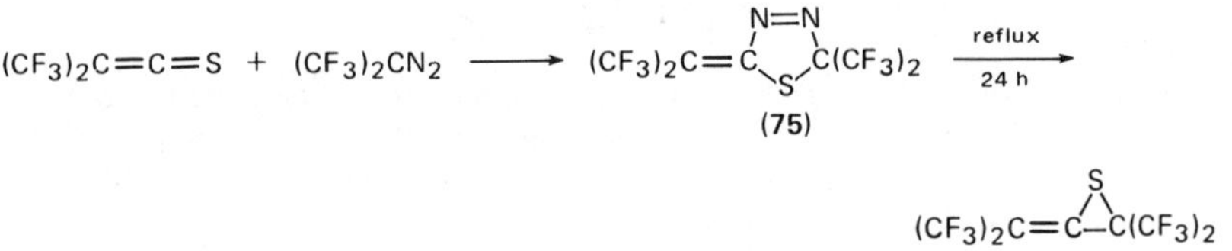

$$(CF_3)_2C=C=S \ + \ (CF_3)_2CN_2 \longrightarrow \ (CF_3)_2C=\overset{\overset{\displaystyle N=N}{|}}{C}\underset{S}{\diagup}C(CF_3)_2 \ \xrightarrow[\text{24 h}]{\text{reflux}}$$

(75)

$$(CF_3)_2C=\overset{S}{\overset{\diagup\diagdown}{C-C}}C(CF_3)_2 \qquad (19)$$

(76)

The tetramethylallene episulphide 80 was prepared by vacuum pyrolysis of 77 as shown in Scheme 15. The carbene 78 and the ylide 79 were proposed[55] as possible intermediates in the pyrolysis of 77 to give 80 as a colourless stable liquid.

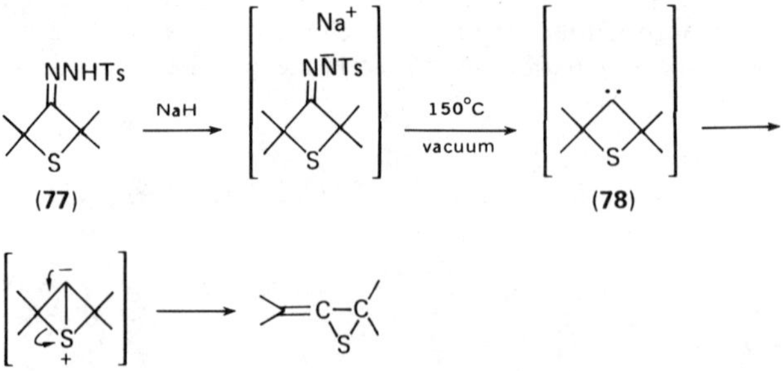

SCHEME 15.

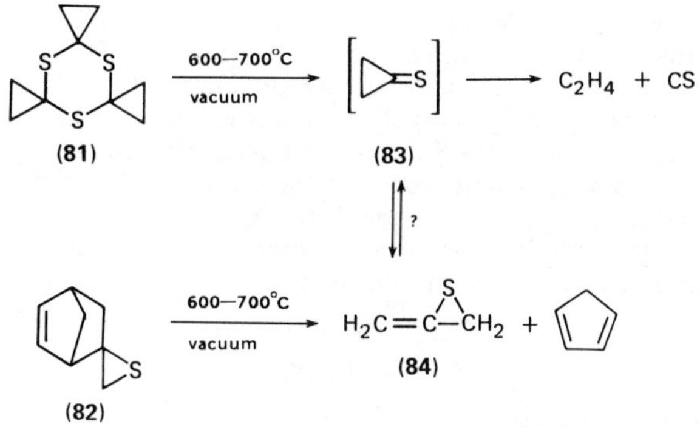

SCHEME 16.

Recently, the parent allene episulphide **84** has been prepared and characterized by flask vacuum pyrolysis of either **81** or **82** as shown in Scheme 16, Decomposition of **81** is proposed to proceed through **83** as evidenced by the formation of ethylene and CS besides **84**, whereas precursor **82** is proposed to give the episulfide directly via a retro-Diels—Alder loss of cyclopentadiene. A microwave determination confirms the structure of **84** with an unsually long $C(sp^3)$—S bond of 1.849 Å. The dipole moment of **84** was found to be 1.36 D^{56}. Allene episulphide was found to have a gas-phase lifetime of about 3 min at room temperature and 20 min at dry-ice temperature at 0.05 Torr[56].

Episulphide **84** can also be prepared by the pyrolyisis of **85** and **86** at 520°C.

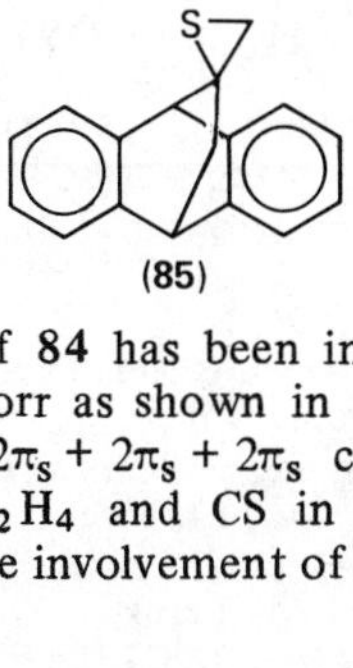

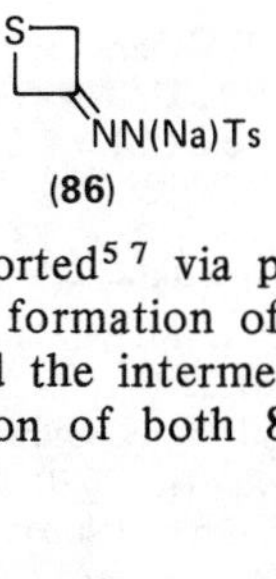

The formation of **84** has been independently reported[57] via pyrolysis of **87** at 500°C and 0.5 Torr as shown in Scheme 17. The formation of **84** from **87** was explained via a $2\pi_s + 2\pi_s + 2\pi_s$ cycloreversion and the intermediacy of **83**. The observation of C_2H_4 and CS in the decomposition of both **81** and **87** seems consistent with the involvement of **83**.

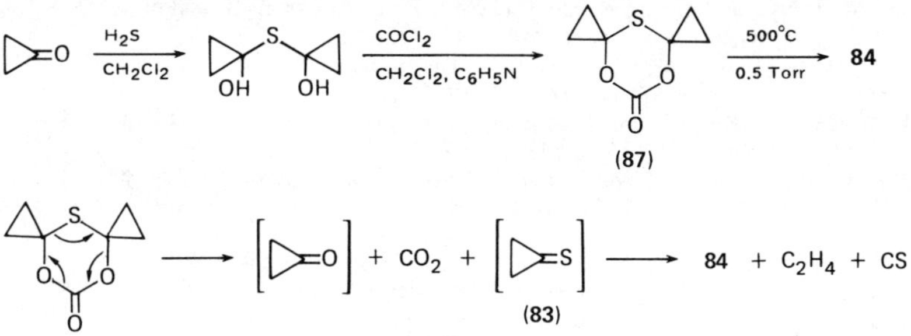

SCHEME 17.

Both experimental observations as well as thermodynamic considerations[56] indicate that the more stable isomer is **84** rather than **83**. Using appropriate bond energies **84** is predicted to be some 7 kcal/mol more stable than **83**[56]. This, of course, is in contrast to the greater stability of the cyclopropanone rather than the allene oxide in the case of the oxygen analogue. The greater stability of cyclopropanone compared to allene oxide is probably partially due to the strong 172 kcal/mol bond strength of a carbonyl, whereas the analogous C=S bond is only 129 kcal/mol thus providing less of a thermodynamic stability to the thiocyclopropanone compared to its isomeric allene episulphide.

VI. ACKNOWLEDGEMENT

Financial support by Public Health Service Research Grant CA 16903-04 from the National Canier Institute is gratefully acknowledged.

VII. REFERENCES

1. R. Hoffman, *J. Amer. Chem. Soc.*, **90**, 1475 (1968).
2. N. Bodor, M. J. S. Dewar, A. Harget and E. Haselbach, *J. Amer. Chem. Soc.*, **92**, 3854 (1970).
3. J. F. Olsen, S. Kang and L. Burnelle, *J. Mol. Struct.*, **9**, 305 (1971).
4. A. Liberles, A. Greenberg and A. Lesk, *J. Amer. Chem. Soc.*, **94**, 8685 (1972).
5. A. Liberles, S. Kang and A. Greenberg, *J. Org. Chem.*, **38**, 1922 (1973).
6. M. E. Zandler, C. E. Choc and C. K. Johnson, *J. Amer. Chem. Soc.*, **96**, 3317 (1974).
7. R. C. Bingham, M. J. S. Dewar and D. H. Lo, *J. Amer. Chem. Soc.*, **97**, 1302 (1975).
8. N. J. Turro, *Acc. Chem. Res.*, **2**, 25 (1969).
9. H. E. Zimmerman, D. Dopp and P. S. Huyffer, *J. Amer. Chem. Soc.*, **88**, 5352 (1966); H. E. Zimmerman and D. S. Crumine, *J. Amer. Chem. Soc.*, **90**, 5612 (1968); T. M. Brennan and R. K. Hill, *J. Amer. Chem. Soc.*, **90**, 5615 (1968); P. Kropp, *Org. Photochem.*, **1**, 1 (1968).
10. M. Fisch and F. E. Richards, *J. Amer. Chem. Soc.*, **90**, 1547, 1553 (1968); R. Noyori, F. Shimizu, K. Fukuta, H. Takaya and Y. Hayakawa, *J. Amer. Chem. Soc.*, **99**, 5196 (1977).
11. E. Block, R. E. Penn, M. D. Ennis, T. A. Owens and S. L. Yu, *J. Amer. Chem. Soc.*, **100**, 7436 (1978).
12. S. W. Benson, *Chem. Rev.*, **78**, 23 (1978); **69**, 279 (1969).
13. J. M. Pochan, J. E. Baldwin and W. H. Flygare, *J. Amer. Chem. Soc.*, **91**, 1896 (1969).
14. P. C. Martino, P. B. Shevlin and S. D. Worley, *J. Amer. Chem. Soc.*, **99**, 8003 (1977).
15. For a summary of early references see A. S. Kende, *Org. Reactions*, **11**, 261 (1960).
16. H. O. House and W. F. Gilmore, *J. Amer. Chem. Soc.*, **83**, 3980 (1961); A. W. Fort, *J. Amer. Chem. Soc.*, **84**, 2620, 2625, 4979 (1962); H. O. House and H. W. Thompson, *J. Org. Chem.*, **28**, 164 (1963); H. O. House and G. A. Frank, *J. Org. Chem.*, **30**, 2948 (1965); H. O. House and F. A. Richey, *J. Org. Chem.*, **32**, 2151 (1967); R. C. Cookson and M. J. Nye, *J. Chem. Soc.*, 2009 (1965); R. C. Cookson, B. Halton, I. D. R. Stevens and C. T. Watts, *J. Chem. Soc. (C)*, **473**, 928 (1967).
17. H. H. Wasserman, G. C. Clark and P. C. Turley, *Fortschr. Chem. Forsch.*, **47**, 73 (1974).
18. J. Boeseken, *Rec. Trav. Chim.*, **54**, 657 (1935).
19. V. I. Pansevich-Kolyada and Z. B. Idelchik, *J. Gen. Chem. USSR*, **24**, 1601 (1954).
20. J. K. Crandall and W. H. Machleder, *Tetrahedron Letters*, 6037 (1966).
21. J. K. Crandall and W. H. Machleder, *J. Amer. Chem. Soc.*, **90**, 7292 (1968).
22. J. K. Crandall, W. H. Machleder and M. J. Thomas, *J. Amer. Chem. Soc.*, **90**, 7346 (1968).
23. J. K. Crandall and W. H. Machleder, *J. Amer. Chem. Soc.*, **90**, 7347 (1968).
24. J. K. Crandall and W. H. Machleder, *J. Heterocycl. Chem.*, **6**, 777 (1969).
25. J. K. Crandall, W. H. Machleder and S. A. Sojka, *J. Org. Chem.*, **38**, 1149 (1973).
26. R. L. Camp and F. D. Greene, *J. Amer. Chem. Soc.*, **90**, 7349 (1968).

27. T. H. Chan, M. P. Li, W. Mychejlowskij and D. N. Harpp, *Tetrahedron Letters*, 3511 (1974).
28. T. H. Chan, B. S. Ong and W. Mychejlowskij, *Tetrahedron Letters*, 3253 (1976).
29. B. S. Ong and T. H. Chan, *Tetrahedron Letters*, 3257 (1976).
30. T. H. Chan and B. S. Ong, *J. Org. Chem.*, **43**, 2994 (1978).
31. For reviews and leading references on dehalosilylation see: A. W. P. Jarvie, *J. Organomet. Rev. (A)*, 176 (1970); T. H. Chan, *Acc. Chem. Rev.*, **10**, 442 (1977).
32. T. H. Chan, W. Mychejlowskij, B. S. Ong and D. N. Harpp, *J. Org. Chem.*, **43**, 1526 (1978).
33. For a recent review and leading references on unsaturated carbenes see: P. J. Stang, *Chem. Rev.*, **78**, 383 (1978).
34. Y. N. Kuo and M. J. Nye, *Can. J. Chem.*, **51**, 1995 (1973).
35. M. S. Newman and W. C. Liang, *J. Org. Chem.*, **38**, 2438 (1973).
36. H. M. R. Hoffmann and R. H. Smithers, *Angew. Chem. (Intern Ed. Engl.)*, **9**, 71 (1970).
37. H. M. R. Hoffmann, K. E. Clemens, E. A. Schmidt and R. H. Smithers, *J. Amer. Chem. Soc.*, **94**, 3201 (1972).
38. W. P. Reeves and G. G. Stroebel, *Tetrahedron Letters*, 2945 (1971).
39. For a previous discussion of the allene oxide–cyclopropanone isomerization and a review of early work on allene oxides see: F. D. Greene, R. L. Camp, L. Kim, J. F. Pazos, D. B. Sclove and C. J. Wilberson, *Proc. Internat. Congr. Pure Appl. Chem.*, **2**, 325 (1971).
40. B. S. Ong and T. H. Chang, *Heterocycles*, **7**, 913 (1977).
41. J. Grimaldi and M. Bertrand, *Tetrahedron Letters*, 3269 (1969).
42. W. Skorianetz, K. H. Schulte-Elke and G. Ohloff, *Helv. Chim. Acta*, **54**, 1913 (1971); N. Harada, S. Suzuki, H. Uda and H. Ueno, *J. Amer. Chem. Soc.*, **94**, 1777 (1972).
43. J. D. Cox, *Tetrahedron*, **19**, 1175 (1963).
44. P. v. R. Schleyer, J. E. Williams and K. R. Blanchard, *J. Amer. Chem. Soc.*, **92**, 2377 (1970).
45. J. K. Crandall and D. R. Paulson, *J. Org. Chem.*, **33**, 991 (1968); *Tetrahedron Letters*, 2751 (1969).
46. J. R. Salaun and J. M. Conia, *Chem. Commun.*, 1579 (1971); *Tetrahedron*, **30**, 1413 (1974).
47. D. H. Aue, M. J. Meshishnek and D. F. Shellhamer, *Tetrahedron Letters*, 4799 (1973).
48. J. K. Crandall and W. W. Conover, *J. Org. Chem.*, **43**, 3533 (1978).
49. M. J. Bogdanowicz and B. M. Trost, *Tetrahedron Letters*, 887 (1972).
50. B. M. Trost and M. J. Bogdanowicz, *J. Amer. Chem. Soc.*, **94**, 4779 (1972).
51. B. M. Trost and M. J. Bogdanowicz, *J. Amer. Chem. Soc.*, **95**, 289, 5311, 5321 (1973).
52. J. R. Wiseman and H. F. Chan, *J. Amer. Chem. Soc.*, **92**, 4749 (1970).
53. C. R. Johnson, G. F. Katekar, R. F. Huxol and E. R. Janiga, *J. Amer. Chem. Soc.*, **93**, 3771 (1971).
54. W. J. Middleton, *J. Org. Chem.*, **34**, 3201 (1969).
55. A. G. Hortmann and A. Bhattacharjya, *J. Amer. Chem. Soc.*, **98**, 7081 (1976).
56. E. Block, R. E. Penn, M. D. Ennis, T. A. Owens and S.-L. Yu, *J. Amer. Chem. Soc.*, **100**, 7436 (1978).
57. E. Jongejan, Th. S. V. Buys, H. Steinberg and Th. J. DeBoer, *Rec. Trav. Chim.*, **97**, 214 (1978).

CHAPTER **20**

Advances in the chemistry of acetals, ketals and ortho esters

R. G. BERGSTROM

California State University, Hayward, California, U.S.A.

I. INTRODUCTION

Acetals and ketals are characterized by the presence of two alkoxy groups (–OR) attached to a carbon atom. Acetals (1) differ from ketals (2) in that they always have at least one hydrogen atom attached to the central carbon atom involved in C–O bond formation. Ketals are obtained by replacing the hydrogen atom of the acetal with an alkyl group (–R). Because of the similarity of acetals and ketals, it is common to find both categorized as acetals. Replacement of the alkyl group of **2** with an alkoxy group leads to an ortho ester (**3**).

881

$$
\begin{array}{ccc}
\underset{\substack{| \\ R}}{\overset{\substack{H \\ |}}{RO-C-OR}} &
\underset{\substack{| \\ R}}{\overset{\substack{R \\ |}}{RO-C-OR}} &
\underset{\substack{| \\ OR}}{\overset{\substack{R \\ |}}{RO-C-OR}} \\
(1) & (2) & (3)
\end{array}
$$

The corresponding sulphur compounds are known as thioacetals, thioketals (mercaptals) and orthothio esters. They are formed by substitution of the alkoxy groups in 1, 2 or 3 with mercapto groups (−SR). Mixed O, S-acetals are also well known.

Previous reviews of the preparation and chemistry of acetals, ketals and ortho esters have appeared. In two earlier volumes of this series Schmitz and Eichhorn[1] have written a chapter on the chemistry of acetals and ketals, and Cordes[2] has contributed a chapter on ortho esters. Ortho esters have also been reviewed by DeWolfe[3] in his monograph on ortho acid derivatives. The mechanism of hydrolysis of acetals and related substances has been the subject of several reviews[4-7] since 1970, the most comprehensive by Cordes[8] appearing in 1974. Since these reviews are so recent and readily accessible, this chapter will deal primarily with material published since 1973.

We begin this review with a discussion of some recent developments in the synthesis of acetals, ortho esters and related substances. Mechanistic considerations are also included whenever they may serve to clarify conditions conducive to the formation of the compounds. It should be noted that during the last few years a good deal of important work on the hydrolysis of acetals has been carried out in a number of laboratories. Consequently, in order to bring the subject up to date, we shall devote a substantial portion of this chapter to the hydrolysis mechanism and its useful implications.

II. FORMATION OF ACETALS, KETALS AND ORTHO ESTERS

A. Introduction

The chief methods for preparing acetals, ketals, ortho esters and their thio analogues have been treated adequately in the forementioned reviews[1-3] and it will suffice in this chapter to give a perfunctory survey of these methods, in particular giving references to more recent work.

The main methods of formation of acetals and ortho esters involve addition and substitution reactions. Simple acid-catalysed additions of alcohols and thiols to aldehydes and ketones are of primary importance due to the wide use of this reaction as a method of protecting the carbonyl group by conversion to an acetal or related compound. Alcohols and thiols also add readily to oxocarbonium ions[9], alkynes[1] and α,β-unsaturated ethers[1] to yield acetals and thioacetals. Ortho esters are products of alcohol additions to ketene acetals[2].

The second type of reaction involves nucleophilic substitution by an alcohol or thiol for a suitable leaving group attached to the central carbon of the substrate (equation 1). For example, addition of excess alcohol to imidate salts (4) gives

$$ROCR_2Y + RXH \longrightarrow ROCR_2XR + YH \tag{1}$$

$$X = O, S$$

simple[10] or mixed[11,12] ortho esters (equation 2). This reaction, known as the Pinner synthesis, is restricted to substitution by primary and secondary alcohols.

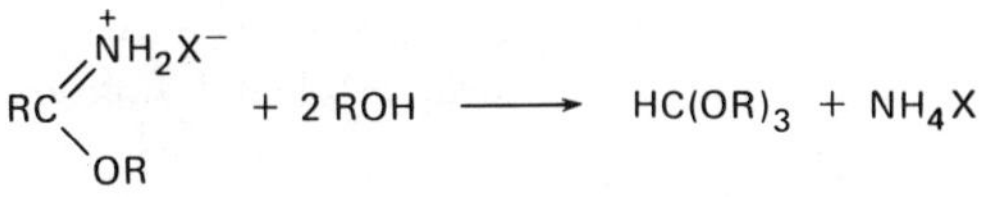

$$RC\overset{\overset{+}{N}H_2X^-}{\underset{OR}{\diagdown}} + 2\,ROH \longrightarrow HC(OR)_3 + NH_4X \tag{2}$$

(4)

Ortho esters may also be obtained from the action of sodium alkoxides on polyhalides[2], as shown in equation (3).

$$R^1CX_3 + 3\,NaOR^2 \longrightarrow R^1C(OR^2)_3 + 3\,NaX \tag{3}$$

B. Some Recent Methods

Direct acetalization (or ketalization) of an aldehyde (or ketone) is not generally an obstacle in synthetic sequences. Sometimes, however, conventional methods fail completely or give low yields when the product is a strained cyclic acetal or an acetal of unusually low stability. Recently, successful syntheses of strained 1,3-dioxacyclanes (5) have been reported involving mixed acetal precursors[13] (equation 4). After initial formation of the mixed acetal, benzene it added and excess alcohol

$$HO-(CH_2)_n-OH + \overset{R^3}{\underset{R^1}{\diagdown}}C{=}O \xrightarrow[2.\ C_6H_6]{1.\ R^2OH/H^+} R^2O-\overset{R^3}{\underset{R^1}{\overset{|}{C}}}-O-(CH_2)_n-O-\overset{R^3}{\underset{R^1}{\overset{|}{C}}}-OR^2 \xrightarrow{\Delta}$$

$$(\overset{.}{C}H_2)_n \overset{O}{\underset{O}{\diagdown}}C\overset{R^3}{\underset{R^1}{\diagup}} \tag{4}$$

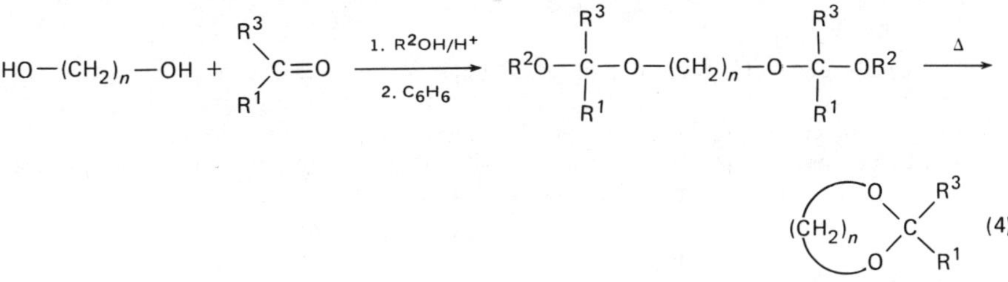

(5)

and water are removed by azeotropic entrainment. Thermal decomposition of the mixed acetal gives rise to the final cyclic acetal.

Monomeric (6) and dimeric (7) 2,2-dimethyl-1,3-dioxacyclanes are formed by

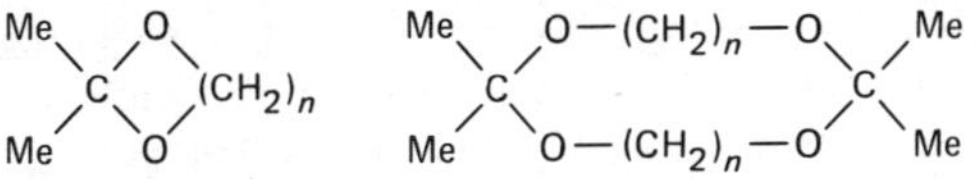

(6) (7)

the reaction of a diol with 2,2-dimethoxypropane under the influence of an acid catalyst[14]. Dimeric cyclic ketals of ring-size 12–22 form readily by this method; however only monomeric cyclic ketals were isolated from 1,3-propane- and 1,4-butane-diol. The dimeric cyclic ketal of butanediol could be prepared from but-2-yn-1,4-diol using this same method and by the oligomerization of 2,2-dimethyl-1,3-dioxopan[15].

Barton, Dawes and Magnus[16] have recently shown that diethylene orthocarbonate (8) is a useful reagent for the conversion of ketones into their corresponding dioxolanes in good yield at room temperature (equation 5). Pyrrole-2- (9a) and

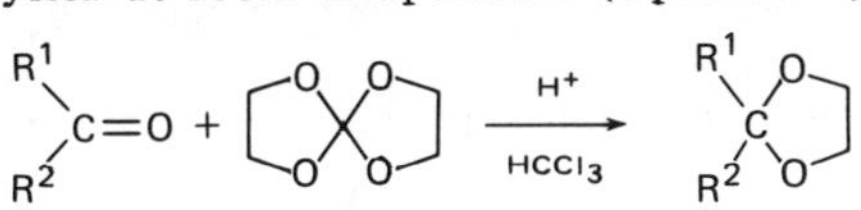

$$\overset{R^1}{\underset{R^2}{\diagdown}}C{=}O + \text{(8)} \xrightarrow[HCCl_3]{H^+} \overset{R^1}{\underset{R^2}{\diagdown}}C\overset{O}{\underset{O}{\diagup}} \tag{5}$$

(8)

pyrrole-3-carbaldehydes (9b) yield interesting acetals on treatment with 2,2-dimethyl-1,3-propanediol in the presence of p-toluenesulphonic acid catalyst and dry benzene[17] (equation 6).

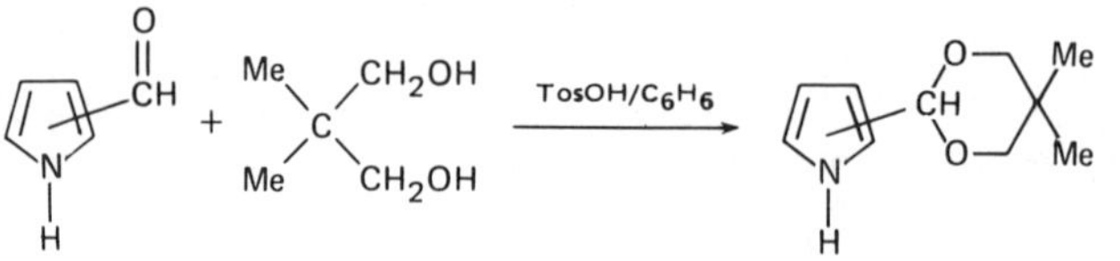

$$\text{(6)}$$

(9) (a) 2-substituted

   (b) 3-substituted

Dimethylformamide–dialkyl sulphate adducts (10) react rapidly with aldehydes and alcohols to give acetals as products in excellent yield[18] (equation 7).

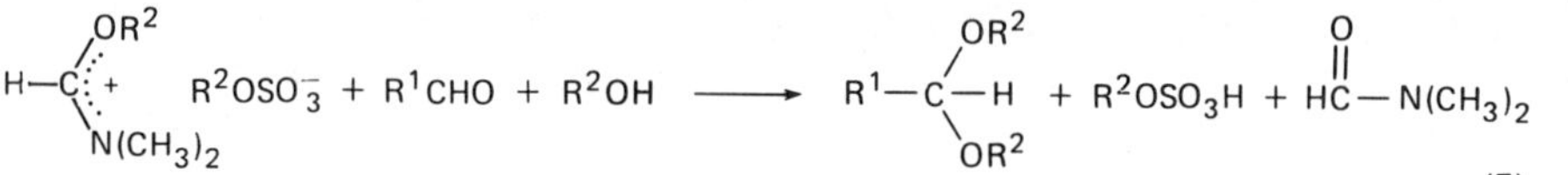

$$H-\underset{\underset{N(CH_3)_2}{|}}{\overset{OR^2}{\overset{|}{C}}} + R^2OSO_3^- + R^1CHO + R^2OH \longrightarrow R^1-\underset{\underset{OR^2}{|}}{\overset{OR^2}{\overset{|}{C}}}-H + R^2OSO_3H + \overset{O}{\overset{||}{HC}}-N(CH_3)_2$$

(10)

$$\text{(7)}$$

Base-catalysed ketalization has also been observed. Newkome, Sauer and McClure[19] showed that di-2-pyridyl ketone (11) could be converted to 2,2-di(2-pyridyl)1,3-dioxolane (12) in 45% yield in refluxing 2-chloroethanol with anhydrous lithium carbonate added (equation 8). The reaction is believed to proceed through initial quaternization of 11 by 2-chloroethanol.

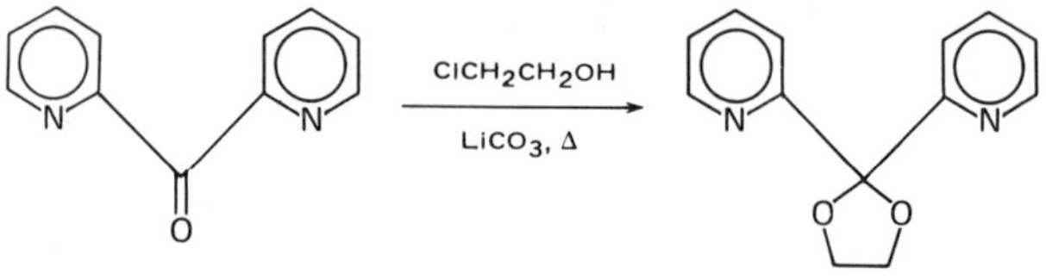

$$\xrightarrow[\text{LiCO}_3, \Delta]{\text{ClCH}_2\text{CH}_2\text{OH}}$$

$$\text{(8)}$$

(11)   (12)

Hall and coworkers[20-23] have recently developed methods of preparing highly reactive bicyclic acetals. The syntheses require diol acetals (13) as intermediates, which undergo intramolecular acid-catalysed acetal exchange to yield bicyclic acetals, as illustrated in the synthesis of 2,6-dioxabicyclo[2.2.2] octane (14) (equation 9).

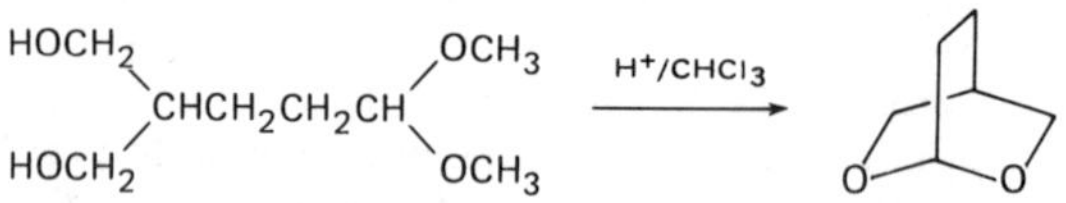

$$\xrightarrow{\text{H}^+/\text{CHCl}_3}$$

$$\text{(9)}$$

(13)   (14)

In a similar reaction, ethyl orthoformate (15) reacts with triols (e.g. glycerol) in the presence of p-toluenesulphonic acid as catalyst to give the corresponding bicyclic ortho esters in good yield[24] (equation 10).

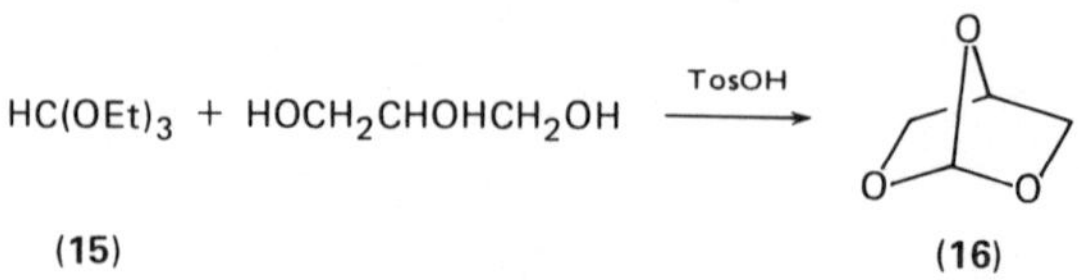

$$HC(OEt)_3 + HOCH_2CHOHCH_2OH \xrightarrow{\text{TosOH}}$$

$$\text{(10)}$$

(15)   (16)

Acetals and ketals have also been recorded as products from the reaction of methyl orthoformate and aldehydes or ketones in the presence of acid catalysts such as sulphuric acid[24-26] ethanolic hydrogen chloride[27,28], p-toluenesulphonic acid[29-31], ferric chloride[32,33], ammonium nitrate[34,35], ammonium chloride[36] or amberlyst-15[37], an acidic ion exchange resin (equation 11).

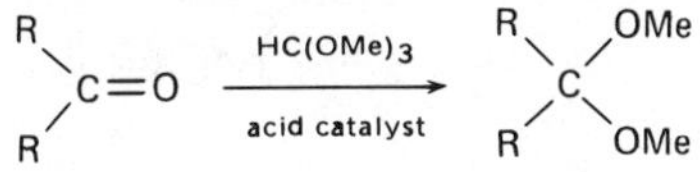

$$
\underset{R}{\overset{R}{>}}C=O \xrightarrow[\text{acid catalyst}]{HC(OMe)_3} \underset{R}{\overset{R}{>}}C\underset{OMe}{\overset{OMe}{<}} \tag{11}
$$

More recently, Taylor and Chiang[38] found that the reaction proceeds most readily and with highest yields ($>90\%$ for all cases reported) when acidic montmorillonite clay K-10 is used as the catalyst.

Evans and coworkers[39] examined a new method for the formation of thioketals: an aldehyde or ketone reacts spontaneously with methylthiotrimethylsilane (17) to give the thioketal (18) in excellent yield in the absence of an acid catalyst (equation 12).

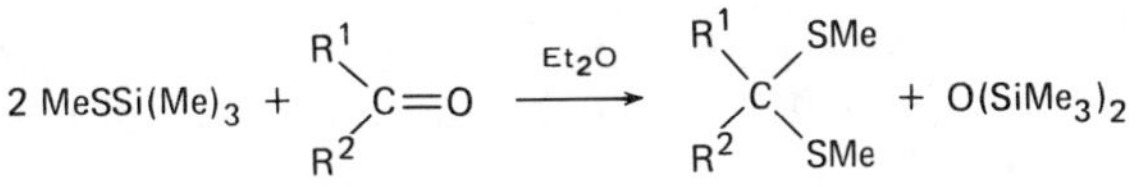

$$
2\,MeSSi(Me)_3 + \underset{R^2}{\overset{R^1}{>}}C=O \xrightarrow{Et_2O} \underset{R^2}{\overset{R^1}{>}}C\underset{SMe}{\overset{SMe}{<}} + O(SiMe_3)_2 \tag{12}
$$

(17)   (18)

C. Miscellaneous Preparations

The following methods are less general, and starting materials may contain functional groups other than carbonyl groups.

1. From olefins

According to Frimer[40], α-hydroxyacetals (21) can be conveniently prepared by the action of a peracid on the corresponding vinyl ether (19) in alcoholic solvents. The proposed mechanism represents formation of an epoxy ether intermediate (20) followed by its rapid solvolysis. The ether oxygen may be either exo- or endo-cyclic as shown in equations (13) and (14). Yields are high and the reaction can be used

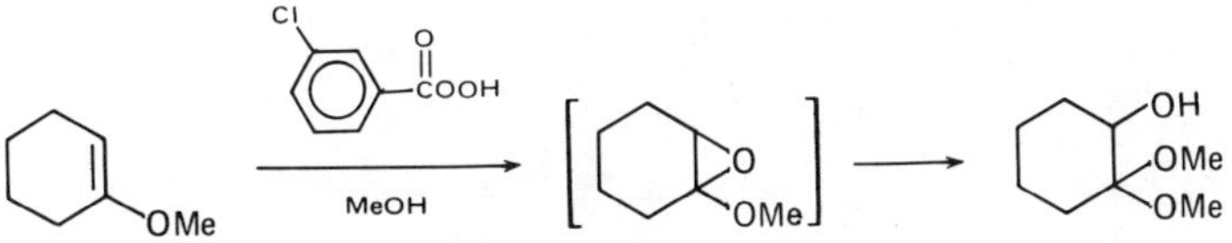

(19)   (20)   (21)   (13)

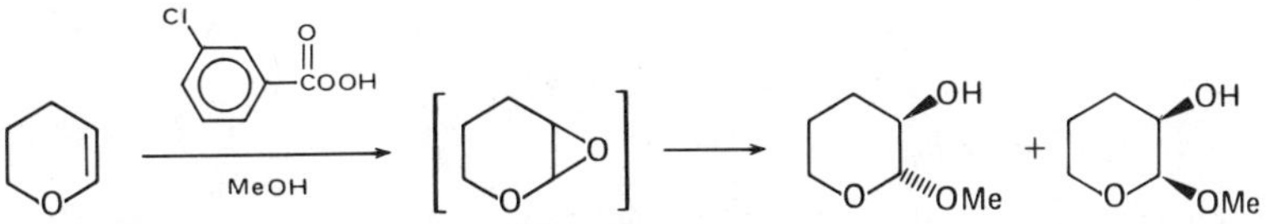

(14)

with acid- and base-sensitive compounds. It is also possible to obtain hydroxy spiroacetals 22 by the reaction of enol acetals with hydroxyketones in the presence of ultraviolet light[41] (equation 15).

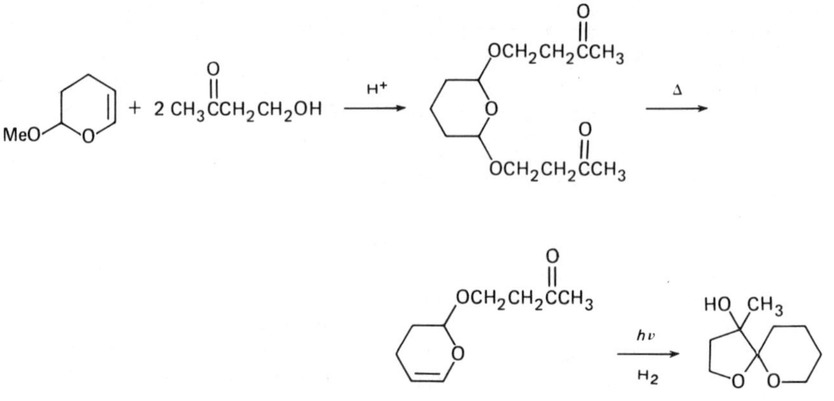

(15)

(22)

Griengl and Bleikolm[42,43] report that 5-alkyl-2,3-dihydrofurans (23) react with 1,3-oxazolidines (24) in dimethyl sulphoxide in the presence of Lewis acids to give cyclic acetals (equation 16).

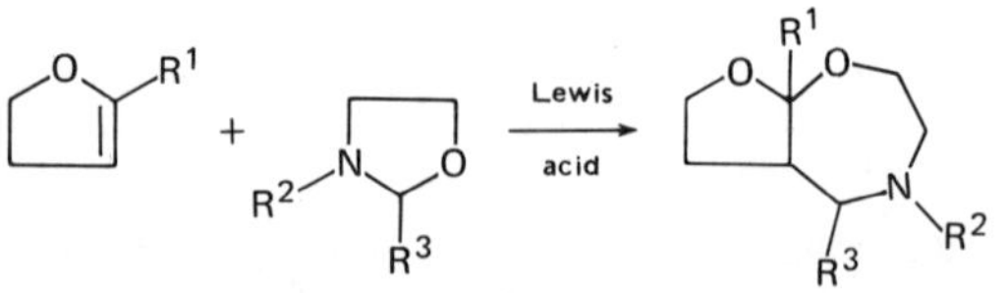

(16)

(23)   (24)

Simple alkenes such as cyclohexene, styrene and 1-phenyl-1-propene (26, R = Me) undergo extremely rapid oxidative rearrangement to give the corresponding dimethyl acetals (25) and (27) by interaction with thallium (III) nitrate absorbed on K-10, a readily available and inexpensive acidic montmorillonite clay, in an inert solvent (heptane, methylene chloride, carbon tetrachloride, toluene, dioxane)[44] (equations 17 and 18).

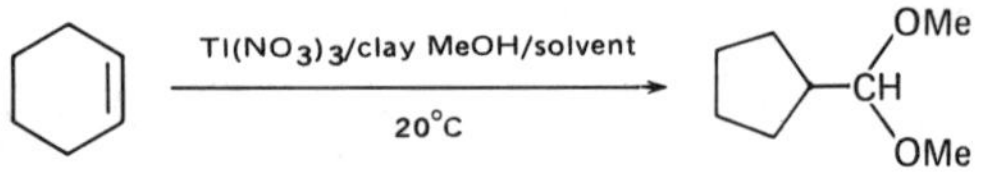

(17)

(25)

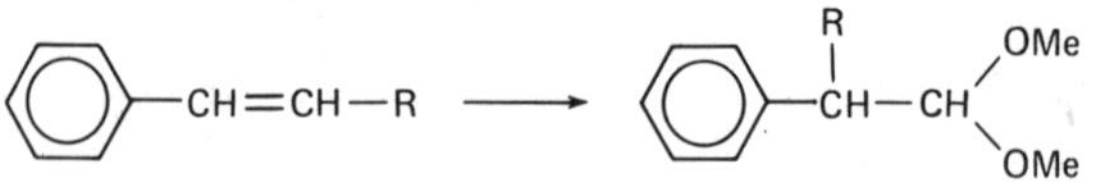

(18)

(26)   (27)

2. From organoborane derivatives

Several preparations of acetals involving boron intermediates have been reported. For example, alkenylboronic acids (28) react with bromine[45] in the presence of sodium methoxide and methanol to form the corresponding α-bromo dimethyl acetals (29) in good yield (equation 19). The reaction apparently proceeds through

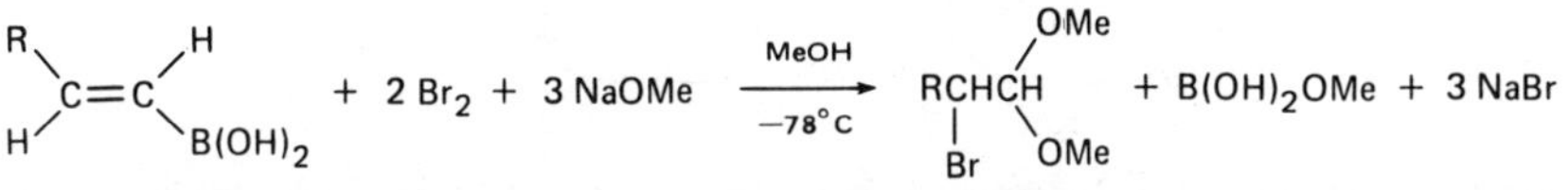

$$\underset{(28)}{\underset{H}{\overset{R}{>}}C=C\overset{H}{\underset{B(OH)_2}{<}}} + 2\,Br_2 + 3\,NaOMe \xrightarrow[-78^\circ C]{MeOH} \underset{(29)}{RCHCH\overset{OMe}{\underset{Br\quad OMe}{<}}} + B(OH)_2OMe + 3\,NaBr \quad (19)$$

a methyl vinyl ether intermediate formed by the *trans* elimination of boron and bromine from **30**.

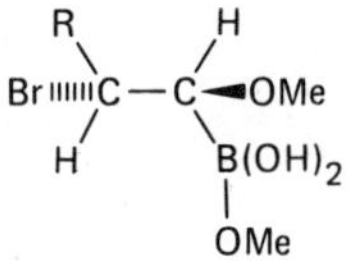

$$\underset{(30)}{Br_{\,\text{\tiny IIIII}}C\overset{R}{\underset{H}{|}}-C\overset{H}{\underset{B(OH)_2}{|}}\blacktriangleleft OMe}$$
$$|$$
$$OMe$$

α-(Phenylthio)alkylboron compounds of the type **31** are efficiently and selectively cleaved by *N*-chlorosuccinimide (NCS) in basic methanol to give the corresponding monothioacetal (**32**) or, in the presence of excess NCS, the acetal[46] (equation 20). The reaction is reported to be compatible with an alkene or acetal

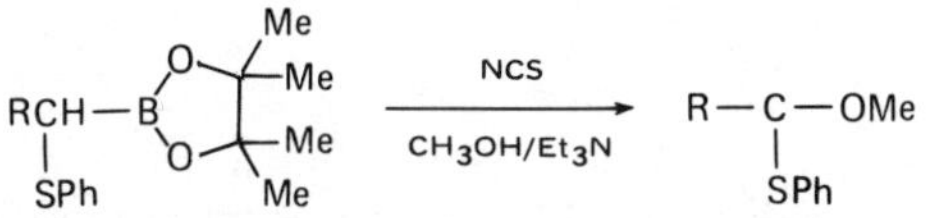

$$\underset{(31)}{RCH\overset{|}{\underset{SPh}{-}}B\overset{O}{\underset{O}{<}}\overset{Me}{\underset{Me}{<}}\overset{-Me}{-Me}} \xrightarrow[CH_3OH/Et_3N]{NCS} \underset{(32)}{R-\overset{|}{\underset{SPh}{C}}-OMe} \quad (20)$$

function elsewhere in the molecule and is useful in that it converts an organoborane directly into a thioacetal under mild basic conditions.

In an isolated example, Clive and Menchen[47] have shown that tris-(phenylseleno)borane (**33**) converts aldehydes and ketones into selenoacetals (**34**) in good yield (equation 21).

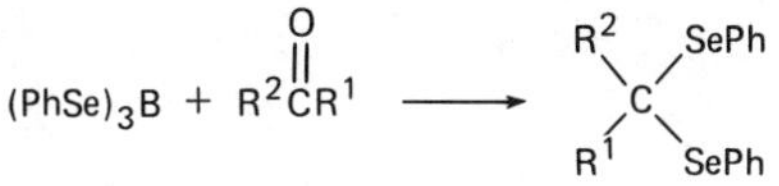

$$\underset{(33)}{(PhSe)_3B + R^2\overset{O}{\overset{||}{C}}R^1} \longrightarrow \underset{(34)}{\overset{R^2}{\underset{R^1}{>}}C\overset{SePh}{\underset{SePh}{<}}} \quad (21)$$

3. From oxidations

Shono and Matsumura[48] have shown that certain aliphatic saturated ethers can be converted to acetals by electrochemical anodic substitution of hydrogen atoms by methoxy groups (equation 22). It was suggested that the reaction involves

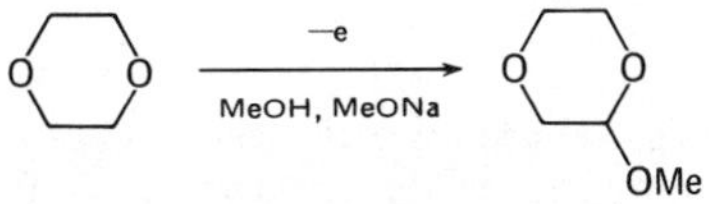

$$(22)$$

hydrogen atom abstraction from the α-position of the ether by an anodically generated radical. Consequently acetal yields are observed to be dependent on the reactivity–selectivity of the α-hydrogen abstraction step.

Extending the foregoing procedure, Scheeren and coworkers[49] showed that acetals can be converted electrochemically into ortho esters (equation 23). Again

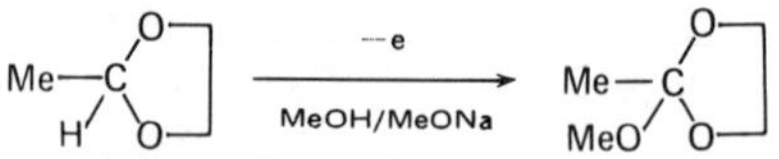
Me
C
O
H
O
−e
MeOH/MeONa
Me
C
O
MeO
O
(23)

the reaction was shown to be dependent on the accessibility of the hydrogen, since acetals with bulky groups at the carbon atom gave low yields.

In addition, 2-methoxy-1,4-dioxanes (**36**) have been obtained electrochemically[50] by anodic oxidation of β-oxocarboxylate ethylene acetals (**35**) (equation 24).

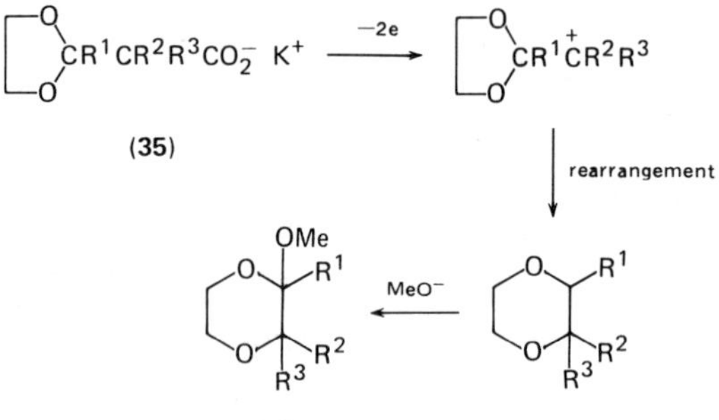
CR¹CR²R³CO₂⁻ K⁺
−2e
CR¹C⁺R²R³
rearrangement
OMe
R¹
MeO⁻
R²
R³
(35)

(24)

(36)

Hewgill and coworkers[51,52] have recently shown that mixtures of phenols with alkoxyphenols can be oxidized by silver oxide or potassium ferricyanide to yield interesting and novel trimeric spiroacetals such as **37** (equation 25). Since one pair

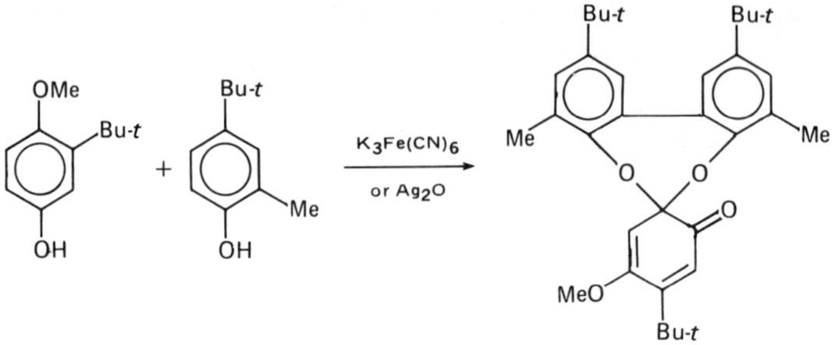
OMe
Bu-t
OH
Bu-t
Me
OH
K₃Fe(CN)₆
or Ag₂O
Bu-t
Bu-t
Me
Me
O
O
O
MeO
Bu-t
(25)

(37)

of phenols can yield up to six trimers, separation of the products can be a formidable task.

III. HYDROLYSIS OF ACETALS, KETALS AND ORTHO ESTERS

A. Introduction

The hydrolysis of acetals, ketals and ortho esters may be generally understood in terms of three basic reaction stages: (1) protonation of the acetal to generate an oxocarbonium ion, (2) hydrolysis of the oxocarbonium ion to a hemiacetal and (3) breakdown of the latter to an alcohol and an aldehyde or ketone (equations 26–28[2,4,8].

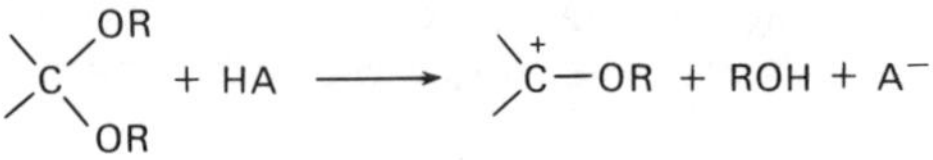
OR
C
OR
+ HA
C⁺—OR + ROH + A⁻
(26)

$$\underset{\diagup}{\overset{\diagdown}{C}}{}^{+}-OR + H_2O \longrightarrow \underset{\diagup}{\overset{\diagdown}{C}}\overset{OH}{\underset{OR}{\diagdown}} + H^+ \qquad (27)$$

$$\underset{\diagup}{\overset{\diagdown}{C}}\overset{OH}{\underset{OR}{\diagdown}} \longrightarrow \underset{\diagup}{\overset{O}{\overset{\|}{C}}}{}_{\diagdown} + ROH \qquad (28)$$

Some mechanistic studies have addressed the problem of ascertaining which stage in the mechanism is rate-determining, while others have investigated the degree to which proton transfer from the catalyst to an ether oxygen of the acetal (equation 26) is sychronous with C—O bond cleavage between this oxygen and the central carbon atom. Investigators have relied primarily on kinetics to elucidate the mechanistic details and for most of the substrates studied the rate-determining step involved C—O bond cleavage[2,4,8] (equation 26). Usually preequilibrium protonation of the acetal occurs much more rapidly than C—O bond cleavage, the hydrolysis being subject to specific acid catalysis. However, general acid catalysis has been observed in a number of acetals, ketals and ortho esters in which either a resonably stable oxocarbonium ion is formed (e.g. tropone diethyl ketal[53,54]) or oxygen basicity is suppressed (e.g. 2-(4-nitrophenoxy)tetrahydropyran[55,56]).

The detection of general acid catalysis implies that proton transfer must be involved in the rate-determining step. The nature of this involvement has presented interesting and challenging mechanistic questions which bear directly on the validity of the currently accepted general mechanism[8] and are of general interest in physical organic chemistry.

Until recently, essentially all kinetic studies inferred that the reaction stage involving formation of the oxocarbonium ion intermediate is the rate-determining step in the hydrolysis[8]. Consequently, direct kinetic studies of the latter stages of the reaction were not possible, although some indirect kinetic investigations have been reported[57-60]. In the remainder of this section, we shall discuss some of the more recent studies which have been carried out on acetal hydrolysis, including those where direct detection and study of the oxocarbonium ion and the hemiacetal intermediates formed in these reactions has been possible.

B. Rate-determining Step

Without apparent exception experimental investigations have shown that acetals, ketals and ortho esters hydrolyse by similar mechanisms at high pH[2,4,8], i.e. rate-limiting formation of the oxocarbonium ion (equation 26). On the other hand, in the pH region near neutrality or below, this conclusion may not be justified. In some recent studies of acetal hydrolysis it has been possible to detect a change in the rate-determining step under certain conditions. The key to the understanding of the changes in the rate-determining step comes from a consideration of the nature of acid catalysis on each step in the hydrolysis mechanism. Discussion of this important aspect of the mechanism will be postponed until the end of this section.

1. Detection of hemiacetal intermediates

Schaleger and coworkers[61,62] thoroughly investigated the kinetics of hydrolysis of 1-methoxy-2-ethyl-1,2-epoxybutane (38) to form methanol and 2-ethyl-2-hydroxybutanal (40) (equation 29). They found that the pH-dependence of the rates of hydrolysis for 38 displayed a maximum at about pH 8, indicative of a

$$Et \overset{O}{\underset{Et}{\diagup\!\!\diagdown}} \overset{OMe}{\underset{H}{}} \longrightarrow (Et)_2\overset{OH}{\underset{|}{C}}-\overset{OH}{\underset{|}{CH}} \longrightarrow (Et)_2\overset{OH}{\underset{|}{C}}CHO + MeOH \qquad (29)$$
$$\qquad\qquad\qquad\qquad\qquad\qquad\qquad OMe$$

(38)   **(39)**   **(40)**

change in the rate-determining step. In the region of the rate maximum the reaction exhibited an induction period which could be accounted for by using the standard rate expression for two consecutive reactions and the rate constants obtained in the high and low acidity regions. The authors argued that these observations lend support to a mechanism in which oxocarbonium ion formation is rate-determining at pH values greater than 8.0, and hydrolysis of a hemiacetal intermediate (39) becomes rate-limiting at low pH values. An alternative explanation for the change in rate-determining step would involve a mechanism where hydrolysis of the oxocarbonium ion has become the slow step at high acidity. However, theoretical and experimental evidence to be discussed below preclude this possibility.

Atkinson and Bruice[63] have similarly observed that during general acid-catalysed hydrolysis of 2-methoxy-3,3-dimethyloxetane (41) (equation 30) an induction period occurs in the pH region 6.1–7.9. As in the preceding example, the authors postulated that the induction period was due to the build-up of hemiacetal.

$$Me\overset{}{\underset{Me}{\diagdown}}\underset{OMe}{\overset{O}{\square}} \longrightarrow HOCH_2CMe_2\overset{OH}{\underset{|}{CH}}-OMe \longrightarrow HOCH_2\overset{Me}{\underset{|}{CC}}-CHO \qquad (30)$$
$$\qquad\qquad\qquad\qquad\qquad\qquad\qquad\qquad\qquad Me$$

(41)   **(42)**

The exceptional behaviour of these two cyclic acetals, **38** and **41**, can be attributed to relief of steric strain in the ground state[63] which facilitates bond breaking and promotes general acid catalysis. Thus one might also expect to detect hemiacetal intermediates during hydrolysis of other acetals in which both alkoxy groups are unusually bulky. In search of such an acetal, Capon[64] reinvestigated the hydrolysis reaction of benzaldehyde di-*t*-butyl acetal (43) (equation 31), originally

$$\underset{OBu\text{-}t}{\overset{OBu\text{-}t}{\bigcirc\!\!-CH}} \overset{-\,t\text{-}BuOH}{\longrightarrow} \underset{OH}{\overset{OBu\text{-}t}{\bigcirc\!\!-CH}} \overset{-\,t\text{-}BuOH}{\longrightarrow} \overset{O}{\underset{}{\bigcirc\!\!-\overset{\|}{C}H}} \qquad (31)$$

(43)

studied by Anderson and Fife[65] and found to be subject to general acid catalysis. He discovered that under the conditions of aqueous buffer concentrations less than 0.025M and in the pH range 4.6–7.0 the reaction of 43 showed an induction period. On the basis of this observation the reaction was postulated to involve hemiacetal intermediates.

Very soon thereafter, Jensen and Lenz[66] showed that hemiacetals could equally well be detected in a number of substituted benzaldehyde diethyl acetals. By means of rapid quenching experiments which utilized the fact that hemiacetal decomposition is acid- and base-catalysed, whereas its formation is only acid-catalysed, these authors were able to determine [hemiacetal]/[acetal] ratios at various reaction times. They concluded that the concentration of hemiacetal can be quite substantial, approaching 40% of the total substrate concentration (for *p*-methoxybenzaldehyde) at optimum times.

Further important evidence for the existence of hemiacetal intermediates in acetal hydrolysis has been gained by means of studies of analogous acylal hydrolysis. Capon and coworkers[67] selected α-acetoxy-α-methoxytoluene (44), an acylal, as a model compound. In 44, the acetoxy function is a much better leaving group than the corresponding alkoxy group in an acetal and consequently its expulsion does not require acid catalysis. Since the authors found that the reaction (equation 32)

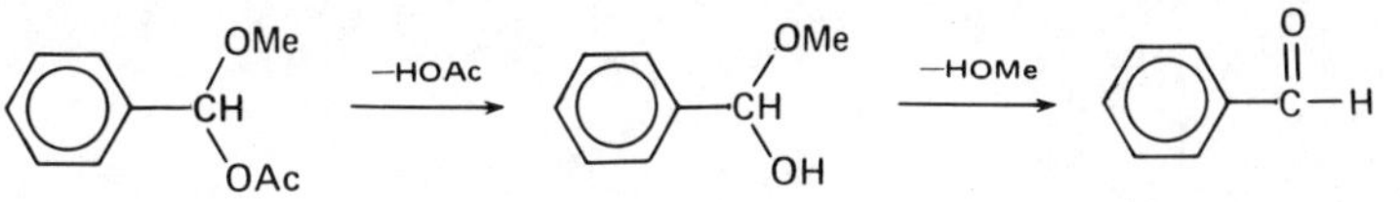

(44) (45) (32)

showed general acid and general base catalysis, they postulated that the rate-determining step in the hydrolysis was decomposition of the hemiacetal (45). This conclusion was further substantiated by the fact that the rate constants for 44 and the α-chloroacetoxy derivative were identical. In a related investigation, Capon and coworkers[68] were able to record the nuclear magnetic resonance spectrum of dimethyl hemiorthoformate (47) derived from the hydrolysis of acetoxydimethoxy methane (46) (equation 33), thus supplying direct spectroscopic evidence for the existence of the hydrogen ortho ester.

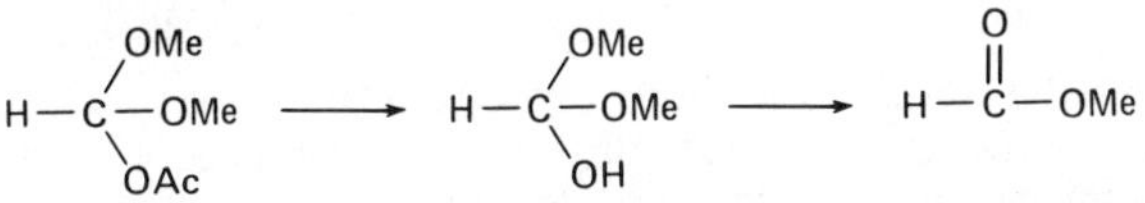

(46) (47) (33)

In an earlier investigation Bladon and Forrest[69] treated cis-3,4-dihydroxytetrahydrofuran with excess trifluoracetic anhydride and obtained a crystalline compound. The cyclic hydrogen ortho ester structure (48), was suggested, since the

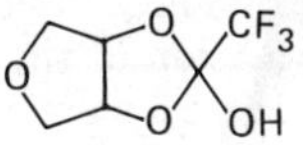

(48)

compound lacked a carbonyl stretching band in the solid infrared spectrum and displayed a proton NMR spectrum characteristic of a cyclic structure.

2. Detection of oxocarbonium ion intermediates

As we have seen in the examples quoted in our preceding discussion, a changeover of the rate-determining step in the overall hydrolysis has allowed the detection and direct measurement of the rate constant for decomposition of the hemiacetal intermediate. In some cases it has also been possible to detect oxocarbonium ions as transient intermediates, again by arranging conditions such that the oxocarbonium ion forms more rapidly than it decays.

Recently, McClelland and Ahmad[70,71] studied the kinetics of hydrolysis of certain ketals and ortho esters, and reported that oxocarbonium ion intermediates could be detected spectroscopically during the reaction. These authors selected as model compounds for the hydrolysis studies ketals known to produce very stable oxocarbonium ions such as tropone diethyl ketal (49)[53,54] trimethyl orthomesitoate (52)[54] and dialkyl ketals of 2,3-diphenylcyclopropenone (55)[54] (equations 34–36).

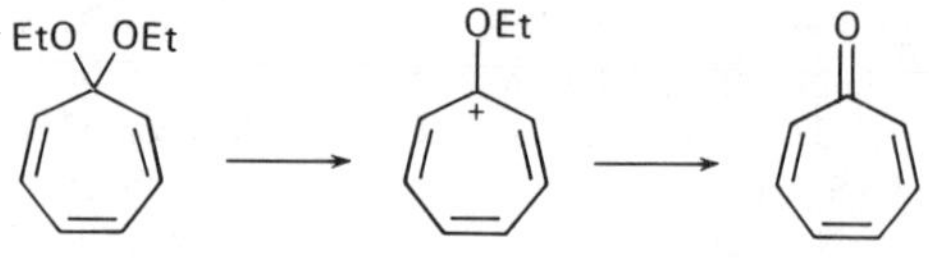

(34)

(49)   (50)   (51)

(35)

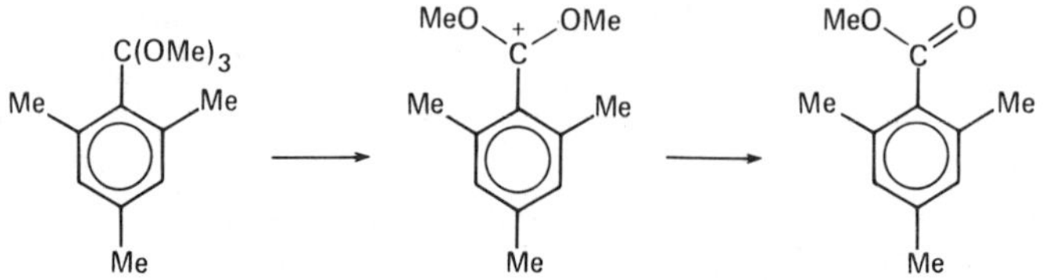

(52)   (53)   (54)

(36)

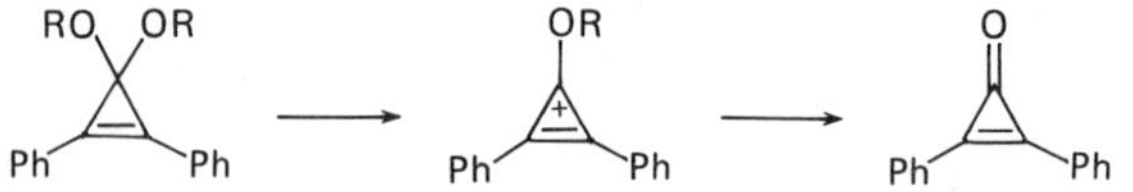

(55)   (56)   (57)

Below pH 5, the initial ultraviolet spectra of aqueous solutions of **49** are identical to the ultraviolet spectrum obtained on dissolving in water the borofluorate salt of the ethoxytropylium ion (**50**). In both cases the spectrum slowly changes to that of tropone (**51**) as the hydrolysis product is formed. The rate constants obtained following this change were identical within experimental error starting with either **49** or the salt **50**. These spectral and kinetic observations were found to be concordant with a mechanism in this pH region involving rapid conversion of the ketal **49** to the oxocarbonium ion **50** and subsequent rate-limiting hydrolysis of **50** to tropone (**51**). Above pH 8.5, formation of the ion, **50**, becomes rate-limiting.

The experimental results for **52** and **55** were analogous to that of **49** and support a similar mechanism for hydrolysis in acidic solutions. Since ions **53** and **56** are much less stabilized than **50**, their rates of decay were found to be significantly faster than that of **50** requiring stopped-flow techniques to obtain rate constants and spectra of the transient oxocarbonium ion intermediates.

For oxocarbonium ions which have very high reactivity in water, i.e. very short life-times, their existence cannot be demonstrated by the direct methods outlined above. One approach to studying these ions has been to follow the hydrolysis in aqueous sulphuric acid solutions where the activity of water is substantially reduced and consequently the reactivity of the ion is decreased[71]. The results obtained in strong acid media are then extrapolated to water.

Recently Young and Jencks[60] have described a different approach for demonstrating the existence of oxocarbonium ions as intermediates in ketal hydrolysis and to estimate the life-time of the free ions. These authors examined the hydrolysis of acetophenone dimethyl ketals (**58**) in the presence of sulphite ion, which acts as a

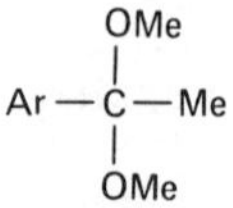

OMe
|
Ar — C — Me
|
OMe

(**58**)

trap for the intermediate oxocarbonium ion derived from **58**. A detailed study of the trapping and partitioning of products obtained from the acid-catalysed cleavage of **58** revealed that the reaction proceeds through a free solvent-equilibriated oxocarbonium ion intermediate. Addition of the sulphite trap did not affect the kinetics of the hydrolysis; therefore, trapping must occur after the rate-determining step. In addition the intermediate was found to have a sufficiently long life-time to react with either sulphite ion or water. This was reflected in the ratios of the rate constants for reaction of the oxocarbonium with 1M sulphite ion (k_S) and with water (k_{H_2O}), which were found to be in the range 1.3×10^{-1} to 7×10^2. The ρ^+ value for the ratio k_{H_2O}/k_S of a series of m- and p-substituted acetophenone dimethyl ketals is 1.6. This suggested that both k_{H_2O} and k_S cannot represent activation-controlled rate constants since the substituent effects on the ratio of rate constants should approximately cancel ($\rho^+ \simeq 0$). This lack of insensitivity of the product ratio to substituent effects taken with the absolute magnitude of the rate ratios, indicated that rate the constant k_S must represent a diffusion-controlled reaction of sulphite ion with the oxocarbonium ion.

Kresge and coworkers[72-74] studied the kinetics of hydrolysis of a series of 2-aryl-(and 2-cyclopropyl-)2-alkoxy-1,3-dioxolanes (**59** and **60**). These compounds

(59)   (60)

are of interest because they represent the only known examples where both oxocarbonium ion and hydrogen ortho ester intermediates can be detected together in the same reacting system.

In dilute acid solutions (pH 4.5–7.5), the first stage of the three-stage mechanism of equations (26)–(28), formation of the dioxolenium ion (**61**), is rate-limiting. Direct evidence for the existence of **61** was provided by the detection of

(61)

N-hydroxybenzimidate ester products[75] upon addition of hydroxylamine as an oxocarbonium ion trapping agent[76,77]. Further evidence for rate-limiting expulsion of the exocyclic alkoxy group was provided by monitoring the reaction using a radiochemical tracer (tritium) in the exocyclic alkoxy group of 2-(2,2-dichloroethoxy)-2-phenyl-1,3-dioxolane. The authors found that the rate of expulsion of the exocyclic group was identical to the rate of formation of the carboxylic ester. When a comparison is made of the rates of acid-catalysed hydrolysis of the substrates containing various exocyclic groups, one finds that the rates depend on the nature of the leaving group. For the series of 2-alkoxy-2-phenyl-1,3-dioxolanes the following relative rates were reported: R = OCH_2CHCl_2 : 1; $OCH_2{\equiv}CH$: 1.34; OCH_2CH_2Cl : 1.48; OCH_2CH_2OMe : 2.11; OMe : 4.36; OEt : 6.60. These data clearly show that loss of the exocyclic group and consequently formation of the dioxolenium ion is involved in the rate-determining steps.

As might be expected from the foregoing discussions, the authors observed that the kinetics for the hydrolysis reaction of **59** underwent a change as the pH of the solution was lowered, and in regions of intermediate acidity a pronounced induction period was observed. In hydrochloric acid solutions of low pH (<3.0), it turns out that the hydronium ion catalytic coefficients (k_{H^+}) become independent of the nature of the exocyclic group. For the series of six substrates which in solutions of high pH gave a sevenfold variation in k_{H^+}, at low pH all give carboxylic acid ester at the same rate ($k_{H^+} = 3.0 \pm 0.13 \times 10^2$ M^{-1} s^{-1}). Evidently at low pH the decomposition of the hydrogen ortho ester (62) has become the slow step.

(62)

Dialkoxycarbonium ions have characteristic ultraviolet spectra with absorption maxima near 300 nm. By using a stopped-flow apparatus as a transient spectrometer, Kresge and coworkers[73] were able to detect dioxolenium ions during the hydrolysis of some cyclic ortho esters. The absorbance of the transient dioxolenium ion present during hydrolysis of 2-methoxy-2-(p-methoxyphenyl)1,3-dioxolane (59; R = Me, X = p-OMe) in 0.5M HClO₄ decayed according to first-order kinetics. The data yielded rate constants identical to those obtained by monitoring the formation of carboxylic acid ester product under the same conditions. At lower acidity the decay of the transient dioxolenium ion (generated from either 59 (R = Me, X = p-OMe) in 0.02M HClO₄ or from the corresponding amide acetal, 2-(N,N-dimethylamine)-2-(p-methoxyphenyl)-1,3-dioxolane), was observed to be biphasic. The initial fast portion of the decay curve could be attributed to reaction between water and the dioxolenium ion since the first-order rate constants which were obtained from the data were of the magnitude expected for reaction of 61 with water[71] ($k = 1.0 \times 10^3$ s^{-1}).

The second slower portion of the biphasic dioxolenium ion decay also yielded first-order rate constants which were identical to those obtained by monitoring the carboxylic acid ester product. This portion of the decay reaction was found to be acid-catalysed, but the relationship between the observed rate constant and the acid concentration was not linear. It was suggested that these experimental results are understandable in terms of a reaction scheme (equation 37) where the dioxolenium ion (61) is in equilibrium with the hydrogen ortho ester (62) plus a proton.

(37)

The rate law required by this mechanism is:

$$k_{obs} = \frac{k_0 + k_{H^+}[H^+]}{1 + [H^+]/K_R}$$

The best values of the three parameters, $k_0 = 1.4$ s^{-1}, $k_{H^+} = 7.5 \times 10^2$ M^{-1} s^{-1} and $pK_R = 1.1$ for 61 (R = Me, X = p-OMe), were obtained by fitting the observed first-order rate constants to this equation.

Dioxolenium ion intermediates could also be detected during hydrolysis of the cyclopropyl derivative 60; however, only the second phase of the decay curve could be discerned. For the other substituted phenyldioxolanes studied[73] (59; R = OMe; X = p-tolyl, H, p-F, p-Cl, p-Br, m-Cl and p-NO₂) only weak transient dioxolenium ion absorbances could be detected and therefore calculation of pK_R values was not possible.

3. Origin of the change in the rate-determining step

In the preceding discussion we have encountered a number of examples of acetals and ortho esters which undergo a change in rate-determining step during hydrolysis as the acidity of the media is varied. However, in general most acetals, ketals and ortho esters are not found to undergo a change in rate-determining step, C—O bond cleavage (equation 26) remaining as the slow step at all pH values. Kresge and coworkers[73] have pointed out that normally, the third stage of the hydrolysis mechanism (equation 28) should always be somewhat faster than the first stage (equation 26), since unstable cationic intermediates (63) like those formed in stage 1, can be avoided in stage 3 (equation 38). Therefore, as the acidity

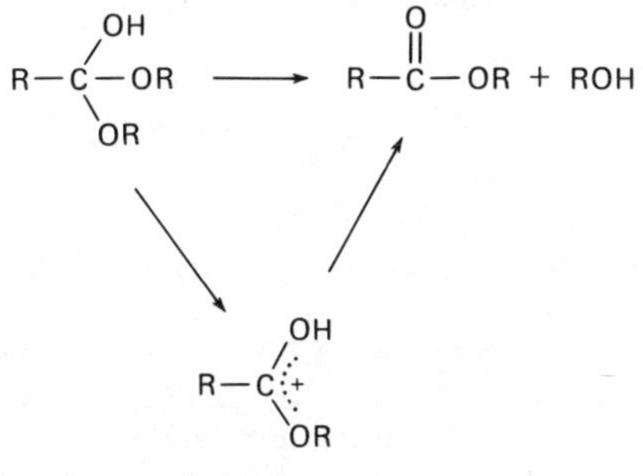

$$
\begin{array}{c}
\text{R—C—OR} \quad \longrightarrow \quad \text{R—C—OR} + \text{ROH} \\
\end{array}
\tag{38}
$$

(63)

of the media is decreased a change in rate-determining step from stage 3 to stage 1 should not occur. This prediction appears to be fully corroborated in the case of acetals, ketals and ortho esters derived from aliphatic substrates. Among the examples which do exhibit a change, various perturbations in the substrates can be recognized that make the first stage of the hydrolysis more rapid than the third stage, and by virtue of the base catalysis of stage 3 allow a change as acidity is decreased. Some of these structural features which can cause an increase in stage 1 have been previously noted: e.g. the highly strained cyclic[61-63] and t-butyl acetals[64,66] and acyloxy ortho esters[68] and acylals[67] which contain very good leaving groups. In the case of aromatic dioxolanes the change in rate-determining step has been ascribed to the phenyl group effect[78].

C. General Acid Catalysis

As we have detailed in the preceding discussion, several examples of acetals, ketals and ortho esters are now known which undergo a change in rate-determining step, providing strong direct evidence for a three-stage mechanism for hydrolysis. We now direct our attention to the first stage of this acid-catalysed reaction — generation of the oxocarbonium ion by loss of an OR group from the substrate.

1. Evidence for concerted C—O bond cleavage

For most substrates studied stage 1 involves rate-determining C—O bond cleavage without accompanying buffer catalysis (A1 mechanism)[4]. The factors which promote general acid catalysis in the hydrolysis of these substrates as well as much of the previous work in this area have recently been reviewed in detail[4,8].

In cases where general acid catalysis has been established unambiguously, the expulsion of the alkoxide ion from the substrate is consistent with a concerted process involving a transition state like **64**.

$$\left[\begin{array}{c} H^{\text{\tiny IIIII}}A^{\delta-} \\ \text{\Large ‖} \\ \overset{\delta+}{\underset{\diagdown}{C}}^{\text{\tiny IIIII}}\overset{..}{O}R \\ \diagup \quad \diagdown OR \end{array} \right]$$

(64)

Alternative mechanisms for general acid catalysis are unattractive. For instance, a stepwise mechanism (equation 39) of general catalysis can be excluded on the

$$\begin{array}{ccc} \diagup OR \\ C \\ \diagup \diagdown OR \end{array} + HA \; \rightleftharpoons \; \begin{array}{c} \diagup H\overset{+}{O}R \\ C \\ \diagup \diagdown OR \end{array} \longrightarrow \begin{array}{c} \diagdown \\ C=\overset{+}{O}R \\ \diagup \end{array} + ROH \qquad (39)$$

basis of several arguments[79]. For this mechanism to satisfactorily account for the observed general acid catalysis, simple proton transfer must be rate-limiting. Acetals, ketals and ortho esters are only weakly basic, the pK_a values of the conjugate acids varying over a range of -3.70 to -8.5^8. Thus, protonation of these compounds in aqueous solution is thermodynamically very unfavourable and the processes should have very late transition states. Assuming that Brønsted exponents, α and β, can be used as a measure of transition state structure[80], it follows that Brønsted plots for the hydrolysis would be expected to yield α-values close to one[81]. However, α-values which have been determined for acetal and ortho ester hydrolysis are generally found to be around 0.5^8.

Secondly, the magnitude of the calculated rate constants for protonation of the substrate is insufficient to account for the observed overall rate constant for hydrolysis. Assuming the rate of deprotonation of the conjugate acid to be diffusion-controlled, 10^{10} M^{-1} s^{-1}, and using known pK_a values of the substrates, the calculated rate constants for protonation are as much as 10^5 times smaller than the observed rate constants.

2. Structure—reactivity relationships

The general problem of concerted versus stepwise reaction pathways, such as the hydrolysis of acetals, ketals and ortho esters considered here, has received considerable attention recently and is still a matter of controversy[82-87]. For reactions which can occur by either a stepwise route or by a concerted route one must analyse reaction paths in terms of motion along more than one dimension of a potential energy surface. This approach, recently popularized by More O'Ferrall[88] and Jencks[82], was first used by Ingold, Hughes and Shapiro[89], recognized by Bunnett[90,91] in his formulation of the theory of the variable E2 transition state and later applied to proton transfer reactions by Albery[92]. Thornton has summarized these arguments as the reacting bond rules[93] which consider the effect of change in structure along the reaction coordinate (parallel effects) and effects perpendicular to it (perpendicular effects). Parallel effects correspond closely to the predictions based on the Leffler—Hammond[94,95] postulate while perpendicular effects lead to conclusions opposite of these predictions.

It is useful to illustrate these structure—reactivity relationships on a three-dimensional potential energy contour diagram. Such a diagram (referred to as a

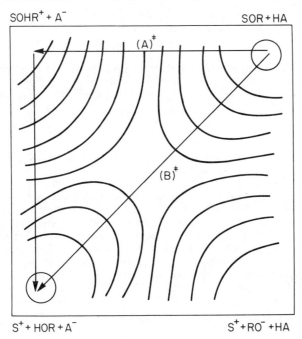

FIGURE 1. A contour map representing the potential energy surface for the first stage of acetal hydrolysis (equation 26). Path (A) represents the stepwise reaction route, while path (B) represents the concerted reaction route.

More O'Ferrall–Jencks plot) for the first stage of hydrolysis of an acetal or related substance is shown in Figure 1. The horizontal axis represents the progress of proton transfer and the vertical axis denotes the progress of C–O bond cleavage. Potential energy is the third dimension, and is represented by the contour lines in the figure. The starting materials, acetal (SOR) and the acid (HA), are in the upper right-hand corner of the diagram, and the products, oxocarbonium ion (S^+), alcohol (ROH) and conjugate base of the catalyst (A^-), are in the lower left-hand corner. Starting in the upper right-hand corner the reaction can proceed via a stepwise reaction mechanism along the edges of the diagram from SOR to $SOHR^+$ through transition state (A), followed by C–O bond cleavage to products S^+. The reaction coordinate for the concerted route would lie near the diagonal from SOR to S^+, avoiding the high-energy intermediates at the corners, and involves passage through transition state (B).

With reference to the two possible pathways for hydrolysis of an acetal presented in Figure 1, we now consider the effect on the system of changing the R group of the substrate. Introduction of an electron-withdrawing substituent into R will stabilize RO^- and destabilize $SOHR^+$. Consequently, the upper left-hand corner of Figure 1 will be raised relative to the lower right-hand corner. This will induce a parallel shift of transition state (A) for the stepwise process toward the destabilized corner (a Hammond effect). The result will be a transition state involving more proton transfer and more positive charge development on the oxygen atom in SOR, corresponding to an increase in the Brønsted exponent α. On the other hand, this same change in R will cause transition state (B) for the concerted pathway to move toward the stabilized corner (anti-Hammond, perpendicular effect). This will result in measurement of lower Brønsted α-values as the electron-withdrawing power of R increases.

It is convenient to express the relationship between the extent of proton transfer (α) and the basicity of the proton accepting site of the leaving group (pK_{lg}) in terms of the interaction coefficient $p_{xy'}$[96] of equation (40). Since hydrolysis via a

$$p_{xy'} = \frac{\partial \alpha}{\partial pK_{lg}} \tag{40}$$

concerted mechanism predicts an increase in α with increasing basicity of the leaving alcohol, this corresponds to a positive $p_{xy'}$ coefficient. A look into the experimental picture clearly shows that anti-Hammond behaviour has been observed in the hydrolysis of acetals and ortho esters and in a number of other systems in which alcohol or water is expelled from a substrate by a concerted acid-catalysed pathway. For instance, Capon and Nimmo[97] obtained an interaction coefficient $p_{xy'} = 0.2$ for the aryl oxide ion expulsion from benzaldehyde aryl methyl acetals, and Kresge and coworkers[74] in a similar study of alcohol expulsion from 2-alkoxy-2-phenyl-1,3-dioxolanes obtained a value of $p_{xy'} = 0.08$. Other studies include alkoxide ion expulsion from addition compounds of phthalimidium ion[79] ($p_{xy'} = 0.07$), from formaldehyde[98] ($p_{xy'} = 0.09$), from tosylhydrazone addition compounds[99] ($p_{xy'} = 0.05$) and from Meisenheimer complexes of the 1,1-dialkoxy-2,6-dinitro-4-X-cyclohexadianate type[100] ($p_{xy'} = 0.12$).

3. Secondary deuterium isotope effects

It is generally believed that the magnitude of secondary deuterium kinetic isotope effects can be used as a probe of transition-state structure. The secondary effects depend on the strengthening or loosening of C–H bonds which are not broken in the rate-determining step. In the hydrolysis of acetals, ketals and ortho esters, the hybridization of the central carbon changes from sp^3 to sp^2 with a concomitant change of the C–H bond force constants. Thus k_H/k_D should reveal the 'product-like' or 'reactant-like' nature of the transition state. Earlier investigations of secondary deuterium isotope effects in acid-catalysed hydrolysis of acetals, ketals and ortho esters have been surveyed in detail by Cordes[8].

Recently, Lamaty and Nguyen[101] determined the α-secondary isotope effect for the hydrolysis of benzaldehyde ethyl phenyl acetal (65) catalysed by a series of

$$\begin{array}{c} Ph \quad\quad OEt \\ \diagdown\;\diagup \\ C \\ \diagup\;\diagdown \\ D \quad\quad OPh \end{array}$$

(65)

acetic and cacodylic acid buffers. The reaction was found to exhibit an α-secondary isotope effect which depended on the strength of the acid catalyst. At $25°C$, k_H/k_D are: for H_3O^+, 1.045; acetic acid, 1.175; cacodylic acid, 1.190; H_2O, 1.243. Thus these data indicate that as the strength of the catalysing acid decreases there is a shift toward a transition state that more closely resembles the carbonium ion.

It is interesting to consider this trend in the α-deuterium isotope effect with reference to a More O'Ferrall–Jencks diagram (Figure 1). If the strength of the acid catalyst is increased, the right-hand side of the diagram will be raised relative to the left-hand side of the diagram. If the reaction coordinate is diagonal this will have the effect of moving the position of the transition state toward the upper right-hand corner (Leffler–Hammond effect[94,95]) and at the same time toward the upper left-hand corner (Thornton effect[93]). The resultant of the vectors for the movements will cause the reaction coordinate to move closer to the top edge of the diagram and in the direction of less C—O bond cleavage, in agreement with the observed isotope effects.

D. Medium Effects

1. Kinetic solvent isotope effects

The kinetic solvent isotope effects resulting from a change in solvent from H_2O to D_2O in the hydrolysis of various acetals and ortho esters can be useful in studying the reaction mechanism. For the A1 mechanism, preequilibrium proton transfer followed by a rate-determining reaction of the protonated substrate, the calculations of Schowen[102] predict, and it is observed experimentally[8], that k_{H_2O}/k_{D_2O} should fall in the range 0.29–0.43. On the other hand, if the first step involves rate-determining proton transfer (A-S$_E$2) the reaction will be influenced by both primary and secondary isotope effects. In a rate-determining proton transfer from hydronium ion to an acetal or ortho ester the maximum value of k_{H_2O}/k_{D_2O} will be around 3 since both primary and secondary effects contribute[103]. Only primary effects are important when the proton transfer is from a molecular acid and values of k_{H_2O}/k_{D_2O} in the neighbourhood of 7 are expected[103]. In cases where general catalysis can be detected in the hydrolysis of acetals and ortho esters, the observed isotope effects fall in the range $k_{H_2O}/k_{D_2O} = 1.4-3.4$[8,104]. Consequently, these results do not support the A1 mechanism, nor are they large enough to be in complete agreement with a true A-S$_E$2 mechanism. These results might be interpreted as supporting evidence for the concerted process involving proton transfer and C—O bond breakage occurring in the same step. This view, however, is in opposition to the rule of Swain, Kuhn and Schowen[105], which states that, for proton transfers between electronegative atoms in a reaction which requires heavy atom reorganization, the proton lies in a completely bonded potential well and should not give rise to primary hydrogen isotope effects. In other words, the hydrogenic motion must take place in a rapid step before or after C—O bond breakage. It follows then, that a Brønsted plot should have a slope α equal to zero or one, contrary to what is observed experimentally (α generally has a value around 0.5).

Recently, Eliason and Kreevoy[106] attempted to resolve the question of the apparent failure of the Swain—Schown rule. They have shown that application of a hydrogenic potential function model that has a double minium and a shallow central maximum leads to the correct prediction of the experimental results. In this model the transferring proton is always in a bound state, retaining zero point energy, while the reaction coordinate consists almost entirely of heavy atom motion. A similar model also has been proposed by Young and Jencks[60].

2. Salt effects

Kubler and coworkers examined kinetic salt effects on the hydrolysis of benzaldehyde dimethyl acetal in water[107] and in 95% methanol–5% water[108]. Neutral salts such as alkali metal and ammonium perchlorates and halides increase the rate of acid hydrolysis. The rate enhancement showed specific cation effects in the order $Li^+ < Na^+ < K^+ < NH_4^+$. According to the authors the observed kinetic salt effects could not be rationalized in terms of the Debye–Huckel–Brønsted approach, indicating that factors other than activity coefficient changes (for example steric effects) are important when considering salt effects on acetal hydrolysis reactions.

Gold and Sghibartz[109] examined the kinetic salt effects on the acid-catalysed hydrolysis of some crown ether acetals (66) in dioxane–water (60 : 40 by volume) at $25°C$.

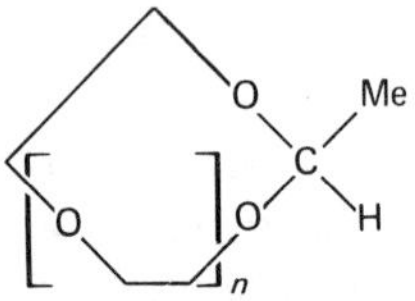

(66)

For the series of compounds (66) with $n = 0–3$, corresponding to acetals containing 5-, 8-, 11- and 14-membered rings and 2,3,4 and 5 oxygen atoms in the ring, respectively, they found that in the presence of $0.25 M$ alkali metal salts only a small increase in the hydrolysis rate was observed. On the other hand, alkali metal salts produced marked rate retardation in acetals of ring-size 17 and 20 ($n = 4–5$). They explained these results by pointing out that these latter acetals have very similar ring-sizes to 18-crown-6 and other cyclic polyethers which are known to be strong chelating agents of alkali metals. It is reasonable then to suppose that cation binding reduces the rate of hydrolysis and accounts for the observed salt effects.

Unlike the crown ether acetals, only small salt effects on the rate of hydrolysis of acyclic ether acetals 67 were found.

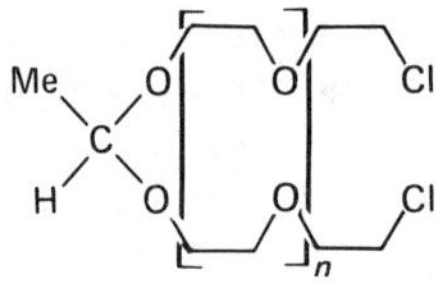

(67)

IV. REFERENCES

1. E. Schmitz and J. Eichhorn in *The Chemistry of the Ether Linkage* (Ed. S. Patai), John Wiley and Sons, London, 1967.
2. E. H. Cordes in *The Chemistry of Carboxylic Acids and Esters*, (Ed. S. Patai), John Wiley and Sons, London, 1969.
3. R. H. DeWolfe, *Carboxylic Ortho Acid Derivatives*, Academic Press, New York, 1970.
4. T. H. Fife, *Acc. Chem. Res.*, **8**, 264 (1972).
5. B. M. Dunn, *Int. J. Chem. Kinet*, **VI**, 143 (1974).
6. B. Capon, *Org. Reac. Mech.*, 1 (1975).
7. A. V. Willi in *Comprehensive Chemical Kinetics*, Vol. 8 (Ed. C. H. Bamford and C. F. H. Tipper), Elsevier, New York, 1977 (review of material to 1970).
8. E. H. Cordes and H. G. Bull, *Chem. Rev.*, **74**, 581 (1974).
9. H. Perst, *Oxonium Ions in Organic Chemistry*, Academic Press, New York, 1971.
10. H. W. Post, *The Chemistry of Aliphatic Ortho Esters*, Rheinhold Publishing Co., New York, 1943.
11. A. Pinner, *Chem. Ber.*, **16**, 352 (1883).

12. A. Pinner, *Chem. Ber.*, **16**, 1643 (1883).
13. M. Anteunis and C. Becu, *Synthesis*, 23 (1974).
14. G. Borgen, *Acta Chem. Scand.*, **B29**, 265 (1975).
15. M. Okada, K. Yagi and H. Sumitomo, *Makromol. Chem.*, **163**, 225 (1973).
16. D. H. R. Barton, C. C. Dawes and P. D. Magnus, *J. Chem. Soc., Chem. Commun.*, 432 (1975).
17. C. E. Loader and H. J. Anderson, *Synthesis*, 295 (1978).
18. W. Kantlehner, H.-D. Gutbrod and P. Gross, *Liebigs Ann. Chem.*, 690 (1974).
19. G. R. Newkome, J. D. Sauer and G. L. McClure, *Tetrahedron Letters*, 1599 (1973).
20. H. K. Hall, Jr., L. J. Carr, R. Kellman and F. DeBlauwe, *J. Amer. Chem. Soc.*, **96**, 7265 (1974).
21. H. K. Hall, Jr. and F. DeBlauwe, *J. Amer. Chem. Soc.*, **97**, 655 (1975).
22. H. K. Hall, Jr. and M. J. Steuck, *J. Polym. Sci. Polym. Chem. Ed.*, **11**, 1035 (1973).
23. H. K. Hall, Jr., F. DeBlauwe, L. J. Carr, V. S. Rao and G. S. Reddy, *J. Polymer. Sci., Symposium No. 56*, 101 (1976).
24. H. K. Hall, Jr., F. DeBlauwe and T. Pyradi, *J. Amer. Chem. Soc.*, **97**, 3854 (1975).
25. R. G. Jones, *J. Amer. Chem. Soc.*, **77**, 4074 (1955).
26. E. Schwenk, G. Fleischer and B. Whitman, *J. Amer. Chem. Soc.*, **60**, 1702 (1938).
27. A. J. Birch, P. Hextall and S. Sternhell, *Australian J. Chem.*, **7**, 256 (1954).
28. L. Claisen, *Ber. Dtsch. Chem. Ges.*, **40**, 3903 (1907).
29. R. Villotti, C. Djerassi and M. J. Ringold, *J. Amer. Chem. Soc.*, **81**, 4566 (1959).
30. C. A. MacKinzie and J. H. Stocker, *J. Org. Chem.*, **20**, 1695 (1955).
31. C. D. Hurd and M. A. Pollack, *J. Amer. Chem. Soc.*, **60**, 1905 (1938).
32. J. Bronstein, S. F. Bedell, P. E. Drummond and C. L. Kosoloski, *J. Amer. Chem. Soc.*, **78**, 83 (1956).
33. L. F. Fieser and M. Fieser, *Reagents for Organic Synthesis*, Vol. 1, John Wiley and Sons, New York, 1967, p. 1206.
34. J. A. Van Allan, *Org. Synth. Coll. Vol. IV*, 21 (1963).
35. J. Klein and E. D. Bergmann, *J. Amer. Chem. Soc.*, **79**, 3452 (1957).
36. L. Claiser, *Ber. Dtsch. Chem. Ges.*, **47**, 3171 (1914).
37. S. A. Patwardham and S. Dev, *Synthesis*, 348 (1974).
38. E. C. Taylor and C.-S. Chiang, *Synthesis*, 467 (1977).
39. D. A. Evans, K. G. Grimm and L. K. Truesdale, *J. Amer. Chem. Soc.*, **97**, 3229 (1975).
40. A. A. Frimer, *Synthesis*, 578 (1977).
41. L. Coltier and G. Descotes, *J. Heterocyclic Chem.*, **14**, 1271 (1977).
42. H. Griengl and A. Bleikolm, *Liebigs Ann. Chem.*, 1783 (1976).
43. H. Griengl and A. Bleikolm, *Liebigs Ann. Chem.*, 1792 (1976).
44. E. C. Taylor, C.-S. Chiang, A. McKillop and J. F. White, *J. Amer. Chem. Soc.*, **98**, 6750 (1976).
45. T. Hamaoka and H. C. Brown, *J. Org. Chem.*, **40**, 1189 (1975).
46. A. Mendoza and D. S. Matteson, *J. Chem. Soc., Chem. Commun.*, 356 (1978).
47. D. L. J. Clive and S. M. Menchen, *J. Chem. Soc., Chem. Commun.*, 356 (1978).
48. T. Shono and Y. Matsumura, *J. Amer. Chem. Soc.*, **91**, 2803 (1969).
49. J. W. Scheeren, H. J. M. Goossens and A. W. H. Top, *Synthesis*, 284 (1978).
50. D. Lelandais, C. Bacquet and J. Einhorn, *J. Chem. Soc., Chem. Commun.*, 195 (1978).
51. F. Hewgill and G. B. Howie, *Australian J. Chem.*, **31**, 1069 (1978).
52. F. Hewgill and D. G. Hewitt, *J. Chem. Soc.*, 3660 (1965).
53. E. Anderson and T. H. Fife, *J. Amer. Chem. Soc.*, **91**, 7163 (1969).
54. T. H. Fife and E. Anderson, *J. Org. Chem.*, **36**, 2357 (1971).
55. T. H. Fife and L. K. Jao, *J. Amer. Chem. Soc.*, **90**, 4081 (1968).
56. T. H. Fife and L. H. Brod, *J. Amer. Chem. Soc.*, **92**, 1681 (1970).
57. J. G. Fullington and E. H. Cordes, *J. Org. Chem.*, **29**, 970 (1964).
58. A. J. Kresge and R. J. Preto, *J. Amer. Chem. Soc.*, **87**, 4593 (1965).
59. K. Koehler and E. H. Cordes, *J. Amer. Chem. Soc.*, **92**, 1576 (1970).
60. P. R. Young and W. P. Jencks, *J. Amer. Chem. Soc.*, **99**, 8238 (1977).
61. A. L. Mori, M. A. Porzio and L. L. Schaleger, *J. Amer. Chem. Soc.*, **94**, 5034 (1972).
62. A. L. Mori and L. L. Schaleger, *J. Amer. Chem. Soc.*, **94**, 5039 (1972).

63. R. F. Atkinson and T. C. Bruice, *J. Amer. Chem. Soc.*, **96**, 819 (1974).
64. B. Capon, *Pure Appl. Chem.*, **49**, 1001 (1977).
65. E. Anderson and T. H. Fife, *J. Amer. Chem. Soc.*, **93**, 1701 (1971).
66. J. L. Jensen and P. A. Lenz, *J. Amer. Chem. Soc.*, **100**, 1291 (1978).
67. B. Capon, K. Nimmo and G. L. Reid, *J. Chem. Soc., Chem. Commun.*, 871 (1976).
68. B. Capon, J. H. Hall and D. McL. A. Grieve, *J. Chem. Soc., Chem. Commun.*, 1034 (1976).
69. P. Bladon and G. C. Forrest, *J. Chem. Soc., Chem. Commun.*, 481 (1966).
70. R. A. McClelland and M. Ahmad, *J. Amer. Chem. Soc.*, **100**, 7027 (1978).
71. R. A. McClelland and M. Ahmad, *J. Amer. Chem. Soc.*, **100**, 7031 (1978).
72. M. Ahmad, R. G. Bergstrom, M. J. Cashen, A. J. Kresge, R. A. McClelland and M. F. Powell, *J. Amer. Chem. Soc.*, **99**, 4827 (1977).
73. M. Ahmad, R. G. Bergstrom, M. J. Cashen, Y. Chiang, A. J. Kresge, R. A. McClelland and M. F. Powell, *J. Amer. Chem. Soc.*, **101**, 2669 (1979).
74. R. G. Bergstrom, M. J. Cashen and A. J. Kresge, *J. Org. Chem.*, **44**, 1639 (1979).
75. Y. Chiang, A. J. Kresge and C. I. Young, *J. Org. Chem.*, **44**, 619 (1979).
76. J. G. Fullington and E. H. Cordes, *J. Org. Chem.*, **29**, 970 (1964).
77. K. Koehler and E. H. Cordes, *J. Amer. Chem. Soc.*, **92**, 1576 (1970).
78. Y. Chiang, A. J. Kresge, S. Salomaa and C. I. Young, *J. Amer. Chem. Soc.*, **96**, 4494 (1974).
79. N. Gravitz and W. P. Jencks, *J. Amer. Chem. Soc.*, **96**, 507, (1974).
80. A. J. Kresge in *Proton Transfer Reactions* (Ed. E. F. Caldin and V. Gold), Chapman and Hall, London, 1975, Chap. 7.
81. M. Eigen, *Angew. Chem. (Intern. Ed.)*, **3**, 1 (1964).
82. W. P. Jencks, *Chem. Rev.*, 72, 705 (1972).
83. J. E. Critchelow, *J. Chem. Soc., Faraday Trans. 1*, **68**, 1774 (1972).
84. R. D. Gandour, G. M. Maggiora and R. L. Schowen, *J. Amer. Chem. Soc.*, **96**, 6967 (1974).
85. W. J. Albery in *Proton Transfer Reactions* (Ed. E. F. Caldin and V. Gold), Chapman and Hall, London, 1975, p. 298.
86. F. G. Bordwell, *Acc. Chem. Res.*, **3**, 281 (1970); **5**, 374 (1972).
87. W. H. Saunders, Jr., *Acc. Chem. Res.*, **9**, 19 (1976).
88. R. A. More O'Ferrall, *J. Chem. Soc. (B)*, 274 (1970).
89. E. D. Hughes, C. K. Ingold and U. G. Shapiro, *J. Chem. Soc.*, 288 (1936).
90. J. F. Bunnett, *Angew. Chem. (Intern. Ed.)*, **1**, 225 (1962).
91. J. F. Bunnett, *Survey Prog. Chem.*, **5**, 53 (1969).
92. W. J. Albery, *Prog. React. Kinet.*, **4**, 355 (1967).
93. E. R. Thornton, *J. Amer. Chem. Soc.*, **89**, 2915 (1967).
94. J. E. Leffler, *Science*, **117**, 340 (1953); J. E. Leffler and E. Grundwald, *Rates and Equilibria of Organic Reactions*, John Wiley and Sons, New York, 1963, p. 162.
95. G. S. Hammond, *J. Amer. Chem. Soc.*, **77**, 334 (1955).
96. D. A. Jencks and W. P. Jencks, *J. Amer. Chem. Soc.*, **99**, 7948 (1977).
97. B. Capon and K. Nimmo, *J. Chem. Soc., Perkin Trans. 2*, 1113 (1975).
98. L. Funderburk, L. Aldwin and W. P. Jencks. *J. Amer. Chem. Soc.*, **100**, 5444 (1978).
99. J. M. Sayer and W. P. Jencks, *J. Amer. Chem. Soc.*, **99**, 464 (1977).
100. C. F. Bernasconi and J. R. Gandler, *J. Amer. Chem. Soc.*, **100**, 8117 (1978).
101. G. Lamaty and M. Nguyen, *IUPAC Fourth International Symposium on Physical Organic Chemistry*, York, England, September, 1978, Abstract number C2.
102. R. L. Schowen, *Prog. Phys. Org. Chem.*, **9**, 275 (1972).
103. J. M. Williams, Jr. and M. M. Kreevoy, *Adv. Phys. Org. Chem.*, **6**, 63 (1968).
104. R. G. Bergstrom and A. J. Kresge, unpublished results.
105. C. G. Swain, D. A. Kuhn and R. L. Schowen, *J. Amer. Chem. Soc.*, **87**, 1553 (1965).
106. R. Eliason and M. M. Kreevoy, *J. Amer. Chem. Soc.*, **100**, 7037 (1978).
107. D. B. Dennison, G. A. Gettys, D. G. Kubler and D. Shepard, *J. Org. Chem.*, **41**, 2344 (1976).
108. T. S. Davis, G. A. Gettys and D. G. Kubler, *J. Org. Chem.*, **41**, 2349 (1976).
109. V. Gold and C. M. Sghibartz, *J. Chem. Soc., Chem. Commun.*, 506 (1978).

CHAPTER **21**

The photochemistry of saturated alcohols, ethers and acetals

CLEMENS VON SONNTAG and
HEINZ-PETER SCHUCHMANN

Institut für Strahlenchemie im Max-Planck-Institut für Kohlenforschung, Stiftstrasse 34–36, D-4330 Mülheim a. d. Ruhr, W. Germany

I. INTRODUCTION

In this review we intend to consider the photochemistry of only those title compounds where the alcohol, ether or acetal function supplies the chromophore. The oxygen lone-pair electrons undergo an n → 3s Rydberg-type transition[1] around 185 nm. The photochemistry of compounds with additional chromophores that are excited at longer wavelengths, such as carbonyl or aryl substituents, is dominated by these chromophores. The 'real' photochemistry of alcohols, ethers and acetals can, strictly speaking, only be studied with the saturated compounds. A less restrictive approach has been taken in two reviews[2,3] in this series.

Since the topic has been reviewed by us recently[4] in some detail, we shall give a briefer and more general account here. Material that has become available in the meantime has been included.

Although the carbonyl-sensitized photolysis[2,3] of the title compounds is not discussed in the present review, we are reporting briefly on the present knowledge of the Hg-sensitized photolysis and the photolysis of O_2-charge-transfer complexes. In both kinds of systems, the alcohol and ether oxygen may be involved as the fulcrum of the interaction.

II. ABSORPTION SPECTRA. ACTINOMETRY AT 185 nm

Saturated alcohols, ethers, and acetals start to absorb noticeably around 200 nm. The maximum of the first absorption band which has been attributed to an $n \rightarrow \sigma^*$[5] or Rydberg-type[1] transition lies near 185 nm. In the gas phase this first absorption band of alcohols is structureless whereas with ethers it usually shows pronounced fine structure[1]. The absorption coefficients of some selected compounds are compiled in Table 1. Formaldehyde dimethyl acetal[16] has a rather low extinction coefficient at 185 nm. Possibly its first absorption maximum lies well below 185 nm. This would correlate with its comparatively high ionization potential[19]. With the remarkable exception of 1,4-dioxane[15], the liquid-phase absorption coefficients of ethers and acetals match those of the gas phase, at least over the range where both can be measured.

With alcohols there is no such matching. Their absorbance at 185 nm is much lower in the neat liquid (e.g. $\epsilon(\text{MeOH}) \approx 7$[20,21], $\epsilon(\text{i-PrOH}) = 32$[22], $\epsilon(t\text{-BuOH}) = 90$[22]) than in the gas phase (see Table 1). This is most likely due to hydrogen bonding in the liquid which causes a blue shift of the absorption band, as is also observed with water[1]. In agreement with this interpretation the extinction coefficient of t-butanol at 185 nm increases on dilution with saturated hydrocarbons[23-25]. At shorter wavelengths other chromophores ($\sigma \rightarrow \bar{\sigma}^*$) are excited. In this wavelength region, fine structure of the absorption bands is observed with alcohols as well[26].

TABLE 1. Molar extinction coefficients (base ten, averaged) of some saturated alcohols, ethers, and acetals at 185 nm in the gas phase

Compound	$\epsilon_{185}(\text{M}^{-1}\,\text{cm}^{-1})$	Reference
Methanol	~ 160	6–8
Isopropanol	~ 240	6–8
t-Butanol	1150	7
Diethyl ether	~2000	7, 9–11
Diisopropyl ether	500	7
Di-t-butyl ether	2200	12
t-Butyl methyl ether	200	9
Tetrahydrofuran	~ 650	13, 14
1,4-Dioxane	3000	14, 15
Formaldehyde dimethyl acetal	50	16
Pivalaldehyde dimethyl acetal	400	17
1,3-Dioxolane	480	18

In the saturated systems considered here, the alcohol or ether chromophore is selectively excited at 185 nm, a major spectral line of the Hg low-pressure arc lamp. The other major spectral line of this lamp, 254 nm, is not absorbed by these systems, or does not contribute significantly to their photolysis. At 185 nm the actinometry of liquid systems is most easily accomplished using the Farkas actinometer, a 5M aqueous ethanol solution which gives H_2 with a quantum yield of 0.4. The Farkas actinometer has been discussed in detail elsewhere[4].

III. PHOTOLYSIS OF ALCOHOLS

Studies have been made of the photochemistry of methanol[26-33], ethanol[27,29,34], isopropanol[27,35-39], t-butanol[24,25,38,40-42] and ethylene glycol[43]. In these systems the quantum yields of the sum of the primary processes leading to products approaches unity. Judging from the work on methanol it appears to make little difference, with respect to the importance of the various primary processes (see Scheme 1), whether the photolysis is carried out in the gas phase[30-33] or in the neat liquid[27-29]. However, considering the strong influence that nonabsorbing solvents exert on the primary processes of t-butanol[24,25] this may not be generally true. Extensive gas-phase studies on the direct photolysis of alcohols other than methanol are lacking.

Primary and secondary alcohols appear to show a similar photolytic behaviour which differs strongly from that of tertiary alcohols if t-butanol is taken as an example which can be generalized.

A. Primary and Secondary Alcohols

The most important process in the photolysis of primary and secondary alcohols is the scission of the O—H bond, a bond which is the strongest in the ground state of these molecules. Two processes are conceivable: (i) the homolytic scission of the O—H bond (reaction 1) or (ii) the elimination of molecular hydrogen (reaction 2).

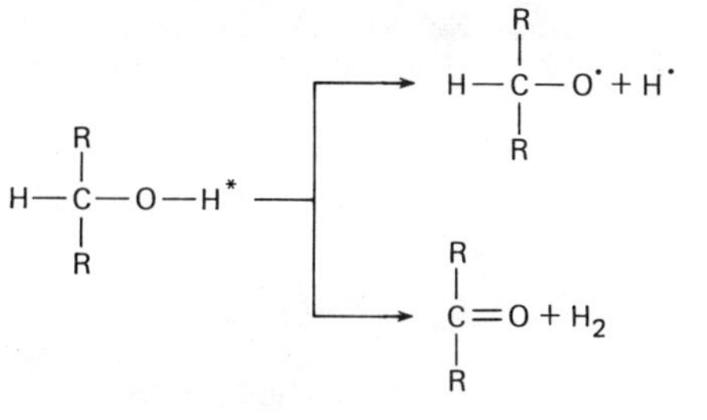

$$H-\overset{\overset{\displaystyle R}{|}}{\underset{\underset{\displaystyle R}{|}}{C}}-O-H^* \quad\begin{cases} \longrightarrow H-\overset{\overset{\displaystyle R}{|}}{\underset{\underset{\displaystyle R}{|}}{C}}-O^{\cdot}+H^{\cdot} & (1) \\[2em] \longrightarrow \overset{\overset{\displaystyle R}{|}}{\underset{\underset{\displaystyle R}{|}}{C}}{=}O+H_2 & (2) \end{cases}$$

	Liquid phase[28,29]	Gas phase[31]
(3) $CH_3O^{\cdot}+H^{\cdot}$ ⩾ 75%	} 88%	} 79%
(4) $H^{\cdot}+{}^{\cdot}CH_2OH$ ⩽ 13%		
(5) CH_2O+H_2	6.5%	20%
(6) ${}^{\cdot}CH_3+{}^{\cdot}OH$	5.5%	1%

CH_3OH^*

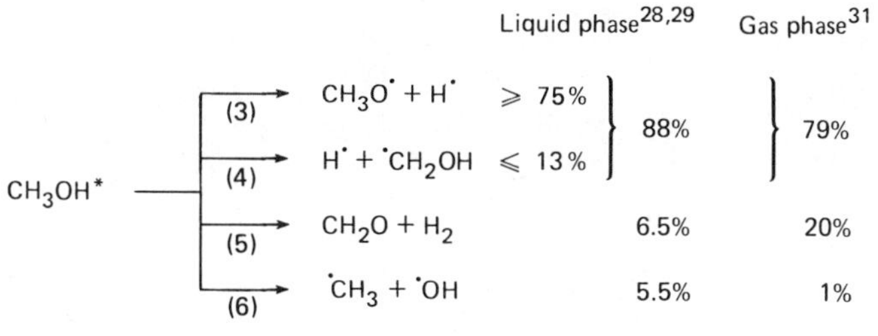

SCHEME 1. 185 nm photolysis of neat methanol.

In methanol process (1) (reaction 3 in Scheme 1) is predominant and process (2) (reaction 5 in Scheme 1) is almost negligible. With increasing methyl substitution process (2) appears to gain at the cost of process (1) (in methanol, $\phi(2)/\phi(1) \approx$ 0.09; in ethanol, $\phi(2)/\phi(1) \approx 1$; in isopropanol, $\phi(2)/\phi(1) \approx 3$). The scission of the C—O bond is of minor importance ($<10\%$) as is the scission of the C—C bond in ethanol ($<2\%$) and isopropanol (5.5%). It has been shown[37] that in isopropanol the C—C bond is preferentially broken via elimination of molecular methane, as depicted in reactions (7) and (8), and that methyl radicals (reaction 9) play only a very minor role. Also of little importance is the homolytic scission of a C—H bond.

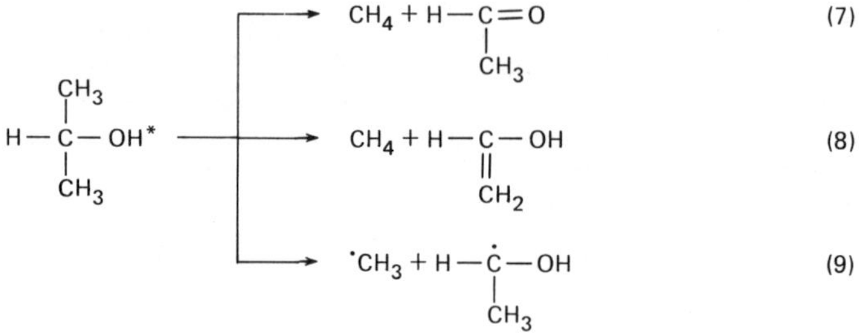

$$CH_4 + H-\underset{\underset{CH_3}{|}}{C}=O \qquad (7)$$

$$H-\underset{\underset{CH_3}{|}}{\overset{\overset{CH_3}{|}}{C}}-OH^* \longrightarrow CH_4 + H-\underset{\underset{CH_2}{||}}{C}-OH \qquad (8)$$

$$^{\bullet}CH_3 + H-\underset{\underset{CH_3}{|}}{\overset{\bullet}{C}}-OH \qquad (9)$$

Such a process does not contribute to more than about 15%, mostly less, of all primary processes in the lower alcohols investigated.

Excitation at wavelengths shorter than 185 nm does not bring about a drastic change in the gas-phase photolysis of methanol[26]. Below a threshold of 130 nm for methanol and 145 nm for C_2-C_4 alcohols the formation of electronically excited OH radicals was observed[44].

B. Tertiary Butanol

It appears that with the primary and secondary alcohols the excitation of the oxygen lone pair activates above all the oxygen—hydrogen bond. However, this is no longer true in neat t-butanol (Scheme 2). Homolytic scission of the O—H bond appears not to occur. Instead O—H bond scission occurs via two molecular modes:

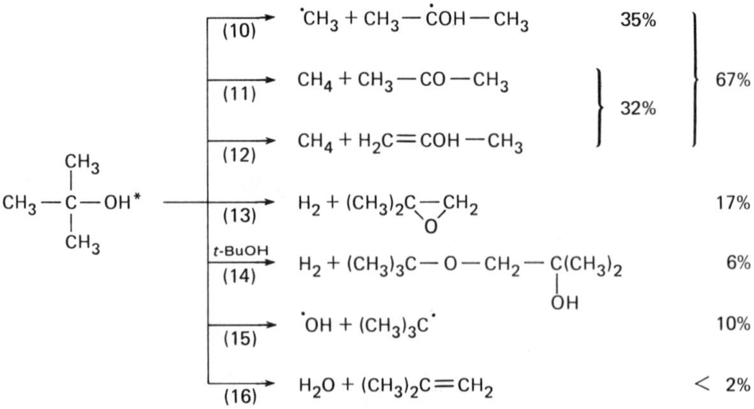

$$CH_3-\underset{\underset{CH_3}{|}}{\overset{\overset{CH_3}{|}}{C}}-OH^* \longrightarrow$$

(10)	$^{\bullet}CH_3 + CH_3-\overset{\bullet}{C}OH-CH_3$	35%
(11)	$CH_4 + CH_3-CO-CH_3$	32% (67%)
(12)	$CH_4 + H_2C=COH-CH_3$	
(13)	$H_2 + (CH_3)_2C-CH_2$ (epoxide, O)	17%
(14) t-BuOH	$H_2 + (CH_3)_3C-O-CH_2-\underset{\underset{OH}{|}}{C}(CH_3)_2$	6%
(15)	$^{\bullet}OH + (CH_3)_3C^{\bullet}$	10%
(16)	$H_2O + (CH_3)_2C=CH_2$	$< 2\%$

SCHEME 2. 185 nm photolysis of neat t-butanol[41].

(*i*) intramolecular epoxide formation (reaction 13), and (*ii*) intermolecular ether formation (reaction 14)[40,41].

A process similar to reaction (13) has also been observed[38] with isopropanol and *s*-butanol but not with ethanol, *n*-propanol, *n*-butanol and isobutanol. The most likely ground-state conformation in the primary alcohols does not favour epoxide formation, whereas the Newman projections show that there is a likelihood of such an interaction in isopropanol and in *t*-butanol. The primary and secondary alcohols

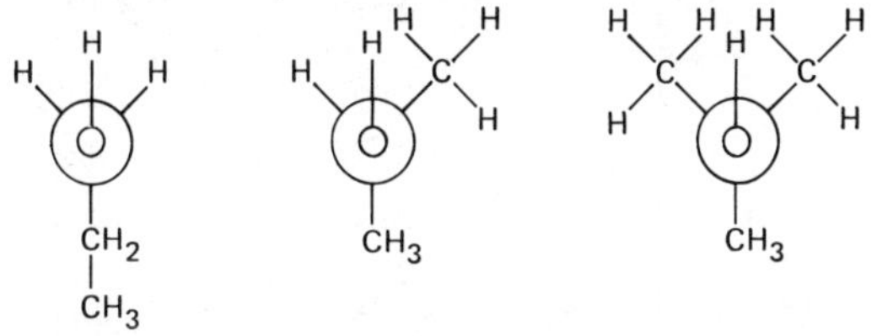

CH_2 | CH_3	CH_3	CH_3
n-Propanol	Isopropanol	*t*-Butanol

could also, in principle, undergo a reaction similar to that depicted in reaction (14). The corresponding products, however, were not observed. If the connection happens to be made to the hydroxyl-bearing carbon atom, one expects not to see the product, which as a hemiacetal is unstable, and the process therefore would be indistinguishable from reaction (6).

In the case of *t*-butanol the scission of a C—C bond is dominant (67%). This is strictly true only for the neat liquid. On dilution with a hydrocarbon such as cyclohexane the importance of C—C bond breakage drops and that of O—H scission rises drastically. An attempt to correlate this effect with changes in the degree of association has been only partially successful[25]. The present state of knowledge is insufficient to theoretically predict the photolytic behaviour of these simple molecules, even when they exist isolated in the gas phase, and there is a still lesser chance to explain such strong solvent effects, considering that the quantum energy is about 200 kJ mol^{-1} above the dissociation energy of any of the bonds involved, and that the energy changes due to hydrogen bonding are only a few kJ. All these strong effects must result from minute alterations in the structure of the excited state.

C. Alkoxide Ions

Alkoxide ions absorb light at longer wavelengths than do the alcohols themselves, and in liquid ammonia electrons are ejected at 254 as well as 316 nm with quantum yields of unity[45] (reaction 17). In liquid ammonia the electrons become solvated and are detected by their blue colour.

$$EtO^- \xrightarrow{h\nu} EtO^\cdot + e^-_{solv} \tag{17}$$

IV. PHOTOLYSIS OF ETHERS

A. Acyclic Ethers

In the photolysis of saturated acyclic ethers at 185 nm in the liquid phase[9,12,46-49] the major process is the scission of a C—O bond. This scission can proceed by homolysis or via a molecular process, the latter being indistinguishable

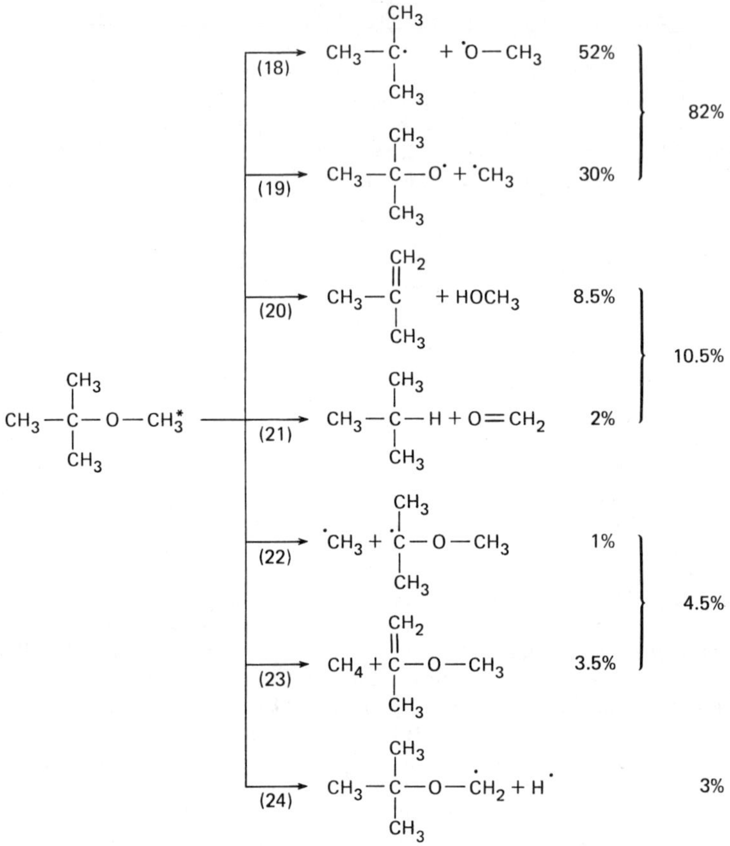

$$
CH_3-\underset{\underset{CH_3}{|}}{\overset{\overset{CH_3}{|}}{C}}-O-CH_3^*
$$

$$
\xrightarrow{(18)} \quad CH_3-\underset{\underset{CH_3}{|}}{\overset{\overset{CH_3}{|}}{C}}\cdot \ + \ \cdot O-CH_3 \qquad 52\%
$$

$$
\xrightarrow{(19)} \quad CH_3-\underset{\underset{CH_3}{|}}{\overset{\overset{CH_3}{|}}{C}}-O^\cdot + \cdot CH_3 \qquad 30\%
$$

$\left.\vphantom{\begin{array}{c}a\\a\\a\end{array}}\right\}$ 82%

$$
\xrightarrow{(20)} \quad CH_3-\underset{\underset{CH_3}{|}}{\overset{\overset{CH_2}{\|}}{C}} \ + \ HOCH_3 \qquad 8.5\%
$$

$$
\xrightarrow{(21)} \quad CH_3-\underset{\underset{CH_3}{|}}{\overset{\overset{CH_3}{|}}{C}}-H + O{=}CH_2 \qquad 2\%
$$

$\left.\vphantom{\begin{array}{c}a\\a\\a\end{array}}\right\}$ 10.5%

$$
\xrightarrow{(22)} \quad \cdot CH_3 + \underset{\underset{CH_3}{|}}{\overset{\overset{CH_3}{|}}{\overset{\cdot}{C}}}-O-CH_3 \qquad 1\%
$$

$$
\xrightarrow{(23)} \quad CH_4 + \underset{\underset{CH_3}{|}}{\overset{\overset{CH_2}{\|}}{C}}-O-CH_3 \qquad 3.5\%
$$

$\left.\vphantom{\begin{array}{c}a\\a\\a\end{array}}\right\}$ 4.5%

$$
\xrightarrow{(24)} \quad CH_3-\underset{\underset{CH_3}{|}}{\overset{\overset{CH_3}{|}}{C}}-O-\overset{\cdot}{C}H_2 + H^\cdot \qquad 3\%
$$

SCHEME 3. 185 nm photolysis of t-butyl methyl ether[9].

from cage disproportionation reactions. A typical example is the photolysis of t-butyl methyl ether. Its primary processes are depicted in Scheme 3. In general, the homolytic scissions predominate over the molecular processes (cf. t-butyl methyl ether[9], diethyl ether[4 6] and methyl n-propyl ether[4 7]). The reverse is the case with di-t-butyl ether[1 2]. Other reactions than those involving the oxygen in one way or other are negligible by comparison (see Scheme 3 and cf. Reference 4).

Asymmetrically substituted ethers split the C—O bonds with different probabilities (reactions 25 and 26). Because of the high hydrogen-abstracting power of the alkoxyl radicals ($R^1O^\cdot$ and $R^2O^\cdot$) these radicals are rapidly converted into the corresponding alcohols (R^1OH and R^2OH). The quantum yields of the alcohols are an approximate measure of the primary processes (25) and (26), approximate only in so far as molecular processes such as reaction (20) in Scheme 3 also contribute to the formation of alcohols. Table 2 gives a compilation of data presently available.

$$R^1-O-R^{2^*}$$

$$\xrightarrow{} R^{1\cdot} + \cdot O-R^2 \qquad (25)$$

$$\xrightarrow{} R^1-O^\cdot + \cdot R^2 \qquad (26)$$

TABLE 2. UV photolysis (λ = 185 nm) of liquid ethers (R^1-O-R^2).
Quantum yields of alcohols

R^1-O-R^2	$\phi(R^1-OH)$		$\phi(R^2-OH)$	Reference
Et–O–Et		0.46		46
Me–O–Pr-n	0.16		0.70	47
Me–O–Pr-i	0.16		0.40	4
Me–O–Bu-n	0.08		0.44	4
Me–O–Bu-t	0.41		0.20	9
t-Bu–O–Bu-t		0.84		12
Et–O–Pr-n	0.31		0.28	4
Et–O–Pr-i	0.26		0.25	4

These data suggest that in the competition between reactions (25) and (26) the smaller alkyl group is split off preferentially, though t-butyl methyl ether presents an exception to the rule. This behaviour of the ether chromophore is in contrast[47] to that of the carbonyl chromophore in aliphatic ketones, where the large alkyl group is preferentially eliminated in the α-cleavage process. No theoretical studies are yet available that could interpret the photolytical behaviour of the ethers.

The photolysis of acyclic ethers in the gas phase[50–52] is probably[4] mechanistically similar to that in the liquid phase. The elucidation of the primary processes on the basis of the products formed is more difficult because of the formation of thermally excited radicals which break down into smaller fragments.

B. Cyclic Ethers

The photolysis of cyclic ethers presents a more complicated picture. Here as well, it is the C—O bond that is mostly cleaved. The intermediacy of a biradical has been suggested in the photolysis of 2,5-dimethyltetrahydrofuran[53] where the *cis* (*trans*) form is converted into the *trans* (*cis*) form (reaction 39, see below). Similar to the acyclic ethers where true molecular processes could not be distinguished from cage disproportionation reactions, the reactive intermediate biradical may undergo disproportionation reactions as well (e.g. reactions 42 and 43, see below). In competition the biradical may, especially in the gas phase at low pressures, undergo a fragmentation by elimination of an unsaturated molecule (e.g. reaction 40, see below) resulting in a smaller biradical. Evidence obtained with tetrahydrofuran[53] indicates that molecular processes also lead to such fragment products.

Table 3 comprises a selection of data obtained in the photolysis of some cyclic ethers in the liquid phase. These data reflect the great differences in the photolytic behaviour of these ethers. For oxiranes no liquid-phase data are available. In oxetanes only breakdown into unsaturated molecules has been observed[54]. In tetrahydrofuran[53], reclosure of the biradical and molecular breakdown into cyclopropane and formaldehyde predominates, whereas in tetrahydropyran[55] mainly the disproportionation products are observed, and breakdown into smaller fragments is negligible (on a further reaction see below). In the oxepane[56] system there is no fragmentation. In 1,4-dioxane[15] only one disproportionation route (or molecular process?), i.e. that leading to the unsaturated alcohol, is observed. Fragmentation, either via the biradical or a molecular process does not halt at the first step (cyclobutane and formaldehyde) but efficiently proceeds to ethylene and further formaldehyde.

The photolysis of the various cyclic ethers is discussed below in more detail.

TABLE 3. Some characteristic products in the photolysis of cyclic ethers in the liquid phase

Ether	Aldehyde	Olefin-alcohol	Simple fragmentation	Double fragmentation	Reference
	CH₃CH₂CHO	H₂C=CHCH₂OH	CH₂O + H₂C=CH₂	CO + H₂ + H₂C=CH₂	
(oxetane, 4-membered ring) — In solution	Absent	Not observed	≥ 95% of products	Little if any	54
(tetrahydrofuran) — Neat	(aldehyde structure: H₃C…CHO), φ = 0.06	(alkenol structure, HO), φ = 0.02	(cyclopropane) + CH₂O, φ = 0.19; H₂C=CH₂ + (oxirane), CH₃CHO, φ = 0.025	None	53
(tetrahydropyran) — Neat	(aldehyde structure: H₃C…CHO), φ = 0.13	(alkenol structure, HO), φ = 0.4	(cyclobutane) + CH₂O, φ = 0.002	2 H₂C=CH₂ + CH₂O, φ = 0.005	55
(oxepane, 7-membered ring) — Neat	(aldehyde structure: H₃C…CHO), φ = 0.09	(alkenol structure, HO), φ = 0.2	(cyclopentane) Absent	None	56
(1,4-dioxane) — Neat	(aldehyde structure: H₃C–O…CHO), Absent	(structure with O and HO), φ = 0.17	(oxetane) + CH₂O, φ = 0.04	H₂C=CH₂ + 2 CH₂O, φ = 0.15	15

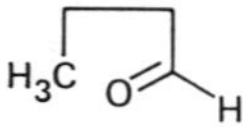

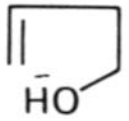

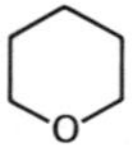

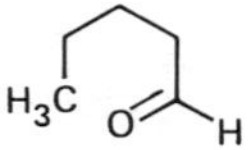

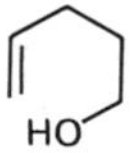

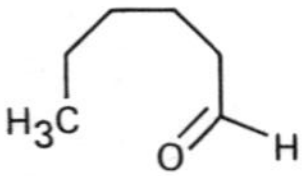

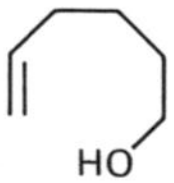

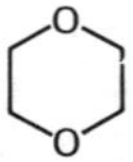

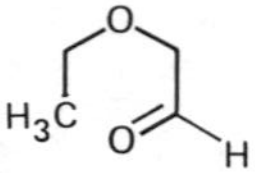

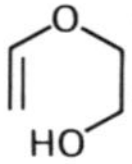

1. Oxiranes

The photolysis of oxirane has been studied in the gas phase[57-60] only. It may be conjectured that the main primary photochemical event is C—O bond scission, which is followed by extensive breakdown into smaller fragments (reaction 27). There is a strong wavelength dependence in the pattern of primary processes[60]. Whereas reaction (28), the extrusion of an oxygen atom and the inverse to epoxide formation[61], is of little importance above 174 nm, it plays a considerable role at 147 nm. At this wavelength two further primary processes (reaction 30 and 31) are believed[60] to set in. Much of the excess energy of reaction (28) is carried off by the ethylene molecule which can break down further into acetylene and hydrogen.

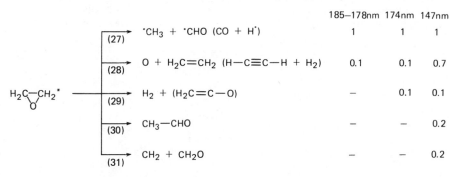

		185–178nm	174nm	147nm
(27)	$\cdot CH_3$ + $\cdot CHO$ (CO + H$\cdot$)	1	1	1
(28)	O + $H_2C=CH_2$ (H—C≡C—H + H_2)	0.1	0.1	0.7
(29)	H_2 + ($H_2C=C—O$)	–	0.1	0.1
(30)	$CH_3—CHO$	–	–	0.2
(31)	CH_2 + CH_2O	–	–	0.2

SCHEME 4. Photolysis of ethylene oxide in the gas phase. Relative importance of primary processes at different wavelengths[60].

The fate of the oxygen atom and the CH_2 remains unclear. If oxygen atoms are generated in the singlet state they might give rise to formaldehyde via insertion into an epoxide C—H bond and subsequent fragmentation. Formaldehyde is the main product at 147 nm and half of it has been ascribed[60] to reaction (32) even at the comparatively low pressures (13 torr) that were employed.

$$\cdot CHO + \cdot H + M \longrightarrow CH_2O + M \qquad (32)$$

Another open question is that of the fate of the oxiranyl radicals which one expects in this system where hydrogen atoms and methyl radicals, possibly hot, are formed with a substantial quantum yield. It is known that such radicals readily undergo fragmentation because of ring strain[62,63].

The 185 nm photolysis of 2-methyloxirane vapour[64] apparently leads to considerable primary rearrangement into propanal. Some acetone and propanal were thought to be produced via rearrangement of the 2- and 3-oxiranyl radicals. It is not known whether there were any hydrogen and hydrocarbons produced.

2. Oxetanes

The photolysis of oxetanes has been studied in the gas phase as well as in isooctane and aqueous solutions[54]. Oxetane has been reported to give exclusively formaldehyde and ethylene (reaction 33) whereas 2,2-dimethyloxetane gives acetone and ethylene (reaction 34) as well as formaldehyde and isobutylene (reaction 35), $\phi(34)/\phi(35)$ being 1.2. The conceivable ring-opened products (see

$$\begin{matrix} CH_2-O^* \\ | \quad\quad | \\ CH_2-CH_2 \end{matrix} \longrightarrow H_2C{=}CH_2 + CH_2O \tag{33}$$

$$\begin{matrix} CH_2-O^* \\ | \quad\quad | \\ CH_2-\underset{\underset{CH_3}{|}}{C}-CH_3 \end{matrix} \quad \left[\begin{array}{l} \longrightarrow H_2C{=}CH_2 + (CH_3)_2CO \tag{34} \\[2mm] \\ \longrightarrow H_2C{=}\underset{\underset{CH_3}{|}}{\overset{\overset{CH_3}{|}}{C}} + CH_2O \tag{35} \end{array} \right.$$

Table 3) were not observed (propionaldehyde) or not looked for (allyl alcohol). At photolysis temperatures above 100°C a chain-reaction sets in (reactions 36–38).

$$\begin{matrix} \overset{\cdot}{C}H-O \\ | \quad\quad | \\ CH_2{-}CH_2 \end{matrix} \longrightarrow H_2C{=}CH_2 + \overset{\cdot}{C}HO \tag{36}$$

$$\overset{\cdot}{C}HO \longrightarrow CO + H^{\cdot} \tag{37}$$

$$H^{\cdot} + \begin{matrix} CH_2-O \\ | \quad\quad | \\ CH_2-CH_2 \end{matrix} \longrightarrow H_2 + \begin{matrix} \overset{\cdot}{C}H-O \\ | \quad\quad | \\ CH_2{-}CH_2 \end{matrix} \tag{38}$$

3. Tetrahydrofurans

The photolysis (λ = 185 nm) of tetrahydrofuran and some of its methyl derivatives has been studied in the liquid phase[53]. The fact that *cis(trans)*-2,5-dimethyltetrahydrofuran gives the *trans(cis)* compound with a quantum yield of 0.2 is good evidence that formation of a biradical by C—O bond scission must play a considerable role (reaction 39). The same biradical may also be considered the

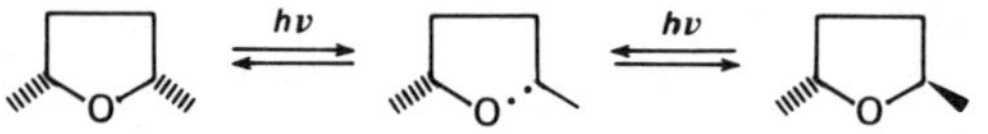

$$\tag{39}$$

precursor of some other products (see reactions 42 and 43, Scheme 5). Substitution of hydrogen by methyl has a strong but as yet unexplained effect on the primary photochemical and some of the subsequent processes. The most noticeable influence is on $\phi(H_2)$ which rises from below 10^{-4} in 2,2,5,5-tetramethyltetrahydrofuran to 0.07 in tetrahydrofuran, 0.17 in 2-methyltetrahydrofuran and ultimately to the high value of 0.29 (0.27) in the case of *trans(cis)*-2,5-dimethyltetrahydrofuran. Although H atoms may be involved (reaction 44) it is not unlikely that in the cases where $\phi(H_2)$ is very high, hydrogen results from a molecular process (reaction 45). A molecular process has also been postulated for the formation of hydrogen in the photolysis of liquid diethyl ether[46].

In the gas-phase photolysis of tetrahydrofuran[65,66] fragmentation dominates the other processes, and the products are not yet thermalized, e.g. the hot cyclopropane from reaction (40) gives largely propene. In the liquid-phase photolysis, however, the cyclopropane: propene ration is 97 : 3[53]. In a recent gas-phase study[67] where some deuterium-labelled tetrahydrofurans were investigated, evidence is presented that not only the hydrocarbon radicals methyl and/or methylene, but also vinyl and allyl are produced in primary processes.

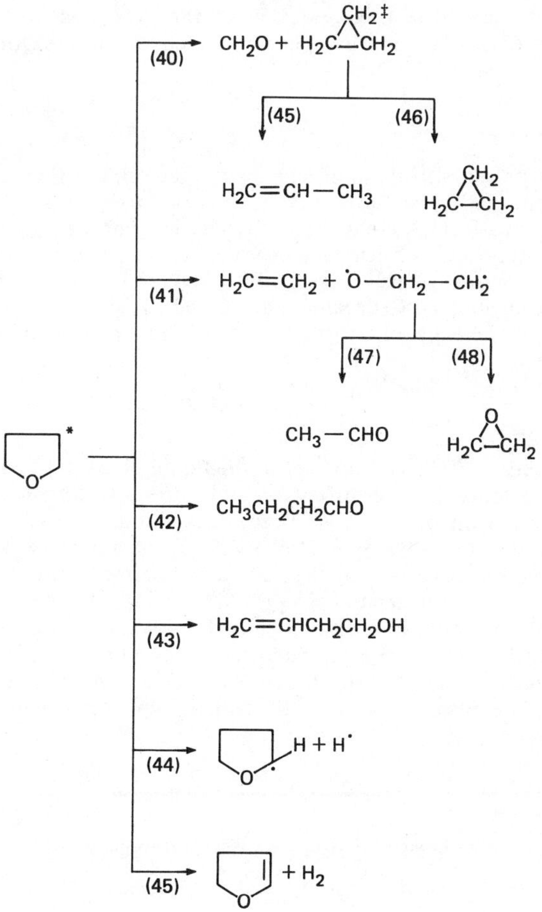

SCHEME 5. 185 nm photolysis of liquid tetrahydrofuran[53].

4. Tetrahydropyran and oxepane

The photolysis of liquid tetrahydropyran[55] at 185 nm resembles that of tetrahydrofuran. Typical products are listed in Table 3. However, there is a major product, 2-(5-hydroxypentyl)tetrahydropyran ($\phi = 0.21$) which appears to be formed without free radicals as intermediates (reaction 46). The mechanism of this

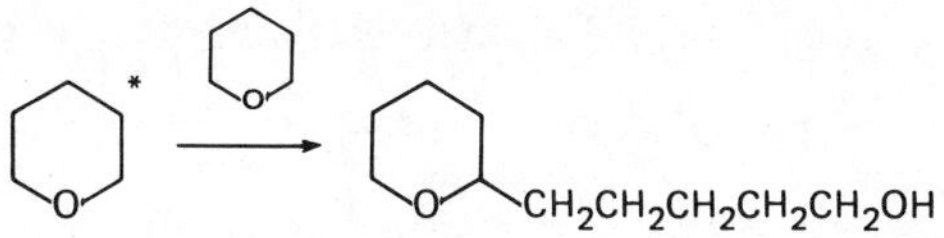

(46)

reaction is not known. Similar compounds also appear to be formed in the photolysis of tetrahydrofuran[53] and oxepane[56], albeit with lower quantum yields.

There is only very little fragmentation of the presumed biradical in the tetra-hydropyran system, and none has been observed in the case of oxepane[56] (see Table 3).

5. 1,4-Dioxane

1,4-Dioxane presents in its photochemistry some interesting features compared to the cyclic ethers discussed hitherto. In the gas phase[68a] it shows a fairly clean decomposition into formaldehyde and ethylene in a ratio of about 2 : 1 (reaction 47), with a quantum yield of ethylene near 0.9. This is also one of the main primary processes in the liquid phase[15] (see Table 3). Similarly, the related compound 1,4,6,9-tetraoxabicyclo[4,4,0]decane photolysed in cyclohexane gives ethylene ($\phi = 0.56$) and ethylene glycol diformate ($\phi = 0.5$) as the only major products[68b].

$$\text{(1,4-dioxane)}^* \longrightarrow 2\,CH_2O + C_2H_4 \tag{47}$$

It has been shown[69,70] to fluoresce in the liquid phase with a quantum yield of 0.03. The fluorescence is blue-shifted on addition of saturated hydrocarbons and red-shifted on addition of water. In both cases the additives decrease the fluorescence quantum yield. N_2O[71,72] also quenches the fluorescence, more strongly than it quenches the formation of the products described above (~85% vs. ~35%). At the same time, nitrogen [$\phi(N_2) \approx 0.6$] and 2-hydroxy-1,4-dioxane are formed[15]. These results have been explained by assuming an excimer state for the fluorescence which is more strongly quenched by N_2O than is the product-forming state[15]. In both cases energy is transferred to N_2O, giving rise to oxygen atoms and nitrogen. The former insert into the C—H bond of 1,4-dioxane giving 2-hydroxy-1,4-dioxane (reactions 48—51).

$$\text{1,4-dioxane} \xrightarrow{h\nu} \text{1,4-dioxane}^* \tag{48}$$

$$\text{1,4-dioxane}^* + N_2O \longrightarrow \text{1,4-dioxane} + N_2O^* \tag{49}$$

$$N_2O^* \longrightarrow N_2 + O \tag{50}$$

$$O + \text{1,4-dioxane} \longrightarrow \text{2-hydroxy-1,4-dioxane} \tag{51}$$

The photolysis of 1,4-dioxane in water appears to be quite different, with hydrogen being a major product. N_2O suppresses the formation of hydrogen, and nitrogen is formed instead with a quantum yield near unity. The corresponding product is bidioxanyl. There are negligible amounts of 2-hydroxy-1,4-dioxane[15]. The proposed mechanism involves the formation of a solvated electron in the first step (reaction 52)[15,71]. The radical cation is considered to rapidly lose a proton (reaction 53). The solvated electron reacts with the proton to give a hydrogen atom (reaction 54), or with N_2O to give a hydroxyl radical (reaction 55). Both will abstract a hydrogen atom from 1,4-dioxane (reaction 56). The resulting dioxanyl

$$\text{1,4-dioxane}^* \longrightarrow (\text{1,4-dioxane})^{\cdot+} + e_{aq}^- \tag{52}$$

$$(\text{1,4-dioxane})^{\cdot+} \longrightarrow (\text{1,4-dioxane}-H)^{\cdot} + H^+ \tag{53}$$

$$e_{aq}^- + H^+ \longrightarrow H^{\cdot} \tag{54}$$

$$e_{aq}^- + N_2O \longrightarrow {}^{\cdot}OH + N_2 + OH^- \tag{55}$$

$$H^{\cdot}(^{\cdot}OH) + 1,4\text{-dioxane} \longrightarrow H_2(H_2O) + (1,4\text{-dioxane} - H)^{\cdot} \tag{56}$$

$$2(1,4\text{-dioxane}-H)^{\cdot} \longrightarrow \text{bidioxanyl} \tag{57}$$

radicals combine to bidioxanyl (reaction 57). Support for the hypothesis of the solvated electron as an intermediate had been drawn from the fact that N_2O and H^+ compete for the same species[71]. However, it has been pointed out[15,71] that the results could also be explained by the assumption that the excited 1,4-dioxane transfers an electron to the proton and to N_2O such that the ratio of the rates of these reactions is the same as the ratio of the rates of the reactions (54) and (55).

A variety of products was found when 1,4-dioxane was irradiated at 254 nm[73,74]. Since 1,4-dioxane is frequently used as a solvent for photochemical reactions in this wavelength region the finding is clearly important. It seems possible, though, that one is dealing here with a decomposition sensitized by traces of carbonyl impurities and oxygen. The latter causes charge-transfer absorptions in ethers (see below). As some of its products are carbonyl compounds the decomposition is self-enhancing.

V. PHOTOLYSIS OF ACETALS

The photolytic behaviour of acyclic saturated acetals resembles that of the ethers. Again, C—O bond cleavage is the major process. Scheme 6 presents the reactions of the simplest compound in this series, formaldehyde dimethyl acetal[16]. Data on acetaldehyde dimethyl acetal[75] and pivalaldehyde dimethyl acetal[17] are also available. Acetaldehyde dimethyl acetal varies in that, to a considerable extent, reaction (64) seems to take place, to the possible exclusion of the molecular route (65), the analogue of which is thought to play a major role in the photolysis of

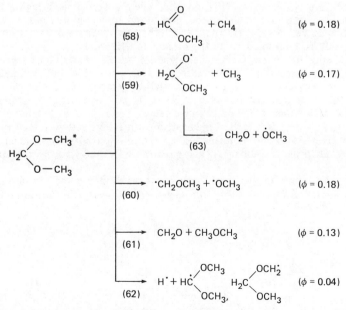

SCHEME 6. Primary processes and their quantum yields in the 185 nm photolysis of liquid formaldehyde dimethyl acetal[16].

$$CH_3-CH(OCH_3)_2{}^* \longrightarrow H_2C=CHOCH_3 + CH_3OH \qquad (64)$$

$$CH_3-CH(OCH_3)_2{}^* \longrightarrow CH_3-COOCH_3 + CH_4 \qquad (65)$$

formaldehyde dimethyl acetal (reaction 58 in Scheme 6). In acetaldehyde dimethyl acetal[75] the scission of the C—C bond in a primary process is only of small importance ($\phi < 0.02$). However, this process appears quite important in pivalaldehyde dimethyl acetal where the processes (66) and (67) together have a quantum yield of 0.16.

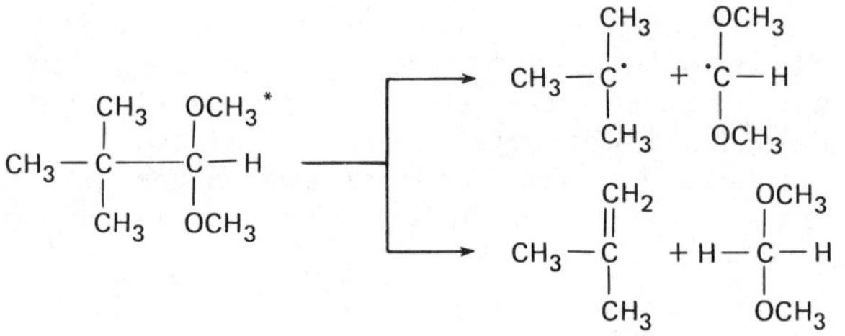

Among the cyclic acetals, 1,3-dioxolane[76,18] and 2,2-dimethyl-1,3-dioxolane[18] have been studied, the former in both the gas phase[76] and the liquid phase[18]. The gas-phase photolysis leads to a nearly complete breakdown into small fragments whereas in the liquid phase some of the intermediates are thermally stabilized so that the reaction paths can be traced with more confidence. The scission of a C—O bond predominates. A mechanism is proposed in Scheme 7. As in the cyclic ethers the intermediacy of biradicals leads to various fragments. CO_2 ($\phi \approx 0.1$), and acetaldehyde and ethylene oxide (together $\phi \approx 0.3$) are important products. The CO_2 may be formed via the dioxirane intermediate or its biradical equivalent (from reaction 68, see Scheme 7). Dioxirane has been detected as a highly unstable product in the ozonization of ethylene, and found to decompose into formic acid, CO and H_2O as well as CO_2, H_2 and 2 H[77,78]. The precursor of acetaldehyde and ethylene oxide is considered to be the biradical ${}^\bullet CH_2-CH_2-O^\bullet$. In 1,3-dioxolane the yield of ethylene oxide is somewhat higher ($\phi = 0.18$) than that of acetaldehyde ($\phi = 0.16$) because reaction (71) can also give ethylene oxide. In 2,3-dimethyl-1,3-dioxolane, ϕ(acetaldehyde) = 0.14 and ϕ(ethylene oxide) = 0.12 has been found, possibly indicating that the biradical ${}^\bullet CH_2-CH_2-O^\bullet$ rearranges to acetaldehyde with a slight preference compared to ring-closure.

In 1,3-dioxolane a free radical-induced chain-reaction sets in (reaction 81 and 82) which gives rise to ethyl formate. The radical-induced rearrangement of 1,3-dioxolanes and other 1,3-dioxacyclanes into esters is well known (cf. References 79—81).

Whereas the quantum yields of primary processes in the 185 nm photolysis of the above aliphatic acetals range from about 0.6 to near unity no products were

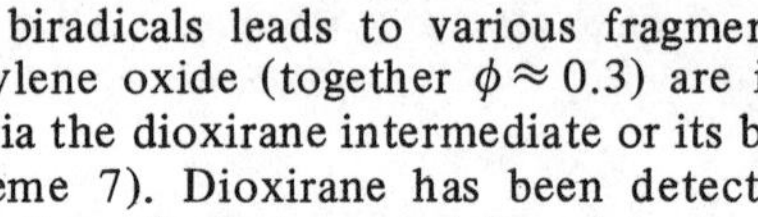

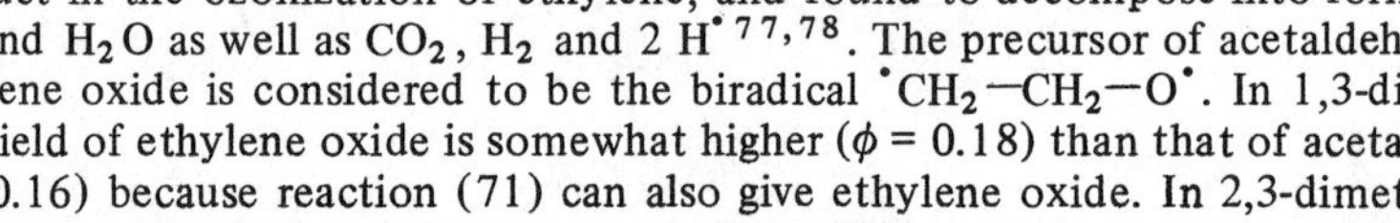

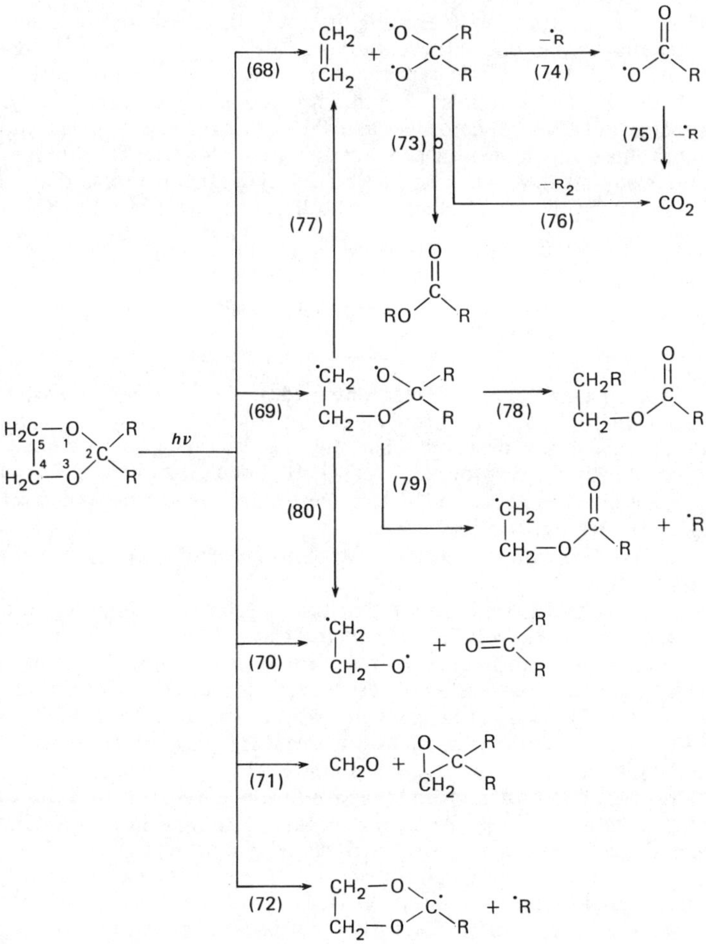

SCHEME 7. Primary processes in the photolysis (λ = 185 nm) of liquid 1,3-dioxolane (R = H) and 2,2-dimethyl-1,3-dioxolane (R = CH_3)[18].

found in the photolysis of 2-phenyl-1,3-dioxolane[82] at 254 nm, where it absorbs strongly. This indicates that primary processes leading to products must have a quantum yield of much less than 10^{-2}, considering that this compound also undergoes radical-induced rearrangement to ethyl benzoate via a chain reaction[83]. Likewise, the photodegradation of polyoxymethylene around 300 nm has been shown[84] to proceed only through sensitization, e.g. if carbonyl groups are present.

VI. Hg-SENSITIZED PHOTOLYSIS OF ALCOHOLS AND ETHERS

The Hg-photosensitized decomposition (λ = 254 nm) of alcohols[85-100] and ethers[88,101-111] has found considerable attention. Two possible primary processes have been envisioned: (i) abstraction of hydrogen by Hg*, and (ii) energy transfer from Hg* to the substrate, with ensuing decomposition. Whereas alcohols (except perhaps t-butanol[94]), acyclic saturated ethers[88,101,103,105,108-110] and alkanes (cf. Reference 112) fit the first hypothesis, the behaviour of vinyl ethers[113-115], epoxides[104,106,107,111], thiols[116] and sulphides (cf. Reference 11) agree more with the second, in that bonds other than those to hydrogen are cleaved, often to the virtual exclusion of hydrogen production. Alcohols suffer O—H bond cleavage, acyclic ethers and alkanes lose a carbon-bound hydrogen. It seems that a complex $(Hg \cdot RH)^*$ might be the intermediate for both paths (equations 83—85). Evidence

$$Hg^* + RH \longrightarrow (Hg \cdot RH)^* \qquad (83)$$

$$(Hg \cdot RH)^* \begin{cases} \longrightarrow R^\cdot + {}^\cdot HgH & (84) \\ \longrightarrow Hg + \text{fragments} & (85) \end{cases}$$

in favour of such complexes has been obtained[99,112,118-120]. In particular, the lifetime of $(Hg \cdot CH_3OH)^*$ has been determined[99] at 14 ns. (A similar complex $(Cd \cdot CH_3OH)^*$ has been observed. The Cd-photosensitized decomposition (λ = 326 nm) apparently also proceeds via O—H bond fission[121]. The transient species HgH has also been observed[97,122]. Depending on whether the substrate was CH_3OH or CH_3OD HgH or HgD was seen[97], supporting conclusions drawn from earlier work that the simple alcohols lose hydrogen from the hydroxyl group in the primary process.

A recent study[100] of the Hg-photosensitized decomposition of liquid methanol and of its aqueous solutions indicated, on the basis of isotopic labelling, that both oxygen- and carbon-bound hydrogen atoms are initially removed. It must be noted that this is a complicated system because Hg* forms complexes with, and decomposes, water as well, even though with a comparatively small quantum yield[100]. In the gas phase, the hydrogen quantum yield of methanol is 30—40 times higher than that of water (cf. Reference 5). One expects, therefore, some decomposition of methanol induced by active species generated from the water.

In acyclic ethers, the case for attack at the C—H bond has been convincingly presented (cf. Reference 101). Epoxides show a more complex behaviour. For instance, in the Hg-sensitized photolysis of $trans$-2,3-epoxybutane[107], methyl radicals play a major role, and some cis isomer was also found. The latter points toward a biradical intermediate. Recently it has been suggested that there may be at least one biphotonic process involved in the Hg-sensitized photolysis of ethylene oxide[111]. A further cause for complexity of the mechanism is the fact that owing to ring strain oxiranyl radicals are prone to ring-opening rearrangement[62,63].

VII. PHOTOLYSIS OF O_2—CHARGE-TRANSFER COMPLEXES

Like many other compounds, ethers[123-130] and alcohols[123,131] on saturation with oxygen show a new absorption in the UV which disappears again when the liquid is purged with an inert gas. This absorption has been attributed to a substrate—oxygen charge-transfer complex[132]. In diethyl ether[127] the maximum of this absorption is at 215 nm. This CT complex is very photoactive [ϕ(primary processes leading to products) $\approx$ 0.5 at 254 nm]. The formation of all products can be accounted for if the primary process is assumed to be the transfer of an electron from the ether to O_2 (reaction 86) followed by a number of subsequent reactions (87—92). The ether—O_2 CT complexes show a considerable absorbance even above 260 nm, and part of the light-induced autoxidation that is observed in ethers may

$$EtOEt\cdots O_2 \xrightarrow{h\nu} EtOEt^{\cdot+} + O_2^{\cdot-} \tag{86}$$

$$EtOEt^{\cdot+} + O_2^{\cdot-} \longrightarrow CH_3\dot{C}HOEt + HO_2^{\cdot} \tag{87}$$

$$CH_3\dot{C}HOEt + HO_2^{\cdot} \longrightarrow \underset{\underset{O_2H}{|}}{CH_3CHOEt} \tag{88}$$

$$CH_3\dot{C}HOEt + O_2 \longrightarrow \underset{\underset{O_2^{\cdot}}{|}}{CH_3CHOEt} \tag{89}$$

$$2\ \underset{\underset{O_2^{\cdot}}{|}}{CH_3CHOEt} \longrightarrow CH_3COOEt + O_2 + \underset{\underset{OH}{|}}{CH_3CHOEt} \tag{90}$$

$$2\ \underset{\underset{O_2^{\cdot}}{|}}{CH_3CHOEt} \longrightarrow 2\ ^{\cdot}CH_3 + 2\ CHOOEt + O_2 \tag{91}$$

$$\underset{\underset{O_2^{\cdot}}{|}}{CH_3CHOEt} + HO_2^{\cdot} \longrightarrow CH_3COOEt + H_2O + O_2 \tag{92}$$

occur by way of such reactions. Since their products are hydroperoxides and carbonyl compounds which are also photoactive, the system is self-enhancing Alcohol–O_2 CT complexes begin to absorb appreciably at shorter wavelengths[123,131] than do the ether–O_2 CT complexes, and are therefore perhaps less likely to interfere with photochemical studies at wavelengths usually employed. It has been pointed out[133] that in cases where ethanol is used as a solvent in dye lasers the products of the reaction, among them acetic acid, acetaldehyde and hydrogen peroxide[131], can impair the functioning of the laser system.

VIII. REFERENCES

1. (a) M. B. Robin, *Higher Excited States of Polyatomic Molecules*, Vol. 1, Academic Press, 1974, p. 254.
   (b) C. Sandorfy, *Top. Curr. Chem.*, **86**, 92 (1979).
2. H.-D. Becker in *The Chemistry of the Hydroxyl Group* (Ed. S. Patai), John Wiley and Sons, London, 1971, p. 835.
3. D. Elad in *The Chemistry of the Ether Linkage* (Ed. S. Patai), John Wiley and Sons, London, 1967, p. 353.
4. C. von Sonntag and H.-P. Schuchmann, *Advan. Photochem.*, **10**, 59 (1977).
5. J. G. Calvert and J. N. Pitts, Jr., *Photochemistry*, John Wiley and Sons, New York, 1966.
6. A. J. Harrison, B. J. Cederholm and M. A. Terwilliger, *J. Chem. Phys.*, **30**, 355 (1959).
7. H. Kaiser, *Doctoral Thesis*, Universität München, 1970.
8. D. R. Salahub and C. Sandorfy, *Chem. Phys. Letters*, **8**, 71 (1971).
9. H.-P. Schuchmann and C. von Sonntag, *Tetrahedron*, **29**, 1811 (1973).
10. A. J. Harrison and D. R. W. Price, *J. Chem. Phys.*, **30**, 357 (1959).
11. H. Tsubomura, K. Kimura, K. Kaya, J. Tanaka and S. Nagakura, *Bull. Chem. Soc. Japan*, **37**, 417 (1964).
12. H.-P. Schuchmann and C. von Sonntag, *Tetrahedron*, **29**, 3351 (1973).
13. J. Doucet, P. Sauvageau and C. Sandorfy, *Chem. Phys. Letters*, **17**, 316 (1972).
14. L. W. Pickett, N. J. Hoeflich and T.-C. Liu, *J. Amer. Chem. Soc.*, **73**, 4865 (1951).
15. H.-P. Schuchmann, H. Bandmann and C. von Sonntag, *Z. Naturforsch.*, **34b**, 327 (1979).
16. H.-P. Schuchmann and C. von Sonntag, *J. Chem. Soc., Perkin Trans. II*, 1408 (1976).

17. P. Naderwitz, H.-P. Schuchmann and C. von Sonntag, Z. Naturforsch., **32b**, 209 (1977).
18. E. Çetinkaya, H.-P. Schuchmann and C. von Sonntag, J. Chem. Soc., Perkin Trans. II, 985 (1978).
19. K. Watanabe, T. Nakayama and J. Mottl, J. Quant. Spectry. Radiat. Transfer, **2**, 369 (1962).
20. J. L. Weeks, G. M. A. C. Meaburn and S. Gordon, Radiat. Res., **19**, 559 (1963).
21. J. Barrett, A. L. Mansell and M. F. Fox, J. Chem. Soc. (B), 173 (1971).
22. D. Sänger, Doctoral Thesis, Universität Karlsruhe, 1969.
23. W. Kaye and R. Poulson, Nature, **193**, 675 (1962).
24. D. Schulte-Frohlinde, D. Sänger and C. von Sonntag, Z. Naturforsch., **27b**, 205 (1972).
25. H.-P. Schuchmann, C. von Sonntag and D. Schulte-Frohlinde, J. Photochem., **4**, 63 (1975).
26. J. Hagège, P. C. Roberge and C. Vermeil, J. Chim. Phys., **65**, 641 (1968); Ber. Bunsenges. Physik. Chem., **72**, 138 (1968).
27. N. C. Yang, D. P. C. Tang, Do-Minh Thap and J. S. Sallo, J. Amer. Chem. Soc., **88**, 2851 (1966).
28. C. von Sonntag, Tetrahedron, **25**, 5853 (1969).
29. C. von Sonntag and D. Schulte-Frohlinde, Z. Physik. Chem. (Frankfurt), **55**, 329 (1967).
30. R. P. Porter and W. A. Noyes, Jr., J. Amer. Chem. Soc., **81**, 2307 (1959).
31. J. Hagège, S. Leach and C. Vermeil, J. Chim. Phys., **62**, 736 (1965).
32. J. Hagège, P. C. Roberge and C. Vermeil, Trans. Faraday Soc., **64**, 3288 (1968).
33. O. S. Herasymowych and A. R. Knight, Can. J. Chem., **51**, 147 (1973).
34. C. von Sonntag, Z. Physik. Chem. (Frankfurt), **69**, 292 (1970).
35. C. von Sonntag, Z. Naturforsch., **27b**, 41 (1972).
36. C. von Sonntag, Tetrahedron, **24**, 117 (1968).
37. C. von Sonntag, Intern. J. Radiat. Phys. Chem., **1**, 33 (1969).
38. C. von Sonntag and D. Sänger, Tetrahedron Letters, 4515 (1968).
39. O. S. Herasymowych and A. R. Knight, Can. J. Chem., **50**, 2217 (1972).
40. D. Sänger and C. von Sonntag, Z. Naturforsch., **25b**, 1491 (1970).
41. D. Sänger and C. von Sonntag, Tetrahedron, **26**, 5489 (1970).
42. H.-P. Schuchmann, C. von Sonntag and D. Schulte-Frohlinde, J. Photochem., **3**, 267 (1974/75).
43. H. J. van der Linde and C. von Sonntag, Photochem. Photobiol., **13**, 147 (1971).
44. I. P. Vinogradov and F. I. Vilesov, High Energy Chem., **11**, 17 (1977).
45. J. Belloni, E. Saito and F. Tissier, J. Phys. Chem., **79**, 308 (1975).
46. C. von Sonntag, H.-P. Schuchmann and G. Schomburg, Tetrahedron, **28**, 4333 (1972).
47. H.-P. Schuchmann and C. von Sonntag, Z. Naturforsch., **30b**, 399 (1975).
48. S. S. Yang, Doctoral Thesis, University of Chicago, 1968.
49. R. Ford, H.-P. Schuchmann and C. von Sonntag, J. Chem. Soc., Perkin Trans. II, 1338 (1975).
50. A. J. Harrison and J. S. Lake, J. Phys. Chem., **63**, 1489 (1959).
51. (a) J. F. Meagher and R. B. Timmons, J. Chem. Phys., **57**, 3175 (1972).
    (b) H. Mikuni, M. Takahasi and S. Tsuchiya, J. Photochem., **9**, 481 (1978).
52. C. A. F. Johnson and W. M. C. Lawson, J. Chem. Soc., Perkin Trans. II, 353 (1974).
53. N. Kizilkiliç, H.-P. Schuchmann and C. von Sonntag, to be published.
54. J. D. Margerum, J. N. Pitts, Jr., J. G. Rutgers and S. Searles, J. Amer. Chem. Soc., **81**, 1549 (1959).
55. H.-P. Schuchmann, P. Naderwitz and C. von Sonntag, Z. Naturforsch., **33b**, 942 (1978).
56. H.-P. Schuchmann and C. von Sonntag, J. Photochem., in the press.
57. G. Fleming, M. M. Anderson, A. J. Harrison and L. W. Pickett, J. Chem. Phys., **30**, 351 (1959).
58. R. Gomer and W. A. Noyes, Jr., J. Amer. Chem. Soc., **72**, 101 (1950).
59. B. C. Roquitte, J. Phys. Chem., **70**, 2699 (1966).
60. M. Kawasaki, T. Ibuki, M. Iwasaki and Y. Takezaki, J. Chem. Phys., **59**, 2076 (1973).
61. R. J. Cvetanović, Advan. Photochem., **1**, 115 (1963).
62. G. Behrens and D. Schulte-Frohlinde, Angew. Chem. (Intern Ed.), **12**, 932 (1973).
63. H. Itzel and H. Fischer, Helv. Chim. Acta, **59**, 880 (1976).

64. D. R. Paulson, A. S. Murray, D. Bennett, E. Mills, Jr., V. O. Terry and S. D. Lopez, *J. Org. Chem.*, **42**, 1252 (1977).
65. B. C. Roquitte, *J. Phys. Chem.*, **70**, 1334 (1966).
66. B. C. Roquitte, *J. Amer. Chem. Soc.*, **91**, 7664 (1969).
67. Z. Diaz and R. D. Doepker, *J. Phys. Chem.*, **82**, 10 (1978).
68. (a) R. R. Hentz and C. F. Parrish, *J. Phys. Chem.*, **75**, 3899 (1971).
    (b) H.-P. Schuchmann and C. von Sonntag, to be published.
69. F. Hirayama, C. W. Lawson and S. Lipsky, *J. Phys. Chem.*, **74**, 2411 (1970).
70. A. M. Halpern and W. R. Ware, *J. Phys. Chem.*, **74**, 2413 (1970).
71. C. von Sonntag and H. Bandmann, *J. Phys. Chem.*, **78**, 2181 (1974).
72. J. Kiwi, *J. Photochem.*, **7**, 237 (1977).
73. P. H. Mazzocchi and M. W. Bowen, *J. Org. Chem.*, **40**, 2689 (1975).
74. J. J. Houser and B. A. Sibbio, *J. Org. Chem.*, **42**, 2145 (1977).
75. H.-P. Schuchmann and C. von Sonntag, *Z. Naturforsch.*, **32b**, 207 (1977).
76. B. C. Roquitte, *J. Phys. Chem.*, **70**, 2863 (1966).
77. 'R. I. Martinez, R. E. Huie and J. T. Herron, *Chem. Phys. Letters*, **51**, 457 (1977).
78. R. D. Suenram and F. J. Lovas, *J. Amer. Chem. Soc.*, **100**, 5117 (1978).
79. M. J. Perkins and B. P. Roberts, *J. Chem. Soc., Perkin Trans. II*, 77 (1975).
80. V. V. Zorin, S. S. Zlotskii, V. P. Nayanov and D. L. Rakhmankulov, *Zh. prikl. Khim.*, **50**, 1131 (1977).
81. M. Ya. Botnikov, S. S. Zlotskii, V. V. Zorin, E. Kh. Kravets, V. M. Zhulin and D. L. Rakhmankulov, *Izv. Akad. Nauk, Ser. Khim*, 690 (1977).
82. H.-P. Schuchmann, P. Naderwitz and C. von Sonntag, unpublished.
83. M. Cazaux, B. Maillard and R. Lalande, *Tetrahedron Letters*, 1487 (1972).
84. D. G. Marsh, *J. Polym. Sci. Polym. Chem. Ed.*, **14**, 3013 (1976).
85. R. S. Juvet, Jr. and L. P. Turner, *Anal. Chem.*, **37**, 1464 (1965).
86. N. Bremer, B. J. Brown, G. H. Morine and J. E. Willard, *J. Phys. Chem.*, **79**, 2187 (1975).
87. M. K. Phibbs and B. de B. Darwent, *J. Chem. Phys.*, **18**, 495 (1950).
88. R. F. Pottie, A. G. Harrison and F. P. Lossing, *Can. J. Chem.*, **39**, 102 (1961).
89. A. R. Knight and H. E. Gunning, *Can. J. Chem.*, **39**, 1231 (1961).
90. A. R. Knight and H. E. Gunning, *Can. J. Chem.*, **39**, 2251 (1961).
91. A. R. Knight and H. E. Gunning, *Can. J. Chem.*, **39**, 2466 (1961).
92. A. R. Knight and H. E. Gunning, *Can. J. Chem.*, **40**, 1134 (1962).
93. A. R. Knight and H. E. Gunning, *Can. J. Chem.*, **41**, 763 (1963).
94. A. R. Knight and H. E. Gunning, *Can. J. Chem.*, **41**, 2849 (1963).
95. A. Kato and R. J. Cvetanović, *Can. J. Chem.*, **45**, 1845 (1967).
96. A. Kato and R. J. Cvetanović, *Can. J. Chem.*, **46**, 235 (1968).
97. S. L. N. G. Krishnamachari and R. Venkatasubramanian, *Mol. Photochem.*, **8**, 419 (1977).
98. T. A. Garibyan, A. A. Mantashyan, A. B. Nalbandyan and A. S. Saakyan, *Arm. Khim. Zhur.*, **24**, 13 (1971).
99. K. Luther, H. R. Wendt and H. E. Hunziker, *Chem. Phys. Letters*, **33**, 146 (1975).
100. J. Y. Morimoto, *Diss. Abstr. Int. B*, **37**, 1272 (1976–77).
101. R. Payette, M. Bertrand and Y. Rousseau, *Can. J. Chem.*, **46**, 2693 (1968).
102. J. C. Y. Tsao, *J. Chinese Chem. Soc. (Taipei)*, **16**, 152 (1969).
103. B. de B. Darwent, E. W. R. Steacie and A. F. van Winckel, *J. Chem. Phys.*, **14**, 551 (1946).
104. M. K. Phibbs, B. deB. Darwent and E. W. R. Steacie, *J. Chem. Phys.*, **16**, 39 (1948).
105. R. A. Marcus, B. deB. Darwent and E. W. R. Steacie, *J. Chem. Phys.*, **16**, 987 (1948).
106. R. J. Cvetanović, *Can. J. Chem.*, **33**, 1684 (1955).
107. R. J. Cvetanović and L. C. Doyle, *Can. J. Chem.*, **35**, 605 (1957).
108. Y. Takezaki, S. Mori and H. Kawasaki, *Bull. Chem. Soc. Japan*, **39**, 1643 (1966).
109. L. F. Loucks and K. J. Laidler, *Can. J. Chem.*, **45**, 2763 (1967).
110. S. V. Filseth, *J. Phys. Chem.*, **73**, 793 (1969).
111. G. R. De Maré, *J. Photochem.*, **7**, 101 (1977).
112. H. E. Gunning, J. M. Campbell, H. S. Sandhu and O. P. Strausz, *J. Amer. Chem. Soc.*, **95**, 746 (1973).

113. E. Murad, *J. Amer. Chem. Soc.*, **83**, 1327 (1961).
114. R. V. Morris and S. V. Filseth, *Can. J. Chem.*, **48**, 924 (1970).
115. J. Castonguay and Y. Rousseau, *Can. J. Chem.*, **49**, 2125 (1971).
116. S. Yamashita and F. P. Lossing, *Can. J. Chem.*, **46**, 2925 (1968).
117. C. S. Smith and A. R. Knight, *Can. J. Chem.*, **54**, 1290 (1976).
118. (a) C. G. Freeman, M. J. McEwan, R. F. C. Claridge and L. F. Phillips, *Trans. Faraday Soc.*, **67**, 67 (1971).
     (b) C. G. Freeman, M. J. McEwan, R. F. C. Claridge and L. F. Phillips, *Trans. Faraday Soc.*, **67**, 2567 (1971).
119. R. H. Newman, C. G. Freeman, M. J. McEwan, R. F. C. Claridge and L. F. Phillips, *Trans. Faraday Soc.*, **67**, 1360 (1971).
120. C. G. Freeman, M. J. McEwan, R. F. C. Claridge and L. F. Phillips, *Trans. Faraday Soc.*, **67**, 3247 (1971).
121. (a) S. Tsunashima, K. Morita and S. Sato, *Bull. Chem. Soc. Japan*, **50**, 2283 (1977).
     (b) S. Yamamoto, K. Tanaka and S. Sato, *Bull. Chem. Soc. Japan*, **48**, 2172 (1975).
122. A. C. Vikis and D. J. LeRoy, *Can. J. Chem.*, **50**, 595 (1972).
123. C. Miyauchi and H. Endo, *Kagaku Keisatsu Kenkyusho Hokoku*, **25**, 99 (1972); *Chem. Abstr.*, **78**, 166592e (1973).
124. L. Horner and W. Jurgeleit, *Liebigs Ann. Chem.*, **591**, 138 (1955).
125. V. I. Stenberg, R. D. Olson, C. T. Wang and N. Kulevsky, *J. Org. Chem.*, **32**, 3227 (1967).
126. N. Kulevsky, C. T. Wang and V. I. Stenberg, *J. Org. Chem.*, **34**, 1345 (1969).
127. C. von Sonntag, K. Neuwald, H.-P. Schuchmann, F. Weeke and E. Janssen, *J. Chem. Soc., Perkin Trans. II*, 171 (1975).
128. K. Maeda, A. Nakane and H. Tsubomura, *Bull. Chem. Soc. Japan*, **48**, 2448 (1975).
129. V. I. Stenberg, C. T. Wang and N. Kulevsky, *J. Org. Chem.*, **35**, 1774 (1970).
130. (a) I. Reimann, G. Hollatz and T. Eckert, *Arch. Pharmaz.*, **307**, 321 (1974).
     (b) I. Reimann, G. Hollatz and T. Eckert, *Arch. Pharmaz.*, **307**, 328 (1974).
131. L. M. Gurdzhiyan, O. L. Kaliyan, O. L. Lebedev and T. N. Fesenko, *Zh. Prikl. Spektrosk.*, **25**, 320 (1976).
132. (a) R. S. Mulliken and W. B. Person, *Molecular Complexes*, Interscience, New York, 1969.
     (b) R. S. Mulliken and W. B. Person, 'Electron donor–acceptor complexes and charge-transfer spectra, in *Physical Chemistry: an Advanced Treatise*, Vol. 3 (Ed. D. Henderson), Academic Press, New York, 1969, p. 537.
133. A. A. Balashova, A. S. Bebchuk, G. A. Matyushin, V. A. Strunkin, V. Ya. Fain, and T. M. Tchernyshova, *Zh. Prikl. Spektrosk.*, **24**, 707 (1976).

CHAPTER **22**

The photolysis of saturated thiols, sulphides and disulphides

CLEMENS VON SONNTAG and HEINZ-PETER SCHUCHMANN

Institut für Strahlenchemie im Max-Planck-Institut für Kohlenforschung, Stiftstrasse 34–36, D-4330 Mülheim a. d. Ruhr, W. Germany

I. INTRODUCTION

This chapter deals with the photochemistry of saturated organic divalent sulphur compounds. In a preceding volume of this series, the photochemistry of thiols has been reviewed[1], and in this respect the present chapter is supplementary. Other reviews touching on the subject have appeared[2-7].

The title compounds start to absorb at considerably longer wavelengths than their oxygen analogues. Their first absorption band is assigned to a transition which has more or less n—σ* nature while at shorter wavelengths Ryberg-type transitions come into play[8]. Spectral data of some compounds of interest are presented in Table 1.

TABLE 1. Molar extinction coefficients (base ten) of some organic divalent sulphur compounds near 254 nm, and some λ_{max} values

Compound	Medium	$\epsilon, M^{-1} cm^{-1} (\lambda, nm)$	$\lambda_{max}{}^a$	$\lambda_{max}{}^b$
MeSH	Vapour	~60 (254)[9]	~230[10]	204[10]
EtSH	Vapour	~80 (254)[10]	~230[10]	202[10]
	n-Heptane	40 (254)[11]	~230[11]	196[11]
Me$_2$S	Vapour	~10 (240)[9]	~220[10]	205[10]
Et$_2$S	Vapour	~30 (240)[9]	~220[10]	205[10]
 (thiirane) S	Vapour	17 (254)[10]	~260[10]	209[10] c
 (thietane) S	Vapour	12 (254)[10]	~260[10]	205[10] c
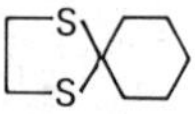 (dithiane)	Ethanol	833 (248)[12]		248[12]
Me$_2$S$_2$	Vapour	~300 (254)[9]		250[9]
	Liquid	316 (254)[13]		
Et$_2$S$_2$	Vapour	~310 (254)[9]		255[9]

[a]First absorption band.
[b]Second absorption band.
[c]Band shows structure.

II. PHOTOLYSIS OF THIOLS

A. The General Reactions

The photolysis of saturated thiols[1,13-25] can be generally described by the primary reactions (1) and (2). The subsequent reactions (3)–(6) explain the major products: hydrogen, disulphides, alkanes and hydrogen sulphide. Although reactions (1)–(6) account well for the general picture there are some variations in

$$RSH \xrightarrow{h\nu} \begin{cases} RS^{\cdot} + H^{\cdot} & (1) \\ R^{\cdot} + {}^{\cdot}SH & (2) \end{cases}$$

$$H^{\cdot} + RSH \longrightarrow H_2 + RS^{\cdot} \qquad (3)$$

$$RS^{\cdot} + RS^{\cdot} \longrightarrow RSSR \qquad (4)$$

$${}^{\cdot}R + RSH \longrightarrow RH + RS^{\cdot} \qquad (5)$$

$${}^{\cdot}SH + RSH \longrightarrow H_2S + RS^{\cdot} \qquad (6)$$

detail, depending on the nature of the substrate, the excitation wavelength and the medium in which the photolysis takes place. Hg photosensitization also leads to both S–H and C–S cleavage[26].

B. Factors Changing the Relative Importance of the Primary Processes

In the gas-phase photolysis of methanethiol[14] and ethanethiol[15] the sum of the quantum yields of reactions (1) and (2) is essentially unity. With increasing quantum energy the contribution of reaction (2) grows. For methanethiol, the ratio $\phi(1)/\phi(2)$ drops from 13 at 254 nm to 3 at 214 nm[14]. At 195 nm it is 1.7[18]. Ethanethiol is similar[15]. In assessing these ratios, the possibility of hot hydrogen atoms H[.]* being present (from reaction 1) has been taken into account[14,15]. H[.]* can mimic reaction (2) through the displacement reaction (7) (see also below).

$$H^{\cdot *} + RSH \longrightarrow R^{\cdot} + H_2S \qquad (7)$$

There is evidence that even thermalized H atoms can bring about such a displacement[23,27]. The substitution of the sulphur-bound hydrogen by deuterium considerably enhances $\phi(2)$ at the cost of $\phi(1)$[17].

A suppression of reaction (2) has been noted in liquid ethanethiol photolysed at 254 nm where $\phi(1)$ was found to be 0.25, which value apparently represents the total quantum yield of primary processes since other products were not detected[13].

C. The Secondary Processes

Reactions (3)–(6) represent the obvious subsequent reactions of thermally equilibrated radicals in the thiol system where the thiol group constitutes an excellent hydrogen donor. However, in reaction (1) hot H atoms and hot RS[.] radicals are initially formed[14-17,20], and there is evidence that reaction (7) is far from negligible[14-17,28]. Such a displacement has also been postulated[23] to occur in the liquid phase where it is thought to involve thermal H atoms[23,27]. The photolysis of liquid thiols yields H_2 and H_2S, their ratio depending on the nature of the thiol. In liquid t-butanethiol the ratio $\phi(H_2S)/\phi(H_2) \approx 1$, in the (secondary) cyclohexanethiol it drops to ~0.25[23], and in the (primary) ethanethiol no H_2S appears to be formed at all[13]. Addition of hydrogen donors led to a decrease in this ratio with increasing donor concentration[23]. The donor (QH) is considered to compete for hydrogen atoms (reactions 8 and 9) thus reducing the H_2S and

$$H^{\cdot} + RSH \longrightarrow R^{\cdot} + H_2S \qquad (8)$$

$$H^{\cdot} + QH \longrightarrow H_2 + Q^{\cdot} \qquad (9)$$

enhancing the H_2 yield. It is reasonable to assume that reaction (8) involves the intermediate RSH_2. Similar complex radicals are formed from sulphides with hydroxyl $(R_2S\dot{O}H)^{29}$, phenyl, hydrogen and $^{\cdot}SH^{30b}$. The formation of these complexes is subject to conformational constraints[30a].

Hot H atoms should be able to abstract carbon-bound hydrogen in competition to reactions (3) and (7). However, no HD has been found in the gas-phase photolysis of CH_3SD^{24}. In contrast, there is e.s.r. spectroscopic evidence that carbon-centred radicals are formed when thiols are irradiated in a rare-gas matrix at 77 K[20]. There seems to be a contradiction between these two facts which is not resolved by the report of the absence of dithiol from the gas-phase photolysis products of thiols[14,15] because it can be argued that the thiol radicals are converted into thiyl radicals in reaction (10).

$$R\dot{C}HSH + RCH_2SH \longrightarrow RCH_2SH + RCH_2S^{\cdot} \qquad (10)$$

There is also evidence for hot thiyl radicals. Ethylthiyl radicals obtained in the 195 nm photolysis of ethanethiol decompose into methyl radicals and thioform-aldehyde. If thiols are photolysed at 254 nm in an organic matrix at 77 K, hot RS[.] radicals as well as hot H atoms are produced[20]. Both mostly generate solvent radicals by hydrogen abstraction, rather than being thermalized. The hotness of the RS[.] is conclusively proved by the fact that at 77 K only a fraction of all radicals in the system are RS[.], most being solvent radicals. On warming the solvent radicals disappear and regenerate $RS^{\cdot}$ radicals. The $RS^{\cdot}$ radicals are stable until the matrix is melted[20]. Even though thermal thiyl radicals at room temperature are practically inert with respect to abstraction of aliphatic carbon-bound hydrogen atoms, they are known to abstract more weakly bound hydrogen[31,32].

The usual fate of the $RS^{\cdot}$ radicals is their dimerization (reaction 4) as the disproportionation/combination ratio for $RS^{\cdot}$ radicals is small (~ 0.04 for $MeS^{\cdot}$[33] and ~ 0.13 for $EtS^{\cdot}$[34] in the gas phase and near zero for $EtS^{\cdot}$ in the liquid phase[13]). Thioformaldehyde, a conceivable disproportionation product of $MeS^{\cdot}$, is produced in the photolysis of methanethiol in an argon matrix, most likely from photolysis of the primarily generated $MeS^{\cdot}$[19].

When thiyl radicals are produced in the presence of trivalent organophosphorus compounds, they are desulphurized to the alkyl radicals (reaction 11)[35]. Thiyl radicals add reversibly to olefins as shown by $cis-trans$ isomerization

$$R^1S^{\cdot} + PR^2_3 \longrightarrow \dot{R}^1 + S{=}PR^2_3 \tag{11}$$

that occurs in the presence of thiyl[36,37] (e.g. reaction 12), and induce a chain-reaction (reactions 12 and 13) which can be of preparative value (cf. Reference 6). Further examples of reactions undergone by thiyl radicals can be found in Chapter 24.

$$RS^{\cdot} + H_2C{=}CH_2 \rightleftharpoons RSCH_2{-}\dot{C}H_2 \tag{12}$$

$$RS{-}CH_2{-}\dot{C}H_2 + RSH \longrightarrow RSCH_2CH_3 + RS^{\cdot} \tag{13}$$

There are conflicting statements in the literature as to the affinity of thiyl radicals towards oxygen (reaction 14). The gas-phase photolysis of methanethiol

$$RS^{\cdot} + O_2 \longrightarrow RS\dot{O}_2 \tag{14}$$

near 230 and 260 nm in the presence of oxygen[25] leads to dimethyl disulphide and a peroxidic compound as the major products. The latter compound was believed to be hydrogen peroxide. These findings were taken to indicate that reaction (14) does not effectively compete with disulphide formation (reaction 4). However $MeS^{\cdot}$ generated at shorter wavelengths has been found to react rapidly with oxygen[18]. Also, the radiolysis of mercaptoethanol[27] and cysteine[38] in oxygenated aqueous solution has shown that reaction (14) is fast, almost diffusion-controlled in these systems[39], where it is in part followed by reaction (15).

$$RS\dot{O}_2 + RSH \longrightarrow RSOOH + RS^{\cdot} \tag{15}$$

D. Photolysis of Thiols in Aqueous Solutions

In aqueous solutions the thiols are in equilibrium with their anions [e.g. $pK(SH$ of cysteine$) = 8.5$[40]]. On photoexcitation the thiolates eject an electron (reaction 16). The electrons become solvated (see Chapter 23) and rapidly react with the thiols to give $R^{\cdot}$ radicals (reaction 17, see Chapter 24). $RS^{\cdot}$ radicals and thiolates

$$RS^- \xrightarrow{h\nu} RS^{\cdot} + e^-_{aq} \tag{16}$$

$$e^-_{aq} + RSH \longrightarrow HS^- + R^{\cdot} \tag{17}$$

$$RS^{\cdot} + RS^- \rightleftharpoons (RSSR)^{\cdot -} \tag{18}$$

form complexes[41] (reaction 18, for details see Chapter 24), which can be easily monitored by their strong optical absorption near 420 nm[42].

III. PHOTOLYSIS OF SULPHIDES

A. Acyclic Alkyl Sulphides

In the photolysis of acyclic alkyl sulphides[18,34,43-49] the main if not the only process is the scission of a carbon–sulphur bond (reactions 19a and 19b). In the

$$R^1-S-R^2 \xrightarrow{h\nu} \begin{cases} R^{1\cdot} + \cdot SR^2 & (19a) \\ R^1S^{\cdot} + \cdot R^2 & (19b) \end{cases}$$

competition between methyl and larger alkyl groups it is the methyl radical which is preferentially eliminated[43]. In the gas-phase photolysis ($\lambda = 229$ nm) of $CH_3-S-C_2H_5$, $\phi(19a)/\phi(19b) = 1.3$ is observed[47]. This preference appears to parallel the photolytic behaviour of ethers. Whereas in the gas-phase photolysis of thiols the sum of the quantum yields of primary decomposition is essentially unity (see above), this seems to be no longer true with dialkyl sulphides, e.g. with dimethyl, ethyl methyl and diethyl sulphide a value of only about 0.5 has been found[34]. The absence of hydrogen cannot entirely rule out C–H bond rupture in view of the possible displacement reaction[30b] analogous to reaction (8). The absence of methane in the photolysis of diethyl sulphide[48] indicates that C–C bond rupture does not occur.

Minor contributions of molecular processes (reactions 20 and 21) in the diethyl sulphide photolysis are possible but not established since the same products could also arise from disproportionation reactions of the radicals formed in reaction (19).

$$CH_3-CH_2-S-C_2H_5 \xrightarrow{h\nu} H_2C{=}CH_2 + C_2H_5SH \qquad (20)$$

$$CH_3-CH_2-S-C_2H_5 \xrightarrow{h\nu} CH_3-CH_3 + S{=}CH-CH_3 \qquad (21)$$

The radicals generated in reaction (19) retain a certain amount of excess energy depending on the wavelength of the exciting light. Particularly in the case of MeS$^{\cdot}$ generated from dimethyl sulphide[46] this excess energy manifests itself by permitting hydrogen abstraction reactions to occur (reaction 22). Because the excess

$$MeS^{\cdot*} + RH \longrightarrow MeSH + R^{\cdot} \qquad (22)$$

energy is spread over more degrees of freedom the radicals formed in the photolysis of diethyl sulphide[48] are less hot. Similar reactions are observed in organic matrices at 77 K[49], their behaviour resembling that of the thiol-containing glasses[20].

The Hg-photosensitized decomposition of acyclic sulphides[33,34,50] leads to the same products that are obtained in the direct photolysis. In contrast with the ethers and hydrocarbons, no hydrogen is observed, and the main primary process is apparently reaction (23). However, one might keep in mind that alkyl displacement by H atoms[30b] can mask C–H bond cleavage. In diethyl sulphide there may be a side-reaction which could amount to at most 20% (reaction 24)[34].

$$R-S-R + Hg^* \longrightarrow RS^{\cdot} + R^{\cdot} + Hg \qquad (23)$$

$$(C_2H_5)_2S + Hg^* \longrightarrow C_2H_5SH + C_2H_4 + Hg \qquad (24)$$

B. Cyclic Sulphides

The photolysis mechanisms of cyclic sulphides (for reviews see also References 3, 5 and 7) strongly differ from their acyclic analogues. Major differences are also observed between thiiranes and thietanes, which will be separately dealt with.

1. Thiiranes

The essential mechanism of the thiirane photolysis is represented[3,51] by the reactions (25)–(30). Direct excitation leads to the singlet excited state (reaction

$$H_2C\!\!-\!\!CH_2 \text{ (S)} \xrightarrow{h\nu} {}^1\!\left(H_2C\!\!-\!\!CH_2 \text{ (S)}\right)^* \tag{25}$$

$$
{}^1\!\left(H_2C\!\!-\!\!CH_2 \text{ (S)}\right)^* \longrightarrow
\begin{cases}
H_2S + CH\!\equiv\!CH & (26)\\[4pt]
{}^3(\dot{\,}CH_2\!\!-\!\!CH_2\!\!-\!\!S\dot{\,}) & (27)
\end{cases}
$$

$$
{}^3(\dot{\,}CH_2\!\!-\!\!CH_2\!\!-\!\!S\dot{\,}) \longrightarrow
\begin{cases}
H_2C\!=\!CH_2 + S(^3P) & (28)\\[4pt]
\text{deactivation} & (29)
\end{cases}
$$

$$S(^3P) + H_2C\!\!-\!\!CH_2 \text{ (S)} \longrightarrow S_2 + H_2C\!=\!CH_2 \tag{30}$$

25) which either decomposes into minor products hydrogen sulphide and acetylene (reaction 26) or mainly crosses over to the triplet state (reaction 27). The triplet species can decompose into ethylene and $S(^3P)$ (reaction 28), or be deactivated (reaction 29). The excited sulphur atoms appear to react efficiently with thiirane, extracting a sulphur atom (reaction 30). The importance of reaction (28) followed by reaction (30) is shown by the high quantum yield of ethylene [$\phi(C_2H_4) = 1.9$]. The existence of sulphur atoms was proved through the formation of thiiranes from added olefins[51]. This reaction is given by both singlet and triplet sulphur atoms[52,53], but thiols which are produced from paraffins and excited singlet sulphur atoms were not detected[51]. Therefore the S atoms must be in the triplet state, which implies the triplet state precursor (reaction 28). This intermediate may have biradical character, since tetrahydrothiophene was found when ethylene had been added[53] (reaction 31). It has been shown in a different system that $S(^3P)$

$$
{}^3(\dot{\,}CH_2\!\!-\!\!CH_2\!\!-\!\!S\dot{\,}) + H_2C\!=\!CH_2 \longrightarrow [{}^3(\dot{\,}CH_2\!\!-\!\!CH_2\!\!-\!\!CH_2\!\!-\!\!CH_2\!\!-\!\!S\dot{\,})] \longrightarrow \text{(S)}
\tag{31}
$$

rapidly reacts with thiirane[54] (reaction 30). Other reactive species such as hydrogen atoms, carbon atoms, and methyl radicals do the same (cf. Reference 55). The S_2 formed in reaction (30) can be identified by its UV absorption spectrum. A further reaction leading to ethylene is possible (reaction 32). Such a reaction operates in the pyrolysis of thiirane[56a]. The system may be even more complex since thermally excited thiirane generated by the addition of $S(^1D_2)$ to ethylene rearranges into vinylthiol with high efficiency[56b].

$$H_2C\!\!-\!\!CH_2 \text{ (S)} + \dot{\,}S\!\!-\!\!CH_2\!\!-\!\!CH_2^{\boldsymbol{\cdot}} \longrightarrow 2\,H_2C\!=\!CH_2 + S_2 \tag{32}$$

The photolysis of thiirane in the liquid state, neat and in hydrocarbon solutions, is explained similarly[57]. $\phi(C_2H_4)$ increases with increasing thiirane concentration. This could be due to a competition between reactions (29) and (32). The maximum value of 0.8 for $\phi(C_2H_4)$ is reached in neat thiirane. Considerable formation of polymeric products was observed in the photolysis of liquid methylthiirane[58].

The photolysis of tetrafluorothiirane has been reported to show very little conversion even after prolonged irradiation[59]. The reason for this might be that the spectrum of the irradiating light and the absorption spectrum did not match sufficiently well. It is known that the UV absorption spectra of perfluorinated compounds often exhibit a marked blue-shift compared to their prototypes (cf. References 8 and 60).

2. Thietanes

The photochemistry of thietane[61-64] (including some alkyl-substituted thietanes[64,65]) has been studied over a wide wavelength range, between 214 and 313 nm. This range straddles two absorption bands. The maximum of the first is near 260 nm, that of the second more structured one is near 206 nm[10]. These two bands lead to remarkably different photochemistries.

The results obtained using 254 and 313 nm light (first absorption band) lend themselves to interpretation more readily. The essential features of the mechanism[64] consist of reactions (33)–(37). The main product is ethylene. Its quantum yield rises with the temperature. At elevated temperatures where the deactivation step (36) is disfavoured the sum of $\phi(34)$ and $\phi(37)$ attains unity[64]. Their ratio is larger than 10 : 1 in all cases[64].

$$\begin{array}{c} CH_2\!-\!S \\ |\qquad | \\ CH_2\!-\!CH_2 \end{array} \xrightarrow{\ h\nu\ } \ ^\bullet CH_2CH_2CH_2S^\bullet \qquad (33)$$

$$^\bullet CH_2CH_2CH_2S^\bullet \longrightarrow H_2C\!=\!CH_2 + H_2C\!=\!S \qquad (34)$$

$$^\bullet CH_2CH_2CH_2S^\bullet + \begin{array}{c} CH_2\!-\!S \\ |\qquad | \\ CH_2\!-\!CH_2 \end{array} \begin{cases} \longrightarrow\ ^\bullet C_3H_6SSC_3H_6^\bullet & (35) \\ \longrightarrow\ \text{deactivation} & (36) \end{cases}$$

$$^\bullet C_3H_6SCC_3H_6^\bullet \longrightarrow \begin{array}{c} CH_2\!-\!S\!-\!\!-\!S\!-\!\!-\!CH_2 \\ |\qquad\qquad\qquad\quad | \\ CH_2\!-\!CH_2\!-\!CH_2\!-\!CH_2 \end{array} \qquad (37)$$

The biradical hypothesis is strongly supported by the finding that suitably substituted thietanes on photolysis undergo cis–trans isomerization[64,65] and that in solution propyl disulphide is produced (reactions 38 and 39). As is to be expected,

$$^\bullet CH_2\!-\!CH_2\!-\!CH_2\!-\!S^\bullet + RH \longrightarrow CH_3\!-\!CH_2\!-\!CH_2\!-\!S^\bullet + R^\bullet \qquad (38)$$

$$2\ C_3H_7S^\bullet \longrightarrow C_3H_7SSC_3H_7 \qquad (39)$$

1,6-hexanedithiol is not formed because the biradical with its alkyl end preferentially abstracts a hydrogen atom from the substrate (reaction 38) and the propanethiyl radicals so formed then combine (reaction 39)[64].

Propylene[61,62,64] and cyclopropane[61] have been found in small amounts and evidence has been presented[62,64] that propylene is formed in a secondary reaction.

The mechanism of the photolysis in the second absorption band appears to be more complex[61]. Irradiation at 214 and 229 nm produces much cyclopropane and propylene beside ethylene. The ratio $\phi(C_2H_4)/\phi(c\text{-}C_3H_6 + C_3H_6)$ is near 0.7 at 214 nm, and near 1.2 at 229 nm[61]. In agreement with the assumption of a trimethylene intermediate species, the cyclopropane to propylene ratio increases with increasing pressure. It appears extremely unlikely that the C_3H_6 hydrocarbon products are secondary here. The conversions reached[61] were less than 1% in the experiments with the higher substrate pressures, and it is there that the C_3H_6

products are relatively most important. The essential process that has been postulated[61] to explain the formation of C_3H_6 is reaction (40). A similar process

$$C_3H_6S^* + \begin{array}{c} CH_2—S \\ | \quad\quad | \\ CH_2—CH_2 \end{array} \longrightarrow 2\,(CH_2)_3^* + S_2 \qquad (40)$$

(reaction 32) may occur in the photolysis of thiirane[3,51]. The postulate of reaction (40) is in accordance with the observation that the sum of $\phi(C_2H_4)$, $\phi(c\text{-}C_3H_6)$ and $\phi(C_3H_6)$ under some conditions exceeds unity, reaching a value near 1.4[61] at a substrate pressure of about 1 torr and a temperature of $236°C$.

The Hg-photosensitized decomposition of thietanes has also been studied[61,66,67]. Its results are similar to those of the direct photolysis in the first UV absorption band. The C–S bond is cleaved to the biradical which then undergoes fragmentation, or *cis–trans* rearrangement and reclosure. There is some evidence[67] for a small contribution by reaction (41). The alternative possibility of process (42) is not excluded.

$$\dot{C}H_2CH_2CH_2\dot{S} \longrightarrow \dot{C}H_2—CH=CH_2 + H\dot{S} \qquad (41)$$

$$Hg^* + C_3H_6S \longrightarrow \dot{H}gSH + \dot{C}_3H_5 \qquad (42)$$

3. Thiolane and higher cyclic sulphides

The photolysis of thiolane resembles that of thietane in that here also there is a strong variation in photolytic behaviour depending on whether the compound is photolysed at 254 nm[3] or at 214 nm[68] and that the biradical (here $\dot{C}H_2—CH_2—CH_2—CH_2—\dot{S}$) plays a major role as an intermediate at both wavelengths. The intermediacy of the thiapentamethylene biradical was proved through addition reactions with olefins[3]. The wavelength dependence is largely expressed in the change of relative abundance of products. Reactions (43) to (51) constitute a plausible mechanism with features similar to those postulated for thiirane and thietane.

$$\begin{array}{c} CH_2—CH_2 \\ | \quad\quad\quad \diagdown S \\ CH_2—CH_2 \end{array} \xrightarrow{h\nu} \dot{C}H_2CH_2CH_2CH_2\dot{S}^* \qquad (43)$$

$$\dot{C}H_2CH_2CH_2CH_2\dot{S}^* \Big\langle \begin{array}{l} \longrightarrow H_2C=CH_2 + \dot{C}H_2CH_2\dot{S} \qquad (44) \\ \\ \longrightarrow \dot{C}H_2CH_2\dot{C}H_2 + CH_2S \qquad (45) \end{array}$$

$$\dot{C}H_2—(CH_2)_m—\dot{S} + \dot{C}H_2—(CH_2)_n—\dot{S} \longrightarrow S_2 + \dot{C}H_2—(CH_2)_m^{\cdot} + \dot{C}H_2—(CH_2)_n^{\cdot} \qquad (46)$$

$$\dot{C}H_2—(CH_2)_n—\dot{S}^* + \begin{array}{c} CH_2—CH_2 \\ | \quad\quad\quad \diagdown S \\ CH_2—CH_2 \end{array} \Big\langle \begin{array}{l} \longrightarrow S_2 + C_4H_8 + \dot{C}H_2—(CH_2)_n^{\cdot} \qquad (47) \\ \\ \longrightarrow \begin{array}{c} CH_2—\dot{C}H \\ | \quad\quad\quad \diagdown S \\ CH_2—CH_2 \end{array} + CH_3—(CH_2)_n S \qquad (48) \\ \\ \longrightarrow \text{deactivation} \qquad (49) \end{array}$$

$$\dot{C}H_2CH_2CH_2\dot{C}H_2 \longrightarrow 2\,H_2C=CH_2,\; c\text{-}C_4H_8 \qquad (50)$$

$$\dot{R}^1 + \dot{R}^2 \longrightarrow \text{products} \qquad (51)$$

$(\dot{R}^1, \dot{R}^2 = \text{any radical}; m, n = 1 \text{ or } 3)$

Ethylene has also been observed as a product in the liquid-phase photolysis of thiacyclopentane and thiacyclohexane[43], and ESR experiments at 77 K have given evidence for $^{\bullet}(CH_2)_n S^{\bullet}$ biradicals from these compounds[69,70].

IV. PHOTOLYSIS OF DITHIAACETALS

The photolysis of a few dithiaacetals[12,71-73] has been studied and is similar to that of the sulphides in so far as here too, an S—C bond is cleaved in the primary process. The products observed in the photolysis of 1,1-bis(methylthio)cyclohexane[71] are cyclohexyl methyl sulphide and dimethyl disulphide. The formation of the latter indicates that the scission of a C—S bond (reaction 52) in an

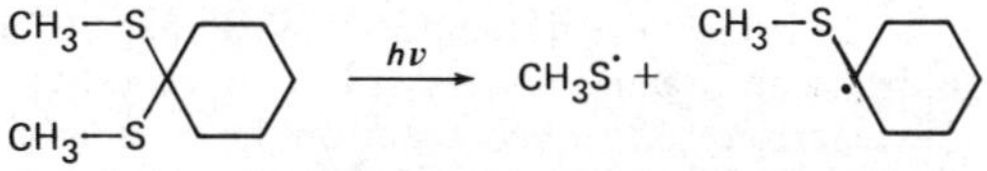

$$\text{(52)}$$

important primary process. The formation of cyclohexyl methyl sulphide is not as straightforward as the route to dimethyl disulphide (reaction 53). One might

$$2\,CH_3-S^{\bullet} \longrightarrow CH_3-S-S-CH_3 \tag{53}$$

consider disproportionation reactions but also a molecular process such as reaction (54). A similar process has been invoked[12] to explain the formation of cyclo-

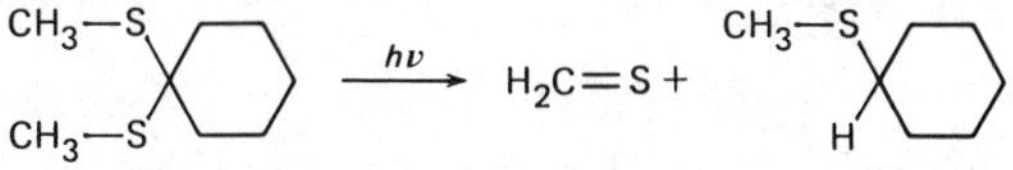

$$\text{(54)}$$

hexanethion which appears to be the precursor of its dimer, the major identified product in the photolysis of 1,3-dithiacyclopentane-2-spiro-1'-cyclohexane (reactions 55 and 56). Possibly by a process similar to reaction (54), the photolysis of

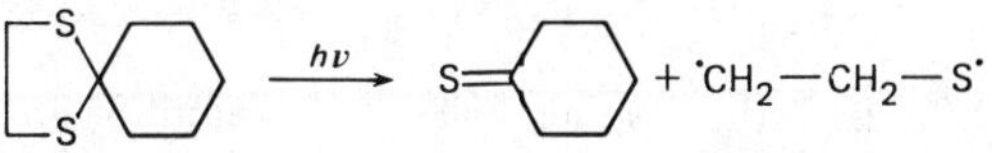

$$\text{(55)}$$

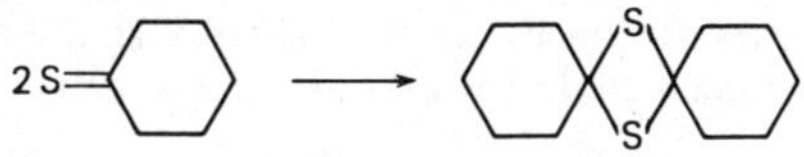

$$\text{(56)}$$

D-galactose diethyl dithioacetal[72] yields 1-S-ethyl-1-thio-D-galactitol in 60% yield (reaction 57).

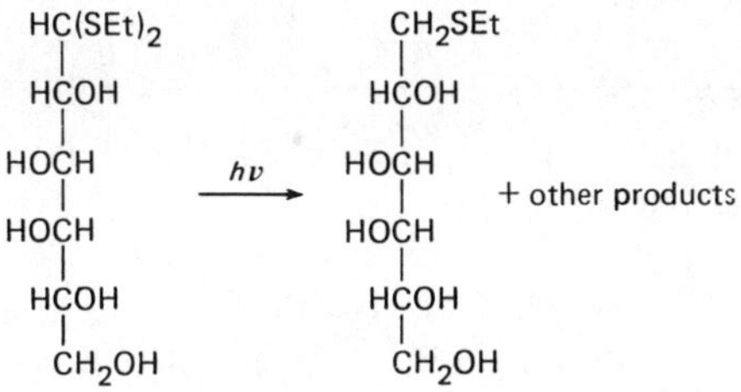

$$\text{(57)}$$

V. PHOTOLYSIS OF DISULPHIDES

The photolysis of disulphides found considerable attention[13,18,23,45,49,69,74-89] and has been the subject of a number of reviews[2-4].

The scission of the S—S bond (reaction 58) and of a C—S bond (reaction 59) are the two major primary processes. There are many investigations reporting that on

$$R^1—S—S—R^2 \xrightarrow{h\nu} \begin{cases} R^1—S^\cdot + {}^\cdot S—R^2 & (58) \\ {}^\cdot R^1 + {}^\cdot S—S—R^2 & (59a) \\ R^1—S—S^\cdot + {}^\cdot R^2 & (59b) \end{cases}$$

excitation at wavelengths above 230 nm only reaction (58) occurs (cf. References 13 and 80–82). However, it has also been reported that under such conditions methane was a product in the photolysis of dimethyl disulphide[86], polarized isobutane and isobutene were detected in a photo-CIDNP study of di-t-butyl disulphide[85], and RSS[·] radicals were detected by ESR spectroscopy in the 254 nm photolysis of disulphides in an organic matrix at low temperature[49].

The rate ratio of reaction (58) over reaction (59) strongly varies with the wavelength of the exciting light, C—S splitting (reaction 59) becoming increasingly important at shorter wavelengths. In the dimethyl disulphide gas-phase system where $\phi(59)/\phi(58)$ has been reported to be practically nil at 254 nm[80], $\phi(59)/\phi(58)$ is around 0.7 at 185 nm[18], whereas in its Hg-photosensitized decomposition[50] this ratio is 0.25. Equally, C—S cleavage is induced by other photosensitizers[90]. The RS[·] radicals formed in reaction (58) can undergo the transposition reaction (60) which is part of a chain-reaction. ϕ(Me-S-S-Et) = 330 was found in the

$$R^1S^\cdot + R^1—S—S—R^2 \longrightarrow R^1—S—S—R^1 + R^2—S^\cdot \qquad (60)$$

cophotolysis ($\lambda \sim 260$ nm) of dimethyl disulphide and diethyl disulphide in the liquid phase[81]. Since any other process is insignificant compared to the transposition reaction a photostationary state can be established (reaction 61). For

$$2R^1—S—S—R^2 \underset{h\nu}{\overset{h\nu}{\rightleftharpoons}} R^1—S—S—R^1 + R^2—S—S—R^2 \qquad (61)$$

R^1 = Me and R^2 = Et the value of the equilibrium constant $K =$ [MeSSEt]^2/[MeSSMe][EtSSEt] has been found to be ~5 in the liquid phase[76,81]. The efficiency of the transposition diminishes rapidly as the alkyls get larger[13,77].

A similar transposition takes place in the presence of thiols[23] (reactions 62–64).

$$R^1—S—S—R^1 \xrightarrow{h\nu} 2R^1—S^\cdot \qquad (62)$$

$$R^1—S^\cdot + R^2—SH \longrightarrow R^1—SH + R^2—S^\cdot \qquad (63)$$

$$R^2—S^\cdot + R^1—S—S—R^1 \longrightarrow R^2—S—S—R^1 + R^1S^\cdot \qquad (64)$$

Cleavage of the disulphide bond by radicals other than thiyl (cf. Reference 2) e.g. OH radicals[91], has also been observed.

VI. REFERENCES

1. A. R. Knight, in *The Chemistry of the Thiol Group* (Ed. S. Patai), John Wiley and Sons, London, 1974, p. 455.
2. W. A. Pryor, *Mechanisms of Sulfur Reactions*, McGraw-Hill, New York, 1962.
3. O. P. Strausz, H. E. Gunning and J. W. Lown, in *Comprehensive Chemical Kinetics*, Vol. 5 (Eds. C. H. Bamford and C. F. H. Tipper), Elsevier, Amsterdam, 1972, p. 697.
4. E. Block, *Quart. Rep. Sulfur Chem.*, **4**, 283 (1969).
5. S. Braslavsky and J. Heicklen, *Chem. Rev.*, **77**, 473 (1977).

6. H. Dürr, in *Houben–Weyl, Methoden der Organischen Chemie*, Vol. 4/5 b (Ed. E. Müller), Thieme, Stuttgart, 1975, p. 1008.
7. A. Padwa, *Int. J. Sulfur Chem. (B)*, **7**, 331 (1972).
8. M. B. Robin, *Higher Excited States of Polyatomic Molecules*, Vol. 1, Academic Press, 1974, p. 276.
9. J. G. Calvert and J. N. Pitts, Jr., *Photochemistry*, John Wiley and Sons, New York, 1966, p. 489 f.
10. L. B. Clark and W. T. Simpson, *J. Chem. Phys.*, **43**, 3666 (1965).
11. *UV Atlas of Organic Compounds, Spectrum I/6*, Butterworths, 1971.
12. J. D. Willett, J. R. Grunwell and G. A. Berchtold, *J. Org. Chem.*, **33**, 2297 (1968).
13. D. D. Carlson and A. R. Knight, *Can. J. Chem.*, **51**, 1410 (1973).
14. L. Bridges and J. M. White, *J. Phys. Chem.*, **77**, 295 (1973).
15. L. Bridges, G. L. Hemphill and J. M. White, *J. Phys. Chem.*, **76**, 2668 (1972).
16. D. Kamra and J. M. White, *J. Photochem.*, **4**, 361 (1975).
17. D. Kamra and J. M. White, *J. Photochem.*, **7**, 171 (1977).
18. A. B. Callear and D. R. Dickson, *Trans. Faraday Soc.*, **66**, 1987 (1970).
19. M. E. Jacox and D. E. Milligan, *J. Mol. Spectry*, **58**, 142 (1975).
20. (a) J. Skelton and F. C. Adam, *Can. J. Chem.*, **49**, 3536 (1971).
    (b) A. J. Elliott and F. C. Adam, *Can. J. Chem.*, **52**, 102 (1974).
21. (a) W. A. Pryor and M. G. Griffith, *J. Amer. Chem. Soc.*, **93**, 1408 (1971).
    (b) W. A. Pryor and J. P. Stanley, *J. Amer. Chem. Soc.*, **93**, 1412 (1971).
22. J. P. Stanley, R. W. Henderson and W. A. Pryor, *Adv. Chem. Ser.*, **110**, 130 (1972).
23. W. A. Pryor and E. G. Olsen, *J. Amer. Chem. Soc.*, **100**, 2852 (1978).
24. T. Inaba and B. deB. Darwent, *J. Phys. Chem.*, **64**, 1431 (1960).
25. D. M. Graham and B. K. T. Sie, *Can. J. Chem.*, **49**, 3895 (1971).
26. S. Yamashita and F. P. Lossing, *Can. J. Chem.*, **46**, 2925 (1968).
27. G. G. Jayson, D. A. Stirling and A. J. Swallow, *Int. J. Radiat. Biol.*, **19**, 143 (1971).
28. L. Bridges and J. M. White, *J. Chem. Phys.*, **59**, 2148 (1973).
29. K.-D. Asmus, D. Bahnemann, M. Bonifačić and H. A. Gillis, *Faraday Discuss. Chem. Soc.*, **63**, 213 (1978).
30. (a) J. A. Kampmeier, R. B. Jordan, M. S. Liu, H. Yamanaka and D. J. Bishop, in *Organic Free Radicals* (Ed. W. A. Pryor), ACS Symposium Series 69, American Chemical Society, Washington, D.C., 1978, p. 275.
    (b) T. Yokata and O. P. Strausz, *J. Phys. Chem.*, **83**, 3196 (1979).
31. C. Walling and R. Rabinovitz, *J. Amer. Chem. Soc.*, **81**, 1137 (1959).
32. (a) W. A. Pryor, G. Gojon and J. P. Stanley, *J. Amer. Chem. Soc.*, **95**, 945 (1973).
    (b) L. Lunazzi, G. Placucci and L. Grossi, *J. Chem. Soc., Chem. Commun.*, 533 (1979).
33. D. R. Tycholiz and A. R. Knight, *J. Amer. Chem. Soc.*, **95**, 1726 (1973).
34. C. S. Smith and A. R. Knight, *Can. J. Chem.*, **54**, 1290 (1976).
35. E. J. Corey and E. Block, *J. Org. Chem.*, **34**, 1233 (1969).
36. C. Sivertz, *J. Phys. Chem.*, **63**, 34 (1959).
37. C. Walling and W. Helmreich, *J. Amer. Chem. Soc.*, **81**, 1144 (1959).
38. A. A. Al-Thannon, J. P. Barton, J. E. Packer, R. J. Sims, C. N. Trumbore and R. V. Winchester, *Int. J. Radiat. Phys. Chem.*, **6**, 233 (1974).
39. (a) J. P. Barton and J. E. Packer, *Int. J. Radiat. Phys. Chem.*, **2**, 159 (1970).
    (b) K. Schäfer, M. Bonifačić, D. Bahnemann and K.-D. Asmus, *J. Phys. Chem.*, **82**, 2777 (1978).
40. R. E. Benesch and R. Benesch, *J. Amer. Chem. Soc.*, **77**, 5877 (1955).
41. T.-L. Tung and J. A. Stone, *J. Phys. Chem.*, **78**, 1130 (1974); T.-L. Tung and J. A. Stone, *Can. J. Chem.*, **53**, 3153 (1975).
42. A. Habersbergerová, I. Janovský and P. Kourim, *Rad. Res. Rev.*, **4**, 123 (1972).
43. W. E. Haines, G. L. Cook and J. S. Ball, *J. Amer. Chem. Soc.*, **78**, 5213 (1956).
44. L. Horner and J. Dörges, *Tetrahedron Letters*, 757 (1963).
45. B. Milligan, D. E. Rivett and W. E. Savige, *Australian J. Chem.*, **16**, 1020 (1963).
46. P. M. Rao and A. R. Knight, *Can. J. Chem.*, **50**, 844 (1972).
47. D. R. Tycholiz and A. R. Knight, *Can. J. Chem.*, **50**, 1734 (1972).
48. C. S. Smith and A. R. Knight, *Can. J. Chem.*, **51**, 780 (1973).

49. F. C. Adam and A. J. Elliot, *Can. J. Chem.*, **55**, 1546 (1977).
50. A. Jones, S. Yamashita and F. P. Lossing, *Can. J. Chem.*, **46**, 833 (1968).
51. P. Fowles, M. DeSorgo, A. J. Yarwood, O. P. Strausz and H. E. Gunning, *J. Amer. Chem. Soc.*, **89**, 1352 (1967).
52. E. M. Lown, E. L. Dedio, O. P. Strausz and H. E. Gunning, *J. Amer. Chem. Soc.*, **89**, 1056 (1967).
53. K. S. Sidhu, E. M. Lown, O. P. Strausz and H. E. Gunning, *J. Amer. Chem. Soc.*, **88**, 254 (1966).
54. R. J. Donovan, D. Husain, R. W. Fair, O. P. Strausz and H. E. Gunning, *Trans. Faraday Soc.*, **66**, 1635 (1970).
55. T. Yokota, M. G. Ahmed, I. Safarik, O. P. Strausz and H. E. Gunning, *J. Phys. Chem.*, **79**, 1758 (1975).
56. (a) E. M. Lown, H. S. Sandhu, H. E. Gunning and O. P. Strausz, *J. Amer. Chem. Soc.*, **90**, 7164 (1968).
    (b) A. G. Sherwood, I. Safarik, B. Verkoczy, G. Almadi, H. A. Wiebe and O. P. Strausz, *J. Amer. Chem. Soc.*, **101**, 3000 (1979).
57. R. Kumar and K. S. Sidhu, *Indian J. Chem.*, **11**, 899 (1973).
58. R. J. Gritter and E. C. Sabatino, *J. Org. Chem.*, **29**, 1965 (1964).
59. W. R. Brasen, H. N. Cripps, C. G. Bottomley, M. W. Farlow and C. G. Krespan, *J. Org. Chem.*, **30**, 4188 (1965).
60. D. R. Salahub and C. Sandorfy, *Chem. Phys. Letters*, **8**, 71 (1971).
61. H. A. Wiebe and J. Heicklen, *J. Amer. Chem.*, **92**, 7031 (1970).
62. D. R. Dice and R. P. Steer, *J. Phys. Chem.*, **77**, 434 (1973).
63. D. R. Dice and R. P. Steer, *Can. J. Chem.*, **52**, 3518 (1974).
64. D. R. Dice and R. P. Steer, *Can. J. Chem.*, **53**, 1744 (1975).
65. D. R. Dice and R. P. Steer, *J. Chem. Soc., Chem. Commun.* 106 (1973).
66. D. R. Dice and R. P. Steer, *J. Amer. Chem. Soc.*, **96**, 7361 (1974).
67. D. R. Dice and R. P. Steer, *Can. J. Chem.*, **56**, 114 (1978).
68. S. Braslavsky and J. Heicklen, *Can. J. Chem.*, **49**, 1316 (1971).
69. P. S. H. Bolman, I. Safarik, D. A. Stiles, W. J. R. Tyerman and O. P. Strausz, *Can. J. Chem.*, **48**, 3872 (1970).
70. G. C. Dismukes and J. E. Willard, *J. Phys. Chem.*, **80**, 2072 (1976).
71. R. E. Kohrman and G. A. Berchtold, *J. Org. Chem.*, **36**, 3971 (1971).
72. D. Horton and J. S. Jewell, *J. Org. Chem.*, **31**, 509 (1966).
73. A. B. Terent'ev and G. N. Shvedova, *Izv. Akad. Nauk SSSR, Ser. Khim.*, 2239 (1968).
74. W. E. Lyons, *Nature*, **162**, 1004 (1948).
75. K. J. Rosengren, *Acta Chem. Scand.*, **16**, 1401 (1962).
76. L. Haraldson, G. J. Olander, S. Sunner and E. Varde, *Acta Chem. Scand.*, **14**, 1509 (1960).
77. S. F. Birch, T. V. Cullum and R. A. Dean, *J. Inst. Petroleum*, **39**, 206 (1953).
78. R. B. Whitney and M. Calvin, *J. Chem. Phys.*, **23**, 1750 (1955).
79. E. E. Smissman and J. R. J. Sorenson, *J. Org. Chem.*, **30**, 4008 (1965).
80. P. M. Rao, J. A. Copeck and A. R. Knight, *Can. J. Chem.*, **45**, 1369 (1967).
81. K. Sayamol and A. R. Knight, *Can. J. Chem.*, **46**, 999 (1968).
82. P. M. Rao and A. R. Knight, *Can. J. Chem.*, **46**, 2462 (1968).
83. J. E. Eager and W. E. Savige, *Photochem. Photobiol.*, **2**, 25 (1963).
84. T. Ueno and Y. Takezaki, *Bull. Inst. Chem. Res. Kyoto Univ.*, **36**, 19 (1958).
85. S. M. Rosenfeld, R. G. Lawler and H. R. Ward, *J. Amer. Chem. Soc.*, **94**, 9255 (1972).
86. M. Ya. Mel'nikov and N. V. Fok, *Khim. Vysok. Energii*, **10**, 466 (1976).
87. J. J. Windle, A. K. Wiersema and A. L. Tappel, *J. Chem. Phys.*, **41**, 1996 (1964).
88. O. Ito and M. Matsuda, *Bull. Chem. Soc. Japan*, **51**, 427 (1978).
89. V. Ramakrishnan, S. D. Thompson and S. P. McGlynn, *Photochem. Photobiol.*, **4**, 907 (1965).
90. G. W. Byers, H. Gruen, H. G. Giles, H. N. Schott and J. A. Kampmeier, *J. Amer. Chem. Soc.*, **94**, 1016 (1972).
91. M. Bonifačić, K. Schäfer, H. Möckel and K.-D. Asmus, *J. Phys. Chem.*, **79**, 1496 (1975).

CHAPTER **23**

Radiation chemistry of alcohols and ethers

CLEMENS VON SONNTAG and
HEINZ-PETER SCHUCHMANN

*Institut für Strahlenchemie im Max-Planck-Institut für Kohlenforschung,
Stiftstrasse 34–36, D-4330 Mülheim a. d. Ruhr, W. Germany*

I. INTRODUCTION

The great interest in the radiation chemistry of alcohols is reflected in the number of reviews that deal with this topic (cf. References 1–5). Alcohols are among the most polar organic compounds. In so far as alcohols as a class are especially closely related to water, which has served as the main substrate for investigating the effect of ionizing radiation on condensed matter in the beginning stages of radiation chemistry, the scope of this interest is easily understandable. In comparison with this, ethers have found little attention[1]. The material on the radiolysis of alcohols and ethers in aqueous solution[6] is at least as extensive as that devoted to these

935

compounds in the neat state, and has contributed much to the present knowledge of their free-radical chemistry.

Most of the kinetic data are obtained using the pulse radiolysis technique[7], where a short pulse of high-energy ($>$1 MeV) electrons is made to penetrate a cell filled with the material to be investigated. Pulse durations of about one microsecond are standard conditions, but equipment delivering pulses on the nanosecond and picosecond time-scale is becoming increasingly widespread. The short pulses of ionizing radiation produce radical and ionic intermediates. Their fate can be monitored by following the change of the optical absorption, or of the conductivity, to the extent that in the course of the reaction charged species are formed or destroyed. It is recalled that a radical $^{\cdot}$QH may be involved in hydrolytic equilibria (1) and (2), and that the differently protonated forms of a radical behave as

$$\text{(1)} \qquad\qquad Q^{\cdot -} \;\overset{H^+}{\underset{}{\rightleftharpoons}}\; QH^{\cdot} \;\overset{H^+}{\underset{}{\rightleftharpoons}}\; QH_2^{\cdot +} \qquad\qquad \text{(2)}$$

chemically distinct species. The conductivity technique has been increasingly used in the recent past and has yielded most interesting results. The accelerator may also be coupled with an ESR spectrometer. This *in situ* technique[8] is usually run under steady-state conditions to identify the radicals but can also be used under pulsed conditions for kinetic measurements. CIDNP studies of the radiolysis of alcohols in aqueous solution have been reported[9], and the combination of pulse radiolysis and polarography has been reviewed[10]. For the investigation of polymer degradation a light-scattering method has been used together with pulse radiolysis[11].

The present review is divided into two sections. The first one deals with the results obtained in the radiolysis of the neat compounds. It consists mainly of material on alcohols, and supplements the picture given by Basson[5] in a previous volume of this series. In the second section some emphasis is placed on the radiation chemistry of aqueous solutions as it is felt that its results might perhaps be of a more general interest.

II. NEAT ALCOHOLS IN THE LIQUID AND SOLID STATE

A. Energy Absorption and Primary Processes

Ionizing radiation absorbed by matter is dissipated by ionization (reaction 3) and excitation processes (reaction 4). The energy is deposited at random along the

$$M \;\xrightarrow{\;\gamma\;}\; M^{\cdot +} + e^- \tag{3}$$

$$M \;\xrightarrow{\;\gamma\;}\; M^* \tag{4}$$

tracks of the energetic charged particles (in γ-radiolysis these are electrons produced mostly through the Compton effect) in small packages called spurs. In these spurs, one or more ion pairs or radical fragment pairs (from reaction 5) are

$$M^* \;\xrightarrow{\hspace{1.5cm}}\; {}^{\cdot}R^1 + {}^{\cdot}R^2 \tag{5}$$

generated from the substrate molecules, with these reactive species existing at first in close proximity so that their concentration within the spurs is much greater than in the bulk of the medium. In this respect, as well as because the spurs are strung along the linear tracks, the concentration of the reactive intermediates is inhomogeneous (for details see References 12–16). Part of the species will react with each other before they can escape into the bulk of the solution. Excitation energy that does not lead to chemical change within the spur may be transferred between substrate molecules under certain conditions and so leave the spur. Quanta of mobile energy of this kind have been termed excitons (cf. Reference 17).

A complete radiolysis mechanism would require that the yields be known of reactions (3) and (4) which precede all the other processes of chemical change. This information is usually not available because part of the excited substrate molecules are superexcited and therefore able to undergo reaction (6). In the gas phase,

$$M^* \longrightarrow M^{\cdot +} + e^- \tag{6}$$

G(ionization)† ≈ 4 has been found for alcohols[18]. Because of a lowering of the ionization potential in the liquid compared to the gaseous state[19], G(ionization) might be somewhat higher in the liquid. As it cannot be determined directly in liquid alcohols, attempts have been made to establish it indirectly by using high concentrations of electron scavengers such as N_2O where $G(N_2)$ has been considered[20] to reflect G(scavenged electrons) (reaction 7). The results[20] are in agreement with the above reasoning. One has to keep in mind, however, that excited states may transfer their energy to N_2O and thus also yield N_2 (reaction 8)[21,22]. As the

$$e^- + N_2O \longrightarrow N_2 + O^{\cdot -} \tag{7}$$

$$M^* + N_2O \longrightarrow M + N_2 + O \tag{8}$$

excited states of alcohols appear to be either very short-lived, or inefficiently to transfer energy to N_2O dissolved therein at atmospheric pressure ($\leqslant 4\%$ transfer from the lowest excited state of t-butanol[23]), $G(N_2)$ might begin to be driven beyond G(reaction 7) at elevated N_2O pressure. On the other hand, sufficiently high electron scavenger concentrations are desirable and necessary to compete successfully against the spur reactions of the electron such as reactions (9)–(11).

$$e^- + ROH_2^+ \longrightarrow ROH + H^{\cdot} \tag{9}$$

$$e^- + RO^{\cdot} \longrightarrow RO^- \tag{10}$$

$$e^- + \overset{|}{\underset{|}{C}}=O \longrightarrow {\cdot}\overset{|}{\underset{|}{C}}-O^- \tag{11}$$

(The carbonyl compound in reaction 11 is formed as a molecular product; see below). The radiation-induced chain-reaction of N_2O with alcohols has been shown to occur only at elevated temperatures[24,25].

The G-value of excitation is more difficult to assess. If the theory of the optical approximation[26,27] holds, higher excited states play a larger role than does the lowest one. Unfortunately, in liquid alcohols only information on the breakdown of the lowest excited state is available[28] (see Chapter 21). Besides straightforward ionization (reaction 3), ionization accompanied by a fragmentation of the radical cation (reaction 12) has often been considered to account for some products, and

$$M \overset{\gamma}{\longrightarrow} N^+ + R^{\cdot} + e^- \tag{12}$$

attempts have been made to correlate product formation in the liquid phase with mass spectra[29-31]. In that approach data obtained at pressures below 10^{-5} bar must be extrapolated to the conditions of the liquid state, where, however, rapid thermalization of a vibrationally excited radical cation can occur. Electronically excited radical cations may behave differently, though.

---

†The radiation-chemical yield G, 'G-value', is defined by the equation $G = N/E$; unit: $(100 \text{ eV})^{-1}$; N = number of species or events of whatever kind, E = radiation energy absorbed causing these events.

B. Solvation of the Electron

The electron ejected in the ionization process (reaction 3) can, after thermalization, become solvated (reaction 13). (Negative solvation clusters in the gas phase

$$e^- + n\,ROH \longrightarrow e^-_{solv} \qquad (13)$$

are also known[31b]). If the electrons are solvated outside the so-called Onsager radius where its potential energy in the field of the geminate positive ion ($e^2/\varepsilon r$) equals its thermal energy (kT) (cf. Reference 32), they are called free electrons. Considerable effort has been spent to determine the yield of free electrons in alcohols (cf. References 33 and 34). Because the Onsager radius depends inversely on the dielectic constant of the medium, the free-ion yield is also a function of the dielectric constant (note that in condensed states the free ion yield is necessarily smaller than the ionization yield). For the lower alcohols G-values between 1 and 2 have been found (for a compilation see References 4 and 35).

It would exceed the scope of this article to extensively review the present knowledge about the properties of the solvated electron in alcohols (for reviews see References 3, 4, 36–43), but a brief account seems in order. In alcohols the solvated electron can be readily detected by its strong optical absorption peaking between 600 and 800 nm, and also by its ESR signal in the glassy state at low temperatures, where one speaks of the trapped electron[3,44–48]. Making use of the picosecond pulse radiolysis technique at room temperature[49–53] or by slowing down the solvation process through lowering the temperature and working at the nanosecond or microsecond time-scale[54–63], the solvation of the electron can be followed spectroscopically. The photodisentrapment of the partially or fully solvated electrons, called photobleaching, has been used to obtain information on the different kinds of electron trap that may exist in a polar medium[63b,64–68]. During photobleaching alcohol radicals are formed[69] via reaction (14) and subsequent

$$e^-_{solv} + ROH \longrightarrow RO^- + H^{\cdot} \qquad (14)$$

hydrogen abstraction by H$^{\cdot}$ (see below). At the early stages where the electron trap is not yet fully established (shallow) a strong absorption in the infrared, due to the 'presolvated' electron, is observed which shifts to the visible as solvation proceeds[64,70–72]. The broadness of the final absorption band of the solvated electron is considered to be due to a distribution of trap depths, or to a superposition of different optical transitions from the same ground state[73–79]. The nature of the solvation shell of the trapped electron in low-temperature glasses has also been studied by ESR spectroscopy[80,81].

The solvated electron is considered to reside within a cavity formed by a shell of polarized solvent molecules. The change of the nature of this cavity with temperature or pressure influences the optical absorption spectrum of the solvated electron, a decrease in temperature[82,83] or an increase in pressure[84,85] causing a blue-shift as the cavity is contracted (cf. References 64 and 86) or compressed[87]. The change with temperature has also been explained on the basis of thermal disorientation of the cavity-forming dipoles[88]. In mixed solvents the electron tends to associate with molecular aggregates of the more polar constituent[89–91] as shown by the fact that its absorption spectrum is essentially the same as in the pure polar compound at concentrations of the latter of 10 mol % or even less[92–95]. As expected, under certain conditions a build-up of the solvation shell has also been observed whereby the less polar neighbour molecules around the electron are progressively replaced by molecules of the more polar compound[82,96–98].

It is thought that in low-temperature glasses the trapped electron reacts with acceptors mostly by tunnelling[99]. The ease of the tunnelling phenomenon seems to depend on the nature of the medium[100]. Presolvated and solvated electrons react at different rates[62,101-108]. The orientation of the acceptor with respect to the tunnelling electron may also influence the reaction rate[109]. If the solid is crystalline instead of glassy, then under otherwise equal conditions the number of electrons becoming trapped is much smaller[110].

In liquid alcohols the reactions of the solvated electron have been monitored by pulse radiolysis, making use of its strong optical absorption[84,85,111-113], and by the salt effect on its reactions with scavengers[114,115]. At room temperature the solvated electron reacts comparatively slowly with alcohols (reaction 14: k_{14}(MeOH, EtOH) $\leqslant 10^5$ M^{-1} s^{-1} (Ref. 113); k_{14}(EtOH) = 7 x 10^3 M^{-1} s^{-1} (Ref. 116)). Reaction (15) predominates over reaction (14) in benzyl[117] and allyl[118]

$$e_{solv}^- + ROH \longrightarrow R^\cdot + OH^-$$ (15)

alcohols where the ensuing radical is resonance-stabilized. It has been suggested[119] that in t-butanol the presolvated electron can undergo a reaction similar to (15), but with a higher specific rate. Data on solvated electron reactions in alcohols have been compiled[120-122].

C. Ion-molecule Reactions

Knowledge about ion-molecule reactions stems from studies in the gas phase[123-127] where it has been shown that the molecular ions of alcohols efficiently (in principle on every encounter) transfer a proton to an alcohol molecule. Oxygen-bound and α-carbon-bound H atoms are transferred with about equal rates (exemplified by reactions 16 and 17). In the condensed state reaction (16) has been considered to be much favoured over reaction (17) because the oxygen-bound hydrogen is involved in hydrogen bonding, in contrast to the carbon-bound one[128a].

$$CH_3OD^{+\cdot} + CH_3OD \longrightarrow CH_3O^\cdot + CH_3OD_2^+$$ (16)

$$CH_3OD^{+\cdot} + CH_3OD \longrightarrow {}^\cdot CH_2OD + CH_3ODH^+$$ (17)

The direct measurement by ESR spectroscopy of the alkoxyl radical in irradiated crystalline methanol has been reported[128b], but its detection in alcohol glasses[129] is difficult because of line broadening and its presence in irradiated liquid alcohols has only been established using the spin labelling technique[130-135]. The G-values obtained by making use of the alkoxyl radical's oxidizing properties reach values between 1.5 and 2.0 for ethanol and methanol[136]. The question as to whether, in methanol, G(CH$_3$O$^\cdot$) and G($^\cdot$CH$_2$OH) (from reaction 17) are roughly equal[130,133,135] or whether essentially only the alkoxyl radical is primary[134a,c] is still being debated[128a,134]. The abundance of α-hydroxyalkyl radicals in the radiolysis of primary and secondary alcohols is no indication of the importance of reaction (17) because these radicals have several different precursors, mainly alkoxyl and H$^\cdot$.

The alkoxyl radicals react rapidly and in primary and secondary alcohols they are converted into α-hydroxyalkyl radicals (reaction 18). By pulse radiolysis k_{18} has been measured[137a] as 2.6 x 10^5 M^{-1} s^{-1}. An intramolecular rearrangement of the methoxyl into the hydroxymethyl radical has also been invoked[128b], a reaction which might be analogous to the same reaction occurring in aqueous solution where it is mediated by the solvent[137b,c]. Some of the reactions of the hydroxymethyl radical in methanol have been studied by pulse radiolysis[137d].

$$CH_3O^\cdot + CH_3OH \longrightarrow CH_3OH + {}^\cdot CH_2OH$$ (18)

D. Formation of Hydrogen

In Table 1 the major products of the radiolysis of some neat alcohols (meth-anol[122], ethanol[121], propanol[138-139], 2-propanol[31,140-143], n-butanol[30], 2-butanol[29], isobutanol[29] and t-butanol[29,144,145] in the liquid phase near room temperature are summarized. Where different values exist in the literature the average has been taken, or preference has been given to work where the applied dose was kept low and a reasonable material balance was obtained. It is seen from Table 1 that in all these alcohols, except t-butanol, hydrogen is the main product.

The predominant reaction is considered to be the reaction of the electron with the protonated alcohol (from the very fast reactions 16 and 17). Reaction (19)

$$H - \overset{|}{\underset{|}{C}} - OH_2^+ + e_{solv}^- \longrightarrow H - \overset{|}{\underset{|}{C}} - OH + H^\cdot \qquad (19)$$

produces an H atom which rapidly reacts with the alcohol by abstracting hydrogen preferentially at the position α to the hydroxyl group (reaction 20). Indeed, if the

$$H^\cdot + H - \overset{|}{\underset{|}{C}} - OH \longrightarrow H_2 + \cdot \overset{|}{\underset{|}{C}} - OH \qquad (20)$$

electrons are removed by electron scavengers $G(H_2)$ is strongly reduced (but not fully suppressed, see below)[142,146,147a]. In competition with this reaction the electron might react with the alkoxyl radicals from reaction (16) (reaction 21: its

$$e_{solv}^- + RO^\cdot \longrightarrow RO^- \qquad (21)$$

reaction with the α-hydroxyalkyl radical has also been considered[147b]), or might be scavenged by carbonyl compounds (reaction 22) present as impurities or formed

$$e_{solv}^- + \overset{|}{C} = O \longrightarrow \cdot \overset{|}{\underset{|}{C}} - O^- \qquad (22)$$

during radiolysis. On addition of acid these reactions are suppressed. $G(H_2)$ rises accordingly and in the series of n-alcohols reaches a value of about $6^{146,148,149}$. In Table 2 are shown the effects of the electron scavenger N_2O and of acid on the relative isotopic composition of the hydrogen evolved in the γ-radiolysis of several deuterated n-butanols[150]. Under strongly acidic conditions the major part (about two thirds to three quarters) of the hydrogen evolved from monodeuterated (at oxygen) alcohols is found as HD. The yield of D_2 is negligible.

It has been suggested[151] that the electron might be chemically trapped by $H^\cdot$ (reaction 23). The hydride is expected to form hydrogen in reaction (24), but the

$$e^- + H^\cdot \longrightarrow H^- \qquad (23)$$

$$H^- + ROH \longrightarrow H_2 + RO^- \qquad (24)$$

smallness of the D_2 yield from O-deuterated alcohols where considerable formation of $D^\cdot$ is expected means that reaction (23) should be rather minor.

The foregoing facts are in agreement with the electron being an important hydrogen precursor, and with a good likelihood of reaction (19) followed by (20). However, the remainder ($G \approx 1.5$; about a quarter) of the hydrogen evolved (consisting of H_2 in the mono(-oxygen-)deuterated alcohols) must have other sources which are not yet fully understood. The fragmentation of the primary ion has been proposed as a possibility (e.g. reaction 27). On the basis of the fact that formyl radical is produced in the radiolysis of methanol at 4 K even after low

TABLE 1. G-values of the major products in the radiolysis of liquid alcohols

Alcohol	H_2	Carbonyl compound	Dehydrodimer	α-Fragmentation products (G-values)
Methanol	5.4	1.95	3.5	—
Ethanol	5.0	3.2	1.7	Methane (0.6)
Propanol	4.4	2.9	1.5	Ethane (2.0), formaldehyde (1.9)
2-Propanol	4.5	4.0	0.6	Methane (1.55), acetaldehyde (0.9)
n-Butanol	4.45	3.15	1.55	Propane (1.9), formaldehyde (1.9)
2-Butanol	3.7	3.6	Not measured	Methane (1.2), propionaldehyde (0.8), ethane (3.5), acetaldehyde (3.4)
Isobutanol	Not measured	Not measured	Not measured	Propane (2.5), propene (1.2), formaldehyde (2.5)
t-Butanol	0.8	—	0.9	Methane (3.7), acetone (3.5)

TABLE 2. H_2, HD, D_2 distribution in the γ-radiolysis of some deuterated n-butanols[150]. Additives: (a) 0.1 M H_2SO_4 (D_2SO_4), (b) ~0.1 M N_2O. Average absolute total hydrogen G-values: (a) 5.8, (b) 1.6

Alcohol	H_2 (%) Acid[a]	H_2 (%) N_2O[b]	HD (%) Acid[a]	HD (%) N_2O[b]	D_2 (%) Acid[a]	D_2 (%) N_2O[b]
(A) $CH_3CH_2CH_2CH_2OD$	30	82	69	18	1	0
(B) $CH_3CH_2CH_2CD_2OD$	12	63	56	29	32	8
(C) $CH_3CH_2CD_2CH_2OH$	95	—	5	—	0	—
(D) $CD_3CD_2CD_2CH_2OH$	86	59	12	34	2	7
(E) $CD_3CH_2CH_2CH_2OH$	97	94	3	6	0	0

exposures, reactions (25) and (26) have been considered[128]. It is certain that hydrogen atoms are formed in the dissociation of excited alcohol molecules (see

$$CH_3O^{\cdot *} \longrightarrow H_2 + H\dot{C}O \tag{25}$$

$$^{\cdot}CH_2OD^* \longrightarrow HD + H\dot{C}O \tag{26}$$

$$CH_3OH^{\cdot +} \xrightarrow{\gamma} CH_2OH^+ + H^{\cdot} \tag{27}$$

Chapter 21). No material is available on the behaviour of liquid alcohols excited at wavelengths below 185 nm. In the 185 nm photolysis of methanol (see Chapter 22) reaction (28) strongly predominates over reaction (29).

In the radiolysis of O-deuterated alcohols some of the HD is expected from reactions such as (28) and (30), whereas reactions such as (29) followed by (20) will lead to the formation of some H_2.

$$CH_3OD^* \longrightarrow CH_3O^{\cdot} + D^{\cdot} \tag{28}$$

$$CH_3OD^* \longrightarrow {}^{\cdot}CH_2OD + H^{\cdot} \tag{29}$$

$$CH_3OD^* \longrightarrow CH_2O + HD \tag{30}$$

The results listed in Table 2 indicate that there is also some primary carbon–hydrogen cleavage from carbons other than $C_{(1)}$, possibly through formation of hydrogen atoms. Molecular hydrogen elimination from vicinal hydrogen-bearing carbon atoms cannot be excluded whereas carbene formation appears unlikely noting the absence of D_2 from the hydrogen produced by the butanols C and E (Table 2), which is in line with expectations based on saturated hydrocarbon radiolysis where carbene formation through geminal molecular hydrogen elimination is considered a minor process[152].

A further uncertainty with respect to the interpretation of the mechanism of hydrogen formation comes from the very low $G(H_2)$ in the case of t-butanol (Table 1). This may be partially due to the low reactivity of hydrogen atoms with t-butanol (cf. Table 5) which could lead to a reaction of the hydrogen atom with another radical in the same spur, whereas the chance that it meets a radical from another spur (randomized radical) is minute (1 : 10,000) at the dose rates commonly used. Fully one third of $G(H_2)$ finds its equivalent in the sum of G(isobutene oxide) and $G(t$-butoxy-2-hydroxy-2-methylpropane)[145]. These two products are also formed in the photolysis of t-butanol at 185 nm[153] where they balance all the H_2 formed (see Chapter 21). A more detailed study on the radiolysis of t-butanol would certainly also help to better understand the radiolysis of primary and secondary alcohols.

E. Fragmentation of the Carbon–Oxygen Skeleton

It is seen from Table 1 that the higher alcohols show considerable C–C bond fragmentation. It is not clear whether the apparent decrease of $G(H_2)$ in the neat higher alcohols perhaps reflects a real decrease of the contribution to the hydrogen yield from nonionic fragmentation, or is due to considerable electron scavenging by impurities or accumulated radiolysis products. On acidification $G(H_2)$ 6 is found[148], at least for all n-alcohols shown in Table 1. It has been proposed (cf. References 29 and 154) that C–C bond rupture may result from the fragmentation of the primary radical cation, e.g. reactions (31) and (32).

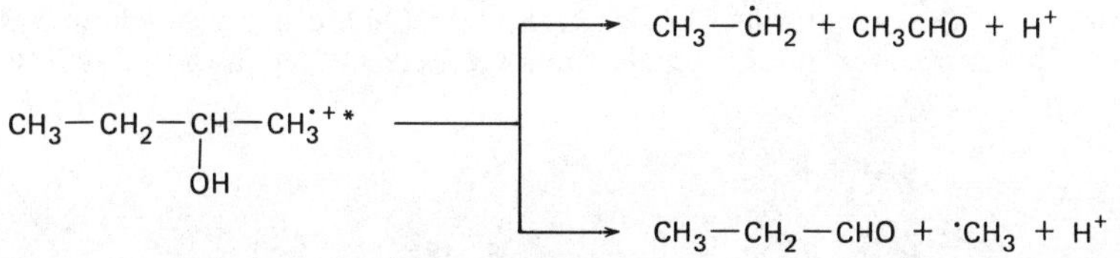

$$CH_3-CH_2-CH-CH_3^{+*} \quad \begin{cases} \longrightarrow \quad CH_3-\dot{C}H_2 + CH_3CHO + H^+ \quad (31) \\ \\ \longrightarrow \quad CH_3-CH_2-CHO + {}^{\cdot}CH_3 + H^+ \quad (32) \end{cases}$$

There is some more detailed material on the C—C bond fragmentation in isopropanol[155,156]. Using deuterated isopropanols and radical scavengers it has been shown that electron scavengers do not influence methane formation, that the major part of the methane has methyl radicals (95%) as precursors (70% scavengable and 25% 'hot'), and only ~5% is formed via the molecular processes (34) and (35). In reaction (33) a hydroxyethyl radical is formed together with the methyl

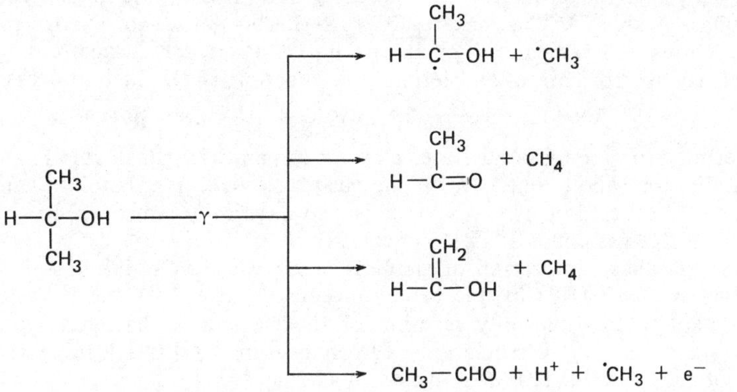

$$H-C-OH \xrightarrow{\gamma} \begin{cases} \longrightarrow \quad H-\underset{CH_3}{\overset{CH_3}{\underset{|}{\overset{|}{C}}}}-OH \cdot + {}^{\cdot}CH_3 \quad (33) \\ \\ \longrightarrow \quad \underset{H-C=O}{\overset{CH_3}{\underset{|}{\overset{|}{}}}} + CH_4 \quad (34) \\ \\ \longrightarrow \quad \underset{H-C-OH}{\overset{CH_2}{\underset{|}{\overset{||}{}}}} + CH_4 \quad (35) \\ \\ \longrightarrow \quad CH_3-CHO + H^+ + {}^{\cdot}CH_3 + e^- \quad (36) \end{cases}$$

$${}^{\cdot}CH_3 + H-\underset{CH_3}{\overset{CH_3}{\underset{|}{\overset{|}{C}}}}-OH \longrightarrow CH_4 + {}^{\cdot}\underset{CH_3}{\overset{CH_3}{\underset{|}{\overset{|}{C}}}}-OH \quad (37)$$

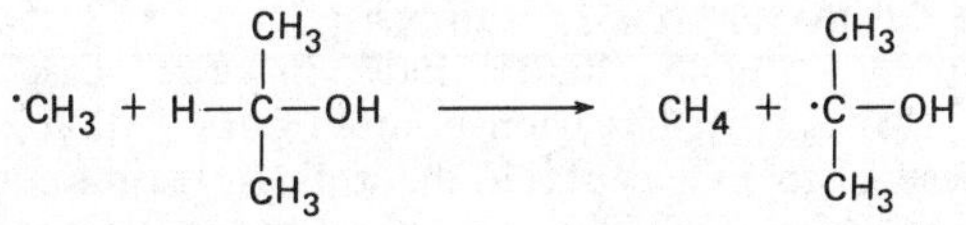

radical. The former may be scavenged by naphthalene and oxidized by benzophenone to acetaldehyde, leaving the methyl radical reactions (methane formation by H abstraction from isopropanol, reaction 37) unaffected. The acetaldehyde results indicate that only about 40% of the acetaldehyde is formed directly (reactions 34—36). There appears to be a major (60%) contribution from reaction (33) with an excited state as the precursor. This excited state must be an upper excited state because the lowest excited state which is reached in the 185 nm photolysis shows, as far as C—C bond cleavage is concerned, essentially merely reactions (34) and (35), and these with only a low quantum yield[157,158].

It has been suggested that, apart from in reaction (15), the 'parent' alkyl radical may be formed from alcohols by the dissociative electron capture of the protonated alcohol[159,160] (reaction 38). However, this reaction does not appear to play a role

$$R-OH_2^+ + e_{solv}^- \longrightarrow R^{\cdot} + H_2O \quad (38)$$

in methane formation from methanol where it has been shown[161] that $G(CH_4)$ is unaffected by addition of either H^+ or N_2O.

In the radiolysis of alcohols at room temperature or below, ethers are formed with low G-values[162]. Besides the trivial reaction (39), reaction (40) has been considered[162]. One would also envisage reaction (41), a reaction which has been shown to be implicated in the formation of ethers in the gas-phase radiolysis of alcohols (see below).

$$R^1O^{\cdot} + {}^{\cdot}R^2 \longrightarrow R^1OR^2 \tag{39}$$

$$R^1OH + {}^+R^2 \longrightarrow R^1OR^2 + H^+ \tag{40}$$

$$R^1OH_2^+ + R^1OH \longrightarrow R^1OR^1 + H_2O + H^+ \tag{41}$$

III. ALCOHOLS IN THE GAS PHASE

In the gas phase the G-values of products (Table 3) are much higher than in the liquid phase (Table 1). This may result from the breakdown of excited molecules, radical cations and radicals which in the liquid phase are thermalized. A typical example is the formation of olefins, e.g. reaction (42). Such processes, being

$$CH_3-CH_2OH \longrightarrow H_2C{=}CH_2 + H_2O, \ \Delta H = 46\,kJ/mol \tag{42}$$

endothermic from the ground state, play a comparatively small role in the liquid-phase radiolysis (about one fifth of the gas-phase yield in ethanol[121] and isopropanol[143,163]). Scission of C—C bonds is also drastically enhanced on going from the liquid to the gas phase (cf. Reference 121).

In the gas phase, ionization of alcohols occurs with a G-value of 4[18]. Electron scavengers reduce $G(H_2)$ by the same amount[164,165,166a], and it has therefore been argued[164] that the only reaction of the electron is that with a protonated alcohol (reaction 19). Dissociative electron capture by ROH leading to the formation of RO$^-$ as well as H$^-$ may play a small role[166b].

There are a number of attempts to correlate mass spectral data with the reactions occurring in the gas-phase radiolysis[123,167-170]. Obviously, such an approach is more justifiable here than in the case of liquid-phase data[29-31,139,154]. However, it has been pointed out[168] that there remain many uncertainties with respect to an acceptable theoretical treatment of this problem.

At elevated temperatures ($>250°C$) chain-reactions set in[163,171-175] (for a review see Reference 176). There are essentially four types of chain-reactions which are depicted by the overall reactions (43)–(46).

The protonated alcohols from reactions (16), (17) and (50) are probably the common precursors in the formation of olefins and ethers (e.g. reactions 47—49). It has been shown[127,177] that extensive clustering (reaction 48) occurs, the number of alcohol molecules within the cluster depending on alcohol pressure.

The chain-reactions leading to hydrogen and carbonyl compound (reaction 45) and to alkane and aldehyde (reaction 46) are considered to be free radical in nature.

TABLE 3. G-values of modes of cleavage in the gas-phase radiolysis of some alcohols

	MeOH[167] [a]	EtOH[164] [b]	i-PrOH[163]
G(C—H and O—H bond cleavage)	10.4	9.9	7.2
G(C—O bond cleavage)	0.3	1.8	2.9
G(C—C bond cleavage)		3.3	5.3

[a]For other work see compilation[122].
[b]For other work see compilation[121].

$$\text{Alcohol} \longrightarrow \begin{cases} \longrightarrow H_2O + \text{olefin} & (43) \\ \longrightarrow H_2O + \text{ether} & (44) \\ \longrightarrow H_2 + \text{carbonyl} & (45) \\ \longrightarrow \text{alkane} + \text{aldehyde} & (46) \end{cases}$$

$$C_2H_5OH_2^+ \longrightarrow H_2C{=}CH_2 + H_3O^+ \qquad (47)$$

$$C_2H_5OH_2^+ + C_2H_5OH \longrightarrow (C_2H_5OH)_2H^+ \qquad (48)$$

$$(C_2H_5OH)_2H^+ \longrightarrow C_2H_5{-}O{-}C_2H_5 + H_3O^+ \qquad (49)$$

$$H_3O^+ + C_2H_5OH \longrightarrow C_2H_5OH_2^+ + H_2O \qquad (50)$$

IV. NEAT ETHERS

The most readily apparent difference between the radiolysis of neat ethers and neat alcohols is based on the fact that the dielectric constants of ethers are smaller and their basicity greater than the corresponding properties in alcohols. Because of the smaller dielectric constant the free ion yield is smaller on account of a higher probability of geminate charge recombination $[G(\text{free ion}) < 1^{178}]$. Higher basicity means that the positive charge, which is formed and stabilized in reactions (51) and (52), remains somewhat more localized because the proton tends to be less mobile in R_2OH^+ than in ROH_2^+.

$$R_2O \xrightarrow{\quad \gamma \quad} R_2O^{+\cdot} + e^- \qquad (51)$$

$$R_2O^{+\cdot} + R_2O \longrightarrow R_2OH^+ + R_2O({-}H)^\cdot \qquad (52)$$

In contrast, the mobility of the solvated[179] or the trapped[180] electron is higher because of the lower polarity of the ether molecule. Direct evidence of the lesser stabilization of the electron is provided by its optical absorption spectrum. Whereas in alcohols its absorption maximum lies between about 600 to 800 nm, in ethers it absorbs near 2000 nm at room temperature[181-184]. There is a relatively larger blue-shift of the absorption maximum with decreasing temperature[184], presumably because the weaker ether molecular dipoles are more easily depolarized as the temperature rises. At about 77 K the blue-shift reaches its maximum value, with the spectrum peaking near 1200 nm[185,186] (cf. Reference 1). At still lower temperatures the maximum is again found at somewhat longer wavelengths but flattened as the dipoles are frozen in[185,187] (for a review see Reference 188a). In 2-methyl-tetrahydrofuran glass an inner solvation shell of three equivalent solvent molecules appears to envelop the electron[188b].

An interesting method, not applicable to protic media such as alcohols, to extend (by a factor of five) the lifetime of the solvated electron in 1,4-dioxane consists of exchanging the oxonium ion against the unreactive Li^+ (reaction 53)[189]. In ethers, alkali metal cations and solvated electrons coexist as ion pairs (M^+, e_{solv}^-) which

$$R_2OH^+ + LiAlH_4 \longrightarrow Li^+ + R_2O + AlH_3 + H_2 \qquad (53)$$

are characterized by a strong blue-shift of the solvated electron absorption spectrum[190-193]. The other alkali ions are not as stable as Li^+ toward the solvated electron. Na^- and K^- were produced in the radiolysis of tetrahydrofuran solutions of the alkali metals, or their boronates[191,192]. Spectra of e^-_{solv}, Na^- and K^- in various ethers have been obtained[194].

As with the alcohols, the radiolysis of ethers is through ionic as well as through excited states. A G-value near 4.3 for total ionization has been measured in the gas phase for various ethers[195], and similar values have been accepted for the liquid phase[196-198].

Apart from undergoing fragmentation, molecules is excited states may also transfer energy to solutes[189], or show luminescence[199]. The latter behaviour is seen to be of particular importance in 1,4-dioxane, in some contrast to other ethers. Dioxane fluoresces (λ_{max} 247 nm) on excitation with 185 nm light[200,201] and also on radiolysis[199,202]. This fluorescence is quenched by N_2O[22,203] and other quenchers[202,203]. Energy transfer to scintillators yields visible light; this property together with the ability of 1,4-dioxane to accommodate aqueous material in homogeneous distribution have earned it a place among the media employed for low-energy β-radiation counting (cf. References 204 and 205).

Hydrogen is a major radiolysis product (cf. Reference 1) in all ethers investigated, including diethyl[206,207], di-n-propyl[196], diisopropyl[208] and dibutyl[208] ethers, tetrahydrofuran[209-211], 2-methyltetrahydrofuran[208,212-215] and 1,4-dioxane[208,216-219]. It is thought that several different modes of hydrogen formation are in operation. Following reactions (51) and (52), the solvated electron neutralizes the oxonium ion (reaction 54). Hydrogen atoms abstract from the ether, predominantly in the α-position (reaction 55). Also, atomic as well as molecular hydrogen is formed from excited molecules (reactions 56 and 57), or in spur reactions irrepressible by electron or radical scavengers.

$$R_2^1OH^+ + e^-_{solv} \longrightarrow R_2^1O + H^\cdot \tag{54}$$

$$H^\cdot + R^2-CH_2-O-R^1 \longrightarrow R^2-\overset{\cdot}{C}H-OR^1 + H_2 \tag{55}$$

$$R_2^1O \overset{\gamma}{\longrightarrow} R_2^1O(-H)^\cdot + H^\cdot \tag{56}$$

$$R^3-CH_2-CH_2-OR^1 \overset{\gamma}{\longrightarrow} R^3-HC=CH-OR^1 + H_2 \tag{57}$$

In most cases few if any other products have been measured. Especially in the cases of the cyclic ethers the radiolysis mechanism is far from clear. Fragmentation of the carbon—oxygen skeleton probably leads to biradical intermediates which may react in a variety of ways, side by side with the different monoradicals. In the case of 2-methyltetrahydrofuran both the tertiary and the secondary α-radical seem now established[220,221].

Diethyl ether presents a case where an extensive product analysis has been carried out for both gas- and liquid-phase radiolysis[206,207]. The product distribution differs in the two phases although G(ether consumption) is nearly the same, at about 11.3. Similar to photolysis[222] and pyrolysis[223], cleavage of the C—O bond is a major event[206,207,210,216] in radiolysis, probably partly through ionic, and partly through excited states (reactions 58—61). There is some evidence that dissociative electron capture (reaction 62) may also occur[224,225]. Fragments resulting from C—C bond rupture such as methane and successor products of CH_3 are of lesser importance and, as expected, more in evidence among the products from the gas-phase radiolysis[207].

$$R_2OH^+ + e^- \longrightarrow ROH + R^{\cdot} \tag{58}$$

$$R_2O \begin{cases} \overset{\gamma}{\longrightarrow} ROH + \text{olefin} & (59) \\ \overset{\gamma}{\longrightarrow} R'CH{=}O + RH & (60) \\ \overset{\gamma}{\longrightarrow} RO^{\cdot} + {}^{\cdot}R & (61) \end{cases}$$

$$R_2O + e^- \longrightarrow RO^- + {}^{\cdot}R \tag{62}$$

V. AQUEOUS SOLUTIONS OF ALCOHOLS AND ETHERS

A. Primary Species in the Radiolysis of Aqueous Solutions

If dilute aqueous solutions are irradiated with ionizing radiation, the radiation energy is largely absorbed by the solvent water leading to OH radicals, hydrated electrons (e$_{aq}^-$), H atoms, the molecular products H_2O_2 and H_2, as well as the ions H^+ and OH^- (reaction 63).

$$H_2O \overset{\gamma}{\longrightarrow} {}^{\cdot}OH, e_{aq}^-, H^{\cdot}, H_2O_2, H_2, H^+, OH^- \tag{63}$$

$$e_{aq}^- + N_2O \longrightarrow {}^{\cdot}OH + N_2 + OH^- \tag{64}$$

$$e_{aq}^- + H^+ \longrightarrow H^{\cdot} \tag{65}$$

$$e_{aq}^- + O_2 \longrightarrow O_2^{\cdot -} \tag{66}$$

$$H^{\cdot} + O_2 \longrightarrow HO_2^{\cdot} \tag{67}$$

The hydrated electrons (for rate constants see References 226 and 227) can be converted into OH radicals by saturating the solution with N_2O (reaction 64; $[N_2O] = 2.2 \times 10^{-2}$ M at $20°C$ and atmospheric pressure, $k_{64} = 5.6 \times 10^9$ M^{-1} s^{-1}). At low pH they are converted into H atoms (reaction 65; $k_{65} = 2.3 \times 10^{10}$ M^{-1} s^{-1}). In the presence of O_2, hydrated electrons can be converted into $O_2^{\cdot -}$ radicals (reaction 66; $k_{66} = 2 \times 10^{10}$ M^{-1} s^{-1}). The H atom (for rate constants see Reference 228) does not react with N_2O but reacts readily with O_2 (reaction 67; $k_{67} = 10^{10}$ M^{-1} s^{-1}). The resulting $HO_2^{\cdot}$ is in equilibrium with its basic form $O_2^{\cdot -}$ [pK_a ($HO_2^{\cdot}$) = 4.75)[229]. Saturation of an aqueous solution with a mixture of N_2O/O_2 (4/1 v/v) converts hydrated electrons into OH radicals whereas the H atoms are scavenged by O_2. The G-values of the molecular products and of the ions are little

TABLE 4. G-values of radicals generated in the γ-radiolysis of neutral water in the presence of inert gases (e.g. He, Ar, N_2), N_2O and O_2

Saturating gas	$G({}^{\cdot}OH)$	$G(e_{aq}^-)$	$G(H^{\cdot})$	$G(O_2^{\cdot -})$
Inert gas	2.7	2.7	0.55	—
N_2O	5.4[a]	—	0.55	—
O_2	2.7	—	—	3.25
N_2O/O_2 (4/1 v/v)	5.4[a]	—	—	0.55

[a]There is evidence[230,231] that under N_2O saturation $G(OH)$ may be as high as 6.

changed by these additives $[(G(H_2O_2) = 0.7, \quad G(H_2) = 0.45, \quad G(H^+) = 3.4,$ $G(OH^-) = 0.6]$. The G-values of the radicals at the various conditions are summarized in Table 4. The values for $O_2^{\cdot-}$ given in Table 4 are only valid as long as other additives are used in concentrations which do not interfere with reactions (66) or (67).

B. Deoxygenated Solutions

1. Saturated alcohols

Hydrated electrons do not react with saturated alcohols at a measurable rate $(k < 10^6 \ \text{M}^{-1}\text{s}^{-1})^{226,232}$. However, OH radicals and H atoms rapidly react with these substrates by hydrogen abstraction. The OH radical (for rate constants see References 233 and 234) reacts considerably faster than the H atom[228]. The reactivity of the $SO_4^{\cdot-}$ radical is somewhere in between[235] (Table 5). The latter can be generated by reaction of the solvated electron with $S_2O_8^{--}$ (reaction 68). The

$$e_{aq}^- + S_2O_8^{--} \longrightarrow SO_4^{--} + SO_4^{\cdot-} \tag{68}$$

preferred site of attack of the OH radical is the position α to the hydroxyl group[236,237]. With increasing chain-length of the alcohol the probability of H abstraction at positions other than α to the hydroxyl group increases. There is always a very low probability of H abstraction at the hydroxyl group (Table 6).

TABLE 5. Rate constants $(\text{M}^{-1}\text{s}^{-1})$ of OH radicals, H atoms and $SO_4^{\cdot-}$ radicals with some alcohols in aqueous solutions (references see text)

Substrate	$^{\cdot}$OH	H$^{\cdot}$	$SO_4^{\cdot-}$
Methanol	9×10^8	2×10^6	3.2×10^6
Methanol-d$_3$	4.2×10^8	$2.5 \times 10^{5\,a}$	1.2×10^6
Ethanol	1.8×10^9	2.6×10^7	1.6×10^7
2-Propanol	2.0×10^9	6.5×10^7	3.2×10^7
2-Methyl-2-propanol	4.5×10^8	8×10^4	4.0×10^5

[a]Value calculated from $k(\text{H}^{\cdot} + \text{methanol})$ on the basis of an H/D isotope effect of $7.5^{235\,b}$.

TABLE 6. Relative yields (%) of H abstraction by OH radicals at different positions from various alcohols[237]

Substrate	α	β,γ,δ etc.	OH
CH_3OH	93.0	–	7.0
C_2H_5OH	84.3	13.2	2.5
$CH_3CH_2CH_2OH$	53.4	46.0	<0.5
$(CH_3)_2CHOH$	85.5	13.3	1.2
$CH_3CH_2CH_2CH_2OH$	41.0	58.5	<0.5
$(CH_3)_3COH$	–	95.7	4.3
CH_2OH-CH_2OH	100	–	<0.1
$CH_2OH-CHOH-CH_3$	79.2	20.7	<0.1
$CH_3-CHOH-CHOH-CH_3$	71.0	29.0	<0.1

TABLE 7. pK values of some α-hydroxyalkyl radicals and their parent alcohols

Radical	pK	pK of parent alcohol	ΔpK
$\cdot CH_2OH$	10.71[a] 10.7[b]	15.09[c]	−4.38
$CH_3\dot{C}HOH$	11.51[a] 11.6[b]	15.93[c]	−4.42
$(CH_3)_2\dot{C}OH$	12.03[a] 12.2[b]	17.1[c]	−5.07
$(CF_3)_2\dot{C}OH$	1.70[a]	9.8[a]	−8.1

[a]From Reference 239.
[b]From Reference 238.
[c]From Reference 240.

Using the pulse radiolysis technique it has been shown that the α-hydroxyalkyl radicals are more acidic by about four pK units than their parent alcohols[238]. This has been confirmed by in situ ESR spectroscopic studies[239] (Table 7).

In their self-termination, the α-hydroxyalkyl radicals disproportionate and dimerize (reactions 69 and 70).

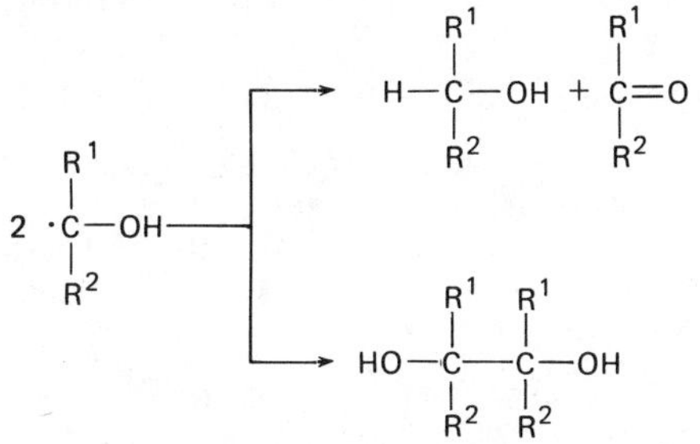

$$2 \cdot\underset{\underset{R^2}{|}}{\overset{\overset{R^1}{|}}{C}}\!\!-OH$$

$$\longrightarrow H-\underset{\underset{R^2}{|}}{\overset{\overset{R^1}{|}}{C}}\!\!-OH + \underset{\underset{R^2}{|}}{\overset{\overset{R^1}{|}}{C}}\!\!=O \qquad (69)$$

$$\longrightarrow HO-\underset{\underset{R^2}{|}}{\overset{\overset{R^1}{|}}{C}}\!\!-\underset{\underset{R^2}{|}}{\overset{\overset{R^1}{|}}{C}}\!\!-OH \qquad (70)$$

The disproportionation/dimerization ratio increases with increasing methyl substitution ($R^1 = R^2 = H$, $k_{69}/k_{70} < 0.1$; $R^1 = H$, $R^2 = CH_3$, $k_{69}/k_{70} = 0.43$; $R^1 = R^2 = CH_3$, $k_{69}/k_{70} \sim 4$)[28]. In the disproportionation of 2-hydroxypropyl-(2) radicals the transfer of a carbon-bound hydrogen atom (reaction 71) has a higher probability than the transfer of the oxygen-bound hydrogen atom (reaction 72)[241].

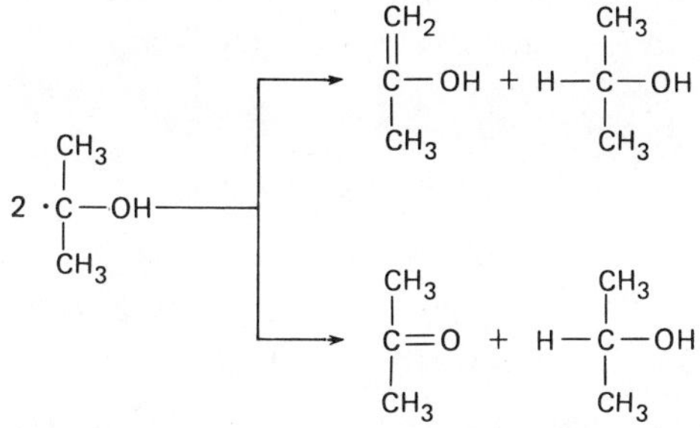

$$2 \cdot\underset{\underset{CH_3}{|}}{\overset{\overset{CH_3}{|}}{C}}\!\!-OH$$

$$\longrightarrow \underset{\underset{CH_3}{|}}{\overset{\overset{CH_2}{||}}{C}}\!\!-OH + H-\underset{\underset{CH_3}{|}}{\overset{\overset{CH_3}{|}}{C}}\!\!-OH \qquad (71)$$

$$\longrightarrow \underset{\underset{CH_3}{|}}{\overset{\overset{CH_3}{|}}{C}}\!\!=O + H-\underset{\underset{CH_3}{|}}{\overset{\overset{CH_3}{|}}{C}}\!\!-OH \qquad (72)$$

An optically active carbon atom which carries an OH group may lose its former optical activity on going from the alcohol through the radical state back to the alcohol. This has been found[242] with scyllo-inositol where the major disproportionation product is myo-inositol (reactions 73 and 74).

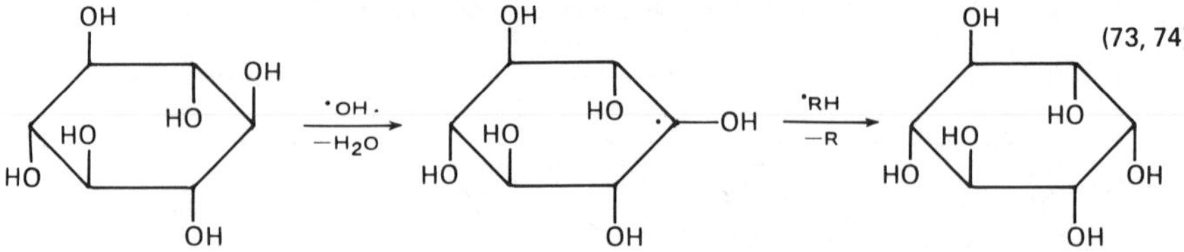

(73, 74)

The α-hydroxyalkyl radicals are rapidly oxidized by $Fe(CN)_6^{3-}$ (reaction 75; $k \approx 4 \times 10^9$ $M^{-1}s^{-1}$), a reaction which has been followed by pulse radiolysis[236].

$$[Fe(CN)_6]^{3-} + R_2\dot{C}OH \longrightarrow [Fe(CN)_6]^{4-} + R_2CO + H^+ \qquad (75)$$

With Fe^{2+} ions these radicals form a complex which can be monitored by its short-lived absorption[243]. However, reduction of these radicals does not take place, and the products are the same as observed in the absence of Fe^{2+} ions. Intermediate complexes of the α-hydroxyalkyl radicals with other metal ions such as Ag^{+2}[244], Ag_2^{+2}[245], Ni^{+2}[246], Cd^{+2}[247] and Pb^{+2}[248 a] were also detected. A compilation of rate constants for the reaction of metal ions in unusual valence states has appeared[248 b].

The reaction of α-hydroxyalkyl radicals with hydrogen peroxide is quite rapid (reaction 76; $k \approx 10^5$ $M^{-1}s^{-1}$)[249-251] and leads to the formation of an OH radical which propagates a chain.

$$R_2\dot{C}OH + H_2O_2 \longrightarrow R_2CO + H_2O + \dot{O}H \qquad (76)$$

The anions of the α-hydroxyalkyl radicals are better electron donors than the α-hydroxyalkyl radicals themselves (for pK values see Table 7). Therefore, electron-transfer reactions are more efficient at high pH where chain-reactions have been observed with alkyl halides and with N_2O[252-256]. Likely propagating steps are the reactions (77) and (78).

$$R_2^1\dot{C}O^{\cdot-} + R^2Br \longrightarrow R_2^1CO + R^{2\cdot} + Br^- \qquad (77)$$

$$R_2^1CO^{\cdot-} + N_2O \longrightarrow R_2^1CO + N_2 + O^{\cdot-} \qquad (78)$$

β-Hydroxyalkyl radicals are also formed in the reactions of OH radicals with primary and secondary alcohols, even though with low yields (see Table 6). They can be generated more conveniently by reacting OH radicals with olefins, for example reaction (79). Further, β-hydroxyalkyl radicals are formed in the reaction of OH radicals with tertiary alcohols, e.g. t-butanol (reaction 80).

$$\dot{O}H + H_2C{=}CH_2 \longrightarrow \dot{C}H_2{-}CH_2OH \qquad (79)$$

$$\dot{O}H + CH_3{-}\underset{\underset{CH_3}{|}}{\overset{\overset{CH_3}{|}}{C}}{-}OH \longrightarrow H_2O + \cdot CH_2{-}\underset{\underset{CH_3}{|}}{\overset{\overset{CH_3}{|}}{C}}{-}OH \qquad (80)$$

In their reaction with Cu^{2+} they are reported to give epoxides (e.g. reaction 81)[257,258].

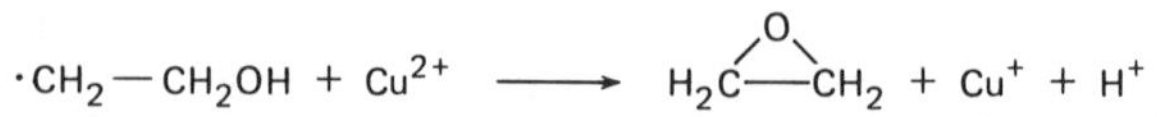

$$\cdot CH_2{-}CH_2OH + Cu^{2+} \longrightarrow H_2\overset{O}{\overset{/\backslash}{C{-}}}CH_2 + Cu^+ + H^+ \qquad (81)$$

From strongly reducing metal ions such as Ni^+ the β-hydroxyalkyl radicals accept an electron, yielding olefins (e.g. reaction 82)[246].

β-Hydroxyalkyl radicals also abstract hydrogen atoms from their parent alcohol

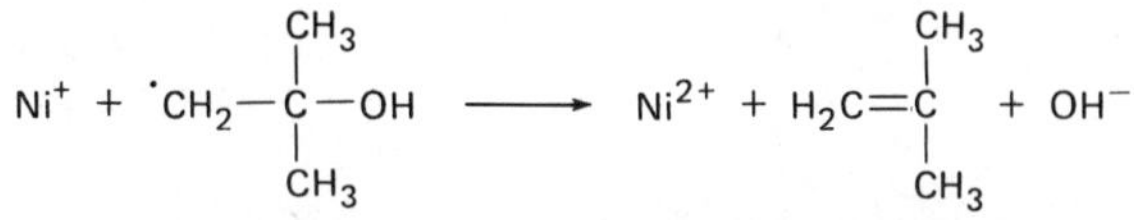

$$Ni^+ + {}^\bullet CH_2 - \underset{\underset{CH_3}{|}}{\overset{\overset{CH_3}{|}}{C}} - OH \longrightarrow Ni^{2+} + H_2C = \underset{\underset{CH_3}{|}}{\overset{\overset{CH_3}{|}}{C}} + OH^- \qquad (82)$$

if derived from a primary or secondary alcohol. Thereby the β-hydroxyalkyl radicals are converted into α-hydroxyalkyl radicals (e.g. reaction 83). The rate

$${}^\bullet CH_2 - CH_2OH + CH_3 - CH_2OH \longrightarrow CH_3 - CH_2OH + CH_3 - {}^\bullet CHOH \qquad (83)$$

constant of this reaction is around 30–50 M $^{-1}$s^{-1} [249,259,260]. A value higher by one order of magnitude has also been reported[261].

2. Polyhydric alcohols and carbohydrates

The radiolysis of polyhydric alcohols and carbohydrates in deoxygenated aqueous solution is characterized by the elimination of water from the original 1,2-dihydroxyalkyl radicals (reaction 84). This reactions has first been observed by

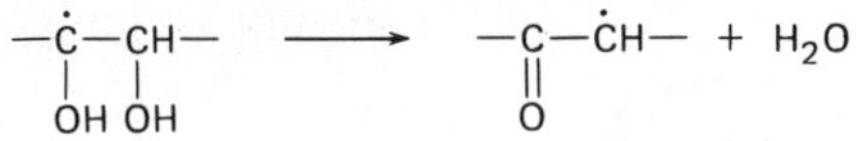

$$-\underset{\underset{OH}{|}}{\overset{\bullet}{C}} - \underset{\underset{OH}{|}}{\overset{}{CH}} - \longrightarrow -\underset{\underset{O}{\|}}{\overset{}{C}} - {}^\bullet CH - + H_2O \qquad (84)$$

ESR spectroscopy[262-265] and was later further investigated by product analysis[266-269] and pulse radiolysis[270]. The elimination of water is acid- and base-catalysed. The acylalkyl radicals ($-CO-{}^\bullet CH-$) have oxidizing properties[270b] and readily abstract hydrogen atoms from the starting material (e.g. reaction 85), thus inducing chain-reactions. In the case of ethylene glycol as a substrate the rate constant of reaction (85) was found to be 75 M $^{-1}$s^{-1} [268]. A typical example of the

$${}^\bullet CH_2 - CHO + CH_2OH - CH_2OH \longrightarrow CH_3 - CHO + {}^\bullet CHOH - CH_2OH \qquad (85)$$

various reactions involved is given in Scheme 1 for the simplest molecule in this series, ethylene glycol.

In the case of erythritol the radical at $C_{(2)}$ has two possible ways to eliminate water. It is noted[271,272] that the elimination towards $C_{(1)}$ (reaction 86) is preferred by a factor of seven over that towards $C_{(3)}$ (reaction 87). The reasons for this unexpected preference are not yet known.

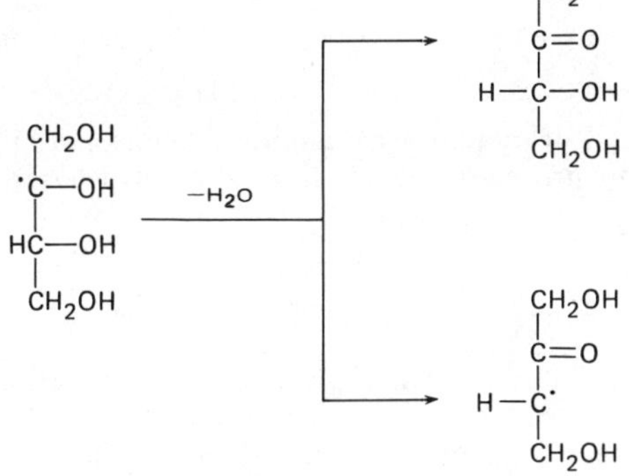

The same type of reaction can proceed with β-alkoxy-α-hydroxyl radicals (reaction 88, X = OR)[270b,272-276]. Reaction (88) is especially fast if X is a good leaving group, e.g. F, Cl, Br, I, CH_3CO_2 and H_2PO_4 [262,265,277-281].

$$CH_2OH-CH_2OH + {}^{\cdot}OH \longrightarrow {}^{\cdot}CHOH-CH_2OH + H_2O$$

(1)   (2)

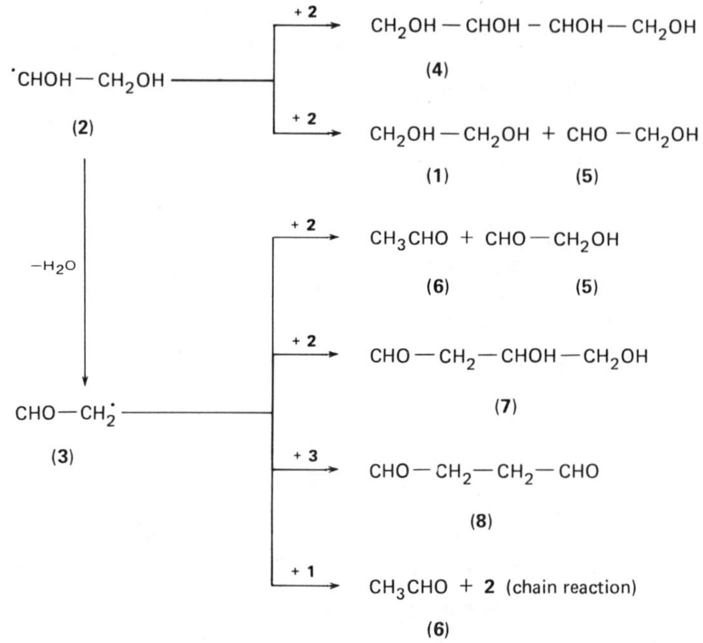

${}^{\cdot}CHOH-CH_2OH$

(2)

+ 2 → $CH_2OH-CHOH-CHOH-CH_2OH$

(4)

+ 2 → $CH_2OH-CH_2OH + CHO-CH_2OH$

(1)   (5)

$-H_2O$

$CHO-CH_2^{\cdot}$

(3)

+ 2 → $CH_3CHO + CHO-CH_2OH$

(6)   (5)

+ 2 → $CHO-CH_2-CHOH-CH_2OH$

(7)

+ 3 → $CHO-CH_2-CH_2-CHO$

(8)

+ 1 → $CH_3CHO + 2$ (chain reaction)

(6)

SCHEME 1. Reactions of radicals derived from ethylene glycol[a]

[a]G-values of products (Reference 267) of a N_2O-saturated 0.1 M solution of ethylene glycol at $20°C$ and at a dose rate of 0.1 W kg^{-1}. G(tetritol, 4),= 0.15, G(glycolaldehyde, 5) = 1.05, G(acetaldehyde, 6) = 1.2, G(2-deoxy-tetrose, 7) = 0.25 and G(succinaldehyde, 8) = 1.7.

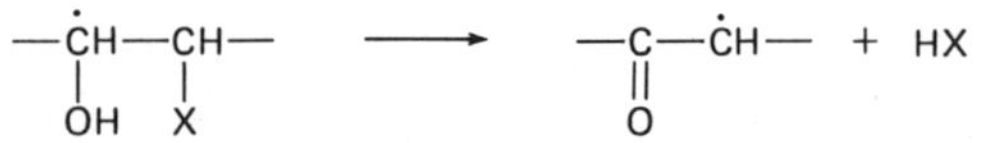

$$-\overset{\cdot}{C}H-\underset{X}{\underset{|}{\underset{OH}{\overset{|}{C}}}}H- \longrightarrow -\overset{\|}{\underset{O}{C}}-\overset{\cdot}{C}H- + HX \qquad (88)$$

The radical-induced deamination of amino alcohols and amino sugars has been considered[282,283] to proceed through the radical zwitterion (reactions 89, 90).

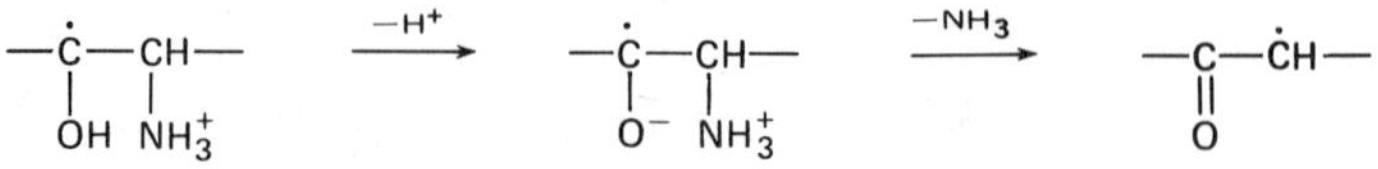

$$-\overset{\cdot}{\underset{OH}{\underset{|}{C}}}-\underset{NH_3^+}{\underset{|}{C}}H- \xrightarrow{-H^+} -\overset{\cdot}{\underset{O^-}{\underset{|}{C}}}-\underset{NH_3^+}{\underset{|}{C}}H- \xrightarrow{-NH_3} -\overset{\|}{\underset{O}{C}}-\overset{\cdot}{C}H- \qquad (89, 90)$$

This pathway may be followed even at pH 5 where deamination is still observed. In the hydrolytic equilibrium, the radical zwitterion might be present at a sufficient concentration. The acidity increase of an OH group that is attached to a carbon atom carrying a free spin is well known (cf. Table 7).

The products which have been identified thus far in the γ-radiolysis of aqueous solutions of D-glucose are listed in Table 8. The importance of the water elimination reaction (reaction 84) and the analogous reaction (reaction 88) is apparent from the number of the deoxy sugars that are formed via these reactions. (About

TABLE 8. Products and their initial G-values from the γ-radiolysis of deoxygenated N_2O-saturated[273] or N_2O/O_2(4 : 1)-saturated[284] aqueous solutions of D-glucose at a dose rate of 0.18 W kg^{-1} at room temperature

Products	G-values	
	N_2O	N_2O/O_2
D-Gluconic acid	0.15	0.90
D-arabino-Hexosulose	0.15	0.90
D-ribo-Hexos-3-ulose	0.10	0.57
D-xylo-Hexos-4-ulose	0.075	0.50
D-xylo-Hexos-5-ulose	0.18	0.60
D-gluco-Hexodialdose	0.22	1.55
2-Deoxy-D-arabino-hexonic acid	0.95	Absent
5-Deoxy-D-threo-hexos-4-ulose		Absent
5-Deoxy-D-xylo-hexonic acid	0.08	Absent
2-Deoxy-D-erythro-hexos-5-ulose		Absent
5-Deoxy-D-xylo-hexodialdose		Absent
3-Deoxy-D-erythro-hexos-4-ulose		Absent
3-Deoxy-D-erythro-hexosulose	0.25	Absent
4-Deoxy-L-threo-hexos-5-ulose		Absent
6-Deoxy-D-xylo-hexos-5-ulose	0.05	Absent
2-Deoxy-D-erythro-hexos-3-ulose	a	Absent
4-Deoxy-D-threo-hexos-3-ulose	a	Absent
D-Arabinose	0.01	} 0.10
D-Arabinonic acid	Absent	
D-Ribose	<0.005	Absent
D-Xylose	<0.005	} 0.08
xylo-Pentodialdose	Absent	
2-Deoxy-D-erythro-pentose	0.04	Absent
D-Erythrose	0.01	} 0.02
D-Erythronic acid	Absent	
Threose	<0.003	Absent
L-threo-Tetrodialdose	Absent	0.20
3-Deoxytetrulose	0.02	Absent
Dihydroxyacetone	0.03	Absent
D-Glyceraldehyde and glyceric acid	Absent	0.13
Glyoxal	b	0.11
Glyoxylic acid and glycolic acid	b	0.4
Formaldehyde	b	0.12
Formic acid	b	0.6
D-Glucose consumption	5.6	5.6

aProducts identified (no G-values given) in Reference 285. They are expected to be included in the G-values of the other deoxy-hexosulose given in the table.
bNot determined, probably absent.

reactions typical for the lactol function see below.) For a detailed discussion of the radiation chemistry of carbohydrates see Reference 286.

3. Saturated ethers and acetals

Solvated electrons do not react with these substrates but OH radicals and H atoms rapidly abstract hydrogen atoms if such are available in the α-position to the ether linkage (reaction 91).

$$R^1CH_2OR^2 + {}^{\cdot}OH \longrightarrow R^1\overset{\cdot|}{C}HOR^2 + H_2O \tag{91}$$

The resulting α-alkoxyalkyl radicals show a number of reactions which resemble those observed with α-hydroxyalkyl radicals. They are readily oxidized by $[Fe(CN)_6]^{3-}$ ($k = 2 \times 10^9$ $M^{-1}s^{-1}$)[236] or hydrogen peroxide ($k = 5.5 \times 10^4$ $M^{-1}s^{-1}$)[287].

In reaction (92), the 1-ethoxyethyl radicals derived from diethyl ether via reaction (91) yield only acetaldehyde and ethanol. A likely intermediate is the carbonium ion (oxonium ion). This must react much more rapidly (>20-fold) with water to give acetaldehyde ethyl hemiacetal (reaction 93) and ultimately acetaldehyde and ethanol (reaction 94) rather than lose a proton to give ethyl vinyl ether (reaction 95)[288]. In the absence of an oxidant the latter is formed as a disproportionation product of two ethoxyethyl radicals. Their dimer, 2,3-diethoxybutane, is

$$CH_3\overset{\cdot}{C}HOC_2H_5 + [Fe(CN)_6]^{-3} \longrightarrow CH_3CH{=}\overset{+}{O}C_2H_5 + [Fe(CN)_6]^{4-} \tag{92}$$

$$CH_3\overset{+}{C}HOC_2H_5 + H_2O \longrightarrow CH_3CHOHOC_2H_5 + H^+ \tag{93}$$

$$CH_3CHOHOC_2H_5 \longrightarrow CH_3CHO + C_2H_5OH \tag{94}$$

$$CH_3CH{=}\overset{+}{O}C_2H_5 \longrightarrow CH_2{=}CH_2OC_2H_5 + H^+ \tag{95}$$

also formed. Because of the high reactivity of H_2O_2 with the ethoxyethyl radicals the radiolytically generated hydrogen peroxide (see above) can only attain very low steady-state concentrations at the usual dose rates of ${}^{60}Co$ γ-irradiation. The products of the reaction of the ethoxyethyl radicals with hydrogen peroxide are acetaldehyde and ethanol. A chain-reaction is induced by the OH radical liberated in this reaction[288].

α-Alkoxyalkyl radicals which carry a good leaving group (e.g. $X = $ halogen or phosphate) in the position β to the free spin rapidly eliminate this group and two new radicals are observed[278,289] by ESR spectroscopy, instead of the original radical (reactions 96 and 97). The most likely intermediate is the radical cation formed in reaction 98. Evidence for this, among other indications, is the fact that

$$RO\overset{\cdot}{C}HCH_2X + H_2O \longrightarrow \begin{cases} RO\overset{\cdot}{C}HCH_2OH + HX & (96) \\ ROCHOH\overset{\cdot}{C}H_2 + HX & (97) \end{cases}$$

$$RO\overset{\cdot}{C}HCH_2X \longrightarrow RO\overset{\cdot}{C}H\overset{+}{C}H_2 + X^- \tag{98}$$

$$\Big\updownarrow$$

$$RO\overset{+}{\underset{}{C}}H\overset{\cdot}{C}H_2$$

$$\Big\updownarrow$$

$$R\overset{+}{O}{=}CH\overset{\cdot}{C}H_2$$

the radical cation which one gets from an acetal (reactions 99, 100) is stable against hydrolysis within its life-time with respect to diffusion-controlled second-order decay[290].

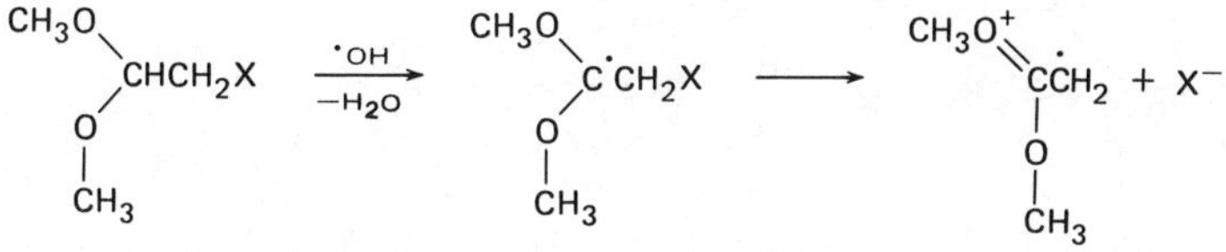

(99, 100)

The rate of the elimination of phosphate from such radicals ($R = CH_3$, X = phosphate; reaction 98) strongly depends on the state of protonation of the phosphate group. Going from the dianion to the neutral form, the rate constant of phosphate elimination increases by three to four orders of magnitude with each protonation step ($X = PO_4^{2-}$, $k_{98} \approx 0.1-1 \text{ s}^{-1}$; $X = HPO_4^-$, $k_{98} \approx 10^3 \text{ s}^{-1}$; $X = H_2PO_4$, $k_{98} \approx 3 \times 10^6$)[289].

The mechanism which has been described here appears also to operate in the formation of OH radical-induced strand breaks of DNA[291-293]. The isolated products are in agreement with the radical at $C_{(4')}$ being their precursor. The DNA strand is broken by the elimination of the phosphate ester group at the 3' and 5' positions (reactions 101 and 102).

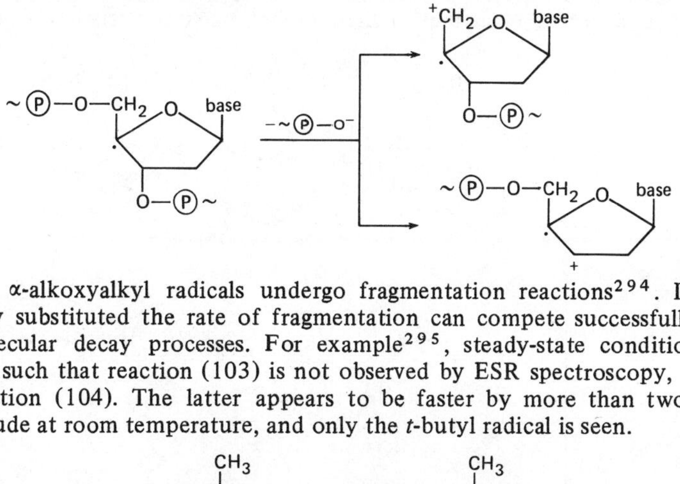

(101)

(102)

The α-alkoxyalkyl radicals undergo fragmentation reactions[294]. If they are suitably substituted the rate of fragmentation can compete successfully with the biomolecular decay processes. For example[295], steady-state conditions can be chosen such that reaction (103) is not observed by ESR spectroscopy, in contrast to reaction (104). The latter appears to be faster by more than two orders of magnitude at room temperature, and only the t-butyl radical is seen.

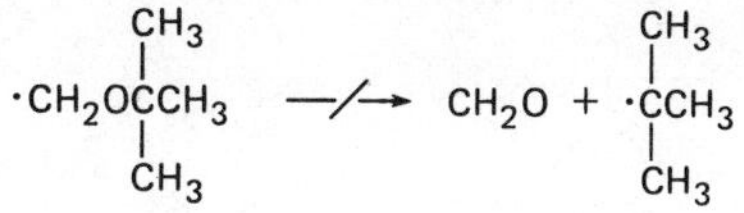

(103)

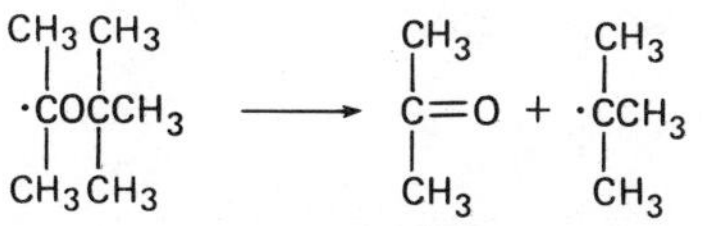

(104)

Similar reactions have been invoked to explain some products in the radiolysis of sugars, e.g. 5-deoxy-D-xylo-hexonic acid from D-glucose[273] (cf. Table 8; reactions 105 and 106).

One of the pathways in the radiation-induced scission of the glycosidic linkage of disaccharides[296,297a] and fragmentation of dioxolanes[297b] is also thought to

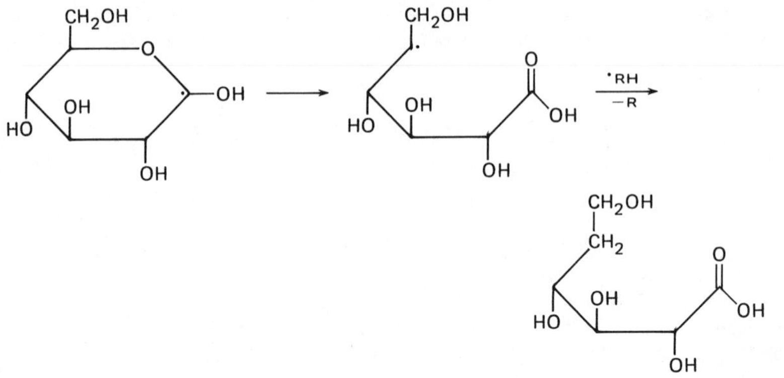

(105, 106)

follow this type of reaction, as do some interesting chain-reactions in crystalline carbohydrates induced by γ-irradiation[298-300].

4. Phenols and aromatic ethers

The radiation chemistry of phenols[301-311] and aromatic ethers[305,309, 310,312-318] is different from that of their aliphatic counterparts largely because

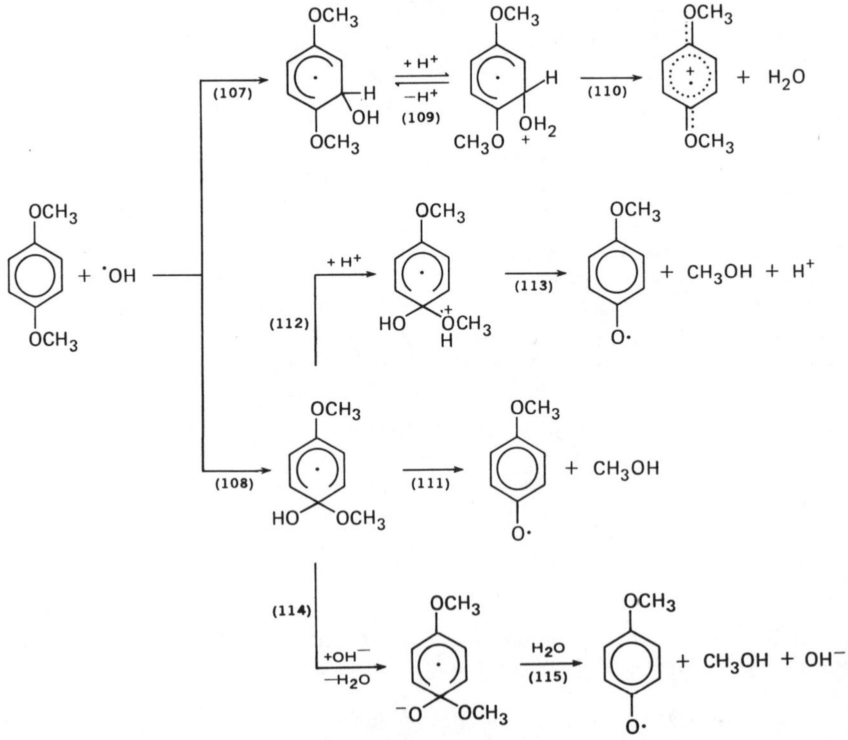

SCHEME 2.

the OH radicals add to the ring but do not abstract hydrogen. Indeed, there appears to be negligible, if any, H abstraction even from the methyl groups of methoxylated benzenes[318]. Because of their electrophilicity the OH radicals add preferentially at positions activated by electron-donating substituents[309-311,317].

Scheme 2 gives an example of the general reaction mechanism. In the chosen case of 1,4-dimethoxybenzene[318] there are two possibilities for OH addition, namely at a free position (reaction 107, see Scheme 2) and at the *ipso* position (reaction 108). The protonation of the OH group and elimination of water leads to the formation of the radical cation (reactions 109 and 110), which is well characterized by its ESR spectrum[318]. The same species is obtained by electron transfer from 1,4-dimethoxybenzene to Tl^{2+}, Ag^{2+} and $SO_4^{\cdot-}$ [316]. The *ipso* OH adduct eliminates methanol in a spontaneous reaction (111) and in an acid- (112 and 113) and base- (114 and 115) catalysed reaction giving a phenoxyl radical. Similar reactions were observed with methoxylated benzoic acids where the formation of the radical cation leads to a zwitterion[315,316].

In the phenol series the radical cations immediately $(t_{1/2} < 1\mu s)$[309] lose a proton and are converted into phenoxyl radicals (reaction 116) which are observed by ESR spectroscopy[308]. Phenoxyl radicals are also formed under basic conditions from the deprotonated dihydroxycyclohexadienyl radicals (reaction 117)[309].

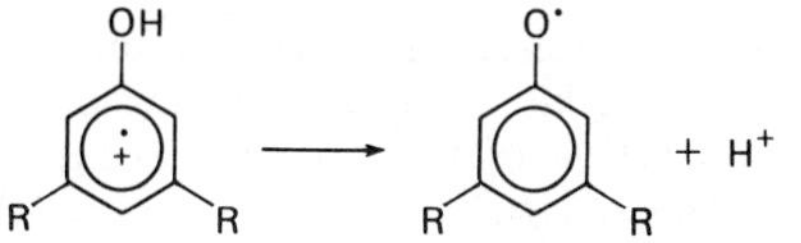

$$\text{(116)}$$

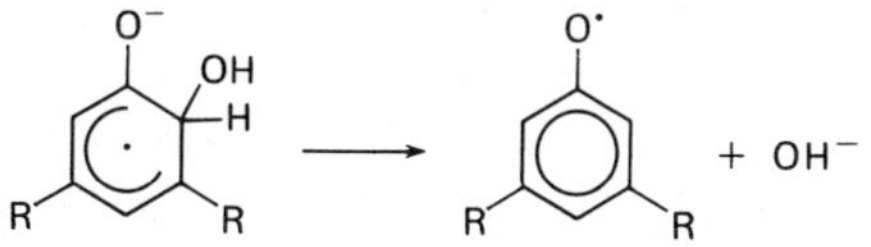

$$\text{(117)}$$

Among the final products of the reaction of OH radicals with phenols are more highly hydroxylated phenols[301-304,306,311b]. Some of these reactions are also of preparative interest[303]. In the presence of HBr the OH radical can be converted into a $Br^{\cdot}$ atom which also adds to the aromatic ring. Under such conditions 2-bromo-4-nitrophenol, among other products, is formed from 4-nitrophenol[307].

C. Oxygenated Solutions of Saturated Alcohols, Ethers and Carbohydrates

The reactions of peroxyl radicals derived from alcohols can be most conveniently studied using radiation techniques. If N_2O/O_2 (4 : 1 v/v) saturated solutions containing aliphatic alcohols are irradiated with ionizing radiation the majority of the primary radicals are OH radicals (cf. Table 4) which rapidly (cf. Table 5) abstract carbon-bound hydrogen atoms (cf. Table 6). These carbon-centred radicals add molecular oxygen at virtually diffusion-controlled rates (e.g. reactions 118 and 119)[236].

The peroxyl radicals derived from α-hydroxyalkyl radicals and from β-hydroxyalkyl radicals show quite a different behaviour.

$$\text{(118)}$$

$$\cdot CH_2-\underset{\underset{R^2}{|}}{\overset{\overset{R^1}{|}}{C}}-OH + O_2 \longrightarrow \cdot O-O-CH_2-\underset{\underset{R^2}{|}}{\overset{\overset{R^1}{|}}{C}}-OH \qquad (119)$$

The α-hydroxyalkylperoxyl radicals undergo a unimolecular elimination of $HO_2^{\cdot}$[284,319-324], most likely[321] via a five-membered transition state (reaction 120). The reaction parameters are given in Table 9. There is also a base-catalysed

$$\underset{R^2}{\overset{R^1}{>}}\!\!C\!\!\underset{O-H}{\overset{O-O\cdot}{<}} \longrightarrow \underset{R^2}{\overset{R^1}{>}}C=O + HO_2^{\cdot} \qquad (120)$$

pathway (reactions 121 and 122) which is nearly diffusion-controlled in the case of hydroxide ion acting as the base (Table 9), but is about three orders of magnitude slower with phosphate.

$$\underset{R^2}{\overset{R^1}{>}}\!\!C\!\!\underset{O-H}{\overset{O-O\cdot}{<}} \xrightarrow[-BH^+]{+B} \underset{R^2}{\overset{R^1}{>}}\!\!C\!\!\underset{O^-}{\overset{O-O\cdot}{<}} \longrightarrow \underset{\underset{R^2}{|}}{\overset{\overset{R^1}{|}}{C}}=O + O_2^{\cdot -} \qquad (121, 122)$$

In competition to the elimination of $HO_2^{\cdot}$ there is the bimolecular decay of the α-hydroxyalkylperoxyl radicals which is near to diffusion-controlled. Because of the comparatively slow elimination of $HO_2^{\cdot}$ at pH 7 in the case of the $HOCH_2O_2^{\cdot}$ radical (cf. Table 9), the bimolecular decay kinetics and its products can be studied more conveniently than in other cases. It has been shown[325] that the major route (> 80%) leads to formic acid and hydrogen peroxide (reactions 123 and 124). A

$$2\ HOCH_2O_2^{\cdot} \longrightarrow \underset{H}{\overset{HO}{>}}C\underset{O\ddagger O\cdots H}{\overset{H\cdots O\ddagger O}{<}}C\underset{H}{\overset{OH}{<}} \longrightarrow H_2O_2 + 2\ HC\underset{OH}{\overset{O}{\diagup\!\!\!\diagdown}} \qquad (123, 124)$$

very short-lived tetroxide and a bicyclic transition state which resembles the monocyclic transition state of the $HO_2^{\cdot}$ elimination has been postulated.

The β-hydroxyalkylperoxyl radicals decay only by second-order reactions which are also near to diffusion-controlled judging from the data obtained with the peroxyl radical derived from *t*-butanol[326]. A very short-lived tetroxide has been considered to decompose along various pathways as indicated in Scheme 3. Reaction (126) is formulated according to the Russell mechanism (cf. Reference 327), a concerted process with a six-membered transition state. Reaction (128) depicts the elimination of O_2 and the formation of two caged oxyl radicals which either combine to the peroxide (reaction 130) or disproportionate (reaction 131) to give

TABLE 9. Rate constants for the first-order formation of H^+ and $O_2^{\cdot -}$ from $R^1R^2C(OH)OO^{\cdot}$ radicals (k_1) and for the OH^--catalysed reaction (k_2) in aqueous solutions

R^1	R^2	k_1 at 22°C (s^{-1})	Activation energy (kJ mol^{-1})	Preexponential factor (s^{-1})	k_2 at 22°C (M^{-1} s^{-1})
H	H	<10			~15 × 10^9
H	CH$_3$	52	60	2 × 10^{12}	8 × 10^9
CH$_3$	CH$_3$	~670	56	6 × 10^{12}	5 × 10^9
OR	OH	>70,000			

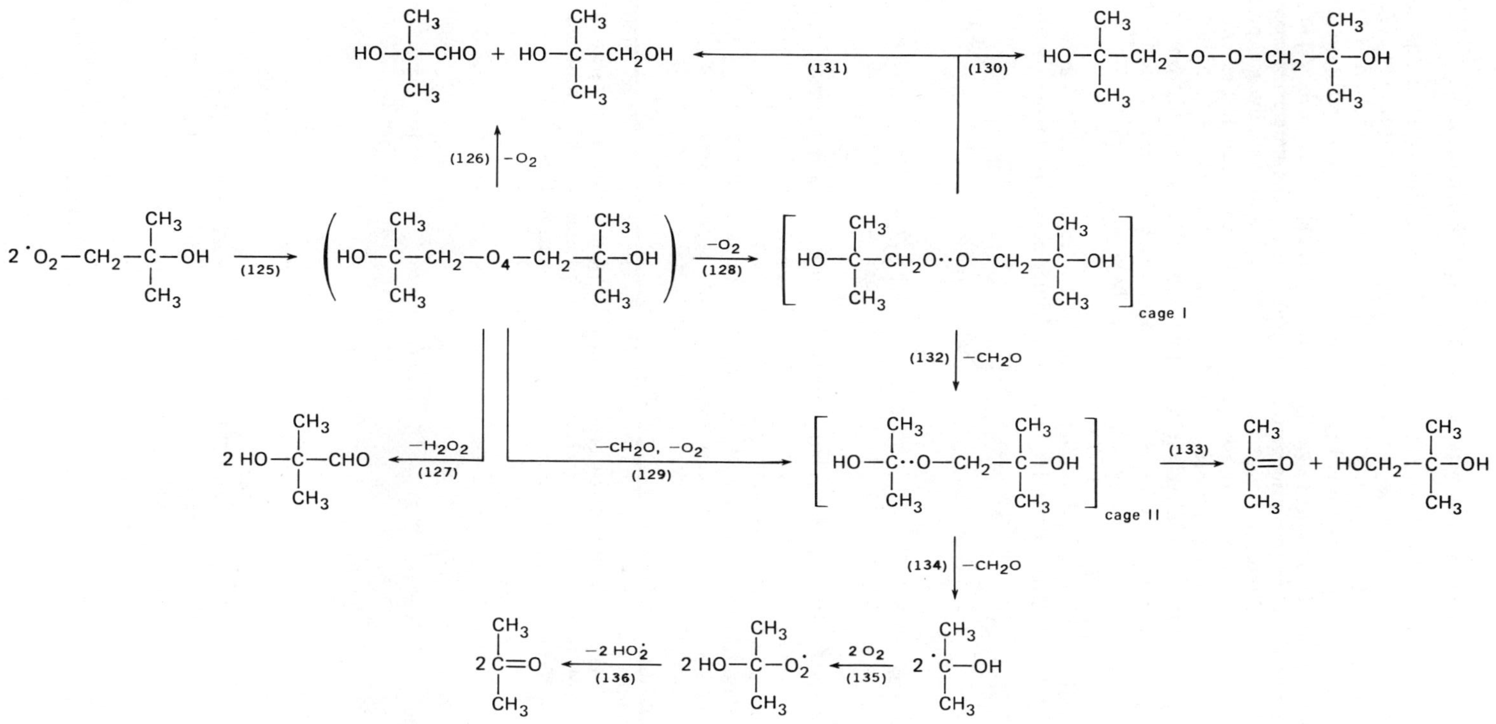

SCHEME 3.

the same products as obtained via the Russell mechanism. The oxyl radicals can also fragment (reactions 132 and 134). Formaldehyde and 2-hydroxypropyl-(2) radicals are the products. Another path to the same products is given by reaction (129). The elimination of H_2O_2 (reaction 127) is similar to the major reaction of two $HOCH_2O_2^{\cdot}$ radicals (reaction 124). The hydroxypropyl-(2) radicals rapidly add oxygen to give the corresponding peroxyl radicals which eliminate $HO_2^{\cdot}$ according to the mechanisms discussed (reactions 120–122). In a pulse radiolysis experiment the kinetics of the overall process have been followed through the change of conductivity caused by the appearance of H^+ and $O_2^{\cdot-}$ $[pK_a(HO_2^{\cdot}) = 4.75)]$.

There is very little material on the fate of peroxyl radicals derived from ethers in aqueous solutions as studied by radiation techniques. The decay kinetics of the α-alkoxyalkylperoxyl radicals generated under these conditions are still open to question (cf. References 322 and 328).

The α-alkoxyalkylperoxyl radicals readily undergo a chain autoxidation reaction[324,329] (e.g. reactions 137 and 138). This reaction is apparently not given

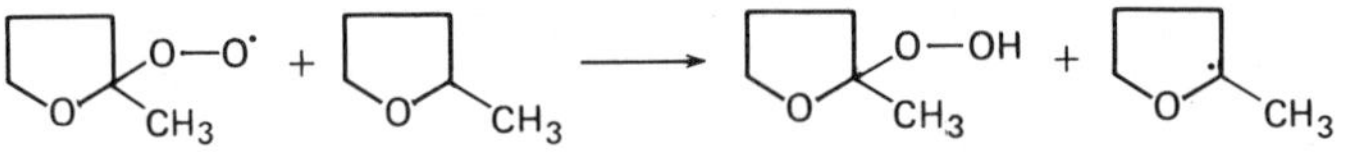

$$(137)$$

$$(138)$$

at neutral pH by the α-hydroxyalkylperoxyl radicals because of their fast $HO_2^{\cdot}$ elimination (reactions 120–122). At neutral pH this leads to $O_2^{\cdot-}$, a species of low H-abstractive power which is incapable of propagating a chain[324,330].

Because of the fast reaction of O_2 with the radicals formed by OH attack on carbohydrates the transformation reactions of the sugar radicals (see above) are fully suppressed in neutral oxygen- or air-saturated solutions. Instead, the reactions of the corresponding peroxyl radicals occur.

As discussed above, the high reactivity of the OH radical leads to an approximately random abstraction of carbon-bound hydrogen atoms from carbohydrates, and the radiolysis of D-glucose in N_2O/O_2-saturated aqueous solutions leads to six different peroxyl radicals with about equal yields (reaction 139)[284].

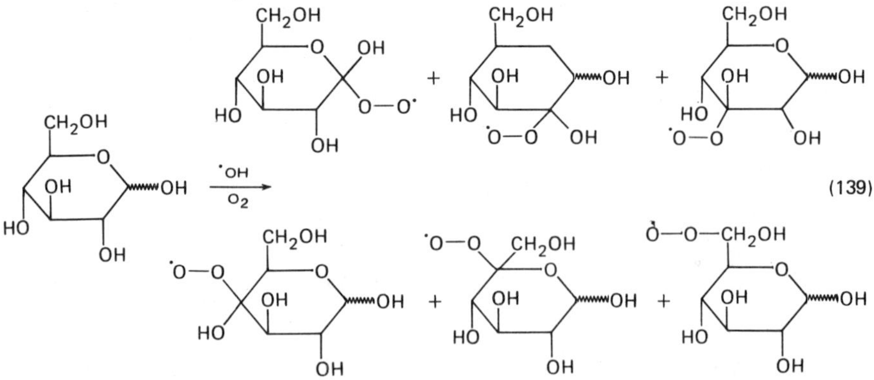

$$(139)$$

Five of these (those at $C_{(1)}$ to $C_{(4)}$ and $C_{(6)}$) are α-hydroxyalkylperoxyl radicals which readily eliminate $HO_2^{\cdot}$ (reactions 120–122). Especially fast $(k > 7 \times 10^4 \text{ s}^{-1})$ is the $HO_2^{\cdot}$ elimination from the peroxyl radical at $C_{(1)}$. But even the peroxyl radical at $C_{(5)}$ may, with base catalysis, eliminate $HO_2^\bullet$ (reactions 140 and 141). The corresponding carbonyl compounds are thus the major products (see

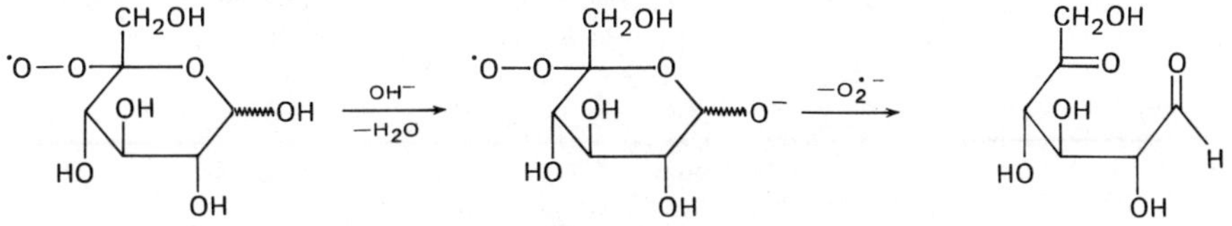

(140, 141)

Table 8). In competition with this $HO_2^\bullet$ elimination the sugar peroxyl radicals undergo reactions second order in peroxyl radicals. The longest-lived peroxyl radical, that at $C_{(5)}$, shows most clearly such a reaction (reactions 142–145). The reaction

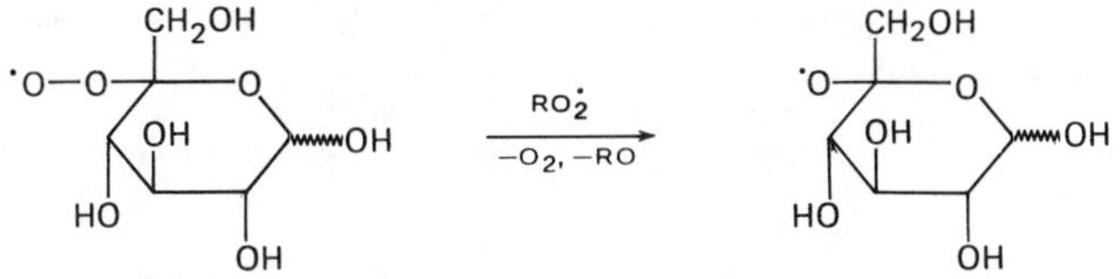

(142)

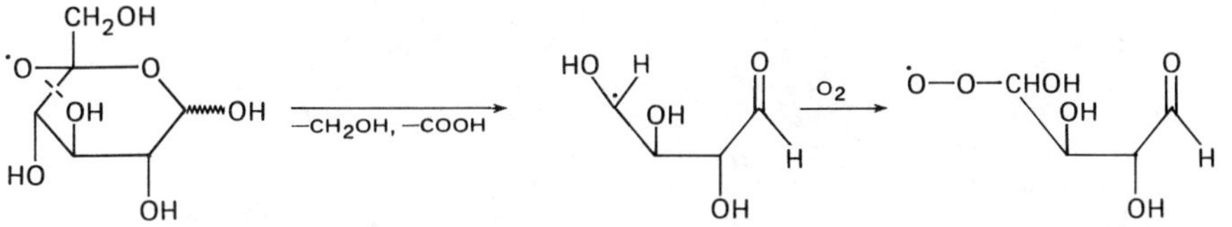

(143, 144)

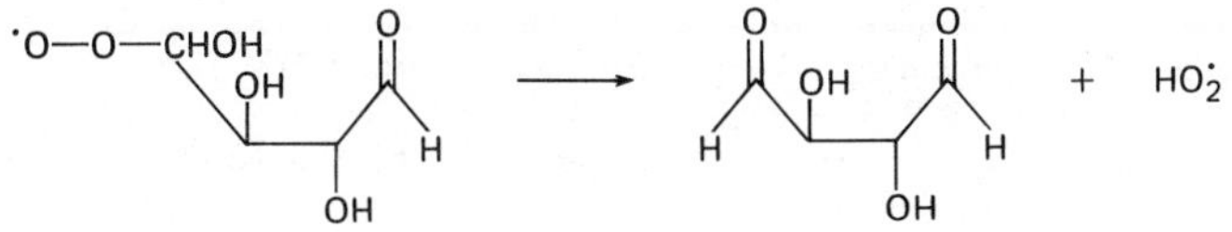

(145)

sequence is similar to that discussed above (cf. Scheme 3). The end-product is L-*threo*-tetrodialdose (see Table 8). As expected the *erythro* isomer is formed from the peroxyl radical at $C_{(5)}$ of D-ribose[331]. Similar reaction sequences have been considered for an explanation of some products from the radiolysis of oxygenated solutions of ribose-5-phosphate[332], N-acetylglucosamine[333] and thymidine[334]. In DNA the peroxyl radical at $C_{(5')}$ has been considered[335] to give rise to DNA strand breaks via such a mechanism, and that at $C_{(2')}$ to an alkali-labile[336] site (for a review see Reference 293).

VI. REFERENCES

1. J. Teplý, *Radiat. Res. Rev.*, **1**, 361 (1968).
2. C. von Sonntag, *Top. Curr. Chem.*, **13**, 333 (1969).
3. L. Kevan in *Actions Chimiques et Biologiques des Radiations* (Ed. M. Haissinsky), Sér. 13, Masson et Cie, Paris, 1969, p. 57.
4. G. R. Freeman in *Actions Chimiques et Biologiques des Radiations* (Ed. M. Haissinsky), Sér. 14, Masson et Cie, Paris, 1970, p. 73.

5. R. A. Basson in *The Chemistry of the Hydroxyl Group* (Ed. S. Patai) Vol. 2, John Wiley and Sons, London, 1971, p. 937.
6. (a) A. J. Swallow, in *MTP International Review of Science: Organic Chemistry*, Vol. 10 (Eds. D. H. Hey and W. A. Waters), *Free Radical Reactions*, Butterworths, London, 1973, p. 263.
    (b) A. J. Swallow, *Progr. React. Kinet.*, **9**, 195 (1978).
7. (a) P. K. Ludwig, *Advan. Radiat. Chem.*, **3**, 1 (1972).
    (b) J. H. Baxendale and M. A. J. Rodgers, *Chem. Soc. Rev.*, **7**, 235 (1978).
8. K. Eiben and R. W. Fessenden, *J. Phys. Chem.*, **75**, 1186 (1971).
9. (a) A. D. Trifunac and D. J. Nelson, *J. Amer. Chem. Soc.*, **99**, 1745 (1977).
    (b) D. J. Nelson, C. Mottley and A. D. Trifunac, *Chem. Phys. Letters*, **55**, 323 (1978).
10. A. Henglein, *Electroanal. Chem.*, **9**, 163 (1976).
11. (a) G. Beck, J. Kiwi, D. Lindenau and W. Schnabel, *European Polym. J.*, **10**, 1069 (1974).
    (b) G. Beck, D. Lindenau and W. Schnabel, *European Polym. J.*, **11**, 761–6 (1975).
12. A. Henglein, W. Schnabel and J. Wendenburg, *Einführung in die Strahlenchemie*, Verlag Chemie, Weinhein, 1969.
13. A. J. Swallow, *Radiation Chemistry*, Longman, London, 1973.
14. J. W. T. Spinks and R. J. Woods, *An Introduction to Radiation Chemistry*, 2nd ed., John Wiley and Sons, New York, 1976.
15. A. Kuppermann, *Nucleonics*, **19**, 38 (1961).
16. A. Mozumder and J. L. Magee in *Physical Chemistry: An Advanced Treatise* (Eds. H. Eyring, D. Henderson and W. Jost), Vol. 7, Academic Press, New York, 1975, p. 699.
17. A. Voltz, *Radiat. Res. Rev.*, **1**, 301 (1968).
18. (a) P. Adler and H.-K. Bothe, *Z. Naturforsch.*, **20a**, 1707 (1965).
    (b) G. G. Meisels and D. R. Ethridge, *J. Phys. Chem.*, **76**, 3842 (1972).
19. A. Bernas, J. Blais, M. Gauthier and D. Grand, *Chem. Phys. Letters*, **30**, 383 (1975).
20. J. C. Russell and G. R. Freeman, *J. Phys. Chem.*, **72**, 816 (1968).
21. T. Wada and Y. Hatano, *J. Phys. Chem.*, **79**, 2210 (1975).
22. H.-P. Schuchmann, H. Bandmann and C. von Sonntag, *Z. Naturforsch.*, **34b**, 327 (1979).
23. C. von Sonntag and H.-P. Schuchmann, unpublished results. 185 nm photolysis of N_2O–saturated *t*-butanol gave a nitrogen quantum yield of 0.04.
24. T. G. Ryan, T. E. M. Sambrook and G. R. Freeman, *J. Phys. Chem.*, **82**, 26 (1978).
25. T. G. Ryan and G. R. Freeman, *J. Phys. Chem.*, **81**, 1455 (1977).
26. R. L. Platzman, *Vortex*, **23**, 372 (1962).
27. R. L. Platzman in *Radiation Research* (Ed. G. Silini), North Holland, Amsterdam, 1966, p. 20.
28. C. von Sonntag and H.-P. Schuchmann, *Advan. Photochem.*, **10**, 59 (1977).
29. H. J. van der Linde and R. A. Basson, *J. S. African Chem. Inst.*, **28**, 115 (1975).
30. L. G. J. Ackerman, R. A. Basson and H. J. van der Linde, *J. Chem. Soc., Faraday Trans. I*, **68**, 1258, (1972).
31. (a) R. A. Basson and H. J. van der Linde, *J. Chem. Soc., Faraday Trans. I*, **70**, 431 (1974).
    (b) H. Knof, V. Hansen and D. Krafft, *Z. Naturforsch.*, **27a**, 162 (1972).
32. G. R. Freeman, *Radiat. Res. Rev.*, **1**, 1 (1968).
33. J. Lilie, S. A. Chaudhri, A. Mamou, M. Graetzel and J. Rabani, *J. Phys. Chem.*, **77**, 597 (1973).
34. (a) R. S. Dixon and V. J. Lopata, *J. Chem. Phys.*, **62**, 4573 (1975).
    (b) R. S. Dixon and V. J. Lopata, *J. Chem. Phys.*, **63**, 3679 (1976).
35. A. O. Allen, *NSRDS-NBS*, 57, U.S. Dept. of Commerce, Washington, D. C., 1976.
36. (a) U. Schindewolf, *Angew. Chem.*, **80**, 165 (1968).
    (b) U. Schindewolf, *Angew. Chem.*, **90**, 939 (1978).
37. K. Eiben, *Angew. Chem.*, **82**, 652 (1970).
38. A. Ekstrom, *Radiat. Res. Rev.*, **2**, 381 (1970).
39. B. C. Webster and G. Howat, *Radiat. Res. Rev.*, **4**, 259 (1972).
40. L. Kevan, *Advan. Radiat. Chem.*, **4**, 181 (1979).

41. F. S. Dainton, *Chem. Soc. Rev.*, **4**, 323 (1975).
42. (a) A. V. Vannikov, *Russ. Chem. Rev.*, **44**, 906 (1975).
 (b) A. K. Pikaev, *High Energy Chem.*, **10**, 95 (1976).
43. M. S. Matheson in *Physical Chemistry: An Advanced Treatise* (Eds. H. Eyring, D. Henderson and W. Jost), Vol. 7, Academic Press, New York, 1975, p. 533.
44. F. S. Dainton, G. A. Salmon and U. F. Zucker, *Chem. Commun.*, 1172 (1968).
45. F. S. Dainton, G. A. Salmon and P. Wardman, *Proc. Roy. Soc. (Lond.)*, A **313**, 1 (1969).
46. F. S. Dainton, G. A. Salmon and U. F. Zucker, *Proc. Roy. Soc. (Lond.)*, A **320**, 1 (1970).
47. F. S. Dainton, G. A. Salmon, P. Wardman and U. F. Zucker, *Proc. Roy. Soc. (Lond.)*, A **325**, 23 (1971).
48. F. S. Dainton, G. A. Salmon and P. Wardman, *Chem. Commun.*, 1174 (1968).
49. (a) L. Gilles, J. E. Aldrich and J. W. Hunt, *Nature Phys. Sci.*, **243**, 70 (1973).
 (b) W. J. Chase and J. W. Hunt, *J. Phys. Chem.*, **79**, 2835 (1975).
50. K. Y. Lam and J. W. Hunt, *J. Phys. Chem.*, **78**, 2414 (1974).
51. G. A. Kenney-Wallace, *Can. J. Chem.*, **55**, 2009 (1977).
52. G. A. Kenney-Wallace and C. D. Jonah, *Chem. Phys. Letters*, **47**, 362 (1977).
53. J. W. Hunt and W. J. Chase, *Can. J. Chem.*, **55**, 2080 (1977).
54. J. H. Baxendale and P. Wardman, *Chem. Commun.*, 429 (1971).
55. (a) L. Kevan, *Chem. Phys. Letters*, **11**, 140 (1971).
 (b) L. Kevan, *J. Chem. Phys.*, **56**, 838 (1972).
56. J. H. Baxendale and P. Wardman, *J. Chem. Soc., Faraday Trans. 1*, **69**, 584 (1973).
57. N. V. Klassen, H. A. Gillis, G. G. Teather and L. Kevan, *J. Chem. Phys.*, **62**, 2474 (1975).
58. J. R. Miller, B. E. Clifft, J. J. Hines, R. F. Runowski and K. W. Johnston, *J. Phys. Chem.*, **80**, 457 (1976).
59. J. H. Baxendale and P. H. G. Sharpe, *Intern. J. Radiat. Phys. Chem.*, **8**, 621 (1976).
60. D.-P. Lin, L. Kevan and H. B. Steen, *Intern. J. Radiat. Phys. Chem.*, **8**, 713 (1976).
61. L. Gilles, M. R. Bono and M. Schmidt, *Can. J. Chem.*, **55**, 2003 (1977).
62. K. Okazaki and G. R. Freeman, *Can. J. Chem.*, **56**, 2313 (1978).
63. (a) J. H. Baxendale and P. H. G. Sharpe, *Chem. Phys. Letters*, **39**, 401 (1976).
 (b) L. M. Perkey and J. F. Smalley, *J. Phys. Chem.*, **83**, 2959 (1979).
 (c) G. V. Buxton, J. Kroh and G. A. Salmon, *Chem. Phys. Letters*, **68**, 554 (1979).
64. H. Hase, T. Warashina, M. Noda, A. Namiki and T. Higashimura, *J. Chem. Phys.*, **57**, 1039 (1972).
65. A. Namiki, M. Noda and T. Higashimura, *Chem. Phys. Letters*, **23**, 402 (1973).
66. L. M. Perkey, Farhataziz and R. R. Hentz, *Chem. Phys. Letters*, **27**, 531 (1974).
67. T. Shida and M. Imamura, *J. Phys. Chem.*, **78**, 232 (1974).
68. H. Hase, L. Kevan and T. Higashimura, *Chem. Phys. Letters*, **55**, 171 (1978).
69. J. Moan, *Acta Chem. Scand. (A)*, **30**, 483 (1976).
70. L. Kevan, *J. Phys. Chem.*, **79**, 2846 (1975).
71. L. Kevan and K. Fueki, *Chem. Phys. Letters*, **49**, 101 (1977).
72. (a) K. Funabashi and W. H. Hamill, *Chem. Phys. Letters*, **56**, 175 (1978).
 (b) G. Dolivo and L. Kevan, *J. Chem. Phys.*, **70**, 2599 (1979).
 (c) M. Ogasawara, K. Shimizu, K. Yoshida, J. Kroh and H. Yoshida, *Chem. Phys. Letters*, **64**, 43 (1979).
 (d) Y. Ito, H. Hase and I. Higashimura, *Radiat. Phys. Chem.*, **13**, 195 (1979).
73. G. R. Freeman, *J. Phys. Chem.*, **77**, 7 (1973).
74. R. R. Hentz and G. A. Kenney-Wallace, *J. Phys. Chem.*, **78**, 514 (1974).
75. L. Kevan, *Intern. J. Radiat. Phys. Chem.*, **6**, 297 (1974).
76. A. V. Vannikov, A. V. Rudnev and G. M. Zimina, *Radiat. Eff.*, **23**, 15 (1974).
77. G. A. Kenney-Wallace and C. D. Jonah, *Chem. Phys. Letters*, **39**, 596 (1976).
78. G. A. Kenney-Wallace, *Chem. Phys. Letters*, **43**, 529 (1976).
79. (a) J. E. Willard, *J. Phys. Chem.*, **79**, 2966 (1975).
 (b) G. A. Kenney-Wallace, *Acc. Chem. Res.*, **11**, 433 (1978).
 (c) F.-Y. Jou and G. R. Freeman, *Can. J. Chem.*, **57**, 591 (1979).

80. B. L. Bales and L. Kevan, *J. Chem. Phys.*, **60**, 710 (1974).
81. R. N. Schwartz, M. K. Bowman and L. Kevan, *J. Chem. Phys.*, **60**, 1690 (1974).
82. R. S. Dixon, V. J. Lopata and C. R. Roy, *Intern. J. Radiat. Phys. Chem.*, **8**, 707 (1976).
83. K. Okazaki and G. R. Freeman, *Can. J. Chem.*, **56**, 2305 (1978).
84. M. G. Robinson, K. N. Jha and G. R. Freeman, *J. Chem. Phys.*, **55**, 4933 (1971).
85. F.-Y. Jou and G. R. Freeman, *J. Phys. Chem.*, **81**, 909 (1977).
86. G. Nilsson, *J. Chem. Phys.*, **56**, 3427 (1972).
87. R. R. Hentz, Farhataziz and E. M. Hansen, *J. Chem. Phys.* **57**, 2595 (1972).
88. K. Fueki, D. F. Feng and L. Kevan, *J. Phys. Chem.*, **78**, 393 (1974).
89. G. R. Freeman, *J. Phys. Chem.*, **76**, 944 (1972).
90. J. H. Baxendale and P. H. G. Sharpe, *Chem. Phys. Letters*, **41**, 440 (1976).
91. J. H. Baxendale, *Can. J. Chem.*, **55**, 1996 (1977).
92. R. R. Hentz and G. A. Kenney-Wallace, *J. Phys. Chem.*, **76**, 2931 (1972).
93. (a) B. J. Brown, N. T. Barker and D. F. Sangster, *Australian J. Chem.*, **26**, 2089 (1973).
    (b) B. J. Brown, N. T. Barker and D. F. Sangster, *Australian J. Chem.*, **27**, 2529 (1974).
94. J. R. Brandon and R. F. Firestone, *J. Phys. Chem.*, **78**, 792 (1974).
95. T. E. Gangwer, A. O. Allen and R. A. Holroyd, *J. Phys. Chem.*, **81**, 1469 (1977).
96. F. S. Dainton and R. J. Whewell, *J. Chem. Soc., Chem. Commun.*, 493 (1974).
97. M. Ogasawara, L. Kevan and H. A. Gillis, *Chem. Phys. Letters*, **49**, 459 (1977).
98. J. Mayer, J. L. Gebicki and J. Kroh, *Radiat. Phys. Chem.*, **11**, 101 (1978).
99. S. A. Rice and M. J. Pilling, *Progr. Reaction Kinet.*, **9**, 93 (1978).
100. J. Kroh, E. Romanovska and Cz. Stradowski, *Chem. Phys. Letters*, **47**, 597 (1977).
101. T. Sasaki and S. Ohno, *Bull. Chem. Soc. Japan*, **44**, 2626 (1971).
102. A. Namiki, M. Noda and T. Higashimura, *J. Phys. Chem.*, **79**, 2975 (1975).
103. T. Ito, K. Fueki and Z. Kuri, *J. Phys. Chem.*, **79**, 1513 (1975).
104. (a) G. L. Bolton, K. N. Jha and G. R. Freeman, *Can. J. Chem.*, **54**, 1497 (1976).
     (b) J. H. Baxendale and P. Wardman, *Can. J. Chem.*, **55**, 3058 (1977).
     (c) G. R. Freeman, *Can. J. Chem.*, **55**, 3059 (1977).
105. G. R. Freeman and G. L. Bolton, *Proceedings Fourth Symposium on Radiation Chemistry, Tihany, Hungary, 1976*, Hungarian Academy of Sciences, Budapest, 1977, p. 999.
106. C. Stradowski, *Radiochem. Radioanal. Letters*, **29**, 267 (1977).
107. J. R. Miller, *J. Phys. Chem.*, **82**, 767 (1978).
108. D. Ražem, W. H. Hamill and F. Funabashi, *Chem. Phys. Letters*, **53**, 84 (1978).
109. E. J. Marshall, M. J. Pilling and S. A. Rice, *J. Chem. Soc., Faraday Trans. II*, **71**, 1555 (1975).
110. H. Barzynski and D. Schulte-Frohlinde, *Z. Naturforsch.*, **22a**, 2131 (1967).
111. K. N. Jha, G. L. Bolton and G. R. Freeman, *J. Phys. Chem.*, **76**, 3876 (1972).
112. G. L. Bolton and G. R. Freeman, *J. Amer. Chem. Soc.*, **98**, 6825 (1976).
113. (a) G. L. Bolton, M. G. Robinson and G. R. Freeman, *Can. J. Chem.*, **54**, 1177 (1976).
     (b) A. M. Afanassiev, K. Okazaki and G. R. Freeman, *J. Phys. Chem.*, **83**, 1244 (1979).
114. G. V. Buxton, F. S. Dainton and M. Hammerli, *Trans. Faraday Soc.*, **63**, 1191 (1967).
115. B. Hickel, *J. Phys. Chem.*, **82**, 1005 (1978).
116. J. W. Fletcher, P. J. Richards and W. A. Seddon, *Can. J. Chem.*, **48**, 1645 (1970).
117. A. Kira and J. K. Thomas, *J. Chem. Phys.*, **60**, 766 (1974).
118. S. Noda, K. Torimoto, K. Fueki and Z. Kuri, *Bull. Chem. Soc. Japan*, **44**, 273 (1971).
119. M. C. R. Symons and K. V. Subba Rao, *Radiat. Phys. Chem.*, **10**, 35 (1977).
120. E. Watson, Jr. and S. Roy, *NSRDS-NBS*, 42, U.S. Dept. of Commerce, Washington, D.C., 1972.
121. G. R. Freeman, *NSRDS-NBS*, 48, U.S. Dept. of Commerce, Washington, D.C., 1974.
122. J. H. Baxendale and P. Wardman, *NSRDS-NBS*, 54, U.S. Dept. of Commerce, Washington, D.C., 1975.
123. E. Lindholm and P. Wilmenius, *Arkiv Kemi*, **20**, 255 (1963).
124. K. R. Ryan, L. W. Sieck and J. H. Futrell, *J. Chem. Phys.*, **41**, 111 (1964).
125. J. C. J. Thynne, F. K. Amenu-Kpodo and A. G. Harrison, *Can. J. Chem.*, **44**, 1655 (1966).
126. D. J. Hyatt, E. A. Dodman and M. J. Henchman, *Advan. Chem. Ser.*, **58**, 131 (1966).
127. M. E. Russell and W. A. Chupka, *J. Phys. Chem.*, **75**, 3797 (1971).

128. (a) M. C. R. Symons and G. W. Eastland, *J. Chem. Res. (M)*, 2901 (1977).
       (b) K. Toriyama and M. Iwasaki, *J. Amer. Chem. Soc.*, **101**, 2516 (1979).
129. M. Iwasaki and K. Toriyama, *J. Amer. Chem. Soc.*, **100**, 1964 (1978).
130. F. P. Sargent, E. M. Gardy and H. R. Falle, *Chem. Phys. Letters*, **24**, 120 (1974).
131. F. P. Sargent and E. M. Gardy, *Can. J. Chem.*, **52**, 3645 (1974).
132. F. P. Sargent and E. M. Gardy, *J. Phys. Chem.*, **80**, 854 (1976).
133. (a) S. W. Mao and L. Kevan, *Chem. Phys. Letters*, **24**, 505 (1974).
       (b) V. E. Zubarev, V. N. Belevskii and L. T. Bugaenko, *High Energy Chem.*, **12**, 178 (1978).
134. (a) M. Shiotani, S. Murabayashi and J. Sohma, *Intern. J. Radiat. Phys. Chem.*, **8**, 483 (1976).
       (b) F. P. Sargent, *Radiat. Phys. Chem.*, **10**, 137 (1977).
       (c) M. Shiotani, S. Murabayashi and J. Sohma, *Radiat. Phys. Chem.*, **11**, 203 (1978).
135. F. P. Sargent, *J. Phys. Chem.*, **81**, 89 (1977).
136. (a) F. Dainton, I. Janovský and G. A. Salmon, *Proc. Roy. Soc. (Lond.)*, A **327**, 305 (1972).
       (b) J. Lind, A. Jowko and T. E. Eriksen, *Radiat. Phys. Chem.*, **13**, 159 (1979).
137. (a) D. H. Ellison, G. A. Salmon and F. Wilkinson, *Proc. Roy. Soc. (Lond.)*, A **328**, 23 (1972).
       (b) B. C. Gilbert, R. G. G. Holmes, H. A. H. Lane and R. O. C. Norman, *J. Chem. Soc., Perkin Trans. II*, 1047 (1976).
       (c) B. C. Gilbert, R. G. G. Holmes and R. O. C. Norman, *J. Chem. Res.*, 1 (1977).
       (d) D. W. Johnson and G. A. Salmon, *J. Chem. Soc., Faraday Trans. I*, 446 (1979).
138. A. M. Afanas'ev and E. P. Kalyazin, *Vestn. Mosk. Univ. Khim.*, **12**, 731 (1971).
139. (a) R. A. Basson and H. J. van der Linde, *J. Chem. Soc. (A)*, 1182 (1967).
       (b) R. A. Basson and H. J. van der Linde, *J. Chem. Soc. (A)*, 662 (1968).
       (c) R. A. Basson and H. J. van der Linde, *J. Chem. Soc. (A)*, 1618 (1969).
140. W. V. Sherman, *J. Phys. Chem.*, **70**, 667 (1966).
141. C. von Sonntag, G. Lang and D. Schulte-Frohlinde in *The Chemistry of Ionization and Excitation* (Eds. G. R. A. Johnson and G. Scholes), Taylor and Francis, London, 1967, p. 123.
142. J. C. Russell and G. R. Freeman, *J. Phys. Chem.*, **72**, 808 (1968).
143. A. M. Afanas'ev and E. P. Kalyazin, *High Energy Chem.*, **7**, 13 (1973).
144. D. Verdin, *Intern. J. Radiat. Phys. Chem.*, **2**, 201 (1970).
145. C. von Sonntag, unpublished results.
146. K. N. Jha and G. R. Freeman, *J. Amer. Chem. Soc.*, **95**, 5891 (1973).
147. (a) K. N. Jha and G. R. Freeman, *J. Chem. Phys.*, **57**, 1408 (1972).
       (b) D. W. Johnson and G. A. Salmon, *Radiat. Phys. Chem.*, **10**, 294 (1977).
148. D. Schulte-Frohlinde, G. Lang and C. von Sonntag, *Ber. Bunsenges. Phys. Chem.*, **72**, 63 (1968).
149. K. N. Jha and G. R. Freeman, *Can. J. Chem.*, **51**, 2033 (1973).
150. D. Röhm and C. von Sonntag, unpublished results.
151. T. B. Truong, *Chem. Phys. Letters*, **35**, 426 (1975).
152. P. Ausloos and S.-G. Lias in *Actions Chimiques et Biologiques des Radiations* (Ed. M. Haissinsky), Ser. 11, Masson et Cie, Paris, 1967, p. 1.
153. D. Sänger and C. von Sonntag, *Tetrahedron*, **26**, 5489 (1970).
154. R. A. Basson, *J. S. African Chem. Inst.*, **22**, 63 (1969).
155. C. von Sonntag and W. Brüning, *Intern. J. Radiat. Phys. Chem.*, **1**, 25 (1969).
156. C. von Sonntag, *Z. Naturforsch.*, **25b**, 654 (1970).
157. C. von Sonntag, *Intern. J. Radiat. Phys. Chem.*, **1**, 33 (1969).
158. C. von Sonntag, *Z. Naturforsch.*, **27b**, 41 (1972).
159. D. R. G. Brimage, J. D. P. Cassell, J. H. Sharp and M. C. R. Symons, *J. Chem. Soc. (A)*, 2619 (1969).
160. V. N. Belevskii, V. E. Zubarev and L. T. Bugaenko, *Vestn. Mosk. Univ., Khim.*, **30**, 184 (1975).
161. H. Seki and M. Imamura, *J. Phys. Chem.*, **71**, 870 (1967).
162. A. M. Afanas'ev and E. P. Kalyazin, *High Energy Chem.*, **7**, 42 (1973).
163. H. J. van der Linde and G. R. Freeman, *J. Phys. Chem.*, **75**, 20 (1971).

164. K. M. Bansal and G. R. Freeman, *J. Amer. Chem. Soc.*, **90**, 7183 (1968).
165. E. Klosová, J. Teplý and Z. Prášil, *Intern. J. Radiat. Phys. Chem.*, **2**, 177 (1970).
166. (a) M. Meaburn and F. W. Mellows, *Trans. Faraday Soc.*, **61**, 1701 (1965).
     (b) R. Large and H. Knof, *Org. Mass Spectrom.*, **11**, 582 (1976).
167. J. H. Baxendale and R. D. Sedgwick, *Trans. Faraday Soc.*, **57**, 2157 (1961).
168. (a) Z. Prášil, *Coll. Czech. Chem. Commun.*, **31**, 3252 (1966).
     (b) Z. Prášil, *Coll. Czech. Chem. Commun.*, **31**, 3263 (1966).
169. (a) P. Wilmenius and E. Lindholm, *Arkiv Fysik*, **21**, 97 (1961).
     (b) B.-Ö. Jonsson and J. Lind, *J. Chem. Soc., Faraday Trans. II*, **72**, 906 (1976).
170. R. Gorden, Jr. and L. W. Sieck, *J. Res. Natl. Bur. Stand.*, **A 76**, 655 (1972).
171. K. M. Bansal and G. R. Freeman, *J. Amer. Chem. Soc.*, **90**, 7190 (1968).
172. K. M. Bansal and G. R. Freeman, *J. Amer. Chem. Soc.*, **92**, 4173 (1970).
173. H. J. van der Linde and G. R. Freeman, *J. Amer. Chem. Soc.*, **92**, 4417 (1970).
174. H. J. van der Linde and G. R. Freeman, *Ind. Eng. Chem. Prod. Res. Develop.*, **11**, 192 (1972).
175. H. J. LeRoux and H. J. van der Linde, *J. S. African Chem. Inst.*, **29**, 40 (1976).
176. K. M. Bansal and G. R. Freeman, *Radiat. Res. Rev.*, **3**, 209 (1971).
177. E. P. Grimsrud and P. Kebarle, *J. Amer. Chem. Soc.*, **95**, 7939 (1973).
178. J.-P. Dodelet and G. R. Freeman, *Can. J. Chem.*, **53**, 1263 (1975).
179. J.-P. Dodelet, F.-Y. Jou and G. R. Freeman, *J. Phys. Chem.*, **79**, 2876 (1975).
180. T. Huang and L. Kevan, *J. Chem. Phys.*, **61**, 4660 (1974).
181. L. M. Dorfman, F.-Y. Jou and R. Wageman, *Ber. Bunsenge. Phys. Chem.*, **75**, 681 (1971).
182. F.-Y. Jou and L. M. Dorfman, *J. Chem. Phys.*, **58**, 4715 (1973).
183. L. M. Dorfman and F.-Y. Jou in *Electrons in Fluids: The Nature of Metal—ammonia Solutions* (Eds. J. Jortner and N. R. Kestner), Springer, New York, 1973, p. 447.
184. F.-Y. Jou and G. R. Freeman, *Can. J. Chem.*, **54**, 3693 (1976).
185. H. Hase, M. Noda and T. Higashimura, *J. Chem. Phys.*, **54**, 2975 (1971).
186. T. Ichikawa, H. Yoshida and K. Hayashi, *Bull. Chem. Soc. Japan*, **46**, 812 (1973).
187. H. Hase, F. Q. H. Nyo and L. Kevan, *J. Chem. Phys.*, **62**, 985 (1975).
188. (a) L. Kevan in *Actions Chimiques et Biologiques des Radiations* (Ed. M. Haissinsky), Sér. 15, Masson et Cie, Paris, 1971, p. 83.
     (b) T. Ichikawa, L. Kevan, M. Bowman, S. A. Dikanov and Yu. D. Tsvetkov, *J. Chem. Phys.*, **71**, 1167 (1979).
189. J. H. Baxendale and M. A. J. Rodgers, *J. Phys. Chem.*, **72**, 3849 (1968).
190. B. Bockrath and L. M. Dorfman, *J. Phys. Chem.*, **77**, 1002 (1973).
191. G. A. Salmon and W. A. Seddon, *Chem. Phys. Letters*, **24**, 366 (1974).
192. G. A. Salmon, W. A. Seddon and J. W. Fletcher, *Can. J. Chem.*, **52**, 3259 (1974).
193. W. A. Seddon, J. W. Fletcher, F. C. Sopchichin and R. Catterall, *Can. J. Chem.*, **55**, 3356 (1977).
194. M. T. Lok, F. J. Tehan and J. I. Dye, *J. Phys. Chem.*, **76**, 2975 (1972).
195. R. M. Leblanc and J. A. Herman, *J. Chim. Phys.*, **63**, 1055 (1966).
196. R. A. Vermeer and G. R. Freeman, *Can. J. Chem.*, **52**, 1181 (1974).
197. J. Teplý and I. Janovský, *Radiochem. Radioanal. Letters*, **22**, 299 (1975).
198. J. Kroh, E. Hankiewicz and I. Zuchowicz, *Bull. Acad. Pol. Sci., Ser. Sci. Chim.*, **21**, 153 (1973).
199. J. H. Baxendale, D. Beaumond and M. A. J. Rodgers, *Chem. Phys. Letters*, **4**, 3 (1969).
200. F. Hirayama, C. W. Lawson and S. Lipsky, *J. Phys. Chem.*, **74**, 2411 (1970).
201. A. M. Halpern and W. R. Ware, *J. Phys. Chem.*, **74**, 2413 (1970).
202. A. Singh, S. P. Vaish and M. J. Quinn, *J. Photochem.*, **5**, 168 (1976).
203. J. Kiwi, *J. Photochem.*, **7**, 237 (1977).
204. F. Spurny, *Proceedings of the 3rd Tihany Symposium on Radiation Chemistry*, Vol. 1, Akademiai Kiado, Budapest, 1972, p. 59.
205. Z. Polacki, *Acta Phys. Polon.*, **A44**, 465 (1973).
206. M. K. M. Ng and G. R. Freeman, *J. Amer. Chem. Soc.*, **87**, 1635 (1965).
207. M. K. M. Ng and G. R. Freeman, *J. Amer. Chem. Soc.*, **87**, 1639 (1965).
208. F. Kiss and J. Teplý, *Intern. J. Radiat. Phys. Chem.*, **3**, 503 (1971).

209. J. H. Baxendale, D. Beaumond and M. A. J. Rodgers, *Intern. J. Radiat. Phys. Chem.*, **2**, 39 (1970).
210. Y. Llabador and J.-P. Adloff, *J. Chim. Phys.*, **61**, 1467 (1964).
211. M. Matsui and M. Imamura, *Bull. Chem. Soc. Japan*, **51**, 2191 (1978).
212. J. Teplý, I. Janovský, F. Kiss and K. Vacek, *Intern. J. Radiat. Phys. Chem.*, **4**, 265 (1972).
213. J. Teplý, *Intern. J. Radiat. Phys. Chem.*, **6**, 379 (1974).
214. C. Chachaty, A. Forchioni, J. Désalos and M. Arvis, *Intern. J. Radiat. Phys. Chem.*, **2**, 69 (1970).
215. F. S. Dainton and G. A. Salmon, *Proc. Roy. Soc. (Lond.)*, **A 285**, 319 (1965).
216. Y. Llabador and J.-P. Adloff, *J. Chim. Phys.*, **61**, 681 (1964).
217. R. R. Hentz and W. V. Sherman, *J. Phys. Chem.*, **72**, 2635 (1968).
218. J. H. Baxendale and M. A. J. Rodgers, *Trans. Faraday Soc.*, **63**, 2004 (1967).
219. E. A. Rojo and R. R. Hentz, *J. Phys. Chem.*, **69**, 3024 (1965).
220. S. Murabayashi, M. Shiotani and J. Sohma, *Chem. Phys. Letters*, **51**, 568 (1977).
221. G. C. Dismukes and J. E. Willard, *J. Phys. Chem.*, **80**, 2072 (1976).
222. C. von Sonntag, H.-P. Schuchmann and G. Schomburg, *Tetrahedron*, **28**, 4333 (1972).
223. (a) K. J. Laidler and D. J. McKenney, *Proc. Roy. Soc. (Lond.)*, **A 278**, 505 (1964).
    (b) K. J. Laidler and D. J. McKenney, *Proc. Roy. Soc. (Lond.)*, **A 278**, 517 (1964).
224. H. Yoshida, M. Irie, O. Shimada and K. Hayashi, *J. Phys. Chem.*, **76**, 3747 (1972).
225. M. Irie, K. Hayashi, S. Okamura and H. Yoshida, *J. Phys. Chem.*, **75**, 476 (1971).
226. M. Anbar, M. Bambenek and A. B. Ross, *NSRDS-NBS*, 43, U.S. Dept. of Commerce, Washington, D.C., 1973.
227. A. B. Ross, *NSRDS-NBS*, 43 (Suppl.), U.S. Dept. of Commerce, Washington, D.C., 1975.
228. M. Anbar, Farhataziz and A. B. Ross, *NSRDS-NBS*, 51, U.S. Dept. of Commerce, Washington, D.C., 1975.
229. B. H. Bielski and A. O. Allen, *J. Phys. Chem.*, **81**, 1048 (1977).
230. G. V. Buxton, *Radiat. Res. Rev.*, **1**, 209 (1968).
231. G. W. Klein and R. H. Schuler, *Radiat. Phys. Chem.*, **11**, 167 (1978).
232. G. G. Teather and N. V. Klassen, *Intern. J. Radiat. Phys. Chem.*, **7**, 475 (1975).
233. Farhataziz and A. B. Ross, *NSRDS-NBS*, 59, U.S. Dept. of Commerce, Washington, D.C., 1977.
234. L. M. Dorfman and G. E. Adams, *NSRDS-NBS*, 46, U.S. Dept. of Commerce, Washington, D.C. 1973.
235. (a) H. Eibenberger, S. Steenken, P. O'Neill and D. Schulte-Frohlinde, *J. Phys. Chem.*, **82**, 749 (1978).
    (b) M. Anbar and D. Meyerstein, *J. Phys. Chem.*, **68**, 3184 (1964).
236. G. E. Adams and R. L. Willson, *Trans. Faraday Soc.*, **65**, 2981 (1969).
237. K.-D. Asmus, H. Möckel and A. Henglein, *J. Phys. Chem.*, **77**, 1218 (1973).
238. K.-D. Asmus, A. Henglein, A. Wigger and G. Beck, *Ber. Bunsenges. Phys. Chem.*, **70**, 756 (1966).
239. G. P. Laroff and R. W. Fessenden, *J. Phys. Chem.*, **77**, 1283 (1973).
240. J. Murto, *Acta Chem. Scand.*, **18**, 1043 (1964).
241. B. Blank, A. Henne, G. P. Laroff and H. Fischer, *Pure Appl. Chem.*, **41**, 475 (1975).
242. A. Kirsch, C. von Sonntag and D. Schulte-Frohlinde, *J. Chem. Soc., Perkin Trans. II*, 1334 (1975).
243. (a) A. G. Pribush, S. A. Brusentseva, V. N. Shubin and P. I. Dolin, *High Energy Chem.*, **9**, 206 (1975).
    (b) A. G. Pribush, S. A. Brusentseva, V. N. Shubin and P. I. Dolin, *High Energy Chem.*, **8**, 217 (1974).
244. R. S. Eachus and M. C. R. Symons, *J. Chem. Soc. (A)*, 1336 (1970).
245. R. Tausch-Treml, A. Henglein and J. Lilie, *Ber. Bunsenges. Phys. Chem.*, **82**, 1335 (1978).
246. M. Kelm, J. Lilie, A. Henglein and E. Janata, *J. Phys. Chem.*, **78**, 882 (1974).
247. M. Kelm, J. Lilie and A. Henglein, *J. Chem. Soc., Faraday Trans. I*, **71**, 1132 (1975).
248. (a) M. Breitenkamp, A. Henglein and J. Lilie, *Ber. Bunsenges. Phys. Chem.*, **80**, 973 (1976).

(b) G. V. Buxton and R. M. Sellers, *NSRDS-NBS*, 62, U.S. Dept. of Commerce, Washington, D.C., 1978.

249. C. E. Burchill and I. S. Ginns, *Can. J. Chem.*, 48, 1232 (1970).

250. C. E. Burchill and P. W. Jones, *Can. J. Chem.*, 49, 4005 (1971).

251. N. S. Kalyazina, K. S. Kalugin and Yu. I. Moskalev, *High Energy Chem.*, 7, 416 (1973).

252. W. V. Sherman, *Chem. Commun.*, 790 (1966).

253. W. V. Sherman, *Chem. Commun.*, 250 (1966).

254. W. V. Sherman, *J. Phys. Chem.*, 71, 4245 (1967).

255. (a) W. V. Sherman, *J. Amer. Chem. Soc.*, 89, 1302 (1967).
(b) W. V. Sherman, *J. Phys. Chem.*, 72, 2287 (1968).

256. R. Backlin and W. V. Sherman, *Chem. Commun.*, 453 (1971).

257. G. V. Buxton and J. C. Green, *J. Chem. Soc., Faraday Trans. I*, 74, 697 (1978).

258. (a) G. V. Buxton, J. C. Green, R. Higgins and S. Kanji, *J. Chem. Soc., Chem. Commun.*, 158 (1976).
(b) T. Söylemez and C. von Sonntag, *J. Chem. Soc. Perkin Trans. II*, in the press.

259. C. E. Burchill and I. S. Ginns, *Can. J. Chem.*, 48, 2628 (1970).

260. C. E. Burchill and G. F. Thompson, *Can. J. Chem.*, 49, 1305 (1971).

261. C. E. Burchill and G. P. Wollner, *Can. J. Chem.*, 50, 1751 (1972).

262. A. L. Buley, R. O. C. Norman and R. J. Pritchett, *J. Chem. Soc. (B)*, 849 (1966).

263. N. M. Bazhin, E. V. Kuznetsov, N. N. Bubnov and V. V. Voevodskii, *Kinetika i Kataliz*, 7, 732 (1966).

264. R. Livingston and H. Zeldes, *J. Amer. Chem. Soc.*, 88, 4333 (1966).

265. S. Steenken, G. Behrens and D. Schulte-Frohlinde, *Intern. J. Radiat. Biol.*, 25, 205 (1974).

266. F. Seidler and C. von Sonntag, *Z. Naturforsch.*, 24b, 780 (1969).

267. C. von Sonntag and E. Thoms, *Z. Naturforsch.*, 25b, 1405 (1969).

268. C. E. Burchill and K. M. Perron, *Can. J. Chem.*, 49, 2382 (1971).

269. A. K. Pikaev and L. I. Kartasheva, *Intern. J. Radiat. Phys. Chem.*, 7, 395 (1975).

270. (a) K. M. Bansal, M. Grätzel, A. Henglein and E. Janata, *J. Phys. Chem.*, 77, 16 (1973).
(b) S. Steenken, *J. Phys. Chem.*, 83, 595 (1979).

271. M. Dizdaroglu, H. Scherz and C. von Sonntag, *Z. Naturforsch.*, 27b, 29 (1972).

272. G. Behrens, unpublished results: ESR spectroscopic work.

273. M. Dizdaroglu, D. Henneberg, G. Schomburg and C. von Sonntag, *Z. Naturforsch.*, 30b, 416 (1975).

274. S. Kawakishi, Y. Kito and M. Namiki, *Agric. Biol. Chem.*, 39, 1897 (1975).

275. M. Dizdaroglu, J. Leitich and C. von Sonntag, *Carbohydr. Res.*, 47, 15 (1976).

276. M. Dizdaroglu, K. Neuwald and C. von Sonntag, *Z. Naturforsch.*, 31b, 227 (1976).

277. G. Behrens and D. Schulte-Frohlinde, *Ber. Bunsenges. Phys. Chem.*, 80, 429 (1976).

278. B. C. Gilbert, J. P. Larkin and R. O. C. Norman, *J. Chem. Soc., Perkin Trans. II*, 794 (1972).

279. T. Matsushige, G. Koltzenburg and D. Schulte-Frohlinde, *Ber. Bunsenges. Phys. Chem.*, 79, 657 (1975).

280. G. Koltzenburg, T. Matsushige and D. Schulte-Frohlinde, *Z. Naturforsch.*, 31b, 960 (1976).

281. A. Samuni and P. Neta, *J. Phys. Chem.*, 77, 2425 (1973).

282. T. Foster and P. R. West, *Can. J. Chem.*, 52, 3589 (1974).

283. A. G. W. Bradbury and C. von Sonntag, *Carbohydr. Res.*, 62, 223 (1978).

284. M. N. Schuchmann and C. von Sonntag, *J. Chem. Soc., Perkin Trans. II*, 1958 (1977).

285. S. Kawakishi, Y. Kito and M. Namiki, *Agric. Biol. Chem.*, 41, 951 (1977).

286. C. von Sonntag, *Advan. Carbohydr. Chem.*, in the press.

287. B. C. Gilbert, R. O. C. Norman and R. C. Sealy, *J. Chem. Soc., Perkin Trans. II*, 824 (1974).

288. H.-P. Schuchmann and C. von Sonntag, unpublished results.

289. G. Behrens, G. Koltzenburg, A. Ritter and D. Schulte-Frohlinde, *Intern. J. Radiat. Biol.*, 33, 163 (1978).

290. G. Behrens, E. Bothe, J. Eibenberger, G. Koltzenburg and D. Schulte-Frohlinde, (a) *Angew. Chem.*, **90**, 639 (1978); (b) *J. Chem. Soc., Perkin Trans. II*, in the press.

291. M. Dizdaroglu, C. von Sonntag and D. Schulte-Frohlinde, *J. Amer. Chem. Soc.*, **97**, 2277 (1975).

292. F. Beesk, M. Dizdaroglu, D. Schulte-Frohlinde and C. von Sonntag, *Intern. J. Radiat. Biol.*, **36**, 565 (1979).

293. C. von Sonntag, U. Hagen, A. Bopp-Schön and D. Schulte-Frohlinde, *Advan. Radiat. Biol.*, in the press.

294. J. K. Kochi (Ed.) in *Free Radicals*, Vol. 2, John Wiley and Sons, New York, 1973, p. 665.

295. S. Steenken, H.-P. Schuchmann and C. von Sonntag, *J. Phys. Chem.*, **79**, 763 (1975).

296. C. von Sonntag, M. Dizdaroglu and D. Schulte-Frohlinde, *Z. Naturforsch.*, **31b**, 857 (1976).

297. (a) H. Zegota and C. von Sonntag, *Z. Naturforsch.*, **32b**, 1060 (1977).
(b) E. P. Petryaev, G. N. Vasil'ev, L. A. Maslovskaya and O. I. Shabyro, *J. Org. Chem. USSR*, **15**, 793 (1979).

298. C. von Sonntag and M. Dizdaroglu, *Z. Naturforsch.*, **28b**, 367 (1973).

299. M. Dizdaroglu, C. von Sonntag, D. Schulte-Frohlinde and W. V. Dahlhoff, *Liebigs Ann. Chem.*, 1592 (1973).

300. C. von Sonntag, K. Neuwald and M. Dizdaroglu, *Radiat. Res.*, **58**, 1 (1974).

301. O. Volkert, G. Termens and D. Schulte-Frohlinde, *Z. Physik. Chem. N.F. (Frankfurt)*, **56**, 261 (1967).

302. D. Grässlin, F. Merger, D. Schulte-Frohlinde and O. Volkert, *Z. physik. Chem. N.F. (Frankfurt)*, **51**, 84 (1966).

303. D. Grässlin, F. Merger, D. Schulte-Frohlinde and O. Volkert, *Chem. Ber.*, **100**, 3077 (1967).

304. K. Omura and T. Matsuura, *Tetrahedron*, **26**, 255 (1970).

305. D. F. Sangster in *The Chemistry of the Hydroxyl Group* (Ed. S. Patai), John Wiley and Sons, London, 1971, p. 133.

306. K. Eiben, D. Schulte-Frohlinde, C. Suarez and H. Zorn, *Intern. J. Radiat. Phys. Chem.*, **3**, 409 (1971).

307. D. Schulte-Frohlinde, G. Reutebuch and C. von Sonntag, *Intern. J. Radiat. Phys. Chem.*, **5**, 331 (1973).

308. (a) P. Neta and R. W. Fessenden, *J. Phys. Chem.*, **78**, 523 (1974).
(b) E. J. Land and M. Ebert, *Trans. Faraday Soc.*, **63**, 1181 (1967).

309. P. O'Neill and S. Steenken, *Ber. Bunsenges. Phys. Chem.*, **81**, 550 (1977).

310. S. Steenken and P. O'Neill, *J. Phys. Chem.*, **81**, 505 (1977).

311. (a) P. O'Neill, S. Steenken, H. J. van der Linde and D. Schulte-Frohlinde, *Radiat. Phys. Chem.*, **12**, 13 (1978).
(b) N. V. Raghavan and S. Steenken, *J. Amer. Chem. Soc.*, in the press.

312. J. H. Fendler and G. L. Gasowski, *J. Org. Chem.*, **33**, 2755 (1968).

313. J. Holcman and K. Sehested, *J. Phys. Chem.*, **80**, 1642 (1976).

314. M. K. Eberhardt, *J. Phys. Chem.*, **81**, 1051 (1977).

315. P. O'Neill, S. Steenken and D. Schulte-Frohlinde, *J. Phys. Chem.*, **81**, 31 (1977).

316. S. Steenken, P. O'Neill and D. Schulte-Frohlinde, *J. Phys. Chem.*, **81**, 26 (1977).

317. N. Latif, P. O'Neill, D. Schulte-Frohlinde and S. Steenken, *Ber. Bunsenges. Phys. Chem.*, **82**, 468 (1978).

318. (a) P. O'Neill, D. Schulte-Frohlinde and S. Steenken, *Faraday Disc. Chem. Soc.*, **63**, 141 (1977).
(b) P. O'Neill, S. Steenken and D. Schulte-Frohlinde, *J. Phys. Chem.*, **79**, 2773 (1975).

319. J. Rabani, D. Klug-Roth and A. Henglein, *J. Phys. Chem.*, **78**, 2089 (1974).

320. Y. Ilan, J. Rabani and A. Henglein, *J. Phys. Chem.*, **80**, 1558 (1976).

321. E. Bothe, G. Behrens and D. Schulte-Frohlinde, *Z. Naturforsch.*, **32b**, 886 (1977).

322. E. Bothe, D. Schulte-Frohlinde and C. von Sonntag, *J. Chem. Soc., Perkin Trans. 2*, 416 (1978).

323. E. Bothe, M. N. Schuchmann, D. Schulte-Frohlinde and C. von Sonntag, *Photochem. Photobiol.*, **28**, 639 (1978).

324. M. N. Schuchmann and C. von Sonntag, *Z. Naturforsch.*, **33b**, 329 (1978).
325. E. Bothe and D. Schulte-Frohlinde, *Z. Naturforsch.*, **33b**, 786 (1978).
326. M. N. Schuchmann and C. von Sonntag, *J. Phys. Chem.*, **83**, 780 (1979).
327. J. A. Howard in *Free Radicals* (Ed. J. K. Kochi), Vol. II, John Wiley and Sons, New York, 1973, p. 3.
328. K. Stockhausen, A. Fojtik and A. Henglein, *Ber. Bunsenges. Phys. Chem.*, **74**, 34 (1970).
329. K. V. S. Rama Rao and A. V. Sapre, *Radiation Effects*, **3**, 183 (1970).
330. G. Hughes and H. A. Makada, *Adv. Chem. Ser.*, **75**, 102 (1968).
331. C. von Sonntag and M. Dizdaroglu, *Carbohydr. Res.*, **58**, 21 (1977).
332. L. Stelter, C. von Sonntag and D. Schulte-Frohlinde, *Z. Naturforsch.*, **30b**, 609 (1975).
333. A. G. W. Bradbury and C. von Sonntag, *Z. Naturforsch.*, **31b**, 1274 (1976).
334. M. Dizdaroglu, K. Neuwald and C. von Sonntag, *Z. Naturforsch.*, **31b**, 227 (1976).
335. M. Dizdaroglu, D. Schulte-Frohlinde and C. von Sonntag, *Z. Naturforsch.*, **30c**, 826 (1975).
336. M. Dizdaroglu, D. Schulte-Frohlinde and C. von Sonntag, *Z. Naturforsch.*, **32c**, 1021 (1977).

CHAPTER 24

Radiation chemistry of thiols, sulphides and disulphides

CLEMENS VON SONNTAG and
HEINZ-PETER SCHUCHMANN

Institut für Strahlenchemie im Max-Planck-Institut für Kohlenforschung,
Stiftstrasse 34–36, D-4330 Mülheim a. d. Ruhr, W. Germany

I. INTRODUCTION

The observation that sulphhydryl compounds can to some extent prevent radiation damage *in vivo*[1,2] has stimulated considerable interest in the radiation chemistry of these compounds. Research in this field has been further motivated by the fact that thiol and disulphide groups, although there are only relatively few of them along the protein chain, are nevertheless crucial to the proper functioning of many enzymes. Thiol and disulphide groups are among the most radiation-sensitive functions in proteins, and it has been suggested that disulphide cleavage can result in enzyme inactivation. In this and other contexts radiation techniques have helped to shed some light on the nature of the active sites[3].

The radiolysis of thiols and disulphides has been reviewed in a previous volume of this series[4a], with an emphasis on aqueous solutions (cf. also Reference 4b). The radiation chemistry of thiols and sulphides has also been studied in nonaqueous systems, neat and in solution, spectroscopically and by product analysis. Some of this work has been discussed in a number of reviews[5-9].

At first glance, the radiation chemistry of thiols may seem deceptively simple. Because the sulphhydryl hydrogen is easily abstracted by most radicals, the alkylthiyl radical is the most frequent radical species in such systems. Their disproportionation/combination ratios tend to be small (cf. Chapter 22), and their main product therefore is the disulphide. However, the scope of thiol radiation chemistry is wide compared to that of alcohols because of several features relating to the sulphur atom. Thiols are more acidic but also seem to undergo protonation more easily than alcohols (at least in the gas phase[10]). With sulphides, this basicity is reflected in the existence of stable trialkylsulphonium compounds such as, for instance, $[SR_3]^+ SR^-$ and $([SR_3]^+)_2 S^{2-}$ which are stoichiometrically a complex of several sulphide molecules[11].

In further contrast to alcohols, ethers and peroxides, the divalent sulphur atom in their sulphur analogues manifests a readiness to acquire a tetravalent nature, in that complex radicals $R_3S^{\bullet}$ often appear as intermediates leading in some cases to radical chain reactions. The easy formation and relatively long life-time of radical cations (e.g. $RSR^{\bullet +}$, $RSSR^{\bullet +}$) is another feature of the sulphur compounds not observed with their oxygen analogues. The variety of radical species often present in such systems in fact seems sometimes to have led to the misassignment of ESR signals to the simplest of these radicals, thiyl[12], whose spectrum is often obscured by the spectra of the other species. However, thiyl has been detected by spin trapping[13-17].

It will be shown that radiation techniques have already considerably expanded the knowledge of the chemistry of carbon-bound sulphur in its lower unstable oxidation states, even though all the complexities are far from being fully understood, especially in nonaqueous media. For this reason mechanisms which are suggested in the section on nonaqueous systems have to be taken with more reservations than those proposed in the aqueous systems where far better kinetic data are available.

II. RADIOLYSIS IN NONAQUEOUS MEDIA

A. Thiols

The present section deals with the radiolysis of thiols (neat liquid and solid[15,17-29], nonaqueous solutions[19-22,26,30-38], gas phase[39-44]), and where it seemed appropriate mass-spectrometric data[44-50] (cf. Reference 10) have been used to interpret the results.

Isotopically labelled thiols have been employed in hydrocarbon radiolysis as a probe to distinguish between the contributions of molecular nonionic primary processes such as reaction (1) and free-radical processes (reactions 2 and 3), in the hydrocarbon (R^1H). The main radical processes involving the thiol and its radicals in saturated hydrocarbon solution are given by reactions (4)–(10). Processes (9)

$$R^1H \xrightarrow{\quad \gamma \quad} H_2 \; + \; \text{olefin} \tag{1}$$

$$R^1H \xrightarrow{\quad \gamma \quad} {}^{\bullet}R^1 \; + \; H^{\bullet} \tag{2}$$

$$R^1H \xrightarrow{\quad \gamma \quad} R^1H^{+\bullet} \; + \; e^- \tag{3}$$

$$^{\bullet}R^1 \; + \; R^2SH \longrightarrow R^1H \; + \; R^2S^{\bullet} \tag{4}$$

$$H^{\cdot} + R^2SH \longrightarrow H_2 + R^2S^{\cdot} \tag{5}$$

$$H^{\cdot} + R^2SH \longrightarrow {}^{\cdot}R^2 + H_2S \tag{6}$$

$$R^2S^{\cdot} + R^2S^{\cdot} \longrightarrow R^2SSR^2 \tag{7}$$

$$R^2S^{\cdot} + {}^{\cdot}R^1 \longrightarrow R^2SR^1 \tag{8}$$

$$R^2S^{\cdot} + R^2SH \underset{(10)}{\overset{(9)}{\rightleftharpoons}} R^2SS^{\cdot}(H)R^2 \tag{9, 10}$$

and (10) are an example[17] of the tendency of complex radicals to be formed in the radiolysis of organosulphur systems. The optical absorption spectrum of a species thought to be EtSS(H)Et has been observed[38]; the existence of the homologous species $RSSR_2{}^{\cdot}$ would not seem improbable[17]. The homolytic displacement reaction (6) has recently been established in the photolysis of thiols[51] and may have to be taken into consideration in radiolytic systems as well.

In olefinic hydrocarbons chain processes occur such as (11) and (12) which have been shown to lead to G-values[†] of the order of 10^5 for the isomerization of *cis*-into *trans*-2-butene[52]. The reverse reaction is also observed and has a G-value three to ten times smaller.

$$RS^{\cdot} + Et{-}CH{=}CH{-}Et \underset{(12)}{\overset{(11)}{\rightleftharpoons}} Et{-}CH(SR){-}\overset{\cdot}{C}H{-}Et \tag{11,12}$$

In the gaseous thiol/carbon monoxide system[53] a chain-reaction of a different kind appears to be operating at elevated temperatures through the addition of the thiyl radical to carbon monoxide (reaction 13). The resulting radical loses carbon oxide sulphide (reaction 14), and the alkyl radical propagates the chain by hydrogen abstraction from the thiol (reaction 4).

$$RS^{\cdot} + CO \longrightarrow (R{-}S{=}C{=}O)^{\cdot} \tag{13}$$

$$(R{-}S{=}C{=}O)^{\cdot} \longrightarrow R^{\cdot} + S{=}C{=}O \tag{14}$$

As discussed so far, the reactions of thiols in the radiolytic systems are the same as in other free-radical generating systems. The formation of charged species (radical cations and electrons) by the absorption of ionizing radiation (cf. Chapter 23) brings about new aspects. Thiols appear to be able to scavenge positive charges (reactions 15 and 16). Their gas-phase ionization potential (I) is lower and their

$$R^1H^{+\cdot} + R^2SH \begin{cases} \longrightarrow {}^{\cdot}R^1 + R^2RSH_2^+ & \text{(15)} \\ \longrightarrow R^1H + R^2SH^{+\cdot} & \text{(16)} \end{cases}$$

gas-phase proton affinity (P) is perhaps slightly higher than the corresponding properties of alcohols; $I(\text{EtSH}) = 9.28$ eV, $I(\text{EtOH}) = 10.48$ eV[54]; $P(\text{MeSH}) = 770$ kJ mol^{-1}, $P(\text{MeOH}) = 760$ kJ mol^{-1} [55]. The same holds with respect to saturated hydrocarbons but may not always be the case with unsaturated ones[56].

The scavenging of the positive charge from a hydrocarbon radical cation, $R^1H^{+\cdot}$ (reactions 15 and 16), may be followed by a proton transfer between thiol radical cation and thiol (reaction 17) or the formation of a complexed radical cation (reaction 18). Complexed radical cations from sulphides, $(RSR)_2^{+\cdot}$, are well estab-

---

[†]The quantity G, called G-value, is defined through $G = N/E$, unit $(100 \text{ eV})^{-1}$, where N is the number of radiolytically generated species or events of whatever kind caused by the absorbed radiation energy E.

$$R^2SH^{+\cdot} + R^2SH \longrightarrow \begin{cases} R^2S^{\cdot} + R^2SH_2^+ & \text{(17)} \\ (R^2SH)_2^{+\cdot} & \text{(18)} \end{cases}$$

lished species[57] (see below). They are the structural analogues of $(RSH)_2^{+\cdot}$. It can be estimated from thermochemical data that reaction (15) tends to be more exothermic than reaction (16) but one might expect that its activation energy be higher. Reaction (17) appears close to thermoneutral (see below). Formation of the complex $(RSH)_2^{+\cdot}$ (reaction 18) is considered[17] as an alternative to proton loss (reaction 17). In fact the latter appears to be unimportant, because in low-temperature glasses the thiyl radical is not seen unless the matrix is bleached or annealed[23,38]. Reaction (17) has also been excluded in the radiolysis of thiophenol[19].

The train of events undergone by the negative charge is not clear. In hydrocarbons containing alcohols it is known that the electron becomes solvated within a solute domain[58]. The smaller polarity of the thiol molecule [dielectric constants: $\epsilon(EtSH) = 6.9$; $\epsilon(EtOH) = 24.3$[59]] would make a similar effect (reaction 19) energetically less rewarding but not impossible. Also, owing to their relatively low polarity thiols have a lower tendency than alcohols to form domains in hydrocarbon solution. Other possibilities could be suggested (reactions 20–22).

$$e^- + n\,RSH \longrightarrow e_{solv}^- \qquad \text{(19)}$$

$$e^- + RSH \longrightarrow \begin{cases} RS^- + H^{\cdot} & \text{(20)} \\ R^{\cdot} + {}^-SH & \text{(21)} \\ RSH^{\cdot-} & \text{(22)} \end{cases}$$

Thermochemical argument indicates that reactions (20) and (21) should be endothermic in the gas phase, reaction (21) to a lesser extent than reaction (20)[49,50] However, the appearance potential of reaction (20) is found to be lower than that of (21)[44]. In methanolic and aqueous glasses reaction (21) has been shown to occur[17], but there is no evidence that it occurs in a hydrocarbon matrix[38] or in the neat thiol[17] at 77 K. This would leave $RSH^{\cdot-}$ as the most likely carrier of negative charge in non-aqueous media. In fact the radical anion $RSH^{\cdot-}$ is supposed to have been observed by ESR spectroscopy in the low-temperature radiolysis of thiols[23, 28,38,60,61] (but cf. Reference 17) whereas the trapped electron was not seen[38,61].

In view of the foregoing, there are many possible neutralization reactions. In particular, reaction (23) has been discussed to explain the growth in the thiyl ESR

$$RSH^{-\cdot} + RSH^{+\cdot} \longrightarrow RSH + RS^{\cdot} + H^{\cdot} \qquad \text{(23)}$$

signal during annealing of γ-irradiated thiol glasses while the signals assigned to $RSH^{\cdot-}$ and $RSH^{+\cdot}$ diminish[26,38].

The radiolysis of neat thiols awaits further investigations, and the mechanisms presented here are largely reasonable extrapolations from data obtained with similar systems. Thus, reactions (24)–(31) should be considered together with the above-mentioned ones. Reactions such as (25) and (26) are observed in the photolysis of thiols, the former predominating by roughly an order of magnitude (see Chapter 22). In the radiolysis[19,39] the situation may be not much different considering the similar ratios of disulphide and H_2S formation.

Reaction (27) is an intriguing one. There seems to be ESR spectroscopic evidence that it does not occur in methyl, ethyl, propyl and butyl thiols, but from pentyl

$$RSH \longrightarrow \gamma \longrightarrow$$

$$\longrightarrow RSH^{+\cdot} + e^- \tag{24}$$

$$\longrightarrow RS^{\cdot} + H^{\cdot} \tag{25}$$

$$\longrightarrow R^{\cdot} + {}^{\cdot}SH \tag{26}$$

$$\longrightarrow {}^{\cdot}R(-H)SH + H^{\cdot} \tag{27}$$

$$RSH^{+\cdot} + RSH \longrightarrow {}^{\cdot}R(-H)SH + RSH_2^+ \tag{28}$$

$$HS^{\cdot} + RSH \longrightarrow RS^{\cdot} + H_2S \tag{29}$$

$${}^{\cdot}R(-H)SH + RSH \longrightarrow RSH + RS^{\cdot} \tag{30}$$

$$RS^{\cdot} + {}^{\cdot}SH \longrightarrow RSSH \tag{31}$$

thiol onward an increasing proportion of the radicals observed appear to be alkyl thiol radicals ${}^{\cdot}R(-H)SH$[24,25,29]. It has been surmised that efficient energy transfer down the alkyl chain to the sulphhydryl or other accepting groups is possible only if the distance to be spanned is less than about five carbon links[24,62,63]. Another implication of the absence of these radicals including the thiol α-radical is that reaction (28) ought to be even less important than reaction (17) (if, indeed, they occur at all in the condensed state). In the gas phase, it has been shown[46] with methanethiol that reactions (17) and (28) occur at a ratio of about 10 : 1; epithermal ions are perhaps involved. This is in contrast to the alcohols where proton transfer is about equally likely from the oxygen and the α-carbon atom[64].

Interesting results have been obtained with 1,4-butanedithiol[21]. In dilute hydrocarbon solutions 1,2-dithiane was formed in high yield (reactions 32 and 33). Its

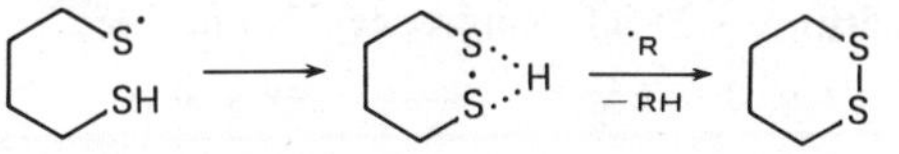

$$\text{(32, 33)}$$

yield was shown to decrease with increasing dithiol concentration while that of disulphidic compounds of higher molecular weight increased. One might suggest a cyclization reaction (reaction 32) to occur in competition with bimolecular addition (reaction 34), the latter being favoured at high dithiol concentrations.

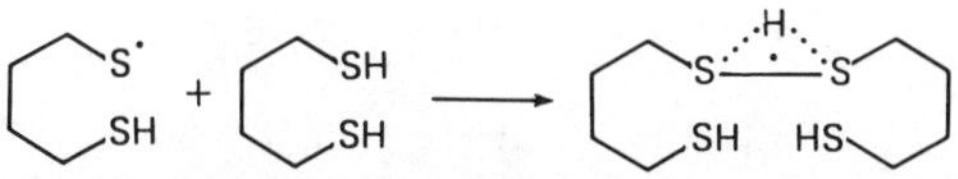

$$\text{(34)}$$

In the presence of oxygen (cf. References[5,6,8,33,65]), initial G-values of thiol consumption rise strongly with falling dose rate and increasing thiol concentration, thus suggesting a chain-reaction. A considerable part of the thiol consumed is transformed into the disulphide, but other more highly oxidized products which are certainly formed have not been measured.

B. Disulphides

In studies on the formation and properties of radicals and radical ions from the radiolysis of organic disulphides in low-temperature glasses[28,37,38,66-70] and in the gas-phase[44,71] it has been shown that disulphides are remarkably good acceptors of various charged and radical species in nonaqueous media. In hydrocarbon solution the efficiency of disulphide as electron scavenger[72] is comparable to that of sulphur hexafluoride and other good electron scavengers[73], and as hydrogen atom scavenger, to that of ethylene. There is no doubt that some of the positive charge is trapped as well, probably by disulphide radical cation, and perhaps also by sulphonium ion formation (reactions 35 and 36).

$$R^1H^{\cdot+} + R^2SSR^2 \quad \begin{cases} \longrightarrow R^1H + R^2SS\dot{R}^{+2} & (35) \\ \longrightarrow {}^\cdot R^1 + R^2SS(H)\overset{+}{R}^2 & (36) \end{cases}$$

In dilute cyclohexane solution diethyl and dipropyl disulphides rapidly equilibrate under the influence of ionizing radiation[72], and it may be inferred[74] that thiyl radicals are also present, generated via reactions (37)–(43). Reactions (40)

$$PrSSPr + e^- \longrightarrow PrSSPr^{\cdot-} \qquad (37)$$

$$PrSSPr^{\cdot-} \underset{(39)}{\overset{(38)}{\rightleftharpoons}} PrS^\cdot + PrS^- \qquad (38, 39)$$

$$PrSSPr + H^\cdot \longrightarrow PrS\dot{S}(H)Pr \qquad (40)$$

$$PrS\dot{S}(H)Pr \longrightarrow PrS^{\cdot\prime} + PrSH \qquad (41)$$

$$PrSSPr + R^\cdot \longrightarrow PrS\dot{S}(R)Pr \qquad (42)$$

$$PrS\dot{S}(R)Pr \longrightarrow PrS^\cdot + RSPr \qquad (43)$$

and (41) explain the formation of thiol[72]. The other product, cyclohexyl propyl sulphide RSPr, is formed[17] in reactions (42) and (43). The mixed disulphide is formed[72] via reactions (44)–(46). In the presence of thiols a similar transposition takes place[72] via reactions (47) (cf. Reference 70) and (48).

$$PrS^\cdot + EtSSEt \longrightarrow EtS\dot{S}(SPr)Et \qquad (44)$$

$$PrS^- + EtSSEt^{\cdot+} \longrightarrow EtS\dot{S}(SPr)Et \qquad (45)$$

$$EtS\dot{S}(SPr)Et \longrightarrow EtSSPr + EtS^\cdot \qquad (46)$$

$$PrS^\cdot + EtSH \longrightarrow PrS\dot{S}(H)Et \qquad (47)$$

$$PrS\dot{S}(H)Et \longrightarrow EtS^\cdot + PrSH \qquad (48)$$

There is a strong reduction of cyclohexane consumption, from $G = 7.3$ in the pure solvent to about half this value in a solution 0.005 M in the disulphide[72]. On the other hand, G(disulphide consumption) is about unity at this concentration. The possibility of the formation of undetected sulphur-containing products has been considered. The apparent discrepancy could also imply a protective action, possibly via processes such as (49)–(51). Radical cation complexes of the type $[RSSR]_2^{+\cdot}$ have been observed in the gas phase[71] and in solid media[38].

$$RSSR^{+\cdot} + RSSR^{-\cdot} \longrightarrow 2\,RSSR \qquad (49)$$

$$RSSR^{+\cdot} + RSSR \rightleftharpoons [RSSR]_2^{+\cdot} \qquad (50)$$

$$[RSSR]_2^{+\cdot} + RSSR^{-\cdot} \longrightarrow 3\,RSSR \qquad (51)$$

C. Sulphides

The present information on the radiolysis of sulphides (nonaqueous liquid[75-80] and solid[20,37,38,70,81-86,87a] conditions; mass-spectrometric studies[45,47,51, 71,87b]) reveals little about the nature of the final products. Apart from studies on thiophene[75,77] the only product that seems to have been measured is hydrogen[20, 81-83]. It is noted that hydrogen formation declines as the atomic fraction of sulphur in the system is increased either intramolecularly by employing lower alkyl homologues [from $G(H_2) = 1.5$ in $(C_{11}H_{23})_2S$ to 0.14 in $(CH_3)_2S)$], or in an alkane/dialkyl sulphide mixture by increasing the sulphide content[20,83]. It is not yet ascertained whether or not the cleavage of the carbon—sulphur bond plays a major role. This reaction has been shown to be the main process in the photolysis of sulphides (see Chapter 22). Carbon-centred radicals have been observed by ESR spectroscopy of glassy radiolysed samples of sulphides where apparently sulphide radicals ˙R(—H)SR of all possible types are being formed[20,81,82]. 2-Methyl-tetrahydrothiophene, in contrast to 2-methyltetrahydrofuran, does not physically trap electrons[84]. Instead, anion radicals are formed which seem to be ring-opened forms of the type $R_2\overset{.}{C}-S^-$. Dissociative electron capture by dimethyl sulphide[44]

$$CH_3SCH_3 + e^- \longrightarrow CH_3S^- + \overset{.}{C}H_3 \tag{52}$$

$$(CH_3)_2S^{+\cdot} + (CH_3)_2S \longrightarrow (CH_3)_2SH^+ + {}^{\cdot}CH_2SCH_3 \tag{53}$$

(reaction 52) is endothermic in the gas phase but occurs in methanolic glass[70]. Proton transfer (reaction 53) appears to be slightly endothermic in the gas phase[71] which would suggest that, until it is neutralized, the positive charge remains in the

$$R_2S^{+\cdot} + R_2S \rightleftharpoons [R_2SSR_2]^{+\cdot} \tag{54}$$

form of the original radical cation of its complex $(R_2S)_2^{+\cdot}$ (reaction 54). Optical, ESR, mass spectrometrical, and product studies have adduced evidence for such complexes[12,38,57,70,79,80a,88,89]. From 1,4-dithiane an intramolecular cation-radical complex (1) is formed by electron removal that absorbs near 600 nm. An intermolecular complex (2) is formed from 1,3-dithiane in nonpolar media which absorbs at the remarkably long wavelength of 750 nm[79,80a].

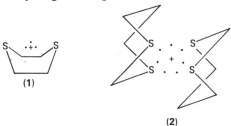

(1)

(2)

III. RADIOLYSIS IN AQUEOUS SOLUTIONS

A. Radiolysis of Water

The primary processes in the radiolysis of aqueous solutions have been discussed in some detail in an earlier review[4] and in a preceding chapter of this volume (Chapter 23). In the latter a compilation of the G-values of the primary species under various conditions can be found. A brief account is given here.

The primary free radical species formed in the radiolysis of water are OH radicals, solvated electrons (e_{aq}^-) and H atoms. Protons and hydroxide ions as well as some molecular hydrogen and hydrogen peroxide are also formed (reaction 55). The solvated electrons can be converted into OH radicals by N_2O (reaction 56). In acidic solutions the solvated electron is converted into a hydrogen atom (reaction 57). Hydroxyl radicals, e_{aq}^-, and H atoms readily react with the title

$$H_2O \xrightarrow{\quad\gamma\quad} {}^{\cdot}OH, e_{aq}^-, H^{\cdot}, H^+, OH^-, H_2, H_2O_2 \tag{55}$$

$$e_{aq}^- + N_2O \longrightarrow {}^{\cdot}OH + N_2 + OH^- \tag{56}$$

$$e_{aq}^- + H^+ \longrightarrow H^{\cdot} \tag{57}$$

compounds. There is now a wealth of rate constants available (for compilations see: OH radicals[90,91], solvated electrons[92,93], H atoms[94]). In the following sections the reactions of the three water-derived radicals with the title compounds and the subsequent free-radical reactions are discussed.

In order to aid the reader, the formulae and trivial names of some sulphur-containing compounds of biochemical interest that are mentioned below are listed in Table 1.

TABLE 1. Trivial names and formulae of sulphur-containing compounds mentioned in the text

Name	Formula
Cysteamine	$HSCH_2CH_2NH_2$
Cysteine	$HSCH_2CH(NH_2)COOH$
Cystine	$(SCH_2CH(NH_2)COOH)_2$
1,4-Dithiothreitol	CH_2SH | $HOCH$ | $HCOH$ | CH_2SH
Gluathione(= glutamylcysteinylglycine)	$HOOCCH(NH_2)CH_2CH_2$ | $C{=}O$ | NH | $HSCH_2CH$ | $C{=}O$ | NH | $HOOCCH_2$
Lipoic acid	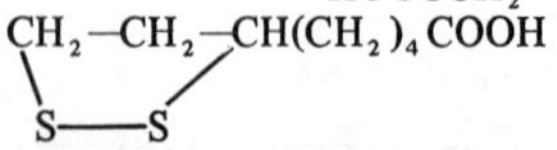 $CH_2{-}CH_2{-}CH(CH_2)_4COOH$ $\diagdown\qquad\diagup$ $S{-\!-}S$
Methional	$CH_3SCH_2CH_2CHO$
Methionine	$CH_3SCH_2CH_2CH(NH_2)COOH$
Penicillamine	$HSC(CH_3)_2CH(NH_2)COOH$

B. Deoxygenated Solutions

1. Thiols

Thiols rapidly react with the hydrated electron. The rate[95,96] is near to diffusion-controlled if the thiol is neutral or positively charged [$k(e_{aq}^- + RSH) \sim 10^{10}$ M^{-1} s^{-1}]. The rate constant drops if the electron reacts with a negatively charged species. It appears not to make much difference whether an adjacent carboxyl group is dissociated, or the sulphydryl group itself. A further strong reduction in the reaction rate is observed with doubly negatively charged species. Under these conditions the rate constants $k(e_{aq}^- + RSH)$ drop to $\sim 3 \times 10^8$ M^{-1} s^{-1}. Two processes are conceivable (reactions 58 and 59). Because of the lower dissociation

$$RSH + e_{aq}^- \begin{cases} \longrightarrow R^\cdot + SH^- & (58) \\ \longrightarrow RS^- + H^\cdot & (59) \end{cases}$$

energy of the C—S bond compared to that of the S—H bond one might expect reaction (58) to be favoured over reaction (59). Indeed, it has been suggested[95] that only reaction (58) occurs and that reaction (59) can be neglected. However, there is evidence that at least in 2-hydroxyethanethiol[97] and in 2-aminoethanethiol(cysteamine)[98] reaction (59) may play a considerable role. This is seen from the fact that $G(H_2S)$ (from reaction 58) does not reach the expected value of 2.7, but only 1.65 in the case of 2-hydroxyethanethiol and 2.0 with 2-aminoethanethiol. There are some more cases which, however, do not show such a strong effect. It is recalled that dissociative electron capture similar to reaction (59) has been observed in the gas phase (see above).

In acidic solutions the hydrated electron is converted into H atoms (reaction 57). Under these conditions the reaction of the H atoms can conveniently be studied. There are two major processes (reactions 60 and 61). The overall rate constant

$$RSH + H^\cdot \begin{cases} \longrightarrow H_2 + RS^\cdot & (60) \\ \longrightarrow R^\cdot + H_2S & (61) \end{cases}$$

$k_{(60+61)}$ is around 10^9 M^{-1} s^{-1} for a number of thiols studied. The ratio k_{60}/k_{61} can be derived from the ratio $G(H_2)/G(H_2S)$ if $G(H_2)$ is corrected for the 'molecular H_2' from reaction (55). The k_{60}/k_{61} ratio is near five[99,100] for primary thiols, but decreases for secondary (1.8[101] and 0.55[99], observed for two different thiols) and tertiary thiols (0.82[101] and 0.44[99], observed for two different thiols). It has been reported[102] that $G(H_2S)$ increases with increasing temperature. This effect has been reinvestigated[101] but could not be reproduced with either cysteine or penicillamine. Thus it appears that k_{60}/k_{61} is not much temperature-dependent. Reaction (60) can be interpreted as a hydrogen abstraction reaction whereas reaction (61) constitutes a displacement reaction. However, it might well be that both reactions have a common precursor, a hydrogen atom adduct radical (reaction 62) in which the sulphur exhibits a three-electron bond. It has already been

$$RSH + H^\cdot \longrightarrow R\dot{S}H_2 \qquad (62)$$

emphasized (and further examples will be shown below) that there is increasing evidence for organic sulphur compounds to be able to complex radicals before decomposing into other free-radical species.

The hydroxyl radical has been shown[103] to react with thiols (reaction 63) at virtually diffusion-controlled rates [$k_{63} = (1-2) \times 10^{10}$ M^{-1} s^{-1}]. The reaction with thiolates (reaction 64) is generally slower by a factor of two[103]. There is good evidence from ESR spectroscopic studies[104] that carbon-centred radicals are also formed from thiols on OH attack (reaction 65).

$$\cdot OH + RSH \longrightarrow H_2O + RS \cdot \tag{63}$$

$$\cdot OH + RS^- \longrightarrow OH^- + RS \cdot \tag{64}$$

$$\cdot OH + RSH \longrightarrow H_2O + \cdot(R-H)SH \tag{65}$$

$$\cdot R + RSH \longrightarrow RH + RS \cdot \tag{66}$$

Rate constants of the reactions of some radicals with various thiols (reaction 66) are summarized in Table 2. These rate constants mostly cluster around 10^8 M^{-1} s^{-1}. However, there are a number of other radicals which show rate constants smaller than 10^7 M^{-1} s^{-1}, among them the OH adduct radicals of uracil and thymidine[105]. This must be borne in mind when the radiation protection of cellular DNA by sulphhydryl compounds is discussed[106] (for a review see Reference 107). It appears worth noting that the 2-hydroxy-2-propyl radical derived from isopropanol reacts considerably faster than the hydroxymethyl radical derived from methanol. The 1-hydroxyethyl radical (derived from ethanol) lies in between. This finding is somewhat surprising. In fact, one might expect the reverse order, because in general hydrogen is more difficult to abstract from methanol than from isopropanol, and therefore the reduction of the corresponding radical should be easier for hydroxymethyl than for 2-hydroxy-2-propyl. Attempts to detect a short-lived complex such as formed by H_2S[108] (reactions 67–69), have failed with thiols[103]. 2-Hydroxy-2-

$$HO-\underset{R^2}{\overset{R^1}{C}}\cdot + H_2S \underset{(68)}{\overset{(67)}{\rightleftharpoons}} HO-\underset{R^2}{\overset{R^1}{C}}-S\overset{H}{\underset{H}{\diagup}} \xrightarrow{(69)} HO-\underset{R^2}{\overset{R^1}{C}}-H + HS\cdot \tag{67-69}$$

propyl is electron-richer than the hydroxymethyl radical and therefore it should undergo formation of the tetravalent complex $RR'\dot{S}H$ more readily, which might help to explain the unexpected behaviour of these alkyl radicals. In this context it is perhaps useful to remember that sulphur tetrahalides are known but not the sulphur tetrahydride.

The reactions of some inorganic radicals with thiols have also been studied[109,110] (see Table 2). It is interesting[111] that the carboxyl anion radical, CO_2^-, can abstract an H atom from thiols (reaction 70), but that the $RS\cdot$ radical also abstracts an H atom from formate (reaction 71). This conclusion has

$$CO_2^- + RSH \underset{(71)}{\overset{(70)}{\rightleftharpoons}} HCO_2^- + RS\cdot \tag{70, 71}$$

been drawn from the fact that tritium-labelled formate solutions exchange with water large amounts of tritium if irradiated in the presence of cysteine. Formation of oxalic acid, the combination product of two CO_2^- entities, is suppressed and the formation of CO_2 is observed instead. This might result from a reaction of the $RS\cdot$ radicals with CO_2^- (reaction 72).

$$CO_2^- + RS\cdot \longrightarrow CO_2 + RS^- \tag{72}$$

A similar equilibrium is observed[112,113] in the phosphite/thiol system (reactions 73 and 74). The equilibrium constant is 800, k_{73} being 3×10^8 and k_{74} 3.8×10^5 M^{-1} s^{-1}.

Thiolate ions readily complex with $RS\cdot$ radicals (reaction 75). The rate constant

TABLE 2. Rate constants for the reaction of some organic and inorganic radicals with various thiols

Radical	Substrate	pH	Rate constant $(M^{-1} s^{-1})$	References
$\cdot CH_3$	CH_3SH	11	7.4×10^7	103
$HOCH_2CH_2\cdot$	$HOCH_2CH_2SH$	10	4.7×10^7	103
$(CH_3)_2C(OH)CH_2\cdot$	$HOCH_2CH_2SH$	10	8.2×10^7	103
	$HSCH_2CHOH-CHOH-CH_2SH$	7	6.8×10^7	110
	$H_2NCH_2CH_2SH$	7.6	1.8×10^7	105
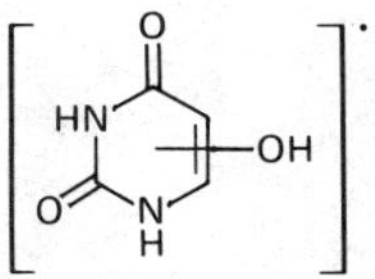	$H_2NCH_2CH_2SH$	7.6	$<10^7$	105
$\cdot CH_2OH$	$HOCH_2CH_2SH$	10	1.3×10^8	103
	$HSCH_2CHOH-CHOH-CH_2SH$	7	6.8×10^7	110
	$H_2NCH_2CH_2SH$	7.6	6.8×10^7	105
$CH_3\dot{C}HOH$	$HOCH_2CH_2SH$	10	2.3×10^8	103
	$H_2NCH_2CH_2SH$	7.6	1.4×10^8	105
$(CH_3)_2\dot{C}OH$	$HOCH_2CH_2SH$	10	5.1×10^8	103
	$HSCH_2CHOH-CHOH-CH_2SH$	7	2.1×10^8	110
	$H_2NCH_2CH_2SH$	7.6	4.2×10^8	105
$Cl_2^{\cdot-}$	$HSCH_2CHOH-CHOH-CH_2SH$	2	3.0×10^9	110
$(CNS)_2^{\cdot-}$	$HSCH_2CHOH-CHOH-CH_2SH$	7	2.1×10^7	110
$I_2^{\cdot-}$	$HSCH_2CHOH-CHOH-CH_2SH$	7	1.9×10^7	110
$CO_3^{\cdot-}$	$HSCH_2CHOH-CHOH-CH_2SH$	11	4.1×10^8	110

$$PO_3^{\cdot\,2-} + RSH \underset{(74)}{\overset{(73)}{\rightleftharpoons}} HPO_3^{2-} + RS^{\cdot} \qquad (73, 74)$$

k_{75} is of the order of 10^9 M^{-1} s^{-1} for a large number of thiols. This behaviour of the RS$^{\cdot}$ is similar to that of halogen and pseudohalogen radicals which readily complex with halogenide and pseudohalogenide ions. The back-reaction (reaction 76) is usually three orders of magnitude slower ($k_{76} \sim 10^6$ s^{-1}) and hence the

$$RS^{\cdot} + RS^{-} \underset{(76)}{\overset{(75)}{\rightleftharpoons}} RSSR^{\cdot-} \qquad (75, 76)$$

equilibrium constants are around 10^3 M[95,114]. In the case of dithiothreitol[110,115] the corresponding RS$^{\cdot}$ radical complexes only with the RS^{-} group within the same molecule (equilibrium 77) but not intermolecularly. The resulting complex has a pK of 5.5.

Whereas the linear disulphide radical anions decay by first-order, the cyclic ones[115,116] decay only by second order (e.g. reaction 78). Because on protonation the corresponding thiyl radicals are formed the decay rate will depend on the pH[115].

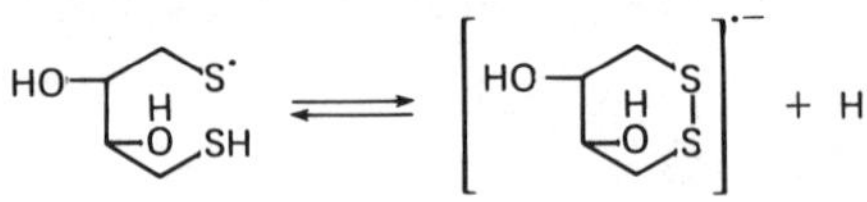

$$(77)$$

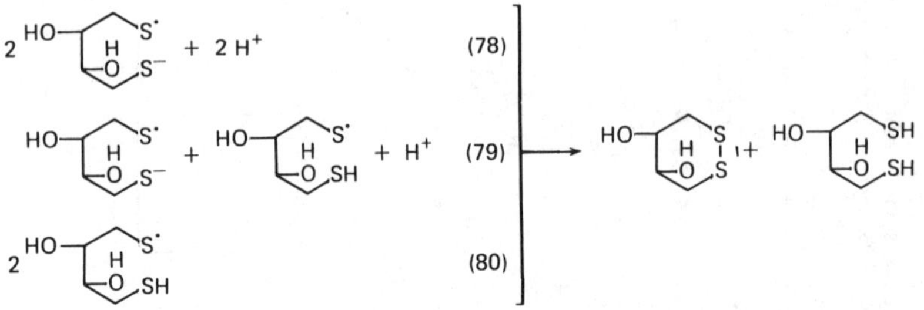

As expected k_{78} is smaller (1.7×10^8 M^{-1} s^{-1}) than k_{80} (1.7×10^9 M^{-1} s^{-1}) whereas the reaction of the anion with the neutral thiyl radical is the fastest ($k_{79} = 2.5 \times 10^9$). A remarkable product from the radiolysis of penicillamine is the trisulphide. It has been proposed[117,118] that it is formed via reactions (81) and (82).

$$RSSR^{\cdot-} + H_2O \longrightarrow RSS^{\cdot} + RH + OH^{-} \qquad (81)$$

$$RSS^{\cdot} + RS^{\cdot} \longrightarrow RSSSR \qquad (82)$$

2. Disulphides

Disulphides react with the solvated electron at virtually diffusion-controlled rates to give radical anions. The latter can dissociate[113] (reaction 75) into thiyl radicals and thiolate ions as discussed above. The disulphide anion radicals are protonated (reaction 83) with rate constants[119,120] between 6×10^8 and 7×10^{10}

$$RSSR^{\cdot-} + H^{+} \longrightarrow RSSRH^{\cdot} \qquad (83)$$

M^{-1} s^{-1}. The resulting H adduct radical is thought[119,120] to decompose rapidly into thiol and a thiyl radical (reaction 84). Thiyl radicals react readily with disulphides (reaction 85) and mixed disulphides are formed via a chain reaction on

$$\text{RSSRH}^{\cdot} \longrightarrow \text{RSH} + \text{RS}^{\cdot} \tag{84}$$

irradiation of a mixture of two different disulphides[112,113,121], just as in non-aqueous media (see above).

$$\text{R}^1\text{SSR}^1 + \text{R}^2\text{S}^{\cdot} \longrightarrow \text{R}^1\text{SSR}^2 + \text{R}^1\text{S}^{\cdot} \tag{85}$$

The reaction of OH radicals with disulphides has been shown[122,123] to give rise to about equal yields of radical cations (reaction 86) and OH radical adducts (reaction 87). The formation of these radical cations which had already been

$$\text{RSSR} + {\cdot}\text{OH} \left\{ \begin{array}{ll} \longrightarrow \text{RSSR}^{\cdot+} + \text{OH}^- & (86) \\ \longrightarrow \text{RSSR(OH)}^{\cdot} & (87) \end{array} \right.$$

postulated earlier[124] has been proven by the appearance of conducting species. The existence of the OH adduct radicals is more indirectly inferred and finds its support by a number of subsequent reactions (see below) that help to explain the data.

The formation of disulphide radical cations is not only brought about by OH radicals but more efficiently ($\sim$100%) by other oxidizing radicals such as the radical cations of 1,3,5-trimethoxybenzene and thio ethers, SO_4^-, $\text{Br}_2^{\cdot-}$, and by metal ions in unstable valency states such as Ag^{2+}, Ag(OH)^+ and Tl^{2+}. Tl(OH)^+ reacts with 80% efficiency and the carbonate radical ion, $\text{CO}_3^{\cdot-}$, with only 10% efficiency[125].

In alkaline solutions the cation radicals decay in a first-order reaction[123] (reaction 88). The rate of reaction (88) is not diffusion-controlled. A good correlation of a log k/k_0 plot against the Taft σ-parameters of the R groups was taken as

$$\text{RSSR}^{\cdot+} + \text{OH}^- \longrightarrow \text{RSSR(OH)}^{\cdot} \tag{88}$$

an indication that the rate of reaction (88) depends on the effective charge at the sulphur bridge. In addition, structural effects may contribute to the observed changes in the rate constants since steric hindrance also increases parallel to the inductive effect[123].

In neutral and slightly acid solution these species decay by second-order kinetics which can be followed using their strong absorption near 420 nm, and it has been shown that the rate is virtually diffusion-controlled. The rate of the disappearance of conductivity is slower than the decay of the optical signal, suggesting that the less-absorbing doubly-charged product of reaction (89) has a certain life-time.

$$2\,\text{RSSR}^{\cdot+} \longrightarrow \text{RSSR}^{2+} + \text{RSSR} \tag{89}$$

The radical cation $\text{RSSR}^{+\cdot}$ itself is an oxidizing species and readily reacts with Fe(CN)_6^{4-} at a diffusion-controlled rate, but about four orders of magnitude slower with $\text{Fe}_{\text{aq}}^{2+}$. In the latter case the variations in the rate of reaction (90), depending on

$$\text{RSSR}^{\cdot+} + \text{Fe}_{\text{aq}}^{2+} \longrightarrow \text{RSSR} + \text{Fe}_{\text{aq}}^{3+} \tag{90}$$

the nature of R in $\text{RSSR}^{\cdot+}$, have been explained to be due to similar effects as in the case of reaction (88).

Pulse radiolysis experiments[123] suggest that at pH > 10 the OH adduct radicals (from reactions 87 and 88) undergo a base-catalysed decomposition (reaction 91). In fact, at pH > 12, $G(\text{EtSH}) = 5.5$ was found in the case of diethyl disulphide[123].

$$\text{RSSR(OH)}^{\cdot} + \text{OH}^- \longrightarrow \text{RS}^- + \text{RS(OH)}_2^{\cdot} \tag{91}$$

In competition with reaction (91) the OH adduct radical may break up according to reaction (92). Sulphenic acid (RSOH) may also be formed from RSSR^{2+} (reaction

93) which is generated by reaction (89). Sulphenic acid, which is a fairly unstable, reducing compound[126], and the $RSO^{\cdot}$ ($RS(OH)_2^{\cdot}$) radical further undergo a number

$$RSSR(OH)^{\cdot} \longrightarrow RS^{\cdot} + RSOH \tag{92}$$

$$RSSR^{2+} + 2\,H_2O \longrightarrow 2\,RSOH + 2\,H^+ \tag{93}$$

of reactions, the products of which have not been fully characterized. In the case of di-t-butyl disulphide, isobutylene and trisulphide is produced[123], and it has been suggested that they may be formed via reaction (94) which is reminiscent of reaction (81).

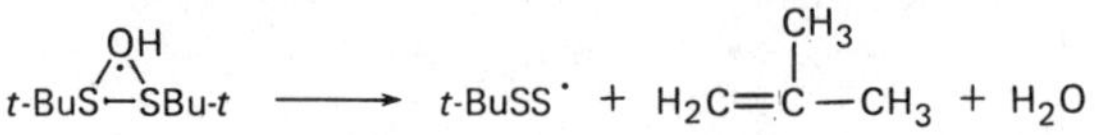

$$t\text{-BuS}\overset{\overset{\displaystyle OH}{\diagup\diagdown}}{-}\text{SBu-}t \longrightarrow t\text{-BuSS}^{\cdot} + H_2C=\overset{\overset{\displaystyle CH_3}{|}}{C}-CH_3 + H_2O \tag{94}$$

Attention is drawn to the possibility that complications could arise with some disulphides on account of hydrolysis when they are investigated in alkaline media (reaction 95)[127].

$$RSSR + OH^- \longrightarrow RSOH + RS^- \tag{95}$$

3. Sulphides

Sulphides appear to react much more slowly ($k_{96} \approx 5 \times 10^7$ M^{-1} s^{-1})[95,128] with hydrated electrons than do thiols and disulphides. In this reaction a C—S bond is cleaved (reaction 96) as has been confirmed by ESR spectroscopic studies

$$R-S-R + e_{aq}^- \longrightarrow R-S^- + {}^{\cdot}R \tag{96}$$

and by product analysis[129]. The subsequent reactions have so far not found much attention.

In the case of thiophene[128] the electron adduct appears to become protonated. 2,2'-Bithienyl has been found as the major reaction product. In acidic solutions, the same optical spectra are observed. However, under these conditions the thiophene ring appears to break down and sulphur is liberated while the yield of 2,2'-bithienyl is drastically reduced.

The OH radical reacts with sulphides [$k_{97} \approx (1-2) \times 10^{10}$ M^{-1} s^{-1}][130]. The first step has been suggested[130] to consist of OH addition to the sulphur (reaction 97). At low sulphide concentrations ($<10^{-4}$ M) the R_2SOH radicals eliminate H_2O (e.g. reaction 98). At sulphide concentrations above 10^{-4} M it was observed[130a]

$$R-S-R + OH \longrightarrow R_2\overset{\cdot}{S}OH \tag{97}$$

$$(CH_3)_2\overset{\cdot}{S}OH \longrightarrow H_2O + CH_2=\overset{\cdot}{S}-CH_3 \tag{98}$$

$$\Big\downarrow$$

$${}^{\cdot}CH_2-S-CH_3$$

that the OH adduct radical $R_2\overset{\cdot}{S}OH$ complexes with another sulphide molecule (reaction 99). The complexed OH-adduct is readily converted into the complexed

$$R_2\overset{\cdot}{S}OH + R_2S \longrightarrow [(R_2S)_2OH]^{\cdot} \tag{99}$$

radical cation either directly (reactions 100—102) or via the molecular species ($R_2\overset{\cdot}{S}OH$, $R_2S^{\cdot+}$) reaction (99) being reversible[130b]. Whereas with a number of

$$[(R_2S)_2OH]^{\cdot} \longrightarrow (R_2S)_2^{\cdot +} + OH^- \tag{100}$$

$$\xrightarrow{H^+} (R_2S)_2^{\cdot +} + H_2O \tag{101}$$

$$\xrightarrow{H_2PO_4^-} (R_2S)_2^{\cdot +} + HPO_4^{2-} + H_2O \tag{102}$$

simple sulphides the formation of the complexed cation radical proceeds even in solutions at pH 10 (reaction 100), methionine is converted under more acidic conditions only (reactions 101 and 102)[130c]. The radical cation complex $(R_2S)_2^{\cdot +}$ is relatively stable but is in equilibrium with its components (reaction 103). There is

$$(R_2S)_2^{\cdot +} \rightleftharpoons R_2S^{+\cdot} + R_2S \tag{103}$$

increasing evidence for cation radicals in these systems, both from ESR spectroscopic studies[14,70,88] and pulse radiolytic investigations[57,89a,130–132a,b].

Intermolecular as well as intramolecular complexes are formed with di- and tri-thianes[57,89a,132c]. Stabilization of the oxidized sulphur atom can be effected by heteroatoms other than sulphur[89a,132a,b]. For example $R_2\overset{\cdot}{S}Br$ or $R_2\overset{\cdot}{S}Cl$ are formed in the reaction of a sulphide with a complexed halogen atom, $Br_2^{\cdot -}$ or $Cl_2^{\cdot -}$ (e.g. reaction 104). At low bromide concentrations where primarily $R_2\overset{\cdot}{S}OH$ is formed the same absorption has been observed[89a] suggesting that reaction (105) can also take place.

$$R_2S + Br_2^{\cdot -} \longrightarrow R_2SBr^{\cdot} + Br^- \tag{104}$$

$$R_2\overset{\cdot}{S}OH + Br^- \longrightarrow R_2SBr^{\cdot} + OH^- \tag{105}$$

The suggestion[133] that thiophene adds OH radicals predominantly at $C_{(2)}$ (reaction 106) has been confirmed by ESR spectroscopic studies[134]. In alkaline

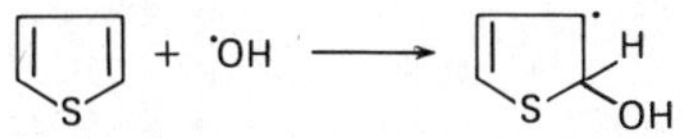

$$\tag{106}$$

solutions the OH adduct radical rearranges and opens the ring[133,134] (reactions 107 and 108). Whereas earlier work[133] had indicated that an equilibrium between the

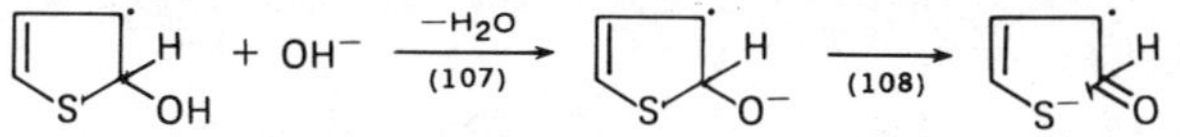

$$\xrightarrow[\text{(107)}]{-H_2O} \xrightarrow{\text{(108)}} \tag{107, 108}$$

OH adduct and its ring-closed anion exists, it was later[134] concluded that deprotonation immediately leads to the ring-opened species. Attempts to identify this species by ESR spectroscopy failed, however[134]. Because of the high tendency of polymerization of hydroxylated thiophenes, product analysis was restricted to the identification of the thiolactone (from the disproportionation reaction 109) and of 2,2'-bithienyl, a product which most likely arises by water elimination (reaction 111) of the combination product formed in reaction (110).

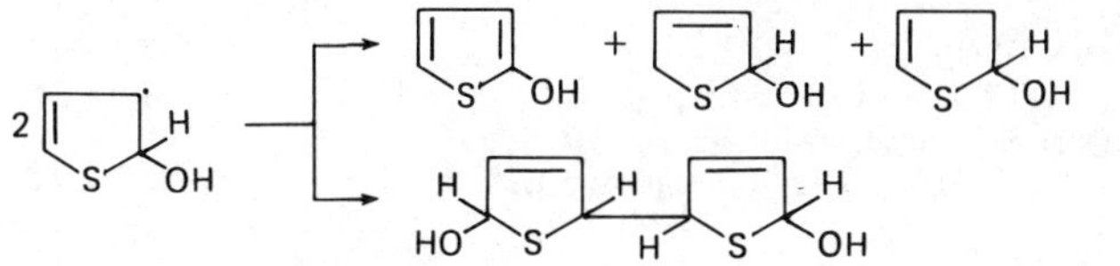

$$\tag{109}$$

$$\tag{110}$$

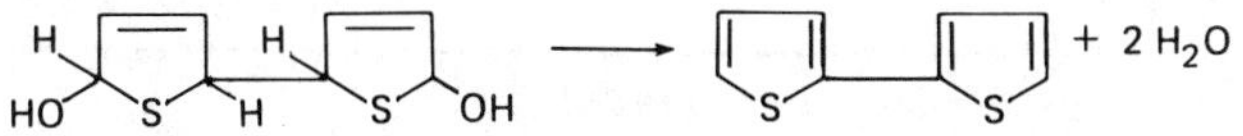

$$\longrightarrow \quad + 2 H_2O \tag{111}$$

C. Oxygenated Solutions

Whereas the free-radical chemistry of deoxygenated solutions of thiols and their derivatives is reasonably well understood, this is not the case with oxygenated solutions. One reason for this may be the relatively low rate of oxygen addition to sulphur-centred radicals (reaction 112). Oxygen adds to carbon-centred radicals at virtually diffusion-controlled rates $[k(R_3C^• + O_2) \approx 2 \times 10^9 \text{ M}^{-1}\text{s}^{-1}]$ while the rate of reaction of thiyl radicals with oxygen appears to be considerably lower (Table 3). Thus at thiolate concentrations higher than those of oxygen the reaction of the thiyl radical with the thiolate anion to give $RSSR^{•-}$ (reaction 75) might successfully compete with reaction (112). This effect is most prominent in compounds which contain two sulphhydryl groups such as dithiothreitol. As a result of this, O_2 reacts with $RSSR^{•-}$ giving the disulphide and $O_2^{•-}$ (reaction 113). At

$$RS^• + O_2 \longrightarrow RSO_2^• \quad (112)$$

$$RSSR^{•-} + O_2 \longrightarrow RSSR + O_2^{•-} \quad (113)$$

low pH where reaction (112) predominates, the resulting $RSO_2^•$ radicals may undergo a number of reactions. Although the system is not yet fully understood some mechanistic aspects can be discussed here.

It is observed that a chain-reaction takes place, the importance of which depends on thiol concentration and on the dose rate. The first step appears to be reaction (114). The resulting hydroperoxide may undergo two competing processes, a rearrangement into sulphinic acid (reaction 115) and hydrolysis (reaction 116). Both reactions may well proceed by proton catalysis, and the substituent R may have an influence on k_{115}/k_{116}. Thus in the case of glutathione[136] both the sulphinic acid

$$RSO_2^• + RSH \longrightarrow RS-O-O-H + RS^• \quad (114)$$

$$RS-O-OH \longrightarrow R-\overset{\overset{\displaystyle O}{\displaystyle \|}}{S}-OH \quad (115)$$

$$RS-O-OH + H_2O \longrightarrow RSOH + H_2O_2 \quad (116)$$

TABLE 3. Rate constants of the reaction of O_2 with some free-radical species derived from thiols and their derivatives[135]

Radical	k ($\text{M}^{-1}\text{s}^{-1}$)
$HOOCCH(NH_2)C(CH_3)_2S^•$	4×10^7
$C_2H_5S^•$	3.4×10^8
$t\text{-BuS}^•$	7.8×10^8
$HOCH_2CH_2S^•$	2.3×10^8
$CH_3S\overset{•}{C}H_2 \longleftrightarrow CH_3\overset{•}{S}=CH_2$	4.4×10^8
$C_2H_5S-\overset{•}{C}HCH_3 \longleftrightarrow C_2H_5\overset{•}{S}=CHCH_3$	6.2×10^8
$(CH_3)_2CHS-\overset{•}{C}(CH_3)_2 \longleftrightarrow (CH_3)_2CH\overset{•}{S}=(CH_3)_2$	1.0×10^9
$RSSR^{•+}$	Unreactive[a]
$R_2S^{•+}$	Unreactive[a]
$(R_2S)_2^{•+}$	Unreactive[a]

[a]Time-scale of pulse radiolysis experiments.

$$RS\text{---}O\text{---}OH + RSH \longrightarrow RSSR + H_2O_2 \qquad (117)$$

$$RSOH + RSH \longrightarrow RSSR + H_2O \qquad (118)$$

and the disulphide are formed, whereas in the case of cysteine[137] only the disulphide and H_2O_2 have been reported as products. However, there appears to be a further reaction (reaction 117) which competes with reactions (115) and (116). Reaction (117) depends on the thiol concentration and therefore should only be noticeable at higher thiol concentrations. Its product is the disulphide, and indeed it has been found[136] that the disulphide/sulphinic acid ratio increases with increasing thiol concentration. The termination of the chain is less clear than the propagation and reactions such as (119) have been suggested[137].

$$2\,RSO_2^- \longrightarrow RSSR + 2\,O_2 \qquad (119)$$

In neutral and alkaline solutions values are reached[137-140] for $G(-RSH)$ which suggest that a chain-reaction must occur under these conditions as well. Because of the fast establishment of the equilibrium leading to $RSSR^{\cdot -}$ reaction (113) must take part. The $O_2^{\cdot -}$ radical must be the chain-carrier as it has been convincingly shown[137] that its conjugated acid $HO_2^{\cdot}$ is not capable of propagating a chain. It has been argued[137] that the $HO_2^{\cdot}$ radical cannot abstract an H atom from the thiol,

$$O_2^{\cdot -} + RSH \xrightarrow{\ H^+\ } RS^{\cdot} + H_2O_2 \qquad (120)$$

but that $O_2^{\cdot -}$ does (reaction 120). This reasoning is somewhat surprising as $O_2^{\cdot -}$ is expected to be a poorer hydrogen abstractor than its conjugated acid $HO_2^{\cdot}$. Evidence for this is given in experiments where it has been shown that $O_2^{\cdot -}$ does not react with alcohols but that $HO_2^{\cdot}$ has sufficient abstractive power to propagate a chain (see Chapter 23). Thus one might have to reformulate the mechanism of this chain-reaction and consider that the thiolate anion could be involved, or that $O_2^{\cdot -}$ could form a labile complex with the thiol, a reaction which might not be undergone by the $HO_2^{\cdot}$ radical. In this context it might be mentioned that the question as to whether $O_2^{\cdot -}$ can react with a sulphide (methional) has been considered[141].

The reaction of oxygen with radicals derived from OH attack on disulphides is far from being understood. Major products are the corresponding sulphonic acids[142-147] The straight disulphides were observed[144] on irradiation of the mixed disulphides, e.g. CySSCy and CyaSSCya from CySSCya.

It has been shown[135] that the radical cations $RSSR^{\cdot +}$ do not react with O_2, at least not on the time-scale of pulse radiolysis experiments. However, it cannot be excluded that such a reaction takes place under $^{60}Co\text{-}\gamma$ conditions where the lifetime of the radical cations would be longer because of the usually much lower dose rates of $^{60}Co\text{-}\gamma$ sources compared to those employed in pulse radiolysis. A similar passivity towards O_2 is also observed with the radical cations derived from sulphides[135].

D. Some Biochemical Aspects

DNA is considered the major target in the radiation-induced deactivation of the living cell[2,148]. It has been found that sulphhydryl compounds can to some extent protect against this damage[2] (cf. Reference 149). In order to rationalize this observation it has been postulated that sulphhydryl compounds can repair radiation-induced DNA radicals. These radicals can be formed by an attack of radicals generated in the neighbourhood of DNA, or by its direct ionization. On hydrogen abstraction in the former case (reaction 121), or proton loss in the latter (reaction 122), a radical is formed which may undergo reactions leading to a damaged site, or may be repaired by sulphhydryl compounds according to reaction (123). The same sort of protection could also be exerted in favour of other vital components of the cell.

$$RH \text{ (i.e. DNA)} + X^{\cdot} \longrightarrow R^{\cdot} + XH \tag{121}$$

$$RH^{+\cdot} \longrightarrow R^{\cdot} + H^{+} \tag{122}$$

$$R^{\cdot} + R^{1}SH \longrightarrow RH + R^{1}S^{\cdot} \tag{123}$$

Another aspect is the radiation-induced deactivation of enzymes, and in the present context this topic is of interest in so far as they contain[150] sulphhydryl, sulphide and disulphide functions. It has been found that in some (e.g. papain[151-154], trypsin[155], ribonuclease[156-158], lactate dehydrogenase[159], yeast alcohol dehydrogenase[160] and glyceraldehyde-3-phosphate dehydrogenase[161]), but not all, enzymes (e.g. α-chymotrypsin[162] and carboxypeptidase A[163]), sulphur-containing functions appear to be critically involved.

Impairment[164,165] of enzymatic activity may be through damage to the active site as well as through disruption of the proper conformation[166]. Inactivation of an enzyme through radiation is complete only after several hits have been scored[167,168] even though transfer of charge and free-radical sites occurs to some extent within the enzyme molecule[169-174]. It has been shown with papain that the degree of inactivation by OH radicals is higher in the presence of oxygen[175].

Other important free-radical targets in proteins are the aromatic amino acids, tyrosine and tryptophane[176]. Even radicals derived from sulphur-containing amino acids bind to proteins through addition to the aromatic constituents[177,178]. The involvement of complexed inorganic [e.g. Br_2^- or $(SCN)_2^{\cdot-}$] and other radicals in these deactivation processes has been studied[151,179-183]. These radicals have been shown to react with more specificity than the highly reactive OH radical.

IV. REFERENCES

1. P. C. Jocelyn, *Biochemistry of the SH Group*, Academic Press, London, 1972, p. 323.
2. H. Dertinger and H. Jung, *Molekulare Strahlenbiologie*, Springer, Berlin, 1969, p. 90.
3. G. E. Adams, *Advan. Radiat. Chem.*, **3**, 125 (1972).
4. (a) J. E. Packer in *The Chemistry of the Thiol Group* (Ed. S. Patai), John Wiley and Sons, London, 1974, p. 481.
   (b) A. J. Swallow, *Progr. React. Kinet.*, **9**, 195 (1978).
5. E. M. Nanobashvili, G. G. Chirakadze, M. Sh. Simonidze, I. G. Bakhtadze and L. V. Ivanitskaya, *Radiolysis of Sulphur Compounds*, Part 1, Metsniereba, Tbilisi, Gruz. SSR, 1967 (Russian).
6. E. M. Nanobashvili, G. G. Chirakadze, M. V. Panchvidze, S. E. Gvilava and G. I. Khidesheli, *Radiolysis of Sulphur Compounds*, Part 2, Metsniereba, Tbilisi, Gruz. SSR, 1973 (Russian).
7. E. M. Nanobashvili and A. D. Bichiashvili, *Radiolysis of Sulphur Compounds*, Part 3, Metsniereba, Tbilisi, Gruz, SSR, 1973 (Russian).
8. E. M. Nanobashvili, M. V. Panchvidze, R. G. Tushurashvili, A. G. Dapkviashvili and G. R. Natroshvili, *Radiolysis of Sulphur Compounds*, Part 4, Metsniereba, Tbilisi, Gruz, SSR, 1975 (Russian).
9. E. M. Nanobashvili, G. G. Chirakadze and M. V. Panchvidze, *Radiolysis of Sulphur Compounds*, Part 5, Metsniereba, Tbilisi, Gruz. SSR, 1977 (Russian).
10. Ch. Lifshitz and Z. V. Zaretskii in *The Chemistry of the Thiol Group* (Ed. S. Patai), John Wiley and Sons, London, 1974, p. 325.

11. F. Klages, *Lehrbuch der organischen Chemie*, Vol. 1, Pt. 2, de Gruyter, Berlin, 1953, p. 674.
12. M. C. R. Symons, *J. Chem. Soc., Perkin Trans. II*, 1618 (1974).
13. I. H. Leaver, G. C. Ramsay and E. Suzuki, *Australian J. Chem.*, **22**, 1891 (1969).
14. B. C. Gilbert, J. P. Larkin and R. O. C. Norman, *J. Chem. Soc., Perkin II*, 272 (1973).
15. J. A. Wargon and F. Williams, *J. Chem. Soc., Chem. Commun.*, 947 (1975).
16. W. H. Davis and J. K. Kochi, *Tetrahedron Letters*, 1761 (1976).
17. D. J. Nelson, R. L. Petersen and M. C. R. Symons, *J. Chem. Soc., Perkin Trans. II*, 2005 (1977).
18. J. J. J. Myron and R. H. Johnsen, *J. Phys. Chem.*, **70**, 2951 (1966).
19. G. Lunde and R. R. Hentz, *J. Phys. Chem.*, **71**, 863 (1967).
20. E. M. Nanobashvili, A. G. Dapkviashvili and R. G. Tushurashvili, *Soobshch. Akad. Nauk Gruz. SSR*, **68**, 353 (1972).
21. G. G. Chirakadze, G. A. Mosashvili and E. M. Nanobashvili, *Soobshch. Akad. Nauk Gruz. SSR*, **75**, 353 (1974).
22. A. D. Bichiashvili, N. N. Tsomaya and E. M. Nanobashvili, *Soobshch. Akad. Nauk Gruz. SSR*, **83**, 629 (1976).
23. R. G. Barsegov, A. D. Bichiashvili, M. V. Panchvidze and E. M. Nanobashvili, *Soobshch. Akad. Nauk Gruz. SSR*, **49**, 91 (1968).
24. A. D. Bichiashvili, R. G. Barsegov and E. M. Nanobashvili, *High Energy Chem.*, **3**, 164 (1969).
25. A. D. Bichiashvili, E. M. Nanobashvili and R. G. Barsegov, *Soobshch. Akad. Nauk Gruz. SSR*, **53**, 337 (1969).
26. G. G. Chirakadze, E. M. Nanobashvili and G. A. Mosashvili, *Soobshch. Akad. Nauk Gruz. SSR*, **57**, 341 (1970).
27. A. Torikai, S. Sawada, K. Fueki and Z.-I. Kuri, *Bull. Chem. Soc. Japan*, **43**, 1617 (1970).
28. T. Gillbro, *Chem. Phys.*, **4**, 476 (1974).
29. Ts. M. Basiladze, A. D. Bichiashvili and E. M. Nanobashvili, *Soobshch. Akad. Nauk Gruz. SSR*, **85**, 89 (1977).
30. M. V. Panchvidze, G. G. Chirakadze and E. M. Nanobashvili, *Soobshch. Akad. Nauk Gruz. SSR*, **43**, 75 (1966).
31. A. Bergdolt and D. Schulte-Frohlinde, *Z. Naturforsch.*, **22b**, 270 (1967).
32. A. Bergdolt and D. Schulte-Frohlinde, *Z. Phys. Chem. (Frankfurt)*, **56**, 254 (1967).
33. E. M. Nanobashvili and M. Sh. Simonidze, *Khim. Seraorg. Soedin. Soderzh. Neftyakh Nefteprod.*, **9**, 168 (1972).
34. J. Skelton and F. C. Adam, *Can. J. Chem.*, **49**, 3536 (1971).
35. N. N. Tsomaya, A. D. Bichiashvili and E. M. Nanobashvili, *Soobshch. Akad. Nauk Gruz. SSR*, **65**, 337 (1972).
36. J. Esser and J. A. Stone, *Can. J. Chem.*, **51**, 192 (1973).
37. J. Wendenburg, H. Möckel, A. Granzow and A. Henglein, *Z. Naturforsch.*, **21b**, 632 (1966).
38. F. C. Adam, G. E. Smith and A. J. Elliot, *Can. J. Chem.*, **56**, 1856 (1978).
39. E. Migdal and M. Forys, *Radiat. Eff.*, **18**, 17 (1973).
40. J. Lind, B. Bjellqvist and T. E. Eriksen, *Intern. J. Radiat. Phys. Chem.*, **5**, 479 (1973).
41. E. Migdal and J. Sobkowski, *Radiat. Eff.*, **23**, 159 (1974).
42. E. Migdal and J. Sobkowski, *Radiat. Eff.*, **23**, 151 (1974).
43. B.-Ö. Jonsson and J. Lind, *Radiat. Eff.*, **32**, 79 (1977).
44. K. Jäger and A. Henglein, *Z. Naturforsch.*, **21a**, 1251 (1966).
45. B. G. Keyes and A. G. Harrison, *J. Amer. Chem. Soc.*, **90**, 5671 (1968).
46. G. P. Nagy, J. C. J. Thynne and A. G. Harrison, *Can. J. Chem.*, **46**, 3609 (1968).
47. W. E. W. Ruska and J. L. Franklin, *Intern. J. Mass Spectr. Ion Phys.*, **3**, 221 (1969).
48. B.-Ö. Jonsson and J. Lind, *J. Chem. Soc., Faraday Trans. II*, **70**, 1399 (1974).
49. R. Large and H. Knof, *Org. Mass Spectrom.*, **11**, 582 (1976).
50. H. Knof, R. Large and G. Albers, *Anal. Chem.*, **48**, 2120 (1976).
51. W. A. Pryor and E. G. Olsen, *J. Amer. Chem. Soc.*, **100**, 2852 (1978).

990   Clemens von Sonntag and Heinz-Peter Schuchmann

52. G. J. Collin, P. M. Perrin and F. X. Garneau, *Can. J. Chem.*, **52**, 2337 (1974).
53. A. S. Berenblum, S. L. Mund, G. A. Kovtun, N. Ya. Usachev, V. G. Sorokin, E. D. Radchenko and I. I. Moiseev in *Tesisy Dokl. 14. Nauchn. Sess. Khim. Tehnol. Org. Soedin. Sery Sernistykh Neftei, 1975* (Ed. I. G. Barkhtalze), Zinatne, Riga, Latv. SSR, 1976, p. 115.
54. K. Watanabe, T. Nakayama and J. Mottl, *J. Quant. Spectrosc. Radiat. Transfer*, **2**, 369 (1962).
55. M. A. Haney and J. L. Franklin, *J. Phys. Chem.*, **73**, 4328 (1969).
56. G. R. Freeman, *Radiat. Res. Rev.*, **1**, 1 (1968).
57. K.-D. Asmus, *Acc. Chem. Res.*, **12**, 436 (1979).
58. J. H. Baxendale, *Can. J. Chem.*, **55**, 1996 (1977).
59. *American Institute of Physics Handbook*, 2nd ed., McGraw-Hill, New York, 1963, pp. 5–127.
60. L. P. Kayushin, V. G. Krivenko and M. K. Pulatova, *Stud. Biophys.*, **33**, 59 (1972).
61. Reference 7, p. 55.
62. A. D. Bichiashvili, *Soobshch. Akad. Nauk Gruz. SSR*, **62**, 73 (1971).
63. Reference 7, pp. 115 and 120.
64. E. Lindholm and P. Wilmenius, *Arkiv Kemi*, **20**, 255 (1963).
65. M. V. Panchvidze, G. G. Chirakadze and E. M. Nanobashvili, *Soobshch. Akad. Nauk Gruz. SSR*, **43**, 75 (1966).
66. F. K. Truby, D. C. Wallace and J. E. Hesse, *J. Chem. Phys.*, **42**, 3845 (1965).
67. T. Shida, *J. Phys. Chem.*, **72**, 2597 (1968).
68. T. Shida, *J. Phys. Chem.*, **74**, 3055 (1970).
69. K. Akasaka, S. Kominami and H. Hatano, *J. Phys. Chem.*, **75**, 3746 (1971).
70. R. C. Petersen, D. J. Nelson and M. C. R. Symons, *J. Chem. Soc., Perkin Trans. II*, 225 (1978).
71. J. E. Kronberg and J. A. Stone, *Intern. J. Mass Spectr. Ion Phys.*, **24**, 373 (1977).
72. J. A. Stone and J. Esser, *Can. J. Chem.*, **52**, 1253 (1974).
73. K.-D. Asmus, J. M. Warman and R. H. Schuler, *J. Phys. Chem.*, **74**, 246 (1970).
74. K. Sayamol and A. R. Knight, *Can. J. Chem.*, **46**, 999 (1968).
75. A. Granzow, J. Wendenburg and A. Henglein, *Z. Naturforsch.*, **19b**, 1015 (1964).
76. Reference 6, p. 108.
77. S. Berk and H. Gisser, *Radiat. Res.*, **56**, 71 (1973).
78. E. S. Brodskii, I. M. Lukashenko, S. V. Voznesenskaya, V. V. Nesterovskii, M. V. Ermolaev and V. V. Tveritneva, *High Energy Chem.*, **9**, 328 (1975).
79. K.-D. Asmus, in *Radicaux libres organiques*, Centre National de la Recherche Scientifique, Paris, 1978, p. 305.
80. (a) K.-D. Asmsu, H. A. Gillis and G. G. Teather, *J. Phys. Chem.*, **82**, 2677 (1978).
    (b) J.-L. Marignier, *Thèse*, Université Paris-Sud, 1979.
81. A. G. Dapkviashvili, M. V. Panchvidze, G. I. Khidesheli and E. M. Nanobashvili, *Soobshch. Akad. Nauk Gruz. SSR*, **59**, 581 (1970).
82. E. M. Nanobashvili, M. V. Panchvidze, A. G. Dapkviashvili and G. I. Khidesheli, *Soobshch. Akad. Nauk Gruz. SSR*, **57**, 81 (1970).
83. V. P. Strunin and B. V. Bol'shakov, *High Energy Chem.*, **5**, 479 (1971).
84. G. C. Dismukes and J. E. Willard, *J. Phys. Chem.*, **80**, 2072 (1976).
85. G. C. Dismukes and J. E. Willard, *J. Phys. Chem.*, **80**, 1435 (1976).
86. E. Andrzejewska, A. Zuk, J. Pietrzak and R. Krzyminiewski, *J. Polym. Sci., Polym. Chem. Ed.*, **16**, 2991 (1978).
87. (a) S. Nagai and T. Gillbro, *J. Phys. Chem.*, **83**, 402 (1979).
    (b) H. J. Möckel, *Z. Analyt. Chem.*, **295**, 241 (1979).
88. B. C. Gilbert, D. K. C. Hodgeman and R. O. C. Norman, *J. Chem. Soc., Perkin Trans. II*, 1748 (1973).
89. (a) K.-D. Asmus, D. Bahnemann, M. Bonifačić and H. A. Gillis, *Faraday Discuss. Chem. Soc.*, **63**, 213 (1977).
    (b) W. B. Gara, J. R. M. Giles and B. P. Roberts, *J. Chem. Soc., Perkin Trans. II*, 1444 (1979).
    (c) W. K. Musker, B. V. Gorewit, P. B. Roush and T. L. Wolford, *J. Org. Chem.*, **43**, 3235 (1978).

(d) W. K. Musker, T. L. Wolford and P. B. Roush, *J. Amer. Chem. Soc.*, **100**, 6416 (1978).
90. L. M. Dorfman and G. E. Adams, *NSRDS-NBS*, Vol. 46, U.S. Dept. of Commerce, Washington, D.C., 1973.
91. Farhataziz and A. B. Ross, *NSRDS-NBS*, Vol. 59, U.S. Dept. of Commerce, Washington, D.C., 1977.
92. M. Anbar, M. Bambenek and A. B. Ross, *NSRDS-NBS*, Vol. 43, U.S. Dept. of Commerce, Washington, D.C., 1973.
93. A. B. Ross, *NSRDS-NBS*, Vol. 43 (Suppl.), U.S. Dept. of Commerce, Washington, D.C., 1975.
94. M. Anbar, Farhataziz and A. B. Ross, *NSRDS-NBS*, Vol. 51, U.S. Dept. of Commerce, Washington, D.C., 1975.
95. M. Z. Hoffman and E. Hayon, *J. Phys. Chem.*, **77**, 990 (1973).
96. T.-L. Tung and R. R. Kuntz, *Radiat. Res.*, **55**, 256 (1973).
97. G. G. Jayson, D. A. Stirling and A. J. Swallow, *Intern. J. Radiat. Biol.*, **19**, 143 (1971).
98. S. A. Grachev, E. V. Kropachev, G. I. Litvyakova and S. P. Orlov, *Bull. Acad. Sci. USSR Chem. Ser.*, 1248 (1976).
99. T.-L. Tung and R. R. Kuntz, *Radiat. Res.*, **55**, 10 (1973).
100. S. A. Grachev, E. V. Kropachev, G. I. Litvyakova and S. P. Orlov, *J. Gen. Chem. USSR*, **46**, 1813 (1976).
101. W. L. Severs, P. A. Hamilton, T.-L. Tung and J. A. Stone, *Intern. J. Radiat. Phys. Chem.*, **8**, 461 (1976).
102. V. G. Wilkening, M. Lal, M. Arends and D. A. Armstrong, *Can. J. Chem.*, **45**, 1209 (1967).
103. W. Karmann, A. Granzow, G. Meissner and A. Henglein, *Intern. J. Radiat. Phys. Chem.*, **1**, 395 (1969).
104. P. Neta and R. W. Fessenden, *J. Phys. Chem.*, **75**, 2277 (1971).
105. G. E. Adams, G. S. McNaughton and B. D. Michael, *Trans. Faraday Soc.* **64**, 902 (1968).
106. J. J. van Hemmen, W. J. A. Meuling, J. de Jong and L. H. Luthjens, *Intern. J. Radiat. Biol.*, **25**, 455 (1974).
107. J. Hähn, U. Prösch and G. Siegel, *Isotopenpraxis*, **6**, 241 (1970).
108. W. Karmann, G. Meissner and A. Henglein, *Z. Naturforsch.*, **22b**, 273 (1967).
109. G. E. Adams, J. E. Aldrich, R. H. Bisby, R. B. Cundall, J. L. Redpath and R. L. Willson, *Radiat. Res.*, **49**, 278 (1972).
110. J. L. Redpath, *Radiat. Res.*, **54**, 364 (1973).
111. M. Morita, K. Sasai, M. Tajima and M. Fujimaki, *Bull. Chem. Soc. Japan*, **44**, 2257 (1971).
112. K. Schäfer, *Thesis*, Technische Universität, Berlin, 1977.
113. K. Schäfer and K.-D. Asmus, to be published.
114. O. I. Mićić, M. T. Nenadović and P. A. Carapellucci, *J. Amer. Chem. Soc.*, **100**, 2209 (1978).
115. P. C. Chan and B. H. J. Bielski, *J. Amer. Chem. Soc.*, **95**, 5504 (1973).
116. T.-L. Tung and J. A. Stone, *Can. J. Chem.*, **53**, 3153 (1975).
117. J. W. Purdie, H. A. Gillis and N. V. Klassen, *Can. J. Chem.*, **51**, 3132 (1973).
118. G. C. Goyal and D. A. Armstrong, *Can. J. Chem.*, **54**, 1938 (1976).
119. M. Z. Hoffman and E. Hayon, *J. Amer. Chem. Soc.*, **94**, 7950 (1972).
120. A. Shafferman, *Israel J. Chem.*, **10**, 725 (1972).
121. T. C. Owen and D. R. Ellis, *Radiat. Res.*, **53**, 24 (1973).
122. H. Möckel, M. Bonifačić and K.-D. Asmus, *J. Phys. Chem.*, **78**, 282 (1974).
123. M. Bonifačić, K. Schäfer, H. Möckel and K.-D. Asmus, *J. Phys. Chem.*, **79**, 1496 (1975). (1975).
124. B. Lalitha and J. P. Mittal, *Radiat. Eff.*, **7**, 159 (1971).
125. M. Bonifačić and K.-D. Asmus, *J. Phys. Chem.*, **80**, 2426 (1976).
126. A. Schöberl and A. Wagner in *Methoden der organischen Chemie* (Houben–Weyl), Vol. 9 Georg Thieme Verlag, Stuttgart, 1955, p. 265.
127. Reference 126, p. 75.
128. B. B. Saunders, *J. Phys. Chem.*, **82**, 151 (1978).
129. F. Shimazu, U. S. Kumta and A. L. Tappel, *Radiat. Res.*, **22**, 276 (1964).

992   Clemens von Sonntag and Heinz-Peter Schuchmann

130. (a) M. Bonifačić, H. Möckel, D. Bahnemann and K.-D. Asmus, J. Chem. Soc., Perkin Trans. II, 675 (1975).
     (b) D. Veltwisch, E. Janata and K.-D. Asmus, Radiat. Phys. Chem., in the press.
     (c) K.-O. Hiller, M. Göbl, B. Masloch and K.-D. Asmus, Radiation, Biology and Chemistry, Research Developments (Eds. H. E. Edwards, S. Navaratnam, B. J. Parsons and G. O. Phillips), Elsevier, Amsterdam, 1979, p. 73.
131. G. Meissner, A. Henglein and G. Beck, Z. Naturforsch., 22b, 13 (1967).
132. (a) D. Bahnemann and K.-D. Asmus, J. Chem. Soc., Chem. Commun., 238 (1975).
     (b) K.-D. Asmus, D. Bahnemann, Ch.-H. Fischer and D. Veltwisch, J. Amer. Chem. Soc., 101, 5322 (1979).
     (c) W. K. Musker, A. S. Hirschon and J. T. Doi, J. Amer. Chem. Soc., 100, 7754 (1978).
133. J. Lilie, Z. Naturforsch., 26b, 197 (1971).
134. B. B. Saunders, P. C. Kaufman and M. S. Matheson, J. Phys. Chem., 82, 142 (1978).
135. K. Schäfer, M. Bonifačić, D. Bahnemann and K.-D. Asmus, J. Phys. Chem., 82, 2777 (1978).
136. M. Lal, Can. J. Chem., 54, 1092 (1976).
137. A. A. Al-Thannon, J. P. Barton, J. E. Packer, R. J. Sims, C. N. Trumbore and R. V. Winchester, Intern. J. Radiat. Phys. Chem., 6, 233 (1974).
138. M. Sh. Simonidze and E. M. Nanobashvili, Issled. v oblasti elektrokhim. i radiatsionnoi khim., No. 5, Metsniereba, Tbilisi, Gruz. SSR, 1965, p. 49.
139. E. M. Nanobashvili and G. G. Chirakadze, Issled. v. oblasti elektrokhim. i radiatsionnoi khim., No. 5, Metsniereba, Tbilisi, Gruz. SSR, 1965, p. 40.
140. M. Quintiliani, R. Badiello, M. Tamba, A. Esfandi and G. Gorin, Intern. J. Radiat. Biol., 32, 195 (1977).
141. W. Bors, E. Lengfelder, M. Saran, C. Fuchs and C. Michel, Biochem. Biophys. Res. Commun., 70, 81 (1976).
142. G. G. Jayson, T. C. Owen and A. C. Wilbraham, J. Chem. Soc. (B), 944 (1967).
143. T. C. Owen, M. Rodriguez, B. G. Johnson and J. A. G. Roach, J. Amer. Chem. Soc., 90, 196 (1968).
144. J. W. Purdie, Can. J. Chem., 49, 725 (1971).
145. T. C. Owen, A. C. Wilbraham, J. A. G. Roach and D. R. Ellis, Radiat. Res., 50, 234 (1972).
146. T. C. Owen and A. C. Wilbraham, Radiat. Res., 50, 253 (1972).
147. S. A. Grachev and E. V. Kropachev, High Energy Chem., 5, 136 (1971).
148. J. Hüttermann, W. Köhnlein, R. Téoule and A. J. Bertinchamps (Eds.), Effects of Ionizing Radiation on DNA, Springer, Berlin, 1978.
149. J. L. Redpath, Radiat. Res., 55, 109 (1973).
150. P. D. Boyer in The Enzymes (Eds. P. D. Boyer, H. Lardy and K. Myrbäck), Vol. 1, Academic Press, New York, 1959, p. 511.
151. G. M. Gaucher, B. L. Mainman and D. A. Armstrong, Can. J. Chem., 51, 2443 (1973).
152. K. R. Lynn and D. Louis, Intern. J. Radiat. Biol., 23, 477 (1973).
153. M. Lal, W. S. Lin, G. M. Gaucher and D. A. Armstrong, Intern. J. Radiat. Biol., 28, 549 (1975).
154. W. S. Lin, J. R. Clement, G. M. Gaucher and D. A. Armstrong, Radiat. Res., 62, 438 (1975).
155. T. Masuda, J. Ovadia and L. I. Grossweiner, Intern. J. Radiat. Biol., 20, 447 (1971).
156. L. K. Mee, S. J. Adelstein and G. Stein, Radiat. Res., 52, 588 (1972).
157. S. J. Adelstein, L. K. Mee and G. Stein, Israel J. Chem., 10, 1059 (1972).
158. L. K. Mee and S. J. Adelstein, Radiat. Res., 60, 422 (1974).
159. J. D. Buchanan and D. A. Armstrong, Intern. J. Radiat. Biol., 30, 115 (1976).
160. R. Badiello, M. Tamba and M. Quintiliani, Intern. J. Radiat. Biol., 26, 311 (1974).
161. J. D. Buchanan and D. A. Armstrong, Intern. J. Radiat. Biol., 33, 409 (1978).
162. G. E. Adams, K. F. Baverstock, R. B. Cundall and J. L. Redpath, Radiat. Res., 54, 375 (1973).
163. P. B. Roberts, Intern. J. Radiat. Biol., 24, 143 (1973).
164. D. J. Marciani and B. M. Tolbert, Biochim. Biophys. Acta, 351, 387 (1974).

165. H. Schüssler, M. Ebert and J. V. Davies, *Intern. J. Radiat. Biol.*, **32**, 391 (1977).
166. K. U. Linderstrøm-Lang and J. A. Schellman in *The Enzymes* (Eds. P. D. Boyer, H. Lardy and K. Myrbäck), Vol. 1, Academic Press, New York, 1959, p. 443.
167. G. E. Adams, R. L. Willson, R. H. Bisby and R. B. Cundall, *Intern. J. Radiat. Biol.*, **20**, 405 (1971).
168. S. M. Herbert and B. M. Tolbert, *Radiat. Res.*, **65**, 268 (1976).
169. N. N. Lichtin, *Israel J. Chem.*, **10**, 1041 (1972).
170. N. N. Lichtin, J. Ogdan and G. Stein, *Biochim. Biophys. Acta*, **276**, 124 (1972).
171. N. Lichtin, J. Ogdan and G. Stein, *Radiat. Res.*, **55**, 69 (1973).
172. J. R. Clement, W. S. Lin, D. A. Armstrong, G. M. Gaucher, N. V. Klassen and H. A. Gillis, *Intern. J. Radiat. Biol.*, **26**, 571 (1974).
173. E. S. Copeland, *Radiat. Res.*, **61**, 63 (1975).
174. E. S. Copeland, *Radiat. Res.*, **68**, 190 (1976).
175. W. S. Lin, D. A. Armstrong and M. Lal, *Intern. J. Radiat. Biol.*, **33**, 231 (1978).
176. J. R. Clement, D. A. Armstrong, N. V. Klassen and H. A. Gillis, *Can. J. Chem.*, **50**, 2833 (1972).
177. O. Yamamoto, *Intern. J. Radiat. Phys. Chem.*, **4**, 227 (1972).
178. O. Yamamoto, *Intern. J. Radiat. Phys. Chem.*, **4**, 335 (1972).
179. G. E. Adams, J. L. Redpath, R. H. Bisby and R. B. Cundall, *Israel J. Chem.*, **10**, 1079 (1972).
180. G. E. Adams, R. H. Bisby, R. B. Cundall, J. L. Redpath and R. L. Willson, *Radiat. Res.*, **49**, 290 (1972).
181. G. E. Adams and J. L. Redpath, *Intern. J. Radiat. Biol.*, **25**, 129 (1974).
182. D. A. Armstrong and J. D. Buchanan, *Photochem. Photobiol.*, **28**, 743 (1978).
183. W. A. Prütz, *Radiat. Environ. Biophys.*, **16**, 43 (1979).

Author Index

This author index is designed to enable the reader to locate an author's name and work with the aid of the reference numbers appearing in the text. The page numbers are printed in normal type in ascending numerical order, followed by the reference numbers in parentheses. The numbers in *italics* refer to the pages on which the references are actually listed.

Subject Index